Areas Under the Normal Curve (*continued*)

z	.00	.01	.02	.03	.04	.05	.06	.07	.08	.09
0.0	0.5000	0.5040	0.5080	0.5120	0.5160	0.5199	0.5239	0.5279	0.5319	0.5359
0.1	0.5398	0.5438	0.5478	0.5517	0.5557	0.5596	0.5636	0.5675	0.5714	0.5753
0.2	0.5793	0.5832	0.5871	0.5910	0.5948	0.5987	0.6026	0.6064	0.6103	0.6141
0.3	0.6179	0.6217	0.6255	0.6293	0.6331	0.6368	0.6406	0.6443	0.6480	0.6517
0.4	0.6554	0.6591	0.6628	0.6664	0.6700	0.6736	0.6772	0.6808	0.6844	0.6879
0.5	0.6915	0.6950	0.6985	0.7019	0.7054	0.7088	0.7123	0.7157	0.7190	0.7224
0.6	0.7257	0.7291	0.7324	0.7357	0.7389	0.7422	0.7454	0.7486	0.7517	0.7549
0.7	0.7580	0.7611	0.7642	0.7673	0.7704	0.7734	0.7764	0.7794	0.7823	0.7852
0.8	0.7881	0.7910	0.7939	0.7967	0.7995	0.8023	0.8051	0.8078	0.8106	0.8133
0.9	0.8159	0.8186	0.8212	0.8328	0.8264	0.8289	0.8315	0.8340	0.8365	0.8389
1.0	0.8413	0.8438	0.8461	0.8485	0.8508	0.8531	0.8554	0.8577	0.8599	0.8621
1.1	0.8643	0.8665	0.8686	0.8708	0.8729	0.8749	0.8770	0.8790	0.8810	0.8830
1.2	0.8849	0.8869	0.8888	0.8907	0.8925	0.8944	0.8962	0.8980	0.8997	0.9015
1.3	0.9032	0.9049	0.9066	0.9082	0.9099	0.9115	0.9131	0.9147	0.9162	0.9177
1.4	0.9192	0.9207	0.9222	0.9236	0.9251	0.9265	0.9278	0.9292	0.9306	0.9319
1.5	0.9332	0.9345	0.9357	0.9370	0.9382	0.9394	0.9406	0.9418	0.9429	0.9441
1.6	0.9452	0.9463	0.9474	0.9484	0.9495	0.9505	0.9515	0.9525	0.9535	0.9545
1.7	0.9554	0.9564	0.9573	0.9582	0.9591	0.9599	0.9608	0.9616	0.9625	0.9633
1.8	0.9641	0.9649	0.9656	0.9664	0.9671	0.9678	0.9686	0.9693	0.9699	0.9706
1.9	0.9713	0.9719	0.9726	0.9732	0.9738	0.9744	0.9750	0.9756	0.9761	0.9767
2.0	0.9772	0.9778	0.9783	0.9788	0.9793	0.9798	0.9803	0.9808	0.9812	0.9817
2.1	0.9821	0.9826	0.9830	0.9834	0.9838	0.9842	0.9846	0.9850	0.9854	0.9857
2.2	0.9861	0.9864	0.9868	0.9871	0.9875	0.9878	0.9881	0.9884	0.9887	0.9890
2.3	0.9893	0.9896	0.9898	0.9901	0.9904	0.9906	0.9909	0.9911	0.9913	0.9916
2.4	0.9918	0.9920	0.9922	0.9925	0.9927	0.9929	0.9931	0.9932	0.9934	0.9936
2.5	0.9938	0.9940	0.9941	0.9943	0.9945	0.9946	0.9948	0.9949	0.9951	0.9952
2.6	0.9953	0.9955	0.9956	0.9957	0.9959	0.9960	0.9961	0.9962	0.9963	0.9964
2.7	0.9965	0.9966	0.9967	0.9968	0.9969	0.9970	0.9971	0.9972	0.9973	0.9974
2.8	0.9974	0.9975	0.9976	0.9977	0.9977	0.9978	0.9979	0.9979	0.9980	0.9981
2.9	0.9981	0.9982	0.9982	0.9983	0.9984	0.9984	0.9985	0.9985	0.9986	0.9986
3.0	0.9987	0.9987	0.9987	0.9988	0.9988	0.9989	0.9989	0.9989	0.9990	0.9990
3.1	0.9990	0.9991	0.9991	0.9991	0.9992	0.9992	0.9992	0.9992	0.9993	0.9993
3.2	0.9993	0.9993	0.9994	0.9994	0.9994	0.9994	0.9994	0.9995	0.9995	0.9995
3.3	0.9995	0.9995	0.9995	0.9996	0.9996	0.9996	0.9996	0.9996	0.9996	0.9997
3.4	0.9997	0.9997	0.9997	0.9997	0.9997	0.9997	0.9997	0.9997	0.9997	0.9998

Statistics for the
Life Sciences

Third Edition

Myra L. Samuels
Purdue University

Jeffrey A. Witmer
Oberlin College

Pearson Education, Inc.
Upper Saddle River, New Jersey 07458

Library of Congress Cataloging-in Publication Data

Samuels, Myra L.
 Statistics for the life sciences / Myra L. Samuels, Jeffrey A. Witmer. - 3rd ed.
 p. cm.
 Includes bibliographical references (p.)
 ISBN 0-13-041316-X
 1. Biometry. 2. Medical statistics. 3. Agriculture-Statistics. I. Witmer, Jeffrey A.
II. Title
 QH323.5 .S23 2003
 570′ .1′ 5195-dc21

                2002074908

Editor in Chief/Acquisitions Editor: *Sally Yagan*
Vice President/Director of Production and Manufacturing: *David W. Riccardi*
Executive Managing Editor: *Kathleen Schiaparelli*
Senior Managing Editor: *Linda Mihatov Behrens*
Assistant Managing Editor: *Bayani Mendoza de Leon*
Production Editor: *Jeanne Audino*
Manufacturing Buyer: *Alan Fischer*
Manufacturing Manager: *Trudy Pisciotti*
Marketing Manager: *Krista M. Bettino*
Marketing Assistant: *Christine Bayeux*
Editorial Assistant/Supplements Editor: *Joanne Wendelken*
Art Director: *Jonathan Boylan*
Assistant to the Art Director: *John Christiana*
Interior and Cover Designer: *Alamini Design*
Art Editor: *Thomas Benfatti*
Assistant Manager of Math Media Production: *John Matthews*
Media Production Editor: *Donna Crilly*
Creative Director: *Carole Anson*
Director of Creative Services: *Paul Belfanti*
Cover Photo: *Paul Katz/Index Stock Imagery*
Art Studio: *Laserwords*
Compositor: *Preparé*

Printed in the United States of America

10 9 8 7 6 5 4 3 2 1

ISBN 0-13-041316-X

Pearson Education LTD., *London*
Pearson Education Australia PTY, Limited, *Sydney*
Pearson Education Singapore, Pte. Ltd
Pearson Education North Asia Ltd, *Hong Kong*
Pearson Education Canada, Ltd., *Toronto*
Pearson Educacion de Mexico, S.A. de C.V.
Pearson Education—Japan, *Tokyo*
Pearson Education Malaysia, Pte. Ltd

Contents

Preface

Statistics for the Life Sciences is an introductory text in statistics, specifically addressed to students specializing in the life sciences. Its primary aims are (1) to show students how statistical reasoning is used in biological, medical, and agricultural research; (2) to enable students confidently to carry out simple statistical analyses and to interpret the results; and (3) to raise students' awareness of basic statistical issues such as randomization, confounding, and the role of independent replication.

Style and Approach

The style of *Statistics for the Life Sciences* is informal and uses only minimal mathematical notation. There are no prerequisites except elementary algebra; anyone who can read a biology or chemistry textbook can read this text. It is suitable for use by graduate or undergraduate students in biology, agronomy, medical and health sciences, nutrition, pharmacy, animal science, physical education, forestry, and other life sciences.

Use of Real Data   Real examples are more interesting and often more enlightening than artificial ones. *Statistics for the Life Sciences* includes hundreds of examples and exercises that use real data, representing a wide variety of research in the life sciences. Each example has been chosen to illustrate a particular statistical issue. The exercises have been designed to reduce computational effort and focus students' attention on concepts and interpretations.

Emphasis on Ideas   The text emphasizes statistical ideas rather than computations or mathematical formulations. Probability theory is included only to support statistics concepts. Throughout the discussion of descriptive and inferential statistics, interpretation is stressed. By means of salient examples, the student is shown why it is important that an analysis be appropriate for the research question to be answered, for the statistical design of the study, and for the nature of the underlying distributions. The student is warned against the common blunder of confusing statistical nonsignificance with practical insignificance, and is encouraged to use confidence intervals to assess the magnitude of an effect. The student is led to recognize the impact on real research of design concepts such as random sampling, randomization, efficiency, and the control of extraneous variation by blocking or adjustment. Numerous exercises amplify and reinforce the student's grasp of these ideas.

The Role of the Computer   The analysis of research data is usually carried out with the aid of a computer. Computer-generated graphs and output, either from the statistical software DataDesk or MINITAB, are shown at several places in the text. MINITAB commands are given in a number of places (although MINITAB output can also be generated from menus while running the software). However, in studying statistics it is desirable for the student to gain experience working directly with data, using paper and pencil and a hand-held calculator, as well as a computer. This experience will help the student appreciate the nature and purpose of the statistical computations. The student is thus prepared to make intelligent use of the computer—to give it appropriate instructions and properly interpret the output. Accordingly, most of the exercises in this text are intended for hand calculation. Selected exercises, identified with the words "computer exercise" are intended to be completed with use of a computer. (Typically, the computer exercises require calculations that would be unduly burdensome if carried out by hand.)

Organization

This text is organized to permit coverage in one semester of the maximum number of important statistical ideas, including power, multiple inference, and the basic principles of design. By including or excluding optional sections, the instructor can also use the text for a one-quarter course or a two-quarter course. It is suitable for a terminal course or for the first course of a sequence.

The following is a brief outline of the text:

Chapter 1: Introduction. The nature and impact of variability in biological data.

Chapter 2: Orientation. Frequency distributions, descriptive statistics, the concept of population versus sample.

Chapters 3, 4, and 5: Theoretical preparation. Probability, binomial and normal distributions, sampling distributions.

Chapter 6: Confidence interval for a mean or for a proportion.

Chapter 7: Comparison of two independent samples. The *t*-test and the Wilcoxon-Mann-Whitney test.

Chapter 8: Design. Randomization, blocking, hazards of observational studies.

Chapter 9: Inference for paired samples. Confidence interval, *t*-test, sign test, and Wilcoxon signed-rank test.

Chapter 10: Categorical data. Chi-square goodness-of-fit test, conditional probability, contingency tables. Optional sections cover Fisher's exact test, McNemar's test, and odds ratios.

Chapter 11: Analysis of variance: one-way layout. Multiple comparison procedures, two-way analysis of variance, contrasts, and interaction in two-factor designs are included in optional sections.

Chapter 12: Regression and correlation. Descriptive and inferential aspects of simple linear regression and correlation and the relationship between them.

Chapter 13: A summary of inference methods.

Statistical tables are provided at the back of the book. The tables of critical values are especially easy to use, because they follow mutually consistent layouts and so are used in essentially the same way.

Optional appendices at the back of the book give the interested student a deeper look into such matters as how the Wilcoxon-Mann-Whitney null distribution is calculated.

Changes to the Third Edition

- A quarter of the problems in the book are new or revised. As before, the majority are based on real data and draw from a variety of subjects of interest to life science majors. Many are keyed to the data disk that accompanies the book.

- New illustrations and graphics have been added in many places, and several sections of the text have been rewritten for greater clarity.

- New material has been added to several chapters. This includes:
  - More complete coverage of probability rules and of random variables in Chapter 3
  - Presentation of the Wilcoxon signed-rank test in Chapter 9
  - An introduction to two-way analysis of variance, including treatment of ANOVA for randomized block designs in Chapter 11
  - An introduction to logistic regression in Chapter 12

Supplements

Instructor's Solutions Manual   (ISBN 0-13-034144-4)

Solutions to the exercises are provided in this manual. Careful attention has been paid to ensure that all methods of solution and notation are consistent with those used in the core text.

Data Disk

The larger data sets used in problems and exercises in the book are saved as ASCII, Excel, Minitab and SPSS files on the data disk that is packaged in the back of every copy of the book. The disk is compatible with both PC and Macintosh platforms. For more information on the web, refer to:

http://www.prenhall.com/samuels

Supplements Available for Purchase by Students

Technology Supplements and Packaging Options

A MINITAB Guide to Statistics   (By Ruth Meyer and David Krueger) (ISBN 0-13-784232-5)

This comprehensive manual assumes no prior knowledge of MINITAB. Organized to correspond to the table of contents of most statistics texts, it provides step-by-step instruction to using MINITAB for statistical analysis.

An Introduction to Data Analysis Using MINITAB for Windows   (by Dorothy Wakefield and Kathleen McLaughlin) (ISBN 0-13-612508-3)

A hands-on guide to using MINITAB 12.0, this spiral-bound workbook provides step-by-step instruction for learning how to perform basic statistical analysis with MINITAB 12.0 for Windows. Each lesson is set up with an activity that is designed to be completed and handed in, making this manual ideal for lab sessions or independent study.

Acknowledgments for the Third Edition

The third edition of *Statistics for the Life Sciences* retains the style and spirit of the writing of Myra Samuels. Prior to her tragic death from cancer, Myra wrote the first edition of the text, based on her experience both as a teacher of statistics and as a statistical consultant. Without her vision and efforts there never would have been a first edition, let alone a third.

Many researchers have contributed sets of data to the text. Among these I am particularly thankful for the help that Yolanda Cruz has given me concerning interpretation of biological data. I have benefited from countless conversations with David Moore, Dick Scheaffer, Murray Clayton, Alan Agresti, Don Bentley, George Cobb, Katherine Halvorsen, Pete Hayslett, Gudmund Iversen, Robin Lock, Tom Moore, Norean Radke Sharpe, Rosemary Roberts, Alan Rossman, and Dex Whittinghill, all of whom have my thanks (although I retain responsibility for failing to make adequate use of the advice they have given).

I am grateful for the sound guidance and encouragement of Jeanne Audino. Steve Samuels, and Chris Andrews have reviewed drafts of many parts of the third edition and have provided many helpful comments and suggestions, which were most welcome.

J. A. W.

CHAPTER 1

Introduction

1.1 STATISTICS AND THE LIFE SCIENCES

Researchers in the life sciences carry out investigations in various settings: in the clinic, in the laboratory, in the greenhouse, in the field. Generally, the resulting data exhibit some *variability*. For instance, patients given the same drug respond somewhat differently; cell cultures prepared identically develop somewhat differently; adjacent plots of genetically identical wheat plants yield somewhat different amounts of grain. Often the degree of variability is substantial even when experimental conditions are held as constant as possible.

   The challenge to the life scientist is to discern the patterns that may be more or less obscured by the variability of responses in living systems. The scientist must try to distinguish the "signal" from the "noise."

   Statistics is the science of understanding data and of making decisions in the face of variability and uncertainty. The discipline of statistics has evolved in response to the needs of scientists and others whose data exhibit variability. The concepts and methods of statistics enable the investigator to describe variability and to plan research so as to take variability into account (i.e., to make the "signal" strong in comparison to the background "noise" in data that are collected). Statistical methods are used to analyze data so as to extract the maximum information and to quantify the reliability of that information.

> ### Objective
>
> - *In this chapter we will look at a series of examples of areas in the life sciences in which statistics is used, with the goal of understanding the scope of the field of statistics.*

1.2 EXAMPLES AND OVERVIEW

In this section we give some examples to illustrate the degree of variability found in biological data and the ways in which variability poses a challenge to the biological researcher. We will briefly mention some of the statistical issues raised by each example and indicate where in this book the issues are addressed.

The first two examples provide a contrast between an experiment that showed no variability and another that showed considerable variability.

Example 1.1

Vaccine for Anthrax. Anthrax is a serious disease of sheep and cattle. In 1881 Louis Pasteur conducted a famous experiment to demonstrate the effect of his vaccine against anthrax. A group of 24 sheep were vaccinated; another group of 24 unvaccinated sheep served as controls. Then, all 48 animals were inoculated with a virulent culture of anthrax bacillus. Table 1.1 shows the results.[1] The data of Table 1.1 show no variability; all the vaccinated animals survived and all the unvaccinated animals died. ∎

TABLE 1.1 Response of Sheep to Anthrax

	Treatment	
Response	Vaccinated	Not vaccinated
Died of anthrax	0	24
Survived	24	0
Total	24	24
Percent survival	100%	0%

Example 1.2

Bacteria and Cancer. To study the effect of bacteria on tumor development, researchers used a strain of mice with a naturally high incidence of liver tumors. One group of mice were maintained entirely germ free, while another group were exposed to the intestinal bacteria *Escherichia coli*. The incidence of liver tumors is shown in Table 1.2.[2]

In contrast to Table 1.1, the data of Table 1.2 show variability; mice given the same treatment did not all respond the same way. Because of this variability, the results in Table 1.2 are equivocal; the data suggest that exposure to *E. coli* increases the risk of liver tumors, but the possibility remains that the observed difference in percentages (62% versus 39%) might reflect only chance variation rather than an effect of *E. coli*. If the experiment were replicated with different animals, the percentages might be substantially changed; note especially that the 62% is based on only 13 animals. ∎

TABLE 1.2 Incidence of Liver Tumors in Mice

	Treatment	
Response	E. coli	Germ Free
Liver tumors	8	19
No liver tumors	5	30
Total	13	49
Percent with liver tumors	62%	39%

In Chapter 10 we will discuss statistical techniques for evaluating data such as those in Tables 1.1 and 1.2. Of course, in some experiments variability is minimal and the message in the data stands out clearly without any special statistical analysis. It is worth noting, however, that absence of variability is itself an experimental result that must be justified by sufficient data. For instance, because Pasteur's anthrax data (Table 1.1) show no variability at all, it is intuitively plausible to conclude that the data provide "solid" evidence for the efficacy of the vaccination. But note that this conclusion involves a judgment; consider how much *less* "solid" the evidence would be if Pasteur had included only 3 animals in each group, rather than 24. In fact, a judgment that variability is negligible can be justified by an appropriate statistical analysis. Thus, a statistical view can be helpful even in the absence of variability.

The next two examples illustrate some of the questions that a statistical approach can help to answer.

Flooding and ATP.   In an experiment on root metabolism, a plant physiologist grew birch tree seedlings in the greenhouse. He flooded four seedlings with water for one day and kept four others as controls. He then harvested the seedlings and analyzed the roots for adenosine triphosphate (ATP). The measured amounts of ATP (nmols per mg tissue) are given in Table 1.3 and displayed in Figure 1.1.[3]

The data of Table 1.3 raise several questions: How should one summarize the ATP values in each experimental condition? How much information do the data provide about the effect of flooding? How confident can one be that the reduced ATP in the flooded group is really a response to flooding rather than just random variation? What size experiment would be required in order to firmly corroborate the apparent effect seen in these data?   ■

| **Example 1.3** | |

TABLE 1.3 ATP Concentration in Birch Tree Roots (nmol/mg)	
Flooded	**Control**
1.45	1.70
1.19	2.04
1.05	1.49
1.07	1.91

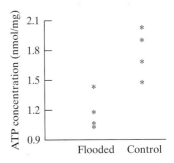

Figure 1.1 ATP concentration in birch tree roots

Chapters 2, 6, 7, and 8 address questions like those posed in Example 1.3.

MAO and Schizophrenia.   Monoamine oxidase (MAO) is an enzyme that is thought to play a role in the regulation of behavior. To see whether different categories of schizophrenic patients have different levels of MAO activity, researchers collected blood specimens from 42 patients and measured the MAO activity in the platelets. The results are given in Table 1.4 and displayed in Figure 1.2. (Values are expressed as nmol benzylaldehyde product per 108 platelets per hour.)[4] Note that it is much easier to get a feeling for the data by looking at the graph (Figure 1.2) than it is to read through the data in the table. The use of graphical displays of data is a very important part of data analysis.

To analyze the MAO data, one would naturally want to make comparisons among the three groups of patients, to describe the reliability of those comparisons, and to characterize the variability within the groups. To go beyond the data to a biological interpretation, one must also consider more subtle issues, such as the following: How were the patients selected? Were they chosen from a common hospital population, or were the three groups obtained at different times or places? Were precautions taken so that the person measuring the MAO was unaware of

Example 1.4

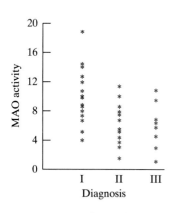

Figure 1.2 MAO activity in schizophrenic patients

TABLE 1.4 MAO Activity in Schizophrenic Patients

Diagnosis	MAO activity				
I:	6.8	4.1	7.3	14.2	18.8
Chronic undifferentiated	9.9	7.4	11.9	5.2	7.8
schizophrenic	7.8	8.7	12.7	14.5	10.7
(18 patients)	8.4	9.7	10.6		
II:	7.8	4.4	11.4	3.1	4.3
Undifferentiated with	10.1	1.5	7.4	5.2	10.0
paranoid features	3.7	5.5	8.5	7.7	6.8
(16 patients)	3.1				
III:	6.4	10.8	1.1	2.9	4.5
Paranoid schizophrenic	5.8	9.4	6.8		
(8 patients)					

the patient's diagnosis? Did the investigators consider various ways of subdividing the patients before choosing the particular diagnostic categories used in Table 1.4? At first glance, these questions may seem irrelevant—can we not let the measurements speak for themselves? We will see, however, that the proper interpretation of data always requires careful consideration of how the data were obtained. ∎

Chapters 2, 3, 8, and 9 include discussions of selection of experimental subjects and of guarding against unconscious investigator bias. In Chapter 11 we will show how sifting through a data set in search of patterns can lead to serious misinterpretations, and we will give guidelines for avoiding the pitfalls in such searches.

The next example shows how the effects of variability can distort the results of an experiment and how this distortion can be minimized by careful design of the experiment.

Example 1.5

Food Choice by Insect Larvae. The clover root curculio, *Sitona hispidulus,* is a root-feeding pest of alfalfa. An entomologist conducted an experiment to study food choice by *Sitona* larvae. She wished to investigate whether larvae would preferentially choose alfalfa roots that were nodulated (their natural state) over roots whose nodulation had been suppressed. Larvae were released in a dish where both nodulated and nonnodulated roots were available. After 24 hours the investigator counted the larvae that had clearly made a choice between root types. The results are shown in Table 1.5.[5]

TABLE 1.5 Food Choice by *Sitona* Larvae

Choice	Number of Larvae
Chose nodulated roots	46
Chose nonnodulated roots	12
Other (no choice, died, lost)	62
Total	120

The data in Table 1.5 appear to suggest rather strongly that *Sitona* larvae prefer nodulated roots. But our description of the experiment has obscured an important point—we have not stated how the roots were arranged. To see the relevance of the arrangement, suppose the experimenter had used only one dish, placing all the nodulated roots on one side of the dish and all the nonnodulated roots on the other side, as shown in Figure 1.3(a), and had then released 120 larvae in the center of the dish. This experimental arrangement would be seriously deficient, because the data of Table 1.5 would then permit several competing interpretations—for instance, (a) perhaps the larvae really do prefer nodulated roots; or (b) perhaps the two sides of the dish were at slightly different temperatures, and the larvae were responding to temperature rather than nodulation; or (c) perhaps one larva chose the nodulated roots just by chance and the other larvae followed its trail. Because of these possibilities, the experimental arrangement shown in Figure 1.3(a) can yield only weak information about larval food preference.

The experiment was actually arranged as in Figure 1.3(b), using six dishes with nodulated and nonnodulated roots arranged in a symmetric pattern. Twenty larvae were released into the center of each dish. This arrangement avoids the pitfalls of the arrangement in Figure 1.3(a). Because of the alternating regions of nodulated and nonnodulated roots, any fluctuation in environmental conditions (such as temperature) would tend to affect the two root types equally. By using several dishes, the experimenter has generated data that can be interpreted even if the larvae do tend to follow each other. To analyze the experiment properly, we would need to know the results in each dish; the condensed summary in Table 1.5 is not adequate. ∎

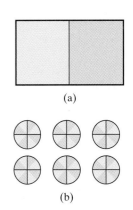

Figure 1.3 Possible arrangements of food choice experiment. The dark-shaded areas contain nodulated roots and the light-shaded areas contain nonnodulated roots. (a) A poor arrangement. (b) A good arrangement.

In Chapter 8 we will describe various ways of arranging experimental material in space and time so as to yield the most informative experiment. In later chapters we will discuss how to analyze the data to extract as much information as possible, and yet to resist the temptation to over interpret patterns that may represent only random variation.

Example 1.6

Sexual Orientation. Some research has suggested that there is a genetic basis for sexual orientation. One such study involved measuring the midsagittal area of the anterior commissure (AC) of the brain for 30 homosexual men, 30 heterosexual men, and 30 heterosexual women. The researchers found that the AC tends to be larger in heterosexual women than in heterosexual men and that it is even larger in homosexual men. These data are summarized in Table 1.6 and are shown graphically in Figure 1.4.

TABLE 1.6 Midsagittal Area of the Anterior Commissure (mm^2)	
Group	**Average midsagittal area (mm^2) of the anterior commissure**
Homosexual men	14.20
Heterosexual men	10.61
Heterosexual women	12.03

The data suggest that the size of the AC in homosexual men is more like that of heterosexual women than that of heterosexual men. When analyzing these data, we should take into account two things: (1) The measurements for two of the

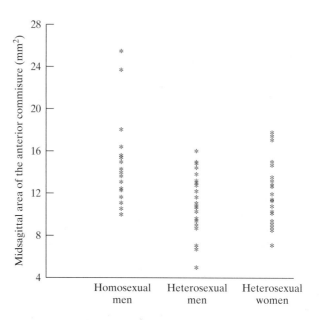

Figure 1.4 Midsagittal area of the anterior commissure (mm²)

homosexual men were much larger than any of the other measurements; sometimes one or two such outliers can have a big impact on the conclusions of a study. (2) Twenty-four of the thirty homosexual men had died of AIDS, as opposed to 6 of the 30 heterosexual men; if AIDS affects the size of the anterior commissure, then this factor could account for some of the difference between the two groups of men.[6]

Note that the *context* in which the data arose is of central importance in statistics. This is quite clear in the present example: The numbers themselves can be used to compute averages or to make graphs, like Figure 1.4, but if we are to understand what the data have to say, we must understand the context in which they arose. This context tells us to be on the alert for the effects of other factors, such as the impact AIDS may have on the size of the anterior commissure. Data analysis without reference to context is meaningless. ■

In Chapter 8 we will consider aspects of data collection and analysis that help to deal with the concerns raised in Example 1.6.

Example 1.7

Toxicity in Dogs.   Before new drugs are given to human subjects, it is common practice to test them first in dogs or other animals. In part of one study, a new investigational drug was given to 4 male and 4 female dogs, at doses 8 mg/kg and 25 mg/kg. Many "endpoints" were measured, such as cholesterol, sodium, and glucose, from blood samples in order to screen for toxicity problems in the dogs before starting studies on humans. One endpoint was alkaline phosphatase level (measured in U/Li). The data are shown in Table 1.7 and plotted in Figure 1.5.[7]

The design of this experiment allows for the investigation of the interaction between two factors: sex of the dog and dose. These factors interacted in the following sense: For females the effect of increasing the dose from 8 to 25 was positive, although small (the average increased from 133.5 to 143), but for males the effect of increasing the dose from 8 to 25 was negative (the average dropped from 143 to 124.5). Techniques for studying such interactions will be considered in Chapter 11. ■

TABLE 1.7 Alkaline Phosphate Level in Units per Liter

Dose (mg/kg)	Male	Female
8	171	150
	154	127
	104	152
	143	105
Ave.	**143**	**133.5**
25	80	101
	149	113
	138	161
	131	197
Ave.	**124.5**	**143**

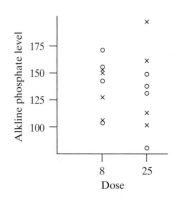

Figure 1.5 Alkaline phosphate level in dogs. Males are shown with circles, females with *x*'s.

The following example is a study of the relationship between two measured quantities.

Body Size and Energy Expenditure. How much food does a person need? To investigate the dependence of nutritional requirements on body size, researchers used underwater weighing techniques to determine the fat-free body mass for each of seven men. They also measured the total 24-hour energy expenditure during conditions of quiet sedentary activity; this was repeated twice for each subject. The results are shown in Table 1.8 and plotted in Figure 1.6.[8]

Example 1.8

TABLE 1.8 Fat-Free Mass and Energy Expenditure

Subject	Fat-free mass (kg)	24-hour energy expenditure (kcal)	
1	49.3	1,851	1,936
2	59.3	2,209	1,891
3	68.3	2,283	2,423
4	48.1	1,885	1,791
5	57.6	1,929	1,967
6	78.1	2,490	2,567
7	76.1	2,484	2,653

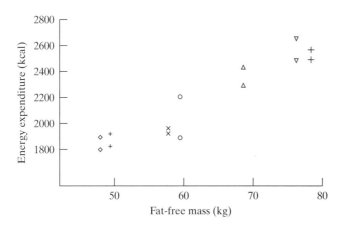

Figure 1.6 Fat-free mass and energy expenditure in seven men. Each man is represented by a different symbol.

A primary goal in the analysis of these data would be to describe the relationship between fat-free mass and energy expenditure—to characterize not only the overall trend of the relationship, but also the degree of scatter or variability in the relationship. (Note also that, to analyze the data, one needs to decide how to handle the duplicate observations on each subject.) ■

The focus of Example 1.8 is on the relationship between two variables: fat-free mass and energy expenditure. Chapter 12 deals with methods for describing such relationships and for quantifying the reliability of the descriptions.

A Look Ahead

Where appropriate, statisticians make use of the computer as a tool in data analysis; computer-generated output and statistical graphics appear throughout this book. The computer is a powerful tool, but it must be used with caution. Using the computer to perform calculations allows us to concentrate on concepts. The danger when using a computer in statistics is that we will jump straight to the calculations without looking closely at the data and asking the right questions about the data. Our goal is to analyze, understand, and interpret data—which are numbers *in a specific context*—not just to perform calculations.

In order to understand a data set, it is necessary to know how and why the data were collected. In addition to considering the most widely used methods in statistical inference, we will consider issues in data collection and experimental design. Together, these topics should provide the reader with the background needed to read the scientific literature and to design and analyze simple research projects.

The preceding examples illustrate the kind of data to be considered in this book. In fact, each of the examples will reappear as an exercise or example in an appropriate chapter. As the examples show, research in the life sciences is usually concerned with the comparison of two or more groups of observations, or with the relationship between two or more variables. We will begin our study of statistics by focusing on a simpler situation—observations of a *single* variable for a *single* group. Many of the basic ideas of statistics will be introduced in this oversimplified context. Two-group comparisons and more complicated analyses will then be discussed in Chapter 7 and later chapters.

CHAPTER 2

Description of Samples and Populations

2.1 INTRODUCTION

Statistics is the science of analyzing and learning from data. In this section we introduce some terminology and notation for dealing with data.

Variables

We begin with the concept of a **variable**. A variable is a characteristic of a person or a thing that can be assigned a number or a category. For example, blood type (A, B, AB, O) and age are two variables we might measure on a person.

Blood type is an example of a **categorical variable**: A categorical variable is a variable that records which of several categories a person or thing is in. Examples of categorical variables are:

Blood type of a person: A, B, AB, O
Sex of a fish: male, female
Color of a flower: red, pink, white
Shape of a seed: wrinkled, smooth

For some categorical variables, the categories can be arrayed in a meaningful rank order. Such a variable is said to be **ordinal**. Examples of ordinal categorical variables are:

Response of a patient to therapy: none, partial, complete
Tenderness of beef: tough, slightly tough, tender, very tender
Cloudiness: overcast, mostly cloudy, partly cloudy, sunny

Age is an example of a **quantitative variable**: A quantitative variable is a variable that records the amount of something. A **continuous variable** is a quantitative variable that is measured on a continuous scale. Examples of continuous variables are

Objectives

In this chapter we will study how to describe populations and samples. In particular, we will

- *learn how frequency distributions are used to make histograms*

- *study the mean and median as measures of center*

- *learn how to read and construct boxplots*

- *study the standard deviation as a measure of variability*

- *consider the relationship between populations and samples*

Weight of a baby
Cholesterol concentration in a blood specimen
Optical density of a solution

A variable such as weight is continuous because, in principle, two weights can be arbitrarily close together. Some types of quantitative variables are not continuous but fall on a discrete scale, with spaces between the possible values. A **discrete variable** is a quantitative variable for which we can list the possible values. For example, the number of eggs in a bird's nest is a discrete variable because only the values $0, 1, 2, 3, \ldots$, are possible. Other examples of discrete variables are

Age of a person (in years)
Number of bacteria colonies in a petri dish
Number of cancerous lymph nodes detected in a patient

The distinction between continuous and discrete variables is not a rigid one. After all, physical measurements are always rounded off. We may measure the weight of a steer to the nearest kilogram, of a rat to the nearest gram, or of an insect to the nearest milligram. The scale of the actual measurements is always discrete, strictly speaking. The continuous scale can be thought of as an approximation to the actual scale of measurement.

In summary, variables can be of the following types:

1. Categorical variables
 (a) Ordinal
 (b) Not ordinal

2. Quantitative variables
 (a) Discrete
 (b) Continuous

We will sometimes find it useful to discuss these types separately when considering methods of data analysis.

Samples

A **sample** is a collection of persons or things on which we measure one or more variables. The number of observations in a sample is called the **sample size** and is denoted by the letter **n**. The following are some examples of samples:

The birthweights of 150 babies born in a certain hospital
The sexes of 73 *Cecropia* moths caught in a trap
The flower colors of 81 plants that are progeny of a single parental cross
The number of bacterial colonies in each of six petri dishes

In conceptualizing a sample, it is helpful to be aware of the following elements:

(a) The observed *variable.* For example,
 birthweight
 sex
 flower color
 number of colonies

(b) The *observational unit* (or *case*). For example,

baby

moth

plant

petri dish

(c) The *sample size.* For example,

$n = 150$

$n = 73$

$n = 81$

$n = 6$

Remark: There is some potential for confusion between the statistical meaning of the term *sample* and the sense in which this word is sometimes used in biology. If a biologist draws blood from 20 people and measures the glucose concentration in each, she might say she has 20 samples of blood. However, the statistician says she has *one* sample of 20 gluco se measurements; the sample size is $n = 20$. In the interest of clarity, throughout this book we will use the term *specimen* where a biologist might prefer *sample.* So we would speak of glucose measurements on 20 specimens of blood.

Notation for Variables and Observations

We will adopt a notational convention to distinguish between a variable and an observed value of that variable. We will denote variables by uppercase letters such as Y. We will denote the observations themselves (that is, the data) by lowercase letters such as y. Thus, we distinguish, for example, between $Y =$ birthweight (the variable) and $y = 7.9$ lb (the observation). This distinction will be helpful in explaining some fundamental ideas concerning variability.

Exercises 2.1–2.3

For each of the following settings in Exercises 2.1–2.3, (i) identify the variable(s) in the study, (ii) for each variable tell the type of variable (e.g., categorical and ordinal, discrete, etc.), (iii) identify the observational unit, and (iv) determine the sample size.

2.1   (a) A paleontologist measured the width (in mm) of the last upper molar in 36 specimens of the extinct mammal *Acropithecus rigidus*.

(b) The birthweight, date of birth, and the mother's race were recorded for each of 65 babies.

2.2   (a) A physician measured the height and weight of each of 37 children.

(b) During a blood drive, a blood bank offered to check the cholesterol of anyone who donated blood. A total of 129 persons denoted blood. For each of them, the blood type and cholesterol levels were recorded.

2.3   (a) A biologist measured the number of leaves on each of 25 plants.

(b) A physician recorded the number of seizures that each of 20 patients with severe epilepsy had during an eight-week period.

2.2 FREQUENCY DISTRIBUTIONS: TECHNIQUES FOR DATA

A first step toward understanding a set of data on a given variable is to explore the data and describe the data in summary form. In this chapter we discuss three mutually complementary aspects of summary data description: frequency distributions, measures of center, and measures of dispersion. These tell us about the shape, center, and spread of the data.

Frequency Distributions

A **frequency distribution** is simply a display of the **frequency**, or number of occurrences, of each value in the data set. The information can be presented in tabular form or, more vividly, with a graph. A **bar chart** is a simple graphic showing the categories that a categorical variable takes on and the number of observations in each category for the data in the sample. Here are two examples of frequency distributions for categorical data.

Example 2.1

Color of Poinsettias. Poinsettias can be red, pink, or white. In one investigation of the hereditary mechanism controlling the color, 182 progeny of a certain parental cross were categorized by color.[1] The bar graph in Figure 2.1 is a visual display of the results given in Table 2.1. ■

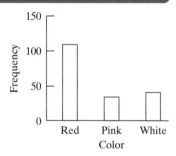

Figure 2.1 Bar chart of color of 182 poinsettias

TABLE 2.1 Color of 182 Poinsettias

Color	Frequency (number of plants)
Red	108
Pink	34
White	40
Total	182

Example 2.2

Clumping of Blood. The strength of reaction of a blood specimen to a certain antigen is categorized into one of six classes according to the degree of clumping of the red blood cells: Class I, complete clumping; Class II, marked clumping; . . . ; Class VI, no clumping. The results for specimens from 70 type-B people are given in Table 2.2 and displayed as a bar graph in Figure 2.2.[2] ■

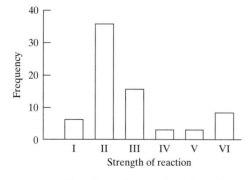

Figure 2.2 Bar chart of strength of clumping reaction of 70 blood specimens

TABLE 2.2 Strength of Clumping Reaction of 70 Blood Specimens

Strength of reaction	Frequency (number of specimens)
I	6
II	35
III	15
IV	3
V	3
VI	8
Total	70

A **dotplot** is a simple graph that can be used to show the distribution of a quantitative variable when the sample size is small. To make a dotplot, we draw a number line covering the range of the data and then put a dot above the number line for each observation, as the following example shows.

Life Expectancy.   Table 2.3 shows the infant mortality rate (infant deaths per 1,000 live births) in each of 12 countries in South America, as of 1999.[3] The distribution is shown in Figure 2.3. ■

Example 2.3

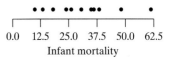

Figure 2.3 Dotplot of infant mortality in 12 South American countries

TABLE 2.3 Infant Mortality in 12 South American Countries	
Country	**Infant Mortality Rate**
Argentina	18.4
Bolivia	62.0
Brazil	35.4
Chile	10.0
Colombia	24.3
Ecuador	30.7
Guyana	48.6
Paraguay	36.4
Peru	39.0
Surinam	26.5
Uruguay	13.5
Venezuela	26.5

When two or more observations take on the same value, we stack the dots in a dotplot on top of each other. This gives an effect similar to the effect of the bars in a bar chart. If we create bars, in place of the stacks of dots, we then have a **histogram**. A histogram is like a bar chart, except that a histogram displays a quantitative variable, which means that there is a natural order and scale for the variable. In a bar chart the amount of space between the bars (if any) is arbitrary, since the data being displayed are categorical. In a histogram the scale of the variable determines the placement of the bars. The following example shows a dotplot and a histogram for a frequency distribution.

Litter Size of Sows.   A group of 36 two-year-old sows of the same breed ($\frac{3}{4}$ Duroc, $\frac{1}{4}$ Yorkshire) were bred to Yorkshire boars. The number of piglets surviving to 21 days of age was recorded for each sow.[4] The results are given in Table 2.4 and displayed as a dotplot in Figure 2.4 and as a histogram in Figure 2.5. ■

Relative Frequency

The frequency scale is often replaced by a **relative frequency** scale:

$$\text{Relative frequency} = \frac{\text{Frequency}}{n}$$

The relative frequency scale is useful if several data sets of different sizes (*n*'s) are to be displayed together for comparison. As another option, a relative frequency can be expressed as a percentage frequency. The shape of the display is not affected by the choice of frequency scale, as the following example shows.

Example 2.4

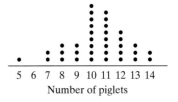

Figure 2.4 Dotplot of number of surviving piglets of 36 sows

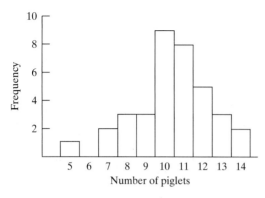

Figure 2.5 Histogram of number of surviving piglets of 36 sows

TABLE 2.4 Number of Surviving Piglets of 36 Sows

Number of piglets	Frequency (number of sows)
5	1
6	0
7	2
8	3
9	3
10	9
11	8
12	5
13	3
14	2
Total	36

Example 2.5

Color of Poinsettias. The poinsettia color distribution of Example 2.1 is expressed as frequency, relative frequency, and percent frequency in Table 2.5 and Figure 2.6. ∎

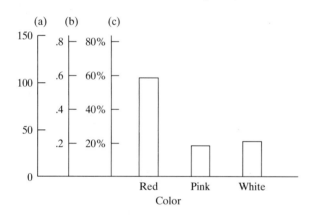

Figure 2.6 Histogram of poinsettia colors on three scales:
(a) Frequency
(b) Relative frequency
(c) Percent frequency

TABLE 2.5 Color of 182 Poinsettias

Color	Frequency	Relative Frequency	Percent Frequency
Red	108	.59	59
Pink	34	.19	19
White	40	.22	22
Total	182	1.00	100

Grouped Frequency Distributions

In the preceding examples, simple ungrouped frequency distributions provided concise summaries of the data. For many data sets, it is necessary to group the data in order to condense the information adequately. (This is usually the case with continuous variables.) The following example shows a grouped frequency distribution.

Example 2.6

Serum CK. Creatine phosphokinase (CK) is an enzyme related to muscle and brain function. As part of a study to determine the natural variation in CK concentration, blood was drawn from 36 male volunteers. Their serum concentrations of CK (measured in u/Li) are given in Table 2.6[5]. Table 2.7 shows these data grouped into **classes**. For instance, the frequency of the class 20–39 is 1, which means that one CK value fell in this range. The grouped frequency distribution is displayed as a histogram in Figure 2.7. ■

TABLE 2.6 Serum CK Values for 36 Men

121	82	100	151	68	58
95	145	64	201	101	163
84	57	139	60	78	94
119	104	110	113	118	203
62	83	67	93	92	110
25	123	70	48	95	42

TABLE 2.7 Frequency Distribution of Serum CK Values for 36 Men

Serum CK (u/Li)	Frequency (number of men)
20–39	1
40–59	4
60–79	7
80–99	8
100–119	8
120–139	3
140–159	2
160–179	1
180–199	0
200–219	2
Total	36

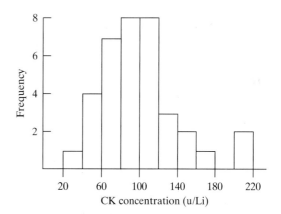

Figure 2.7 Histogram of serum CK concentrations for 36 men

A grouped frequency distribution should display the essential features of the data. For instance, the histogram of Figure 2.7 shows that the average CK value is about 100 U/Li, with the majority of the values falling between 60 and 140 U/Li. In addition, the histogram shows the *shape* of the distribution. Note that the CK values are piled up around a central peak, or **mode**. On either side of this mode, the frequencies decline and ultimately form the **tails** of the distribution. These shape features are labeled in Figure 2.8. The CK distribution is not symmetric but is **skewed to the right**, which means that the right tail is more stretched out than the left.*

Computer note: Computer software is often used to make a histogram. For example, if the data have been entered into the statistical package MINITAB as column C1, then the following command will produce a histogram:

```
MTB > HISTOGRAM C1
```

---

* To help remember which tail of a skewed distribution is the longer tail, think of a skewer. The peak of the distribution corresponds to the handle of the skewer and the tail corresponds to the pointed end. Thus, a distribution that is skewed to the right is one in which the right tail stretches out, like the pointed end of a skewer.

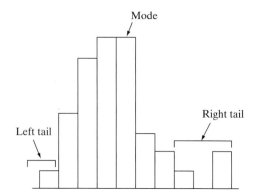

Figure 2.8 Shape features of the CK distribution

When making a histogram, we need to decide how many classes to have and how wide the classes should be. If we use computer software to generate a histogram, the program will choose the number of classes and the class width for us, but most software allows the user to change the number of classes and to specify the class width. If a data set is large and is quite spread out, it is a good idea to look at more than one histogram of the data, as is done in Example 2.7.

Example 2.7

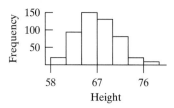

Figure 2.9 Heights of students, using 7 classes (class width = 3)

Heights of Students. A sample of 510 college students were asked how tall they were. Note that they were not measured; rather, they just reported their heights. Figure 2.9 shows the distribution of the self-reported values, using 7 classes and a class width of 3 (inches). By using only 7 classes, the distribution appears to be reasonably symmetric, with a single peak around 66 inches.

Figure 2.10 shows the height data, but in a histogram that uses 18 classes and a class width of 1.1. This view of the data shows two modes—one for women and one for men.

Figure 2.11 shows the height data again, this time using 37 classes, each of width .5. Using such a large number of classes makes the distribution look jagged. In this case, we see an alternating pattern between classes with lots of observations and classes with few observations. In the middle of the distribution we see that there were many students who reported a height of 63 inches, few who reported a height of 63.5 inches, many who reported a height of 64 inches, and so on. It seems that most students round off to the nearest inch! ■

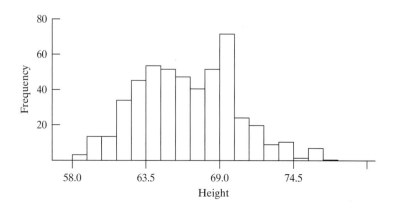

Figure 2.10 Heights of students, using 18 classes (class width = 1.1)

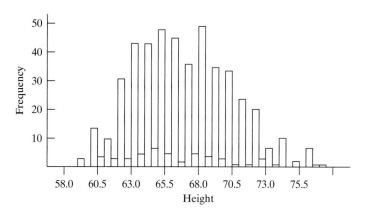

Figure 2.11 Heights of students, using 37 classes (class width = .5)

Computer note: To make a histogram with 37 classes within the MINITAB system, use the command

```
MTB > HISTOGRAM C1;
SUBC > NINTERVAL 37.
```

The semicolon at the end of the first line tells the computer that a subcommand follows. In this case, the subcommand tells the computer that the number of intervals (NINTERVAL) is 37.

Interpreting Areas in a Histogram

A histogram can be looked at in two ways. The tops of the bars sketch out the shape of the distribution. But the *areas* within the bars also have a meaning. The area of each bar is proportional to the corresponding frequency. Consequently, the area of one or several bars can be interpreted as expressing the number of observations in the classes represented by the bars. For example, Figure 2.12 shows a histogram of the CK distribution of Example 2.6. The shaded area is 42% of the total area in all the bars. Accordingly, 42% of the CK values are in the corresponding classes; that is, 15 of 36 or 42% of the values are between 60 u/Li and 100 u/Li.*

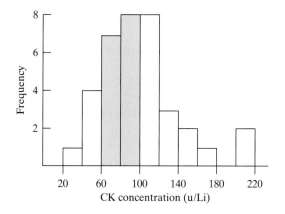

Figure 2.12 Histogram of CK distribution. The shaded area is 42% of the total area and represents 42% of the observations.

---

* Strictly speaking, between 60 u/Li and 99 u/Li, inclusive.

The area interpretation of histograms is a simple but important idea. In our later work with distributions we will find the idea to be indispensable.

Frequency Distributions with Unequal Class Widths

When a grouped frequency distribution is formed, classes are usually chosen to be of equal width. Occasionally classes of unequal width are used, for example, to smooth the distribution in a region where the data are sparse. If the classes are of unequal width, the method for drawing the histogram must be modified. To take an exaggerated example, suppose the last four classes of the CK grouping of Table 2.7 were coalesced into one class: 140–219. This class would have a frequency of 5. If the resulting distribution were plotted using raw frequencies, the histogram would be distorted in shape, as illustrated in Figure 2.13(a). Furthermore, the areas of the bars would no longer be proportional to the frequencies of the corresponding classes. The distortion can be removed by dividing the frequency of the coalesced class by 4, since it is 4 times as wide as the other classes. This gives the histogram of Figure 2.13(b). Notice that in this modified histogram the height of the wide bar is the *average* of the heights of the four narrow bars that it has replaced. This averaging process tends to retain the approximate shape of the original histogram; also, the proportionality between area and frequency is preserved. [Of course, the vertical axis in Figure 2.13(b) can no longer be labeled "frequency"; this will be discussed further in Section 3.5.]

Even if you are not actually drawing a histogram, it is important to check the class widths when interpreting a tabulated distribution. If they are unequal, the frequencies do not indicate the shape of the distribution.

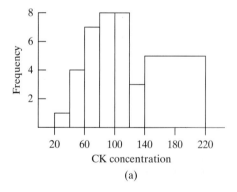 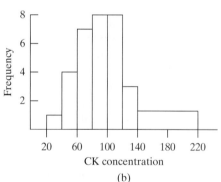

Figure 2.13 Histograms of CK distribution with unequal class widths. (a) Distorted; (b) appropriate.

Stem-and-Leaf Diagrams

Another graphic that is useful for small data sets is a **stem-and-leaf diagram**. The construction of a stem-and-leaf diagram is illustrated in the following example.

Example 2.8 **Radish Growth.** A common biology experiment involves growing radish seedlings under various conditions. In one version of this experiment, a moist paper towel is put into a plastic bag. Staples are put in the bag about one-third of the way from the bottom of the bag; then radish seeds are placed along the staple seam. One group of students kept their radish seed bags in total darkness for three days and then measured the length, in mm, of each radish shoot at the end of the three days. They collected 14 observations; the data are shown in Table 2.8.[6]

TABLE 2.8 Radish Growth, in mm, After 3 Days in Total Darkness				
15	20	11	30	33
20	29	35	8	10
22	37	15	25	

A natural way to organize the data is by putting the observations into groups:

0's:	8			
10's:	15	11	10	15
20's:	20	20	29	22	25
30's:	30	33	35	37

We can then split each data value into a "stem" and a "leaf" as follows:

		Stem	Leaf
8	$\longrightarrow$	0	8
15	$\longrightarrow$	1	5
20	$\longrightarrow$	2	0
30	$\longrightarrow$	3	0

and so on. The smallest observation is an 8, for which the stem is 0—that is, we think of "8" as the two-digit number 08. If we continue adding leaves to the stems as we work through the data, the result is the stem-and-leaf diagram of Figure 2.14.

In the diagram, each stem is accompanied by all of its leaves. It helps to arrange the leaves in order, from smallest to largest, on each stem. Figure 2.15 is an ordered stem-and-leaf diagram of the radish growth data.   ■

Notice that a stem-and-leaf diagram can be viewed as a histogram by turning it sideways. Unlike a histogram, however, the stem-and-leaf diagram retains the original data values.

To construct a stem-and-leaf diagram, simply read through the data values and write down each leaf next to its stem. In general, the last digit of an observation is the leaf and the rest is the stem. For example, if the data values are 123, 137, 142, 125, and so on, then we would use the ones digits (the 3, 7, 2, and 5) as leaves and the hundreds and tens digits together (i.e., 12, 13, 14) as the stems. It may be necessary to round the data so that this principle will produce a satisfactory display. Suppose, for instance, that the radish growth data had been measured to the nearest .1 mm, and the values were 15.3, 20.2, 10.8, . . . ; then we would want to round the data to one decimal place before constructing the stem-and-leaf diagram.

Note that the construction of a stem-and-leaf diagram does not depend on the location of the decimal point in the data. For instance, if the radish growth data

```
0 | 8
1 | 5 1 0 5
2 | 0 0 9 2 5
3 | 0 3 5 7
```
Key: 1|5 means 15 mm.

Figure 2.14 Stem-and-leaf diagram for radish growth in darkness

```
0 | 8
1 | 0 1 5 5
2 | 0 0 2 5 9
3 | 0 3 5 7
```
Key: 1|5 means 15 mm.

Figure 2.15 Stem-and-leaf diagram for radish growth in darkness with the leaves arranged in order

```
0 | 8
1 | 0 1
1 | 5 5
2 | 0 0 2
2 | 5 9
3 | 0 3
3 | 5 7
```
Key: 1|5 means 15 mm.

Figure 2.16 Stem-and-leaf diagram for radish growth in darkness using split stems

of Table 2.8 were expressed in cm rather than in mm, then the observations would be 1.5, 2.0, 1.1, ... but the (ordered) stem-and-leaf diagram would be exactly the same as Figure 2.15; the key indicates the scale of measurement.

It is sometimes helpful to stretch out the scale in a stem-and-leaf diagram by splitting the stems in half, with leaves 0–4 going in the lower half and leaves 5–9 going in the upper half of each stem. Figure 2.16 shows this technique applied to the radish data.

Computer note: To make a stem-and-leaf diagram within the MINITAB system, for data stored in column 1, use the command

```
MTB > STEM C1
```

MINITAB will choose how to split the stems. This choice can be overridden with the subcommand "INCREMENT."

Another type of stem-and-leaf diagram is a back-to-back stem-and-leaf diagram, which allows us to compare two distributions, as in Example 2.9.

Example 2.9

```
0 | 4 9
1 | 0 0 1 5 5
2 | 0 0 0 1 2 5 7
```
Key: 1|5 means 15 mm.

Figure 2.17 Stem-and-leaf diagram for radish growth in 12 light/12 dark with the leaves arranged in order

```
    9 4 | 0 | 8
  5 5 1 0 0 | 1 | 0 1 5 5
7 5 2 1 0 0 0 | 2 | 0 0 2 5 9
            | 3 | 0 3 5 7
```
Key: 1|5 means 15 mm.

Figure 2.18 Back-to-back stem-and-leaf diagram for radish growth: light/dark versus total darkness

```
        4 | 0 |
        9 | 0 | 8
      1 0 0 | 1 | 0 1
        5 5 | 1 | 5 5
    2 1 0 0 0 | 2 | 0 0 2
        7 5 | 2 | 5 9
            | 3 | 0 3
            | 3 | 5 7
```
Key: 1|5 means 15 mm.

Figure 2.19 Back-to-back stem-and-leaf diagram with split stems for radish growth: light/dark versus total darkness

Radish Growth.   The data shown in Table 2.8 are for radish seedlings that were kept in total darkness for three days. In a second part of that experiment, the students moved some seedlings back and forth between light for 12 hours and darkness for 12 hours, over the same three-day period. The data for the "12 light/12 dark" seedlings are shown in Table 2.9. Figure 2.17 shows the distribution of these data. Figure 2.18 shows the two distributions in a back-to-back stem-and-leaf diagram. The stems are in the middle of the diagram, with the "darkness" distribution building out to the right and the "12 light/12 dark" distribution building out to the left. Figure 2.19 uses split stems in a back-to-back stem-leaf-diagram to help us see the difference between the two distributions.

TABLE 2.9 Radish Growth, in mm, After 3 Days of 12 Hours of Light Followed by 12 Hours of Darkness				
10	15	22	25	9
11	20	21	27	20
15	4	10	20	

We can see that there is considerable overlap between the two distributions. Nonetheless, radish seedlings grown in total darkness tend to grow more than do seedlings grown in light and darkness. The "light and darkness" distribution is shifted roughly 10 mm, toward lower values, in comparison to the "total darkness" distribution. ∎

In a research report, a frequency distribution would usually be presented as a table or a histogram. However, the stem-and-leaf diagram is a useful working tool during the analysis of data and gives a quick and convenient way to display small data sets.

2.3 FREQUENCY DISTRIBUTIONS: SHAPES AND EXAMPLES

When discussing a set of data, we want to describe the shape, center, and spread of the distribution. In this section we concentrate on the shapes of frequency distributions and illustrate some of the diversity of distributions encountered in the life sciences. The shape of a distribution can be indicated by a smooth curve that approximates the histogram, as shown in Figure 2.20.

Figure 2.20 Approximation of a histogram by a smooth curve

Some distributional shapes are shown in Figure 2.21. A common shape for biological data is **unimodal** (has one mode) and is somewhat skewed to the right, as in (c). Approximately bell-shaped distributions, as in (a), also occur. Sometimes a distribution is symmetric but differs from a bell in having long tails; an exaggerated version is shown in (b). Left-skewed (d) and exponential (e) shapes are less common. **Bimodality** (two modes), as in (f), can indicate the existence of two distinct subgroups of observational units.

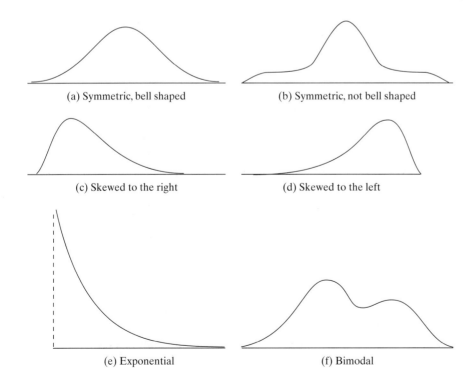

(a) Symmetric, bell shaped

(b) Symmetric, not bell shaped

(c) Skewed to the right

(d) Skewed to the left

(e) Exponential

(f) Bimodal

Figure 2.21 Shapes of distributions

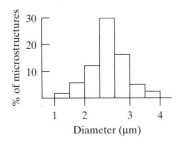

Figure 2.22 Sizes of microfossils

Notice that the shape characteristics we are emphasizing, such as number of modes and degree of symmetry, are *scale free;* that is, they are not affected by the arbitrary choices of vertical and horizontal scale in plotting the distribution. By contrast, a characteristic such as whether the distribution appears short and fat, or tall and skinny, is affected by how the distribution is plotted and so is not an inherent feature of the biological variable.

The following three examples illustrate biological frequency distributions with various shapes. In the first example, the shape provides evidence that the distribution is in fact biological rather than nonbiological.

Example 2.10

Microfossils. In 1977 paleontologists discovered microscopic fossil structures, resembling algae, in rocks 3.5 billion years old. A central question was whether these structures were biological in origin. One line of argument focused on their size distribution, which is shown in Figure 2.22. This distribution, with its unimodal and rather symmetric shape, resembles that of known microbial populations but not that of known nonbiological structures.[7] ■

Example 2.11

Cell Firing Times. A neurobiologist observed discharges from rat muscle cells grown in culture together with nerve cells. The time intervals between 308 successive discharges were distributed as shown in Figure 2.23. Note the exponential shape of the distribution.[8] ■

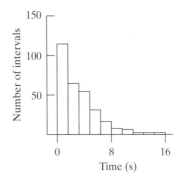

Figure 2.23 Time intervals between electrical discharges in rat muscle cells

Example 2.12

Brain Weight. In 1888 P. Topinard published data on the brain weights of hundreds of French men and women. The data for males and females are shown in Figure 2.24(a) and (b). The male distribution is fairly symmetric and bell shaped; the female distribution is somewhat skewed to the right. Part (c) of the figure shows the brain weight distribution for males and females combined. This combined distribution is slightly bimodal.[9] ■

Sources of Variation

In interpreting biological data, it is helpful to be aware of sources of variability. The variation among observations in a data set often reflects the combined effects of several underlying factors. The following two examples illustrate such situations.

Example 2.13

Weights of Beans. In a classic experiment to distinguish environmental from genetic influence, a geneticist weighed seeds of the princess bean *Phaseolus vulgaris.* Figure 2.25 shows the weight distributions of (a) 5,494 seeds from a com-

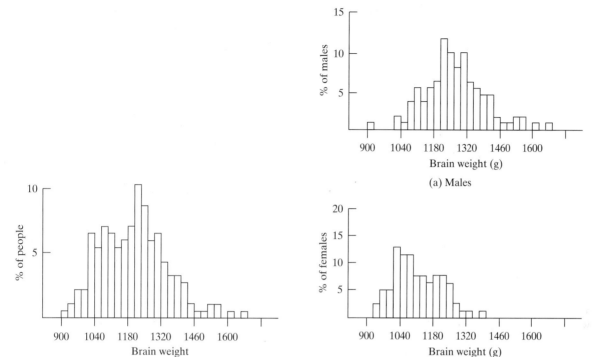

Figure 2.24 Brain weights

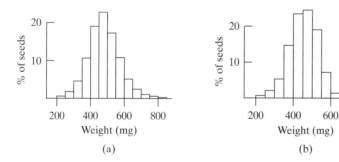

Figure 2.25 Weights of princess beans. (a) From an open-bred population; (b) From an inbred line.

mercial seed lot, and (b) 712 seeds from a highly inbred line that was derived from a single seed from the original lot. The variability in (a) is due to both environmental and genetic factors; in (b), because the plants are nearly genetically identical, the variation in weights is due largely to environmental influence.[10] Thus, there is less variability in the inbred line. ∎

Serum ALT. Alanine aminotransferase (ALT) is an enzyme found in most human tissues. Part (a) of Figure 2.26 shows the serum ALT concentrations for 129 adult volunteers. The following are potential sources of variability among the measurements:

Example 2.14

   1. Interindividual
      (a) Genetic
      (b) Environmental

2. Intraindividual
   (a) Biological: changes over time
   (b) Analytical: imprecision in assay

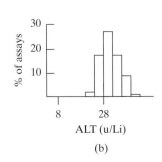

Figure 2.26 Distribution of serum ALT measurements (a) for 129 volunteers; (b) for 109 assays of the same specimen

The effect of the last source—analytical variation—can be seen in Figure 2.26(b), which shows the frequency distribution of 109 assays of the *same* specimen of serum; the figure shows that the ALT assay is fairly imprecise.[11] ■

Exercises 2.4–2.13

2.4 A paleontologist measured the width (in mm) of the last upper molar in 36 specimens of the extinct mammal *Acropithecus rigidus*. The results were as follows:[12]

6.1	5.7	6.0	6.5	6.0	5.7
6.1	5.8	5.9	6.1	6.2	6.0
6.3	6.2	6.1	6.2	6.0	5.7
6.2	5.8	5.7	6.3	6.2	5.7
6.2	6.1	5.9	6.5	5.4	6.7
5.9	6.1	5.9	5.9	6.1	6.1

(a) Construct a frequency distribution and display it as a table and as a histogram.
(b) Describe the shape of the distribution.

2.5 In a study of schizophrenia, researchers measured the activity of the enzyme monoamine oxidase (MAO) in the blood platelets of 18 patients. The results (expressed as nmols benzylaldehyde product per 108 platelets) were as follows:[13]

6.8	8.4	8.7	11.9	14.2	18.8
9.9	4.1	9.7	12.7	5.2	7.8
7.8	7.4	7.3	10.6	14.5	10.7

Construct a dotplot of the data.

2.6 Consider the data presented in Exercise 2.5. Construct a frequency distribution and display it as a table and as a histogram.

2.7 A dendritic tree is a branched structure that emanates from the body of a nerve cell. As part of a study of brain development, 36 nerve cells were taken from the brains of newborn guinea pigs. The investigators counted the number of dendritic branch segments emanating from each nerve cell. The numbers were as follows:[14]

23	30	54	28	31	29	34	35	30
27	21	43	51	35	51	49	35	24
26	29	21	29	37	27	28	33	33
23	37	27	40	48	41	20	30	57

(a) Construct a stem-and-leaf diagram of the data.
(b) Construct a dotplot of the data.

2.8 Consider the data presented in Exercise 2.7. Construct a frequency distribution and display it as a table and as a histogram.

2.9 The total amount of protein produced by a dairy cow can be estimated from periodic testing of her milk. The following are the total annual protein production values (lb) for 28 two-year-old Holstein cows. Diet, milking procedures, and other conditions were the same for all the animals.[15]

425	481	477	434	410	397	438
545	528	496	502	529	500	465
539	408	513	496	477	445	546
471	495	445	565	499	508	426

Construct a frequency distribution and display it as a table and as a histogram.

2.10 For each of 31 healthy dogs, a veterinarian measured the glucose concentration in the anterior chamber of the right eye, and also in the blood serum. The following data are the anterior chamber glucose measurements, expressed as a percentage of the blood glucose.[16]

81	85	93	93	99	76	75	84
78	84	81	82	89	81	96	82
74	70	84	86	80	70	131	75
88	102	115	89	82	79	106	

Construct a frequency distribution and display it as a table and as a histogram.

2.11 Refer to the glucose data of Exercise 2.10. Construct a stem-and-leaf display of the data.

2.12 In a behavioral study of the fruitfly *Drosophila melanogaster*, a biologist measured, for individual flies, the total time spent preening during a six-minute observation period. The following are the preening times (s) for 20 flies:[17]

34	24	10	16	52
76	33	31	46	24
18	26	57	32	25
48	22	48	29	19

(a) Construct a stem-and-leaf display for these data.
(b) Construct a dotplot of the data.
(c) Describe the shape of the distribution.

2.13 *(Computer problem)* Trypanosomes are parasites that cause disease in humans and animals. In an early study of trypanosome morphology, researchers measured the lengths of 500 individual trypanosomes taken from the blood of a rat. The results are summarized in the accompanying frequency distribution.[18]

Length (μm)	Frequency (number of individuals)	Length (μm)	Frequency (number of individuals)
15	1	27	36
16	3	28	41
17	21	29	48
18	27	30	28
19	23	31	43
20	15	32	27
21	10	33	23
22	15	34	10
23	19	35	4
24	21	36	5
25	34	37	1
26	44	38	1

(a) Construct a histogram of the data using 24 classes (i.e., one class for each integer length, from 15 to 38).

(b) What feature of the histogram suggests the interpretation that the 500 individuals are a mixture of two distinct types?

(c) Construct a histogram of the data using only six classes. Discuss how this histogram gives a qualitatively different impression than the histogram from part (a).

2.4 DESCRIPTIVE STATISTICS: MEASURES OF CENTER

For categorical data, the frequency distribution provides a concise and complete summary of a sample. For quantitative variables, the frequency distribution can be usefully supplemented by a few numerical measures. A numerical measure calculated from data is called a **statistic**. **Descriptive statistics** are statistics that describe a set of data. Usually the descriptive statistics for a sample are calculated in order to provide information about a population of interest (see Section 2.8). In this section we discuss measures of the center of the data. There are several different ways to define the "center" or "typical value" of the observations in a sample. We will consider the two most widely used measures of center: the mean and the median.

The Mean

The most familiar measure of center is the ordinary average or **mean** (sometimes called the arithmetic mean). The mean of a sample (or "the sample mean") is the sum of the observations divided by the number of observations. If we denote a variable by Y, then we denote the observations in a sample by $y_1, y_2, \ldots, y_n$ and we denote the mean of the sample by the symbol $\bar{y}$ (read "y-bar). Example 2.15 illustrates this notation.

Example 2.15

Weight Gain of Lambs.   The following are the two-week weight gains (lb) of six young lambs of the same breed who had been raised on the same diet:[19]

11   13   19   2   10   1

Here $y_1 = 11$, $y_2 = 13$, and so on, and $y_6 = 1$. The sum of the observations is $11 + 13 + \ldots + 1 = 56$. We can write this using "summation notation" as $\Sigma\, y_i = 56$. The symbol $\Sigma\, y_i$ means to "add up the y_i's." Thus, when $n = 6$, $\Sigma\, y_i = y_1 + y_2 + y_3 + y_4 + y_5 + y_6$. In this case we get $\Sigma\, y_i = 11 + 13 + 19 + 2 + 10 + 1 = 56$.

The mean weight gain of the 6 lambs in this sample is

$$\bar{y} = \frac{11 + 13 + 19 + 2 + 10 + 1}{6}$$

$$= \frac{56}{6}$$

$$= 9.33 \text{ lb*}$$

> ### The Sample Mean
> The general definition of the sample mean is
> $$\bar{y} = \frac{\Sigma\, y_i}{n}$$
> where the y_i's are the observations in the sample and n is the sample size (that is, the number of y_i's).

The mean is the "point of balance" of the data. Figure 2.27 shows a dotplot of the lamb weight-gain data, along with the location of $\bar{y}$. If the data points were children on a weightless seesaw, then the seesaw would exactly balance if supported at $\bar{y}$. ■

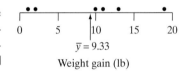

Figure 2.27 Plot of the lamb weight-gain data

The difference between a data point and the mean is called a deviation: deviation$_i = y_i - \bar{y}$. The mean has the property that the sum of the deviations from the mean is zero—that is, $\Sigma(y_i - \bar{y}) = 0$. In this sense, the mean is a center of the distribution.

Weight Gain of Lambs. For the lamb weight-gain data, the deviations are as follows:

Example 2.16

$$\text{deviation}_1 = y_1 - \bar{y} = 11 - 9.33 = 1.67$$
$$\text{deviation}_2 = y_2 - \bar{y} = 13 - 9.33 = 3.67$$
$$\text{deviation}_3 = y_3 - \bar{y} = 19 - 9.33 = 9.67$$
$$\text{deviation}_4 = y_4 - \bar{y} = 2 - 9.33 = -7.33$$
$$\text{deviation}_5 = y_5 - \bar{y} = 10 - 9.33 = 0.67$$
$$\text{deviation}_6 = y_6 - \bar{y} = 1 - 9.33 = -8.33$$

The sum of the deviations is $\Sigma(y_i - \bar{y}) = 1.67 + 3.67 + 9.67 - 7.33 + 0.67 - 8.33 = 0$. ■

---

* We will sometimes round values for clarity of presentation. Thus, we write $56/6 = 9.33$, rather than 9.33333 or $9.3\bar{3}$.

The Median

The sample **median** is the value that most nearly lies in the middle of the sample. To find the median, first arrange the observations in increasing order. In the array of ordered observations, the median is the middle value (if n is odd) or midway between the two middle values (if n is even). Example 2.17 illustrates these definitions.

Example 2.17 **Weight Gain of Lambs.**

(a) For the weight-gain data of Example 2.15, the ordered observations are

$$1 \quad 2 \quad 10 \quad 11 \quad 13 \quad 19$$

The median weight gain is

$$\text{Median} = \frac{10 + 11}{2} = 10.5 \, \text{lb}$$

(b) Suppose the sample contained one more lamb, with the seven ranked observations as follows:

$$1 \quad 2 \quad 10 \quad 10 \quad 11 \quad 13 \quad 19$$

For this sample, the median weight gain is

$$\text{Median} = 10 \, \text{lb}$$

(Notice that in this example there are two lambs whose weight gain is equal to the median. The fourth observation—the second 10—is the median.) ■

A more formal way to define the median is in terms of rank position in the ordered array (counting the smallest observation as rank 1, the next as 2, and so on). The rank position of the median is equal to

$$(.5)(n + 1)$$

Thus, if $n = 7$, we calculate $(.5)(n + 1) = 4$, so that the median is the fourth largest observation; if $n = 6$, we have $(.5)(n + 1) = 3.5$, so that the median is midway between the third and fourth largest observations. Note that the formula $(.5)(n + 1)$ does not give the median; it gives the location of the median within the ordered list of the data.

Robustance. A statistic is said to be **robust** or **resistant** if the value of the statistic is relatively unaffected by changes in a small portion of the data, even if the changes are dramatic ones. The median is a robust statistic, but the mean is not robust because it can be greatly shifted by changes in even one observation. Example 2.18 illustrates this behavior.

Example 2.18 **Weight Gain of Lambs.** Recall that for the lamb weight-gain data

$$1 \quad 2 \quad 10 \quad 11 \quad 13 \quad 19$$

we found

$$\bar{y} = 9.3 \text{ and Median} = 10.5$$

Suppose now that the observation 19 is changed, or even omitted. How would the mean and median be affected? You can visualize the effect by imagining moving or removing the right-hand dot in Figure 2.27. Clearly the mean could change a great deal; the median would generally be less affected. For instance,

> If the 19 is changed to 12, the mean becomes 8.2 and the median does not change.
> If the 19 is omitted, the mean becomes 7.4 and the median becomes 10.

The preceding changes are not wild ones; that is, the changed samples might well have arisen from the same feeding experiment. Of course, a huge change, such as changing the 19 to 100, would shift the mean drastically; note that it would not shift the median at all. ■

Visualizing the Mean and Median

We can visualize the mean and the median in relation to the histogram of a distribution. The median divides the area under the histogram roughly in half because it divides the observations roughly in half ["roughly" because some observations may be tied at the median, as in Example 2.17(b), and because the observations within each class are not uniformly distributed across the class]. The mean can be visualized as the point of balance of the histogram: If the histogram were made out of plywood, it would roughly balance if supported at the mean.

If the frequency distribution is symmetric, the mean and the median are equal and fall in the center of the distribution. If the frequency distribution is skewed, both measures are pulled toward the longer tail, but the mean is usually pulled farther than the median. The effect of skewness is illustrated by the following example.

Cricket Singing Times.   Male Mormon crickets *(Anabrus simplex)* sing to attract mates. A field researcher measured the duration of 51 unsuccessful songs—that is, the time until the singing male gave up and left his perch.[20] Figure 2.28 shows the histogram of the 51 singing times. Table 2.10 gives the raw data. The median is 3.7 min and the mean is 4.3 min. The discrepancy between these measures is due largely to the long straggly tail of the distribution; the few unusually long singing times influence the mean but not the median. ■

Example 2.19

Figure 2.28 Histogram of cricket singing times

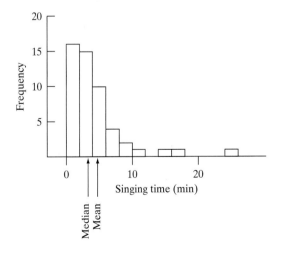

TABLE 2.10 51 Cricket Singing Times (min)							
4.3	3.9	17.4	2.3	.8	1.5	.7	3.7
24.1	9.4	5.6	3.7	5.2	3.9	4.2	3.5
6.6	6.2	2.0	.8	2.0	3.7	4.7	
7.3	1.6	3.8	.5	.7	4.5	2.2	
4.0	6.5	1.2	4.5	1.7	1.8	1.4	
2.6	.2	.7	11.5	5.0	1.2	14.1	
4.0	2.7	1.6	3.5	2.8	.7	8.6	

Mean Versus Median

Both the mean and the median are usually reasonable measures of the center of a data set. The mean is related to the sum; for example, if the mean weight gain of 100 lambs is 9 lb, then the total weight gain is 900 lb, and this total may be of primary interest since it translates more or less directly into profit for the farmer. In some situations the mean makes very little sense. Suppose, for example, that the observations are survival times of cancer patients on a certain treatment protocol, and that most patients survive less than 1 year, while a few respond well and survive for 5 or even 10 years. In this case, the mean survival time might be greater than the survival time of most patients; the median would more nearly represent the experience of a "typical" patient. Note also that the mean survival time cannot be computed until the last patient has died; the median does not share this disadvantage. Situations in which the median can readily be computed, but the mean cannot, are not uncommon in bioassay, survival, and toxicity studies.

We have noted that the median is more resistant than the mean. If a data set contains a few observations rather distant from the main body of the data—that is, a long "straggly" tail—then the mean may be unduly influenced by these few unusual observations. Thus, the "tail" may "wag the dog"—an undesirable situation. In such cases, the resistance of the median may be advantageous.

An advantage of the mean is that in some circumstances it is more efficient than the median. Efficiency is a technical notion in statistical theory; roughly speaking, a method is efficient if it takes full advantage of all the information in the data. Partly because of its efficiency, the mean has played a major role in classical methods in statistics.

Exercises 2.14–2.29

2.14 Invent a sample of size 5 for which the sample mean is 20 and not all the observations are equal.

2.15 Invent a sample of size 5 for which the sample mean is 20 and the sample median is 15.

2.16 A researcher applied the carcinogenic (cancer-causing) compound benzo(a)pyrene to the skin of five mice and measured the concentration in the liver tissue after 48 hours. The results (nmol/g) were as follows:[21]

6.3   5.9   7.0   6.9   5.9

Determine the mean and the median.

2.17 Consider the data from Exercise 2.16. Do the calculated mean and median support the claim that, in general, liver tissue concentration after 48 hours is 6.3 nmol/g?

2.18 Six men with high serum cholesterol participated in a study to evaluate the effects of diet on cholesterol level. At the beginning of the study their serum cholesterol levels (mg/dLi) were as follows:[22]

366   327   274   292   274   230

Determine the mean and the median.

2.19 Consider the data from Exercise 2.18. Suppose an additional observation equal to 400 were added to the sample. What would be the mean and the median of the seven observations?

2.20   The weight gains of beef steers were measured over a 140-day test period. The average daily gains (lb/day) of 9 steers on the same diet were as follows:[23]

3.89   3.51   3.97   3.31   3.21
3.36   3.67   3.24   3.27

Determine the mean and median.

2.21   Consider the data from Exercise 2.20. Do the calculated mean and median support the claim that, in general, steers gain 3.5 lb/day? Do the data support a claim of 4.0 lb/day?

2.22   Consider the data from Exercise 2.20. Suppose an additional observation equal to 2.46 were added to the sample. What would be the mean and the median of the 10 observations?

2.23   As part of a classic experiment on mutations, ten aliquots of identical size were taken from the same culture of the bacterium *E. coli*. For each aliquot, the number of bacteria resistant to a certain virus was determined. The results were as follows:[24]

14   15   13   21   15
14   26   16   20   13

(a) Construct a frequency distribution of these data and display it as a histogram.
(b) Determine the mean and the median of the data and mark their locations on the histogram.

2.24   The accompanying table gives the litter size (number of piglets surviving to 21 days) for each of 36 sows (as in Example 2.4). Determine the median litter size.

Number of piglets	Frequency (Number of sows)
5	1
6	0
7	2
8	3
9	3
10	9
11	8
12	5
13	3
14	2
Total	36

2.25   Consider the data from Exercise 2.24. Determine the mean of the 36 observations.

[*Hint*: Note that there is one 5 but there are two 7's, three 8's, and so on. Thus, $\Sigma y_i = 5 + 7 + 7 + 8 + 8 + 8 + \ldots = 5 + 2(7) + 3(8) + \ldots$.]

2.26   Here is a histogram.

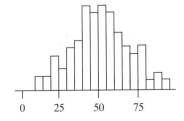

(a) Estimate the median of the distribution.

(b) Estimate the mean of the distribution.

2.27 Consider the histogram from Exercise 2.26. By "reading" the histogram, estimate the percentage of observations that are less than 40. Is this percentage closest to 15%, 25%, 35%, or 45%? *Note:* The frequency scale is not given for this histogram, because there is no need to calculate the number of observations in each class. Rather, the percentage of observations that are less than 40 can be estimated by looking at area.

2.28 Here is a histogram.

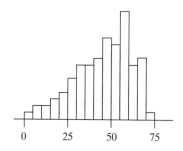

(a) Estimate the median of the distribution.

(b) Estimate the mean of the distribution.

2.29 Consider the histogram from Exercise 2.28. By "reading" the histogram, estimate the percentage of observations that are greater than 55. Is this percentage closest to 15%, 25%, 35%, or 45%? *Note:* The frequency scale is not given for this histogram, because there is no need to calculate the number of observations in each class. Rather, the percentage of observations that are greater than 55 can be estimated by looking at area.

2.5 BOXPLOTS

One of the most efficient graphics, both for examining a single distribution and for making comparisons between distributions, is known as a boxplot, which is the topic of this section. Before discussing boxplots, however, we need to discuss quartiles.

Quartiles and the Interquartile Range

The median of a distribution splits the distribution into two parts, a lower part and an upper part. The **quartiles** of a distribution divide each of these parts in half, thereby dividing the distribution into four quarters. The **first quartile**, denoted by Q_1, is the median of the data values in the lower half of the data set. The **third quartile**, denoted by Q_3, is the median of the data values in the upper half of the data set.* The following example illustrates these definitions.

_____

* Some authors use other definitions of quartiles, as does some computer software. A common alternative definition is to say that the first quartile has rank position $(.25)(n + 1)$ and that the third quartile has rank position $(.75)(n + 1)$. Thus, if $n = 10$, the first quartile would have rank position $(.25)(11) = 2.75$—that is, to find the first quartile we would have to interpolate between the second and third largest observations. If n is large, then there is little practical difference between the definitions that various authors use.

Blood Pressure.   The systolic blood pressures (mm Hg) of seven middle-aged men were as follows:[25]

Example 2.20

$$151 \quad 124 \quad 132 \quad 170 \quad 146 \quad 124 \quad 113$$

Putting these values in rank order, the sample is

$$113 \quad 124 \quad 124 \quad 132 \quad 146 \quad 151 \quad 170$$

The median is the fourth largest observation, which is 132. There are three data points in the lower part of the distribution: 113, 124, and 124. The median of these three values is 124. Thus, the first quartile, Q_1, is 124.

Likewise, there are three data points in the upper part of the distribution: 146, 151 and 170. The median of these three values is 151. Thus, the third quartile, Q_3, is 151.

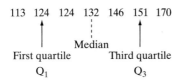

Note that the median is not included in either the lower part nor the upper part of the distribution. If the sample size, n, is even, then exactly half of the observations are in the lower part of the distribution and half are in the upper part.

The **interquartile range** is the difference between the first and third quartiles and is abbreviated as **IQR**: IQR $= Q_3 - Q_1$. For the blood pressure data in Example 2.20, the IQR is $151 - 124 = 27$.

Pulse.   The pulses of twelve college students were measured.[26] Here are the data, arranged in order, with the position of the median indicated by a dashed line:

Example 2.21

$$62 \quad 64 \quad 68 \quad 70 \quad 70 \quad 74 \ \vdots \ 74 \quad 76 \quad 76 \quad 78 \quad 78 \quad 80$$

The median is $\dfrac{74 + 74}{2} = 74$. There are 6 observations in the lower part of the distribution: 62, 64, 68, 70, 70, 74. Thus, the first quartile is the average of the third and fourth largest data values:

$$Q_1 = \frac{68 + 70}{2} = 69$$

There are 6 observations in the upper part of the distribution: 74, 76, 76, 78, 78, 80. Thus, the third quartile is the average of the ninth and tenth largest data values (the third and fourth values in the upper part of the distribution):

$$Q_3 = \frac{76 + 78}{2} = 77$$

Thus, the interquartile range is

$$\text{IQR} = 77 - 69 = 8$$

We have

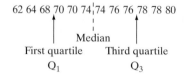

The minimum pulse value is 62 and the maximum is 80. ■

The minimum, the maximum, the median, and the quartiles, taken together, are referred to as the **five-number summary** of the data.

Boxplots

A **boxplot** is a visual representation of the five-number summary. To make a box-plot, we first make a number line; then we mark the positions minimum, Q_1, the median, Q_3, and the maximum:

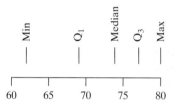

Next, we make a box connecting the quartiles:

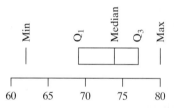

Note that the interquartile range is equal to the length of the box. Finally, we extend "whiskers" from Q_1 down to the minimum and from Q_3 up to the maximum:

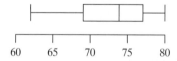

A boxplot gives a quick visual summary of the distribution. We can immediately see where the center of the data is, from the line within the box that locates the median. We see the spread of the total distribution, from the minimum up to the maximum, as well as the spread of the middle half of the distribution—the interquartile range—from the length of the box. The boxplot also gives an indication of the shape of the distribution; the preceding boxplot has a long lower whisker, indicating that the distribution is skewed to the left. Example 2.22 shows a boxplot for the radish growth data considered earlier.

Radish Growth. The stem-and-leaf diagram of Figure 2.29 represents the data on radish growth in darkness from Example 2.8. The quartiles have been circled; they are $Q_1 = 15$ and $Q_3 = 30$. The median, 21, is represented with a dashed line. Figure 2.30 shows a boxplot of the same data. Figure 2.31 shows a vertical boxplot of the same data. ∎

Example 2.22

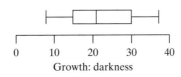

Key: 1|5 means 15 mm.

Figure 2.29 Ordered stem-and-leaf diagram of data on radish growth in darkness. The quartiles are circled and the median is represented with a dashed line.

Figure 2.30 Boxplot of data on radish growth in darkness

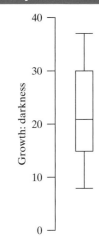

Figure 2.31 Boxplot of data on radish growth in darkness

Parallel Boxplots

One of the advantages to using boxplots is that it is easy to compare distributions by creating parallel boxplots. With boxplots that are drawn on the same scale, we can compare two or more distributions quickly. We get a visual impression of how the medians of the distributions compare, as well as how the spreads of the distributions compare.

Radish Growth. In Example 2.9 we used back-to-back stem-and-leaf diagrams to compare radish growth in total darkness to growth in 12 hours of light followed by 12 hours of darkness. There were actually three parts to the experiment described in Example 2.9. In the third part of the experiment, the students grew radish seedlings in constant light. Figure 2.32 shows three parallel boxplots, one for each of the three data sets. From these boxplots we can see how light inhibits growth of the radish seedlings by noting both that the distributions shift downward as more light is added. Also, the interquartile range of the light distribution is much smaller than the IQRs of the other distributions. The third quartile of the "light" distribution is equal to the first quartile of the "12 light/12 dark" distribution and is less than the first quartile of the "darkness" distribution. ∎

Outliers

Sometimes a data point differs so much from the rest of the data that it doesn't seem to belong with the other data. Such a point is called an **outlier**. An outlier might occur because of a recording error or typographical error when the data are recorded, because of an equipment failure during an experiment, or for many other reasons. Outliers are the most interesting points in a data set. Sometimes outliers tell us about a problem with the experimental protocol (e.g., an equipment failure or a failure of a patient to take his or her medication consistently during a medical trial). At other times an outlier might alert us to the fact that a special circumstance has happened (e.g., an abnormally high or low value on a medical test could indicate the presence of a disease in a patient).

Example 2.23

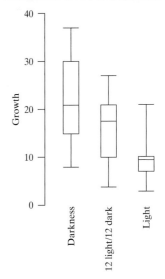

Figure 2.32 Boxplots of data on radish growth under three conditions: constant darkness, half light and half darkness, and constant light

People often use the term *outlier* informally. There is, however, a common definition of *outlier* in statistical practice. To give a definition of outlier, we first discuss what are known as fences. The **lower fence** of a distribution is

$$\text{lower fence} = Q_1 - 1.5 * \text{IQR}$$

The **upper fence** of a distribution is

$$\text{upper fence} = Q_3 + 1.5 * \text{IQR}$$

This means that the fences are located 1.5 IQRs (i.e., 1.5 * the length of the box) beyond the end of the box in a boxplot.

Note that the fences need not be data values; indeed, there might be no data near the fences. The fences just locate limits within the sample distribution. These limits give us a way to define outliers. *An outlier is a data point that falls outside of the fences.* That is, if

$$\text{data point} < Q_1 - 1.5 * \text{IQR}$$

or

$$\text{data point} > Q_3 + 1.5 * \text{IQR}$$

then we call the point an outlier.

Example 2.24

Pulse. In Example 2.21 we saw that $Q_1 = 69$, $Q_3 = 77$, and IQR $= 8$. Thus, the lower fence is $69 - 1.5 * 8 = 69 - 12 = 57$. Any point less than 57 would be an outlier. The upper fence is $77 + 1.5 * 8 = 77 + 12 = 89$. Any point greater than 89 would be an outlier. Since there are no points less than 57 nor greater than 89, there are no outliers in this data set. ■

Example 2.25

Radish Growth in Light. Figure 2.32 shows the distribution of growth for radish seedlings under three conditions. One of the three conditions was constant light. There are 14 seedlings in this set of data. The observations, in order, are

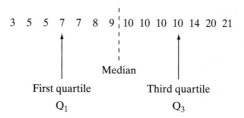

Thus, the median is $\dfrac{9 + 10}{2} = 9.5$, Q_1 is 7, and Q_3 is 10. The interquartile range is IQR $= 10 - 7 = 3$. The lower fence is $7 - 1.5 * 3 = 7 - 4.5 = 2.5$, so any point less than 2.5 would be an outlier. The upper fence is $10 + 1.5 * 3 = 10 + 4.5 = 14.5$, so any point greater than 14.5 is an outlier. Thus, the two largest observations in this data set are outliers: 20 and 21. ■

The method we have defined for identifying outliers allows the bulk of the data to determine how extreme an observation must be before we consider it to be an outlier, since the quartiles and the IQR are determined from the data themselves. Thus, a point that is an outlier in one data set might not be an outlier in another data set. For example, the observations of 20 and 21 are outliers in the "light" distribution, but they would not be outliers in the "12 light/12 dark" distribution. We label a point as an outlier if it is unusual relative to the inherent variability in the entire data set.

After an outlier has been identified, people are often tempted to remove the outlier from the data set. In general, this is not a good idea. If we can identify that an outlier occurred due to an equipment error, for example, then we have good reason to remove the outlier before analyzing the rest of the data. However, quite often outliers appear in data sets without any identifiable, external reason for them. In such cases, we simply proceed with our analysis, aware that there is an outlier present. In some cases, we might want to calculate the mean, for example, with and without the outlier and then report both calculations, to show the effect of the outlier in the overall analysis. This is preferable to removing the outlier, which obscures the fact that there was an unusual data point present. In presenting data graphically, we can draw attention to outliers by using modified boxplots, which we now introduce.

Modified Boxplot

A standard variation on the idea of a boxplot is what is known as a modified boxplot. A **modified boxplot** is a boxplot in which the outliers, if any, are graphed as separate points. The advantage of a modified boxplot is that it lets us quickly see where the outliers are, if there are any.

To make a modified boxplot, we proceed as we did when first making a boxplot, except for the last step. After drawing the box for the boxplot, we check to see if there are outliers. If there are no outliers, then we extend whiskers from the box out to the extremes (the minimum and the maximum). However, if there are outliers in the upper part of the distribution, then we identify them with asterisks. We then extend a whisker from Q_3 up to the largest data point that is not an outlier. Likewise, if there are outliers in the lower part of the distribution, we identify them with asterisks and extend a whisker from Q_1 down to the smallest observation that is not an outlier. Figure 2.33 shows a boxplot and a modified boxplot of the data on radish seedlings grown in constant light.

Most often, when people make boxplots they make modified boxplots. Computer software is typically programmed to produce a modified boxplot when the user asks for a boxplot. Thus, we will use the term *boxplot* to mean "modified boxplot."

Example 2.26 shows the power of boxplots to give us a visual comparison of several distributions.

Temperature.   The high temperature in Oberlin, Ohio varies quite a bit over the course of a year. Figure 2.34 shows 12 parallel boxplots of the daily high temperature for one year, with one boxplot for each month.

These plots allow us to compare months quickly and to see how the high temperature varies as the year progresses. Note that there is more variability in the winter months than in the summer, as indicated by the lengths of the boxes and the whiskers. The only high outliers occurred in November, when there were two days that were unusually warm for November, with temperatures well above 60 degrees. These would have been average days in September, however. There were two low outliers in December. These two cold days would not have been outliers in January, February, or March. ∎

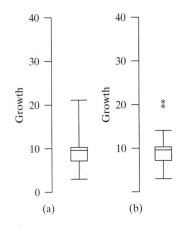

Figure 2.33
(a) Boxplot of data on radish growth in constant light;
(b) modified boxplot of radish growth data

Example 2.26

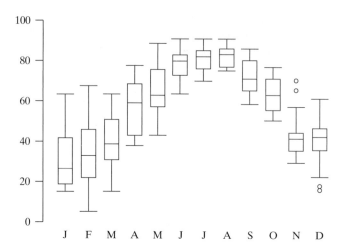

Figure 2.34 Daily high temperature in Oberlin, Ohio for one year

Computer note: To make a (modified) boxplot within the MINITAB system, use the command

```
MTB > BOXPLOT C1
```

Suppose the data are stored in column 1 and that column 2 holds an indicator variable (for example, if we are comparing men and women, then column 2 might have a 1 for men and a 2 for women). Then the command

```
MTB > BOXPLOT C1*C2
```

will produce parallel boxplots of the C1 data, one for each level of the variable in C2 (e.g., a boxplot for the men and a parallel boxplot for the women).

Exercises 2.30–2.39

2.30 Here are the data from Exercise 2.23 on the number of virus-resistant bacteria in each of 10 aliquots:

$$14 \quad 15 \quad 13 \quad 21 \quad 15$$
$$14 \quad 26 \quad 16 \quad 20 \quad 13$$

(a) Determine the median and the quartiles.
(b) Determine the interquartile range.
(c) How large would an observation in this data set have to be in order to be an outlier?

2.31 Here are the 18 measurements of MAO activity reported in Exercise 2.5:

$$6.8 \quad 8.4 \quad 8.7 \quad 11.9 \quad 14.2 \quad 18.8$$
$$9.9 \quad 4.1 \quad 9.7 \quad 12.7 \quad 5.2 \quad 7.8$$
$$7.8 \quad 7.4 \quad 7.3 \quad 10.6 \quad 14.5 \quad 10.7$$

(a) Determine the median and the quartiles.

(b) Determine the interquartile range.

(c) Construct a (modified) boxplot of the data.

2.32   In a study of milk production in sheep (for use in making cheese), a researcher measured the three-month milk yield for each of 11 ewes. The yields (liters) were as follows:[27]

$$56.5 \quad 89.8 \quad 110.1 \quad 65.6 \quad 63.7 \quad 82.6$$
$$75.1 \quad 91.5 \quad 102.9 \quad 44.4 \quad 108.1$$

(a) Determine the median and the quartiles.

(b) Determine the interquartile range.

(c) Construct a (modified) boxplot of the data.

2.33   A group of college students were asked how many hours per week they exercise.[28] The answers given by 12 men were as follows:

$$6 \quad 0 \quad 2 \quad 1 \quad 2 \quad 4.5 \quad 8 \quad 3 \quad 17 \quad 4.5 \quad 4 \quad 5$$

The answers given by 13 women were as follows:

$$5 \quad 13 \quad 3 \quad 2 \quad 6 \quad 14 \quad 3 \quad 1 \quad 1.5 \quad 1.5 \quad 3 \quad 8 \quad 4$$

Construct parallel boxplots of the male and female distributions.

2.34   Consider the data from Exericise 2.33. Describe the two boxplots, including how they compare to each other.

2.35   For each of the following histograms, use the histogram to estimate the median and the quartiles; then construct a boxplot for the distribution.

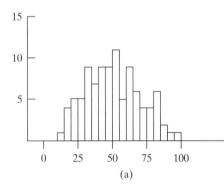

(a)

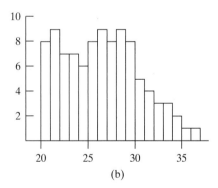

(b)

2.36   The histogram below shows the same data that are shown in one of the four boxplots. Which boxplot goes with the histogram? Explain your answer.

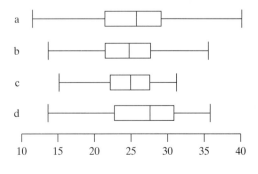

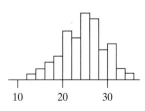

2.37 The boxplot shows the five-number summary for a data set. For these data the minimum is 35, Q_1 is 42, the median is 49, Q_3 is 56, and the maximum is 65. Is it possible that no observation in the data set equals 42? Explain your answer.

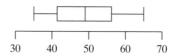

```
      30      40      50      60      70
```

2.38 Statistics software can be used to find the five-number summary of a data set. For example, if data are stored within the MINITAB system in column 1, then the DE-SCRIBE command produces the following:

```
MTB > Describe C1
   Variable     N      Mean    Median    TrMean    StDev    SEMean
      C1        75    119.94    118.40    119.98     9.98     1.15
   Variable    Min     Max       Q1        Q3
      C1      95.16   145.11    113.59    127.42
```

(a) Use the MINITAB output to calculate the interquartile range.
(b) Are there any outliers in this set of data?

2.39 Consider the data from Exercise 2.37. Use the five-number summary that is given to create a boxplot of the data.

2.6 MEASURES OF DISPERSION

We have considered the shapes and centers of distributions, but a good description of a distribution should also characterize how spread out the distribution is—are the observations in the sample all nearly equal, or do they differ substantially? In Section 2.5 we defined the interquartile range, which is one measure of dispersion. We will now consider other measures of dispersion: the range, the standard deviation, and the coefficient of variation.

The Range

The sample **range** is the difference between the largest and smallest observations in a sample. Here is an example.

Example 2.27 **Blood Pressure.** The systolic blood pressures (mm Hg) of seven middle-aged men was given in Example 2.20 as follows:

$$113 \quad 124 \quad 124 \quad 132 \quad 146 \quad 151 \quad 170$$

For these data, the sample range is

$$170 - 113 = 57 \text{ mm Hg} \qquad \blacksquare$$

The range is easy to calculate, but it is very sensitive to extreme values (i.e., it is not robust). If the maximum in the blood pressure sample had been 190 rather than 170, the range would have been changed from 57 to 77.

We defined the interquartile range (IQR) in Section 2.5 as the difference between the quartiles. Unlike the range, the IQR is robust. The IQR of the blood pressure data is $151 - 124 = 27$. If the maximum in the blood pressure sample had been 190 rather than 170, the IQR would not have changed; it would still be 27.

The Standard Deviation

The standard deviation is the classical and most widely used measure of dispersion. Recall that a *deviation* is the difference between an observation and the sample mean:

$$\text{deviation} = \text{observation} - \bar{y}$$

The standard deviation of the sample, or sample **standard deviation**, is determined by combining the deviations in a special way, as described in the accompanying box.

The Sample Standard Deviation

The sample standard deviation is denoted by s and is defined by the following formula:

$$s = \sqrt{\frac{\sum (y_i - \bar{y})^2}{n - 1}}$$

In this formula, the expression $\sum (y_i - \bar{y})^2$ denotes the sum of the squared deviations.

So, to find the standard deviation of a sample, first find the deviations. Then

(a) square

(b) add

(c) divide by $n - 1$

(d) take the square root

To illustrate the use of the formula, we have chosen a data set that is especially simple to handle because the mean happens to be an integer.

Growth of Chrysanthemums. In an experiment on chrysanthemums, a botanist measured the stem elongation (mm in 7 days) of five plants grown on the same greenhouse bench. The results were as follows:[29]

> | Example 2.28 |

$$76 \quad 72 \quad 65 \quad 70 \quad 82$$

The data are tabulated in the first column of Table 2.11. The sample mean is

$$\bar{y} = \frac{365}{5} = 73 \text{ mm}$$

The deviations $(y_i - \bar{y})$ are tabulated in the second column of Table 2.11; the first observation is 3 mm above the mean, the second is 1 mm below the mean, and so on.

The third column of Table 2.11 shows that the sum of the squared deviations is

$$\Sigma(y_i - \bar{y})^2 = 164$$

Since $n = 5$, the standard deviation is

TABLE 2.11 Illustration of the Formula for the Sample Standard Deviation

Observation y_i	Deviation $y_i - \bar{y}$	Squared deviation $(y_i - \bar{y})^2$
76	3	9
72	−1	1
65	−8	64
70	−3	9
82	9	81
Sum 365 $= \Sigma\, y_i$	0	$164 = \Sigma(y_i - \bar{y})^2$

$$s = \sqrt{\frac{164}{4}}$$
$$= \sqrt{41}$$
$$= 6.4 \text{ mm}$$

Note that the units of s (mm) are the same as the units of Y. This is because we have squared the deviations and then later taken the square root. ■

The sample **variance**, denoted by s^2, is simply the standard deviation squared: variance $= s^2$. Thus, $s = \sqrt{\text{variance}}$.

Example 2.29   **Chrysanthemum Growth.**   The variance of the chrysanthemum growth data is

$$s^2 = 41 \text{ mm}^2$$

Note that the units of the variance (mm^2) are not the same as the units of Y. ■

An Abbreviation. We will frequently abbreviate "standard deviation" as SD; the symbol s will be used in formulas.

Interpretation of the Definition of s

The magnitude (disregarding sign) of each deviation $(y_i - \bar{y})$ can be interpreted as the *distance* of the corresponding observation from the sample mean $\bar{y}$. Figure 2.35 shows a plot of the chrysanthemum growth data (Example 2.28) with each distance marked.

From the formula for s, you can see that each deviation contributes to the SD. Thus, a sample of the same size but with less dispersion will have a smaller SD, as illustrated in the following example.

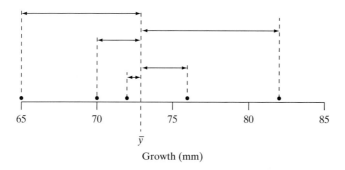

Figure 2.35 Plot of chrysanthemum growth data with deviations indicated as distances

Chrysanthemum Growth.   If the chrysanthemum growth data of Example 2.28 are changed to

Example 2.30

$$75 \quad 72 \quad 73 \quad 75 \quad 70$$

then the mean is the same ($\bar{y} = 73$ mm), but the SD is smaller ($s = 2.1$ mm), because the observations lie closer to the mean. The relative dispersion of the two samples can be easily seen from Figure 2.36.   ∎

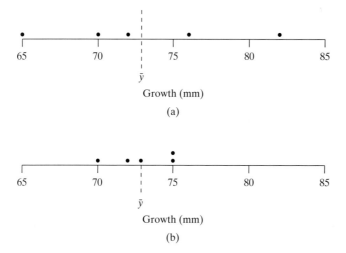

Figure 2.36 Two samples of chrysanthemum growth data with the same mean but different standard deviations. (a) $s = 6.3$ mm; (b) $s = 2.1$ mm.

   Let us look more closely at the way in which the deviations are combined to form the SD. The formula calls for dividing by $(n - 1)$. If the divisor were n instead of $(n - 1)$, then the quantity inside the square root sign would be the average (the mean) of the squared deviations. Unless n is very small, the inflation due to dividing by $(n - 1)$ instead of n is not very great, so that the SD can be interpreted approximately as

$$s \approx \sqrt{\text{sample average value of } (y_i - \bar{y})^2}$$

Thus, it is roughly appropriate to think of the SD as a "typical" distance of the observations from their mean.

Why n−1? Since dividing by n seems more natural, you may wonder why the formula for the SD specifies dividing by $(n - 1)$. Note that the sum of the deviations $y_i - \bar{y}$ is always zero. Thus, once the first $n - 1$ deviations have been calculated, the last deviation is constrained. This means that in a sample with n

observations there are only $n - 1$ units of information concerning deviation from the average. The quantity $n - 1$ is called the **degrees of freedom** of the standard deviation or variance. We can also give an intuitive justification of why $n - 1$ is used by considering the extreme case when $n = 1$, as in the following example.

Example 2.31

Chrysanthemum Growth.   Suppose the chrysanthemum growth experiment of Example 2.28 had included only one plant, so that the sample consisted of the single observation

$$73$$

For this sample, $n = 1$ and $\bar{y} = 73$. However, the SD formula breaks down (giving $\frac{0}{0}$), so the SD cannot be computed. This is reasonable, because the sample gives no information about variability in chrysanthemum growth under the experimental conditions. If the formula for the SD said to divide by n, we would obtain an SD of zero, suggesting that there is little or no variability; such a conclusion hardly seems justified by observation of only one plant.   ■

The Coefficient of Variation

The **coefficient of variation** is the standard deviation expressed as a percentage of the mean: coefficient of variation $= \frac{s}{\bar{y}} \cdot 100\%$. Here is an example.

Example 2.32

Chrysanthemum Growth.   For the chrysanthemum growth data of Example 2.28, we have $\bar{y} = 73.0$ mm and $s = 6.4$ mm. Thus,

$$\frac{s}{\bar{y}} \cdot 100\% = \frac{6.4}{73.0} \cdot 100\% = .088 \cdot 100\% = 8.8\%$$

The sample coefficient of variation is 8.8%. Thus, the standard deviation is 8.8% as large as the mean.   ■

Note that the coefficient of variation is not affected by multiplicative changes of scale. For example, if the chrysanthemum data were expressed in inches instead of mm, then both $\bar{y}$ and s would be in inches, and the coefficient of variation would be unchanged. Because of its imperviousness to scale change, the coefficient of variation is a useful measure for comparing the dispersions of two or more variables that are measured on different scales.

Example 2.33

Girls Height and Weight.   As part of the Berkeley Guidance Study,[30] the heights (in cm) and weights (in kg) of 13 girls were measured at age 2. At age 2, the average height was 86.6 cm and the SD was 2.9 cm. Thus, the coefficient of variation of height at age 2 is

$$\frac{s}{\bar{y}} \cdot 100\% = \frac{2.9}{86.6} \cdot 100\% = .033 \cdot 100\% = 3.3\%$$

For weight at age 2 the average was 12.6 kg and the SD was 1.4 kg. Thus, the coefficient of variation of weight at age 2 is

$$\frac{s}{\bar{y}} \cdot 100\% = \frac{1.4}{12.6} \cdot 100\% = .111 \cdot 100\% = 11.1\%$$

There is considerably more variability in weight than there is in height, when we express each measure of variability as a percentage of the mean. The SD of weight is a fairly large percentage of the average weight, but the SD of height is a rather small percentage of the average height. ■

Visualizing Measures of Dispersion

The range and the interquartile range are easy to interpret. The range is the spread of all the observations and the interquartile range is the spread of (roughly) the middle 50% of the observations. In terms of the histogram of a data set, the range can be visualized as (roughly) the width of the histogram. The quartiles are (roughly) the values that divide the area into four equal parts and the interquartile range is the distance between the first and third quartiles. The following example illustrates these ideas.

Daily Gain of Cattle.   The performance of beef cattle was evaluated by measuring their weight gain during a 140-day testing period on a standard diet. Table 2.12 gives the average daily gains (kg/day) for 39 bulls of the same breed (Charolais); the observations are listed in increasing order.[31] The values range from 1.18 kg/day to 1.92 kg/day. The quartiles are 1.29, 1.41, and 1.58 kg/day. Figure 2.37 shows a histogram of the data, the range, the quartiles, and the interquartile range (IQR). The shaded area represents the middle 50% (approximately) of the observations. ■

Example 2.34

TABLE 2.12 Average Daily Gain (kg/day) of 39 Charolais Bulls							
1.18	1.24	1.29	1.37	1.41	1.51	1.58	1.72
1.20	1.26	1.33	1.37	1.41	1.53	1.59	1.76
1.23	1.27	1.34	1.38	1.44	1.55	1.64	1.83
1.23	1.29	1.36	1.40	1.48	1.57	1.64	1.92
1.23	1.29	1.36	1.41	1.50	1.58	1.65	

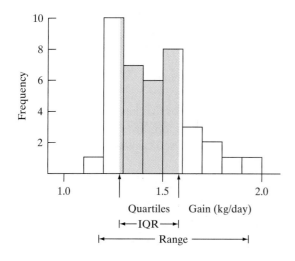

Figure 2.37 Histogram of 39 daily gain measurements, showing the range, the quartiles, and the interquartile range (IQR). The shaded area represents about 50% of the observations.

Visualizing the Standard Deviation

We have seen that the SD is a combined measure of the distances of the observations from their mean. It is natural to ask how many of the observations are within ± 1 SD of the mean, within ± 2 SDs of the mean, and so on. The following example explores this question.

Example 2.35 | **Daily Gain of Cattle.** For the daily-gain data of Example 2.34, the mean is $\bar{y} = 1.445$ kg/day and the SD is $s = .183$ kg/day. In Figure 2.38 the intervals $\bar{y} \pm s$, $\bar{y} \pm 2s$, and $\bar{y} \pm 3s$ have been marked on a histogram of the data. The interval $\bar{y} \pm s$ is

$$1.445 \pm .183 \text{ or } 1.262 \text{ to } 1.628$$

You can verify from Table 2.12 that this interval contains 25 of the 39 observations. Thus, $\frac{25}{39}$ or 64% of the observations are within ± 1 SD of the mean; the corresponding area is shaded in Figure 2.38. The interval $\bar{y} \pm 2s$ is

$$1.445 \pm .366 \text{ or } 1.079 \text{ to } 1.811$$

This interval contains $\frac{37}{39}$ or 95% of the observations. You may verify that the interval $y \pm 3s$ contains all of the observations. ■

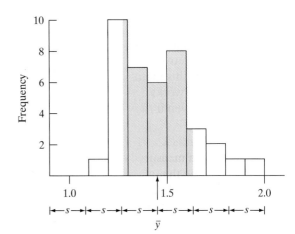

Figure 2.38 Histogram of daily-gain data showing intervals 1, 2, and 3 standard deviations from the mean. The shaded area represents about 64% of the observations.

It turns out that the percentages found in Example 2.35 are fairly typical of distributions that are observed in the life sciences.

> **Typical Percentages: The Empirical Rule**
> For "nicely shaped" distributions—that is, unimodal distributions that are not too skewed and whose tails are not overly long or short—we usually expect to find
>
> about 68% of the observations within ± 1 SD of the mean;
>
> about 95% of the observations within ± 2 SDs of the mean;
>
> $>99\%$ of the observations within ± 3 SDs of the mean.

The typical percentages enable us to construct a rough mental image of a frequency distribution if we know just the mean and SD. (The value 68% may seem to come from nowhere. Its origin will become clear in Chapter 4.)

Estimating the SD from a Histogram

The empirical rule gives us a way to construct a rough mental image of a frequency distribution if we know just the mean and SD: We can envision a histogram centered at the mean and extending out a bit more than 2 SDs in either directions. Of course, the actual distribution might not be symmetric, but our rough mental image will often be fairly accurate.

Thinking about this the other way around, we can look at a histogram and estimate the SD. To do this, we need to estimate the endpoints of an interval that is centered at the mean and that contains about 95% of the data. The empirical rule implies that this interval is roughly the same as $(\bar{y} - 2s, \bar{y} + 2s)$, so the length of the interval should be about 4 times the SD:

$$(\bar{y} - 2s, \bar{y} + 2s) \text{ has length of } 2s + 2s = 4s$$

This means

$$\text{length of interval} = 4s$$

so

$$\text{estimate of } s = \frac{\text{length of interval}}{4}$$

Of course, our visual estimate of the interval that covers the middle 95% of the data could be off. Moreover, the empirical rule works best for distributions that are symmetric. Thus, this method of estimating the SD will only give a general estimate. The method works best when the distribution is fairly symmetric, but it works reasonably well even if the distribution is somewhat skewed.

Pulse after Exercise. A group of 28 adults did some moderate exercise for five minutes and then measured their pulses. Figure 2.39 shows the distribution of the data.[32] We can see that about 95% of the observations are between about 75 and 125. Thus, an interval of length 50 $(125 - 75)$ covers the middle 95% of the data. From this, we can estimate the SD to be $\frac{50}{4} = 12.5$. The actual SD is 13.4, which is not far off from our estimate. ∎

> **Example 2.36**

The typical percentages given by the empirical rule may be grossly wrong if the sample is small or if the shape of the frequency distribution is not "nice." For instance, the cricket singing-time data (Table 2.10 and Figure 2.28) has $s = 4.4$ mm, and the interval $\bar{y} \pm s$ contains 90% of the observations. This is much higher than the "typical" 68% because the SD has been inflated by the long straggly tail of the distribution.

Comparison of Measures of Dispersion

The dispersion, or spread, of the data in a sample can be described by the standard deviation, the range, or the interquartile range. The range is simple to understand, but it can be a poor descriptive measure because it depends only on the extreme tails of the distribution. The interquartile range, by contrast, describes the spread in the central "body" of the distribution. The standard deviation takes account of

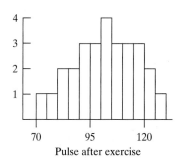

Figure 2.39 Pulse after moderate exercise for a group of adults

all the observations and can be roughly interpreted in terms of the spread of the observations around their mean. However, the SD can be inflated by observations in the extreme tails. The interquartile range is a resistant measure, while the SD is nonresistant. Of course, the range is highly nonresistant.

The descriptive interpretation of the SD is less straightforward than that of the range and the interquartile range. Nevertheless, the SD is the basis for most standard classical statistical methods. The SD enjoys this classic status for various technical reasons, including efficiency in certain situations.

The developments in later chapters will emphasize classical statistical methods, in which the mean and SD play a central role. Consequently, in this book we will rely primarily on the mean and SD rather than other descriptive measures.

Computer note: Calculating a sample standard deviation by hand is tedious. Statistics software can be used to find summary statistics, such as the mean, median, and standard deviation. For example, if the daily-gain data of Example 2.34 are stored within the MINITAB system in column 1, then the DESCRIBE command produces the following:

```
MTB > Describe C1
Descriptive Statistics
Variable     N      Mean   Median TrMean  StDev   SEMean
    C1      39     1.4446  1.4100 1.4346  0.1831  0.0293
Variable    Min     Max      Q1     Q3
    C1     1.1800  1.9200  1.2900 1.5800
```

Exercises 2.40–2.55

2.40 Calculate the standard deviation of each of the following fictitious samples:

(a) 16, 13, 18, 13   (b) 38, 30, 34, 38, 35
(c) 1, −1, 5, −1   (d) 4, 6, −1, 4, 2

2.41 Calculate the standard deviation of each of the following fictitious samples:

(a) 8, 6, 9, 4, 8   (b) 4, 7, 5, 4   (c) 9, 2, 6, 7, 6

2.42 (a) Invent a sample of size 5 for which the deviations $(y_i - \bar{y})$ are $-3, -1, 0, 2, 2$.
(b) Compute the standard deviation of your sample.
(c) Should everyone get the same answer for part (b)? Why?

2.43 Four plots of land, each 346 square feet, were planted with the same variety ("Beau") of wheat. The plot yields (lb) were as follows:[33]

35.1   30.6   36.9   29.8

(a) Calculate the mean and the standard deviation.
(b) Calculate the coefficient of variation.

2.44 A plant physiologist grew birch seedlings in the greenhouse and measured the ATP content of their roots. (See Example 1.3.) The results (nmol ATP/mg tissue) were as follows for four seedlings that had been handled identically.[34]

1.45   1.19   1.05   1.07

(a) Calculate the mean and the standard deviation.
(b) Calculate the coefficient of variation.

2.45 Ten patients with high blood pressure participated in a study to evaluate the effectiveness of the drug Timolol in reducing their blood pressure. The accompanying table shows systolic blood pressure measurements taken before and after two weeks of treatment with Timolol.[35] Calculate the mean and standard deviation of the *change* in blood pressure (note that some values are negative).

Blood Pressure (mm Hg)

Patient	Before	After	Change
1	172	159	−13
2	186	157	−29
3	170	163	−7
4	205	207	2
5	174	164	−10
6	184	141	−43
7	178	182	4
8	156	171	15
9	190	177	−13
10	168	138	−30

2.46 Dopamine is a chemical that plays a role in the transmission of signals in the brain. A pharmacologist measured the amount of dopamine in the brain of each of seven rats. The dopamine levels (nmol/g) were as follows:[36]

6.8   5.3   6.0   5.9   6.8   7.4   6.2

(a) Calculate the mean and standard deviation.
(b) Determine the median and the interquartile range.
(c) Calculate the coefficient of variation.
(d) Replace the observation 7.4 by 10.4 and repeat parts (a) and (b). Which of the descriptive measures display resistance and which do not?

2.47 In a study of the lizard *Sceloporus occidentalis*, biologists measured the distance (m) run in two minutes for each of 15 animals. The results (listed in increasing order) were as follows:[37]

18.4   22.2   24.5   26.4   27.5   28.7   30.6   32.9
32.9   34.0   34.8   37.5   42.1   45.5   45.5

(a) Determine the quartiles and the interquartile range.
(b) Determine the range.

2.48 Refer to the running-distance data of Exercise 2.47. The sample mean is 32.23 m and the SD is 8.07 m. What percentage of the observations are within
(a) 1 SD of the mean?   (b) 2 SDs of the mean?

2.49 Compare the results of Exercise 2.48 with the predictions of the empirical rule.

2.50 Listed in increasing order are the serum creatine phosphokinase (CK) levels (u/Li) of 36 healthy men (these are the data of Example 2.6):

25	62	82	95	110	139
42	64	83	95	113	145
48	67	84	100	118	151
57	68	92	101	119	163
58	70	93	104	121	201
60	78	94	110	123	203

The sample mean CK level is 98.3 u/Li and the SD is 40.4 u/Li. What percentage of the observations are within

(a) 1 SD of the mean?   (b) 2 SDs of the mean?   (c) 3 SDs of the mean?

2.51   Compare the results of Exercise 2.50 with the predictions of the empirical rule.

2.52   The girls in the Berkeley Guidance Study (Example 2.33) who were measured at age two were measured again at age nine. Of course, the average height and weight were much greater at age nine than at age two. Likewise, the SDs of height and of weight were much greater at age nine than they were at age two. But what about the coefficient of variation of height and the coefficient of variation of weight? It turns out that one of these went up a moderate amount from age two to age nine, but for the other variable the increase in the coefficient of variation was fairly large. For which variable, height or weight, would you expect the coefficient of variation to change more between age two and age nine? Why? (*Hint:* Think about how genetic factors influence height and weight and how environmental factors influence height and weight.)

2.53   Consider the 13 girls mentioned in Example 2.33. At age 18 their average height was 166.3 cm and the SD of their heights was 6.8 cm. Calculate the coefficient of variation.

2.54   Here is a histogram. Estimate the mean and the SD of the distribution.

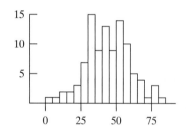

2.55   Here is a histogram. Estimate the mean and the SD of the distribution.

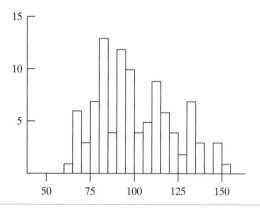

2.7 EFFECT OF TRANSFORMATION OF VARIABLES (OPTIONAL)

Sometimes when we are working with a data set, we find it convenient to transform a variable. For example, we might convert from inches to centimeters or from °F to °C. Transformation, or reexpression, of a variable Y means replacing Y by a new variable, say Y'. To be more comfortable working with data, it is helpful to know how the features of a distribution are affected if the observed variable is transformed.

The simplest transformations are **linear** transformations, so called because a graph of Y against Y' would be a straight line. A familiar reason for linear transformation is a change in the scale of measurement, as illustrated in the following two examples.

Weight.   Suppose Y represents the weight of an animal in kg, and we decide to reexpress the weight in lb. Then

$$Y = \text{Weight in kg}$$

$$Y' = \text{Weight in lb}$$

so

$$Y' = 2.2Y$$

This is a **multiplicative** transformation, because Y' is calculated from Y by multiplying by the constant value 2.2.   ■

Example 2.37

Body Temperature.   Measurements of basal body temperature (temperature on waking) were made on 47 women.[38] Typical observations Y, in °C, were

$$Y: \quad 36.23, \quad 36.41, \quad 36.77, \quad 36.15, \quad \ldots$$

Suppose we convert these data from °C to °F, and call the new variable Y':

$$Y': \quad 97.21, \quad 97.54, \quad 98.19, \quad 97.07, \quad \ldots$$

The relation between Y and Y' is

$$Y' = 1.8Y + 32$$

The combination of **additive** ($+32$) and multiplicative ($\times 1.8$) changes indicates a linear relationship.   ■

Example 2.38

Another reason for linear transformation is **coding**, which means transforming the data for convenience in handling the numbers. The following is an example.

Body Temperature.   Consider the temperature data of Example 2.38. If we subtract 36 from each observation, the data become

$$.23, \quad .41, \quad .77, \quad .15, \quad \ldots$$

This is additive coding, since we added a constant value (-36) to each observation. Now suppose we further transform the data to the form

$$23, \quad 41, \quad 77, \quad 15, \ldots$$

This step of the coding is multiplicative, since each observation is multiplied by a constant value (100).   ■

Example 2.39

As the foregoing examples illustrate, a linear transformation consists of (1) multiplying all the observations by a constant, or (2) adding a constant to all the observations, or (3) both.

How Linear Transformations Affect the Frequency Distribution

A linear transformation of the data does not change the essential shape of its frequency distribution; by suitably scaling the horizontal axis, you can make the transformed histogram identical to the original histogram. Example 2.40 illustrates this idea.

Example 2.40

Body Temperature.   Figure 2.40 shows the distribution of 47 temperature measurements that have been transformed by first subtracting 36 from each observation and then multiplying by 100 (as in Examples 2.38 and 2.39). That is, $Y' = (Y - 36) * 100$. The figure shows that the two distributions can be represented by the same histogram with different horizontal scales. ∎

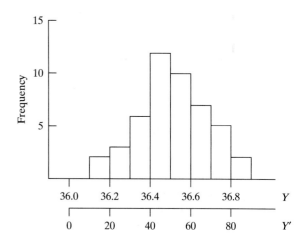

Figure 2.40 Distribution of 47 temperature measurements showing original and linearly transformed scales

How Linear Transformations Affect $\bar{y}$ and s

The effect of a linear transformation on $\bar{y}$ is "natural"; that is, **under a linear transformation**, $\bar{y}$ changes like Y. For instance, if temperatures are converted from °C to °F, then the mean is similarly converted:

$$Y' = 1.8Y + 32 \quad \text{so} \quad \bar{y}' = 1.8\bar{y} + 32$$

The effect of multiplying Y by a positive constant on s is "natural"; if $Y' = c * Y$, with $c > 0$, then $s' = c * s$. For instance, if weights are converted from kg to lb, the SD is similarly converted: $s' = 2.2s$. If $Y' = c * Y$ and $c < 0$, then $s' = -c * s$. In general, if $Y' = c * Y$, then $s' = |c| * s$.

However, an additive transformation does not affect s. If we add or subtract a constant, we do not change how spread out the distribution is, so s does not change. Thus, for example, we would *not* convert the SD of temperature data from °C to °F in the same way as we convert each observation; we would multiply the SD by 1.8 but we would *not* add 32. The fact that the SD is unchanged by additive transformation will appear less surprising if you recall (from the definition) that s depends only on the deviations $(y_i - \bar{y})$, and these are not changed by an additive transformation. The following example illustrates this idea.

Consider a simple set of fictitious data, coded by subtracting 20 from each observation. The original and transformed observations are shown in Table 2.13.

Example 2.41

TABLE 2.13 Effect of Additive Transformation

	Original observations y	Deviations $y_i - \bar{y}$	Transformed observations y'	Deviations $y_i' - \bar{y}$
	25	−1	5	−1
	26	0	6	0
	28	2	8	2
	25	−1	5	−1
Mean	26		6	

The SD for the original observations is

$$s = \sqrt{\frac{(-1)^2 + (0)^2 + (2)^2 + (-1)^2}{3}}$$

$$= 1.4$$

Because the deviations are unaffected by the transformation, the SD for the transformed observations is the same:

$$s' = 1.4 \qquad \blacksquare$$

An additive transformation effectively picks up the histogram of a distribution and moves it to the left or to the right on the number line. The shape of the histogram does not change and the deviations do not change, so the SD does not change. A multiplicative transformation, on the other hand, stretches or shrinks the distribution, so the SD gets larger or smaller accordingly.

Other Statistics. Under linear transformations, other measures of center (for instance, the median) change like $\bar{y}$, and other measures of dispersion (for instance, the interquartile range) change like s. The quartiles themselves change like $\bar{y}$.

Nonlinear Transformations

Data are sometimes reexpressed in a nonlinear way. Examples of nonlinear transformations are

$$Y' = \sqrt{Y}$$
$$Y' = \log(Y)$$
$$Y' = \frac{1}{Y}$$
$$Y' = Y^2$$

These transformations are termed "nonlinear" because a graph of Y' against Y would be a curve rather than a straight line. Computers make it easy to use nonlinear transformations. The logarithmic transformation is especially common in biology because many important relationships can be simply expressed in terms of logs. For instance, there is a phase in the growth of a bacterial colony when log(colony size) increases at a constant rate with time. [Note that logarithms are used in some familiar scales of measurement, such as pH measurement or earthquake magnitude (Richter scale).]

Nonlinear transformations can affect data in complex ways. For example, the mean does not change "naturally" under a log transformation; the log of the mean is *not* the same as the mean of the logs. Furthermore, nonlinear transformations (unlike linear ones) *do* change the essential shape of a frequency distribution.

In future chapters we will see that if a distribution is skewed to the right, such as the singing time distribution shown in Figure 2.41, then we may wish to apply a transformation that makes the distribution more symmetric, by pulling in the right-hand tail. Using $Y' = \sqrt{Y}$ will pull in the right-hand tail of a distribution and push out the left-hand tail. The transformation $Y' = \log(Y)$ is more severe than $\sqrt{Y}$ in this regard. The following example shows the effect of these transformations.

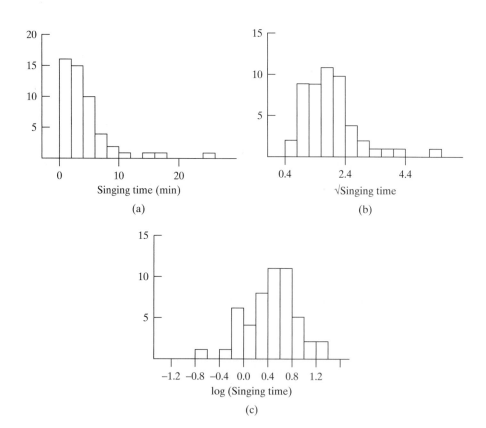

Figure 2.41 Distribution of Y, of $\sqrt{Y}$, and of log (Y) for 51 observations of Y = singing time

Example 2.42 **Cricket Singing Times.** Figure 2.41(a) shows the distribution of the cricket singing-time data of Table 2.10. If we transform these data by taking square roots, the transformed data have the distribution shown in Figure 2.41(b). Taking logs (base 10) yields the distribution shown in Figure 2.41(c). Notice that the transformations have the effect of "pulling in" the straggly upper tail and "stretching out" the clumped values on the lower end of the original distribution. ∎

Computer note: Without the aid of technology, transforming data would be very tedious. However, if the data are stored on a computer or graphing calculator, then it is fairly easy to transform the data, so that one can try a variety of transformations, as was done in Example 2.42. For example, suppose the cricket singing times data of Example 2.42 are stored within the MINITAB system in column 1. Then to transform the data by taking square roots, we use the command

MTB > Sqrt C1 C2.

This puts the transformed data (the square root values) in column 2.

To transform the data by taking logs (base 10), we use the command

MTB > Log Ten C1 C2.

If we want to take the natural logarithm of each observation, we use the command

MTB > LogE C1 C2.

We can also create other transformations by typing in expressions. For example, to create a new variable in column 2 that contains the reciprocals of the square roots of data in column 1, we use the command

MTB > Let C2 = 1/sqrt (C1).

Exercises 2.56–2.61

2.56 A biologist made a certain pH measurement in each of 24 frogs; typical values were[39]

$$7.43, \quad 7.16, \quad 7.51, \ldots$$

She calculated a mean of 7.373 and a standard deviation of .129 for these original pH measurements. Next, she transformed the data by subtracting 7 from each observation and then multiplying by 100. For example, 7.43 was transformed to 43. The transformed data are

$$43, \quad 16, \quad 51, \ldots$$

What are the mean and standard deviation of the transformed data?

2.57 The mean and SD of a set of 47 body temperature measurements were as follows:[40]

$$\bar{y} = 36.497°C \qquad s = .172°C$$

If the 47 measurements were converted to °F,

(a) What would be the new mean and SD?

(b) What would be the new coefficient of variation?

2.58   A researcher measured the average daily gains (in kg/day) of 20 beef cattle; typical values were[41]

$$1.39, \quad 1.57, \quad 1.44, \ldots$$

The mean of the data was 1.461 and the standard deviation was .178.

(a) Express the mean and standard deviation in lb/day. (*Hint*: 1 kg = 2.20 lb.)
(b) Calculate the coefficient of variation when the data are expressed (i) in kg/day; (ii) in lb/day.

2.59   Consider the data from Exercise 2.58. The mean and SD were 1.461 and .178. Suppose we transformed the data from

$$1.39, \quad 1.57, \quad 1.44, \ldots$$

to

$$39, \quad 57, \quad 44, \ldots$$

What would be the mean and standard deviation of the transformed data?

2.60   The following histogram shows the distribution for a sample of data:

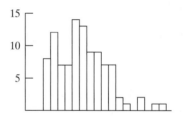

One of the following histrograms is the result of applying a square root transformation and the other is the result of applying a log transformation. Which is which? How do you know?

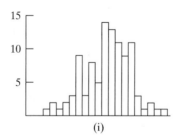

(i)

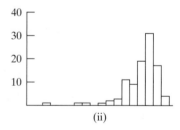

(ii)

2.61   *(Computer problem)*   The file 'dendnewb' is included on the data disk packaged with this text. This file contains 36 observations on the number of dendritic branch segments emanating from nerve cells taken from the brains of newborn guinea pigs. (These data were used in Exercise 2.7.) Open the file and enter the data into a statistics package, such as MINITAB. Make a histogram of the data, which are skewed to the right. Now consider the following possible transformations: sqrt(Y), log(Y), and 1/sqrt(Y). Which of these transformations does the best job of meeting the goal of making the resulting distribution reasonably symmetric?

2.8 SAMPLES AND POPULATIONS: STATISTICAL INFERENCE

In the preceding sections we have examined several ways of describing a set of observations. We have called the data set a "sample." Now we discuss the reason for this terminology.

The description of a data set is sometimes of interest for its own sake. Usually, however, the researcher hopes to generalize, to extend the findings beyond the limited scope of the particular group of animals, plants, or other units that were actually observed. Statistical theory provides a rational basis for this process of generalization, a basis that takes into account the variability of the data. The key idea of the statistical approach is to view the particular data in an experiment as a sample from a larger population; the population is the real focus of scientific and/or practical interest. The following example illustrates this idea.

Blood Types. In an early study of the ABO blood-typing system, researchers determined blood types of 3,696 persons in England. The results are given in Table 2.14.[42]

Example 2.43

TABLE 2.14 Blood Types of 3,696 Persons	
Blood type	**Frequency**
A	1,634
B	327
AB	119
O	1,616
Total	3,696

These data were not collected for the purpose of learning about the blood types of those particular 3,696 people. Rather, they were collected for their scientific value as a source of information about the distribution of blood types in a larger population. For instance, one might presume that the blood type distribution of all English people should resemble the distribution for these 3,696 people. In particular, the observed relative frequency of Type A blood was

$$\frac{1,634}{3,696} \text{ or } 44\% \text{ Type A}$$

One might conclude from this that approximately 44% of the people in England have Type A blood. ∎

Statistical Inference

The process of drawing conclusions about a population, based on observations in a sample from that population, is called **statistical inference**. For instance, in Example 2.43 the conclusion that approximately 44% of the people in England have Type A blood would be a statistical inference. The inference is shown schematically in Figure 2.42. Of course, such an inference might be entirely wrong—perhaps the 3,696 people are not at all representative of English people in general. We might be worried about two possible sources of difficulty: (1) The 3,696 people might have been selected in a way that was systematically biased for (or against) Type A people, and (2) the number of people examined might have been too small to permit generalization to a population of many millions. In general, it turns out that the population size being in the millions is *not* a problem, but bias in the way people are selected is a big concern.

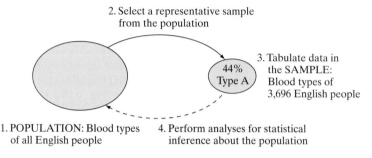

Figure 2.42 Schematic representation of inference from sample to population regarding prevalence of blood Type A

In making a statistical inference, we would prefer that the sample resemble the population closely—that the sample be *representative* of the population. However, we must ask about the likelihood of this happening. In other words, we must ask the important question: *How representative (of the population) is a sample likely to be?* We will see in Chapters 3 and 5 how statistical theory can help to answer this question. But the question itself becomes meaningful only if the population has been defined, a process that we now discuss in more detail.

Defining the Population

Ideally, the population should be defined in such a way that it is plausible to believe that a sufficiently large sample *would* be representative of the population. The first step in defining the population is to ask how the observations were obtained. Two important issues are, How were the observational units selected? and What was the observed variable? The following example illustrates the reasoning involved in defining the population.

Example 2.44   **Blood Types.**   How were the 3,696 English people of Example 2.43 actually chosen? It appears from the original paper that this was a "sample of convenience," that is, friends of the investigators, employees, and sundry unspecified sources. There is little basis for believing that the *people* themselves would be representative of the entire English population. Nevertheless, one might argue that their *blood types* might be (more or less) representative of the population. The argument would be that the biases that entered into the selection of those particular people were probably not related to blood type (although an objection might be made on the basis of race). The argument for representativeness would be much less plausible if the observed variable were blood pressure rather than blood type; we know that blood pressure tends to increase with age, and the selection procedure was undoubtedly biased against certain age groups (for example, elderly people).   ■

As Example 2.44 shows, whether a sample is likely to be representative of a population depends not only on how the observational units (in this case people) were chosen, but also on the variable that was observed. Generally, therefore, it is most appropriate to think of the population as consisting of observations, rather than of people or other observational units. We can conceptualize the population as an indefinitely large extension of the sample. In other words, **in order to try to define the population from which our data came, we try to describe the set of observations that we would obtain if the process generating the data were repeated indefinitely**. The following is another example.

Alcohol and MOPEG. The biochemical MOPEG (3-methoxy-4-hydrox-yphenylethylene) plays a role in brain function. Seven healthy male volunteers participated in a study to determine whether drinking alcohol might elevate the concentration of MOPEG in the cerebrospinal fluid. The MOPEG concentration was measured twice for each man—once at the start of the experiment, and again after he drank 80 g of ethanol. The results (in pmol/mLi) are given in Table 2.15.[43]

Example 2.45

TABLE 2.15 Effect of Alcohol on MOPEG

	MOPEG concentration		
Volunteer	Before	After	Change
1	46	56	10
2	47	52	5
3	41	47	6
4	45	48	3
5	37	37	0
6	48	51	3
7	58	62	4

Let us focus on the rightmost column, which shows the change in MOPEG concentration (that is, the difference between the "after" and the "before" measurements). In thinking of these values as a sample from a population, we need to specify all the details of the experimental conditions—how the cerebrospinal specimens were obtained, the exact timing of the measurements and the alcohol consumption, and so on—as well as relevant characteristics of the volunteers themselves. Thus, the definition of the population might be something like this:

> **Population** Change in cerebrospinal MOPEG concentration in healthy young men when measured before and after drinking 80 g of ethanol, both measurements being made at 8:00 A.M., ... (other relevant experimental conditions are specified here).

There is no single "correct" definition of a population for an experiment like this. A scientist reading a report of the experiment might find the above definition too narrow (for instance, perhaps it does not matter that the volunteers were measured at 8:00 A.M.) or too broad. She might use her knowledge of alcohol and brain chemistry to formulate her own definition, and she would then use that definition as a basis for interpreting these seven observations. ∎

A Dynamic Example

The concept of obtaining precise statements about populations from samples is at the heart of statistical thinking. In the following example we dramatize this concept by looking at larger and larger samples from the same population. (Of course, in practice, one usually takes only one sample from a population rather than samples of various sizes.)

Example 2.46

Sucrose Consumption. An entomologist is interested in the mechanism controlling feeding behavior in the black blowfly *(Phormia regina)*. One variable of interest to him is the amount of sucrose (sugar) solution a fly will drink in 30 minutes. The measurement procedure is such that a given fly can be measured only once. To study the inherent variability of the system, the researcher has measured hundreds of flies under standardized conditions. Figure 2.43 shows histograms of sucrose consumption values (mg) for samples of various numbers of individuals.[44] The means and standard deviations of the samples are as follows:

n	20	40	100	400	900
$\bar{y}$	15.5	14.7	14.3	15.0	14.9
s	6.5	5.9	5.0	5.4	5.4

Notice that, as the sample size is increased, the frequency distribution tends to stabilize and, similarly, the mean and the SD tend to stabilize.

It is natural to define a population from which the samples came, as follows:

Population Sucrose consumption values for all *P. regina* individuals under the standardized conditions ∎

Remark: As noted in Example 2.46, the SD tends to stabilize as the sample size is increased. To see intuitively why this should happen, recall from Section 2.6 that

$$s \approx \sqrt{\text{Sample average value of } (y_i - \bar{y})^2}$$

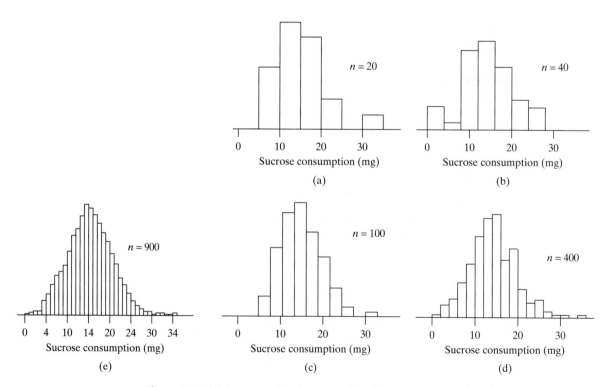

Figure 2.43 Histograms of various samples of sucrose consumption data

The right-hand side of this expression depends only on the *composition* of the sample, not on its size; thus, **samples of different sizes but with similar compositions (relative frequency distributions) will have similar SDs**. Increasingly larger samples from the same population will tend to have compositions increasingly similar to the population, and so also to have means and SDs increasingly similar to the mean and SD of the population.

Describing a Population

Because observations are made only on a sample, characteristics of biological populations are almost never known exactly. Typically, our knowledge of a population characteristic comes from a sample. In statistical language, we say that the sample characteristic is an estimate of the corresponding population characteristic. Thus, estimation is a type of statistical inference.

Just as each sample has a distribution, a mean, and an SD, so also we can envision a population distribution, a population mean, and a population SD. In order to discuss inference from a sample to a population, we will need a language for describing the population. This language parallels the language that describes the sample. A sample characteristic is called a **statistic**; a population characteristic is called a **parameter**.

Proportions

For a categorical variable, we can describe a population by simply stating the proportion, or relative frequency, of the population in each category. The following is a simple example.

Oat Plants.   In a certain population of oat plants, resistance to crown rust disease is distributed as shown in Table 2.16.[45]   ■

Example 2.47

TABLE 2.16 Disease Resistance in Oats	
Resistance	**Proportion of Plants**
Resistant	.47
Intermediate	.43
Susceptible	.10
Total	1.00

Remark: The population described in Example 2.47 is realistic, but it is not a specific real population; the exact proportions for any real population are not known. For similar reasons, we will use fictitious but realistic populations in several other examples, here and in Chapters 3, 4, and 5.

For categorical data, the sample proportion of a category is an estimate of the corresponding population proportion. Because these two proportions are not necessarily the same, it is essential to have a notation that distinguishes between them. We denote the population proportion of a category by p and the sample proportion by $\hat{p}$ (read "p-hat"):

$$p = \text{Population proportion}$$

$$\hat{p} = \text{Sample proportion}$$

The symbol "$\hat{}$" can be interpreted as "estimate of." Thus,

$$\hat{p} \text{ is an estimate of } p.$$

We illustrate this notation with an example.

Example 2.48

Lung Cancer.   Eleven patients suffering from adenocarcinoma (a type of lung cancer) were treated with the chemotherapeutic agent Mitomycin. Three of the patients showed a positive response (defined as shrinkage of the tumor by at least 50%).[46] Suppose we define the population for this study as "responses of all adenocarcinoma patients." Then we can represent the sample and population proportions of the category "positive response" as follows:

p = Proportion of positive responders among all adenocarcinoma patients
$\hat{p}$ = Proportion of positive responders among the 11 patients in the study

$$\hat{p} = \frac{3}{11} = .27$$

Note that p is unknown, and $\hat{p}$, which is known, is an estimate of p. ∎

We should emphasize that an "estimate," as we are using the term, may or may not be a *good* estimate. For instance, the estimate $\hat{p}$ in Example 2.48 is based on very few patients; estimates based on a small number of observations are subject to considerable uncertainty. Of course, the question of whether an estimation procedure is good or poor is an important one, and we will show in later chapters how this question can be answered.

Other Descriptive Measures

If the observed variable is quantitative, one can consider descriptive measures other than proportions—the mean, the quartiles, the SD, and so on. Each of these quantities can be computed for a sample of data, and each is an estimate of its corresponding population analog. For instance, the sample median is an estimate of the population median. In later chapters, we will focus especially on the mean and the SD, and so we will need a special notation for the population mean and SD. **The population mean is denoted by μ (mu), and the population SD is denoted by σ (sigma).** We may define these as follows for a quantitative variable Y:

$$\mu = \text{Population average value of } Y$$
$$\sigma = \sqrt{\text{Population average value of } (Y - \mu)^2}$$

The following example illustrates this notation.

Example 2.49

Tobacco Leaves.   An agronomist counted the number of leaves on each of 150 tobacco plants of the same strain (Havana). The results are shown in Table 2.17.[47]
        The sample mean is

$$\bar{y} = 19.78 = \text{Mean number of leaves on the 150 plants}$$

The population mean is

$$\mu = \text{Mean number of leaves on Havana tobacco}$$
$$\text{plants grown under these conditions}$$

We do not know μ, but we can regard $\bar{y} = 19.78$ as an estimate of μ. The sample SD is

$$s = 1.38 = \text{SD of number of leaves on the 150 plants}$$

The population SD is

$$\sigma = \text{SD of number of leaves on Havana tobacco}$$
$$\text{plants grown under these conditions}$$

We do not know σ but we can regard $s = 1.38$ as an estimate of σ.* ∎

TABLE 2.17 Number of Leaves on Tobacco Plants	
Number of Leaves	**Frequency (Number of Plants)**
17	3
18	22
19	44
20	42
21	22
22	10
23	6
24	1
Total	150

2.9 PERSPECTIVE

In this chapter we have considered various ways of describing a set of data. We have also introduced the notion of regarding a data set as a sample from a suitably defined population, and regarding features of the sample as estimates of corresponding features of the population.

Parameters and Statistics

Some features of a distribution—for instance, the mean—can be represented by a single number, while some—for instance, the shape—cannot. We have noted that a numerical measure that describes a sample is called a statistic. Correspondingly, a numerical measure that describes a population is called a parameter. For the most important numerical measures, we have defined notations to distinguish between the statistic and the parameter. These notations are summarized in Table 2.18 for convenient reference.

A Look Ahead

It is natural to view a sample characteristic (for instance, $\bar{y}$) as an estimate of the corresponding population characteristic (for instance, μ). But in taking such a view one must guard against unjustified optimism. Of course, if the sample were perfectly

---

* You may wonder why we use $\bar{y}$ and s instead of $\hat{\mu}$ and $\hat{\sigma}$. One answer is "tradition." Another answer is that since "ˆ" means estimate, you might have other estimates in mind.

TABLE 2.18 Notation for Some Important Statistics and Parameters

Measure	Sample value (Statistic)	Population value (Parameter)
Proportion	$\hat{p}$	p
Mean	$\bar{y}$	μ
Standard deviation	s	σ

representative of the population, then the estimate would be perfectly accurate. But this raises the central question: How representative (of the population) is a sample likely to be? Intuition suggests that, if the observational units are appropriately selected, then the sample should be more or less representative of the population. Intuition also suggests that larger samples should tend to be more representative than smaller samples. These intuitions are basically correct, but they are too vague to provide practical guidance for research in the life sciences. Practical questions that need to be answered are as follows:

1. How can an investigator judge whether a sample can be viewed as "more or less" representative of a population?

2. How can an investigator quantify "more or less" in a specific case?

In Chapter 3 we will describe a theoretical model—the random sampling model—that provides a framework for the judgment in question (1), and in Chapter 6 we will see how this model can provide a concrete answer to question (2). Specifically, in Chapter 6 we will see how to analyze a set of data so as to quantify how closely the sample mean ($\bar{y}$) estimates the population mean (μ). But before returning to data analysis in Chapter 6, we will need to lay some groundwork in Chapters 3, 4, and 5; the developments in these chapters are an essential prelude to understanding the techniques of statistical inference.

Supplementary Exercises 2.62–2.80

2.62 A botanist grew 15 pepper plants on the same greenhouse bench. After 21 days, she measured the total stem length (cm) of each plant, and obtained the following values:[48]

$$
\begin{array}{ccc}
12.4 & 12.2 & 13.4 \\
10.9 & 12.2 & 12.1 \\
11.8 & 13.5 & 12.0 \\
14.1 & 12.7 & 13.2 \\
12.6 & 11.9 & 13.1 \\
\end{array}
$$

(a) Construct a stem-and-leaf display for these data, and use it to determine the quartiles.
(b) Calculate the interquartile range.

2.63 Here are the 20 measurements of preening time reported in Exercise 2.12:

$$
\begin{array}{ccccc}
34 & 24 & 10 & 16 & 52 \\
76 & 33 & 31 & 46 & 24 \\
18 & 26 & 57 & 32 & 25 \\
48 & 22 & 48 & 29 & 19 \\
\end{array}
$$

(a) Determine the median and the quartiles.
(b) Determine the interquartile range.
(c) Construct a (modified) boxplot of the data.

2.64 To calibrate a standard curve for assaying protein concentrations, a plant pathologist used a spectrophotometer to measure the absorbance of light (wavelength 500 nm) by a protein solution. The results of 27 replicate assays of a standard solution containing 60 μg protein per mLi water were as follows:[49]

MINITAB: Modified boxplot.

Construct a frequency

2.65 Refer to the absorban

(a) Prepare a stem-an
(b) Use the stem-and-
    tiles, and the interc
(c) How large must an

2.66 Twenty patients with sev   re
the numbers of major sei   .[50]

5
5

(a) Determine the median number of seizures.
(b) Determine the mean number of seizures.
(c) Construct a histogram of the data. Mark the positions of the mean and the median on the histogram.
(d) What feature of the frequency distribution suggests that neither the mean nor the median is a meaningful summary of the experience of these patients?

2.67 Calculate the standard deviation of each of the following fictitious samples:

(a) $11, 8, 4, 10, 7$ (b) $23, 29, 24, 21, 23$ (c) $6, 0, -3, 2, 5$

2.68 To study the spatial distribution of Japanese beetle larvae in the soil, researchers divided a 12×12-foot section of a cornfield into 144 one-foot squares. They counted the number of larvae Y in each square, with the results shown in the following table.[51]

Number of Larvae	Frequency (Number of Squares)
0	13
1	34
2	50
3	18
4	16
5	10
6	2
7	1
Total	144

(a) The mean and standard deviation of Y are $\bar{y} = 2.23$ and $s = 1.47$. What percentage of the observations are within

I. 1 standard deviation of the mean?
II. 2 standard deviations of the mean?

(b) Determine the total number of larvae in all 144 squares. How is this number related to $\bar{y}$?

(c) Determine the median value of the distribution.

2.69 One measure of physical fitness is maximal oxygen uptake, which is the maximum rate at which a person can consume oxygen. A treadmill test was used to determine the maximal oxygen uptake of nine college women before and after participation in a ten-week program of vigorous exercise. The accompanying table shows the before and after measurements and the change (after−before); all values are in mLi O_2 per mm per kg body weight.[52]

Maximal Oxygen Uptake

Participant	Before	After	Change
1	48.6	38.8	−9.8
2	38.0	40.7	2.7
3	31.2	32.0	.8
4	45.5	45.4	−.1
5	41.7	43.2	1.5
6	41.8	45.3	3.5
7	37.9	38.9	1.0
8	39.2	43.5	4.3
9	47.2	45.0	−2.2

The following computations are to be done on the *change* in maximal oxygen uptake (the right-hand column).

(a) Calculate the mean and the standard deviation.
(b) Determine the median.
(c) Eliminate participant 1 from the data and repeat parts (a) and (b). Which of the descriptive measures display resistance and which do not?

2.70 A veterinary anatomist investigated the spatial arrangement of the nerve cells in the intestine of a pony. He removed a block of tissue from the intestinal wall, cut the block into many equal sections, and counted the number of nerve cells in each of 23 randomly selected sections. The counts were as follows.[53]

$$35 \quad 19 \quad 33 \quad 34 \quad 17 \quad 26 \quad 16 \quad 40$$
$$28 \quad 30 \quad 23 \quad 12 \quad 27 \quad 33 \quad 22 \quad 31$$
$$28 \quad 28 \quad 35 \quad 23 \quad 23 \quad 19 \quad 29$$

Construct a stem-and-leaf diagram of the data.

2.71 Refer to the nerve-cell data of Exercise 2.70.

(a) Use the stem-and-leaf display of part (a) to determine the median, the quartiles, and the interquartile range.
(b) Construct a boxplot of the data.

2.72 Part (a) of Exercise 2.71 asks for a stem-and-leaf display of the nerve-cell data. Does this graphic support the claim that the data came from a reasonably symmetric and mound-shaped distribution?

2.73   A geneticist counted the number of bristles on a certain region of the abdomen of the fruitfly *Drosophila melanogaster*. The results for 119 individuals were as shown in the table.[54]

Number of Bristles	Number of Flies	Number of Bristles	Number of Flies
29	1	38	18
30	0	39	13
31	1	40	10
32	2	41	15
33	2	42	10
34	6	43	2
35	9	44	2
36	11	45	3
37	12	46	2

(a) Find the median number of bristles.
(b) Find the first and third quartiles of the sample.
(c) Make a boxplot of the data.
(d) The sample mean is 38.45 and the standard deviation is 3.20. What percentage of the observations fall within 1 standard deviation of the mean?

2.74   The carbon monoxide in cigarettes is thought to be hazardous to the fetus of a pregnant woman who smokes. In a study of this theory, blood was drawn from pregnant women before and after smoking a cigarette. Measurements were made of the percent of blood hemoglobin bound to carbon monoxide as carboxyhemoglobin (COHb). The results for ten women are shown in the table.[55]

	Blood COHb (%)		
Subject	*Before*	*After*	*Increase*
1	1.2	7.6	6.4
2	1.4	4.0	2.6
3	1.5	5.0	3.5
4	2.4	6.3	3.9
5	3.6	5.8	2.2
6	.5	6.0	5.5
7	2.0	6.4	4.4
8	1.5	5.0	3.5
9	1.0	4.2	3.2
10	1.7	5.2	3.5

(a) Calculate the mean and standard deviation of the *increase* in COHb.
(b) Calculate the mean COHb before and the mean after. Is the mean increase equal to the increase in means?
(c) Construct a stem-and-leaf diagram of the increase in COHb. Use the diagram to determine the median increase.
(d) Repeat part (c) for the before measurements and for the after measurements. Is the median increase equal to the increase in medians?

2.75   *(Computer problem)*   A medical researcher in India obtained blood specimens from 31 young children, all of whom were infected with malaria. The following data, listed in increasing order, are the numbers of malarial parasites found in 1 ml of blood from each child.[56]

100	140	140	271	400	435	455	770
826	1,400	1,540	1,640	1,920	2,280	2,340	3,672
4,914	6,160	6,560	6,741	7,609	8,547	9,560	10,516
14,960	16,855	18,600	22,995	29,800	83,200	134,232	

(a) Construct a frequency distribution of the data, using a class width of 10,000; display the distribution as a histogram.
(b) Transform the data by taking the logarithm (base 10) of each observation. Construct a frequency distribution of the transformed data and display it as a histogram. How does the log transformation affect the shape of the frequency distribution?
(c) Determine the mean of the original data and the mean of the log-transformed data. Is the mean of the logs equal to the log of the mean?
(d) Determine the median of the original data and the median of the log-transformed data. Is the median of the logs equal to the log of the median?

2.76 Rainfall, measured in inches, for the month of June in Cleveland, Ohio, was recorded for each of 41 years.[57] The values had a minimum of 1.2, an average of 3.6, and a standard deviation of 1.6. Which of the following is a rough histogram for the data? How do you know?

2.77 The following histograms (a, b, and c) show three distributions.

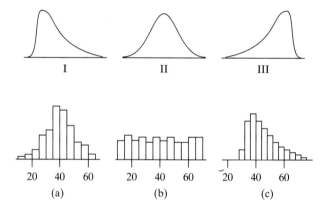

The computer output given below shows the mean, median, and standard deviation of the three distributions, plus the mean, median, and standard deviation for a fourth distribution. Match the histograms with the statistics. Explain your reasoning. (One set of statistics will not be used.)

1. Count	100		2. Count	100
Mean	41.3522		Mean	39.6761
Median	39.5585		Median	39.5377
StdDev	13.0136		StdDev	10.0476
3. Count	100		4. Count	100
Mean	37.7522		Mean	39.6493
Median	39.5585		Median	39.5448
StdDev	13.0136		StdDev	17.5126

2.78   The following boxplots show mortality rates (deaths within one year per 100 patients) for heart transplant patients at various hospitals. The low-volume hospitals are those that perform between 5 and 9 transplants per year. The high-volume hospitals perform 10 or more transplants per year.[58] Describe the distributions, paying special attention to how they compare to one another. Be sure to note the shape, center, and spread of each distribution.

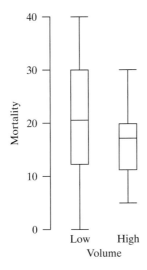

2.79   *(Computer problem)*   Physicians measured the concentration of calcium (n*M*) in blood samples from 38 healthy persons. The data are as follows.[59]

95	110	135	120	88	125
112	100	130	107	86	130
122	122	127	107	107	107
88	126	125	112	78	115
78	102	103	93	88	110
104	122	112	80	121	126
90	96				

Calculate appropriate measures of the center and spread of the distribution. Describe the shape of the distribution and any unusual features in the data.

2.80   The boxplot shows the same data that are shown in one of the three histograms. Which histogram goes with the boxplot? Explain your answer.

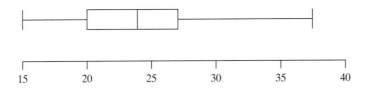

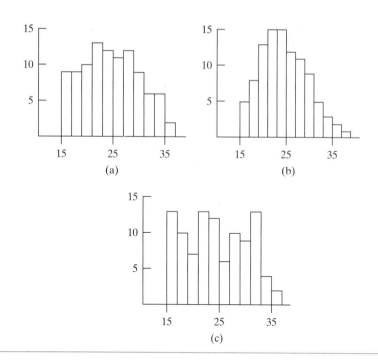

CHAPTER 3

Random Sampling, Probability, and the Binomial Distribution

3.1 PROBABILITY AND THE LIFE SCIENCES

Probability, or chance, plays an important role in scientific thinking about living systems. Some biological processes are affected directly by chance. A familiar example is the segregation of chromosomes in the formation of gametes; another example is the occurrence of mutations.

Even when the biological process itself does not involve chance, the results of an experiment are always somewhat affected by chance: chance fluctuations in environmental conditions, chance variation in the genetic makeup of experimental animals, and so on. Often, chance also enters directly through the design of an experiment; for instance, varieties of wheat may be randomly allocated to plots in a field. (Random allocation is discussed in Chapter 8.)

The conclusions of a statistical data analysis are often stated in terms of probability. Probability enters statistical analysis not only because chance influences the results of an experiment, but also because of theoretical frameworks, or *models*, that are used as a basis for statistical inference. In this chapter we describe the most fundamental of these theoretical models, the random sampling model. In addition, we introduce the language of probability and develop some simple tools for manipulating probabilities.

3.2 RANDOM SAMPLING

The first step in developing a basis for statistical inference is to define what is meant by random sampling.

Definition of Random Sampling

Informally, the process of random sampling can be visualized in terms of labeled tickets, such as those used in a lottery or raffle. Suppose that

---

Objectives

In this chapter we will study the basic ideas of probability, including

- *the role of random sampling in statistics*

- *the "limiting-frequency" definition of probability*

- *the use of probability trees*

- *the concept of a random variable*

- *rules for finding means and standard deviations of random variables*

- *the use of the binomial distribution*

each member of the population is represented by one ticket, and that the tickets are placed in a large box and thoroughly mixed. Then n tickets are drawn from the box by a blindfolded assistant, with new mixing after each ticket is removed; these n tickets constitute the sample. (Equivalently, we may visualize that n assistants reach in the box simultaneously, each assistant drawing one ticket.)

More abstractly, we may define random sampling as follows:

> **A Simple Random Sample**
> A **simple random sample** of n items is a sample in which (a) every member of the population has the same chance of being included in the sample; and (b) the members of the sample are chosen independently of each other. [Requirement (b) means that the chance of a given member of the population being chosen does not depend on which other members are chosen.]*

Simple random sampling can be thought of in other, equivalent, ways. We may envision the sample members being chosen one at a time from the population; under simple random sampling, at each stage of the drawing every remaining member of the population is equally likely to be the next one chosen. Another view is to consider the totality of possible samples of size n; if all possible samples are equally likely to be obtained, then the process gives a simple random sample.

There are other kinds of sampling that are random in a sense but that are not simple. For example, consider sampling from a human population as follows: First choose some families at random, and then include in the sample all members of those families. With this kind of sampling, which is called *cluster sampling*, all members of the population have the same chance of being in the sample, but the various members of the sample are not chosen independently of each other.

A sample chosen by random sampling is often called a *random sample*. But note that it is actually the *process* of sampling rather than the sample itself that is defined as random; randomness is not a property of the particular sample that happens to be chosen.

Choosing a Random Sample

The technique of actually choosing a random sample from a concrete population has two types of application in biological studies: (1) choosing a sample of units for study from a large population that is available; and (2) random allocation of units to treatment groups (as explained in Chapter 8). In addition, some of the exercises

---

* Technically, requirement (b) is that every pair of members of the population has the same chance of being selected for the sample, every group of three members of the population has the same chance of being selected for the sample, and so on. In contrast, suppose we had a population with 30 persons in it and we wrote the names of three persons on each of 10 tickets. Then we could then choose one ticket in order to get a sample of size $n = 3$, but this would not be a simple random sample, since the pair $(1, 2)$ could end up in the sample but the pair $(1, 4)$ could not. Here the selections of members of the sample are not independent of each other. (This kind of sampling is known as "cluster sampling," with 10 clusters of size 3.) If the population is infinite, then the technical definition, that all subsets of a given size are equally likely to be selected as part of the sample, is equivalent to the requirement that the members of the sample are chosen independently.

("sampling exercises") in this book require random sampling; by giving you some experience with drawing random samples and looking at the results, these exercises are designed to help you feel more comfortable with statistical reasoning.

The technique of random sampling is easy to learn. First, you need a source of random digits. A calculator or computer can supply random digits. Alternatively, you can use a table of random digits, such as Table 1 at the end of this book.

How to Read Random Digits from Your Calculator or Computer. Many calculators and computer programs generate random numbers. Sometimes these numbers are expressed as decimal numbers between 0 and 1; to convert these to random digits, simply ignore the decimal and just read the individual digits in each random number. If you need single-digit numbers, read only the first digit; if you need two-digit numbers, read the first two digits; and so on.

How to Use the Table of Random Digits. For ease of reading, the rows and columns of Table 1 are numbered, and the digits in the table are grouped into 5×5 blocks. To use Table 1, begin reading at a random place in the table.* If you need single-digit numbers, just read down the table; if you need two-digit numbers, read two columns across, down the table; and so on. When you get to the bottom, go back to the top, move over an appropriate number of columns so that you will not use the same column twice, and continue reading.

Remark: In calling the digits in Table 1 or your calculator or computer *random* digits, we are using the term *random* loosely. Strictly speaking, random digits are digits produced by a random *process*—for example, tossing a ten-sided die. The digits in Table 1 or in your calculator or computer are actually *pseudorandom* digits; they are generated by a deterministic (although possibly very complex) process that is designed to produce sequences of digits that mimic randomly generated sequences. For those readers who are curious about this, a simple example of a procedure for generating pseudorandom digits is given in Appendix 3.1.

How to Choose a Random Sample. The following is a simple procedure for choosing a random sample of *n* items from a finite population of items.

(a) Label the members of the population with identification numbers. All identification numbers must have the same number of digits; for instance, if the population contains 75 items, the identification numbers could be $01, 02, \ldots, 75$.

(b) Read numbers from Table 1 or your calculator or computer. Reject any numbers that do not correspond to any population member. (For example, if the population has 75 items that have been assigned identification numbers $01, 02, \ldots, 75$, then skip over the numbers $76, 77, \ldots, 99$ and 00.) Continue until *n* numbers have been acquired. (Ignore any repeated occurrence of the same number.)

(c) The population members with the chosen identification numbers constitute the sample.

The following example illustrates this procedure.

---

* There are various ways to choose a random starting place. One simple method is to close your eyes and drop a paper clip onto Table 1; start reading at the digit closest to the outer end of the paper clip wire.

Example 3.1

Suppose we are to choose a random sample of size 6 from a population of 75 members. Label the population members $01, 02, \ldots, 75$. Suppose we dropped a paper clip on Table 1 and selected a starting point at row 04, column 12; we obtain the numbers shaded in Table 3.1, which is a reproduction of part of Table 1.

We ignore the numbers greater than 75, we ignore 00, and we ignore the second occurrence of 23. Thus, the population members with the following identification numbers will constitute the sample:

23 38 59 21 08 09

TABLE 3.1 **Reproduction of Part of Table 1**				
	01	**06**	**11**	**16**	**21**
01	06048	96063	22049	86532	75170
02	25636	73908	85512	78073	19089
03	61378	45410	43511	54364	97334
04	15919	71559	12310	00727	54473
05	47328	20405	88019	82276	33679
06	72548	80667	53893	64400	81955
07	87154	04130	55985	44508	37515
08	68379	96636	32154	94718	22845
09	89391	54041	70806	36012	30833
10	15816	60231	28365	61924	66934
11	29618	55219	18394	11625	27673
12	30723	42988	30002	95364	45473
13	54028	04975	92323	53836	76128
14	40376	02036	48087	05216	26684
15	64439	37357	90935	57330	79738

Remark: If the population is large, then computer software can be quite helpful in generating a sample. If you need a random sample of size 15 from a population with 2,500 members, have the computer (or calculator) generate 15 random numbers between 1 and 2,500. (If there are duplicates in the set of 15, then go back and get more random numbers.)

The Random Sampling Model

We saw in Chapter 2 that, in order to generalize beyond a particular set of data, an investigator may view the data as a sample from a population. But how can we provide a rationale for inference from a limited sample to a very much larger population? The approach of statistical theory is to refer to an idealized model of the sample-population relationship. In this model, which is called the **random sampling model**, the sample is chosen from the population by random sampling. The model is represented schematically in Figure 3.1.

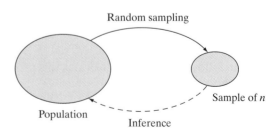

Figure 3.1 The random sampling model

In Chapter 2 we posed a central question of statistical inference: How representative (of the population) is a sample likely to be? The random sampling model is useful because it provides a basis for answering this question. The model can be used to determine how much an inference might be influenced by chance, or "luck of the draw." More explicitly, a randomly chosen sample usually will not exactly resemble the population from which it was drawn. The discrepancy between the sample and the population is called **chance error due to sampling**. We will see in later chapters how statistical theory derived from the random sampling model enables us to set limits on the likely amount of error due to sampling in an experiment. The quantification of such error is a major contribution that statistical theory has made to scientific thinking.

How does the random sampling model relate to reality? In some studies in the life sciences, the observational units are literally chosen by random sampling. In much biological research, however, the observations in a data set are not chosen by an actual random sampling procedure. Before applying the random sampling model to a real study, it is necessary to ask. Can the data in this study reasonably be viewed *as if* they were obtained by random sampling from some population? The first step in answering this question is to define the population. As discussed in detail in Section 2.8, in defining the population one tries to identify those factors that are relevant to the observed variable Y. The next step is to scrutinize the procedure by which the observational units were selected and to ask, Could the *observations* have been chosen at random?

The most clear-cut kind of nonrandomness is **sampling bias**, which is a systematic tendency for some values of Y to be selected more readily than others. The following two examples illustrate sampling bias.

Lengths of Fish.   A biologist plans to study the distribution of body length in a certain population of fish in the Chesapeake Bay. The sample will be collected using a fishing net. Smaller fish can more easily slip through the holes in the net. Thus, smaller fish are less likely to be caught than larger ones, so the sampling procedure is biased.   ■

Example 3.2

Sizes of Nerve Cells.   A neuroanatomist plans to measure the sizes of individual nerve cells in cat brain tissue. In examining a tissue specimen, the investigator must decide which of the hundreds of cells in the specimen should be selected for measurement. Some of the nerve cells are incomplete because the microtome cut through them when the tissue was sectioned. If the size measurement can be made only on complete cells, a bias arises because the smaller cells had a greater chance of being missed by the microtome blade.   ■

Example 3.3

When the sampling procedure is biased, the sample mean is a poor estimate of the population mean because it is systematically distorted. For instance, in Example 3.2 smaller fish will tend to be underrepresented in the sample, so the sample mean length will be an overestimate of the population mean length.

The following example illustrates a kind of nonrandomness that is different from bias.

Sucrose in Beet Roots.   An agronomist plans to sample beet roots from a field in order to measure their sucrose content. Suppose she were to take all her specimens from a randomly selected small area of the field. This sampling

Example 3.4

procedure would not be biased but would tend to produce *too homogeneous* a sample because environmental variation across the field would not be reflected in the sample. ∎

Example 3.4 illustrates an important principle that is sometimes overlooked in the analysis of data: In order to check applicability of the random sampling model, one needs to ask not only whether the sampling procedure might be biased, but also whether the sampling procedure will adequately reflect the variability inherent in the population. Faulty information about variability can distort scientific conclusions just as seriously as bias can.

We now consider some examples where the random sampling model might reasonably be applied.

Example 3.5

Fungus Resistance in Corn.   A certain variety of corn is resistant to fungus disease. To study the inheritance of this resistance, an agronomist crossed the resistant variety with a nonresistant variety and measured the degree of resistance in the progeny plants. The actual progeny in the experiment can be regarded as a random sample from a conceptual population of all *potential* progeny of that particular cross. ∎

When the purpose of a study is to *compare* two or more experimental conditions, a very narrow definition of the population may be satisfactory, as illustrated in the next example.

Example 3.6

Nitrite Metabolism.   To study the conversion of nitrite to nitrate in the blood, researchers injected four New Zealand White rabbits with a solution of radioactively labeled nitrite molecules. Ten minutes after injection, they measured for each rabbit the percentage of the nitrite that had been converted to nitrate.[1] Although the four animals were not literally chosen at random from a specified population, nevertheless it might be reasonable to view the measurements of nitrite metabolism as a random sample from similar measurements made on all New Zealand White rabbits. (This formulation assumes that age and sex are irrelevant to nitrite metabolism.) ∎

Example 3.7

Treatment of Ulcerative Colitis.   A medical team conducted a study of two therapies, A and B, for treatment of ulcerative colitis. All the patients in the study were referral patients in a clinic in a large city. Each patient was observed for satisfactory "response" to therapy. In applying the random sampling model, the researchers might want to make an inference to the population of all ulcerative colitis patients in urban referral clinics. First consider inference about the actual probabilities of response; such an inference would be valid if the probability of response to each therapy is the same at all urban referral clinics. However, this assumption might be somewhat questionable, and the investigators might believe that the population should be defined very narrowly—for instance, as "the type of ulcerative colitis patients who are referred to this clinic." Even such a narrow population can be of interest in a comparative study. For instance, if treatment A is better than treatment B for the narrow population, it might be reasonable to infer that A would be better than B for a broader population (even if the actual response probabilities might be different in the broader population). In fact, it might even be argued that the broad population should include all ulcerative colitis patients, not merely those in urban referral clinics. ∎

      It often happens in research that, for practical reasons, the population actually studied is narrower than the population that is of real interest. In order to apply the kind of rationale illustrated in Example 3.7, one must argue that the results in the narrowly defined population (or, at least, some aspects of those results) can be meaningfully extrapolated to the population of interest. This extrapolation is not a *statistical* inference; it must be defended on biological, not statistical, grounds.

Exercises 3.1–3.2

3.1   *(Sampling exercise)* Refer to the collection of 100 ellipses shown in the accompanying figure, which can be thought of as representing a natural population of the mythical organism *C. ellipticus*. The ellipses have been given identification numbers 00, 01, 99 for convenience in sampling. Certain individuals of *C. ellipticus* are mutants and have two tail bristles.

    (a) Use your *judgment* to choose a sample of size 10 from the population that you think is representative of the entire population. Note the number of mutants in the sample.

    (b) Use *random digits* (from Table 1 or your calculator or computer) to choose a random sample of size 10 from the population and note the number of mutants in the sample.

3.2   *(Sampling exercise)* Refer to the collection of 100 ellipses.

    (a) Use random digits (from Table 1 or your calculator or computer) to choose a random sample of size 5 from the population and note the number of mutants in the sample.

    (b) Repeat part (a) nine more times, for a total of ten samples. (Some of the ten samples may overlap.)

To facilitate pooling of results from the entire class, report your results in the following format:

Number of		Frequency
Mutants	*Nonmutants*	*(No. of Samples)*
0	5	
1	4	
2	3	
3	2	
4	1	
5	0	
		Total: 10

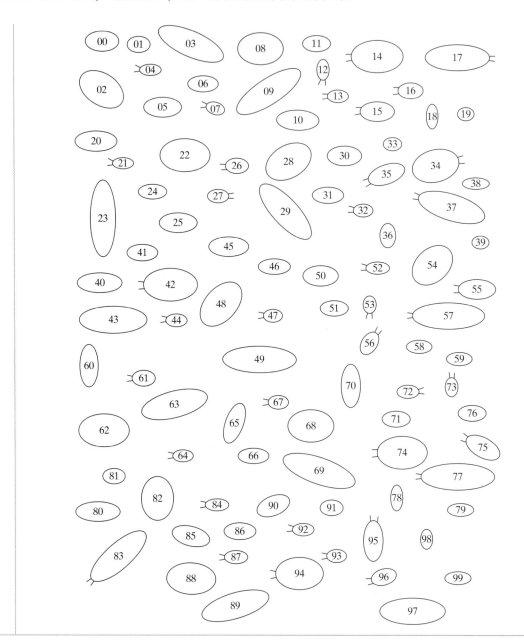

3.3 INTRODUCTION TO PROBABILITY

In this section we introduce the language of probability and its interpretation.

Basic Concepts

A **probability** is a numerical quantity that expresses the likelihood of an event. The probability of an event E is written as

$$\Pr\{E\}$$

The probability $\Pr\{E\}$ is always a number between 0 and 1, inclusive.

We can speak meaningfully about a probability $\Pr\{E\}$ only in the context of a chance operation—that is, an operation whose outcome is determined at least partially by chance. The chance operation must be defined in such a way that *each time the chance operation is performed, the event E either occurs or does not occur.* The following two examples illustrate these ideas.

Coin Tossing.   Consider the familiar chance operation of tossing a coin, and define the event

Example 3.8

$$E\text{: Heads}$$

Each time the coin is tossed, either it falls heads or it does not. If the coin is equally likely to fall heads or tails, then

$$\Pr\{E\} = \frac{1}{2} = .5$$

Such an ideal coin is called a "fair" coin. If the coin is not fair (perhaps because it is slightly bent), then $\Pr\{E\}$ will be some value other than .5, for instance

$$\Pr\{E\} = .6 \qquad \blacksquare$$

Coin Tossing.   Consider the event

Example 3.9

$$E\text{: 3 heads in a row}$$

The chance operation "toss a coin" is *not* adequate for this event because we cannot tell from one toss whether E has occurred. A chance operation that would be adequate is

   Chance operation: Toss a coin 3 times

Another chance operation that would be adequate is

   Chance operation: Toss a coin 100 times

with the understanding that E occurs if there is a run of 3 heads anywhere in the 100 tosses. Intuition suggests that E would be more likely with the second definition of the chance operation (100 tosses) than with the first (3 tosses). This intuition is correct and serves to underscore the importance of the chance operation in interpreting a probability. $\qquad \blacksquare$

The language of probability can be used to describe the results of random sampling from a population. The simplest application of this idea is a sample of size $n = 1$—that is, choosing one member at random from a population. The following is an illustration.

Sampling Fruitflies.   A large population of the fruitfly *Drosophila melanogaster* is maintained in a lab. In the population, 30% of the individuals are black because of a mutation, while 70% of the individuals have the normal gray body color. Suppose one fly is chosen at random from the population. Then the probability that a black fly is chosen is .3. More formally, define

Example 3.10

$$E\text{: Sampled fly is black}$$

Then

$$\Pr\{E\} = .3 \qquad \blacksquare$$

The preceding example illustrates the basic relationship between probability and random sampling: *The probability that a randomly chosen individual has a certain characteristic is equal to the proportion of population members with the characteristic.*

Frequency Interpretation of Probability

The **frequency interpretation** of probability provides a link between probability and the real world by relating the probability of an event to a measurable quantity, namely, the long-run relative frequency of occurrence of the event.*

According to the frequency interpretation, the probability of an event E is meaningful only in relation to a chance operation that can in principle be repeated indefinitely often. Each time the chance operation is repeated, the event E either occurs or does not occur. *The probability $Pr\{E\}$ is interpreted as the relative frequency of occurrence of E in an indefinitely long series of repetitions of the chance operation.*

Specifically, suppose that the chance operation is repeated a large number of times, and that for each repetition the occurrence or nonoccurrence of E is noted. Then we may write

$$\Pr\{E\} \leftrightarrow \frac{\text{\# of times } E \text{ occurs}}{\text{\# of times chance operation is repeated}}$$

The arrow in the preceding expression indicates "approximate equality in the long run"; that is, if the chance operation is repeated many times, the two sides of the expression will be approximately equal. Here is a simple example.

Example 3.11 | **Coin Tossing.** Consider again the chance operation of tossing a coin and the event

$$E: \text{Heads}$$

If the coin is fair, then

$$\Pr\{E\} = .5 \leftrightarrow \frac{\text{\# of heads}}{\text{\# of tosses}}$$

The arrow in the preceding expression indicates that, in a long series of tosses of a fair coin, we expect to get heads about 50% of the time. ∎

The following two examples illustrate the relative frequency interpretation for more complex events.

Example 3.12 | **Coin Tossing.** Suppose that a fair coin is tossed twice. For reasons that will be explained in Section 3.4, the probability of getting heads both times is .25. This probability has the following relative frequency interpretation.

Chance operation: Toss a coin twice

$$E: \text{Both tosses are heads}$$

---

* Some statisticians prefer a different view, namely that the probability of an event is a subjective quantity expressing a person's "degree of belief" that the event will happen. Statistical methods based on this "subjectivist" interpretation are rather different from those presented in this book.

$$\Pr\{E\} = .25 \leftrightarrow \frac{\text{\# of times both tosses are heads}}{\text{\# of pairs of tosses}}$$

Sampling Fruitflies.   In the *Drosophila* population of Example 3.10, 30% of the flies are black and 70% are gray. Suppose that two flies are randomly chosen from the population. We will see in Section 3.4 that the probability that both flies are the same color is .58. This probability can be interpreted as follows:

> **Example 3.13**

Chance operation: Choose a random sample of size $n = 2$

E: Both flies in the sample are the same color

$$\Pr\{E\} = .58 \leftrightarrow \frac{\text{\# of times both flies are same color}}{\text{\# of times a sample of } n = 2 \text{ is chosen}}$$

We can relate this interpretation to a concrete sampling experiment. Suppose that the *Drosophila* population is in a very large container and that we have some mechanism for choosing a fly at random from the container. We choose one fly at random, and then another; these two constitute the first sample of $n = 2$. After recording their colors, we put the two flies back into the container, and we are ready to repeat the sampling operation once again. Such a sampling experiment would be tedious to carry out physically, but it can be readily simulated using a computer. Table 3.2 shows a partial record of the results of choosing 10,000 random samples of size $n = 2$ from a simulated *Drosophila* population. After each repetition of the chance operation (that is, after each sample of $n = 2$), the cumulative relative frequency of occurrence of the event E was updated, as shown in the rightmost column of the table.

Figure 3.2 shows the cumulative relative frequency plotted against the number of samples. Notice that, as the number of samples becomes large, the relative frequency of occurrence of E approaches .58 (which is $\Pr\{E\}$). In other words, the percentage of color-homogeneous samples among all the samples approaches 58% as the number of samples increases. It should be emphasized, however, that the *absolute* number of color-homogeneous samples generally does *not* tend to get closer to 58% of the total number. For instance, if we compare the results shown in Table 3.2 for the first 100 samples and the first 1,000 samples, we find the following:

	Color-Homogenous		Deviation from 58% of Total	
First 100 samples:	54	or 54 %	− 4	or −4 %
First 1,000 samples:	596	or 59.6%	+16	or +1.6%

Note that the deviation from 58% is larger in absolute terms, but smaller in relative terms (i.e., in percentage terms), for 1,000 samples than for 100 samples. Likewise, for 10,000 samples the deviation from 58% is rather larger (a deviation of −30), but the percentage deviation is quite small (30/10,000 is 0.3%). The deficit of 4 color-homogeneous samples among the first 100 samples is not *canceled* by a corresponding excess in later samples, but rather is *swamped*, or overwhelmed, by a larger denominator.

TABLE 3.2 **Partial Results of Simulated Sampling from a *Drosophila* Population**

Sample Number	Color 1st Fly	Color 2nd Fly	Did *E* Occur?	Relative Frequency of *E* (Cumulative)
1	G	B	No	.000
2	B	B	Yes	.500
3	B	G	No	.333
4	G	B	No	.250
5	G	G	Yes	.400
6	G	B	No	.333
7	B	B	Yes	.429
8	G	G	Yes	.500
9	G	B	No	.444
10	B	B	Yes	.500
⋮	⋮	⋮	⋮	⋮
20	G	B	No	.450
⋮	⋮	⋮	⋮	⋮
100	G	B	No	.540
⋮	⋮	⋮	⋮	⋮
1,000	G	G	Yes	.596
⋮	⋮	⋮	⋮	⋮
10,000	B	B	Yes	.577

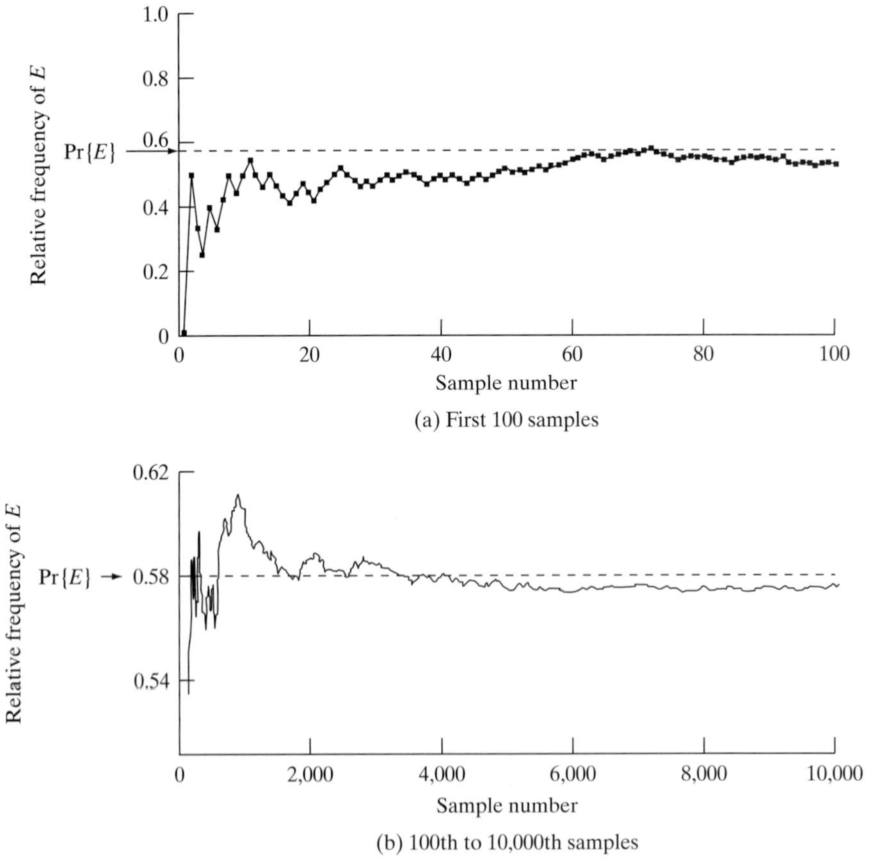

(a) First 100 samples

Figure 3.2 Results of sampling from Fruitfly population. Note that the axes are scaled differently in (a) and (b)

(b) 100th to 10,000th samples

Exercises 3.3–3.5

3.3　*(Sampling exercise)* Consider a string of five randomly generated digits; that is, each digit is equally likely to be any of $0, 1, 2, \ldots, 9$, regardless of the other digits. Let E be the event that all five digits are different. It can be shown that $\Pr\{E\} = .30$. Use Table 1 (or your calculator or computer) to generate 20 strings of 5 random digits each. Keep a record of your results, and tabulate the cumulative relative frequency of occurrence of E (as in Table 3.2). To facilitate pooling the results from the entire class, also report the total number of occurrences of E.

3.4　*(Sampling exercise)* Proceed as in Exercise 3.3, but generate 50 strings of 5 random digits each. Calculate the cumulative relative frequency of E only after every tenth string.

3.5　In a certain population of the freshwater sculpin, *Cottus rotheus*, the distribution of the number of tail vertebrae is as shown in the table.[2]

No. of Vertebrae	Percent of Fish
20	3
21	51
22	40
23	6
Total	100

Find the probability that the number of tail vertebrae in a fish randomly chosen from the population

(a) equals 21

(b) is less than or equal to 22

(c) is greater than 21

(d) is no more than 21

3.4 PROBABILITY TREES

Often it is helpful to use a **probability tree** to analyze a probability problem. A probability tree provides a convenient way to break a problem into parts and to organize the information available. The following examples show some applications of this idea.

Coin Tossing.　If a fair coin is tossed twice, then the probability of heads is .5 on each toss. The first part of a probability tree for this scenario shows that there are two possible outcomes for the first toss and that they have probability .5 each.

Example 3.14

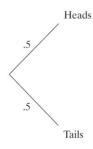

　　　　　　Heads

　　.5

.5

　　　Tails

Then the tree shows that, for either outcome of the first toss, the second toss can be either heads or tails, again with probabilities .5 each.

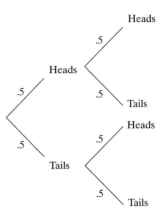

To find the probability of getting heads on both tosses, we consider the path through the tree that produces this event. We multiple together the probabilities that we encounter along the path. Figure 3.3 summarizes this example and shows that

$$\Pr\{\text{heads on both tosses}\} = .5 \times .5 = .25 \qquad ■$$

Combination of Probabilities

If an event can happen in more than one way, the relative frequency interpretation of probability can be a guide to the appropriate combination of the probabilities of subevents. The following example illustrates this idea.

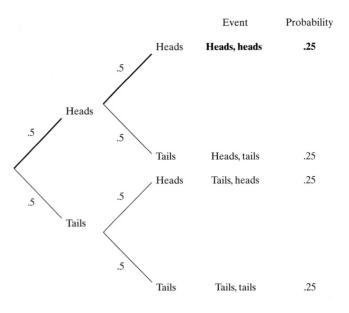

Figure 3.3 Probability tree for two coin tosses

Example 3.15

Sampling Fruitflies.   In the *Drosophila* population of Examples 3.10 and 3.13, 30% of the flies are black and 70% are gray. Suppose that two flies are randomly chosen from the population. Suppose we wish to find the probability that both flies are the same color. A probability tree shown in Figure 3.4 shows the four possible outcomes from sampling two flies. From the tree, we can see that the probability of getting two black flies is $.3 \times .3 = .09$. Likewise, the probability of getting two gray flies is $.7 \times .7 = .49$.

To find the probability of the event

E: Both flies in the sample are the same color

we add the probability of black, black to the probability of gray, gray to get

$$.09 + .49 = .58.$$ ■

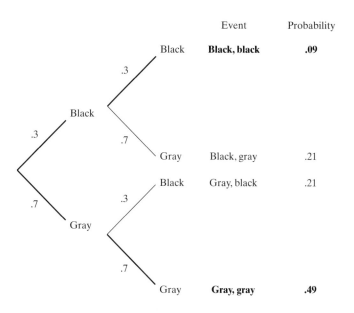

Figure 3.4 Probability tree for sampling two flies

In the coin tossing setting of Example 3.14, the second part of the probability tree had the same structure as the first part—namely, a .5 chance of heads and a .5 chance of tails—because the outcome of the first toss does not affect the probability of heads on the second toss. Likewise, in Example 3.15 the probability of the second fly being black was .3, regardless of the color of the first fly, because the population was assumed to be very large, so that removing one fly from the population would not affect the proportion of flies that are black. However, in some situations we need to treat the second part of the probability tree different from the first part.

Example 3.16

Nitric Oxide.   Hypoxic respiratory failure is a serious condition that affects some newborns. If a newborn has this condition, it is often necessary to use extracorporeal membrane oxygenation (ECMO) to save the life of the child. However, ECMO is an invasive procedure that involves inserting a tube into a vein or artery near the heart, so physicians hope to avoid the need for it. One treatment for hypoxic respiratory failure is to have the newborn inhale nitric oxide. To test the effectiveness of this treatment, newborns suffering hypoxic respiratory failure were assigned at random to either be given nitric oxide or a control group.[3] In the treatment group, 45.6% of the newborns had a negative outcome, meaning that either they needed ECMO or that they died. In the control group, 63.6% of the newborns had a negative outcome. Figure 3.5 shows a probability tree for this experiment.

If we choose a newborn at random from this group, there is a .5 probability that the newborn will be in the treatment group and, if so, a probability of .456 of getting a negative outcome. Likewise, there is a .5 probability that the newborn will be in the control group and, if so, a probability of .636 of getting a negative outcome. Thus, the probability of a negative outcome is

$$.5 \times .456 + .5 \times .636 = .228 + .318 = .546 \qquad \blacksquare$$

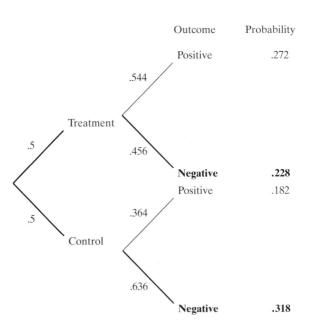

Figure 3.5 Probability tree for nitric oxide example

Example 3.17   **Medical Testing.**   Suppose a medical test is conducted on someone to try to determine whether or not the person has a particular disease. If the test indicates that the disease is present, we say the person has "tested positive." If the test indicates that the disease is not present, we say the person has "tested negative." However, there are two types of mistakes that can be made. It is possible that the test indicates that the disease is present, but the person does not really have the disease; this is known as a false positive. It is also possible that the person has the disease but the test does not detect it; this is known as a false negative.

Suppose that a particular test has a 95% chance of detecting the disease if the person has it (this is called the sensitivity of the test) and a 90% chance of correctly indicating that the disease is absent if the person really does not have the disease (this is called the specificity of the test). Suppose 8% of the population has the disease. What is the probability that a randomly chosen person will test positive?

Figure 3.6 shows a probability tree for this situation. The first split in the tree shows the division between those who have the disease and those who don't. If someone has the disease, then we use .95 as the chance of the person testing positive. If the person doesn't have the disease, then we use .10 as the chance of the person testing positive. Thus, the probability of a randomly chosen person testing positive is

$$.08 \times .95 + .92 \times .10 = .076 + .092 = .168$$ ■

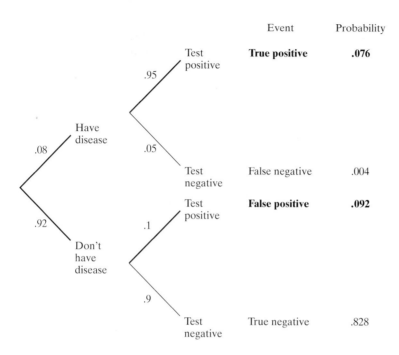

	Event	Probability
Test positive	**True positive**	**.076**
Test negative	False negative	.004
Test positive	**False positive**	**.092**
Test negative	True negative	.828

Figure 3.6 Probability tree for medical testing example

Example 3.18

False Positives.   Consider the medical testing scenario of Example 3.17. If someone tests positive, what is the chance the person really has the disease? In Example 3.17 we found that .168 (16.8%) of the population will test positive. The "true positives" make up .076 of this .168, which is to say that the probability that someone really has the disease, given that the person tests positive, is $\dfrac{.076}{.168} \approx .452.$

This probability is quite a bit smaller than most people expect it to be, given that the sensitivity and specificity of the test are .95 and .90.   ■

Exercises 3.6–3.11

3.6   In a certain college, 55% of the students are women. Suppose we take a sample of two students. Use a probability tree to find the probability

   (a) that both chosen students are women.
   (b) that at least one of the two students is a woman.

3.7   Suppose that a disease is inherited via a sex-linked mode of inheritance, so that a male offspring has a 50% chance of inheriting the disease, but a female offspring has no chance of inheriting the disease. Further suppose that 51.3% of births are male. What is the probability that a randomly chosen child will be affected by the disease?

3.8   Suppose that a student who is about to take a multiple choice test has only learned 40% of the material covered by the exam. Thus, there is a 40% chance that she will know the answer to a question. However, even if she does not know the answer to a question, she still has a 20% chance of getting the right answer by guessing. If we choose a question at random from the exam, what is the probability that she will get it right?

3.9   If a woman takes an early pregnancy test, she will either test positive, meaning that the test says she is pregnant, or test negative, meaning that the test says she is not pregnant. Suppose that if a woman really is pregnant, there is a 98% chance that she will test positive. Also, suppose that if a woman really is *not* pregnant, there is a 99% chance that she will test negative.

(a) Suppose that 1,000 women take early pregnancy tests and that 100 of them really are pregnant. What is the probability that a randomly chosen woman from this group will test positive?

(b) Suppose that 1,000 women take early pregnancy tests and that 50 of them really are pregnant. What is the probability that a randomly chosen woman from this group will test positive?

3.10   (a) Consider the setting of Exercise 3.9, part (a). Suppose that a woman tests positive. What is the probability that she really is pregnant?

(b) Consider the setting of Exercise 3.9, part (b). Suppose that a woman tests positive. What is the probability that she really is pregnant?

3.11   Suppose that a medical test has a 92% chance of detecting a disease if the person has it (i.e., 92% sensitivity) and a 94% chance of correctly indicating that the disease is absent if the person really does not have the disease (i.e., 94% specificity). Suppose 10% of the population has the disease.

(a) What is the probability that a randomly chosen person will test positive?

(b) Suppose that a randomly chosen person does test positive. What is the probability that this person really has the disease?

3.5 PROBABILITY RULES (OPTIONAL)

We have defined the probability of an event, $\Pr\{E\}$, as the long-run relative frequency with which the event occurs. In this section we will briefly consider a few rules that help determine probabilities. We begin with three basic rules.

Basic Rules

Rule 1: The probability of an event E is always between 0 and 1. That is, $0 \le \Pr\{E\} \le 1$.

Rule 2: The sum of the probabilities of all possible events equals 1. That is, if the set of possible events is $E_1, E_2, \ldots, E_k$, then $\Sigma \Pr\{E_i\} = 1$.

Rule 3: The probability that an event E does not happen, denoted by E^C, is one minus the probability that the event happens. That is, $\Pr\{E^C\} = 1 - \Pr\{E\}$. (We refer to E^C as the *complement* of E.)

We illustrate these rules with an example.

Example 3.19   **Blood Type.**   In the United States, 44% of the population has type O blood, 42% are type A, 10% are type B, and 4% are type AB. Consider choosing someone at random and determining the person's blood type. The probability of a given blood type will correspond to the population percentage.

(a) The probability that the person will have type O blood = $Pr\{O\}$ = .44.

(b) $Pr\{O\} + Pr\{A\} + Pr\{B\} + Pr\{AB\}$ = .44 + .42 + .10 + .04 = 1.

(c) The probability that the person will *not* have type O blood = $Pr\{O^C\}$ = 1 − .44 = .56. This could also be found by adding the probabilities of the other blood types: $Pr\{O^C\} = Pr\{A\} + Pr\{B\} + Pr\{AB\}$ = .42 + .10 + .04 = .56.   ∎

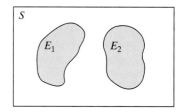

Figure 3.7 Venn diagram showing two disjoint events

We often want to discuss two or more events at once; to do this we will find some terminology to be helpful. We say that two events are *disjoint** if they cannot occur simultaneously. Figure 3.7 is a *Venn diagram* that depicts a *sample space* S of all possible outcomes as a rectangle with two disjoint events depicted as nonoverlapping regions.

The *union* of two events is the event that one or the other occurs or both occur. The *intersection* of two events is the event that they both occur. Figure 3.8 is a Venn diagram that shows the union of two events as the total shaded area, with the intersection of the events being the overlapping region in the middle.

If two events are disjoint, then the probability of their union is the sum of their individual probabilities. If the events are not disjoint, then to find the the probability of their union we take the sum of their individual probabilities and subtract the probability of their intersection (the part that was "counted twice").

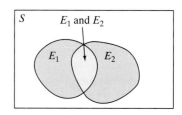

Figure 3.8 Venn diagram showing union (total shaded area) and intersection (middle area) of two events

Addition Rules

Rule 4: If two events E_1 and E_2 are disjoint, then
$Pr\{E_1 \text{ or } E_2\} = Pr\{E_1\} + Pr\{E_2\}$.

Rule 5: For any two events E_1 and E_2, $Pr\{E_1 \text{ or } E_2\} = Pr\{E_1\} + Pr\{E_2\} - Pr\{E_1 \text{ and } E_2\}$.

We illustrate these rules with an example.

Hair Color and Eye Color.   Table 3.3 shows the relationship between hair color and eye color for a group of 1,770 German men.[4]

Example 3.20

TABLE 3.3 Hair Color and Eye Color

		Brown	Black	Red	Total
		Hair color			
Eye color	Brown	400	300	20	720
	Blue	800	200	50	1,050
	Total	1,200	500	70	1,770

(a) Because events "black hair" and "red hair" are disjoint, if we choose someone at random from this group, then $Pr\{\text{black hair or red hair}\}$ = $Pr\{\text{black hair}\} + Pr\{\text{red hair}\}$ = 500/1,770 + 70/1,770 = 570/1,770.

(b) If we choose someone at random from this group, then $Pr\{\text{black hair}\}$ = 500/1,770.

---

* Another term for disjoint events is "mutually exclusive" events.

(c) If we choose someone at random from this group, then Pr{blue eyes} = 1,050/1,770.

(d) The events "black hair" and "blue eyes" are not disjoint, since there are 200 men with both black hair and blue eyes. Thus, Pr{black hair or blue eyes} = Pr{black hair} + Pr{blue eyes} − Pr{black hair and blue eyes} = 500/1,770 + 1,050/1,770 − 200/1,770 = 1,350/1,770. ■

Two events are said to be *independent* if knowing that one of them occurred does not change the probability of the other one occurring. For example, if a coin is tossed twice, the outcome of the second toss is independent of the outcome of the first toss, since knowing whether the first toss resulted in heads or in tails does not change the probability of getting heads on the second toss.

Events that are not independent are said to be *dependent*. When events are dependent, we need to consider the *conditional probability* of one event, given that the other event has happened. We use the notation

$$\Pr\{E_2|E_1\}$$

to represent the probability of E_2 happening, given that E_1 happened.

Example 3.21   **Hair Color and Eye Color.** Consider choosing a man at random from the group shown in Table 3.3. Overall, the probability of blue eyes is 1,050/1,770, or about 59.3%. However, if the man has black hair, then the conditional probability of blue eyes is only 200/500, or 40%; that is, Pr{blue eyes|black hair} = .40. Because the probability of blue eyes depends on hair color, the events "black hair" and "blue eyes" are dependent. ■

Refer again to Figure 3.8, which shows the intersection of two regions (for E_1 and E_2). If we know that the event E_1 has happened, then we can restrict our attention to the E_1 region in the Venn diagram. If we now want to find the chance that E_2 will happen, we need to consider the intersection of E_1 and E_2 relative to the entire E_1 region. In the case of Example 3.21, this corresponds to knowing that a randomly chosen man has black hair, so that we restrict our attention to the 500 men (out of 1,770 total in the group) with black hair. Of these men, 200 have blue eyes. The 200 are in the intersection of "black hair" and "blue eyes." The fraction 200/500 is the conditional probability of having blue eyes, given that the man has black hair. This leads to the following formal definition of the conditional probability of E_2 given E_1:

> **Definition**
> The conditional probability of E_2, given E_1, is
>
> $$\Pr\{E_2|E_1\} = \frac{\Pr\{E_1 \text{ and } E_2\}}{\Pr\{E_1\}}$$
>
> provided that $\Pr\{E_1\} > 0$.

Example 3.22   **Hair Color and Eye Color.** Consider choosing a man at random from the group shown in Table 3.3. The probability of the man having blue eyes given that he has black hair is

$$Pr\{\text{blue eyes|black hair}\} = Pr\{\text{black hair and blue eyes}\}/Pr\{\text{black hair}\}$$

$$= \frac{200/1{,}770}{500/1{,}770} = \frac{200}{500} = .40$$ ■

In Section 3.4 we used probability trees to study compound events. In doing so, we implicitly used multiplication rules that we now make explicit.

Multiplication Rules

Rule 6: If two events E_1 and E_2 are independent, then
$Pr\{E_1 \text{ and } E_2\} = Pr\{E_1\} \times Pr\{E_2\}$.

Rule 7: For any two events E_1 and
$E_2, Pr\{E_1 \text{ and } E_2\} = Pr\{E_1\} \times Pr\{E_2|E_1\}$.

Coin Tossing.   If a fair coin is tossed twice, the two tosses are independent of each other. Thus, the probability of getting heads on both tosses is

$$Pr\{\text{heads twice}\} = Pr\{\text{heads on first toss}\} \times Pr\{\text{heads on second toss}\}$$

$$= .5 \times .5 = .25$$ ■

| **Example 3.23** |

Blood Type.   In Example 3.19 we stated that 44% of the U.S. poplulation has type O blood. It is also true that 15% of the population is Rh negative and that this is independent of blood group. Thus, if someone is chosen at random, the probability that the person has type O, Rh negative blood is

$$Pr\{\text{group O and Rh negative}\} = Pr\{\text{group O}\} \times Pr\{\text{Rh negative}\}$$

$$= .44 \times .15 = .066$$ ■

| **Example 3.24** |

Hair Color and Eye Color.   Consider choosing a man at random from the group shown in Table 3.3. What is the probability that the man will have red hair and brown eyes? Hair color and eye color are dependent, so finding this probability involves using a conditional probability. The probability that the man will have red hair is 70/1,770. Given that the man has red hair, the conditional probability of brown eyes is 20/70. Thus,

$$Pr\{\text{red hair and brown eyes}\} = Pr\{\text{red hair}\} \times Pr\{\text{brown eyes|red hair}\}$$
$$= 70/1{,}770 \times 20/70 = 20/1{,}770$$ ■

| **Example 3.25** |

Hand Size.   Consider choosing someone at random from a population that is 60% female and 40% male. Suppose that for the women the average hand size, in cm^2, is 110, the standard deviation is 20, and the probability of having a hand size smaller than $100 \, cm^2$ is .31.[5] Suppose that for the men the average hand size, in cm^2, is 135, the standard deviation is 25, and the probability of having a hand size smaller than $100 \, cm^2$ is .08.* What is the probability that the randomly chosen person will have a hand size smaller than $100 \, cm^2$?

| **Example 3.26** |

---

* The probabilities follow from the use of a "normal distribution model." The normal curve is presented in detail in Chapter 4.

We are given that if the person is a woman, then the probability of a "small" hand size is .31 and that if the person is a man, then the probability of a "small" hand size is .08.

Thus,

$$\Pr\{\text{hand size} < 100\}$$
$$= \Pr\{\text{woman}\} \times \Pr\{\text{hand size} < 100|\text{woman}\}$$
$$+ \Pr\{\text{man}\} \times \Pr\{\text{hand size} < 100|\text{man}\}$$
$$= .6 \times .31 + .4 \times .08 = .186 + .032 = .218 \qquad ■$$

Exercises 3.12–3.14

3.12   In a study of the relationship between health risk and income, a large group of people living in Massachusetts were asked a series of questions.[6] Some of the results are shown in the following table.

	Income			
	Low	Medium	High	Total
Smoke	634	332	247	1213
Don't smoke	1846	1622	1868	5336
Total	2480	1954	2115	6549

(a) What is the probability that someone in this study smokes?
(b) What is the conditional probability that someone in this study smokes, given that the person has high income?
(c) Is being a smoker independent of having a high income? Why or why not?

3.13   The following data table is taken from the study reported in Exercise 3.12. Here "stressed" means that the person reported that most days are extremely stressful or quite stressful; "not stressed" means that the person reported that most days are a bit stressful, not very stressful, or not at all stressful.

	Income			
	Low	Medium	High	Total
Stressed	526	274	216	1016
Not stressed	1954	1680	1899	5533
Total	2480	1954	2115	6549

(a) What is the probability that someone in this study is stressed?
(b) What is the probability that someone in this study has low income?
(c) What is the probability that someone in this study either is stressed or has low income (or both)?
(d) What is the probability that someone in this study either is stressed and has low income?

3.14   Suppose that in a certain population of married couples 30% of the husbands smoke, 20% of the wives smoke, and in 8% of the couples both the husband and the wife smoke. Is the smoking status (smoker or nonsmoker) of the husband independent of that of the wife? Why or why not?

3.6 DENSITY CURVES

The examples presented in Sections 3.3 and 3.4 dealt with probabilities for discrete variables. In this section we consider probability when the variable is continuous.

Relative Frequency Histograms and Density Curves

In Chapter 2 we discussed the use of a histogram to represent a frequency distribution for a variable. A relative frequency histogram is a histogram in which we indicate the proportion (i.e., the relative frequency) of observations in each category, rather than the count of observations in the category. We can think of the relative frequency histogram as an approximation of the underlying true population distribution from which the data came.

It is often desirable, especially when the observed variable is continuous, to describe a population frequency distribution by a smooth curve. We may visualize the curve as an idealization of a relative frequency histogram with very narrow classes. The following example illustrates this idea.

Blood Glucose.   A glucose tolerance test can be useful in diagnosing diabetes. The blood level of glucose is measured one hour after the subject has drunk 50 mg of glucose dissolved in water. Figure 3.9 shows the distribution of responses to this test for a certain population of women.[7] The distribution is represented by histograms with class widths equal to (a) 10 and (b) 5, and by (c) a smooth curve. ■

Example 3.27

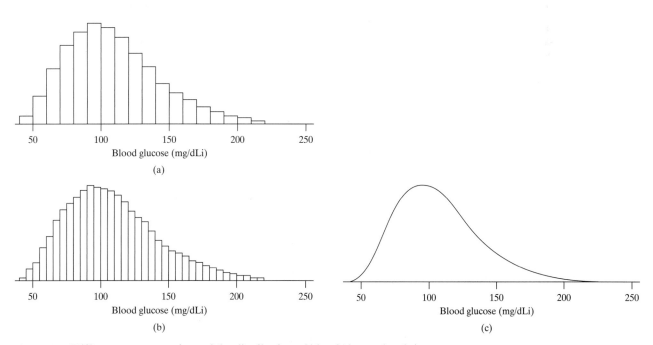

Figure 3.9 Different representations of the distribution of blood glucose levels in a population of women

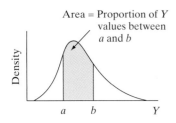

Figure 3.10 Interpretation of area under a density curve

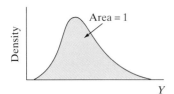

Figure 3.11 The area under an entire density curve must be 1.

A smooth curve representing a frequency distribution is called a **density curve**. The vertical coordinates of a density curve are plotted on a scale called a **density scale**. When the density scale is used, relative frequencies are represented as areas under the curve. Formally, the relation is as follows:

> **Interpretation of Density**
> For any two numbers a and b,
>
> $$\begin{array}{c}\text{Area under density curve} \\ \text{between } a \text{ and } b\end{array} = \begin{array}{c}\text{Proportion of } Y \text{ values} \\ \text{between } a \text{ and } b\end{array}$$
>
> This relation is indicated in Figure 3.10 for an arbitrary distribution.

Because of the way the density curve is interpreted, the density curve is entirely above (or equal to) the x-axis and the area under the entire curve must be equal to 1, as shown in Figure 3.11.

The interpretation of density curves in terms of areas is illustrated concretely in the following example.

Example 3.28

Blood Glucose.   Figure 3.12 shows the density curve for the blood glucose distribution of Example 3.27, with the vertical scale explicitly shown. The shaded area is equal to .42, which indicates that about 42% of the glucose levels are between 100 mg/dLi and 150 mg/dLi. The area under the density curve to the left of 100 mg/dLi is equal to .50; this indicates that the population median glucose level is 100 mg/dLi. The area under the entire curve is 1.   ■

The Continuum Paradox.   The area interpretation of a density curve has a paradoxical element. If we ask for the relative frequency of a single specific Y value, the answer is zero. For example, suppose we want to determine from Figure 3.12 the relative frequency of blood glucose levels *equal* to 150. The area interpretation gives an answer of zero. This seems to be nonsense—how can every value of Y have a relative frequency of zero? Let us look more closely at the question. If blood glucose is measured to the nearest mg/dLi, then we are really asking for the relative frequency of glucose levels between 149.5 and 150.5 mg/dLi, and the corresponding area is not zero. On the other hand, if we are thinking of blood glucose as an *idealized* continuous variable, then the relative frequency of any particular

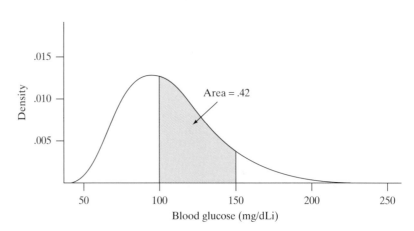

Figure 3.12 Interpretation of an area under the blood glucose density curve

value (such as 150) *is* zero. This is admittedly a paradoxical situation. It is similar to the paradoxical fact that an idealized straight line can be 1 centimeter long, and yet each of the idealized points of which the line is composed has length equal to zero. In practice, the continuum paradox does not cause any trouble; we simply do not discuss the relative frequency of a single Y value (just as we do not discuss the length of a single point).

Probabilities and Density Curves

If a variable has a continuous distribution, then we find probabilities by using the density curve for the variable. A probability for a continuous variable equals the area under the density curve for the variable between two points.

Blood Glucose.   Consider the blood glucose level, in mg/dLi, of a randomly chosen subject from the population described in Example 3.28. We saw in Example 3.28 that 42% of the population glucose levels are between 100 mg/dLi and 150 mg/dLi. Thus, $\Pr\{100 \leq \text{glucose level} \leq 150\} = .42$.

     We are modeling blood glucose level as being a continuous variable, which means that $\Pr\{\text{glucose level} = 100\} = 0$, as we noted previously. Thus,

$$\Pr\{100 \leq \text{glucose level} \leq 150\} = \Pr\{100 < \text{glucose level} < 150\} = .42 \quad \blacksquare$$

Example 3.29

Tree Diameters.   The diameter of a tree trunk is an important variable in forestry. The density curve shown in Figure 3.13 represents the distribution of diameters (measured 4.5 feet above the ground) in a population of 30-year-old Douglas fir trees; areas under the curve are shown in the figure.[8] Consider the diameter, in inches, of a randomly chosen tree. Then, for example, $\Pr\{4 < \text{diameter} < 6\} = .33$. If we want to find the probability that a randomly chosen tree has a diameter greater than 8 inches, we must add the last two areas under the curve in Figure 3.11: $\Pr\{\text{diameter} > 8\} = .12 + .07 = .19$. $\blacksquare$

Example 3.30

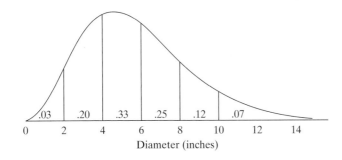

Figure 3.13 Diameters of 30-year-old Douglas fir trees

Exercises 3.15–3.17

3.15   Consider the density curve shown in Figure 3.13, which represents the distribution of diameters (measured 4.5 feet above the ground) in a population of 30-year-old Douglas fir trees. Areas under the curve are shown in the figure. What percentage of the trees have diameters

(a) between 4 inches and 10 inches?
(b) less than 4 inches?
(c) more than 6 inches?

3.16   In a certain population of the parasite *Trypanosoma*, the lengths of individuals are distributed as indicated by the density curve shown here. Areas under the curve are shown in the figure.[9]

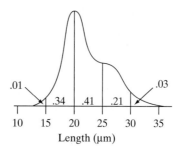

Consider the length of an individual trypanosome chosen at random from the population. Find

(a) $\Pr\{20 < \text{length} < 30\}$
(b) $\Pr\{\text{length} > 20\}$
(c) $\Pr\{\text{length} < 20\}$

3.17   Consider the distribution of *Trypanosoma* lengths shown by the density curve in Exercise 3.16. Suppose we take a sample of two trypanosomes. What is the probability that

(a) Both trypanosomes will be shorter than 20 μm?
(b) The first trypanosome will be shorter than 20 μm and the second trypanosome will be longer than 25 μm?
(c) Exactly one of the trypanosomes will be shorter than 20 μm and one trypanosome will be longer than 25 μm?

3.7 RANDOM VARIABLES

A **random variable** is simply a variable that takes on numerical values that depend on the outcome of a chance operation. The following examples illustrate this idea.

Example 3.31

Dice.   Consider the chance operation of tossing a die. Let the random variable Y represent the number of spots showing. The possible values of Y are $Y = 1, 2, 3, 4, 5,$ or 6. We do not know the value of Y until we have tossed the die. If we know how the die is weighted, then we can specify the probability that Y has a particular value, say $\Pr\{Y = 4\}$, or a particular set of values, say $\Pr\{2 \leq Y \leq 4\}$. For instance, if the die is perfectly balanced so that each of the six faces is equally likely, then

$$\Pr\{Y = 4\} = \frac{1}{6} \approx .17$$

and

$$\Pr\{2 \leq Y \leq 4\} = \frac{3}{6} = .5$$

∎

Family Size.   Suppose a family is chosen at random from a certain population, and let the random variable Y denote the number of children in the chosen family. The possible values of Y are $0, 1, 2, 3, \ldots$ The probability that Y has a particular value is equal to the percentage of families with that many children. For instance, if 23% of the families have 2 children, then

$$\Pr\{Y = 2\} = .23$$

Example 3.32

Medications.   After someone has heart surgery, the person is usually given several medications. Let the random variable Y denote the number of medications that a patient is given following cardiac surgery. If we know the distribution of the number of medications per patient for the entire population, then we can specify the probability that Y has a certain value or falls within a certain interval of values. For instance, if 52% of all patients are given 2, 3, 4, or 5 medications, then

$$\Pr\{2 \le Y \le 5\} = .52$$

Example 3.33

Heights of Men.   Let the random variable Y denote the height of a man chosen at random from a certain population. If we know the distribution of heights in the population, then we can specify the probability that Y falls in a certain range. For instance, if 46% of the men are between 65.2 and 70.4 inches tall, then

$$\Pr\{65.2 \le Y \le 70.4\} = .46$$

Example 3.34

Each of the variables in Examples 3.31–3.33 is a *discrete random variable*, because in each case we can list the possible values that the variable can take on. In contrast, the variable in Example 3.34, height, is a *continuous random variable*: Height, at least in theory, can take on any of an infinite number of values in an interval. Of course, when we measure and record a person's height, we generally measure to the nearest inch or half inch. Nonetheless, we can think of true height as being a continuous variable. We use density curves to model the distributions of continuous random variables, such as blood glucose level or tree diameter as discussed in Section 3.6.

Mean and Variance of a Random Variable

In Chapter 2 we briefly considered the concepts of population mean and population standard deviation. For the case of a discrete random variable, we can calculate the population mean and standard deviation if we know the probability distribution for the random variable. We begin with the mean.

The mean of a discrete random variable Y is defined as

$$\mu_Y = \Sigma \, y_i \Pr(Y = y_i)$$

where the y_i's are the values that the variable takes on and the sum is taken over all possible values.

The mean of a random variable is also known as the *expected value* and is often written as $E(Y)$; that is, $E(Y) = \mu_Y$.

Example 3.35

Fish Vertebrae.   In a certain population of the freshwater sculpin, *Cottus rotheus*, the distribution of the number of tail vertebrae, Y, is as shown in the Table 3.4.[2]

TABLE 3.4 Distribution of Vertebrae	
No. of Vertebrae	Percent of Fish
20	3
21	51
22	40
23	6
Total	100

The mean of Y is

$$\mu_Y = 20 \times \Pr\{Y = 20\} + 21 \times \Pr\{Y = 21\}$$
$$+ 22 \times \Pr\{Y = 22\} + 23 \times \Pr\{Y = 23\}$$
$$= 20 \times .03 + 21 \times .51 + 22 \times .40 + 23 \times .06$$
$$= .6 \qquad + 10.71 \quad + 8.8 \qquad + 1.38$$
$$= 21.49$$

$\blacksquare$

Example 3.36

Dice.   Consider rolling a die that is perfectly balanced so that each of the six faces is equally likely to come up and let the random variable Y represent the number of spots showing. The expected value, or mean, of Y is

$$E(Y) = \mu_Y = 1 \times \frac{1}{6} + 2 \times \frac{1}{6} + 3 \times \frac{1}{6} + 4 \times \frac{1}{6} + 5 \times \frac{1}{6} + 6 \times \frac{1}{6}$$
$$= \frac{21}{6} = 3.5$$

$\blacksquare$

To find the standard deviation of a random variable, we first find the variance, σ^2, of the random variable and then take the square root of the variance to get the the standard deviation, σ.

The variance of a discrete random variable Y is defined as

$$\sigma_Y^2 = \Sigma (y_i - \mu_Y)^2 \Pr(Y = y_i)$$

where the y_i's are the values that the variable takes on and the sum is taken over all possible values.
We often write $\mathrm{VAR}(Y)$ to denote the variance of Y.

Example 3.37

Fish Vertebrae.   Consider the distribution of vertebrae given in Table 3.4. In Example 3.35 we found that the mean of Y is $\mu_Y = 21.49$. The variance of Y is

$$\mathrm{VAR}(Y) = \sigma_Y^2 = (20 - 21.49)^2 \times \Pr\{Y = 20\}$$
$$+ (21 - 21.49)^2 \times \Pr\{Y = 21\}$$
$$+ (22 - 21.49)^2 \times \Pr\{Y = 22\}$$
$$+ (23 - 21.49)^2 \times \Pr\{Y = 23\}$$

$$= (-1.49)^2 \times .03 + (-.49)^2 \times .51$$
$$+ (.51)^2 \times .40 + (1.51)^2 \times .06$$
$$= 2.2201 \times .03 + .2401 \times .51 + .2601 \times .40 + 2.2801 \times .06$$
$$= .066603 + .122451 + .10404 + .136806$$
$$= .4299$$

The standard deviation of Y is $\sigma_Y = \sqrt{.4299} \approx .6557$. ■

Dice. In Example 3.36 we found that the mean number obtained from rolling a fair die is 3.5 (i.e., $\mu_Y = 3.5$). The variance of the number obtained from rolling a fair die is

Example 3.38

$$\sigma_Y^2 = (1 - 3.5)^2 \times \Pr\{Y = 1\} + (2 - 3.5)^2$$
$$\times \Pr\{Y = 2\} + (3 - 3.5)^2 \times \Pr\{Y = 3\}$$
$$+ (4 - 3.5)^2 \times \Pr\{Y = 4\} + (5 - 3.5)^2$$
$$\times \Pr\{Y = 5\} + (6 - 3.5)^2 \times \Pr\{Y = 6\}$$
$$= (-2.5)^2 \times \frac{1}{6} + (-1.5)^2 \times \frac{1}{6} + (-.5)^2 \times \frac{1}{6}$$
$$+ (.5)^2 \times \frac{1}{6} + (1.5)^2 \times \frac{1}{6} + (2.5)^2 \times \frac{1}{6}$$
$$= (6.25) \times \frac{1}{6} + (2.25) \times \frac{1}{6} + (.25) \times \frac{1}{6}$$
$$+ (.25) \times \frac{1}{6} + (2.25) \times \frac{1}{6} + (6.25) \times \frac{1}{6}$$
$$= 17.5 \times \frac{1}{6}$$
$$\approx 2.9167$$

The standard deviation of Y is $\sigma_Y = \sqrt{2.9167} \approx 1.708$. ■

The definitions just given are appropriate for discrete random variables. There are analogous definitions for continuous random variables, but they involve integral calculus and are not presented here.

Adding and Subtracting Random Variables (Optional)

If we add two random variables, it makes sense that we add their means. Likewise, if we create a new random variable by subtracting two random variables, then we subtract the individual means to get the mean of the new random variable. If we multiply a random variable by a constant (for example, if we are converting feet to inches, so that we are multiplying by 12), then we multiply the mean of the random variable by the same constant. If we add a constant to a random variable, then we add that constant to the mean.

The following rules summarize the situation:

Rules for Means of Random Variables

Rule 1: If X and Y are two random variables, then

$$\mu_{X+Y} = \mu_X + \mu_Y$$

$$\mu_{X-Y} = \mu_X - \mu_Y$$

Rule 2: If Y is a random variable and a and b constants, then
$\mu_{a+bY} = a + b\mu_Y$.

Example 3.39

Temperature.   The average summer temperature, μ_Y, in a city is 81°F. To convert °F to °C, we use the formula °C = (°F − 32) × (5/9) or °C = (5/9) × °F − (5/9) × 32. Thus, the mean in degrees Celsius is (5/9) × (81) − (5/9) × 32 = 45 − 17.78 = 27.22. ∎

Dealing with standard deviations of functions of random variables is a bit more compicated. We work with the variance first and then take the square root, at the end, to get the standard deviation we want. If we *multiply* a random variable by a constant (for example, if we are converting inches to centimeters by multiplying by 2.54), then we multiply the variance by the square of the constant. This has the effect of multiplying the standard deviation by the constant. If we *add* a constant to a random variable, then we are not changing the relative spread of the distribution, so the variance does not change.

Example 3.40

Feet to Inches.   Let Y denote the height, in feet, of a person in a given population; suppose the standard deviation of Y is $\sigma_Y = .35$ (feet). If we wish to convert from feet to inches, we can define a new variable X as $X = 12Y$. The variance of Y is $.35^2$ (the square of the standard deviation). The variance of X is $12^2 \times .35^2$, which means that the standard deviation of X is $\sigma_X = 12 \times .35 = 4.2$ (inches). ∎

If we add two random variables *that are independent of one another*, then we add their variances.* Moreover, if we subtract two random variables *that are independent of one another*, then we *add* their variances. If we want to find the standard deviation of the sum (or difference) of two independent random variables, we first find the variance of the sum (or difference) and then take the square root to get the standard deviation of the sum (or difference).

---

* If we add two random variables that are not independent of one another, then the variance of the sum depends on the degree of dependence between the variables. To take an extreme case, suppose that one of the random variables is the negative of the other. Then the sum of the two random variables will always be zero, so that the variance of the sum will be zero. This is quite different from what we would get by adding the two variances together. As another example, suppose Y is the number of questions correct on a 20-question exam and X is the number of questions wrong. Then $Y + X$ is always equal to 20, so that there is no variability at all. Hence, the variance of $Y + X$ is zero, even though the variance of Y is positive, as is the variance of X.

Example 3.41

Mass.   Consider finding the mass of a 10-mLi graduated cylinder. If several measurements are made, using an analytical balance, then in theory we would expect the measurements to all be the same. In reality, however, the readings will vary from one measurement to the next. Suppose that a given balance produces readings that have a standard deviation of .03g; let X denote the value of a reading made using this balance. Suppose that a second balance produces readings that have a standard deviation of .04g; let Y denote denote the value of a reading made using this second balance.[10]

If we use each balance to measure the mass of a graduated cylinder, we might be interested in the difference, $X - Y$, of the two measurements. The standard deviation of $X - Y$ is positive. To find the standard deviation of $X - Y$, we first find the variance of the difference. The variance of X is $.03^2$ and the variance of Y is $.04^2$. The variance of the difference is $.03^2 + .04^2 = .0025$. The standard deviation of $X - Y$ is the square root of .0025, which is .05.   ■

The following rules summarize the situation for variances:

Rules for Variances of Random Variables

Rule 3: If Y is a random variable and a and b constants, then $\sigma^2_{a+bY} = b^2\sigma^2_Y$.

Rule 4: If X and Y are two *independent* random variables, then

$$\sigma^2_{X+Y} = \sigma^2_X + \sigma^2_Y$$
$$\sigma^2_{X-Y} = \sigma^2_X + \sigma^2_Y$$

Exercises 3.18–3.25

3.18   In a certain population of the European starling, there are 5,000 nests with young. The distribution of brood size (number of young in a nest) is given in the accompanying table.[11]

Brood Size	Frequency (No. of Broods)
1	90
2	230
3	610
4	1,400
5	1,760
6	750
7	130
8	26
9	3
10	1
Total	5,000

Suppose one of the 5,000 broods is to be chosen at random, and let Y be the size of the chosen brood. Find

(a) $\Pr\{Y = 3\}$ (b) $\Pr\{Y \geq 7\}$ (c) $\Pr\{4 \leq Y \leq 6\}$

3.19 In the starling population of Exercise 3.18, there are 22,435 young in all the broods taken together. (There are 90 young from broods of size 1, there are 460 from broods of size 2, etc.) Suppose one of the *young* is to be chosen at random, and let Y' be the size of the chosen individual's brood.

(a) Find $\Pr\{Y' = 3\}$.
(b) Find $\Pr\{Y' \geq 7\}$.
(c) Explain why choosing a young at random and then observing its brood is not equivalent to choosing a brood at random. Your explanation should show why the answer to part (b) is greater than the answer to part (b) of Exercise 3.18.

3.20 Calculate the mean, μ_Y, of the random variable Y from Exercise 3.18.

3.21 Consider a population of the fruitfly *Drosophila melanogaster* in which 30% of the individuals are black because of a mutation, while 70% of the individuals have the normal gray body color. Suppose three flies are chosen at random from the population; let Y denote the number of black flies out of the three. Then the probability distribution for Y is given by the following table:

Y (No. Black)	Probability
0	.343
1	.441
2	.189
3	.027
Total	1.000

(a) Find $\Pr\{Y \geq 2\}$.
(b) Find $\Pr\{Y \leq 2\}$.

3.22 Calculate the mean, μ_Y, of the random variable Y from Exercise 3.21.

3.23 Calculate the standard deviation, σ_Y, of the random variable Y from Exercise 3.21.

3.24 A group of college students were surveyed to learn how many times they had visited a dentist in the previous year.[12] The probability distribution for Y, the number of visits, is given by the following table:

Y (No. Visits)	Probability
0	.15
1	.50
2	.35
Total	1.00

Calculate the mean, μ_Y, of the number of visits.

3.25 Calculate the standard deviation, σ_Y, of the random variable Y from Exercise 3.24.

3.8 THE BINOMIAL DISTRIBUTION

To add some depth to the notion of probability and random variables, we now consider a special type of random variable, the **binomial**. The distribution of a binomial random variable is a probability distribution associated with a special kind of chance operation. The chance operation is defined in terms of a set of conditions called the independent-trials model.

The Independent-Trials Model

The **independent-trials model** relates to a sequence of chance "trials." Each trial is assumed to have two possible outcomes, which are arbitrarily labeled "success" and "failure." The probability of success on each individual trial is denoted by the letter p and is assumed to be constant from one trial to the next. In addition, the trials are required to be independent, which means that the chance of success or failure on each trial is independent of what happens on the other trials. The total number of trials is denoted by n. These conditions are summarized in the following definition of the model.

> **Independent-Trials Model**
> A series of n independent trials is conducted. Each trial results in success or failure. The probability of success is equal to the same quantity, p, for each trial, regardless of the outcomes of the other trials.

The following examples illustrate situations that can be described by the independent-trials model.

Albinism. If two carriers of the gene for albinism marry, each of their children has probability 1/4 of being albino. The chance that the second child is albino is the same (1/4) whether or not the first child is albino; similarly, the outcome for the third child is independent of the first two, and so on. Using the labels "success" for albino and "failure" for nonalbino, the independent-trials model applies with $p = 1/4$ and $n =$ the number of children in the family. ∎

Example 3.42

Mutants. Suppose that 39% of the individuals in a large population have a certain mutant trait and that a random sample of individuals is chosen from the population. As each individual is chosen for the sample, the probability is .39 that the chosen individual will be mutant. This probability is the same as each choice is made, regardless of the results of the other choices, because the percentage of mutants in the large population remains equal to .39 even when a few individuals have been removed. Using the labels "success" for mutant and "failure" for nonmutant, the independent-trials model applies with $p = .39$ and $n =$ the sample size. ∎

Example 3.43

An Example of the Binomial Distribution

The binomial distribution specifies the probabilities of various numbers of successes and failures when the basic chance operation consists of n independent trials. Before giving the general formula for the binomial distribution, we consider a simple example.

Albinism. Suppose two carriers of the gene for albinism marry (see Example 3.42) and have two children. Then the probability that both of their children are albino is

Example 3.44

$$\Pr\{\text{both children are albino}\} = \left(\frac{1}{4}\right)\left(\frac{1}{4}\right) = \frac{1}{16}$$

The reason for this probability can be seen by considering the relative frequency interpretation of probability. Of a great many such families with two children, $\frac{1}{4}$

would have the first child albino; furthermore, $\frac{1}{4}$ *of these* would have the second child albino; thus, $\frac{1}{4}$ of $\frac{1}{4}$, or $\frac{1}{16}$, of all the couples would have both albino children. A similar kind of reasoning shows that the probability that both children are not albino is

$$\text{Pr}\{\text{both children are not albino}\} = \left(\frac{3}{4}\right)\left(\frac{3}{4}\right) = \frac{9}{16}$$

A new twist enters if we consider the probability that one child is albino and the other is not. There are two possible ways this can happen:

$$\text{Pr}\{\text{first child is albino, second is not}\} = \left(\frac{1}{4}\right)\left(\frac{3}{4}\right) = \frac{3}{16}$$

$$\text{Pr}\{\text{second child is albino, first is not}\} = \left(\frac{3}{4}\right)\left(\frac{1}{4}\right) = \frac{3}{16}$$

To see how to combine these possibilities, we again consider the relative frequency interpretation of probability. Of a great many such families with two children, the fraction of families with one albino and one nonalbino child would be the total of the two possibilities, or

$$\left(\frac{3}{16}\right) + \left(\frac{3}{16}\right) = \frac{6}{16}$$

Thus, the corresponding probability is

$$\text{Pr}\{\text{one child is albino, the other is not}\} = \frac{6}{16}$$

Another way to see this is to consider a probability tree. The first split in the tree represents the birth of the first child; the second split represents the birth of the second child. The four possible outcomes and their associated probabilities are shown in Figure 3.14. These probabilities are collected in Table 3.5. ■

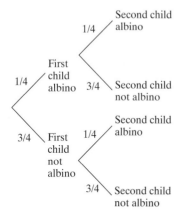

Figure 3.14 Probability tree for albinism among two children of carriers of the gene for albinism

TABLE 3.5 Probability Distribution for Number of Albino Children

Number of		
Albino	*Nonalbino*	**Probability**
0	2	$\frac{9}{16}$
1	1	$\frac{6}{16}$
2	0	$\frac{1}{16}$
		$\frac{}{1}$

The probability distribution in Table 3.5 is called the binomial distribution with $p = \frac{1}{4}$ and $n = 2$. Note that the probabilities add to 1. This makes sense because all possibilities have been accounted for: We expect $\frac{9}{16}$ of the families to have no albino children, $\frac{6}{16}$ to have one albino child, and $\frac{1}{16}$ to have two albino children; there are no other possible compositions for a two-child family. The number of albino children, out of the two children, is an example of a binomial random variable. A **binomial random variable** is a random variable that satisfies the following four conditions, abbreviated as **BInS**:

Binary outcomes: There are two possible outcomes for each trial (success and failure).

Independent trials: The outcomes of the trials are independent of each other.

n is fixed: The number of trials, n, is fixed in advance.

Same value of p: The probability of a success on a single trial is the same for all trials.

The Binomial Distribution Formula

A general formula is available which can be used to calculate probabilities associated with a binomial random variable for any values of n and p. This formula can be proved using logic similar to that in Example 3.44. (The formula is discussed further in Appendix 3.2.) The formula is given in the accompanying box.

The Binomial Distribution Formula

For a binomial random variable Y, the probability that the n trials result in j successes (and $n - j$ failures) is given by the following formula

$$\Pr\{j \text{ successes}\} = \Pr\{Y = j\} = {}_nC_j p^j (1 - p)^{n-j}$$

The quantity ${}_nC_j$ appearing in the formula is called a **binomial coefficient**. Each binomial coefficient is an integer depending on n and on j. Values of binomial coefficients are given in Table 2 at the end of this book and can be found by the formula

$$_nC_j = \frac{n!}{j!(n-j)!}$$

where $x!$ ("x-factorial") is defined for any positive integer x as

$$x! = x(x - 1)(x - 2)\ldots(2)(1)$$

and $0! = 1$. For more details, see Appendix 3.2.

For example, for $n = 5$ the binomial coefficients are as follows:

j:	0	1	2	3	4	5
$_5C_j$:	1	5	10	10	5	1

Thus, for $n = 5$ the binomial probabilities are as indicated in Table 3.6. Notice the pattern in Table 3.6: The powers of p ascend (0, 1, 2, 3, 4, 5) and the powers of $(1 - p)$ descend (5, 4, 3, 2, 1, 0). (In using the binomial distribution formula, remember that $x^0 = 1$ for any nonzero x.)

The following example shows a specific application of the binomial distribution with $n = 5$.

TABLE 3.6 Binomial Probabilities for $n = 5$

Number of Successes j	Number of Failures $n - j$	Probability
0	5	$1p^0(1 - p)^5$
1	4	$5p^1(1 - p)^4$
2	3	$10p^2(1 - p)^3$
3	2	$10p^3(1 - p)^2$
4	1	$5p^4(1 - p)^1$
5	0	$1p^5(1 - p)^0$

Example 3.45

Mutants. Suppose we draw a random sample of five individuals from a large population in which 39% of the individuals are mutants (as in Example 3.43). The probabilities of the various possible samples are then given by the binomial distribution formula with $n = 5$ and $p = .39$; the results are displayed in Table 3.7. For instance, the probability of a sample containing 3 mutants and 2 nonmutants is

$$10(.39)^3(.61)^2 \approx .22$$

Thus, $\Pr\{Y = 3\} \approx .22$. This means that about 22% of random samples of size 5 will contain three mutants and two nonmutants.

TABLE 3.7 Binomial Distribution with $n = 5$ and $p = .39$

Number of		
Mutants	*Nonmutants*	**Probability**
0	5	.08
1	4	.27
2	3	.35
3	2	.22
4	1	.07
5	0	.01
		1.00

Notice that the probabilities in Table 3.7 add to 1. The probabilities in a probability distribution must always add to 1, because they account for 100% of the possibilities. ■

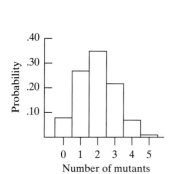

Figure 3.15 Binomial distribution with $n = 5$ and $p = .39$

The binomial distribution of Table 3.7 is pictured graphically in Figure 3.15. Such a graphical display of a probability distribution is called a **probability histogram**.

Remark: In applying the independent-trials model and the binomial distribution, we assign the labels "success" and "failure" arbitrarily. For instance, in Example 3.45, we could say "success" = "mutant" and $p = .39$; or, alternatively, we could say "success" = "nonmutant" and $p = .61$. Either assignment of labels is all right; it is only necessary to be consistent.

Notes on Table 2: The following features in Table 2 are worth noting:

(a) The first and last entries in each row are equal to 1. This will be true for any row; that is, $_nC_0 = 1$ and $_nC_n = 1$ for any value of n.

(b) Each row of the table is symmetric; that is, $_nC_j$ and $_nC_{n-j}$ are equal.

(c) The bottom rows of the table are left incomplete to save space, but you can easily complete them using the symmetry of the $_nC_j$'s; if you need to know $_nC_j$ you can look up $_nC_{n-j}$ in Table 2. For instance, consider $n = 18$; if you want to know $_{18}C_{15}$ you just look up $_{18}C_3$; both $_{18}C_3$ and $_{18}C_{15}$ are equal to 816.

Computational note: Computer and calculator technology make it fairly easy to handle the binomial distribution formula for small or moderate values of n. For example, suppose we want to find $\Pr\{Y = 2\}$ when Y is a binomial random

variable with $n = 5$ and $p = .39$. In the MINITAB system this probability can be found with the following command:

```
MTB > PDF 2;
SUBC> Binomial 5 .39.
```

This command returns the following output, which agrees with Table 3.7 (although MINITAB uses the letter X to denote a random variable, whereas we have used the letter Y):

```
Probability Density Function
Binomial with n = 5 and p = 0.390000
     x          P(X = x)
   2.00           0.3452
```

If a list of possible values is stored in a column in MINITAB, then the Binomial command can be used to find several probabilities at once. Thus, if we want to recreate Table 3.7, we enter the values 0, 1, 2, 3, 4, 5 in column 1 and enter the command

```
MTB > PDF C1;
SUBC> Binomial 5 .39.
```

MINITAB returns the following output:

```
Probability Density Function
Binomial with n = 5 and p = 0.390000
     x          P(X = x)
   0.00           0.0845
   1.00           0.2700
   2.00           0.3452
   3.00           0.2207
   4.00           0.0706
   5.00           0.0090
```

For large values of n, the use of the binomial formula gets to be tedious and even a computer will balk at being asked to calculate a binomial probability. However, the binomial formula can be approximated by other methods. One of these is discussed in the optional Section 5.5.

Sometimes a binomial probability question involves combining two or more possible outcomes. The following example illustrates this idea.

Sampling Fruitflies. In a large *Drosophila* population, 30% of the flies are black (B) and 70% are gray (G). Suppose two flies are randomly chosen from the population (as in Example 3.13). The binomial distribution with $n = 2$ and $p = .3$

Example 3.46

gives probabilities for the possible outcomes as shown in Table 3.8. (Using the binomial formula agrees with the results given by probability tree shown in Figure 3.4.)

TABLE 3.8

Sample Composition	Y	Probability
Both G	0	.49
One B, one G	1	.42
Both B	2	.09
		1.00

Let E be the event that both flies are the same color. Then E can happen in two ways: Both flies are gray or both are black. To find the probability of E, consider what would happen if we repeated the sampling procedure many times: 49% of the samples would have both flies gray, and 9% would have both flies black. Consequently, the percentage of samples with both flies the same color would be 49% + 9% = 58%. Thus, we have shown that the probability of E is

$$\Pr\{E\} = .58$$

as we claimed in Example 3.13.   ∎

Whenever an event E can happen in two or more mutually exclusive ways, a rationale such as that of Example 3.46 can be used to find $\Pr\{E\}$.

Example 3.47

Blood Type.   In the United States, 85% of the population has Rh positive blood. Suppose we take a random sample of 6 persons and count the number with Rh positive blood. The binomial model can be applied here, since the BInS conditions are met: There is a binary outcome on each trial (Rh positive or Rh negative blood), the trials are independent (due to the random sampling), n is fixed at 6, and the same probability of Rh positive blood applies to each person ($p = .85$).

Let Y denote the number of persons, out of 6, with Rh positive blood. The probabilities of the possible values of Y are given by the binomial distribution formula with $n = 6$ and $p = .85$; the results are displayed in Table 3.9. For instance, the probability that $Y = 4$ is

$$_6C_4(.85)^4(.15)^2 \approx 15(.522)(.0225) \approx .1762$$

TABLE 3.9 Binomial Distribution with $n = 6$ and $p = .85$

Number of Successes	Probability
0	<.0001
1	.0004
2	.0055
3	.0415
4	.1762
5	.3993
6	.3771
	1

If we want to find the probability that at least 4 persons (out of the 6 sampled) will have Rh positive blood, we need to find $\Pr\{Y \geq 4\} = \Pr\{Y = 4\} + \Pr\{Y = 5\} + \Pr\{Y = 6\} = .1762 + .3993 + .3771 = .9526$. This means that the probability of getting at least 4 persons with Rh positive blood in a sample of size 6 is .9526.   ■

The probability of an event happening is 1 minus the probability that the event does not happen: $\Pr\{E\} = 1 - \Pr\{E \text{ does not happen}\}$. In some problems, such as in the following example, the easiest way to find $\Pr\{E\}$ is to first find $\Pr\{E \text{ does not happen}\}$ and then to subtract this probability from 1.

Blood Type.   As in Example 3.47, let Y denote the number of persons, out of 6, with Rh positive blood. Suppose we want to find the probability that Y is less than 6 (i.e., the probability that there is *at least 1* person in the sample who has Rh *negative* blood). We could find this directly as $\Pr\{Y = 0\} + \Pr\{Y = 1\} + \cdots + \Pr\{Y = 5\}$. However, it is easier to find $\Pr\{Y \neq 6\}$ and subtract this from 1:

$$\Pr\{Y < 6\} = 1 - \Pr\{Y = 6\} = 1 - .3771 = .6229 \qquad ■$$

Example 3.48

Mean and Standard Deviation of a Binomial

If we toss a fair coin 10 times, then we expect to get 5 heads, on average. This is an example of a general rule: *For a binomial random variable, the mean (that is, the average number of successes) is equal to np*. This is an intuitive fact: The probability of success on each trial is p, so if we conduct n trials, then np is the expected number of successes. In Appendix 3.3, we show that this result is consistent with the rule given in Section 3.7 for finding the mean of the sum of random variables. *The standard deviation for a binomial random variable is given by* $\sqrt{np(1 - p)}$. This formula is not intuitively clear; a derivation of the result is given in Appendix 3.3. For the example of tossing a coin 10 times, the standard deviation of the number of heads is $\sqrt{10 \times .5 \times (1 - .5)} = \sqrt{2.5} \approx 1.58$.

Blood Type.   As discussed in Example 3.47, if Y denotes the number of persons with Rh positive blood in a sample of size 6, then a binomial model can be used to find probabilities associated with Y. The single most likely value of Y is 5 (which has probability .3993). The average value of Y is $6 \times .85 = 5.1$, which means that if we take many samples, each of size 6, and count the number of Rh positive persons in each sample, and then average those counts, we expect to get 5.1. The standard deviation of those counts is $\sqrt{6 \times .85 \times .15} \approx .87$.   ■

Example 3.49

Applicability of the Binomial Distribution

A number of statistical procedures are based on the binomial distribution. We will study some of these procedures in later chapters. Of course, the binomial distribution is applicable only in experiments where the BInS conditions are satisfied in the real biological situation. We briefly discuss some aspects of these conditions.

Application to Sampling. The most important application of the independent-trials model and the binomial distribution is to describe random sampling from a population when the observed variable is dichotomous—that is, a categorical variable with two categories (for instance, black and gray in Example 3.46). This application is valid if the sample size is a negligible fraction of the population size, so that the population composition is not altered appreciably by the removal of the individuals in the sample (thus the S part of BInS is satisfied: The probability of a success remains the same from trial to trial). However, if the sample is *not* a negligibly small part of the population, then the population composition may be altered by the sampling process, so that the "trials" involved in composing the sample are not independent and the probability of a success changes as the sampling progresses. In this case, the probabilities given by the binomial formula are not correct. In most biological studies, the population is so large that this kind of difficulty does not arise.

Contagion. In some applications the phenomenon of contagion can invalidate the condition of independence between trials. The following is an example.

Example 3.50 **Chickenpox.** Consider the occurrence of chickenpox in children. Each child in a family can be categorized according to whether he or she had chickenpox during a certain year. One can say that each child constitutes a "trial" and that "success" is having chickenpox during the year, but the trials are *not* independent because the chance of a particular child catching chickenpox depends on whether his or her sibling caught chickenpox. As a specific example, consider a family with five children, and suppose that the chance of an individual child catching chickenpox during the year is equal to .10. The binomial distribution gives the chance of all five children getting chickenpox as

$$\Pr\{5 \text{ children get chickenpox}\} = (.10)^5 = .00001$$

However, this answer is not correct; because of contagion, the correct probability would be much larger. There would be many families in which one child caught chickenpox and then the other four children got chickenpox from the first child, so that all five children would get chickenpox. ∎

Exercises 3.26–3.34

3.26 The seeds of the garden pea *(Pisum sativum)* are either yellow or green. A certain cross between pea plants produces progeny in the ratio 3 yellow:1 green.[13] If four randomly chosen progeny of such a cross are examined, what is the probability that

(a) three are yellow and one is green?
(b) all four are yellow?
(c) all four are the same color?

3.27 In the United States, 42% of the population has type A blood. Consider taking a sample of size 4. Let Y denote the number of persons in the sample with type A blood. Find

(a) $\Pr\{Y = 0\}$
(b) $\Pr\{Y = 1\}$
(c) $\Pr\{Y = 2\}$
(d) $\Pr\{0 \le Y \le 2\}$
(e) $\Pr\{0 < Y \le 2\}$

3.28 A certain drug treatment cures 90% of cases of hookworm in children.[14] Suppose that 20 children suffering from hookworm are to be treated, and that the children can be regarded as a random sample from the population. Find the probability that

(a) all 20 will be cured
(b) all but one will be cured
(c) exactly 18 will be cured
(d) exactly 90% will be cured

3.29 The shell of the land snail *Limocolaria martensiana* has two possible color forms: streaked and pallid. In a certain population of these snails, 60% of the individuals have streaked shells.[15] Suppose that a random sample of 10 snails is to be chosen from this population. Find the probability that the percentage of streaked-shelled snails in the *sample* will be

(a) 50%   (b) 60%   (c) 70%

3.30 Consider taking a sample of size 10 from the snail population in Exercise 3.29.

(a) What is the mean number of streaked-shelled snails?
(b) What is the standard deviation of the number of streaked-shelled snails?

3.31 The sex ratio of newborn human infants is about 105 males : 100 females.[16] If four infants are chosen at random, what is the probability that

(a) two are male and two are female?
(b) all four are male?
(c) all four are the same sex?

3.32 Neuroblastoma is a rare, serious, but treatable disease. A urine test, the vanilly mandelic acid test, has been developed that gives a positive diagnosis in about 70% of cases of neuroblastoma.[17] It has been proposed that this test be used for large-scale screening of children. Assume that 300,000 children are to be tested, of whom 8 have the disease. We are interested in whether or not the test detects the disease in the 8 children who have the disease. Find the probability that

(a) all 8 cases will be detected
(b) only one case will be missed
(c) two or more cases will be missed [*Hint:* Use parts (a) and (b) to answer part (c).]

3.33 If two carriers of the gene for albinism marry, each of their children has probability $\frac{1}{4}$ of being albino (see Example 3.42). If such a couple has six children, what is the probability that

(a) none will be albino?
(b) at least one will be albino? [*Hint:* Use part (a) to answer part (b); note that "at least one" means "one or more."]

3.34 Childhood lead poisoning is a public health concern in the United States. In a certain population, one child in eight has a high blood lead level (defined as 30 μg/dLi or more).[18] In a randomly chosen group of 16 children from the population, what is the probability that

(a) none has high blood lead?
(b) one has high blood lead?
(c) two have high blood lead?
(d) three or more have high blood lead? [*Hint:* Use parts (a)–(c) to answer part (d).]

3.9 FITTING A BINOMIAL DISTRIBUTION TO DATA (OPTIONAL)

Occasionally it is possible to obtain data that permit a direct check of the applicability of the binomial distribution. One such case is described in the next example.

Example 3.51

Sexes of Children.   In a classic study of the human sex ratio, families were categorized according to the sexes of the children. The data were collected in Germany in the nineteenth century, when large families were common. Table 3.10 shows the results for 6,115 families with 12 children.[19]

TABLE 3.10 Sex Ratios in 6,115 Families with 12 Children

Number of		Observed Frequency
Boys	*Girls*	**(Number of Families)**
0	12	3
1	11	24
2	10	104
3	9	286
4	8	670
5	7	1,033
6	6	1,343
7	5	1,112
8	4	829
9	3	478
10	2	181
11	1	45
12	0	7
		6,115

It is interesting to consider whether the observed variation among families can be explained by the independent-trials model. We will explore this question by fitting a binomial distribution to the data.

The first step in fitting the binomial distribution is to determine a value for $p = \Pr\{\text{boy}\}$. One possibility would be to assume that $p = .50$. However, since it is known that the human sex ratio at birth is not exactly $1 : 1$ (in fact, it favors boys slightly), we will not make this assumption. Rather, we will "fit" p to the data; that is, we will determine a value for p that fits the data best. We observe that the total number of children in all the families is

$$(12)(6{,}115) = 73{,}380 \text{ children}$$

Among these children, the number of boys is

$$(3)(0) + (24)(1) + \cdots + (12)(7) = 38{,}100 \text{ boys}$$

Therefore, the value of p that fits the data best is

$$p = \frac{38{,}100}{73{,}380} = .519215$$

The next step is to compute probabilities from the binomial distribution formula with $n = 12$ and $p = .519215$. For instance, the probability of 3 boys and 9 girls is computed as

$$_{12}C_3(p)^3(1 - p)^9 = 220(.519215)^3(.480785)^9$$
$$\approx .042269$$

For comparison with the observed data, we convert each probability to a theoretical or "expected" frequency by multiplying by 6,115 (the total number of families). For instance, the expected number of families with 3 boys and 9 girls is

$$(6,115)(.042269) \approx 258.5$$

The expected and observed frequencies are displayed together in Table 3.11. Table 3.11 shows reasonable agreement between the observed frequencies and the predictions of the binomial distribution. But a closer look reveals that the discrepancies, although not large, follow a definite pattern. The data contain more unisexual, or preponderantly unisexual, sibships than expected. In fact, the observed frequencies are higher than the expected frequencies for nine types of families in which one sex or the other predominates, while the observed frequencies are lower than the expected frequencies for four types of more "balanced" families. This pattern is clearly revealed by the last column of Table 3.11, which shows the sign of the difference between the observed frequency and the expected frequency. Thus, the observed distribution of sex ratios has heavier "tails" and a lighter "middle" than the best-fitting binomial distribution.

The systematic pattern of deviations from the binomial distribution suggests that the observed variation among families cannot be entirely explained by the independent-trials model.* What factors might account for the discrepancy?

TABLE 3.11 Sex-Ratio Data and Binomial Expected Frequencies

Number of Boys	Girls	Observed Frequency	Expected Frequency	Sign of (OBS. − EXP.)
0	12	3	.9	+
1	11	24	12.1	+
2	10	104	71.8	+
3	9	286	258.5	+
4	8	670	628.1	+
5	7	1,033	1,085.2	−
6	6	1,343	1,367.3	−
7	5	1,112	1,265.6	−
8	4	829	854.3	−
9	3	478	410.0	+
10	2	181	132.8	+
11	1	45	26.1	+
12	0	7	2.3	+
		6,115	6,115.0	

* A chi-square goodness-of-fit test of the binomial model shows that there is strong evidence that the differences between the observed and expected frequencies did not happen due to chance error in the sampling process. We explore the topic of goodness-of-fit tests in Chapter 10.

This intriguing question has stimulated several researchers to undertake more detailed analysis of these data. We briefly discuss some of the issues.

One explanation for the excess of predominantly unisexual families is that the probability of producing a boy may vary among families. If p varies from one family to another, then sex will appear to "run" in families in the sense that the number of predominantly unisexual families will be inflated. In order to visualize this effect, consider the fictitious data set shown in Table 3.12.

TABLE 3.12 Fictitious Sex-Ratio Data and Binomial Expected Frequencies

Number of		Observed	Expected	Sign of
Boys	*Girls*	**Frequency**	**Frequency**	**(OBS. − EXP.)**
0	12	2,940	.9	+
1	11	0	12.1	−
2	10	0	71.8	−
3	9	0	258.5	−
4	8	0	628.1	−
5	7	0	1,085.2	−
6	6	0	1,367.3	−
7	5	0	1,265.6	−
8	4	0	854.3	−
9	3	0	410.0	−
10	2	0	132.8	−
11	1	0	26.1	−
12	0	3,175	2.3	+
		6,115	6,115.0	

In the fictitious data set, there are $(3,175)(12) = 38,100$ males among 73,380 children, just as there are in the real data set. Consequently, the best-fitting p is the same ($p = .519215$) and the expected binomial frequencies are the same as in Table 3.11. The fictitious data set contains only unisexual sibships and so is an extreme example of sex "running" in families. The real data set exhibits the same phenomenon more weakly. One explanation of the fictitious data set would be that some families can have only boys ($p = 1$) and other families can have only girls ($p = 0$). In a parallel way, one explanation of the real data set would be that p varies slightly among families. Variation in p is biologically plausible, even though the mechanism causing the variation has not yet been discovered.

An alternative explanation for the inflated number of sexually homogeneous families would be that the sexes of the children in a family are literally dependent on one another, in the sense that the determination of an individual child's sex is somehow influenced by the sexes of the previous children. This explanation is implausible on biological grounds because it is difficult to imagine how the biological system could "remember" the sexes of previous offspring. ∎

Example 3.51 shows that poorness of fit to the independent-trials model can be biologically interesting. We should emphasize, however, that most statistical applications of the binomial distribution proceed from the assumption that the independent-trials model is applicable. In a typical application, the data are regarded as resulting from a *single* set of n trials. Data such as the family sex-ratio data, which refer to *many* sets of $n = 12$ trials, are not often encountered.

Exercises 3.35–3.37

3.35 The accompanying data on families with 6 children are taken from the same study as the families with 12 children in Example 3.51. Fit a binomial distribution to the data. (Round the expected frequencies to one decimal place.) Compare with the results in Example 3.51; what features do the two data sets share?

Number of		Number of
Boys	*Girls*	**Families**
0	6	1,096
1	5	6,233
2	4	15,700
3	3	22,221
4	2	17,332
5	1	7,908
6	0	1,579
		72,069

3.36 An important method for studying mutation-causing substances involves killing female mice 17 days after mating and examining their uteri for living and dead embryos. The classical method of analysis of such data assumes that the survival or death of each embryo constitutes an independent binomial trial. The accompanying table, which is extracted from a larger study, gives data for 310 females, all of whose uteri contained 9 embryos; all of the animals were treated alike (as controls).[20]

Number of Embryos		Number of
Dead	*Living*	**Female Mice**
0	9	136
1	8	103
2	7	50
3	6	13
4	5	6
5	4	1
6	3	1
7	2	0
8	1	0
9	0	0
		310

(a) Fit a binomial distribution to the observed data. (Round the expected frequencies to one decimal place.)

(b) Interpret the relationship between the observed and expected frequencies. Do the data cast suspicion on the classical assumption?

3.37 Students in a large botany class conducted an experiment on the germination of seeds of the Saguaro cactus. As part of the experiment, each student planted five seeds in a small cup, kept the cup near a window, and checked every day for germination (sprouting). The class results on the seventh day after planting were as displayed in the table.[21]

Number of Seeds		Number
Germinated	*Not Germinated*	of Students
0	5	17
1	4	53
2	3	94
3	2	79
4	1	33
5	0	4
		280

(a) Fit a binomial distribution to the data. (Round the expected frequencies to one decimal place.)

(b) Two students, Fran and Bob, were talking before class. All of Fran's seeds had germinated by the seventh day, whereas none of Bob's had. Bob wondered whether he had done something wrong. With the perspective gained from seeing all 280 students' results, what would you say to Bob? (*Hint:* Can the variation among the students be explained by the hypothesis that some of the seeds were good and some were poor, with each student receiving a randomly chosen five seeds?)

(c) Invent a fictitious set of data for 280 students, with the same overall percentage germination as the observed data given in the table but with all the students getting either Fran's results (perfect) or Bob's results (nothing). How would your answer to Bob differ if the actual data had looked like this fictitious data set?

Supplementary Exercises 3.38–3.47

3.38 In the United States, 10% of adolescent girls have iron deficiency.[22] Suppose two adolescent girls are chosen at random. Find the probability that

(a) both girls have iron deficiency
(b) one girl has iron deficiency and the other does not

3.39 In preparation for an ecological study of centipedes, the floor of a beech woods is divided into a large number of one-foot squares.[23] At a certain moment, the distribution of centipedes in the squares is as shown in the table.

Number of Centipedes	Percent Frequency (% of Squares)
0	45
1	36
2	14
3	4
4	1
	100

Suppose that a square is chosen at random, and let Y be the number of centipedes in the chosen square. Find

(a) $\Pr\{Y = 1\}$ (b) $\Pr\{Y \geq 2\}$

3.40 Refer to the distribution of centipedes given in Exercise 3.39. Suppose five squares are chosen at random. Find the probability that three of the squares contain centipedes and two do not.

3.41 Refer to the distribution of centipedes given in Exercise 3.39. Suppose five squares are chosen at random. Find the expected value (i.e., the mean) of the number of squares that contain at least one centipede.

3.42 Wavy hair in mice is a recessive genetic trait. If mice with wavy hair are mated with straight-haired (heterozygous) mice, each offspring has probability $\frac{1}{2}$ of having wavy hair.[24] Consider a large number of such matings, each producing a litter of five offspring. What percentage of the litters will consist of

(a) two wavy-haired and three straight-haired offspring?
(b) three or more straight-haired offspring?
(c) all the same type (either all wavy- or all straight-haired) offspring?

3.43 A certain drug causes kidney damage in 1% of patients. Suppose the drug is to be tested on 50 patients. Find the probability that

(a) none of the patients will experience kidney damage
(b) one or more of the patients will experience kidney damage [*Hint:* Use part (a) to answer part (b).]

3.44 Refer to Exercise 3.43. Suppose now that the drug is to be tested on *n* patients, and let *E* represent the event that kidney damage occurs in one or more of the patients. The probability $\Pr\{E\}$ is useful in establishing criteria for drug safety.

(a) Find $\Pr\{E\}$ for $n = 100$.
(b) How large must *n* be in order for $\Pr\{E\}$ to exceed .95?

3.45 To study people's ability to deceive lie detectors, researchers sometimes use the "guilty knowledge" technique.[25] Certain subjects memorize six common words; other subjects memorize no words. Each subject is then tested on a polygraph machine (lie detector), as follows. The experimenter reads, in random order, 24 words: the six "critical" words (the memorized list) and, for each critical word, three "control" words with similar or related meanings. If the subject has memorized the six words, he or she tries to conceal that fact. The subject is scored a "failure" on a critical word if his or her electrodermal response is higher on the critical word than on any of the three control words. Thus, on each of the six critical words, even an innocent subject would have a 25% chance of failing. Suppose a subject is labeled "guilty" if the subject fails on four or more of the six critical words. If an innocent subject is tested, what is the probability that he or she will be labeled "guilty"?

3.46 The density curve shown here represents the distribution of systolic blood pressures in a population of middle-aged men.[26] Areas under the curve are shown in the figure. Suppose a man is selected at random from the population, and let *Y* be his blood pressure. Find

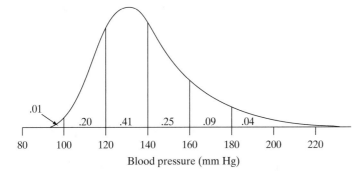

(a) $\Pr\{120 < Y < 160\}$
(b) $\Pr\{Y < 120\}$
(c) $\Pr\{Y > 140\}$

3.47 Refer to the blood pressure distribution of Exercise 3.46. Suppose four men are selected at random from the population. Find the probability that

(a) all four have blood pressures higher than 140 mm Hg
(b) three have blood pressures higher than 140, and one has blood pressure 140 or less

CHAPTER 4

The Normal Distribution

4.1 INTRODUCTION

In Chapter 2 we introduced the idea of regarding a set of data as a sample from a population. In Section 3.6 we saw that the population distribution of a quantitative variable Y can be described by its mean μ and its standard deviation σ and also by a density curve, which represents relative frequencies as areas under the curve. In this chapter we study the most important type of density curve: the **normal curve**. The normal curve is a symmetric bell-shaped curve whose exact form we describe in this chapter. A distribution represented by a normal curve is called a **normal distribution**.

The family of normal distributions plays two roles in statistical applications. Its more straightforward use is as a convenient approximation to the distribution of an observed variable Y. The second role of the normal distribution is more theoretical and will be explored in Chapter 5.

An example of a natural population distribution that can be approximated by a normal distribution follows.

Serum Cholesterol. The relationship between the concentration of cholesterol in the blood and the occurrence of heart disease has been the subject of much research. As part of a government health survey, researchers measured serum cholesterol levels for a large sample of Americans. The distribution for 17-year-olds can be fairly well approximated by a normal curve with mean $\mu = 176$ mg/dLi and standard deviation $\sigma = 30$ mg/dLi. Figure 4.1 shows a histogram based on a sample of 953 17-year-olds, with the normal curve superimposed.[1]   ∎

Objectives

In this chapter we will study the normal distribution, including

- *the use of the normal curve in modeling distributions*

- *finding probabilities using the normal curve*

- *assessing normality of data sets with the use of normal probability plots*

- *applying "continuity correction" to improve normal curve approximations*

Example 4.1

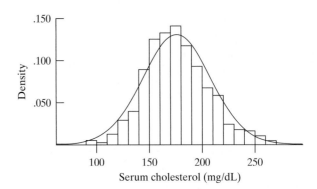

Figure 4.1 Distribution of serum cholesterol in 17-year-olds

To indicate how the mean μ and standard deviation σ relate to the normal curve, Figure 4.2 shows the normal curve for the serum cholesterol distribution of Example 4.1, with tick marks at 1, 2, and 3 standard deviations from the mean.

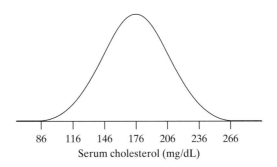

Figure 4.2 Normal distribution of serum cholesterol, with $\mu = 176$ mg/dLi and $\sigma = 30$ mg/dLi

The normal curve can be used to describe the distribution of an observed variable Y in two ways: (1) as a smooth approximation to a histogram based on a sample of Y values; and (2) as an idealized representation of the population distribution of Y. The normal curve in Figure 4.1 could be interpreted either way. For simplicity, in the remainder of this chapter we consider the normal curve as representing a population distribution.

Further Examples

We now give three more examples of normal curves that approximately describe real populations. In each figure, the horizontal axis is scaled with tick marks centered at the mean and one standard deviation apart.

Example 4.2

Eggshell Thickness.   In the commercial production of eggs, breakage is a major problem. Consequently, the thickness of the eggshell is an important variable. In one study, the shell thicknesses of the eggs produced by a large flock of White Leghorn hens were observed to follow approximately a normal distribution with mean $\mu = .38$ mm and standard deviation $\sigma = .03$ mm. This distribution is pictured in Figure 4.3.[2]   ∎

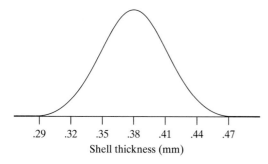

Figure 4.3 Normal distribution of eggshell thickness, with $\mu = .38$ mm and $\sigma = .03$ mm

Interspike Times in Nerve Cells. In certain nerve cells, spontaneous electrical discharges are observed that are so rhythmically repetitive that they are called "clock-spikes." The timing of these spikes, even though remarkably regular, does exhibit variation. In one study, the interspike-time intervals (in milliseconds) for a single housefly *(Musca domestica)* were observed to follow approximately a normal distribution with mean $\mu = 15.6$ ms and standard deviation $\sigma = .4$ ms; this distribution is shown in Figure 4.4.[3]   ■

Example 4.3

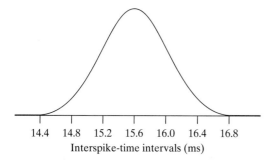

Figure 4.4 Normal distribution of interspike-time intervals, with $\mu = 15.6$ ms and $\sigma = .4$ ms

The preceding examples have illustrated very different kinds of populations. In Example 4.3, the entire population consists of measurements on only one fly. Still another type of population is a *measurement error* population, consisting of repeated measurements of exactly the same quantity. The deviation of an individual measurement from the "correct" value is called measurement error. Measurement error is not the result of a mistake, but rather is due to lack of perfect precision in the measuring process or measuring instrument. Measurement error distributions are often approximately normal; in this case the mean of the distribution of repeated measurements of the same quantity is the true value of the quantity (assuming that the measuring instrument is correctly calibrated), and the standard deviation of the distribution indicates the precision of the instrument. One measurement error distribution was described in Example 2.14. The following is another example.

Measurement Error. When a certain electronic instrument is used for counting particles such as white blood cells, the measurement error distribution is approximately normal. For white blood cells, the standard deviation of repeated counts based on the same blood specimen is about 1.4% of the true count. Thus, if the true count of a certain blood specimen were 7,000 cells/mm³, then the standard deviation would be about 100 cells/mm³ and the distribution of repeated counts on that specimen would resemble Figure 4.5.[4]   ■

Example 4.4

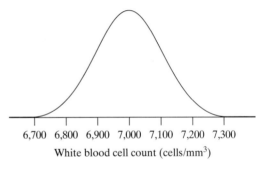

Figure 4.5 Normal distribution of repeated white blood cell counts of a blood specimen whose true value is $\mu = 7{,}000$ cells/mm³. The standard deviation is $\sigma = 100$ cells/mm³.

4.2 THE NORMAL CURVES

As the examples in Section 4.1 show, there are many normal curves; each particular normal curve is characterized by its mean and standard deviation. If a variable Y follows a normal distribution with mean μ and standard deviation σ, then it is common to write $Y \sim N(\mu, \sigma)$. All of the normal curves can be described by a single formula. Even though we do not make any direct use of the formula in this book, we present it here, both as a matter of interest and also to emphasize that a normal curve is not just any symmetric curve but rather a *specific* kind of symmetric curve.

If a variable Y follows a normal distribution with mean μ and standard deviation σ, then the density curve of the distribution of Y is given by the following formula:

$$f(y) = \frac{1}{\sigma\sqrt{2\pi}} e^{-\frac{1}{2}\left(\frac{y-\mu}{\sigma}\right)^2}$$

This function, $f(y)$, is called the *density function* of the distribution and expresses the height of the curve as a function of the position y along the y-axis. The quantities e and π that appear in the formula are constants, with e approximately equal to 2.72 and π approximately equal to 3.14.

Figure 4.6 shows a graph of a normal curve. The shape of the curve is like a symmetric bell, centered at $y = \mu$. The direction of curvature is downward (like an inverted bowl) in the central portion of the curve, and upward in the tail portions. The points where the curvature changes direction are $y = \mu - \sigma$ and $y = \mu + \sigma$; notice that the curve is almost linear near these points. In principle the curve extends to $+\infty$ and $-\infty$, never actually reaching the y-axis; however, the height of the curve is very small for y values more than three standard deviations

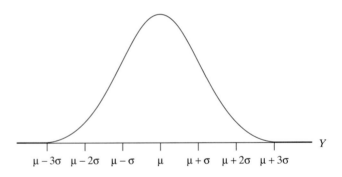

Figure 4.6 A normal curve with mean μ and standard deviation σ

from the mean. The area under the curve is exactly equal to 1. (*Note*: It may seem paradoxical that a curve can enclose a finite area, even though it never descends to touch the *y*-axis. This apparent paradox is clarified in Appendix 4.1.)

All normal curves have the same essential shape, in the sense that they can be made to look identical by suitable choice of the vertical and horizontal scales for each. (For instance, notice that the curves in Figures 4.2–4.5 look identical.) But normal curves with different values of μ and σ will not look identical if they are all plotted to the same scale, as illustrated by Figure 4.7. The location of the normal curve along the *y*-axis is governed by μ since the curve is centered at $y = \mu$; the width of the curve is governed by σ. The height of the curve is also determined by σ: Since the area under each curve must be equal to 1, a curve with a smaller value of σ must be taller. This reflects the fact that the values of Y are more highly concentrated near the mean when the standard deviation is smaller.

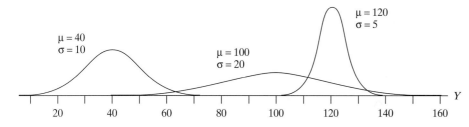

Figure 4.7 Three normal curves with different means and standard deviations

---

4.3 AREAS UNDER A NORMAL CURVE

As explained in Section 3.6, a density curve can be quantitatively interpreted in terms of areas under the curve. While areas can be roughly estimated by eye, for some purposes it is desirable to have fairly precise information about areas.

The Standardized Scale

The areas under a normal curve have been computed mathematically and tabulated for practical use. The use of this tabulated information is much simplified by the fact that all normal curves can be made equivalent with respect to areas under them by suitable rescaling of the horizontal axis. The rescaled variable is denoted by Z; the relationship between the two scales is shown in Figure 4.8.

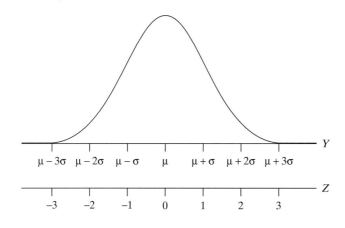

Figure 4.8 A normal curve, showing the relationship between the natural scale (Y) and the standardized scale (Z)

As Figure 4.8 indicates, the Z scale measures standard deviations from the mean: $z = 1.0$ corresponds to 1.0 standard deviation above the mean, $z = -2.5$ corresponds to 2.5 standard deviations below the mean, and so on. The Z scale is referred to as a **standardized scale**.

The correspondence between the Z scale and the Y scale can be expressed by the formula given in the box.

Standardization Formula

$$Z = \frac{Y - \mu}{\sigma}$$

The variable Z is referred to as the **standard normal**; the distribution of Z follows a normal curve with mean zero and standard deviation one. Table 3 at the end of this book gives areas under the standard normal curve, with distances along the horizontal axis measured in the Z scale. Each area tabled in Table 3 is the area under the standard normal curve below a specified value of z. For example, for $z = 1.53$ the tabled area is .9370; this area is shaded in Figure 4.9.

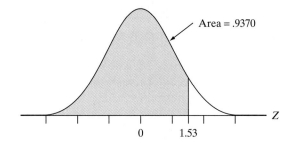

Figure 4.9 Illustration of the use of Table 3

If we want to find the area above a given value of z, we subtract the tabulated area from 1. For example, the area above $z = 1.53$ is $1.0000 - .9370 = .0630$ (Figure 4.10).

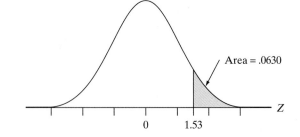

Figure 4.10 Area under a standard normal curve above 1.53

To find the area between two z numbers, we can subtract the areas given in Table 3. For example, to find the area under the Z curve between $z = -1.2$ and $z = 0.8$ (Figure 4.11), we take the area below 0.8, which is .7881, and subtract the area below -1.2, which is .1151, to get $.7881 - .1151 = .6730$.

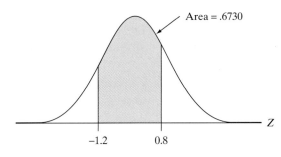

Figure 4.11 Area under a standard normal curve between −1.2 and 0.8

Using Table 3, we see that the area under the normal curve between $z = -1$ and $z = +1$ is $.8413 - .1587 = .6826$. Thus, for any normal distribution, about 68% of the observations are within ± 1 standard deviation of the mean. Likewise, the area under the normal curve between $z = -2$ and $z = +2$ is $.9772 - .0228 = .9544$ and the area under the normal curve between $z = -3$ and $z = +3$ is $.9987 - .0013 = .9974$. This means that for any normal distribution about 95% of the observations are within ± 2 standard deviations of the mean and about 99.7% of the observations are within ± 3 standard deviations of the mean. (see Figure 4.12.) For example, about 68% of the serum cholesterol values in the idealized distribution of Figure 4.2 are between 146 mg/dLi and 206 mg/dLi, about 95% are between 116 mg/dLi and 236 mg/dLi, and virtually all are between 86 mg/dLi and 266 mg/dLi. Figure 4.13 shows the percentages

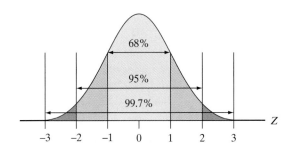

Figure 4.12 Areas under a standard normal curve between −1 and +1, between −2 and +2, and between −3 and +3

If the variable Y follows a normal distribution, then

about 68% of the y's are within ± 1 SD of the mean;

about 95% of the y's are within ± 2 SDs of the mean;

about 99.7% of the y's are within ± 3 SDs of the mean.

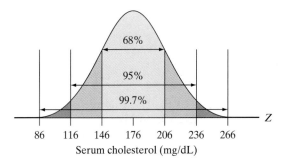

Figure 4.13 The 68/95/99.7 rule and the serum cholesterol distribution

These statements provide a definite interpretation of the standard deviation in cases where a distribution is approximately normal. (In fact, the statements are often approximately true for moderately nonnormal distributions; that is why, in Section 2.6, these percentages—68%, 95%, and >99%—were described as "typical" for "nicely shaped" distributions.)

Determining Areas for a Normal Curve

By taking advantage of the standardized scale, we can use Table 3 to answer detailed questions about any normal population when the population mean and standard deviation are specified. The following example illustrates the use of Table 3. (Of course, the population described in the example is an idealized one, since no actual population follows a normal distribution *exactly*.)

Example 4.5 **Lengths of Fish.**   In a certain population of the herring *Pomolobus aestivalis*, the lengths of the individual fish follow a normal distribution. The mean length of the fish is 54.0 mm, and the standard deviation is 4.5 mm.[5] We use Table 3 to answer various questions about the population.

   (a) What percentage of the fish are less than 60 mm long?
       Figure 4.14 shows the population density curve, with the desired area indicated by shading. In order to use Table 3, we convert the limits of the area from the Y scale to the Z scale, as follows:
       For $y = 60$, the value of z is

$$z = \frac{y - \mu}{\sigma} = \frac{60 - 54}{4.5} = 1.33$$

   Thus, the question "What percentage of the fish are less than 60 mm long?" is equivalent to the question "What is the area under the standard normal curve below the z value of 1.33?" Looking up $z = 1.33$ in Table 3, we find that the area is .9082; thus, 90.82% of the fish are less than 60 mm long.

   (b) What percentage of the fish are more than 51 mm long?
       The standardized value for $y = 51$ is

$$z = \frac{y - \mu}{\sigma} = \frac{51 - 54}{4.5} = -.67$$

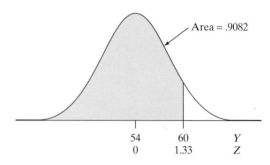

Figure 4.14 Area under the
normal curve in Example 4.5(a)

Thus, the question "What percentage of the fish are more than 51 mm long?" is equivalent to the question "What is the area under the standard normal curve above the z value of $-.67$?" Figure 4.15 shows this relationship. Looking up $z = -.67$ in Table 3, we find that the area is below

−.67 is .2514. This means that the area above −.67 is 1 − .2514 = .7486. Thus, 74.86% of the fish are more than 51 mm long.

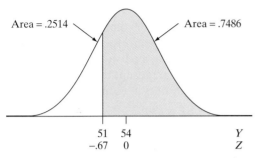

Area = .2514   Area = .7486

| 51 | 54 | Y |
| −.67 | 0 | Z |

Figure 4.15 Area under the normal curve in Example 4.5(b)

(c) What percentage of the fish are between 51 and 60 mm long?
Figure 4.16 shows the desired area. This area can be expressed as a difference of two areas found from Table 3. The area below $y = 60$ is .9082, as found in part (a), and the area below $y = 51$ is .2514, as found in part (b). Consequently, the desired area is computed as

$$.9082 − .2514 = .6568$$

Thus, 65.68% of the fish are between 51 and 60 mm long.

(d) What percentage of the fish are between 58 and 60 mm long?

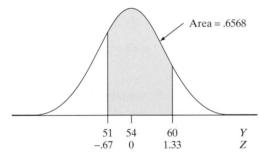

Area = .6568

| 51 | 54 | 60 | Y |
| −.67 | 0 | 1.33 | Z |

Figure 4.16 Area under the normal curve in Example 4.5(c)

Figure 4.17 shows the desired area. This area can be expressed as a difference of two areas found from Table 3. The area below $y = 60$ is .9082, as was found in part (a). To find the area below $y = 58$, we first calculate the z value that corresponds to $y = 58$:

$$z = \frac{y - \mu}{\sigma} = \frac{58 - 54}{4.5} = .89$$

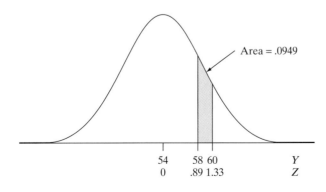

Area = .0949

| 54 | 58 | 60 | Y |
| 0 | .89 | 1.33 | Z |

Figure 4.17 Area under the normal curve in Example 4.5(d)

The area under the Z curve below .89 is .8133. Consequently, the desired area is computed as

$$.9082 - .8133 = .0949$$

Thus, 9.49% of the fish are between 58 and 60 mm long. ∎

Each of the percentages found in Example 4.5 can also be interpreted in terms of probability. Let the random variable Y represent the length of a fish randomly chosen from the population. Then the results in Example 4.5 imply that

$$\Pr\{Y < 60\} = .9082$$
$$\Pr\{Y > 51\} = .7486$$
$$\Pr\{51 < Y < 60\} = .6568$$

and

$$\Pr\{58 < Y < 60\} = .0949$$

Thus, the normal distribution can be interpreted as a continuous probability distribution.

Note that because the idealized normal distribution is perfectly continuous, probabilities such as

$$\Pr\{Y > 48\} \text{ and } \Pr\{Y \geq 48\}$$

are equal (see Section 3.6). That is,

$$\Pr\{Y \geq 48\} = \Pr\{Y > 48\} + \Pr\{Y = 48\}$$
$$= \Pr\{Y > 48\} + 0(\text{since } Y \text{ is taken to be continuous})$$
$$= \Pr\{Y > 48\}$$

If, however, the length were measured only to the nearest mm, then the measured variable would actually be discrete, so that $\Pr\{Y > 48\}$ and $\Pr\{Y \geq 48\}$ would differ somewhat from each other. In cases where this discrepancy is important, the computation can be refined to take into account the discontinuity of the measured distribution (see the optional Section 4.5).

Inverse Reading of Table 3

In determining facts about a normal distribution, it is sometimes necessary to read Table 3 in an "inverse" way—that is, to find the value of z corresponding to a given area rather than the other way around. For example, suppose we want to find the value on the Z scale that cuts off the top 2.5% of the distribution. This number is 1.96, as shown in Figure 4.18.

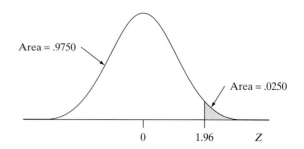

Figure 4.18 Area under the normal curve above 1.96

Area = .9750

Area = .0250

0   1.96   Z

We will find it helpful, for future reference, to introduce some notation. We use the notation Z_α to denote the number such that $\Pr\{Z < Z_\alpha\} = 1 - \alpha$ and $\Pr\{Z > Z_\alpha\} = \alpha$, as shown in Figure 4.19. Thus, $Z_{.025} = 1.96$.

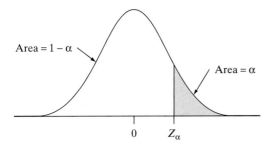

Figure 4.19 Area under the normal curve above α

We often need to determine a Z_α value when we want to determine a *percentile* of a normal distribution. The percentiles of a distribution divide the distribution into 100 equal parts, just as the quartiles divide it into four equal parts [from the Latin roots *centum* (hundred) and *quartus* (fourth)]. For example, suppose we want to find the 70th percentile of a standard normal distribution. That means that we want to find the number $Z_{.30}$ that divides the standard normal distribution into two parts: the bottom 70% and the top 30%. As Figure 4.20 illustrates, we need to look in Table 3 for an area of .7000. The closest value is an area of .6985, corresponding to a z value of .52. Thus, $Z_{.30} = .52$.

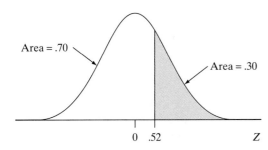

Figure 4.20 Determining the 70th percentile of a normal distribution

Lengths of Fish.

Example 4.6

(a) Suppose we want to find the 70th percentile of the fish length distribution of Example 4.5. Let us denote the 70th percentile by y^*. By definition, y^* is the value such that 70% of the fish lengths are less than y^* and 30% are greater, as illustrated in Figure 4.21.

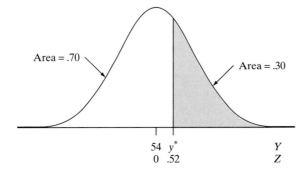

Figure 4.21 Determining the 70th percentile of a normal distribution, Example 4.6(a)

To find y^* we use the value of $Z_{.30} = .52$ that we just determined. Next we convert this z value to the Y scale. We know that the if we were given the value of y^*, we could convert it to a standard normal (z scale) and the result would be .52. Thus, from the standardization formula we obtain the equation

$$.52 = \frac{y^* - 54}{4.5}$$

which can be solved to give $y^* = (.52)(4.5) + 54 = 56.3$. The 70th percentile of the fish length distribution is 56.3 mm.

(b) Suppose we want to find the 20th percentile of the fish length distribution of Example 4.5. Let us denote the 20th percentile by y^*. By definition, y^* is the value such that 20% of the fish lengths are less than y^* and 80% are greater, as illustrated in Figure 4.22.

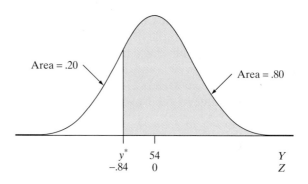

Figure 4.22 Determining the 20th percentile of a normal distribution, Example 4.6(b)

To find y^* we first determine the value of $Z_{.80}$, which is the 20th percentile in the Z scale. As Figure 4.22 illustrates, we need to look in Table 3 for an area of .2000. The closest value is an area of .2005, corresponding to $z = -.84$. The next step is to convert this z value to the Y scale. From the standardization formula we obtain the equation

$$-.84 = \frac{y^* - 54}{4.5}$$

which can be solved to give $y^* = (-.84)(4.5) + 54 = 50.2$. The 20th percentile of the fish length distribution is 50.2 mm.   ∎

Problem-Solving Tip. In solving problems that require the use of Table 3, a sketch of the distribution (as in Figures 4.14–4.17 and 4.20–4.22) is a very handy aid to straight thinking.

Computer note: Computer software can be used to find normal probablities. For example, in Example 4.5, part (a), we found that the percentage of fish less than 60 mm long, for a population with mean length 54 mm and standard deviation 4.5 mm, is 90.82%. The statistical package MINITAB has a built-in version of a standard normal table (Table 3), which can be used to find this percentage. The following command, which makes use of a "cumulative distribution function" (cdf), will produce the percentage

```
MTB > CDF 60;
SUBC > NORMAL 54 4.5.
```

(Note that MINITAB returns an answer of .9088, which differs slightly from the answer of .9082 found in Example 4.5. This is due to the fact that MINITAB carries out calculations to four decimal places, whereas we rounded off to the second decimal place when calculating the value of z in Example 4.5.)

MINITAB can also be used to find percentiles. In Example 4.6, part (a), we found that the 70th percentile of the fish length distribution is 56.3 mm. To find this value using MINITAB, we use the "inverse cumulative distribution function" (invcdf), as follows:

```
MTB > INVCDF .7;
SUBC > NORMAL 54 4.5.
```

Exercises 4.1–4.16

4.1   Suppose a certain population of observations is normally distributed. What percentage of the observations in the population

(a) are within ±1.5 standard deviations of the mean?
(b) are more than 2.5 standard deviations above the mean?
(c) are more than 3.5 standard deviations away from (above or below) the mean?

4.2   (a) The 90th percentile of a normal distribution is how many standard deviations above the mean?
(b) The 10th percentile of a normal distribution is how many standard deviations below the mean?

4.3   The brain weights of a certain population of adult Swedish males follow approximately a normal distribution with mean 1,400 g and standard deviation 100 g.[6] What percentage of the brain weights are

(a) 1,500 g or less?
(b) between 1,325 and 1,500 g?
(c) 1,325 g or more?
(d) 1,475 g or more?
(e) between 1,475 and 1,600 g?
(f) between 1,200 and 1,325 g?

4.4   Let Y represent a brain weight randomly chosen from the population of Exercise 4.3. Find

(a) $\Pr\{Y \leq 1{,}325\}$
(b) $\Pr\{1{,}475 \leq Y \leq 1{,}600\}$

4.5   In an agricultural experiment, a large uniform field was planted with a single variety of wheat. The field was divided into many plots (each plot being 7×100 ft) and the yield (lb) of grain was measured for each plot. These plot yields followed approximately a normal distribution with mean 88 lb and standard deviation 7 lb.[7] What percentage of the plot yields were

(a) 80 lb or more?
(b) 90 lb or more?

(c) 75 lb or less?
(d) between 75 and 90 lb?
(e) between 90 and 100 lb?
(f) between 75 and 80 lb?

4.6   Refer to Exercise 4.5. Let Y represent the yield of a plot chosen at random from the field. Find

(a) $\Pr\{Y > 90\}$
(b) $\Pr\{75 < Y < 90\}$

4.7   Consider a standard normal distribution, Z. Find

(a) $Z_{.10}$
(b) $Z_{.25}$
(c) $Z_{.05}$
(d) $Z_{.01}$

4.8   For the wheat-yield distribution of Exercise 4.5 find

(a) the 65th percentile
(b) the 35th percentile

4.9   The serum cholesterol levels of 17-year-olds follow a normal distribution with mean 176 mg/dLi and standard deviation 30 mg/dLi. What percentage of 17-year-olds have serum cholesterol values

(a) 186 or more?
(b) 156 or less?
(c) 216 or less?
(d) 121 or more?
(e) between 186 and 216?
(f) between 121 and 156?
(g) between 156 and 186?

4.10   Refer to Exercise 4.9. Suppose a 17-year-old is chosen at random and let Y be the person's serum cholesterol value. Find

(a) $\Pr\{Y \geq 180\}$
(b) $\Pr\{180 < Y < 210\}$

4.11   For the serum cholesterol distribution of Exercise 4.9, find

(a) the 80th percentile
(b) the 20th percentile

4.12   When red blood cells are counted using a certain electronic counter, the standard deviation (SD) of repeated counts of the same blood specimen is about .8% of the true value, and the distribution of repeated counts is approximately normal.[8] For example, this means that if the true value is 5,000,000 cells/mm^3, then the SD is 40,000.

(a) If the true value of the red blood count for a certain specimen is 5,000,000 cells/mm^3, what is the probability that the counter would give a reading between 4,900,000 and 5,100,000?
(b) If the true value of the red blood count for a certain specimen is μ, what is the probability that the counter would give a reading between $.98\mu$ and 1.02μ?
(c) A hospital lab performs counts of many specimens every day. For what percentage of these specimens does the reported blood count differ from the correct value by 2% or more?

4.13   The amount of growth, in a 15-day period, for a population of sunflower plants was found to follow a normal distribution with mean 3.18 cm and standard deviation 0.53 cm.[9] What percentage of plants grow

(a) 4 cm or more?
(b) 3 cm or less?
(c) between 2.5 and 3.5 cm?

4.14   Refer to Exercise 4.13. In what range do the middle 90% of all growth values lie?

4.15   For the sunflower plant growth distribution of Exercise 4.13, what is the 25th percentile?

4.16   Many cities sponsor marathon races each year. The following histogram shows the distribution of times that it took for 3,700 runners to complete the Rome marathon in 1996, with a normal curve superimposed. The fastest runner completed the 26.3-mile course in 2 hours and 12 minutes, which is 132 minutes. The average time was 230 minutes, and the standard deviation was 36 minutes. Use the normal curve to answer the following questions.

(a) What percentage of times were greater than 200 minutes?
(b) What is the 60th percentile of the times?
(c) Notice that the normal curve approximation is fairly good except around the 240 minute mark. How can we explain this anomalous behavior of the distribution?

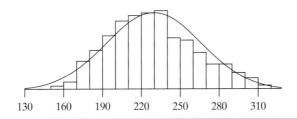

4.4 ASSESSING NORMALITY

Many statistical procedures are based on having data from a normal population. In this section we consider ways to assess whether it is reasonable to use a normal curve model for a set of data and, if not, how we might proceed.

Recall from Section 4.3 that if the variable Y follows a normal distribution, then

about 68% of the y's are within ± 1 SD of the mean;

about 95% of the y's are within ± 2 SDs of the mean;

about 99.7% of the y's are within ± 3 SDs of the mean.

We can use these facts as a check of how closely a normal curve model fits a set of data.

Serum Cholesterol.   For the serum cholesterol data of Example 4.1 the sample mean is 176 and the sample SD is 30. The interval "mean $\pm$ SD" is

$$(176 - 30, 176 + 30) \text{ or } (146, 206)$$

Example 4.7

This interval contains 659 of the 953 observations, or 69.2% of the data. Likewise, the interval

$$(176 - 2 \cdot 30, 176 + 2 \cdot 30) \text{ is } (116, 236)$$

which contains 901, or 94.5%, of the 953 observations. Finally, the interval

$$(176 - 3 \cdot 30, 176 + 3 \cdot 30) \text{ is } (86, 266)$$

which contains 951, or 99.8%, of the 953 observations. The three observed percentages

$$69.2\%, 94.5\%, \text{ and } 99.8\%$$

agree quite well with the theoretical percentages of

$$68\%, 95\%, \text{ and } 99.7\%.$$

This agreement supports the claim that serum cholesterol levels for 17-year-olds have a normal distribution. This reinforces the visual evidence of Figure 4.1. ■

Example 4.8

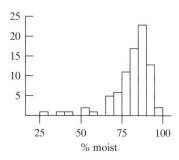

Figure 4.23 Moisture content in freshwater fruit

Moisture Content.   Moisture content was measured in each of 83 freshwater fruit.[10] Figure 4.23 shows that this distribution is strongly skewed to the left. The sample mean of these data is 80.7 and the sample SD is 12.7. The interval

$$(80.7 - 12.7, 80.7 + 12.7)$$

contains 70, or 84.3%, of the 83 observations. The interval

$$(80.7 - 2 \cdot 12.7, 80.7 + 2 \cdot 12.7)$$

contains 78, or 93.8%, of the 83 observations. Finally, the interval

$$(80.7 - 3 \cdot 12.7, 80.7 + 3 \cdot 12.7)$$

contains 80, or 96.4%, of the 83 observations. The three percentages

$$84.3\%, 93.8\%, \text{ and } 96.4\%$$

differ from the theoretical percentages of

$$68\%, 95\%, \text{ and } 99.7\%$$

because the distribution is far from being bell-shaped. This reinforces the visual evidence of Figure 4.23. ■

Normal Probability Plots

A **normal probability plot** is a special statistical graph that is used to assess normalilty. We present this statistical tool with an example using the heights (in inches) of a sample of 11 women, sorted from smallest to largest:

$$61, 62.5, 63, 64, 64.5, 65, 66.5, 67, 68, 68.5, 70.5$$

Based on these data, does it make sense to use a normal curve to model the distribution of women's heights? Figure 4.24 shows a histogram of the data with a normal curve superimposed, using the sample mean of 65.5 and the sample standard deviation of 2.9 as the parameters of the normal curve. This histogram is fairly symmetric, but when we have a small sample it can be hard to tell the shape of the population distribution by looking at a histogram.

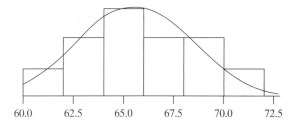

Figure 4.24 Histogram of the heights of 11 women

A normal probability plot is a tool to help assess whether a population is normal. Most statistical computer packages provide normal probability plots. Figure 4.25 shows a normal probability plot for the height data.

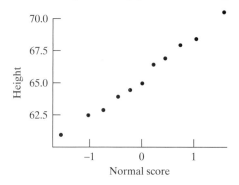

Figure 4.25 Normal probability plot of the height data

When we look at a normal probability plot, we hope to see a straight line. If the points fall along a straight line, then we infer that the population distribution is normal. Many statistical procedures are based on the condition that the data came from a normal population, so it is important to be able to assess normality. It is easier to assess whether or not a graph of points is straight than whether or not a histogram is bell-shaped.

How Normal Probability Plots Work

In our sample the median height is 65 inches and the sample mean height is 65.5 inches. If the population distribution of heights is $N(65.5, 2.9)$ then the population median is 65.5 (the same as the population mean of 65.5). If we were to take several random samples of size 11 from a $N(65.5, 2.9)$ distribution, then we would expect the average of the sample medians to be 65.5.

The shortest woman in our sample is 61 inches tall. If we were to take several random samples of size 11 from a $N(65.5, 2.9)$ distribution, on average how small would the smallest value be? That is, if heights of women really follow a normal distribution, with mean 65.5 and standard deviation 2.9, then how short would we expect the shortest woman in a sample of size 11 to be? Unlike the case of the median, this is not a simple question to answer.

One way to think about this issue is to consider what would happen if we took repeated samples from a $N(0, 1)$ distribution. We know that if $Y \sim N(\mu, \sigma)$, then $Z = \dfrac{Y - \mu}{\sigma} \sim N(0, 1)$, so that Y and Z are related by the linear relationship $Y = \mu + \sigma Z$.

The expected values of the ordered observations in a sample of size 11 from a $N(0, 1)$ distribution are called "normal scores." Using computer software, we can find that the first normal score—the expected value of the smallest observation—

is approximately −1.56.* This means that if we take repeated samples of size 11 from a $N(0, 1)$ distribution and find the smallest value in each sample, these smallest values average approximately −1.56. By symmetry, the largest normal score—the expected value of the largest observation from a $N(0, 1)$ distribution—is 1.56. The only normal score that we can easily find without using computer software is the 6th normal score—the expected value of the median of 11 observations from a $N(0, 1)$ distribution—which is 0 [since the $N(0, 1)$ curve is symmetric about 0].

To make a normal probability plot, we find all 11 normal scores and match them with the 11 data values, creating 11 ordered pairs of the form (normal score, observed height), which we then graph.[11] Of course, we would want to use a computer (or a graphing calculator) to carry out this process. If the points in the plot show a linear pattern, then we infer that there is a linear relationship between Y and Z, of the form $Y = \mu + \sigma Z$. Since we know that Z has a $N(0, 1)$ distribution, we infer that Y also has a normal distribution, with mean μ and standard deviation σ.

Of course, even when we sample from a perfectly normal distribution, we have to expect that there will be some variability between the sample we obtain and the theoretical normal scores. Figure 4.26 shows six normal probability plots based on samples taken from a $N(0, 1)$ distribution. Notice that all six plots show

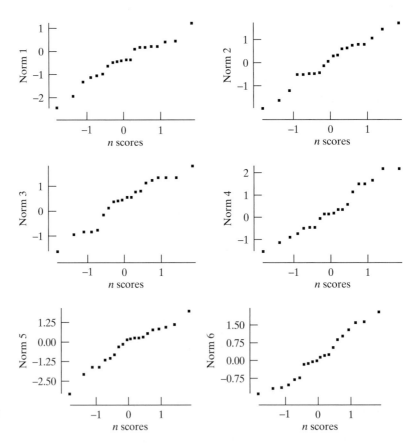

Figure 4.26 Normal probability plots for normal data

---

* This value was found using the program Data Desk. Data Desk calculates the ith normal score as $Z_{1-\alpha}$, where $\alpha = (i - 1/3)/(n + 1/3)$. That is, the ith normal score is the value on the Z scale that cuts off the bottom area under the Z curve of $(i - 1/3)/(n + 1/3)$. Some other software programs use slightly different conventions for calculating normal scores.

a general linear pattern. It is true that there is a fair amount of "wiggle" in some of the plots, but the important feature of each of these plots is that we can draw a line that follows the majority of the data points.

If the points in the normal probability plot do not fall more or less along a straight line, then we infer that the population is not normal. For example, if the top of the plot bends up, that means the y values at the upper end of the distribution are too large for the distribution to be bell shaped (i.e., the distribution is skewed to the right or has large outliers, as in Figure 4.27).

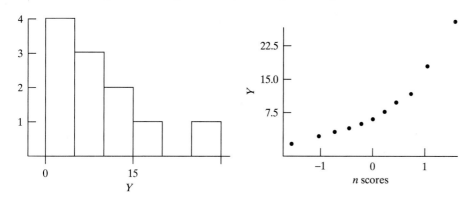

Figure 4.27 Histogram and normal probability plot of a distribution that is skewed to the right

If the bottom of the plot bends down, that means the y values at the lower end of the distribution are too small for the distribution to be bell shaped (i.e., the distribution is skewed to the left or has small outliers). Figure 4.28 shows the distribution of moisture content in freshwater fruit, from Example 4.8, which is strongly skewed to the left.

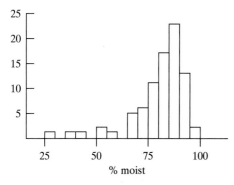

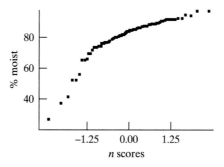

Figure 4.28 Histogram and normal probability plot of a distribution that is skewed to the left

If a distribution has a very long left-hand tail and a long right-hand tail, when compared with a normal curve, then the normal probability plot will have something of an S shape. Figure 4.29 shows such a distribution.

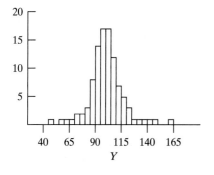

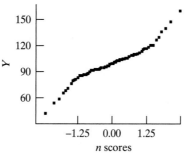

Figure 4.29 Histogram and normal probability plot of a distribution that has long tails

Sometimes the same value shows up repeatedly in a sample, due to rounding in the measurement process. This leads to *granularity* in the normal probability plot, as in Figure 4.30, but this does not stop us from inferring that the underlying distribution is normal.

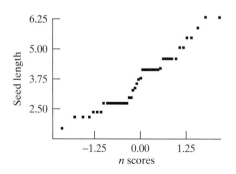

Figure 4.30 Normal probability plot of lengths of 48 seeds from freshwater fruit [10]

Computer note: Creating a normal probability plot requires a great deal of computation and thus is almost always done with the aid of technology. In the statistical package MINITAB, if the data are stored in column 1, then the following command will produce a normal probability plot:

```
MTB >% NormPlot C1
```

Note that MINITAB puts the data, Y, on the horizontal axis and normal probabilities on the verical axis, rather than using the approach presented here (with the data on the vertical axis). Nonetheless, the basic idea remains the same: Create the plot and look to see if there is a linear pattern.

Transformations for Nonnormal Data

A normal probability plot can help us assess whether or not the data came from a normal distribution. Sometimes a histogram or normal probability plot shows that our data are nonnormal, but a transformation of the data gives us a symmetric, bell-shaped curve. In such a situation, we may wish to transform the data and continue our analysis in the new (transformed) scale.

Example 4.9

Lentil Growth. Figures 4.31(a) and (b) show the distribution of the growth rate, in cm per day, for a sample of 47 lentil plants.[12] This distribution is skewed to the right. If we take the natural logarithm of each observation, we get a distribution that is much more nearly symmetric. Figures 4.32(a) and (b) show that in log scale the growth rate distribution is approximately normal. (In Figure 4.32 the natural logarithm, $\log_e$, is used, but we could use any base, such as $\log_{10}$, and the effect on the shape of the distribution would be the same.) ∎

In general, if the distribution is skewed to the right, then one of the following transformations should be considered:

$$\sqrt{Y}, \log Y, \frac{1}{\sqrt{Y}}, \frac{1}{Y}.$$

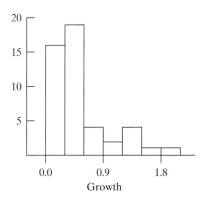

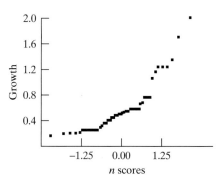

Figures 4.31 (a) and 4.31 (b) Histogram and normal probability plot of growth rates of 47 lentil plants[12]

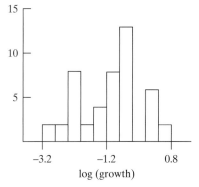

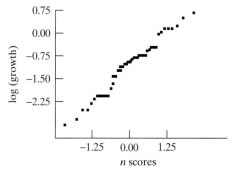

Figures 4.32 (a) and 4.32 (b) Histogram and normal probability plot of the logarithms of the growth rates of 47 lentil plants

These transformations will pull in the long right-hand tail and push out the short left-hand tail, making the distribution more nearly symmetric. Each of these is more drastic than the one before. Thus, a square root transformation will change a mildly skewed distribution into a symmetric distribution, but a log transformation may be needed if the distribution is more heavily skewed. For example, we saw in Example 2.42 (in Section 2.7) how a square root transformation pulls in a long right-hand tail and how a log transformation pulls in the right-hand tail even more. If the distribution of a variable Y is skewed to the left, then raising Y to a power greater than 1 can be helpful.

Exercises 4.17–4.21

4.17 In Example 4.2 it was stated that shell thicknesses in a population of eggs follow a normal distribution with mean $\mu = .38$ mm and standard deviation $\sigma = .03$ mm. Use the 68%–95%–99.7% rule to determine intervals, centered at the mean, that include 68%, 95%, and 99.7% of the shell thicknesses in the distribution.

4.18 The following three normal probability plots, (a), (b), and (c), were generated from the distributions shown by histograms I, II, and III. Which normal probability plot goes with which histogram? How do you know?

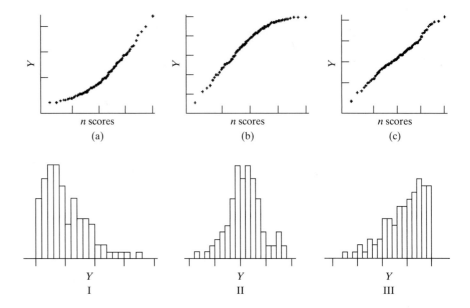

4.19   The June precipitation totals, in inches, for the city of Cleveland, Ohio for the years 1964–1978 are given in the following table together with the corresponding normal scores.[13] (Note that the data are given in chronological order, so the normal scores are not listed in increasing order.) Use these values to create a normal probability plot of the data. Do you conclude that the distribution is normal?

Year	Rainfall	Normal score
1964	2.06	−0.94
1965	3.05	−0.52
1966	1.83	−1.23
1967	1.17	−1.71
1968	2.32	−0.71
1969	4.61	0.52
1970	4.98	0.94
1971	3.79	0.16
1972	9.06	1.71
1973	6.72	1.23
1974	3.57	−0.16
1975	4.10	0.33
1976	3.64	0.00
1977	4.91	0.71
1978	3.30	−0.33

4.20   For each of the following normal probability plots, sketch the corresponding histogram of the data.

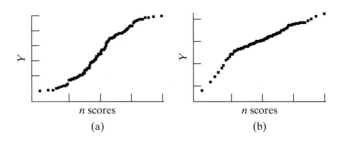

4.21 The following normal probability plot was created from the times that it took 166 bicycle riders to complete the Stage 11 Time Trial, from Grenoble to Chamrousse, France, in the 2001 Tour de France cycling race.

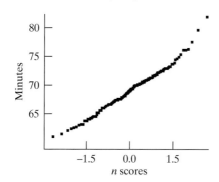

(a) Consider the fastest riders. Are their times better than, worse than, or roughly equal to the times one would expect the fastest riders to have if the data came from a truly normal distribution?
(b) Consider the slowest riders. Are their times better than, worse than, or roughly equal to the times one would expect the slowest riders to have if the data came from a truly normal distribution?

4.5 THE CONTINUITY CORRECTION (OPTIONAL)

Although a normal curve theoretically represents a continuous distribution, it is common practice to use a normal curve to describe approximately the distribution of a discrete variable. Often the discreteness of the variable can be ignored without introducing any serious error; indeed, in the computations of Section 4.3 we did not distinguish between discrete and continuous variables. For greater accuracy, however, we can take account of discreteness by applying a correction, known as the **continuity correction**, when calculating areas under the normal curve. The following example illustrates the use of the continuity correction.

Litter Size. Table 4.1 shows the distribution of litter size (defined as the number of live young in the first litter) for a population of female mice; the population mean is 7.8 and the standard deviation is 2.3.[14] Figure 4.33 shows a normal curve with $\mu = 7.8$ and $\sigma = 2.3$, superimposed on the litter size distribution; the normal curve fits the distribution quite well.

Example 4.10

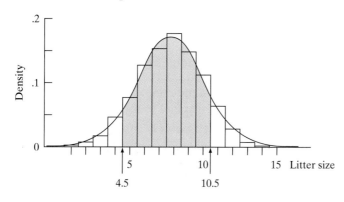

Figure 4.33 Litter size distribution and approximating normal curve

TABLE 4.1

Litter Size	Percent Frequency	Litter Size	Percent Frequency
1	.4	9	15.0
2	.7	10	11.5
3	1.8	11	6.6
4	4.8	12	3.1
5	8.0	13	1.0
6	13.0	14	.5
7	15.5	15	.3
8	17.8		

(a) Let us compare an actual population relative frequency with the relative frequency predicted by the normal curve. From Table 4.1, the percentage of litters with sizes between 5 and 10, inclusive, is

$$8.0 + 13.0 + 15.5 + 17.8 + 15.0 + 11.5 = 80.8\%$$

In other words, if Y represents the size of a randomly selected litter, then

$$\Pr\{5 \le Y \le 10\} = .808$$

This is the sum of the areas of the six histogram bars from $y = 5$ to $y = 10$. What is the corresponding area under the normal curve? If we were to take the approach shown in Section 4.3, we would find that

$$\Pr\{5 \le Y \le 10\} = \Pr\left\{\frac{5 - 7.8}{2.3} < \frac{Y - \mu}{\sigma} < \frac{10 - 7.8}{2.3}\right\}$$
$$\approx \Pr\{-1.22 < Z < .96\}$$
$$= .8315 - .1112 = .7203$$

However, this calculation gives the area between 5.0 and 10.0, which means that it excludes the area for half of the histogram bar for $y = 5$ and half of the histogram bar for $y = 10$. To adjust for the discreteness of Y, we should calculate the area under the curve between $y = 4.5$ and $y = 10.5$, which is shaded in Figure 4.33; the use of 4.5 instead of 5, and 10.5 instead of 10, represents the continuity correction. Using Table 3, we have

$$\Pr\{5 \le Y \le 10\} = \Pr\{4.5 < Y < 10.5\}$$
$$= \Pr\left\{\frac{4.5 - 7.8}{2.3} < \frac{Y - \mu}{\sigma} < \frac{10.5 - 7.8}{2.3}\right\}$$
$$\approx \Pr\{-1.43 < Z < 1.17\} = .8790 - .0764 = .8026$$

Thus, the value found from the normal approximation with the continuity correction (.8026) is quite close to the actual value (.808).

(b) The continuity correction is especially important if we want to consider the probability of a single Y value. For example, the normal approximation to $\Pr\{Y = 10\}$ is the area from $y = 9.5$ to $y = 10.5$, which is shaded in Figure 4.34. This area can be calculated to be

$$\Pr\{9.5 < Y < 10.5\} = \Pr\left\{\frac{9.5 - 7.8}{2.3} < \frac{Y - \mu}{\sigma} < \frac{10.5 - 7.8}{2.3}\right\}$$

$$\approx \Pr\{.74 < Z < 1.17\} = .8790 - .7704$$

$$= .1086 \text{ or } 10.9\%$$

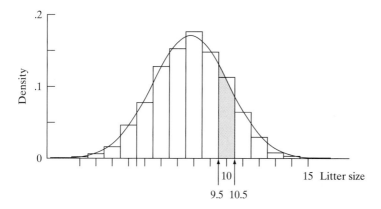

Figure 4.34 Litter size distribution and approximating normal curve

This is reasonably close to the actual value of 11.5% (from Table 4.1). The normal approximation without the continuity correction would be the area from $y = 10$ to $y = 10$, which is zero—not a sensible answer.

(c) Suppose we want to find the probability that Y is at least 9, that is, $\Pr\{Y \geq 9\}$. Thinking about the raw data and the histogram, we see that $\Pr\{Y \geq 9\} = \Pr\{Y = 9\} + \Pr\{Y = 10\} + \ldots + \Pr\{Y = 15\}$. That is, we want to include the histogram bar for $Y = 9$ in our calculation, but we want to leave out the histogram bar for $Y = 8$. Thus we draw the line halfway between 8 and 9, at 8.5 (see Figure 4.35):

$$\Pr\{Y \geq 9\} = \Pr\{Y > 8.5\} = \Pr\left\{\frac{Y - \mu}{\sigma} > \frac{8.5 - 7.8}{2.3}\right\}$$

$$\approx \Pr\{Z > .30\} = 1 - .6179 = .3821$$

This agrees quite well with the actual value of 38% found by adding the percent frequencies in Table 4.1 from 9 through 15. ∎

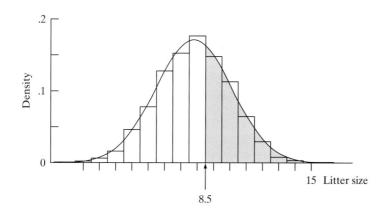

Figure 4.35 Litter size distribution and approximating normal curve

In Example 4.10, the continuity correction always involved an adjustment of ±.5, for example from 5 to 4.5 or from 10 to 10.5. This was due to the fact that the raw data were integer valued, so that when a histogram was made using the smallest bin width possible for the data (a width of 1), one-half of the bin width was .5. This will often be the case, but not always. For example, if we have data that are recorded in units of 100 (e.g., 100, 200, 300, . . .), then applying a continuity correction would involve an adjustment of ±50.

Whenever a discrete distribution is approximated by a normal curve, the continuity correction can be used to obtain more accurate values for predicted relative frequencies. This applies not only to variables that are inherently discrete (such as litter size) but also to variables that are actually continuous but are measured on a discrete scale because of rounding. For instance, blood pressure is a continuous variable, but it is usually measured to the nearest mm Hg, so that the actual measurements fall on a discrete scale. If the "spaces" between the possible values of a variable are small compared with the standard deviation of the distribution, then the continuity correction has little effect. For instance, if the standard deviation of a distribution of blood pressures is 20 mm Hg, then (because 1 is small compared to 20) for most purposes the continuity correction for this distribution could be ignored.

Exercises 4.22–4.25

4.22 In genetic studies of the fruitfly *Drosophila melanogaster*, one variable of interest is the total number of bristles on the ventral surface of the fourth and fifth abdominal segments. For a certain *Drosophila* population,[15] the bristle count follows approximately a normal distribution with mean 38.5 and standard deviation 2.9. Find (using the continuity correction)

(a) the percentage of flies with 40 or more bristles
(b) the percentage of flies with exactly 40 bristles
(c) the percentage of flies whose bristle count is between 35 and 40, inclusive

4.23 Refer to the fruitfly population of Exercise 4.22. Let Y be the bristle count of a fly chosen at random from the population.

(a) Use the continuity correction to calculate $\Pr\{35 \leq Y \leq 40\}$.
(b) Calculate $\Pr\{35 \leq Y \leq 40\}$ without the continuity correction and compare with the result of part (a).

4.24 The litter sizes of a certain population of female mice follow approximately a normal distribution with mean 7.8 and standard deviation 2.3 (as in Example 4.10). Let Y represent the size of a randomly chosen litter. Use the continuity correction to find approximate values for each of the following probabilities:

(a) $\Pr\{Y \leq 6\}$
(b) $\Pr\{Y = 6\}$
(c) $\Pr\{8 \leq Y \leq 11\}$

4.25 In a certain population of healthy people the mean total protein concentration in the blood serum is 6.85 g/dLi, the standard deviation is .42 g/dLi, and the distribution is approximately normal.[16] Let Y be the total protein value of a randomly selected person, as given by an instrument that reports the value to the nearest .1 g/dLi. Use the continuity correction to calculate

(a) $\Pr\{Y = 6.5\}$
(b) $\Pr\{6.5 \leq Y \leq 8.0\}$

4.6 PERSPECTIVE

The normal distribution is also called the Gaussian distribution, after the German mathematician K. F. Gauss. The term *normal*, with its connotations of "typical" or "usual," can be seriously misleading. Consider, for instance, a medical context, where the primary meaning of "normal" is "not abnormal." Thus, confusingly, the phrase "the normal population of serum cholesterol levels" may refer to cholesterol levels in ideally "healthy" people, or it may refer to a Gaussian distribution such as the one in Example 4.1. In fact, for many variables the distribution in the normal (nondiseased) population is decidedly not normal (i.e., not Gaussian).

The examples of this chapter have illustrated one use of the normal distribution—as an approximation to naturally occurring biological distributions. If a natural distribution is well approximated by a normal distribution, then the mean and standard deviation provide a complete description of the distribution: The mean is the center of the distribution, about 68% of the values are within 1 standard deviation of the mean, about 95% are within 2 standard deviations of the mean, and so on.

As noted in Section 2.6, the 68% and 95% benchmarks can be roughly applicable even to distributions that are rather skewed. (But if the distribution is skewed, then the 68% is not symmetrically divided on both sides of the mean, and similarly for the 95%.) However, the benchmarks do not apply to a distribution (even a symmetric one) for which one or both tails are long and thin [see Figures 2.13(b) and 2.20].

We will see in later chapters that many classical statistical methods are specifically designed for, and function best with, data that have been sampled from normal populations. We will further see that in many practical situations these methods work very well also for samples from nonnormal populations.

The normal distribution is of central importance in spite of the fact that many, perhaps most, naturally occurring biological distributions could be described better by a skewed curve than by a normal curve. A major use of the normal distribution is not to describe natural distributions, but to describe certain theoretical distributions, called sampling distributions, that are used in the statistical analysis of data. We will see in Chapter 5 that many sampling distributions are approximately normal; it is this property that makes the normal distribution so important in the study of statistics.

Supplementary Exercises 4.26–4.45

4.26 The activity of a certain enzyme is measured by counting emissions from a radioactively labeled molecule. For a given tissue specimen, the counts in consecutive 10-second time periods may be regarded (approximately) as repeated independent observations from a normal distribution.[17] Suppose the mean 10-second count for a certain tissue specimen is 1,200 and the standard deviation is 35. Let Y denote the count in a randomly chosen 10-second time period. Find

(a) $\Pr\{Y \geq 1{,}250\}$
(b) $\Pr\{Y \leq 1{,}175\}$
(c) $\Pr\{1{,}150 \leq Y \leq 1{,}250\}$
(d) $\Pr\{1{,}150 \leq Y \leq 1{,}175\}$

4.27   The shell thicknesses of the eggs produced by a large flock of hens follow approximately a normal distribution with mean equal to .38 mm and standard deviation equal to .03 mm (as in Example 4.2). Find the 95th percentile of the thickness distribution.

4.28   Refer to the eggshell thickness distribution of Exercise 4.27. Suppose an egg is defined as thin-shelled if its shell is .32 mm thick or less.

(a) What percentage of the eggs are thin shelled?
(b) Suppose a large number of eggs from the flock are randomly packed into boxes of 12 eggs each. What percentage of the boxes will contain at least one thin-shelled egg? (*Hint*: First find the percentage of boxes that will contain no thin-shelled egg.)

4.29   The heights of a certain population of corn plants follow a normal distribution with mean 145 cm and standard deviation 22 cm.[18] What percentage of the plant heights are

(a) 100 cm or more?
(b) 120 cm or less?
(c) between 120 and 150 cm?
(d) between 100 and 120 cm?
(e) between 150 and 180 cm?
(f) 180 cm or more?
(g) 150 cm or less?

4.30   Suppose four plants are to be chosen at random from the corn plant population of Exercise 4.29. Find the probability that none of the four plants will be more than 150 cm tall.

4.31   Refer to the corn plant population of Exercise 4.29. Find the 90th percentile of the height distribution.

4.32   For the corn plant population described in Exercise 4.29, find the quartiles and the interquartile range.

4.33   Suppose a certain population of observations is normally distributed. Find the value of z^* such that 95% of the observations in the population are between $-z^*$ and $+z^*$, on the Z scale.

4.34   In the nerve-cell activity of a certain individual fly, the time intervals between "spike" discharges follow approximately a normal distribution with mean 15.6 ms and standard deviation .4 ms (as in Example 4.3). Let Y denote a randomly selected interspike interval. Find

(a) $\Pr\{Y > 15\}$
(b) $\Pr\{Y > 16.5\}$
(c) $\Pr\{15 < Y < 16.5\}$
(d) $\Pr\{15 < Y < 15.5\}$

4.35   For the distribution of interspike-time intervals described in Exercise 4.34, find the quartiles and the interquartile range.

4.36   Among American women aged 20–29 years, 10% are less than 60.8 inches tall, 80% are between 60.8 and 67.6 inches tall and 10% are more than 67.6 inches tall.[19] Assuming that the height distribution can be adequately approximated by a normal curve, find the mean and standard deviation of the distribution.

4.37   The intelligence quotient (IQ) score, as measured by the Stanford-Binet IQ test, is normally distributed in a certain population of children. The mean IQ score is 100 points, and the standard deviation is 16 points.[20] What percentage of children in the population have IQ scores

(a) 140 or more?
(b) 80 or less?
(c) between 80 and 120?
(d) between 80 and 140?
(e) between 120 and 140?

4.38   Refer to the IQ distribution of Exercise 4.37. Let Y be the IQ score of a child chosen at random from the population. Find $\Pr\{80 \le Y \le 140\}$.

4.39   Refer to the IQ distribution of Exercise 4.37. Suppose five children are to be chosen at random from the population. Find the probability that exactly one of them will have an IQ score of 80 or less and four will have scores higher than 80. (*Hint:* First find the probability that a randomly chosen child will have an IQ score of 80 or less.)

4.40   A certain assay for serum alanine aminotransferase (ALT) is rather imprecise. The results of repeated assays of a single specimen follow a normal distribution with mean equal to the true ALT concentration for that specimen and standard deviation equal to 4 U/Li (see Example 2.14). Suppose that a certain hospital lab measures many specimens every day, performing one assay for each specimen, and that specimens with ALT readings of 40 U/Li or more are flagged as "unusually high." If a patient's true ALT concentration is 35 U/Li, what is the probability that his specimen will be flagged as "unusually high"?

4.41   Resting heart rate was measured for a group of subjects; the subjects then drank 6 ounces of coffee. Ten minutes later their heart rates were measured again. The change in heart rate followed a normal distribution, with a mean increase of 7.3 beats per minute and a standard deviation of 11.1.[21] Let Y denote the change in heart rate for a randomly selected person. Find

(a) $\Pr\{Y > 10\}$
(b) $\Pr\{Y > 20\}$
(c) $\Pr\{5 < Y < 15\}$

4.42   Refer to the heart rate distribution of Exercise 4.41. The fact that the standard deviation is greater than the average and that the distribution is normal tells us that some of the data values are negative, meaning that the person's heart rate went down, rather than up. Find the probability that a randomly chosen person's heart rate will go down. That is, find $\Pr\{Y < 0\}$.

4.43   Refer to the heart rate distribution of Exercise 4.41. Suppose we take a random sample of size 400 from this distribution. How many observations do we expect to obtain that fall between 0 and 15?

4.44   Refer to the heart rate distribution of Exercise 4.41. If we use the $1.5 \cdot \text{IQR}$ rule, from Chapter 2, to identify outliers, how large would an observation need to be in order to be labeled an outlier?

4.45   The following four normal probability plots, (a), (b), (c), and (d), were generated from the distributions shown by histograms I, II, and III and another histogram that is not shown. Which normal probability plot goes with which histogram? How do you know? (There will be one normal probability plot that is not used.)

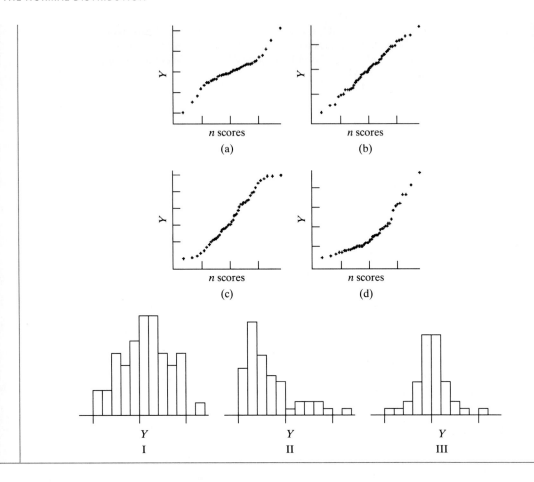

CHAPTER 5

Sampling Distributions

5.1 BASIC IDEAS

An important goal of data analysis is to distinguish between features of the data that reflect real biological facts and features that may reflect only chance effects. As explained in Sections 2.8 and 3.2, the random sampling model provides a framework for making this distinction. The underlying reality is visualized as a population, the data are viewed as a random sample from the population, and chance effects are regarded as sampling error—that is, discrepancy between the sample and the population.

　　In this chapter we develop the theoretical background that will enable us to place specific limits on the degree of sampling error to be expected in an experiment. (Although in later chapters we will distinguish between an experimental study and an observational study, for the present we will call any scientific investigation an *experiment*.) As in Chapters 2 and 3, we continue to confine the discussion to the simple context of an experiment with only one experimental group (one sample).

Sampling Variability

The variability among random samples from the same population is called **sampling variability**. A probability distribution that characterizes some aspect of sampling variability is termed a **sampling distribution**. Usually a random sample will resemble the population from which it came. Of course, we have to expect a certain amount of discrepancy between the sample and the population. A sampling distribution tells us how close the resemblance between the sample and the population is likely to be. We begin with a small (and artificial) example to illustrate the idea of a sampling distribution.

Objectives

In this chapter we will develop the idea of a sampling distribution, which is central to classical statistical inference. In particular, we will

- *study sampling distributions for dichotomous populations*

- *see how the sample size is related to the accuracy of the sample mean*

- *explore the Central Limit Theorem*

- *see how the normal distribution can be used to approximate the binomial distribution*

149

Example 5.1

Weights of Dogs.   Consider a small population of four dogs that have weights 42, 48, 52, and 58 pounds; label the dogs A, B, C, and D (in order). It is highly unusual that we would study such a small population by taking a sample of size $n = 2$, but for sake of illustration, suppose we were to sample two of the dogs, without replacement. Because there are only six possible samples that can be obtained, we can list each possible sample and its sample mean. For example, one possible sample contains dogs A and B, with weights 42 and 48 pounds; this sample has a sample mean of 45 pounds.

The complete list of possible samples is given in Table 5.1, along with the sample mean in each case. We can see from Table 5.1 that there only are five possible values for the sample mean, $\bar{y}$, in this setting. One of these values, 50, is more likely than the others, since there are two possible samples that result in a sample mean of 50. Table 5.2 gives the sampling distribution of the sample mean. ■

TABLE 5.1 Possible Samples of Size $n = 2$ from $N = 4$ Dogs

Sample	Data	Sample mean
A,B	42,48	45
A,C	42,52	47
A,D	42,58	50
B,C	48,52	50
B,D	48,58	53
C,D	52,58	55

TABLE 5.2 Sampling Distribution of Mean Dog Weight for Samples of Size $n = 2$

Sample mean	Probability
45	1/6
47	1/6
50	1/3
53	1/6
55	1/6

In this chapter we will discuss several aspects of sampling variability and study two important sampling distributions. From this point forward, we will assume that the sample size is a negligibly small fraction of the population size. This assumption simplifies the theory because it guarantees that the process of drawing the sample does not change the population composition.

The Meta-Experiment

According to the random sampling model, we regard the data in an experiment as a random sample from a population. Generally we obtain only a single random sample, which comes from a very large population. However, to visualize sampling variability we must broaden our frame of reference to include not merely one sample, but all the possible samples that might be drawn from the population. This wider frame of reference we will call the **meta-experiment**. A meta-experiment consists of indefinitely many repetitions, or replications, of the same experiment.* Thus, if the experiment consists of drawing a random sample of size n from some population, the corresponding meta-experiment involves drawing *repeated* random samples of size n from the same population. The process of repeated drawing is carried on indefinitely, with the members of each sample being replaced before the next sample is drawn. The experiment and the meta-experiment are schematically represented in Figure 5.1.

The following two examples illustrate the notion of a meta-experiment.

Rat Blood Pressure. An experiment consists of measuring the change in blood pressure in each of $n = 10$ rats after administering a certain drug. The corresponding meta-experiment would consist of repeatedly choosing groups of $n = 10$ rats from the same population and making blood pressure measurements under the same conditions. ■

Example 5.2

Bacterial Growth. An experiment consists of observing bacterial growth in $n = 5$ petri dishes that have been treated identically. The corresponding meta-experiment would consist of repeatedly preparing groups of five petri dishes and observing them in the same way. ■

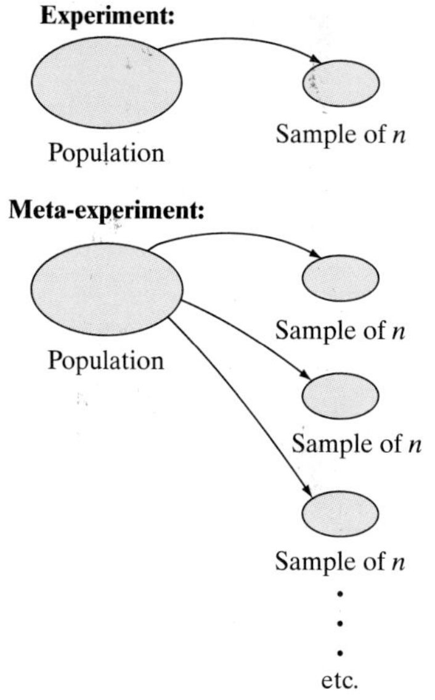

Example 5.3

Note that a meta-experiment is a theoretical construct rather than an operation that is actually performed by an experimenter.

The meta-experiment concept provides a link between sampling variability and probability. Recall from Chapter 3 that the probability of an event can be interpreted as the long-run relative frequency of occurrence of the event. Choosing a random sample is a chance operation; the meta-experiment consists of many repetitions of this chance operation, and so *probabilities concerning a random sample can be interpreted as relative frequencies in a meta-experiment*. Thus, the meta-experiment is a device for explicitly visualizing a sampling distribution: The sampling distribution describes the variability among the many random samples in a meta-experiment.

Figure 5.1 Schematic representation of experiment and meta-experiment

5.2 DICHOTOMOUS OBSERVATIONS

Consider an experiment in which the observed variable is dichotomous. As in Chapters 2 and 3, we will let p represent the population proportion of one category, while $\hat{p}$ represents the corresponding sample proportion. Then the question of how closely the sample resembles the population becomes "How close to p is $\hat{p}$ likely to be?" We will see how to answer this question through the **sampling distribution of $\hat{p}$**—that is, the collection of probabilities of all the various possible values of $\hat{p}$.

The Sampling Distribution of $\hat{p}$

For random sampling from a large dichotomous population, we saw in Chapter 3 how to use the binomial distribution to calculate the probabilities of all the various

---

* The term *meta-experiment* is not a standard term. It is unrelated to the term *meta-analysis*, which denotes a particular type of statistical analysis.

possible sample compositions. These probabilities in turn determine the sampling distribution of $\hat{p}$. The following is an example.

Example 5.4

Superior Vision. In a certain human population, 30% of the individuals have "superior" distance vision, in the sense of scoring 20/15 or better on a standardized vision test without glasses.[1] If we were to examine a random sample of two persons from the population, then we would get either zero, one, or two persons with superior vision. The probability that neither person will have superior vision is $.7 \times .7 = .49$. The probability that both persons will have superior vision is $.3 \times .3 = .09$. There are two ways to get a sample in which one person will have superior vision and one will not: The first could have superior vision and the second not have superior vision, or vice versa. Thus, the probability that exactly one person will have superior vision is $.3 \times .7 + .7 \times .3 = .42$.

If we let $\hat{p}$ represent the sample proportion of individuals with superior vision, then a sample that contains no individuals with superior vision has $\hat{p} = \frac{0}{2} = 0$; this happens with probability .49. A sample that contains one individual with superior vision has $\hat{p} = \frac{1}{2} = .5$; this happens with probability .42. A sample that contains two individuals with superior vision has $\hat{p} = \frac{2}{2} = 1$; this happens with probability .09. Thus, there is a 49% chance that $\hat{p}$ will equal 0, a 42% chance that $\hat{p}$ will equal .5, and a 9% chance that $\hat{p}$ will equal 1. This sampling distribution is given in Table 5.3.

TABLE 5.3 Sampling Distribution of Y (the Number with Superior Vision) and of $\hat{p}$ (the Proportion With Superior Vision) for Samples of Size $n = 2$

Y	$\hat{p}$	Probability
0	0	.49
1	.5	.42
2	1	.09

The probability that $\hat{p}$ will equal .5 can be interpreted in terms of a meta-experiment. The experiment consists of observing the distance vision of two randomly chosen people. The meta-experiment, as pictured in Figure 5.2, consists of repeatedly choosing two people and observing their distance vision. Each sample of size 2 has its own value of $\hat{p}$. In the long run, 42% of the $\hat{p}$'s will be equal to .5.

In the context of this setting, we would describe the *sampling distribution of the sample proportion of persons with superior vision* as the distribution that $\hat{p}$, the proportion of persons with superior vision, takes on in repeated samples of size 2. ∎

Example 5.5

Superior Vision and a Larger Sample. Suppose we were to examine a sample of 20 people from a population in which 30% have superior vision. How many people with superior vision might we expect to find in the sample? As was true in Example 5.4, this question can be answered in the language of probability. However,

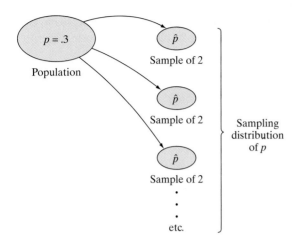

Figure 5.2 Meta-experiment for Example 5.4

since $n = 20$ is rather large, we will not list each possible sample. Rather, we will make calculations using the binomial distribution with $n = 20$ and $p = .30$. For instance, let us calculate the probability that 5 members of the sample would have superior vision and 15 would not:

$$\Pr\{5 \text{ superior}, 15 \text{ not superior}\} = {}_{20}C_5(.3)^5(.7)^{15}$$
$$= 15{,}504(.3)^5(.7)^{15}$$
$$= .179$$

Letting $\hat{p}$ represent the sample proportion of individuals with superior vision, a sample that contains 5 individuals with superior vision has $\hat{p} = \dfrac{5}{20} = .25$. Thus, we have found that

$$\Pr\{\hat{p} = .25\} = .179$$

As before, this probability can be interpreted in terms of a meta-experiment. Now the experiment consists of observing the distance vision of 20 randomly chosen people. The meta-experiment, as pictured in Figure 5.3, consists of repeatedly choosing 20 people, observing their distance vision, and calculating a value of $\hat{p}$. About 17.9% of the $\hat{p}$'s will be equal to .25.

In the context of this setting, we would describe the *sampling distribution of the sample proportion of persons with superior vision* as the distribution that $\hat{p}$, the proportion of persons with superior vision, takes on in repeated samples of size 20.

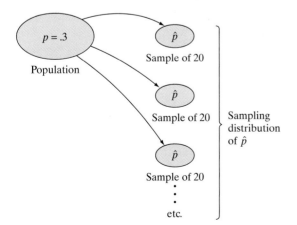

Figure 5.3 Meta-experiment for Example 5.5

The binomial distribution can be used to determine the entire sampling distribution of $\hat{p}$. The distribution is displayed in Table 5.4 and as a probability histogram in Figure 5.4.

We can use the binomial distribution to answer questions such as, "If we take a random sample of size $n = 20$, what is the probability that no more than 4 will have superior vision?" Notice that this question can be asked in two equivalent ways: "What is $\Pr\{Y \leq 4\}$?" and "What is $\Pr\{\hat{p} \leq .20\}$?" The answer to either question is found by adding the first 5 probabilities in Table 5.4: $\Pr\{Y \leq 4\} = \Pr\{\hat{p} \leq .20\} = .001 + .007 + .028 + .072 + .130 = .238.$ ∎

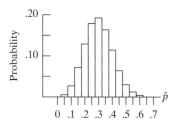

Figure 5.4 Sampling distribution of $\hat{p}$ when $n = 20$ and $p = .3$

TABLE 5.4 Sampling Distribution of Y, the Number of Successes, and of $\hat{p}$, the Proportion of Successes, When $n = 20$ and $p = .3$

Y	$\hat{p}$	Probability	Y	$\hat{p}$	Probability
0	.00	.001	11	.55	.012
1	.05	.007	12	.60	.004
2	.10	.028	13	.65	.001
3	.15	.072	14	.70	.000
4	.20	.130	15	.75	.000
5	.25	.179	16	.80	.000
6	.30	.192	17	.85	.000
7	.35	.164	18	.90	.000
8	.40	.114	19	.95	.000
9	.45	.065	20	1.00	.000
10	.50	.031			

Relationship to Statistical Inference

In making a statistical inference from a sample to the population, it is reasonable to use $\hat{p}$ as our estimate of p. The sampling distribution of $\hat{p}$ can be used to predict how much sampling error to expect in this estimate. For example, suppose we want to know whether the sampling error will be less than 5 percentage points—in other words, whether $\hat{p}$ will be within $\pm.05$ of p. We cannot predict for certain whether this event will occur, but we can find the probability of it happening, as illustrated in the following example.

Example 5.6

Superior Vision.   In the vision example with $n = 20$, we see from Table 5.4 that

$$\Pr\{.25 \leq \hat{p} \leq .35\} = .179 + .192 + .164$$
$$= .535 \approx .53$$

Thus, there is a 53% chance that, for a sample of size 20, $\hat{p}$ will be within $\pm.05$ of p. ∎

It may have occurred to you that the preceding discussion seems to have a fatal flaw. In order to specify the sampling distribution of $\hat{p}$ we need to know p. But if we know p, we do not need to take a sample in order to get information about p. How, then, can the sampling distribution of $\hat{p}$ be a basis for statistical inference in real experiments? It turns out that there is an escape from this apparently circular reasoning. We will see in later chapters that it is not necessary to know p in order to estimate the amount of sampling error associated with $\hat{p}$.

Dependence on Sample Size

It is interesting to consider how the sampling distribution of $\hat{p}$ depends on n. You may think intuitively that a larger n should provide a more informative sample, and indeed this is true. If n is larger, then $\hat{p}$ is more likely to be close to p.* The following example illustrates this effect.

Superior Vision.   Figure 5.5 shows the sampling distribution of $\hat{p}$, for three different values of n, for the vision population of Example 5.5. (Each sampling distribution is determined by a binomial distribution with $p = .3$. The figures are scaled so that their areas are equal.) You can see from the figure that as n increases

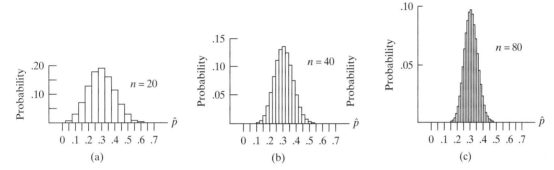

Figure 5.5 Sampling distributions of $\hat{p}$ for $p = .30$ and various values of n

the sampling distribution becomes more compressed around the value $p = .3$; thus, the probability that $\hat{p}$ is close to p tends to increase as n increases. For example, consider the probability that $\hat{p}$ is within ± 5 percentage points of p. We saw in Example 5.6 that for $n = 20$ this probability is equal to .53; Table 5.5 shows how the probability depends on n.

TABLE 5.5	
n	$\Pr\{.25 \le \hat{p} \le .35\}$
20	.53
40	.61
80	.73
400	.97

Note: A larger sample improves the probability that $\hat{p}$ will be close to p. We should be mindful, however, that the probability that $\hat{p}$ is exactly *equal* to p is very small for large n. In fact,

$$\Pr\{\hat{p} = .3\} = .097 \text{ for } n = 80$$

The value $\Pr\{.25 \le \hat{p} \le .35\} = .73$ is the sum of many small probabilities, the largest of which is .097; you can see this effect clearly in Figure 5.5(c).   ■

---

* This statement should not be interpreted too literally. As a function of n, the probability that $\hat{p}$ is close to p has an overall increasing trend, but it can fluctuate somewhat.

Exercises 5.1–5.10

5.1   Consider taking a random sample of size 3 from a population of persons who smoke and recording how many of them, if any, have lung cancer. Let $\hat{p}$ represent the proportion of persons in the sample with lung cancer. What are the possible values in the sampling distribution of $\hat{p}$?

5.2   Suppose we are to draw a random sample of three individuals from a large population in which 39% of the individuals are mutants (as in Example 3.45). Let $\hat{p}$ represent the proportion of mutants in the sample. Calculate the probability that $\hat{p}$ will be equal to

(a) 0
(b) 1/3

5.3   Suppose we are to draw a random sample of five individuals from a large population in which 39% of the individuals are mutants (as in Example 3.45). Let $\hat{p}$ represent the proportion of mutants in the sample.

(a) Use the results in Table 3.7 to determine the probability that $\hat{p}$ will be equal to
     (i) 0
     (ii) .2
     (iii).4
     (iv) .6
     (v) .8
     (vi) 1.0
(b) Display the sampling distribution of $\hat{p}$ in a histogram.

5.4   A new treatment for acquired immune deficiency syndrome (AIDS) is to be tested in a small clinical trial on 15 patients. The proportion $\hat{p}$ who respond to the treatment will be used as an estimate of the proportion p of (potential) responders in the entire population of AIDS patients. If in fact $p = .2$, and if the 15 patients can be regarded as a random sample from the population, find the probability that

(a) $\hat{p} = .2$
(b) $\hat{p} = 0$

5.5   In a certain forest, 25% of the white pine trees are infected with blister rust. Suppose a random sample of four white pine trees is to be chosen, and let $\hat{p}$ be the sample proportion of infected trees.

(a) Compute the probability that $\hat{p}$ will be equal to
     (i) 0
     (ii) .25
     (iii).50
     (iv) .75
     (v) 1.0
(b) Display the sampling distribution of $\hat{p}$ as a histogram.

5.6   Refer to Exercise 5.5.

(a) Determine the sampling distribution of $\hat{p}$ for samples of size $n = 8$ white pine trees from the same forest.
(b) Construct histograms of the sampling distributions of $\hat{p}$ for $n = 4$ and for $n = 8$, using the same horizontal scale for both, but doubling the vertical scale for the distribution with $n = 8$ (so that the two histograms have the same area). Compare the two distributions visually. How do they differ?

5.7   The shell of the land snail *Limocolaria marfensiana* has two possible color forms: streaked and pallid. In a certain population of these snails, 60% of the individuals have streaked shells (as in Exercise 3.29). Suppose a random sample of ten snails is to be chosen from the population; let $\hat{p}$ be the sample proportion of streaked snails. Find

(a) $\Pr\{\hat{p} = .5\}$
(b) $\Pr\{\hat{p} = .6\}$
(c) $\Pr\{\hat{p} = .7\}$
(d) $\Pr\{.5 \le \hat{p} \le .7\}$
(e) the percentage of samples for which $\hat{p}$ is within $\pm.10$ of p

5.8   In a certain human population, 30% of the people have "superior" vision (as in Example 5.4). Suppose a random sample of five people is to be chosen and their vision examined. Let $\hat{p}$ represent the sample proportion of people with superior vision.

(a) Compute the sampling distribution of $\hat{p}$.
(b) Construct a histogram of the distribution found in part (a) and compare it visually with Figure 5.4. How do the two distributions differ?

5.9   Consider random sampling from a dichotomous population; let E be the event that $\hat{p}$ is within $\pm.05$ of p. In Example 5.6, we found that $\Pr\{E\} = .53$ for $n = 20$ and $p = .3$. Calculate $\Pr\{E\}$ for $n = 20$ and $p = .4$. (Perhaps surprisingly, the two probabilities are roughly equal.)

5.10   Consider taking a random sample of size 10 from the population of students at a certain college and asking each of the 10 students whether or not they smoke. In the context of this setting, explain what is meant by the sampling distribution of the sample percentage.

5.3 QUANTITATIVE OBSERVATIONS

If the observed variable is quantitative, then the question of similarity between sample and population is more complex than for dichotomous data. For a quantitative variable, the sample and the population can be described in various ways—by the frequency distribution, the mean, the median, the standard deviation, and so on—and the question must be answered separately for each of these descriptive measures. In this section we will focus primarily on the mean.

The Sampling Distribution of $\overline{Y}$

The sample mean $\bar{y}$ can be used not only as a description of the data in the sample but also as an estimate of the population mean μ. It is natural to ask, "How close to μ is $\bar{y}$?" We cannot answer this question for the mean $\bar{y}$ of a particular sample, but we can answer it if we think in terms of the random sampling model and regard the sample mean as a random variable $\overline{Y}$. The question then becomes, "How close to μ is $\overline{Y}$ *likely* to be?" and the answer is provided by the **sampling distribution of** $\overline{Y}$—that is, the probability distribution that describes sampling variability in $\overline{Y}$.

    To visualize the sampling distribution of $\overline{Y}$, imagine the meta-experiment as follows: Random samples of size n are repeatedly drawn from a fixed population with mean μ and standard deviation σ; each sample has its own mean $\bar{y}$. The variation of the $\bar{y}$'s among the samples is specified by the sampling distribution of $\overline{Y}$. This relationship is indicated schematically in Figure 5.6.

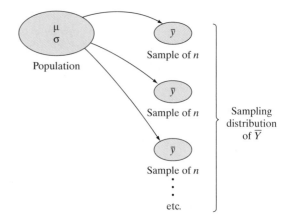

Figure 5.6 Schematic representation of the sampling distribution of $\bar{Y}$

When we think of $\bar{Y}$ as a random variable, we need to be aware of two basic facts. The first of these is intuitive: On average, the sample mean equals the population mean. That is, the average of the sampling distribution of $\bar{Y}$ is μ. The second fact is not obvious: The standard deviation of $\bar{Y}$ is equal to the standard deviation of Y divided by the square root of the sample size. That is, the standard deviation of $\bar{Y}$ is $\sigma / \sqrt{n}$.

Example 5.8

Serum Cholesterol. The serum cholesterol levels of 17-year-olds follow a normal distribution with mean $\mu = 176$ mg/dLi and standard deviation $\sigma = 30$ mg/dLi. If we take a random sample of size $n = 9$, then the standard deviation of the sample mean is $\dfrac{30}{\sqrt{9}} = \dfrac{30}{3} = 10$. This means, loosely speaking, that the sample mean, $\bar{Y}$, will vary from one sample to the next by about 10*; on average the sample mean will be 176. If we took random samples of size $n = 25$, then the standard deviation of the sample mean would be $\dfrac{30}{\sqrt{25}} = \dfrac{30}{5} = 6$, which means that $\bar{Y}$ would vary from one sample to the next by about 6. As the sample size goes up, the variability in the sample mean $\bar{Y}$ goes down. ∎

We now state as a theorem the basic facts about the sampling distribution of $\bar{Y}$. The theorem can be proved using the methods of mathematical statistics; we will state it without proof. The theorem describes the sampling distribution of $\bar{Y}$ in terms of its mean (denoted by $\mu_{\bar{Y}}$), its standard deviation (denoted by $\sigma_{\bar{Y}}$), and its shape.[†]

---

* Strictly speaking, the standard deviation measures deviation from the mean, not the difference between consecutive observations.

[†] We are assuming here that the population is infinitely large or, equivalently, that we are sampling with replacement, so that we never exhaust the population. If we sample without replacement from a finite population, then an adjustment is needed to get the right value for $\sigma_{\bar{Y}}$. Here $\sigma_{\bar{Y}}$ is given by $\dfrac{\sigma}{\sqrt{n}} \times \sqrt{\dfrac{N - n}{N - 1}}$. The term $\sqrt{\dfrac{N - n}{N - 1}}$ is called the **finite population correction factor**. Note that if the sample size n is 10% of the population size N, then the correction factor is $\sqrt{\dfrac{.9N}{N - 1}} \approx .95$, so the adjustment is small. Thus, if n is small, in comparison to N, then the finite population correction factor is close to 1 and can be ignored.

THEOREM 5.1: THE SAMPLING DISTRIBUTION OF $\overline{Y}$

1. *Mean* The mean of the sampling distribution of $\overline{Y}$ is equal to the population mean. In symbols,

$$\mu_{\overline{Y}} = \mu$$

2. *Standard deviation* The standard deviation of the sampling distribution of $\overline{Y}$ is equal to the population standard deviation divided by the square root of the sample size. In symbols,

$$\sigma_{\overline{Y}} = \frac{\sigma}{\sqrt{n}}$$

3. *Shape*
   (a) If the population distribution of Y is normal, then the sampling distribution of $\overline{Y}$ is normal, regardless of the sample size n.
   (b) *Central Limit Theorem* If n is large, then the sampling distribution of $\overline{Y}$ is approximately normal, even if the population distribution of Y is not normal.

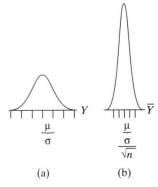

(a)　　　　(b)

Figure 5.7 (a) The population distribution of a normally distributed variable Y. (b) The sampling distribution of $\overline{Y}$ in samples from the population of part (a).

Parts 1 and 2 of Theorem 5.1 specify the relationship between the mean and standard deviation of the population being sampled, and the mean and standard deviation of the sampling distribution of $\overline{Y}$. Part 3a of the theorem states that, if the observed variable Y follows a normal distribution in the population being sampled, then the sampling distribution of $\overline{Y}$ is also a normal distribution. These relationships are indicated in Figure 5.7.

The following example illustrates the meaning of parts 1, 2, and 3(a) of Theorem 5.1.

Weights of Seeds. A large population of seeds of the princess bean *Phaseotus vulgaris* is to be sampled. The weights of the seeds in the population follow a normal distribution with mean $\mu = 500$ mg, and standard deviation $\sigma = 120$ mg.[2] Suppose now that a random sample of four seeds is to be weighed, and let $\overline{Y}$ represent the mean weight of the four seeds. Then, according to Theorem 5.1, the sampling distribution of $\overline{Y}$ will be a normal distribution with mean and standard deviation as follows:

Example 5.9

$$\mu_{\overline{Y}} = \mu = 500 \text{ mg}$$

and

$$\sigma_{\overline{Y}} = \frac{\sigma}{\sqrt{n}} = \frac{120}{\sqrt{4}} = 60 \text{ mg}$$

Thus, on average the sample mean will equal 500, but the variability from one sample of size 4 to the next sample of size 4 is such that about two-thirds of the time $\overline{Y}$ will be between $500 - 60$ and $500 + 60$ (i.e., between 440 and 560). Likewise, allowing for 2 standard deviations, we expect that $\overline{Y}$ will be between $500 - 120$ and $500 + 120$ about 95% of the time. The sampling distribution of $\overline{Y}$ is shown in Figure 5.8; the ticks are 1 standard deviation apart. ∎

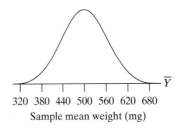

320　380　440　500　560　620　680
Sample mean weight (mg)

Figure 5.8 Sampling distribution of $\overline{Y}$ for Example 5.9

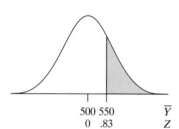

Figure 5.9 Calculation of $\Pr\{\bar{Y} > 550\}$ for Example 5.9

The sampling distribution of $\bar{Y}$ expresses the relative likelihood of the various possible values of $\bar{Y}$. For example, suppose we want to know the probability that the mean weight of the four seeds will be greater than 550 mg. This probability is shown as the shaded area in Figure 5.9. Notice that the value of $\bar{y} = 550$ must be converted to the Z scale using the standard deviation $\sigma_{\bar{Y}} = 60$, not $\sigma = 120$:

$$z = \frac{\bar{y} - \mu_{\bar{Y}}}{\sigma_{\bar{Y}}} = \frac{550 - 500}{60} = .83$$

From Table 3, $z = .83$ corresponds to an area of .7967. Thus,

$$\Pr\{\bar{Y} > 550\} = \Pr\{Z > .83\} = 1 - .7967$$
$$= .2033 \approx .20$$

This probability can be interpreted in terms of a meta-experiment as follows: If we were to choose many random samples of four seeds each from the population, then about 20% of the samples would have a mean weight exceeding 550 mg.

Part 3(b) of Theorem 5.1 is known as the **Central Limit Theorem**. The Central Limit Theorem states that, *no matter what distribution Y may have in the population,** if the sample size is large enough, then the sampling distribution of $\bar{Y}$ will be approximately a normal distribution.

The Central Limit Theorem is of fundamental importance because it can be applied when (as often happens in practice) the form of the population distribution is not known. It is because of the Central Limit Theorem (and other similar theorems) that the normal distribution plays such a central role in statistics.

It is natural to ask how large a sample size is required by the Central Limit Theorem: How large must n be in order that the sampling distribution of $\bar{Y}$ be well approximated by a normal curve? The answer is that the required n depends on the shape of the population distribution. If the shape is normal, any n will do. If the shape is moderately nonnormal, a moderate n is adequate. If the shape is highly nonnormal, then a rather large n will be required. (Some specific examples of this phenomenon are given in the optional Section 5.4.)

Remark: We stated in Section 5.1 that the theory of this chapter is valid if the sample size is small compared with the population size. But the Central Limit Theorem is a statement about large samples. This may seem like a contradiction: How can a large sample be a small sample? In practice, there is no contradiction. In a typical biological application, the population size might be 10^6; a sample of size $n = 100$ would be a small fraction of the population but would nevertheless be large enough for the Central Limit Theorem to be applicable (in most situations).

Dependence on Sample Size

Consider the possibility of choosing random samples of various sizes from the same population. The sampling distribution of $\bar{Y}$ will depend on the sample size n in two ways. First, its standard deviation is

$$\sigma_{\bar{Y}} = \frac{\sigma}{\sqrt{n}}$$

---

* Technically, the Central Limit Theorem requires that the distribution of Y have a standard deviation. In practice this condition is always met.

and this is inversely proportional to $\sqrt{n}$. Second, if the population distribution is not normal, then the *shape* of the sampling distribution of $\overline{Y}$ depends on n, being more nearly normal for larger n. However, if the population distribution is normal, then the sampling distribution of $\overline{Y}$ is always normal, and only the standard deviation depends on n.

The more important of the two effects of sample size is the first: Larger n gives a smaller value of $\sigma_{\overline{Y}}$ and consequently a smaller expected sampling error if $\overline{y}$ is used as an estimate of μ. The following example illustrates this effect for sampling from a normal population.

Weights of Seeds.   Figure 5.10 shows the sampling distribution of $\overline{Y}$ for samples of various sizes from the princess bean population of Example 5.9. Notice that for larger n the sampling distribution is more concentrated around the population mean $\mu = 500$ mg. As a consequence, the probability that $\overline{Y}$ is close to it is larger for larger n. For instance, consider the probability that $\overline{Y}$ is within ±50 mg of μ, that is, $\Pr\{450 \leq \overline{Y} \leq 550\}$. Table 5.6 shows how this probability depends on n.   ■

Example 5.10

TABLE 5.6

n	$\Pr\{450 \leq \overline{Y} \leq 550\}$
4	.59
9	.79
16	.91
64	.999

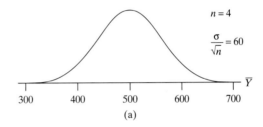

(a)

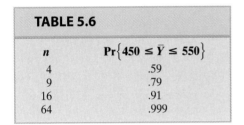

(b)

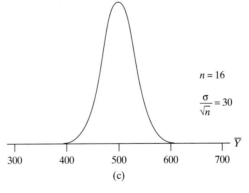

(c)

Figure 5.10 Sampling distribution of $\overline{Y}$ for various sample sizes n

Example 5.10 illustrates how the closeness of to $\overline{Y}$ to μ depends on sample size. The mean of a larger sample is not *necessarily* closer to it than the mean of a smaller sample, but it has a *greater probability* of being close. It is in this sense that a larger sample provides more information about the population mean than a smaller sample.

Populations, Samples, and Sampling Distributions

In thinking about Theorem 5.1, it is important to distinguish clearly among three different distributions related to a quantitative variable Y: (1) the distribution of Y in the population; (2) the distribution of Y in a sample of data, and (3) the sampling distribution of $\overline{Y}$. The means and standard deviations of these distributions are summarized in Table 5.7.

TABLE 5.7

Distribution	Mean	Standard Deviation
Y in population	μ	σ
Y in sample	$\bar{y}$	s
$\bar{Y}$ (in meta-experiment)	μ	$\dfrac{\sigma}{\sqrt{n}}$

The following example illustrates the distinction among the three distributions.

Example 5.11

Weights of Seeds. For the princess bean population of Example 5.9, the population mean and standard deviation are $\mu = 500$ mg and $\sigma = 120$ mg; the population distribution of Y = weight is represented in Figure 5.11(a). Suppose we weigh a random sample of $n = 25$ seeds from the population and obtain the data in Table 5.8.

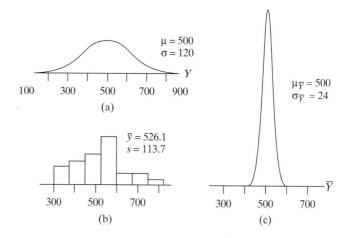

Figure 5.11 Three distributions related to Y = seed weight of princess beans. (a) Population distribution of Y; (b) Distribution of 25 observations of Y; (c) Sampling distribution of $\bar{Y}$ for $n = 25$

TABLE 5.8 Weights of 25 Princess Bean Seeds

Weight (mg)						
343	755	431	480	516	469	694
659	441	562	597	502	612	549
348	469	545	728	416	536	581
433	583	570	334			

For the data in Table 5.8, the sample mean is $\bar{y} = 526.1$ mg and the sample standard deviation is $s = 113.7$ mg. Figure 5.11(b) shows a histogram of the data; this histogram represents the distribution of Y in the sample. The sampling distribution of $\bar{Y}$ is a theoretical distribution that relates not to the particular sample shown in the histogram but rather to the meta-experiment of repeated samples of size $n = 25$. The mean and standard deviation of the sampling distribution are

$$\mu_{\bar{Y}} = 500 \text{ mg} \quad \text{and} \quad \sigma_{\bar{Y}} = 120/\sqrt{25} = 24 \text{ mg}$$

The sampling distribution is represented in Figure 5.11(c). Notice that the distributions in Figures 5.11(a) and (b) are more or less similar; in fact, the distribution in (b) is an estimate (based on the data in Table 5.8) of the distribution in (a). By contrast, the distribution in (c) is much narrower because it represents a distribution of *means* rather than of individual observations. ∎

Other Aspects of Sampling Variability

The preceding discussion has focused on sampling variability in the sample mean, $\bar{Y}$. Two other important aspects of sampling variability are (1) sampling variability in the sample standard deviation, s; and (2) sampling variability in the *shape* of the sample, as represented by the sample histogram. Rather than discuss these aspects formally, we illustrate them with the following example.

Weights of Seeds.   In Figure 5.11(b) we displayed a random sample of 25 observations from the princess bean population of Example 5.9; now we display in Figure 5.12 eight additional random samples from the same population. (All nine samples were actually simulated using a computer.) Notice that, even though the samples were drawn from a normal population [pictured in Figure 5.11(a)], there is very substantial variation in the forms of the histograms. Notice also that there is considerable variation in the sample standard deviations. Of course, if the sample size were larger (say, $n = 100$ rather than $n = 25$), there would be less sampling variation; the histograms would tend to resemble a normal curve more closely, and the standard deviations would tend to be closer to the population value ($\sigma = 120$). ∎

Example 5.12

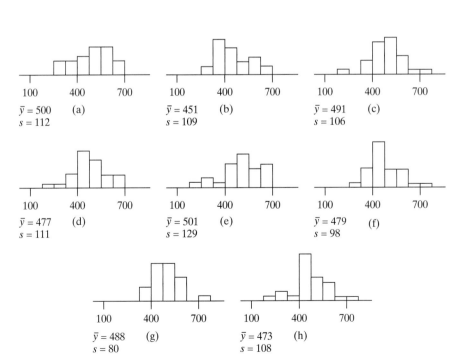

Figure 5.12 Eight random samples, each of size $n = 25$, from a normal population with $\mu = 500$ and $\sigma = 120$

Exercises 5.11–5.28

5.11   *(Sampling exercise)* Refer to Exercise 3.1. The collection of 100 ellipses shown there can be thought of as representing a natural population of the organism *C. ellipticus*. Use your judgment to choose a sample of five ellipses that you think should be reasonably representative of the population. (In order to best simulate the analogous judgment in a real-life setting, you should make your choice intuitively, without any detailed preliminary study of the population.) With a metric ruler, measure the length of each ellipse in your sample. Measure only the body, excluding any tail bristles; measurements to the nearest millimeter will be adequate. Compute the mean and standard deviation of the five lengths. To facilitate the pooling of results from the entire class, express the mean and standard deviation in millimeters, keeping two decimal places.

5.12   *(Sampling exercise)* Proceed as in Exercise 5.11, but use random sampling rather than "judgment" sampling. To do this, choose 10 random digits (from Table 1 or your calculator). Let the first 2 digits be the number of the first ellipse that goes into your sample, etc. The 10 random digits will give you a random sample of five ellipses.

5.13   *(Sampling exercise)* Proceed as in Exercise 5.12, but choose a random sample of 20 ellipses.

5.14   Refer to Exercise 5.12. The following scheme is proposed for choosing a sample of 5 ellipses from the population of 100 ellipses. (i) Choose a point at random in the ellipse "habitat" (that is, the figure); this could be done crudely by dropping a pencil point on the page, or much better by overlaying the page with graph paper and using random digits. (ii) If the chosen point is inside an ellipse, include that ellipse in the sample, otherwise start again at step (i). (iii) Continue until five ellipses have been selected. Explain why this scheme is not equivalent to random sampling. In what direction is the scheme biased—that is, would it tend to produce a $\bar{y}$ that is too large or a $\bar{y}$ that is too small?

5.15   The serum cholesterol levels of a population of 17-year-olds follow a normal distribution with mean 176 mg/dLi and standard deviation 30 mg/dLi (as in Example 4.1).

(a) What percentage of the 17-year-olds have serum cholesterol values between 166 and 186 mg/dLi?

(b) Suppose we were to choose at random from the population a large number of groups of nine 17-year-olds each. In what percentage of the groups would the group mean cholesterol value be between 166 and 186 mg/dLi?

(c) If $\bar{Y}$ represents the mean cholesterol value of a random sample of nine 17-year-olds from the population, what is $\Pr\{166 \le \bar{Y} \le 186\}$?

5.16   An important indicator of lung function is forced expiratory volume (FEV), which is the volume of air that a person can expire in one second. Dr. Jones plans to measure FEV in a random sample of n young women from a certain population, and to use the sample mean $\bar{y}$ as an estimate of the population mean. Let E be the event that Jones's sample mean will be within ± 100 mLi of the population mean. Assume that the population distribution is normal with mean 3,000 mLi and standard deviation 400 mLi.[3] Find $\Pr\{E\}$ if

(a) $n = 15$

(b) $n = 60$

(c) How does $\Pr\{E\}$ depend on the sample size? That is, as n increases, does $\Pr\{E\}$ increase, decrease, or stay the same?

5.17   Refer to Exercise 5.16. Assume that the population distribution of FEV is normal with standard deviation 400 mLi.

(a) Find $\Pr\{E\}$ if $n = 15$ and the population mean is 2,800 mLi.

(b) Find $\Pr\{E\}$ if $n = 15$ and the population mean is 2,600 mLi.

(c) How does $\Pr\{E\}$ depend on the population mean?

5.18 The heights of a certain population of corn plants follow a normal distribution with mean 145 cm and standard deviation 22 cm (as in Exercise 4.29).

(a) What percentage of the plants are between 135 and 155 cm tall?

(b) Suppose we were to choose at random from the population a large number of samples of 16 plants each. In what percentage of the samples would the sample mean height be between 135 and 155 cm?

(c) If $\bar{Y}$ represents the mean height of a random sample of 16 plants from the population, what is $\Pr\{135 \le \bar{Y} \le 155\}$?

(d) If $\bar{Y}$ represents the mean height of a random sample of 36 plants from the population, what is $\Pr\{135 \le \bar{Y} \le 155\}$?

5.19 The basal diameter of a sea anemone is an indicator of its age. The density curve shown here represents the distribution of diameters in a certain large population of anemones; the population mean diameter is 4.2 cm, and the standard deviation is 1.4 cm.[4] Let $\bar{Y}$ represent the mean diameter of 25 anemones randomly chosen from the population.

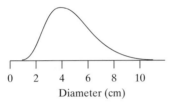

Diameter (cm)

(a) Find the approximate value of $\Pr\{4 \le \bar{Y} \le 5\}$.

(b) Why is your answer to part (a) approximately correct even though the population distribution of diameters is clearly not normal? Would the same approach be equally valid for a sample of size 2 rather than 25? Why or why not?

5.20 In a certain population of fish, the lengths of the individual fish follow approximately a normal distribution with mean 54.0 mm and standard deviation 4.5 mm. We saw in Example 4.5 that in this situation 65.68% of the fish are between 51 and 60 mm long. Suppose a random sample of four fish is chosen from the population. Find the probability that

(a) all four fish are between 51 and 60 mm long

(b) the mean length of the four fish is between 51 and 60 mm

5.21 In Exercise 5.20, the answer to part (b) was larger than the answer to part (a). Argue that this must necessarily be true, no matter what the population mean and standard deviation might be. [*Hint:* Can it happen that the event in part (a) occurs but the event in part (b) does not?]

5.22 Professor Smith conducted a class exercise in which students ran a computer program to generate random samples from a population that had a mean of 50 and a standard deviation of 9 mm. Each of Smith's students took a random sample of size n and calculated the sample mean. Smith found that about 68% of the students had sample means between 48.5 and 51.5 mm. What was n? (Assume that n is large enough that the Central Limit Theorem is applicable.)

5.23 A certain assay for serum alanine aminotransferase (ALT) is rather imprecise. The results of repeated assays of a single specimen follow a normal distribution with mean equal to the ALT concentration for that specimen and standard deviation equal to 4 U/Li (as in Exercise 4.40). Suppose a hospital lab measures many specimens every day, and specimens with reported ALT values of 40 or more are flagged as "unusually high." If a patient's true ALT concentration is 35 U/Li, find the probability that his specimen will be flagged as "unusually high"

(a) if the reported value is the result of a single assay
(b) if the reported value is the mean of three independent assays of the same specimen

5.24 The mean of the distribution shown in the histogram is 41.5 and the standard deviation is 4.7. Consider taking random samples of size $n = 4$ from this distribution and calculating the sample mean, $\bar{y}$, for each sample.

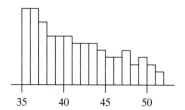

(a) What is the mean of the sampling distribution of $\bar{Y}$?
(b) What is the standard deviation of the sampling distribution of $\bar{Y}$?

5.25 Refer to the histogram in Exercise 5.24. Suppose that 100 random samples are taken from this population and the sample mean is calculated for each sample. If we were to make a histogram of the distribution of the sample means from 100 samples, what kind of shape would we expect the histogram to have for each of the following?

(a) if $n = 2$ for each random sample
(b) if $n = 25$ for each random sample

5.26 Refer to the histogram in Exercise 5.24. Suppose that 100 random samples are taken from this population and the sample mean is calculated for each sample. If we were to make a histogram of the distribution of the sample means from 100 samples, what kind of shape would we expect the histogram to have if $n = 1$ for each random sample? That is, what does the sampling distribution of the mean look like when the sample size is $n = 1$?

5.27 A medical researcher measured systolic blood pressure in 100 middle-aged men.[5] The results are displayed in the accompanying histogram; note that the distribution is rather skewed. According to the Central Limit Theorem, would we expect the distribution of blood pressure readings to be less skewed (and more bell shaped) if it were based on $n = 400$ rather than $n = 100$ men? Explain.

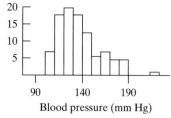

Blood pressure (mm Hg)

5.28 The partial pressure of oxygen, PaO_2, is a measure of the amount of oxygen in the blood. Assume that the distribution of PaO_2 levels among newborns has an average of 38 (mm Hg) and a standard deviation of 9.[6] If we take a sample of size $n = 25$,

(a) what is the probability that the sample average will be greater than 36?
(b) what is the probability that the sample average will be greater than 41?

5.4 ILLUSTRATION OF THE CENTRAL LIMIT THEOREM (OPTIONAL)

The importance of the normal distribution in statistics is due largely to the Central Limit Theorem and related theorems. In this section we take a closer look at the Central Limit Theorem.

According to the Central Limit Theorem, the sampling distribution of $\bar{Y}$ is approximately normal if n is large. If we consider larger and larger samples from a fixed nonnormal population, then the sampling distribution of $\bar{Y}$ will be more nearly normal for larger n. The following examples show the Central Limit Theorem at work for two nonnormal distributions: a moderately skewed distribution (Example 5.13) and a highly skewed distribution (Example 5.14).

Eye Facets. The number of facets in the eye of the fruitfly *Drosophila melanogaster* is of interest in genetic studies. The distribution of this variable in a certain *Drosophila* population can be approximated by the density function shown in Figure 5.13. The distribution is moderately skewed; the population mean and standard deviation are $\mu = 64$ and $\sigma = 22$.[7]

Figure 5.14 shows the sampling distribution of $\bar{Y}$ for samples of various sizes from the eye-facet population. In order to clearly show the shape of these distributions, we have plotted them to different scales; the horizontal scale is stretched more for larger n. Notice that the distributions are somewhat skewed to the right, but the skewness is diminished for larger n; for $n = 32$ the distribution looks very nearly normal. ∎

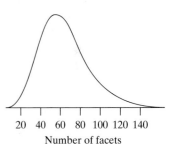

Example 5.13

Figure 5.13 Distribution of eye facet number in a *Drosophila* population

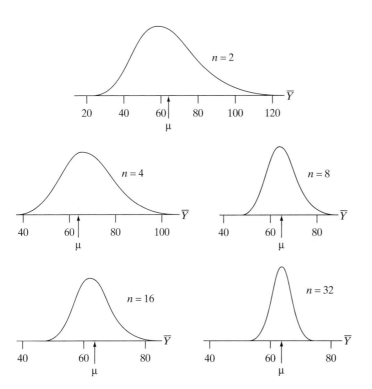

Figure 5.14 Sampling distributions of $\bar{Y}$ for samples from the *Drosophila* eye-facet population

Example 5.14

Reaction Time.   A psychologist measured the time required for a person to reach up from a fixed position and operate a pushbutton with his or her forefinger. The distribution of time scores (in milliseconds) for a single person is represented by the density shown in Figure 5.15. About 10% of the time, the subject fumbled, or missed the button on the first thrust; the resulting delayed times appear as the second peak of the distribution.[8] The first peak is centered at 115 ms and the second at 450 ms; because of the two peaks, the overall distribution is violently skewed. The population mean and standard deviation are $\mu = 148$ ms and $\sigma = 105$ ms, respectively.

Figure 5.16 shows the sampling distribution of $\overline{Y}$ for samples of various sizes from the time-score distribution. To show the shape clearly, the Y scale has been stretched more for larger n. Notice that for small n the distribution has several modes. As n increases, these modes are reduced to bumps and finally disappear, and the distribution becomes increasingly symmetric.   ■

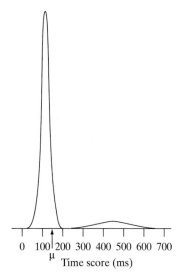

Figure 5.15 Distribution of time scores in a button-pushing task

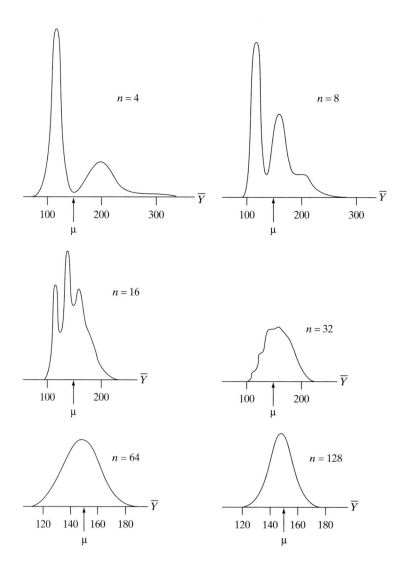

Figure 5.16 Sampling distributions of $\overline{Y}$ for samples from the time-score population

Examples 5.13 and 5.14 illustrate the fact, mentioned in Section 5.3, that the meaning of the requirement "*n* is large" in the Central Limit Theorem depends on the shape of the population distribution. Approximate normality of the sampling distribution of $\bar{Y}$ will be achieved for a moderate *n* if the population distribution is only moderately nonnormal (as in Example 5.13), while a highly nonnormal population (as in Example 5.14) will require a larger *n*. Note, however, that Example 5.14 indicates the remarkable strength of the Central Limit Theorem. The skewness of the time-score distribution is so extreme that we might be reluctant to consider the mean as a summary measure. Even in this worst case, you can see the effect of the Central Limit Theorem in the relative smoothness and symmetry of the sampling distribution for $n = 64$.

The Central Limit Theorem may seem rather like magic. To demystify it somewhat, we look at the time-score sampling distributions in more detail in the following example.

Reaction Time.   Consider the sampling distributions of $\bar{Y}$ displayed in Figure 5.16. Consider first the distribution for $n = 4$, which is the distribution of the mean of four button-pressing times. The high peak at the left of the distribution represents cases in which the subject did not fumble any of the four thrusts, so that all four times were about 115 ms; such an outcome would occur about 66% of the time [from the binomial distribution, because $(.9)^4 = .66$]. The next lower peak represents cases in which three thrusts took about 115 ms each, while one was fumbled and took about 450 ms. (Notice that the average of three 115's and one 450 is about 200, which is the center of the second peak.) Similarly, the third peak (which is barely visible) represents cases in which the subject fumbled two of the four thrusts. The peaks representing three and four fumbles are too low to be visible in the plot.

Now consider the plot for $n = 8$. The first peak represents eight good thrusts (no fumbles), the second represents seven good thrusts and one fumble, the third represents six good thrusts and two fumbles, and so on. The fourth and later peaks are blended together. For $n = 16$ the first peak is lower than the second because the occurrence of 16 good thrusts is less likely than 15 good thrusts and one fumble (as you can verify from the binomial distribution). For larger *n*, the first peaks are lower still and the later peaks are higher. For $n = 32$ the most likely outcome is three fumbles (about 10%) and 29 good thrusts; this outcome gives a mean time of about

$$\frac{(3)(450) + (29)(115)}{32} \approx 146 \text{ ms}$$

which is the location of the central peak. For similar reasons, the distribution for larger *n* is centered at about 148 ms, which is the population mean. ∎

> **Example 5.15**

Exercises 5.29–5.31

5.29   Refer to Example 5.15. In the sampling distribution of $\bar{Y}$ for $n = 4$ (Figure 5.16), approximately what is the area under

(a) the first peak?

(b) the second peak?

(*Hint:* Use the binomial distribution.)

5.30 Refer to Example 5.15. Consider the sampling distribution of $\overline{Y}$ for $n = 2$ (which is not shown in Figure 5.16).

(a) Make a rough sketch of the sampling distribution. How many peaks does it have? Show the location (on the Y-axis) of each peak.

(b) Find the approximate area under each peak. (*Hint:* Use the binomial distribution.)

5.31 Refer to Example 5.15. Consider the sampling distribution of $\overline{Y}$ for $n = 1$ (which is not shown in Figure 5.16). Make a rough sketch of the sampling distribution. How many peaks does it have? Show the location (on the Y-axis) of each peak.

5.5 THE NORMAL APPROXIMATION TO THE BINOMIAL DISTRIBUTION (OPTIONAL)

In Section 5.2 we saw that, for random sampling from a large dichotomous population, the sampling distribution of $\hat{p}$ is governed by the binomial distribution. Probabilities for the binomial distribution can be calculated from the formula

$$_nC_j p^j (1 - p)^{n-j}$$

However, this formula can be burdensome if n is not small. Fortunately, a convenient approximation is available. In this section we show how the binomial distribution can be approximated by a normal distribution, if n is large.

The Normal Approximation

The normal approximation to the binomial distribution can be expressed in two equivalent ways: in terms of the binomial distribution itself, or in terms of the sampling distribution of $\hat{p}$. We state both forms in the following theorem. In this theorem, n represents the sample size (or, more generally, the number of independent trials) and p represents the population proportion (or, more generally, the probability of success in each independent trial).

THEOREM 5.2: NORMAL APPROXIMATION TO BINOMIAL DISTRIBUTION

(a) If n is large, then the binomial distribution can be approximated by a normal distribution with

$$\text{Mean} = np$$

and

$$\text{Standard deviation} = \sqrt{np(1 - p)}$$

(b) If n is large, then the sampling distribution of $\hat{p}$ can be approximated by a normal distribution with

$$\text{Mean} = p$$

and

$$\text{Standard deviation} = \sqrt{\frac{p(1 - p)}{n}}$$

Remarks:

1. It is true, but not obvious, that the normal approximation to the binomial distribution is an application of the Central Limit Theorem (Section 5.3). The relationship is explained more fully in Appendix 5.1.

2. As shown in Appendix 5.2, for a population of 0's and 1's, where the proportion of 1's is given by p, the standard deviation is $\sigma = \sqrt{p(1-p)}$. Theorem 5.1 (Section 5.3) stated that the standard deviation of a mean is given by $\dfrac{\sigma}{\sqrt{n}}$. We can think of $\hat{p}$ in part (b) of Theorem 5.2 as a special kind of sample average, for the setting in which all of the data are 0's and 1's. Thus, Theorem 5.1 tells us that the standard deviation of $\hat{p}$ should be $\dfrac{\sqrt{p(1-p)}}{\sqrt{n}}$, or $\sqrt{\dfrac{p(1-p)}{n}}$, which agrees with the result stated in Theorem 5.2(b).

The following two examples illustrate the use of Theorem 5.2.

We consider a binomial distribution with $n = 20$ and $p = .3$. Figure 5.17(a) shows this binomial distribution; superimposed is a normal curve with

$$\text{Mean} = np = (20)(.3) = 6$$

and

$$SD = \sqrt{np(1-p)} = \sqrt{(20)(.3)(.7)} = 2.049$$

Note that the curve fits the distribution fairly well. Figure 5.17(b) shows the sampling distribution of $\hat{p}$ for $n = 20$ and $p = .3$ (the same distribution was shown in Figure 5.4); superimposed is a normal curve with

$$\text{Mean} = p = .3$$

and

$$SD = \sqrt{\dfrac{p(1-p)}{n}} = \sqrt{\dfrac{(.3)(.7)}{20}} = .1025$$

Note that Figure 5.17(b) is just a relabeled version of Figure 5.16(a).

Example 5.16

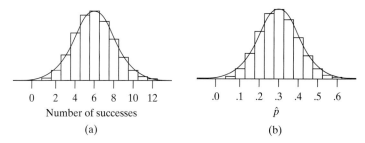

Figure 5.17 The normal approximation to the binomial distribution with $n = 20$ and $p = .3$

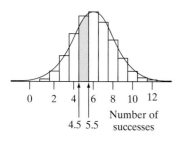

Figure 5.18 Normal approximation to the probability of five successes

To illustrate the use of the normal approximation, let us consider the event that 20 independent trials result in 5 successes and 15 failures. In Example 5.5 we found that the exact probability of this event is .179; this probability can be visualized as the area of the bar above the 5 in Figure 5.18. The normal approximation to the probability is the corresponding area under the normal curve, which is shaded in Figure 5.18. The boundaries of the shaded area are 4.5 and 5.5, which correspond in the Z scale to

$$z = \frac{4.5 - 6}{2.049} = -.73$$

and

$$z = \frac{5.5 - 6}{2.049} = -.24$$

From Table 3, we find that the area is $.4052 - .2327 = .1725$, which is fairly close to the exact value of .179. ∎

Example 5.17

To illustrate part (b) of Theorem 5.2, we again assume that $n = 20$ and $p = .3$. In Example 5.6 we found that

$$\Pr\{.25 \le \hat{p} \le .35\} = .535$$

The normal approximation to this probability is the shaded area in Figure 5.19. The boundaries of the area are $\hat{p} = .225$ and $\hat{p} = .375$, which correspond on the Z scale to

$$z = \frac{.225 - .3}{.1025} = -.73$$

and

$$z = \frac{.375 - .3}{.1025} = .73$$

The resulting approximation (from Table 3) is then

$$\Pr\{.25 \le \hat{p} \le .35\} \approx .7673 - .2327 = .5346$$

which agrees very well with the exact value. ∎

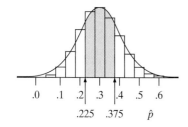

Figure 5.19 Normal approximation to $\Pr\{.25 \le \hat{p} \le .35\}$

The Continuity Correction

Notice that the calculation in Example 5.17 used the boundaries $\hat{p} = .225$ and .375 rather than $\hat{p} = .25$ and .35; this is an example of a continuity correction.* The reason for the continuity correction can be seen from Figure 5.19. The exact probability is the area of the three rectangles corresponding to $\hat{p} = .25, .30$, and .35; the boundaries of this region are .225 and .375. Without the continuity correction, we would calculate the area between $\hat{p} = .25$ and $\hat{p} = .35$, which is equal to .3758; this value is too small because it omits half of the $\hat{p} = .25$ rectangle and half of the $\hat{p} = .35$ rectangle.

In general, when using a continuity correction, the first step is to calculate the half-width of a histogram bar; the desired area is then extended by this amount

---

* The continuity correction was also discussed in the optional Section 4.5.

in each direction. For instance, in Example 5.17 the half-width of a histogram bar is equal to

$$\left(\frac{1}{2}\right)\left(\frac{1}{20}\right) = .025$$

and the boundaries of the shaded region in Figure 5.19 can be calculated as

$$.250 - .025 = .225 \quad \text{and} \quad .350 + .025 = .375$$

If the area to be calculated includes many bars of the probability histogram, then the continuity correction can be omitted without causing much error. If the area includes only a few bars, then the continuity correction greatly improves the accuracy of the approximation; as an extreme example, if the area includes only one bar (as in Figure 5.18), then omitting the continuity correction would give a probability of zero, which is not at all a useful approximation. (In Example 5.16, we applied the continuity correction by using the boundaries 4.5 and 5.5.)

    Remark: Any problem involving the normal approximation to the binomial can be solved in two ways: in terms of Y, using part (a) of Theorem 5.2, or in terms of $\hat{p}$, using part (b) of the theorem. Although it is natural to state questions in terms of proportions (e.g., "What is $\Pr\{\hat{p} > .70\}$?"), it is often easier to solve problems in terms of the binomial count Y (e.g., "What is $\Pr\{Y > 70\}$?"), particularly when using continuity correction. The following example illustrates the approach of converting a question about a sample proportion into a question about the number of successes for a binomial random variable.

Consider a binomial distribution with $n = 20$ and $p = .3$. The sample proportion of successes, out of the 20 trials, is $\hat{p}$. Figure 5.17(b) shows the sampling distribution of $\hat{p}$ with a normal curve superimposed.

Example 5.18

    Suppose we wish to find the probability that $.25 \leq \hat{p} \leq .35$. Since $\hat{p} = Y/20$, this is the probability that $.25 \leq Y/20 \leq .35$, which is the same as the probability that $5 \leq Y \leq 7$. That is, $\Pr\{.25 \leq \hat{p} \leq .35\} = \Pr\{5 \leq Y \leq 7\}$.

    We know that Y has a binomial distribution with mean $= np = (20)(.3) = 6$ and $SD = \sqrt{np(1 - p)} = \sqrt{(20)(.3)(.7)} = 2.049$. Using continuity correction, we would find the Z-scale values of

$$z = \frac{4.5 - 6}{2.049} = -.73$$

and

$$z = \frac{7.5 - 6}{2.049} = .73$$

Then, using Table 3, we have $\Pr\{.25 \leq \hat{p} \leq .35\} = \Pr\{5 \leq Y \leq 7\} \approx .7673 - .2327 = .5346.$ ∎

How Large Must n Be?

Theorem 5.2 states that the binomial distribution can be approximated by a normal distribution if n is "large." It is helpful to know how large n must be in order for the approximation to be adequate. The required n depends on the value of p. If $p = .5$, then the binomial distribution is symmetric and the normal approximation is quite good even for n as small as 10. However, if $p = .1$, the binomial distribution for $n = 10$ is quite skewed and is poorly fitted by a normal curve; for larger n the skewness is diminished and the normal approximation is better. A simple rule of thumb is the following:

> The normal approximation to the binomial distribution is fairly good if both np and $n(1 - p)$ are at least equal to 5.

For example, if $n = 20$ and $p = .3$, as in Example 5.16, then $np = 6$ and $n(1 - p) = 14$; since $6 \geq 5$ and $14 \geq 5$, the rule of thumb indicates that the normal approximation is fairly good.

Exercises 5.32–5.41

5.32 A fair coin is to be tossed 20 times. Find the probability that 10 of the tosses will fall heads and 10 will fall tails

   (a) using the binomial distribution formula
   (b) using the normal approximation with the continuity correction

5.33 In the United States, 44% of the population has type O blood. Suppose a random sample of 12 persons is taken. Find the probability that 6 of the persons will have type O blood (and 6 will not)

   (a) using the binomial distribution formula
   (b) using the normal approximation with the continuity correction

5.34 An epidemiologist is planning a study on the prevalence of oral contraceptive use in a certain population.[9] She plans to choose a random sample of n women and to use the sample proportion of oral contraceptive users ($\hat{p}$) as an estimate of the population proportion (p). Suppose that in fact $p = .12$. Use the normal approximation (with the continuity correction) to determine the probability that $\hat{p}$ will be within $\pm.03$ of p if

   (a) $n = 100$
   (b) $n = 200$
   [*Hint:* If you find using part (b) of Theorem 5.2 to be difficult here, try using part (a) of the theorem instead.]

5.35 In a study of how people make probability judgments, college students (with no background in probability or statistics) were asked the following question.[10] A certain town is served by two hospitals. In the larger hospital about 45 babies are born each day, and in the smaller hospital about 15 babies are born each day. As you know, about 50% of all babies are boys. The exact percentage of baby boys, however, varies from day to day. Sometimes it may be higher than 50%, sometimes lower.
For a period of one year, each hospital recorded the days on which at least 60% of the babies born were boys. Which hospital do you think recorded more such days?

   • The larger hospital
   • The smaller hospital
   • About the same (i.e., within 5% of each other)

   (a) Imagine that you are a participant in the study. Which answer would you choose, based on intuition alone?
   (b) Determine the correct answer by using the normal approximation (without the continuity correction) to calculate the appropriate probabilities.

5.36 Consider random sampling from a dichotomous population with $p = .3$, and let E be the event that $\hat{p}$ is within $\pm.05$ of p. Use the normal approximation (without the continuity correction) to calculate $\Pr\{E\}$ for a sample of size $n = 400$. Your answer should agree with the value given in Table 5.5.

5.37 Refer to Exercise 5.36. Calculate $\Pr\{E\}$ for $n = 40$ (rather than 400)

(a) with the continuity correction
(b) without the continuity correction
Your answer to part (a) should agree with the value given in Table 5.5.

5.38 A certain cross between sweet-pea plants will produce progeny that are either purple flowered or white flowered;[11] the probability of a purple-flowered plant is $p = \dfrac{9}{16}$. Suppose n progeny are to be examined, and let $\hat{p}$ be the sample proportion of purple-flowered plants. It might happen, by chance, that $\hat{p}$ would be closer to $\dfrac{1}{2}$ than to $\dfrac{9}{16}$. Find the probability that this misleading event would occur if

(a) $n = 1$
(b) $n = 64$
(c) $n = 320$
(Use the normal approximation without the continuity correction.)

5.39 A fair coin is to be tossed 10 times. Find the probability that between 30% and 40% (inclusive) of the tosses will fall heads

(a) using the binomial distribution formula
(b) using the normal approximation with the continuity correction

5.40 In a certain population of mussels (*Mytilus edulis*), 80% of the individuals are infected with an intestinal parasite.[12] A marine biologist plans to examine 100 randomly chosen mussels from the population. Find the probability that 85% or more of the sampled mussels will be infected, using the normal approximation

(a) without the continuity correction
(b) with the continuity correction

5.41 Refer to Exercise 5.40. Suppose that the biologist takes a random sample of size 50. Find the probability that fewer than 35 of the sampled mussels will be infected, using the normal approximation

(a) without the continuity correction
(b) with the continuity correction

5.6 PERSPECTIVE

In this chapter we have presented two important sampling distributions—the sampling distribution of $\hat{p}$ and the sampling distribution of $\overline{Y}$. Of course, there are many other important sampling distributions, such as are the sampling distribution of the sample standard deviation and the sampling distribution of the sample median.

The ethereal concept of a sampling distribution is linked to the solid reality of data through the random sampling model. Let us take another look at this model in the light of Chapter 5. As we have seen, a *random* sample is not necessarily a *representative* sample.* But using sampling distributions, we can specify the degree of representativeness to be expected in a random sample. For instance, it is intuitively plausible that a larger sample is likely to be more representative than a smaller sample from the same population. In Sections 5.2 and 5.3 we saw how a sampling distribution can make this vague intuition precise by specifying the probability that a specified degree of representativeness will be achieved by a random sample. Thus, sampling distributions provide what has been called "certainty about uncertainty."[13]

In Chapter 6 we will see for the first time how the theory of sampling distributions can be put to practical use in the analysis of data. We will find that, although the calculations of Chapter 5 seem to require the knowledge of unknowable quantities (such as μ and σ), nevertheless when analyzing data we can estimate the probable magnitude of sampling error using only information contained in the sample itself.

In addition to their application to data analysis, sampling distributions provide a basis for comparing the relative merits of different methods of analysis. For example, consider sampling from a normal population with mean μ. Of course, the sample mean $\overline{Y}$ is an estimator of μ. But since a normal distribution is symmetric, it is also the population median, so the sample *median* is also an estimator of μ. How, then, can we decide which estimator is better? This question can be answered in terms of sampling distributions, as follows: Statisticians have determined that, if the population is normal, the sample median is inferior to the sample mean in the sense that its sampling distribution, while centered at μ, has a standard deviation larger than $\frac{\sigma}{\sqrt{n}}$. Consequently, the sample median is less efficient (as an estimator of μ) than the sample mean; for a given sample size n, the sample median provides less information about μ than does the sample mean. (If the population is not normal, however, the sample median can be much more efficient than the mean.)

---

* It is true, however, that sometimes the investigator can force the sample to be representative with respect to some variable (not the one under study) whose population distribution is known. For example, suppose we are sampling from a human population in order to study $Y =$ blood pressure; since blood pressure is age related, we might want to construct the sample so that it matches the population in age distribution. This kind of sampling is not *simple* random sampling, and the methods of analysis given in this book cannot be applied without suitable modification.

Supplementary Exercises 5.42–5.55

[*Note: Exercises preceded by an asterisk refer to optional sections.*]

5.42 In an agricultural experiment, a large field of wheat was divided into many plots (each plot being 7×100 ft) and the yield of grain was measured for each plot. These plot yields followed approximately a normal distribution with mean 88 lb and standard deviation 7 lb (as in Exercise 4.5). Let $\overline{Y}$ represent the mean yield of five plots chosen at random from the field. Find $\Pr\{\overline{Y} > 90\}$.

5.43 In a certain population, 83% of the people have Rh-positive blood type.[14] Suppose a random sample of $n = 10$ people is to be chosen from the population and let $\hat{p}$ represent the proportion of Rh-positive people in the sample. Find

(a) $\Pr\{\hat{p} = .8\}$
(b) $\Pr\{\hat{p} = .9\}$

5.44   The heights of men in a certain population follow a normal distribution with mean 69.7 inches and standard deviation 2.8 inches.[15]

(a) If a man is chosen at random from the population, find the probability that he will be more than 72 inches tall.

(b) If two men are chosen at random from the population, find the probability that (i) both of them will be more than 72 inches tall; (ii) their mean height will be more than 72 inches.

5.45   Suppose a botanist grows many individually potted eggplants, all treated identically and arranged in groups of four pots on the greenhouse bench. After 30 days of growth, she measures the total leaf area Y of each plant. Assume that the population distribution of Y is approximately normal with mean $= 800\,\text{cm}^2$ and $SD = 90\,\text{cm}^2$.[16]

(a) What percentage of the plants in the population will have leaf area between $750\,\text{cm}^2$ and $850\,\text{cm}^2$?

(b) Suppose each group of four plants can be regarded as a random sample from the population. What percentage of the groups will have a group mean leaf area between $750\,\text{cm}^2$ and $850\,\text{cm}^2$?

5.46   Refer to Exercise 5.45. In a real greenhouse, what factors might tend to invalidate the assumption that each group of plants can be regarded as a random sample from the same population?

5.47   In a population of flatworms (*Planaria*) living in a certain pond, one in five individuals is adult and four are juvenile.[17] An ecologist plans to count the adults in a random sample of 20 flatworms from the pond; she will then use $\hat{p}$, the proportion of adults in the sample, as her estimate of p, the proportion of adults in the pond population. Find

(a) $\Pr\{\hat{p} = p\}$
(b) $\Pr\{p - .05 \le \hat{p} \le p + .05\}$

*5.48   Refer to Exercise 5.47. Use the normal approximation (with the continuity correction) to calculate the probabilities.

*5.49   Consider taking a random sample of size 25 from a population in which 42% of the people have type A blood. What is the probability that the sample proportion with type A blood will be greater than .44? Use the normal approximation to the binomial with continuity correction.

5.50   The activity of a certain enzyme is measured by counting emissions from a radioactively labeled molecule. For a given tissue specimen, the counts in consecutive 10-second time periods may be regarded (approximately) as repeated independent observations from a normal distribution (as in Exercise 4.26). Suppose the mean 10-second count for a certain tissue specimen is 1,200 and the standard deviation is 35. For that specimen, let Y represent a 10-second count and let $\bar{Y}$ represent the mean of six 10-second counts. Find $\Pr\{1{,}175 \le Y \le 1{,}225\}$ and $\Pr\{1{,}175 \le \bar{Y} \le 1{,}225\}$, and compare the two. Does the comparison indicate that counting for one minute and dividing by 6 would tend to give a more precise result than merely counting for a single 10-second time period? How?

5.51   In a certain lab population of mice, the weights at 20 days of age follow approximately a normal distribution with mean weight $= 8.3\,\text{g}$ and standard deviation $= 1.7\,\text{g}$.[18] Suppose many litters of 10 mice each are to be weighed. If each litter can be regarded as a random sample from the population, what percentage of the litters will have a total weight of 90 g or more? (*Hint:* How is the total weight of a litter related to the mean weight of its members?)

5.52   Refer to Exercise 5.51. In reality, what factors would tend to invalidate the assumption that each litter can be regarded as a random sample from the same population?

5.53   A certain drug causes drowsiness in 20% of patients. Suppose the drug is to be given to five randomly chosen patients, and let $\hat{p}$ be the proportion who experience drowsiness.

(a) Compute the sampling distribution of $\hat{p}$.
(b) Display the distribution of part (a) as a histogram.

5.54   Consider taking a random sample of size 28 from the population of plants and measuring the height of each plant. In the context of this setting, explain what is meant by the sampling distribution of the sample average.

5.55   Refer to the setting of Exercise 5.54. Suppose that the population mean is 18 cm and the population standard deviation is 4 cm. If the sample size is 28, what is the standard deviation of the sampling distribution of the sample average?

5.56   The skull breadths of a certain population of rodents follow a normal distribution with a standard deviation of 10 mm. Let $\bar{Y}$ be the mean skull breadth of a random sample of 64 individuals from this population, and let μ be the population mean skull breadth.

(a) Suppose $\mu = 50$ mm. Find $\Pr\{\bar{Y}$ is within ± 2 mm of $\mu\}$.
(b) Suppose $\mu = 100$ mm. Find $\Pr\{\bar{Y}$ is within ± 2 mm of $\mu\}$.
(c) Suppose μ is unknown. Can you find $\Pr\{\bar{Y}$ is within ± 2 mm of $\mu\}$? If so, do it. If not, explain why not.

CHAPTER 6

Confidence Intervals

6.1 STATISTICAL ESTIMATION

In this chapter we undertake our first adventure into statistical inference. Recall that statistical inference is based on the random sampling model: We view our data as a random sample from some population, and we use the information in the sample to infer facts about the population. Statistical estimation is a form of statistical inference in which we use the data to (1) determine an estimate of some feature of the population; and (2) assess the precision of the estimate. Let us consider an example.

Soybean Growth. As part of a study on plant growth, a plant physiologist grew 13 individually potted soybean seedlings of the type called Wells II. She raised the plants in a greenhouse under identical environmental conditions (light, temperature, soil, and so on). She measured the total stem length (cm) for each plant after 16 days of growth. The data are given in Table 6.1.[1]

TABLE 6.1 Stem Length of Soybean Plants

Stem Length (cm)				
20.2	22.9	23.3	20.0	19.4
22.0	22.1	22.0	21.9	21.5
19.7	21.5	20.9		

For these data, the mean and standard deviation are

$$\bar{y} = 21.3385 \approx 21.34 \, \text{cm} \quad \text{and} \quad s = 1.2190 \approx 1.22 \, \text{cm}$$

Suppose we regard the 13 observations as a random sample from a population; the population could be described by (among other things) its mean, μ, and its standard deviation, σ. We might define μ and σ verbally as follows:

μ = the (population) mean stem length of Wells II soybean plants grown under the specified conditions

σ = the (population) SD of stem lengths of Wells II soybean plants grown under the specified conditions

Example 6.1

Objectives

In this chapter we will begin a formal study of statistical inference. We will

- *introduce the concept of the standard error and compare it with the standard deviation*

- *learn how to make and interpret confidence intervals for means*

- *learn how to make and interpret confidence intervals for proportions*

- *learn how to determine the sample size that is needed in order to achieve a desired level of accuracy*

- *consider the conditions under which the use of a confidence interval is valid*

It is natural to estimate μ by the sample mean and σ by the sample standard deviation. Thus, from the data on the 13 plants,

21.34 is an estimate of μ;

1.22 is an estimate of σ.

We know that these estimates are subject to sampling error. Note that we are not speaking merely of measurement error; no matter how accurately each individual plant was measured, the sample information is imperfect due to the fact that only 13 plants were measured, rather than the entire infinite population of plants. ■

In general, for a sample of observations on a quantitative variable Y, the sample mean and SD are estimates of the population mean and SD:

$\bar{y}$ is an estimate of μ;

s is an estimate of σ.

The notation for these means and SDs is summarized schematically in Figure 6.1. Our goal is to estimate μ. We will see how to assess the reliability or precision of this estimate, and how to plan a study large enough to attain a desired precision.

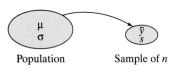

Population   Sample of n

Figure 6.1 Notation for means and SDs of sample and population

6.2 STANDARD ERROR OF THE MEAN

It is intuitively reasonable that the sample mean $\bar{y}$ should be an estimate of μ. It is not so obvious how to determine the reliability of the estimate. As an estimate of μ, the sample mean $\bar{y}$ is imprecise to the extent that it is affected by sampling error. In Section 5.3 we saw that the magnitude of the sampling error—that is, the amount of discrepancy between $\bar{y}$ and μ—is described (in a probability sense) by the sampling distribution of $\bar{Y}$. The standard deviation of the sampling distribution of $\bar{Y}$ is

$$\sigma_{\bar{Y}} = \frac{\sigma}{\sqrt{n}}$$

Since s is an estimate of σ, a natural estimate of $\dfrac{\sigma}{\sqrt{n}}$ would be $\dfrac{s}{\sqrt{n}}$; this quantity is called the **standard error of the mean**. We will denote it as $\text{SE}_{\bar{y}}$ or sometimes simply SE.*

> **Definition**
> The **standard error of the mean** is defined as
>
> $$\text{SE}_{\bar{y}} = \frac{s}{\sqrt{n}}$$

The following example illustrates the definition.

* Some statisticians prefer to reserve the term *standard error* for $\sigma/\sqrt{n}$ and to call $s/\sqrt{n}$ the *estimated standard error*.

Soybean Growth.   For the soybean growth data of Example 6.1, we have $n = 13$, $\bar{y} = 21.3385 \approx 21.34$ cm, and $s = 1.2190 \approx 1.22$ cm. The standard error of the mean is

Example 6.2

$$SE_{\bar{y}} = \frac{s}{\sqrt{n}}$$

$$= \frac{1.2190}{\sqrt{13}} = .338 \text{ cm, which we will round to } .34 \text{ cm*}$$ ∎

As we have seen, the SE is an estimate of $\sigma_{\bar{Y}}$. On a more practical level, the SE can be interpreted in terms of the expected sampling error: Roughly speaking, the difference between $\bar{y}$ and μ is rarely more than a few standard errors. Indeed, we expect $\bar{y}$ to be within about one standard error of μ quite often. Thus, the standard error is a measure of the reliability or precision of $\bar{y}$ as an estimate of μ; the smaller the SE, the more precise the estimate. Notice how the SE incorporates the two factors that affect reliability: (1) the inherent variability of the observations (expressed through s), and (2) the sample size (n).

Standard Error Versus Standard Deviation

The terms *standard error* and *standard deviation* are sometimes confused. It is extremely important to distinguish between standard error (SE) and standard deviation (s, or SD). These two quantities describe entirely different aspects of the data. The SD describes the dispersion of the data, while the SE describes the uncertainty (due to sampling error) in the *mean* of the data. Let us consider a concrete example.

Lamb Birthweights.   A geneticist weighed 28 female lambs at birth. The lambs were all born in April, were all the same breed (Rambouillet), and were all single births (no twins). The diet and other environmental conditions were the same for all the parents. The birthweights are shown in Table 6.2.[2]

Example 6.3

TABLE 6.2 Birthweights of 28 Rambouillet Lambs

Birthweight (kg)						
4.3	5.2	6.2	6.7	5.3	4.9	4.7
5.5	5.3	4.0	4.9	5.2	4.9	5.3
5.4	5.5	3.6	5.8	5.6	5.0	5.2
5.8	6.1	4.9	4.5	4.8	5.4	4.7

---

* Rounding Summary Statistics

For reporting the mean, standard deviation, and standard error of the mean, the following procedure is recommended:

1. Round the SE to two significant digits.
2. Round $\bar{y}$ and s to match the SE with respect to the decimal position of the last significant digit. (The concept of significant digits is reviewed in Appendix 6.1.) For example, if the SE is rounded to two decimal places, then $\bar{y}$ and s are also rounded to two decimal places.

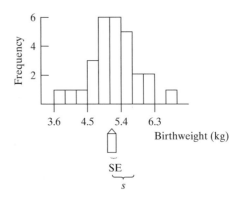

Figure 6.2 Birthweights of 28 lambs

For these data, the mean is $\bar{y} = 5.17$ kg, the standard deviation is $s = .65$ kg, and the standard error is SE $= .12$ kg. The SD, s, describes the variability from one lamb to the next, while the SE indicates the variability associated with the sample mean (5.17 kg), viewed as an estimate of the population mean birthweight. This distinction is emphasized in Figure 6.2, which shows a histogram of the lamb birthweight data; the SD is indicated as a deviation from $\bar{y}$, while the SE is indicated as variability associated with $\bar{y}$ itself. ■

Another way to highlight the contrast between the SE and the SD is to consider samples of various sizes. As the sample size increases, the sample mean and SD tend to approach more closely the population mean and SD; indeed, the distribution of the data tends to approach the population distribution. The standard error, by contrast, tends to decrease as n increases; when n is very large the SE is very small and so the sample mean is a very precise estimate of the population mean. The following example illustrates this effect.

Example 6.4 **Lamb Birthweights.** Suppose we regard the birthweight data of Example 6.3 as a sample of size $n = 28$ from a population, and consider what would happen if we were to choose larger samples from the same population—that is, if we were to measure the birthweights of additional female Rambouillet lambs born under the specified conditions. Figure 6.3 shows the kind of results we might expect; the values given are fictitious but realistic. For very large n, $\bar{y}$ and s would be very close to μ and σ, where

μ = Mean birthweight of female Rambouillet lambs born under the conditions described

	Sample size			
	$n = 28$	$n = 280$	$n = 2800$	$n \longrightarrow \infty$
$\bar{y}$	5.17	5.19	5.14	$\bar{y} \longrightarrow \mu$
s	.65	.67	.65	$s \longrightarrow \sigma$
SE	.12	.040	.012	SE $\longrightarrow 0$

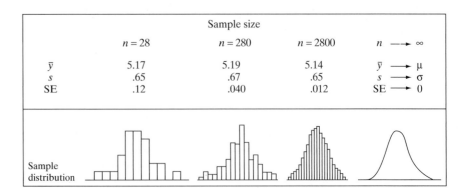

Figure 6.3 Samples of various sizes from the lamb birthweight population

and

σ = Standard deviation of birthweights of female Rambouillet lambs born under the conditions described ■

Graphical Presentation of the SE and the SD

The clarity and impact of a scientific report can be greatly enhanced by well-designed displays of the data. Data can be displayed graphically or in a table. We briefly discuss some of the options.

Let us first consider graphical presentation of data. Here is an example.

MAO and Schizophrenia.   The enzyme monoamine oxidase (MAO) is of interest in the study of human behavior. Figures 6.4 and 6.5 display measurements of MAO activity in the blood platelets in five groups of people: Groups I, II, and III are three diagnostic categories of schizophrenic patients (see Example 1.4), and groups IV and V are healthy male and female controls.[3] The MAO activity values are expressed as nmol benzylaldehyde product per 10^8 platelets per hour. In both Figures 6.4 and 6.5, the dots represent the group means; the vertical lines represent $\pm$ SE in Figure 6.4 and $\pm$ SD in Figure 6.5.

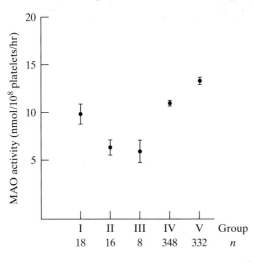

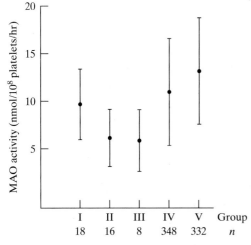

Example 6.5

Figure 6.4 MAO data displayed as $\bar{y} \pm$ SE

Figure 6.5 MAO data displayed as $\bar{y} \pm$ SD

Figures 6.4 and 6.5 convey very different information. Figure 6.4 conveys (1) the mean MAO value in each group, and (2) the reliability of each group mean, viewed as an estimate of its respective population mean. Figure 6.5 conveys (1) the mean MAO value in each group, and (2) the variability of MAO within each group. For instance, group V shows greater variability of MAO than group I (Figure 6.5) but has a much smaller standard error (Figure 6.4) because it is a much larger group.

Figure 6.4 invites the viewer to compare the means and gives some indication of the reliability of the comparisons. (A full discussion of comparison of two or more means must wait until Chapter 7 and later chapters.) Figure 6.5 invites the viewer to compare the means and also to compare the standard deviations. Furthermore, Figure 6.5 gives the viewer some information about the extent of overlap of the MAO values in the various groups. For instance, consider groups IV and V; whereas they appear quite "separate" in Figure 6.4, we can easily see from Figure 6.5 that there is considerable overlap of individual MAO values in the two groups. ■

In some scientific reports, data are summarized in tables rather than graphically. Table 6.3 shows a tabular summary for the MAO data of Example 6.5.

TABLE 6.3 MAO Activity in Five Groups of People

| Group | n | MAO Activity (nmol/10^8 platelets/hr) | | |
		Mean	SE	SD
I	18	9.81	.85	3.62
II	16	6.28	.72	2.88
III	8	5.97	1.13	3.19
IV	348	11.04	.30	5.59
V	332	13.29	.30	5.50

Exercises 6.1–6.7

6.1 A pharmacologist measured the concentration of dopamine in the brains of several rats. The mean concentration was 1,269 ng/g and the standard deviation was 145 ng/g.[4] What was the standard error of the mean if

(a) 8 rats were measured?
(b) 30 rats were measured?

6.2 An agronomist measured the heights of n corn plants.[5] The mean height was 220 cm and the standard deviation was 15 cm. Calculate the standard error of the mean if

(a) $n = 25$
(b) $n = 100$

6.3 In evaluating a forage crop, it is important to measure the concentration of various constituents in the plant tissue. In a study of the reliability of such measurements, a batch of alfalfa was dried, ground, and passed through a fine screen. Five small (.3 g) aliquots of the alfalfa were then analyzed for their content of insoluble ash.[6] The results (g/kg) were as follows:

10.0   8.9   9.1   11.7   7.9

For these data, calculate the mean, the standard deviation, and the standard error of the mean.

6.4   A zoologist measured tail length in 86 individuals, all in the 1-year age group, of the deermouse *Peromyscus*. The mean length was 60.43 mm and the standard deviation was 3.06 mm. The table presents a frequency distribution of the data.[7]

Tail Length (mm)	Number of Mice
52–53	1
54–55	3
56–57	11
58–59	18
60–61	21
62–63	20
64–65	9
66–67	2
68–69	1
Total	86

(a) Calculate the standard error of the mean.
(b) Construct a histogram of the data and indicate the intervals $\bar{y} \pm$ SD and $\bar{y} \pm$ SE on your histogram. (See Figure 6.2.)

6.5   Refer to the mouse data of Exercise 6.4. Suppose the zoologist were to measure 500 additional animals from the same population. Based on the data in Exercise 6.4,

(a) What would you predict would be the standard deviation of the 500 new measurements?
(b) What would you predict would be the standard error of the mean for the 500 new measurements?

6.6   In a report of a pharmacological study, the experimental animals were described as follows:[8] "Rats weighing 150 $\pm$ 10 g were injected . . . " with a certain chemical, and then certain measurements were made on the rats. If the author intends to convey the degree of homogeneity of the group of experimental animals, then should the 10 g be the SD or the SE? Explain.

6.7   For each of the following, decide whether the description fits the SD or the SE.

(a) This quantity is a measure of the accuracy of the sample mean as an estimate of the population mean.
(b) This quantity tends to stay the same as the sample size goes up.
(c) This quantity tends to go down as the sample size goes up.

6.3 CONFIDENCE INTERVAL FOR μ

In Section 6.2 we said that the standard error of the mean (the SE) measures how far $\bar{y}$ is likely to be from the population mean μ. In this section we make that idea precise.

Confidence Interval for μ: Basic Idea

Figure 6.6 is a drawing of an invisible man walking his dog. The dog, which is visible, is on a spring-loaded leash. The tension on the spring is such that the dog is within one SE of the man about two-thirds of the time. The dog is within 2 standard

Figure 6.6 Invisible man walking his dog

errors of the man 95% of the time. Only 5% of the time is the dog more than two SEs from the man—unless the leash breaks, in which case the dog could be anywhere. We can see the dog, but we would like to know where the man is. Since the man and the dog are usually within two SEs of each other, we can take the interval "dog $\pm$ 2 · SE" as an interval that typically would include the man. Indeed, we could say that we are 95% confident that the man is in this interval.

This is the basic idea of a confidence interval. We would like to know the value of the population mean μ—which corresponds to the man—but we cannot see it directly. What we *can* see is the sample mean $\bar{y}$—which corresponds to the dog. We use what we can see, $\bar{y}$, together with the standard error, which we can calculate from the data, as a way of constructing an interval that we hope will include what we cannot see, the population mean μ. We call the interval "position of the dog $\pm$ 2 · SE" a 95% confidence interval for the position of the man. (This all depends on having a model that is correct: We said that if the leash breaks, then knowing where the dog is doesn't tell us much about where the man is. Likewise, if our statistical model is wrong [for example, if we have a biased sample], then knowing $\bar{y}$ doesn't tell us much about μ!)

Confidence Interval for μ: Mathematics

In the invisible man analogy,* we said that the dog is within 1 SE of the man about two-thirds of the time and within 2 SEs of the man 95% of the time. This is based on the idea of the sampling distribution of $\bar{Y}$ when we have a random sample from a normal distribution. If Z is a standard normal random variable, then the probability that Z is between ± 2 is about 95%. More precisely, $\Pr\{-1.96 < Z < 1.96\} = .95$. From Chapter 5 we know that if Y has a normal distribution, then $\dfrac{\bar{Y} - \mu}{\sigma/\sqrt{n}}$ has a standard normal (Z) distribution, so

$$\Pr\left\{-1.96 < \frac{\bar{Y} - \mu}{\sigma/\sqrt{n}} < 1.96\right\} = .95 \tag{6.1}$$

Thus,

$$\Pr\left\{-1.96 \cdot \sigma/\sqrt{n} < \bar{Y} - \mu < 1.96 \cdot \sigma/\sqrt{n}\right\} = .95$$

and

$$\Pr\left\{-\bar{Y} - 1.96 \cdot \sigma/\sqrt{n} < -\mu < -\bar{Y} + 1.96 \cdot \sigma/\sqrt{n}\right\} = .95$$

so

$$\Pr\left\{\bar{Y} - 1.96 \cdot \sigma/\sqrt{n} < \mu < \bar{Y} + 1.96 \cdot \sigma/\sqrt{n}\right\} = .95$$

That is, the interval

$$\bar{y} \pm 1.96 \frac{\sigma}{\sqrt{n}} \tag{6.2}$$

will contain μ for 95% of all samples.

The interval (6.2) cannot be used for data analysis because it contains a quantity—namely, σ—that cannot be determined from the data. If we replace σ by

---

* Credit for this analogy is due to Geoff Jowett.

its estimate—namely, s—then we can calculate an interval from the data, but what happens to the 95% interpretation? Fortunately, it turns out that there is an escape from this dilemma. The escape was discovered by a British scientist named W. S. Gosset, who was employed by the Guinness Brewery; he published his findings in 1908 under the pseudonym "Student," and the method has borne his name ever since.[9] "Student" discovered that *if the data come from a normal population* and if we replace σ in the interval (6.2) by the sample SD, s, then the 95% interpretation can be preserved if the multiplier of $\dfrac{\sigma}{\sqrt{n}}$ (that is, 1.96) is replaced by a suitable quantity; the new quantity is denoted $t_{.025}$ and is related to a distribution known as Student's t distribution.

Student's t Distribution

The **Student's t distributions** are theoretical continuous distributions that are used for many purposes in statistics, including the construction of confidence intervals. The exact shape of a Student's t distribution depends on a quantity called degrees of freedom, abbreviated df. Figure 6.7 shows the density curves of two Student's t distributions with df = 3 and df = 10, and also a normal curve. A t curve is symmetric and bell shaped like the normal curve, but has a larger standard deviation. As the df increase, the t curves approach the normal curve; thus, the normal curve can be regarded as a t curve with infinite df (df = ∞).

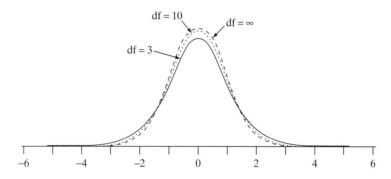

Figure 6.7 Two Student's t curves and a normal curve (df = ∞)

The quantity $t_{.025}$ is called the two-tailed 5% critical value of Student's t distribution and is defined to be the value such that the interval between $-t_{.025}$ and $+t_{.025}$ contains 95% of the area under the curve, as shown in Figure 6.8.* That is, the combined area in the two tails—below $-t_{.025}$ and above $+t_{.025}$—is 5%. The total shaded area in Figure 6.8 is equal to .05; note that the shaded area consists of two "pieces" of area .025 each.

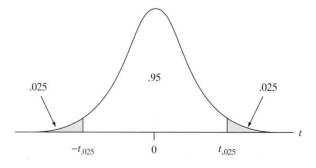

Figure 6.8 Definition of the critical value $t_{.025}$

Critical values of Student's t distribution are tabulated in Table 4. The values of $t_{.025}$ are shown in the column headed "Two-Tailed Area .05." If you glance down this column, you will see that the values of $t_{.025}$ decrease as the df increase; for df $= \infty$ (that is, for the normal distribution) the value is $t_{.025} = 1.960$. You can confirm from Table 3 that the interval ± 1.96 (on the Z scale) contains 95% of the area under a normal curve.

Other columns of Table 4 show other critical values, which are defined analogously; for instance, the interval $\pm t_{.05}$ contains 90% of the area under a Student's t curve.

Confidence Interval for μ: Method

We describe Student's method for constructing a confidence interval for μ, based on a random sample from a normal population. First, suppose we have chosen a confidence level equal to 95% (i.e., we wish to be 95% confident). To construct a 95% confidence interval for μ, we compute the lower and upper limits of the interval as

$$\bar{y} - t_{.025}\text{SE}_{\bar{y}} \quad \text{and} \quad \bar{y} + t_{.025}\text{SE}_{\bar{y}}$$

that is,

$$\bar{y} \pm t_{.025}\frac{s}{\sqrt{n}}$$

where the critical value $t_{.025}$ is determined from Student's t distribution with

$$\text{df} = n - 1$$

The following example illustrates the construction of a confidence interval.

Example 6.6

Soybean Growth. For the soybean stem length data of Example 6.1, we have $n = 13$, $\bar{y} = 21.3385$ cm, and $s = 1.2190$ cm. Figure 6.9 shows a histogram and a normal probability plot of the data; these support the belief that the data came from a normal population. We have 13 observations, so the value of df is

$$\text{df} = n - 1 = 13 - 1 = 12$$

From Table 4 we find

$$t_{.025} = 2.179$$

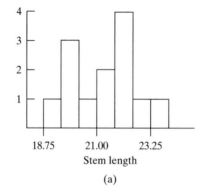

(a)

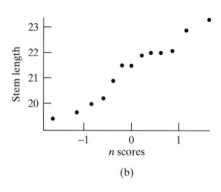

(b)

Figure 6.9 Histogram (a) and normal probability plot (b) of soybean growth data

---

* In some statistics textbooks, you may find other notations, such as $t_{.05}$ or $t_{.975}$, rather than $t_{.025}$.

The 95% confidence interval for μ is

$$21.3385 \pm 2.179\frac{1.2190}{\sqrt{13}}$$

$$21.3385 \pm 2.179(.3381)$$

$$21.3385 \pm .7367$$

or approximately

$$21.34 \pm .74$$

The confidence interval may be left in this form. Alternatively, the endpoints of the interval may be explicitly calculated as

$$21.34 - .74 = 20.60 \text{ and } 21.34 + .74 = 22.08$$

and the interval may be written compactly as

$$(20.6, 22.1)$$

or in a more complete form as the following confidence statement:

$$20.6 \text{ cm} < \mu < 22.1 \text{ cm}$$

The confidence statement asserts that the population mean stem length of Wells II soybean plants, grown under the specified conditions, is between 20.6 cm and 22.1 cm. ∎

The interpretation of the "95% confidence" will be discussed after the next example.

Confidence coefficients other than 95% are used analogously. For instance, a 90% confidence interval for μ is constructed using $t_{.05}$ instead of $t_{.025}$ as follows:

$$\bar{y} \pm t_{.05}\frac{s}{\sqrt{n}}$$

The following is an example.

Soybean Growth. From Table 4, we find that $t_{.05} = 1.782$ with df $= 12$. Thus, the 90% confidence interval for μ from the soybean growth data is

Example 6.7

$$21.3385 \pm 1.782\frac{1.2190}{\sqrt{13}}$$

$$21.3385 \pm .6025$$

or

$$20.7 < \mu < 21.9$$ ∎

As you see, the choice of a confidence level is somewhat arbitrary. For the soybean growth data, the 95% confidence interval is

$$21.34 \pm .74$$

and the 90% confidence interval is

$$21.34 \pm .60$$

Thus, the 90% confidence interval is narrower than the 95% confidence interval. If we want to be 95% confident that our interval contains μ, then we need a wider interval than we would need if we only wanted to be 90% confident: The higher the confidence level, the wider the confidence interval.

Remark: The quantity $(n - 1)$ is referred to as degrees of freedom because the deviations $(y_i - \bar{y})$ must sum to zero, and so only $(n - 1)$ of them are free to vary. A sample of size n provides only $(n - 1)$ independent pieces of information about variability; that is, about σ. This is particularly clear if we consider the case $n = 1$; a sample of size 1 provides some information about μ, but no information about σ, and so no information about sampling error. It makes sense, then, that when $n = 1$ we cannot use Student's t method to calculate a confidence interval: The sample standard deviation does not exist (see Example 2.31) and there is no critical value with df $= 0$. A sample of size 1 is sometimes called an anecdote; for instance, an individual medical case history is an anecdote. Of course, a case history can contribute greatly to medical knowledge, but it does not (in itself) provide a basis for judging how closely the individual case resembles the population at large.

Confidence Intervals and Randomness

In what sense can we be confident in a confidence interval? To answer this question, let us assume that we are dealing with a random sample from a normal population. Consider, for instance, a 95% confidence interval. One way to interpret the confidence level (95%) is to refer to the meta-experiment of repeated samples from the same population. If a 95% confidence interval for μ is constructed for each sample, then 95% of the confidence intervals will contain μ. Of course, the observed data in an experiment comprise only *one* of the possible samples; we can hope confidently that this sample is one of the lucky 95%, but we will never know.

The following example provides a more concrete visualization of the meta-experiment interpretation of a confidence level.

Example 6.8 | **Eggshell Thickness.** In a certain large population of chicken eggs (described in Example 4.2), the distribution of eggshell thickness is normal with mean $\mu = .38$ mm and standard deviation $\sigma = .03$ mm. Figure 6.10 shows some typical samples from this population; plotted on the right are the associated 95% confidence intervals. The sample sizes are $n = 5$ and $n = 20$. Notice that the second confidence interval with $n = 5$ does not contain μ. In the totality of potential confidence intervals, the percentage that would contain μ is 95% for either sample size; as Figure 6.10 shows, the larger samples tend to produce narrower confidence intervals. ∎

A confidence level can be interpreted as a probability, but caution is required. If we consider 95% confidence intervals, for instance, then the following statement is correct:

Pr{the next sample will give us a confidence interval that contains μ} = .95

However, we should realize that it is *the confidence interval* that is the random item in this statement, and it is not correct to replace this item with its value from the data. Thus, for instance, we found in Example 6.6 that the 95% confidence interval for the mean soybean growth is

$$20.6 \, \text{cm} < \mu < 22.1 \, \text{cm} \tag{6.3}$$

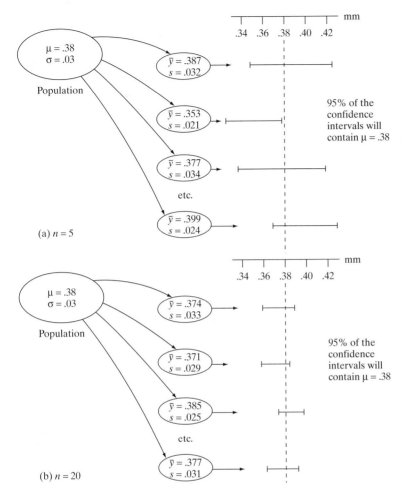

Figure 6.10 Confidence intervals for mean eggshell thickness

Nevertheless, it is *not* correct to say that

$$\Pr\{20.6\,\text{cm} < \mu < 22.1\,\text{cm}\} = .95$$

because this statement has no chance element; either μ is between 20.6 and 22.1 or it is not. If $\mu = 21$, then $\Pr\{20.6\,\text{cm} < \mu < 22.1\,\text{cm}\} = \Pr\{20.6\,\text{cm} < 21 < 22.1\,\text{cm}\} = 1$ (not .95). The following analogy may help to clarify this point. Suppose we let Y represent the number of spots showing when a balanced die is tossed; then

$$\Pr\{Y = 2\} = \frac{1}{6}$$

On the other hand, if we now toss the die and observe five spots, it is obviously *not* correct to substitute this datum in the probability statement to conclude that

$$\Pr\{5 = 2\} = \frac{1}{6}$$

As the preceding discussion indicates, the confidence level (for instance, 95%) is a property of the *method* rather than of a particular interval. An individual statement—such as (6.3)—is either true or false; but in the long run, if the researcher constructs 95% confidence intervals in various experiments, each time producing a statement such as (6.3), then 95% of the statements will be true.

Interpretation of a Confidence Interval

Example 6.9

Bone Mineral Density.   Low bone mineral density often leads to hip fractures in the elderly. In an experiment to assess the effectiveness of hormone replacement therapy, researchers gave conjugated equine estrogen (CEE) to a sample of 94 women between the ages of 45 and 64.[10] After taking the medication for 36 months, the bone mineral density was measured for each of the 94 women. The average density was .878 g/cm^2, with a standard deviation of .126 g/cm^2.

The standard error of the mean is thus $\frac{.126}{\sqrt{94}} = .013$. It is not clear that the distribution of bone mineral density is a normal distribution, but as we will see in Section 6.5, when the sample size is large, the condition of normality is not crucial. There were 94 observations, so there are 93 degrees of freedom. To find the t multiplier for a 95% confidence interval, we will use 80 degrees of freedom (since Table 4 doesn't list 93 degrees of freedom); the t multiplier is $t_{.025} = 1.990$. A 95% confidence interval for μ is

$$.878 \pm 1.990(.013)$$

or approximately

$$.878 \pm .026$$

or

$$(.852, .904)$$

Thus, *we are 95% confident that the average hip bone mineral density of all women age 45 to 64 who take CEE for 36 months is between .852 g/cm^2 and .904 g/cm^2.* ∎

Example 6.10

Seeds per Fruit.   The number of seeds per fruit for the freshwater plant *Vallisneria Americana* varies considerably from one fruit to another. A researcher took a random sample of 12 fruit and found that the average number of seeds was 320, with a standard deviation of 125.[11] The researcher expected the number of seeds to follow, at least approximately, a normal distribution. A normal probability plot of the data is shown in Figure 6.11. This supports the use of a normal distribution model for these data.

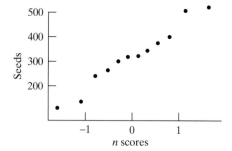

Figure 6.11 Normal probability plot of seeds per fruit for *Vallisneria Americana*

The standard error of the mean is $\frac{125}{\sqrt{12}} = 36$. There are 11 degrees of freedom. The t multiplier for a 90% confidence interval is $t_{.05} = 1.796$. A 90% confidence interval for μ is

$$320 \pm 1.796(36)$$

or approximately

$$320 \pm 65$$

or

$$(255,385)$$

Thus, *we are 90% confident that the (population) average number of seeds per fruit for Vallisneria Americana is between 255 and 385.* ■

    Computer note: Statistical software can be used to calculate confidence intervals. For example, in the MINITAB system the command

```
MTB > TInterval 90 C1
```

will produce a 90% confidence interval for the population mean, using whatever data are stored in column 1. If the seeds per fruit data from Example 6.10 are stored in column 1, then the output of this command is

```
Variable   N    Mean    StDev   SE Mean     90.0 % C.I.
C1        12   319.5    125.2     36.1    (254.6,  384.4)
```

which, except for rounding off, agrees with the calculations shown in Example 6.10.

Relationship to Sampling Distribution of $\bar{Y}$

At this point it may be helpful to look back and see how a confidence interval for μ is related to the sampling distribution of $\bar{Y}$. Recall from Section 5.3 that the mean of the sampling distribution is μ and its standard deviation is $\dfrac{\sigma}{\sqrt{n}}$. Figure 6.12 shows a particular sample mean $(\bar{y})$ and its associated 95% confidence interval for μ, superimposed on the sampling distribution of $\bar{Y}$. Notice that the particular confidence interval does contain μ; this will happen for 95% of samples.

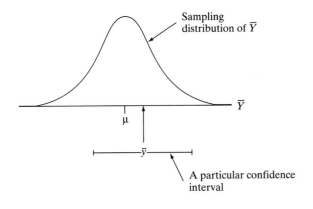

Figure 6.12 Relationship between a particular confidence interval for μ and the sampling distribution of $\bar{Y}$

Exercises 6.8–6.26

6.8   *(Sampling exercise)*   Refer to Exercise 5.11. Use your sample of five ellipse lengths to construct an 80% confidence interval for μ, using the formula $\bar{y} \pm (1.533)s/\sqrt{n}$. To facilitate the pooling of results from the entire class, compute the two endpoints explicitly.

6.9   *(Sampling exercise)*   Refer to Exercise 5.13. Use your sample of 20 ellipse lengths to construct an 80% confidence interval for μ using the formula $\bar{y} \pm (1.328)s/\sqrt{n}$. To facilitate the pooling of results from the entire class, compute the two endpoints explicitly.

6.10   As part of a study of the development of the thymus gland, researchers weighed the glands of five chick embryos after 14 days of incubation. The thymus weights (mg) were as follows:[12]

$$29.6 \quad 21.5 \quad 28.0 \quad 34.6 \quad 44.9$$

For these data, the mean is 31.7 and the standard deviation is 8.7.

(a) Calculate the standard error of the mean.
(b) Construct a 90% confidence interval for the population mean.

6.11   Consider the data from Exercise 6.10.

(a) Construct a 95% confidence interval for the population mean.
(b) Interpret the confidence interval you found in part (a). That is, explain what the numbers in the interval mean. (See Examples 6.9 and 6.10.)

6.12   Six healthy three-year-old female Suffolk sheep were injected with the antibiotic Gentamicin, at a dosage of 10 mg/kg body weight. Their blood serum concentrations (μg/mLi) of Gentamicin 1.5 hours after injection were as follows:[13]

$$33 \quad\quad 26 \quad\quad 34 \quad\quad 31 \quad\quad 23 \quad\quad 25$$

For these data, the mean is 28.7 and the standard deviation is 4.6.

(a) Construct a 95% confidence interval for the population mean.
(b) Define in words the population mean that you estimated in part (a). (See Example 6.1.)
(c) The interval constructed in part (a) nearly contains all of the observations; will this typically be true for a 95% confidence interval? Explain.

6.13   A zoologist measured tail length in 86 individuals, all in the one-year age group, of the deermouse *Peromyscus*. The mean length was 60.43 mm and the standard deviation was 3.06 mm. A 95% confidence interval for the mean is (59.77, 61.09).

(a) True or false (and say why): We are 95% confident that the average tail length of the 86 individuals in the sample is between 59.77 mm and 61.09 mm.
(b) True or false (and say why): We are 95% confident that the average tail length of all the individuals in the population is between 59.77 mm and 61.09 mm.

6.14   Researchers measured the bone mineral density of the spines of 94 women who had taken the drug CEE. (See Example 6.9, which dealt with hip bone mineral density.) The mean was $1.016 \, \text{g/cm}^2$ and the standard deviation was $.155 \, \text{g/cm}^2$. A 95% confidence interval for the mean is (.984, 1.048). True or false (and say why): 95% of the data are between .984 and 1.048.

6.15   There was a control group in the study described in Example 6.9. The 124 women in the control group were given a placebo, rather than an active medication. At the end of the study they had an average bone mineral density of $.840 \, \text{g/cm}^2$. The following are three confidence intervals, one of which is a 90% confidence interval, one of which is an 85% confidence interval, and the other of which is an 80% confidence interval. Without doing any calculations, match the intervals with the confidence levels and explain how you determined which interval goes with which level.

Confidence levels: 90% 85% 80%

Intervals (in scrambled order): (.826, .854) (.824, .856) (.822, .858)

6.16 Human beta-endorphin (HBE) is a hormone secreted by the pituitary gland under conditions of stress. A researcher conducted a study to investigate whether a program of regular exercise might affect the resting (unstressed) concentration of HBE in the blood. He measured blood HBE levels, in January and again in May, in ten participants in a physical fitness program. The results were as shown in the table.[14]

(a) Construct a 95% confidence interval for the population mean difference in HBE levels between January and May. (*Hint*: You need to use only the values in the right-hand column.)

	HBE Level (pg/mLi)		
Participant	*January*	*May*	*Difference*
1	42	22	20
2	47	29	18
3	37	9	28
4	9	9	0
5	33	26	7
6	70	36	34
7	54	38	16
8	27	32	−5
9	41	33	8
10	18	14	4
Mean	37.8	24.8	13.0
SD	17.6	10.9	12.4

(b) Interpret the confidence interval from part (a). That is, explain what the interval tells you about HBE levels. (See Examples 6.9 and 6.10.)

6.17 Consider the data from Exercise 6.16. If the sample size is small, as it is in this case, then in order for a confidence interval based on Student's *t* distribution to be valid, the data must come from a normally distributed population. Is it reasonable to think that difference in HBE level is normally distributed? How do you know?

6.18 Invertase is an enzyme that may aid in spore germination of the fungus *Colletotrichum graminicola*. A botanist incubated specimens of the fungal tissue in petri dishes and then assayed the tissue for invertase activity. The specific activity values for nine petri dishes incubated at 90% relative humidity for 24 hours are summarized as follows:[15]

Mean = 5,111 units   SD = 818 units

(a) Assume that the data are a random sample from a normal population. Construct a 95% confidence interval for the mean invertase activity under these experimental conditions.

(b) Interpret the confidence interval you found in part (a). That is, explain what the numbers in the interval mean. (See Examples 6.9 and 6.10.)

(c) If you had the raw data, how could you check the condition that the data are from a normal population?

6.19 As part of a study of the treatment of anemia in cattle, researchers measured the concentration of selenium in the blood of 36 cows who had been given a dietary supplement of selenium (2 mg/day) for one year. The cows were all the same breed (*Santa Gertrudis*) and had borne their first calf during the year. The mean selenium concentration was 6.21 μg/dLi and the standard deviation was 1.84 μg/dLi.[16] Construct a 95% confidence interval for the population mean.

6.20 In a study of larval development in the tufted apple budmoth *(Platynota idaeusalis)*, an entomologist measured the head widths of 50 larvae. All 50 larvae had been reared under identical conditions and had moulted six times. The mean head width was 1.20 mm and the standard deviation was .14 mm. Construct a 90% confidence interval for the population mean.[17]

6.21 In a study of the effect of aluminum intake on the mental development of infants, a group of 92 infants who had been born prematurely were given a special aluminum-depleted intravenous-feeding solution.[18] At age 18 months the neurologic development of the infants was measured using the Bayley Mental Development Index. (The Bayley Mental Development Index is similar to an IQ score, with 100 being the average in the general population.) A 95% confidence interval for the mean is (93.8, 102.1). Interpret this interval. That is, what does the interval tell us about neurologic development in the population of prematurely born infants who receive intravenous-feeding solutions?

6.22 A group of 101 patients with end-stage renal disease were given the drug epoetin.[19] The mean hemoglobin level of the patients was 10.3 (g/dLi), with an SD of 0.9. Construct a 95% confidence interval for the population mean.

6.23 In Table 4 we find that $t_{.025} = 1.960$ when df $= \infty$. Show how this value can be verified using Table 3.

6.24 Use Table 3 to find the value of $t_{.0025}$ when df $= \infty$. (Do not attempt to interpolate in Table 4.)

6.25 Data are often summarized in this format: $\bar{y} \pm$ SE. Suppose this interval is interpreted as a confidence interval. If the sample size is large, what would be the confidence level of such an interval? That is, what is the chance that an interval computed as

$$\bar{y} \pm (1.00)\text{SE}$$

will actually contain the population mean? [*Hint*: Recall that the confidence level of the interval $\bar{y} \pm (1.96)$SE is 95%.]

6.26 *(Continuation of Exercise 6.25)*

(a) If the sample size is small but the population distribution is normal, is the confidence level of the interval $\bar{y} \pm$ SE larger or smaller than the answer to Exercise 6.25? Explain.

(b) How is the answer to Exercise 6.25 affected if the population distribution of Y is not approximately normal?

6.4 PLANNING A STUDY TO ESTIMATE μ

In planning an experiment, it is wise to consider in advance whether the estimates generated from the data will be sufficiently precise. It can be painful indeed to discover after a long and expensive study that the standard errors are so large that the primary questions addressed by the study cannot be answered.

The precision with which a population mean can be estimated is determined by two factors: (1) the population variability of the observed variable Y, and (2) the sample size.

In some situations the variability of Y cannot, and perhaps should not, be reduced. For example, a wildlife ecologist may wish to conduct a field study of a natural population of fish; the heterogeneity of the population is not controllable, and in fact is a proper subject of investigation. As another example, in a medical investigation, in addition to knowing the average response to a treatment, it may also be important to know how much the response varies from one patient to another, and so it may not be appropriate to use an overly homogeneous group of patients.

On the other hand, it is often appropriate, especially in comparative studies, to reduce the variability of Y by holding *extraneous* conditions as constant as possible. For example, physiological measurements may be taken at a fixed time of day; tissue may be held at a controlled temperature; all animals used in an experiment may be the same age.

Suppose, then, that plans have been made to reduce the variability of Y as much as possible, or desirable. What sample size will be sufficient to achieve a desired degree of precision in estimation of the population mean? If we use the standard error as our measure of precision, then this question can be approached in a straightforward manner. Recall that the SE is defined as

$$\text{SE}_{\bar{y}} = \frac{s}{\sqrt{n}}$$

In order to decide on a value of n, we must (1) specify what value of the SE is considered desirable to achieve, and (2) have available a preliminary guess of the SD, either from a pilot study or other previous experience, or from the scientific literature. The required sample size is then determined from the following equation:

$$\text{Desired SE} = \frac{\text{Guesed SD}}{\sqrt{n}}$$

The following example illustrates the use of this equation.

Soybean Growth.   The soybean stem-length data of Example 6.1 yielded the following summary statistics:

$$\bar{y} = 21.34 \text{ cm}$$
$$s = 1.22 \text{ cm}$$
$$\text{SE} = .34 \text{ cm}$$

Example 6.11

Suppose the researcher is now planning a new study of soybean growth and has decided that it would be desirable that the SE be no more than .2 cm. As a preliminary guess of the SD, she will use the value from the old study, namely 1.22 cm. Thus, the desired n must satisfy the following relation:

$$\text{SE} = \frac{1.22}{\sqrt{n}} \le .2$$

This equation is easily solved to give $n \ge 37.2$. Since we cannot have 37.2 plants, the new experiment should include 38 plants. ∎

You may wonder how a researcher would arrive at a value such as .2 cm for the desired SE. Such a value is determined by considering how much error we are willing to tolerate in the estimate of μ. For example, suppose the researcher in Example 6.11 has decided that she would like to be able to estimate the population mean, μ, to within $\pm.4$ with 95% confidence. That is, she would like her 95% confidence interval for μ to be $\bar{y} \pm .4$. The "$\pm$ part" of the confidence interval, which is sometimes called the margin of error, is $t_{.025} \cdot \text{SE}$. The precise value of $t_{.025}$ depends on the degrees of freedom, but typically $t_{.025}$ is approximately 2. Thus, the researcher wants $2 \cdot \text{SE}$ to be no more than .4. This means that the SE should be no more than .2 cm.

In comparative experiments, the primary consideration is usually the size of anticipated treatment effects. For instance, if we are planning to compare two experimental groups, the anticipated SE for each experimental group should be substantially smaller than (preferably less than one-fourth of) the anticipated difference between the two group means.* Thus, the soybean researcher of Example 6.11 might arrive at the value .2 cm if she were planning to compare two environmental conditions that she expected to produce stem lengths differing (on the average) by about .8 cm. She would then plan to grow 38 plants in each of the two environmental conditions.

To see how the required n depends on the specified precision, suppose the soybean researcher specified the desired SE to be .1 cm rather than .2 cm. Then the relation would be

$$\text{SE} = \frac{1.22}{\sqrt{n}} \le .1$$

which yields $n = 148.84$, so that she would plan to include 149 plants in each group. Thus, to double the precision (by cutting the SE in half) requires not twice as many, but four times as many observations. This phenomenon of diminishing returns is due to the square root in the SE formula.

---

* This is a rough guideline for obtaining adequate sensitivity to discriminate between treatments. Such sensitivity, technically called *power*, is discussed in Chapter 7.

Exercises 6.27–6.30

6.27 An experiment is being planned to compare the effects of several diets on the weight gain of beef cattle, measured over a 140-day test period.[20] In order to have enough precision to compare the diets, it is desired that the standard error of the mean for each diet should not exceed 5 kg.

(a) If the population standard deviation of weight gain is guessed to be about 20 kg on any of the diets, how many cattle should be put on each diet in order to achieve a sufficiently small standard error?

(b) If the guess of the standard deviation is doubled, to 40 kg, does the required number of cattle double? Explain.

6.28 A medical researcher proposes to estimate the mean serum cholesterol level of a certain population of middle-aged men, based on a random sample of the population. He asks a statistician for advice. The ensuing discussion reveals that the researcher wants to estimate the population mean to within ±6 mg/dLi or less, with 95% confidence. Thus, the standard error of the mean should be 3 mg/dLi or less. Also, the researcher believes that the standard deviation of serum cholesterol in the population is probably about 40 mg/dLi.[21] How large a sample does the researcher need to take?

6.29 Suppose you are planning an experiment to test the effects of various diets on the weight gain of young turkeys. The observed variable will be Y = weight gain in three weeks (measured over a period starting one week after birth and ending three weeks later). Previous experiments suggest that the standard deviation of Y under a standard diet is approximately 80 g.[22] Using this as a guess of σ, determine how many turkeys you should have in a treatment group, if you want the standard error of the group mean to be no more than

(a) 20 g
(b) 15 g

6.30 A researcher is planning to compare the effects of two different types of lights on the growth of bean plants. She expects that the means of the two groups will differ by about 1 inch and that in each group the standard deviation of plant growth will be around 1.5 inches. Consider the guideline that the anticipated SE for each experimental group should no more than be one-fourth of the anticipated difference between the two group means. How large should the sample be (for each group) in order to meet this guideline?

6.5 CONDITIONS FOR VALIDITY OF ESTIMATION METHODS

For any sample of quantitative data, we can use the methods of this chapter to compute the mean, its standard error, and various confidence intervals; indeed, computers can make this rather easy to carry out. However, the *interpretations* that we have given for these descriptions of the data are valid only under certain conditions.

Conditions for Validity of the SE Formula

First, the very notion of regarding the sample mean as an estimate of a population mean requires that the data be viewed as if they had been generated by random sampling from some population. To the extent that this is not possible, any inference beyond the actual data is questionable. The following example illustrates the difficulty.

Marijuana and Intelligence. Ten people who used marijuana heavily were found to be quite intelligent; their mean IQ was 128.4, whereas the mean IQ for the general population is known to be 100. The ten people belonged to a religious group that uses marijuana for ritual purposes; since their decision to join the group

Example 6.12

might very well be related to their intelligence, it is not clear that the ten can be regarded (with respect to IQ) as a random sample from any particular population, and therefore there is no apparent basis for thinking of the sample mean (128.4) as an estimate of the mean IQ of a particular population (such as, for instance, all heavy marijuana users). An inference about the *effect* of marijuana on IQ would be even more implausible, especially because data were not available on the IQs of the ten people *before* they began marijuana use.[23]   ■

Second, the use of the standard error formula $SE = s/\sqrt{n}$ requires two further conditions:

1. The population size must be large compared with the sample size. This requirement is rarely a problem in the life sciences; the sample can be as much as 5% of the population without seriously invalidating the SE formula.*

2. The observations must be independent of each other. This requirement means that the n observations actually give n independent pieces of information about the population.

Data often fail to meet the independence requirement if the experiment has a **hierarchical structure**, in which observational units are nested within sampling units, as illustrated by the following example.

Example 6.13

Canine Anatomy. The coccygeus muscle is a bilateral muscle in the pelvic region of the dog. As part of an anatomical study, the left side and the right side of the coccygeus muscle were weighed for each of 21 female dogs. There were thus $2 \cdot 21 = 42$ observations, but only 21 units chosen from the population of interest (female dogs). Because of the symmetry of the coccygeus, the information contained in the right and left sides is largely redundant, so that the data contain not 42, but only 21, independent pieces of information about the coccygeus muscle of female dogs. It would therefore be incorrect to apply the SE formula as if the data comprised a sample of size $n = 42$. The hierarchical nature of the data set is indicated in Figure 6.13.[24]   ■

Figure 6.13 Hierarchical data structure of Example 6.13

Hierarchical data structures are rather common in the life sciences. For instance, observations may be made on 90 nerve cells that come from only three different cats; on 80 kernels of corn that come from only four ears; on 60 young mice who come from only 10 litters. A particularly clear example of nonindependent observations is replicated measurements on the same individual; for instance, if a physician makes triplicate blood pressure measurements on each of 10 patients,

---

* If the sample size, n, is a substantial fraction of the population size, N, then the finite population correction factor should be applied. This factor is $\sqrt{\dfrac{N-n}{N-1}}$. The standard error of the mean then becomes $\dfrac{s}{\sqrt{n}} \cdot \sqrt{\dfrac{N-n}{N-1}}$.

she clearly does not have 30 independent observations. In some situations a correct treatment of hierarchical data is obvious; for instance, the triplicate blood pressure measurements could be averaged to give a single value for each patient. In other situations, however, lack of independence can be more subtle. For instance, suppose 60 young mice from 10 litters are included in an experiment to compare two diets. Then the choice of a correct analysis depends on the *design* of the experiment—on such aspects as whether the diets are fed to the young mice themselves or to the mothers, and how the animals are allocated to the two diets. The subject of design of experiments will be discussed in detail in Chapter 8.

Conditions for Validity of a Confidence Interval for μ

A confidence interval for μ provides a definite quantitative interpretation for $\text{SE}_{\bar{y}}$. Note that the data must be a random sample from the population of interest. If there is bias in the sampling process, then the sampling distribution concepts on which the confidence interval method is based do not hold: Knowing the average of a biased sample does not provide information about the population mean μ. The validity of Student's t method for constructing confidence intervals also depends on the form of the population distribution of the observed variable Y. If Y follows a normal distribution in the population, then Student's t method is exactly **valid**—that is to say, the probability that the confidence interval will contain μ is actually equal to the confidence level (for example, 95%). By the same token, this interpretation is approximately valid if the population distribution is approximately normal. Even if the population distribution is not normal, the Student's t confidence interval is approximately valid *if* the sample size is large. This fact can often be used to justify the use of the confidence interval even in situations where the population distribution cannot be assumed to be approximately normal.

From a practical point of view, the important question is, How large must the sample be in order for the confidence interval to be approximately valid? Not surprisingly, the answer to this question depends on the *degree* of nonnormality of the population distribution: If the population is only moderately nonnormal, then n need not be very large. Table 6.4 shows the actual probability that a Student's t

TABLE 6.4 Actual Probability that Confidence Intervals Will Contain the Population Mean

(a) 95% Confidence Interval

	2	4	8	16	32	64	Very Large
				SAMPLE SIZE			
Population 1	.95	.95	.95	.95	.95	.95	.95
Population 2	.94	.93	.94	.94	.95	.95	.95
Population 3	.87	.53	.57	.80	.88	.92	.95

(b) 99% Confidence Interval

	2	4	8	16	32	64	Very Large
				SAMPLE SIZE			
Population 1	.99	.99	.99	.99	.99	.99	.99
Population 2	.99	.98	.98	.98	.99	.99	.99
Population 3	.97	.82	.60	.81	.93	.96	.99

confidence interval will contain μ, for samples from three different populations.[25] The forms of the population distributions are shown in Figure 6.14. Population 1 is a normal population, population 2 is moderately skewed, and population 3 is a violently skewed, L-shaped distribution. (Populations 2 and 3 were discussed in optional Section 5.4.)

For population 1, Table 6.4 shows that the confidence interval method is exactly valid for all sample sizes, even $n = 2$. For population 2, the method is approximately valid even for fairly small samples. For population 3, the approximation is very poor for small samples and is only fair for samples as large as $n = 64$. In a sense, population 3 is a worst case; it could be argued that the mean is not a meaningful measure for population 3, because of its bizarre shape.

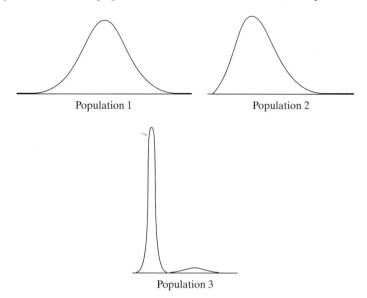

Figure 6.14 The three populations of Example 6.13

Summary of Conditions

In summary, Student's t method of constructing a confidence interval for μ is appropriate if the conditions stated in the box hold.

---

1. Conditions on the design of the study

(a) It must be reasonable to regard the data as a random sample from a large population.

(b) The observations in the sample must be independent of each other.

2. Conditions on the form of the population distribution

(a) If n is small, the population distribution must be approximately normal.

(b) If n is large, the population distribution need not be approximately normal.

The requirement that the data are a random sample is the most important condition.

The required "largeness" in condition 2(b) depends (as shown in Example 5.14) on the degree of nonnormality of the population. In many practical situations, moderate sample sizes (say, $n = 20$ to 30) are large enough.

Verification of Conditions

In practice, the preceding conditions are often assumptions rather than known facts. However, it is always important to check whether the conditions are reasonable in a given case.

To determine whether the random sampling model is applicable to a particular study, the design of the study should be scrutinized, with particular attention to possible biases in the choice of experimental material and to possible nonindependence of the observations due to hierarchical data structures.

As to whether the population distribution is approximately normal, information on this point may be available from previous experience with similar data. If the only source of information is the data at hand, then normality can be roughly checked by making a histogram (or stem-and-leaf display) and normal probability plot of the data. Unfortunately, for small or moderate sample size, this check is fairly crude; for instance, if you look back at Figure 5.12, you will see that even samples of size 25 from a normal population often do not appear particularly normal. Of course, if the sample is large, then the sample histogram gives us good information about the population shape; however, if n is large, the requirement of normality is less important anyway.

In any case, a crude check is better than none, and *every* data analysis should begin with inspection of a graph of the data, with special attention to any observations that lie very far from the center of the distribution.

Sometimes a histogram or normal probability plot of the data indicates that the data did not come from a normal population. If the sample size is small, then Student's t method will not give valid results. However, it may be possible to transform the data to achieve approximate normality and then analyze the data in the transformed scale.

Sediment Yield. Sediment yield, which is a measure of the amount of suspended sediment in water, is a measure of water quality for a river. The distribution of sediment yield often has a skewed distribution. However, taking the logarithm of each observation can produce a distribution that follows a normal curve quite well. Figure 6.15 shows normal probability plots of sediment yields of water samples from the Black River in Northern Ohio for $n = 9$ days (a) in mg/Li and (b) in log scale (i.e., log(mg/Li)).[26]

Example 6.14

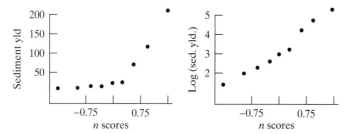

Figure 6.15 Normal probability plots of sediment yields of water samples from the Black River for 9 days (a) in mg/Li and (b) after taking the logarithm of each observation

The logarithms of the sediment yields have an average of $\bar{y} = 3.21$ and a standard deviation of $s = 1.33$. Thus, the standard error of the mean is $\dfrac{1.33}{\sqrt{9}} = .44$. The t multiplier for a 95% confidence interval is $t(8)_{.025} = 2.306$. A 95% confidence interval for μ is

$$3.21 \pm 2.306(.44)$$

or approximately

$$3.21 \pm 1.01$$

or

$$(2.20, 4.22)$$

Thus, *we are 95% confident that the average logarithm of sediment yield for the Black River is between 2.20 and 4.22.*

Note that we have constructed a confidence interval for the population average logarithm of sediment yield. Because the logarithm transformation is not linear, the mean of the logarithms is not the logarithm of the mean, so we cannot convert this confidence interval into a confidence interval for the population mean in the original scale of mg/Li. ∎

Exercises 6.31–6.35

6.31 Serum Glutamic-Oxaloacetic Transamiase (SGOT) is an enzyme that shows elevated activity when the heart muscle is damaged. In a study of 31 patients who underwent heart surgery, serum levels of SGOT were measured 18 hours after surgery.[27] The mean was 49.3 U/Li and the standard deviation was 68.3 U/Li. If we regard the 31 observations as a sample from a population, what feature of the data would cause us to doubt that the population distribution is normal?

6.32 A dendritic tree is a branched structure that emanates from the body of a nerve cell. In a study of brain development, researchers examined brain tissue from seven adult guinea pigs. The investigators randomly selected nerve cells from a certain region of the brain and counted the number of dendritic branch segments emanating from each selected cell. A total of 36 cells were selected, and the resulting counts were as follows:[28]

38	42	25	35	35	33	48	53	17
24	26	26	47	28	24	35	38	26
38	29	49	26	41	26	35	38	44
25	45	28	31	46	32	39	59	53

The mean of these counts is 35.67 and the standard deviation is 9.99.

Suppose we want to construct a 95% confidence interval for the population mean. We could calculate the standard error as

$$SE_{\bar{y}} = \frac{9.99}{\sqrt{36}} = 1.67$$

and obtain the confidence interval as

$$35.67 \pm (2.042)(1.67)$$

or

$$32.3 < \mu < 39.1$$

(a) On what grounds might the preceding analysis be criticized? (*Hint*: Are the observations independent?)

(b) Using the classes 15–19, 20–24, and so on, construct a histogram of the data. Does the shape of the distribution support the criticism you made in part (a)? If so, explain how.

6.33 In an experiment to study the regulation of insulin secretion, blood samples were obtained from seven dogs before and after electrical stimulation of the vagus nerve. The following values show, for each animal, the increase (after minus before) in the immunoreactive insulin concentration (μU/mLi) in pancreatic venous plasma.[29]

30   100   60   30   130   1,060   30

For these data, Student's t method yields the following 95% confidence interval for the population mean:

$$-145 < \mu < 556$$

Is Student's t method appropriate in this case? Why or why not?

6.34 In a study of parasite-host relationships, 242 larvae of the moth *Ephestia* were exposed to parasitization by the Ichneumon fly. The following table shows the number of Ichneumon eggs found in each of the *Ephestia* larva.[30]

Number of Eggs (Y)	Number of Larvae
0	21
1	77
2	52
3	41
4	23
5	13
6	9
7	1
8	2
9	0
10	2
11	0
12	0
13	0
14	0
15	1
Total	242

For these data, $\bar{y} = 2.368$ and $s = 1.950$. Student's t method yields the following 95% confidence interval for μ, the population mean number of eggs per larva:

$$2.12 < \mu < 2.61$$

(a) Does it appear reasonable to assume that the population distribution of Y is approximately normal? Explain.

(b) In view of your answer to part (a), on what grounds can you defend the application of Student's t method to these data?

6.35 The following normal probability plot shows the distribution of the diameters, in cm, of each of nine American sycamore trees.[31]

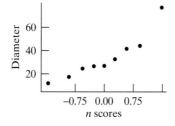

The normal probability plot is not linear, which suggests that a transformation of the data is needed before a confidence interval can be constructed using Student's *t* method. The raw data are

$$12.4 \quad 44.8 \quad 28.2 \quad 77.6 \quad 34 \quad 17.5 \quad 41.5 \quad 25.5 \quad 27.5$$

(a) Take the square root of each observation, and then construct a 90% confidence interval for the mean.
(b) Interpret the confidence interval from part (a). That is, explain what the interval tells you about the square root of the diameters of these trees.

6.6 CONFIDENCE INTERVAL FOR A POPULATION PROPORTION

Up to this point in Chapter 6 we have described confidence intervals when the observed variable is quantitative. Now we will turn our attention to situations in which the variable is *categorical* and the parameter of interest is a population *proportion.* We assume that the data can be regarded as a random sample from some population. The population distribution of a categorical variable can be described in terms of the population proportion, or probability, of each category. In this section we discuss construction of a confidence interval for a population proportion.

Consider a random sample of *n* categorical observations, and let us fix attention on one of the categories. For instance, suppose a geneticist observes *n* guinea pigs whose coat color can be either black, sepia, cream, or albino; let us fix attention on the category "black." Let *p* denote the population proportion of the category, and let $\hat{p}$ denote the corresponding sample proportion. (This is the same notation used in Chapter 3 and in Section 5.2.). The notation is schematically represented in Figure 6.16.

Under the random sampling model, a natural estimate of the population proportion, *p*, is the sample proportion, $\hat{p}$. How close to *p* is $\hat{p}$ likely to be? Recall from Chapter 5 that this question can be answered in terms of the sampling distribution of *p* (which in turn is computed from the binomial distribution).

In Section 6.3 we showed how to use sample data on a quantitative variable to construct a confidence interval for the population mean, μ; the rationale for the method was based on the sampling distribution of $\overline{Y}$. In a similar way, sample data on the relative abundance of a category can be used to construct a confidence interval for the population proportion, *p*.

A confidence interval for *p* can be constructed directly from the binomial distribution. However, for many practical situations a simple approximate method can be used instead. When the sample size, *n*, is large, the sampling distribution of $\hat{p}$ is approximately normal; this approximation is related to the Central Limit Theorem. If you review Figure 5.5, you will see that the sampling distributions resemble normal curves, especially the distribution with *n* = 80. (The approximation is described in detail in optional Section 5.5.) If the sample size is small, then the normal approximation can be quite inadequate. However, there is a method available for constructing approximate confidence intervals that is based on a modification of $\hat{p}$ and that is related to the normal approximation. We present that method here.

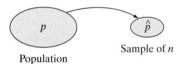

Population   Sample of *n*

Figure 6.16 Notation for population and sample proportion

In Section 6.3 we stated that when the data come from a normal populations, a 95% confidence interval for a population mean μ is constructed as

$$\bar{y} \pm t_{.025}SE_{\bar{y}}$$

A confidence interval for a population proportion p is constructed analogously.

The first step is to calculate an estimate of p from the data. Recall that the sample proportion, $\hat{p}$, is defined as $\hat{p} = \dfrac{y}{n}$, where y is the number of observations, out of n, that fell into the category in question. Related to the sample proportion is the estimate $\tilde{p}$ ("p tilde") given by

$$\tilde{p} = \frac{y + 2}{n + 4}$$

We will use $\tilde{p}$ as the center of a 95% confidence interval for p. (Note that if n is large, then $\hat{p}$ and $\tilde{p}$ are very nearly equal.)

Next, we need to calculate a standard error for $\tilde{p}$.

Standard Error of $\tilde{p}$

The standard error of the estimate is found using the formula given in the box.

> **Standard Error of $\tilde{p}$ (for a 95% Confidence Interval)**
>
> $$SE_{\tilde{p}} = \sqrt{\frac{\tilde{p}(1 - \tilde{p})}{n + 4}}$$

This formula for the standard error of the estimate looks similar to the formula for the standard error of a mean, but with $\sqrt{\tilde{p}(1 - \tilde{p})}$ playing the role of s and with $n + 4$ in place of n.

Iron Deficiency.   As part of the National Health and Nutrition Examination Survey (NHANES), iron levels were checked for a sample of 786 girls aged 12 to 15.[32] Iron deficiency was detected in 71 of those sampled, which is 9% ($71/786 = .09$ or 9%). Thus, $\tilde{p}$ is $\dfrac{71 + 2}{786 + 4} = \dfrac{73}{790} = .092$; the standard error is $\sqrt{\dfrac{.092(1 - .092)}{790}} = .010$ or 1%. A sample value $\tilde{p}$ is typically within ± 2 standard errors of the population proportion p. Based on this standard error, we can expect that the proportion, p, of all girls aged 12 to 15 who have iron deficiency is in the interval $(.07, .11)$ or $(7\%, 11\%)$. A confidence interval for p makes this idea more precise. ■

Example 6.15

95% Confidence Interval for p

Once we have the standard error of $\tilde{p}$, we need to know how likely it is that $\tilde{p}$ will be close to p. The general process of constructing a confidence interval for a proportion is similar to that used in Section 6.3 to construct a confidence interval for a mean. However, when constructing a confidence interval for a mean we multiplied the standard error by a t multiplier. This was based on having a sample from a normal distribution. When dealing with proportion data we know that the population is not normal—there only are two values in the population!—but the Central Limit Theorem tells us that the sampling distribution of $\widetilde{p}$ is approximately normal if the sample size, n, is large. Moreover, it turns out that even for moderate or small samples, intervals based on $\widetilde{p}$ and Z multipliers do a very good job of estimating the population proportion, p.[33]

For a 95% confidence interval, the appropriate Z multiplier is $Z_{.025} = 1.960$. Thus, the approximate 95% confidence interval for a population proportion p is constructed as shown in the box.*

95% Confidence Interval for p

$$95\% \text{ confidence interval: } \widetilde{p} \pm 1.96SE_{\widetilde{p}}$$

Critical values for the confidence interval are obtained from the normal distribution; these can be found most easily from Table 4 with df $= \infty$. (Recall from Section 6.3 that the t distribution with df $= \infty$ is a normal $[Z]$ distribution.) The following example illustrates the confidence interval method.

Example 6.16   **Breast Cancer.**   *BRCA1* is a gene that has been linked to breast cancer. Researchers used DNA analysis to search for *BRCA1* mutations in 169 women with family histories of breast cancer. Of the 169 women tested, 27 (16%) had *BRCA1* mutations.[34] Let p denote the probability that a woman with a family history of breast cancer will have a *BRCA1* mutation. For these data, $\widetilde{p} = \dfrac{29}{173} = .168$. The standard error for $\hat{p}$ is $\sqrt{\dfrac{.168(1 - .168)}{173}} = .028$. Thus, a 95% confidence interval for p is

$$.168 \pm (1.96)(.028)$$

or

$$.168 \pm .055$$

or

$$.113 < p < .223$$

Thus, we are 95% confident that the probability of a BRCA1 mutation in a woman with a family history of breast cancer is between .113 and .223 (i.e., between 11.3% and 22.3%). ∎

---

* Most statistics books present the confidence interval for a proportion as $\hat{p} \pm 1.96\sqrt{\dfrac{\hat{p}(1 - \hat{p})}{n}}$. This commonly used interval is similar to the interval we present, particularly if n is large. For small or moderate sample sizes, the interval we present is more likely to cover the population proportion p. The value $\widetilde{p}$ is sometimes called the Wilson estimate of p, in honor of Edwin B. Wilson, who first proposed its use. A technical discussion of the Wilson estimate is given in Appendix 6.2.

ECMO. Extracorporeal membrane oxygenation (ECMO) is a potentially life-saving procedure that is used to treat newborn babies who suffer from severe respiratory failure. An expirement was conducted in which 11 babies were treated with ECMO; none of the 11 babies died.[35] Let p denote the probability of death for a baby treated with ECMO. For these data, the sample proportion of deaths is $\hat{p} = 0/11 = 0$. However, the fact that none of the babies died should not lead us to believe that the probability of death, p, is precisely zero—only that it is close to zero. The estimate given by $\tilde{p}$ is $2/15 = .133$. The standard error of $\tilde{p}$ is

Example 6.17

$$\sqrt{\frac{.133(.867)}{15}} = .088*$$

Thus, a 95% confidence interval for p is

$$.133 \pm (1.96)(.088)$$

or

$$.133 \pm .172$$

or

$$-.039 < p < .305$$

We know that p cannot be negative, so we state the confidence interval as $(0, .305)$. *Thus, we are 95% confident that the probability of death in a newborn with severe respiratory failure who is treated with ECMO is between 0 and .305 (i.e., between 0% and 30.5%).* ■

Other Confidence Levels

The procedure outlined previously can be used to construct 95% confidence intervals. In order to construct intervals with other confidence coefficients, some modifications to the procedure are needed. The first modification concerns $\tilde{p}$. For a 95% confidence interval we defined $\tilde{p}$ to be $\dfrac{y + 2}{n + 4}$. In general, for a confidence interval of level $100(1 - \alpha)\%$, $\tilde{p}$ is defined as

$$\tilde{p} = \frac{y + .5(Z_{\alpha/2}^2)}{n + Z_{\alpha/2}^2}$$

For a 95% confidence interval $Z_{\alpha/2}$ is 1.96, so that $\tilde{p} = \dfrac{y + .5(1.96^2)}{n + 1.96^2}$. This is equal to $\dfrac{y + 1.92}{n + 3.84}$, which we rounded off as $\dfrac{y + 2}{n + 4}$. However, any confidence level can

---

* Note that if we used the commonly presented method of $\hat{p} \pm 1.96\sqrt{\dfrac{\hat{p}(1 - \hat{p})}{n}}$ we would find that the standard error is zero, leading to a confidence interval of 0 ± 0. Such an interval would not seem to be very useful in practice!

be used. As an example, for a 90% confidence interval, $\widetilde{p} = \dfrac{y + .5(1.645^2)}{n + 1.645^2}$; this is equal to $\dfrac{y + 1.35}{n + 2.7}$.

The second modification concerns the standard error. For a 95% confidence interval we used $\sqrt{\dfrac{\widetilde{p}(1 - \widetilde{p})}{n + 4}}$ as the standard error term. In general, we use $\sqrt{\dfrac{\widetilde{p}(1 - \widetilde{p})}{n + Z_{\alpha/2}^2}}$ as the standard error term.

Finally, the Z multiplier must match the confidence level (1.645 for a 90% confidence interval, etc.). The following example illustrates these modifications.

Example 6.18 | **Left-Handedness.** In a survey of English and Scottish college students, 40 of 400 male students were found to be left-handed. Let us construct a 90% confidence interval for the proportion, p, of left-handed individuals in the population.[36]

The sample estimate of the proportion is

$$\widetilde{p} = \frac{40 + .5(1.645^2)}{400 + 1.645^2} = \frac{40 + 1.35}{400 + 2.7} \approx .103$$

and the SE is

$$\sqrt{\frac{.103(.897)}{402.7}} = .015$$

A 90% confidence interval for p is

$$.103 \pm (1.645)(.015)$$

or

$$.078 < p < .128$$

Thus, we are 90% confident that between 7.8% and 12.8% of the sampled population are left-handed. ∎

Note that the size of the standard error is inversely proportional to $\sqrt{n}$, as illustrtated in the following example.

Example 6.19 | **Left-Handedness.** Suppose, as in Example 6.18, that a sample of n individuals contains approximately 10% left-handers. Then $\widetilde{p} \approx .10$ and

$$SE_{\widetilde{p}} \approx \sqrt{\frac{.10(.90)}{n + 4}}$$

We saw in Example 6.18 that if $n = 400$, then

$$SE_{\widetilde{p}} = .015$$

If $n = 1,600$, then

$$SE_{\widetilde{p}} = .0075$$

Thus, a sample with the same composition (that is, 10% left-handers) but four times as large would yield twice as much precision in the estimation of p. ∎

Planning a Study to Estimate p

In Section 6.4 we discussed a method for choosing the sample size n so that a proposed study would have sufficient precision for its intended purpose. The approach depended on two elements: (1) a specification of the desired $SE_{\bar{y}}$; and (2) a preliminary guess of the SD. In the present context, when the observed variable is categorical, a similar approach can be used. If a desired value of $SE_{\tilde{p}}$ is specified, and if a rough informed guess of $\tilde{p}$ is available, then the required sample size n can be determined from the following equation:

$$\text{Desired SE} = \sqrt{\frac{(\text{Guessed } \tilde{p})(1 - \text{Guessed } \tilde{p})}{n + 4}}$$

The following example illustrates the use of the method.

Left-Handedness. Suppose we regard the left-handedness data of Example 6.18 as a pilot study, and we now wish to plan a new study large enough to estimate p with a standard error of one percentage point; that is, .01. Our guessed value of p from the pilot study is .10, so the required n must satisfy the following relation:

> **Example 6.20**

$$\sqrt{\frac{.10(.90)}{n + 4}} \leq .01$$

This equation is easily solved to give $n + 4 \geq 900$. We should plan to examine 896 students. ∎

Planning in Ignorance. Suppose no preliminary informed guess of p is available. Remarkably, in this situation it is still possible to plan an experiment to achieve a desired value of $SE_{\tilde{p}}$.* Such a "blind" plan depends on the fact that the crucial quantity $\sqrt{\tilde{p}(1 - \tilde{p})}$ is *largest* when $\tilde{p} = .5$; you can see this in the graph of Figure 6.17. It follows that a value of n calculated using "guessed $\tilde{p}$" $= .5$ will be *conservative*—that is, it will certainly be large enough. (Of course, it will be much larger than necessary if $\tilde{p}$ is really very different from .5.) The following example shows how such worst-case planning is used.

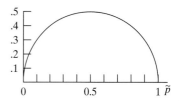

Figure 6.17 How $\sqrt{\tilde{p}(1 - \tilde{p})}$ depends on $\tilde{p}$

Left-Handedness. Suppose, as in Example 6.20, that we are planning a study of left-handedness and that we want $SE_{\tilde{p}}$ to be .01, but suppose that we have no preliminary information whatsoever. We can proceed as in Example 6.20, but using a guessed value of $\tilde{p}$ of .5. Then we have

> **Example 6.21**

$$\sqrt{\frac{.5(.5)}{n + 4}} \leq .01$$

which means that $n + 4 \geq 2,500$, so we need $n = 2,496$. Thus, a sample of 2,496 persons would be adequate to estimate p with a standard error of .01, regardless of the actual value of $\tilde{p}$. (Of course, if $\tilde{p} = .1$, this value of n is much larger than is necessary.) ∎

---

* By contrast, it would not be possible if we were planning a study to estimate a population mean μ and we had no information whatsoever about the value of the SD.

Exercises 6.36–6.50

6.36   A series of patients with bacterial wound infections were treated with the antibiotic Cefotaxime. Bacteriologic response (disappearance of the bacteria from the wound) was considered satisfactory in 84% of the patients.[37] Determine the standard error of the observed proportion of satisfactory responses if the series contained

(a) 50 patients   (b) 200 patients

6.37   In an experiment with a certain mutation in the fruitfly *Drosophila*, n individuals were examined; of these, 20% were found to be mutants. Determine the standard error of the sample proportion of mutants if

(a) $n = 100$ (b) $n = 400$

6.38   Refer to Exercise 6.37. In each case ($n = 100$ and $n = 400$) construct a 95% confidence interval for the population proportion of mutants.

6.39   In a natural population of mice (*Mus musculus*) near Ann Arbor, Michigan, the coats of some individuals are white-spotted on the belly. In a sample of 580 mice from the population, 28 individuals were found to have white-spotted bellies.[38] Construct a 95% confidence interval for the population proportion of this trait.

6.40   To evaluate the policy of routine vaccination of infants for whooping cough, adverse reactions were monitored in 339 infants who received their first injection of vaccine. Reactions were noted in 69 of the infants.[39]

(a) Construct a 95% confidence interval for the probability of an adverse reaction to the vaccine.
(b) Interpret the confidence interval from part (a). What does the interval say about whooping cough vaccinations?

6.41   Researchers tested patients with cardiac pacemakers to see if use of a cellular telephone interferes with the operation of the pacemaker. There were 959 tests conducted for one type of cellular telephone; interference with the pacemaker (detected with electrocardiographic monitoring) was found in 15.7% of these tests.[40]

(a) Use these data to construct an appropriate 90% confidence interval.
(b) The confidence interval from part (a) is a confidence interval for what quantity? Answer in the context of the setting.

6.42   In a study of human blood types in nonhuman primates, a sample of 71 orangutans were tested and 14 were found to be blood type B.[41] Construct a 95% confidence interval for the relative frequency of blood type B in the orangutan population.

6.43   In populations of the snail *Cepaea*, the shells of some individuals have dark bands, while other individuals have unbanded shells.[42] Suppose that a biologist is planning a study to estimate the percentage of banded individuals in a certain natural population, and that she wants to estimate the percentage—which she anticipates will be in the neighborhood of 60%—with a standard error not to exceed 4 percentage points. How many snails should she plan to collect?

6.44   (*Continuation of Exercise 6.43*)   What would the answer be if the anticipated percentage of banded snails were 50% rather than 60%?

6.45   The ability to taste the compound phenylthiocarbamide (PTC) is a genetically controlled trait in humans. In Europe and Asia, about 70% of people are "tasters."[43] Suppose a study is being planned to estimate the relative frequency of tasters in a certain Asian population, and it is desired that the standard error of the estimated relative frequency should be .01. How many people should be included in the study?

6.46   Refer to Exercise 6.45. Suppose a study is being planned for a part of the world for which the percentage of tasters is completely unknown, so that the 70% figure used in Exercise 6.45 is not applicable. What sample size is needed so that the standard error will be no larger than .01?

6.47 Refer to Exercise 6.45. Suppose the SE requirement is relaxed by a factor of 2—from .01 to .02. Would this reduce the required sample size by a factor of 2? Explain.

6.48 A group of 1,438 sexually active patients were counseled on condom use and the risk of contracting a sexually transmitted disease (STD). After six months 103 of the patients had new STDs.[44] Construct a 95% confidence interval for the probability of contracting an STD within six months after being part of a counseling program like the one used in this study.

6.49 The Luso variety of wheat is resistant to the Hessian fly. In order to understand the genetic mechanism controlling this resistance, an agronomist plans to examine the progeny of a certain cross involving Luso and a nonresistant variety. Each progeny plant will be classified as resistant or susceptible and the agronomist will estimate the proportion of progeny that are resistant.[45] How many progeny does he need to classify in order to guarantee that the standard error of his estimate of this proportion will not exceed .05?

6.50 *(Continuation of Exercise 6.49)* Suppose the agronomist is considering two possible genetic mechanisms for the inheritance of resistance; the population ratio of resistant to susceptible progeny would be $1:1$ under one mechanism and $3:1$ under the other. If the agronomist uses the sample size determined in Exercise 6.49, can he be sure that a 95% confidence interval will exclude at least one of the mechanisms? That is, can he be sure that the confidence interval will *not* contain both .50 and .75? Explain.

6.7 PERSPECTIVE AND SUMMARY

In this section we place Chapter 6 in perspective by relating it to other chapters and also to other methods for analyzing a single sample of data. We also present a condensed summary of the methods of Chapter 6.

Sampling Distributions and Data Analysis

The theory of the sampling distribution of $\overline{Y}$ seemed to require knowledge of quantities—μ and σ—that in practice are unknown. In Chapter 6, however, we have seen how to make an inference about μ, including an assessment of the precision of that inference, using only information provided by the sample. Likewise, the sampling distribution of $\widetilde{p}$ depends on the unknown population proportion p. However, we have seen how to use $\widetilde{p}$ to assess the precision of an inference concerning p. Thus, the theory of sampling distributions has led to a practical method of analyzing data.

In later chapters we will study more complex methods of data analysis. Each method is derived from an appropriate sampling distribution; in most cases, however, we will not study the sampling distribution in detail.

Choice of Confidence Level

In illustrating the confidence interval methods, we have often chosen a confidence level equal to 95%. However, it should be remembered that the confidence level is arbitrary. It is true that in practice the 95% level is the confidence level that is most widely used; however, there is nothing wrong with an 80% confidence interval, for example.

Characteristics of Other Measures

This chapter has primarily discussed estimation of a population average—μ for continuous distributions and p for dichotomous distributions. In some situations, we may wish to estimate other parameters of a population. For example, in evaluating a measurement technique, interest may focus on the repeatability of the technique, as indicated by the standard deviation of repeated determinations. As another example, in defining the limits of health, a medical researcher might want to estimate the 95th percentile of serum cholesterol levels in a certain population. Just as the precision of the mean can be indicated by a standard error or a confidence interval, statistical techniques are also available to specify the precision of estimation of parameters such as the population standard deviation or 95th percentile.

Summary of Estimation Methods

For convenient reference, we summarize in the box the confidence interval methods presented in this chapter.

Standard error of the mean

$$\text{SE}_{\bar{y}} = \frac{s}{\sqrt{n}}$$

Confidence interval for μ

$$\text{95\% confidence interval: } \bar{y} \pm t_{.025}\text{SE}_{\bar{y}}$$

Critical value $t_{.025}$ from Student's t distribution with df $= n - 1$.
Intervals with other confidence levels (such as 90%, 99%, etc.) are constructed analogously (using $t_{.05}, t_{.005}$, etc.).
The confidence interval formula is valid if (1) the data can be regarded as a random sample from a large population, (2) the observations are independent, and (3) the population is normal. If n is large, then condition (3) is less important.

95% Confidence interval for p

$$\tilde{p} \pm 1.96\text{SE}_{\tilde{p}}$$

where $\tilde{p} = \dfrac{Y + 2}{n + 4}$ and

$$\text{SE}_{\tilde{p}} = \sqrt{\frac{\tilde{p}(1 - \tilde{p})}{n + 4}}$$

General confidence interval for p

$$\tilde{p} \pm Z_{\alpha/2}\text{SE}_{\tilde{p}}$$

where $\tilde{p} = \dfrac{Y + .5(Z_{\alpha/2}^2)}{n + Z_{\alpha/2}^2}$

$$\text{SE}_{\tilde{p}} = \sqrt{\frac{\tilde{p}(1 - \tilde{p})}{n + Z_{\alpha/2}^2}}$$

The confidence interval formulas are valid if (1) the data can be regarded as a random sample from a large population and (2) the observations are independent.

Supplementary Exercises 6.51–6.71

6.51 To study the conversion of nitrite to nitrate in the blood, researchers injected four rabbits with a solution of radioactively labeled nitrite molecules. Ten minutes after injection, they measured for each rabbit the percentage of the nitrite that had been converted to nitrate. The results were as follows:[46]

51.1   55.4   48.0   49.5

(a) For these data, calculate the mean, the standard deviation, and the standard error of the mean.
(b) Construct a 95% confidence interval for the population mean percentage.
(c) Without doing any calculations, would a 99% confidence interval be wider, narrower, or the same width as the confidence interval you found in part (b)? Why?

6.52 The diameter of the stem of a wheat plant is an important trait because of its relationship to breakage of the stem, which interferes with harvesting the crop. An agronomist measured stem diameter in eight plants of the Tetrastichon cultivar of soft red winter wheat. All observations were made three weeks after flowering of the plant. The stem diameters (mm) were as follows:[47]

2.3   2.6   2.4   2.2   2.3   2.5   1.9   2.0

The mean of these data is 2.275 and the standard deviation is .238.

(a) Calculate the standard error of the mean.
(b) Construct a 95% confidence interval for the population mean.
(c) Define in words the population mean that you estimated in part (b). (See Example 6.1.)

6.53 Refer to Exercise 6.52.

(a) What conditions are needed for the confidence interval to be valid?
(b) Are these conditions met? How do you know?
(c) Which of these conditions is most important?

6.54 Refer to Exercise 6.52. Suppose that the data on the eight plants are regarded as a pilot study, and that the agronomist now wishes to design a new study for which he wants the standard error of the mean to be only .03 mm. How many plants should be measured in the new study?

6.55 A sample of 20 fruitfly (*Drosophila melanogaster*) larva were incubated at 37°C for 30 minutes. It is theorized that such exposure to heat causes polytene chromosomes located in the salivary glands of the fly to unwind, creating puffs on the chromosome arm that are visible under a microscope. The following normal probability plot supports the use of a normal curve to model the distribution of puffs.[48]

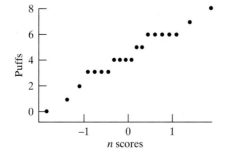

The average number of puffs for the 20 observations was 4.30, with a standard deviation of 2.03.

(a) Construct a 95% confidence interval for μ.
(b) In the context of this problem, describe what μ represents. That is, the confidence interval from part (a) is a confidence interval for what quantity?

6.56 Over a period of about nine months, 1,353 women reported the timing of each of their menstrual cycles. For the first cycle reported by each woman, the mean cycle time was 28.86 days, and the standard deviation of the 1,353 times was 4.24 days.[49]

(a) Construct a 99% confidence interval for the population mean cycle time.
(b) Because environmental rhythms can influence biological rhythms, we might hypothesize that the population mean menstrual cycle time is 29.5 days, the length of the lunar month. Is the confidence interval of part (a) consistent with this hypothesis?

6.57 Refer to the menstrual cycle data of Exercise 6.56.

(a) Over the entire time period of the study, the women reported a total of 12,247 cycles. When all of these cycles are included, the mean cycle time is 28.22 days. Explain why we would expect that this mean would be smaller than the value 28.86 given in Exercise 6.50. (*Hint*: If each woman reported for a fixed time period, which women contributed more cycles to the total of 12,247 observations?)
(b) Instead of using only the first reported cycle as in Exercise 6.56, we could use the first four cycles for each woman, thus obtaining $1,353 \cdot 4 = 5,412$ observations. We could then calculate the mean and standard deviation of the 5,412 observations and divide the SD by $\sqrt{5412}$ to obtain the SE; this would yield a much smaller value than the SE found in Exercise 6.51. Why would this approach not be valid?

6.58 For the 28 lamb birthweights of Example 6.3, the mean is 5.1679 kg, the SD is .6544 kg, and the SE is .1237 kg. Construct

(a) a 95% confidence interval for the population mean
(b) a 99% confidence interval for the population mean
(c) Interpret the confidence interval you found in part (a). That is, explain what the numbers in the interval mean. (*Hint*: See Examples 6.9 and 6.10.)

6.59 Refer to Exercise 6.58.

(a) What conditions are required for the validity of the confidence intervals?
(b) Which of the conditions of part (a) can be checked (roughly) from the histogram of Figure 6.2?
(c) Twin births were excluded from the lamb birthweight data. If twin births had been included, would the confidence intervals be valid? Why or why not?

6.60 Researchers measured the number of tree species in each of 69 vegetational plots in the Lama Forest of Benin, West Africa.[50] The number of species ranged from a low of 1 to a high of 12. The sample mean was 6.8 and the sample SD was 2.4, which results in a 95% confidence interval of (6.2, 7.4). However, the number of tree species in a plot takes on only integer values. Does this mean that the confidence interval should be (7, 7)? Or does it mean that we should round off the endpoints of the confidence interval and report it as (6, 7)? Or should the confidence interval really be (6.2, 7.4)? Explain.

6.61 As part of a study of natural variation in blood chemistry, serum potassium concentrations were measured in 84 healthy women. The mean concentration was 4.36 mEq/Li, and the standard deviation was .42 mEq/Li. The table presents a frequency distribution of the data.[51]

Serum Potassium (mEq/Li)	Number of Women
3.1–3.3	1
3.4–3.6	2
3.7–3.9	7
4.0–4.2	22
4.3–4.5	28
4.6–4.8	16
4.9–5.1	4
5.2–5.4	3
5.5–5.7	1
Total	84

(a) Calculate the standard error of the mean.

(b) Construct a histogram of the data and indicate the intervals $\bar{y} \pm SD$ and $\bar{y} \pm SE$ on the histogram. (See Figure 6.2.)

(c) Construct a 95% confidence interval for the population mean.

(d) Interpret the confidence interval you found in part (c). That is, explain what the numbers in the interval mean. (*Hint*: See Examples 6.9 and 6.10.)

6.62 Refer to Exercise 6.61. In medical diagnosis, physicians often use reference limits for judging blood chemistry values; these are the limits within which we would expect to find 95% of healthy people. Would a 95% confidence interval for the mean be a reasonable choice of reference limits for serum potassium in women? Why or why not?

6.63 Refer to Exercise 6.61. Suppose a similar study is to be conducted next year, to include serum potassium measurements on 200 healthy women. Based on the data in Exercise 6.60, what would you predict would be

(a) the SD of the new measurements?

(b) the SE of the new measurements?

6.64 An agronomist selected six wheat plants at random from a plot, and then, for each plant, selected 12 seeds from the main portion of the wheat head; by weighing, drying, and reweighing, she determined the percent moisture in each batch of seeds. The results were as follows:[52]

$$62.7 \quad 63.6 \quad 60.9 \quad 63.0 \quad 62.7 \quad 63.7$$

(a) Calculate the mean, the standard deviation, and the standard error of the mean.

(b) Construct a 90% confidence interval for the population mean.

6.65 In a study of environmental effects upon reproduction, 123 adult white-tailed deer from the central Adirondack area were captured and 97 were found to be pregnant.[53] Construct a 95% confidence interval for the proportion of females pregnant in this deer population.

6.66 Refer to Exercise 6.65. Which of the conditions for validity of the confidence interval might have been violated in this study?

6.67 Gene mutations have been found in patients with muscular dystrophy. In one study, it was found that there were defects in the gene coding of sarcoglycan proteins in 23 of 180 patients with limb-girdle muscular dystrophy.[54]

(a) Use these data to construct an appropriate 90% confidence interval.

(b) What conditions are necessary for the confidence interval from part (a) to be valid?

(c) Interpret your confidence interval from part (a) in the context of this setting. That is, what do the numbers in the confidence interval mean?

6.68   As part of the National Health and Nutrition Examination Survey (NHANES), hemoglobin levels were checked for a sample of 1139 men age 70 and over.[55] The sample mean was 145.3 g/Li and the standard deviation was 12.87 g/Li.

(a) Use these data to construct a 95% confidence interval for μ.
(b) Does the confidence interval from part (a) give limits in which we expect 95% of the sample data to lie? Why or why not?
(c) Does the confidence interval from part (a) give limits in which we expect 95% of the population to lie? Why or why not?

6.69   At a certain university there are 25,000 students. Suppose you want to estimate the proportion of those students who are nearsighted. The prevalence of near-sightedness in the general population is 45%.[56] Using this as a preliminary guess of p, how many students would need to be included in a random sample if you want the standard error of your estimate to be less than or equal to 2 percentage points?

6.70   Refer to Exercise 6.69. Suppose you do not trust that the 45% nearsightedness rate for the general population is a useful guess for the university population. How many students would need to be included in a random sample if you want the standard error of your estimate to be less than or equal to 2 percentage points, no matter what the value of p is?

6.71   The blood pressure (average of systolic and diastolic measurements) of each of 38 persons was measured.[57] The average was 94.5 (mm Hg). A histogram of the data is shown.

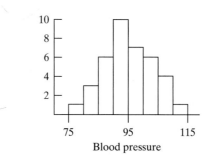

Which of the following is an approximate 95% confidence interval for the population mean blood pressure? Explain.

(a) 94.5 ± 16
(b) 94.5 ± 8
(c) 94.5 ± 2.6
(d) 94.5 ± 1.3

Comparison of Two Independent Samples

7.1 INTRODUCTION

In Chapter 6 we considered the analysis of a single sample of quantitative data. In practice, however, much scientific research involves the comparison of two or more samples from different populations. In the present chapter we introduce methods for comparing two samples.

Two-sample comparisons can arise in a variety of ways. Here are two examples.

Hematocrit in Males and Females.   Hematocrit level is a measure of the concentration of red cells in blood. Figure 7.1 shows the relative frequency distributions of hematocrit values for two samples of 17-year-old American youths—489 males and 469 females.[1] The sample means and standard deviations are given in Table 7.1.

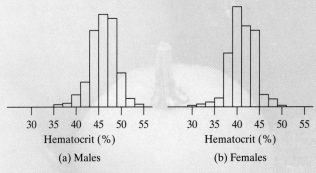

Figure 7.1 Hematocrit values in 17-year-old youths.
(a) 489 males, (b) 469 females

Example 7.1

TABLE 7.1 Hematocrit (Percent)		
	Males	**Females**
Mean	45.8	40.6
SD	2.8	2.9

The following features can be seen from Figure 7.1 and Table 7.1. First, the males tend to have higher levels than the females. (Nevertheless, the two distributions do overlap quite a bit, so that many females have higher levels than many males.) Second, in spite of the substantial difference in means, the two distributions have very nearly the same standard deviation and are quite similar in shape. Thus, the main difference between the two distributions is a *shift* along the Y-axis. ■

Example 7.2

Pargyline and Sucrose Consumption. A study was conducted to determine the effect of the psychoactive drug Pargyline on feeding behavior in the black blowfly *Phormia regina*. The response variable was the amount of sucrose (sugar) solution a fly would drink in 30 minutes. The experimenters used two separate groups of flies: a group injected with Pargyline (905 flies) and a control group injected with saline (900 flies). Comparing the responses of the two groups provides an indirect assessment of the effect of Pargyline. (One might propose that a more *direct* way to determine the effect of the drug would be to measure each fly twice—on one occasion after injecting Pargyline and on another occasion after injecting saline. However, this direct method is not practical because the measurement procedure disturbs the fly so much that each fly can be measured only once.) Figure 7.2 shows the data for the two groups, and Table 7.2 shows the means and standard deviations.[2]

Figure 7.2 Sucrose consumption of flies. (a) 900 control flies; (b) 905 Pargyline-treated flies.

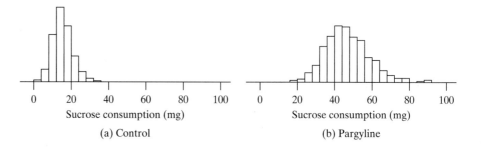

(a) Control

(b) Pargyline

TABLE 7.2 Sucrose Consumption (mg)		
	Control	**Pargyline**
Mean	14.9	46.5
SD	5.4	11.7

It is clear from Figure 7.2 that the two distributions differ in two distinct ways: First, the Pargyline distribution is shifted to the right, and second, the Pargyline distribution is more dispersed. This impression is confirmed by Table 7.2, which shows that the Pargyline distribution has both a larger mean and a larger standard deviation than the control distribution. ■

Examples 7.1 and 7.2 both involve two-sample comparisons. But notice that the two studies differ in a fundamental way. In Example 7.1 the samples come from populations that occur naturally; the investigator is merely an observer:

Population 1: Hematocrit values of 17-year-old males living in the United States

Population 2: Hematocrit values of 17-year old females living in the United States

By contrast, the two populations in Example 7.2 do not actually exist, but rather are defined in terms of specific experimental conditions; in a sense, the populations are created by experimental intervention:

Population 1: Sucrose consumptions of blowflies when injected with saline

Population 2: Sucrose consumptions of blowflies when injected with Pargyline

These two types of two-sample comparisons—the observational and the experimental—are both widely used in research. The formal methods of analysis are often the same for the two types, but the interpretation of the results may be somewhat different. For instance, in Example 7.2 it might be reasonable to say that Pargyline *causes* the increase in sucrose consumption, while no such notion applies in Example 7.1. (We will discuss these two study designs further in Chapter 8.)

When the observed variable is quantitative, the comparison of two samples can include—as we saw in Examples 7.1 and 7.2—several aspects, notably (1) comparison of means, (2) comparison of standard deviations, and (3) comparison of shapes. In this chapter, and indeed throughout this book, the primary emphasis will be on comparison of means and on other comparisons related to shift.

Notation

Figure 7.3 presents our notation for comparison of two samples. The notation is exactly parallel to that of Chapter 6, but now a subscript (1 or 2) is used to differentiate between the two samples. The two "populations" can be naturally occurring populations (as in Example 7.1) or they can be conceptual populations defined by certain experimental conditions (as in Example 7.2). In either case, the data in each sample are viewed as a random sample from the corresponding population.

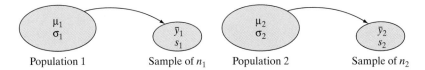

| Population 1 | Sample of n_1 | Population 2 | Sample of n_2 |

Figure 7.3 Notation for comparison of two samples

A Look Ahead

In this chapter we will discuss two different but complementary approaches to the comparison of two means:

1. The confidence interval approach

2. The hypothesis testing approach

The confidence interval approach (Section 7.3) is a natural extension of the technique of Chapter 6. The hypothesis testing approach involves new concepts. We will first (Sections 7.4–7.9) introduce the basic ideas of hypothesis testing in the context of comparing two means. We will then (Section 7.10) discuss these ideas in more generality, and (Section 7.11) consider another hypothesis testing procedure for comparing two samples.

We begin by describing, in the next section, some simple computations that are used both for confidence intervals and for hypothesis testing.

7.2 STANDARD ERROR OF $(\bar{y}_1 - \bar{y}_2)$

In this section we introduce a fundamental quantity for comparing two samples: the standard error of the difference between two sample means.

Basic Ideas

We saw in Chapter 6 that the precision of a sample mean $\bar{y}$ can be expressed by its standard error, which is equal to

$$SE_{\bar{y}} = \frac{s}{\sqrt{n}}$$

To compare two sample means, it is natural to consider the difference between them:

$$\bar{y}_1 - \bar{y}_2$$

which is an estimate of the quantity $(\mu_1 - \mu_2)$. To characterize the sampling error of estimation, we need to be concerned with the standard error of the difference $(\bar{y}_1 - \bar{y}_2)$. We illustrate this idea with an example.

Example 7.3

Vital Capacity. Vital capacity is a measure of the amount of air that someone can exhale after taking a deep breath. One might expect that musicians who play brass instruments would have greater vital capacities, on average, than would other persons of the same age, sex, and height. In one study the vital capacities of eight brass players were compared to the vital capacities of seven control subjects; Table 7.3 shows the data.[3]

TABLE 7.3 Vital Capacity (liters)	
Brass Player	**Control**
4.7	4.2
4.6	4.7
4.3	5.1
4.5	4.7
5.5	5.0
4.9	
5.3	
n 7	5
$\bar{y}$ 4.83	4.74
s .435	.351

The difference between the sample means is

$$\bar{y}_1 - \bar{y}_2 = 4.83 - 4.74 = 0.09$$

We know that both $\bar{y}_1$ and $\bar{y}_2$ are subject to sampling error, and consequently the difference (0.09) is subject to sampling error. The standard error of $\bar{y}_1 - \bar{y}_2$ tells us how much precision to attach to this difference between $\bar{y}_1$ and $\bar{y}_2$. ∎

> **Definition**
> The **standard error of $\bar{y}_1 - \bar{y}_2$** is defined as
> $$SE_{(\bar{y}_1 - \bar{y}_2)} = \sqrt{\frac{s_1^2}{n_1} + \frac{s_2^2}{n_2}}$$

The following alternative form of the formula shows how the SE of the difference is related to the individual SEs of the means:

$$SE_{(\bar{y}_1 - \bar{y}_2)} = \sqrt{SE_1^2 + SE_2^2}$$

where

$$SE_1 = SE_{\bar{y}_1} = \frac{s_1}{\sqrt{n_1}}$$

$$SE_2 = SE_{\bar{y}_2} = \frac{s_2}{\sqrt{n_2}}$$

Notice that this version of the formula shows that "SEs add like Pythagorus." When we have two independent samples, we take the SE of each mean, square them, add them, and then take the square root of the sum. Figure 7.4 illustrates this idea.

It may seem odd that in calculating the SE of a difference we *add* rather than subtract within the formula $SE_{(\bar{y}_1 - \bar{y}_2)} = \sqrt{SE_1^2 + SE_2^2}$. However, as was discussed in Section 3.7, the variability of the difference depends on the variability of each part. Whether we add $\bar{y}_2$ to $\bar{y}_1$ or subtract $\bar{y}_2$ from $\bar{y}_1$, the "noise" associated with $\bar{y}_2$ (i.e., SE_2) adds to the overall uncertainty. The greater the variability in $\bar{y}_2$, the greater the variability in $\bar{y}_1 - \bar{y}_2$. The formula $SE_{(\bar{y}_1 - \bar{y}_2)} = \sqrt{SE_1^2 + SE_2^2}$ accounts for this variability.

We illustrate the formulas in the following example.

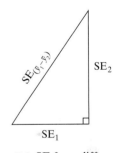

Figure 7.4 SE for a difference

Vital Capacity. For the vital capacity data, preliminary computations yield the results in Table 7.4.

Example 7.4

> **TABLE 7.4**
>
> | | Brass Player | Control |
> |-------|:------------:|:-------:|
> | s^2 | .1892 | .1232 |
> | n | 7 | 5 |
> | SE | .164 | .157 |

The SE of $(\bar{y}_1 - \bar{y}_2)$ is

$$SE_{(\bar{y}_1 - \bar{y}_2)} = \sqrt{\frac{.1892}{7} + \frac{.1232}{5}} = .227 \approx .23$$

Note that

$$.227 = \sqrt{(.164)^2 + (.157)^2}$$

Notice that the SE of the difference is greater than either of the individual SEs but less than their sum. ∎

Example 7.5

Hematocrit Levels.   The data in Table 7.1 showed that the standard deviation of hematocrit levels in 489 males was 2.8. Thus, the SE for the male mean is $2.8/\sqrt{489} = .1266$. For 469 females the SD was 2.9, which gives an SE of $2.9/\sqrt{469} = .1339$. The SE for the difference in the two means is $\sqrt{.1266^2 + .1339^2} = .1843 \approx .18$. ∎

The Pooled Standard Error (Optional)

The standard error just presented is known as the "unpooled" standard error. Many statistics software packages allow the user to specify use of what is known as the "pooled" standard error, which we will discuss briefly.

Recall that the square of the standard deviation, s, is the sample variance, s^2, defined as

$$s^2 = \frac{\Sigma(y_i - \bar{y})^2}{n - 1}$$

The pooled variance is a weighted average of s_1^2, the variance of the first sample, and s_2^2, the variance of the second sample, with weights equal to the degrees of freedom from each sample, $n_i - 1$:

$$s_{pooled}^2 = \frac{(n_1 - 1)s_1^2 + (n_2 - 1)s_2^2}{(n_1 - 1) + (n_2 - 1)} = \frac{(n_1 - 1)s_1^2 + (n_2 - 1)s_2^2}{(n_1 + n_2 - 2)}$$

The pooled standard error is defined as

$$SE_{pooled} = \sqrt{s_{pooled}^2 \left(\frac{1}{n_1} + \frac{1}{n_2}\right)}$$

We illustrate with an example.

Example 7.6

Vital Capacity.   For the vital capacity data we found that $s_1^2 = .1892$ and $s_2^2 = .1232$. The pooled variance is

$$s_{pooled}^2 = \frac{(7 - 1).1892 + (5 - 1).1232}{(7 + 5 - 2)} = .1628$$

and the pooled SE is

$$SE_{pooled} = \sqrt{.1628\left(\frac{1}{7} + \frac{1}{5}\right)} = .236$$

Recall from Example 7.4 that the unpooled SE for the same data was .227. ∎

If the sample sizes are equal ($n_1 = n_2$) or if the sample standard deviations are equal ($s_1 = s_2$), then the unpooled and the pooled method will give the same answer for $SE_{(\bar{y}_1 - \bar{y}_2)}$. The two answers will not differ substantially unless both the sample sizes and the sample SDs are quite discrepant.

To show the analogy between the two SE formulas, we can write them as follows:

$$SE_{(\bar{y}_1-\bar{y}_2)} = \sqrt{\frac{s_1^2}{n_1} + \frac{s_2^2}{n_2}}$$

and

$$SE_{pooled} = \sqrt{\frac{s_{pooled}^2}{n_1} + \frac{s_{pooled}^2}{n_2}}$$

In the pooled method, the separate variances—s_1^2 and s_2^2—are replaced by the single variance s_{pooled}^2, which is calculated from both samples.

Both the unpooled and the pooled SE have the same purpose—to estimate the standard deviation of the sampling distribution of $(\bar{Y}_1 - \bar{Y}_2)$. In fact, it can be shown that the standard deviation is

$$\sigma_{(\bar{Y}_1-\bar{Y}_2)} = \sqrt{\frac{\sigma_1^2}{n_1} + \frac{\sigma_2^2}{n_2}}$$

Note the resemblance between this formula and the formula for $SE_{(\bar{y}_1-\bar{y}_2)}$.

In analyzing data when the sample sizes are unequal ($n_1 \neq n_2$), we need to decide whether to use the pooled or unpooled method for calculating the standard error. The choice depends on whether we are willing to assume that the population SDs (σ_1 and σ_2) are equal. It can be shown that if $\sigma_1 = \sigma_2$, then the pooled method should be used, because in this case s_{pooled} is the best estimate of the population SD. However, in this case the unpooled method will typically give an SE that is quite similar to that given by the pooled method. If $\sigma_1 \neq \sigma_2$, then the unpooled method should be used, because in this case s_{pooled} is not an estimate of either σ_1 or σ_2, so that pooling would accomplish nothing. Because the two methods substantially agree when $\sigma_1 = \sigma_2$ and the pooled method is not valid when $\sigma_1 \neq \sigma_2$, most statisticians prefer the unpooled method. There is little to be gained by pooling when pooling is appropriate and there is much to be lost when pooling is not appropriate. Many software packages use the unpooled method by default; the user must specify use of the pooled method if she or he wishes to pool the variances.

Exercises 7.1–7.9

7.1   Data from two samples gave the following results:

	Sample 1	Sample 2
n	6	12
$\bar{y}$	40	50
s	4.3	5.7

Compute the standard error of $(\bar{y}_1 - \bar{y}_2)$.

7.2   Compute the standard error of $(\bar{y}_1 - \bar{y}_2)$ for the following data:

	Sample 1	Sample 2
n	10	10
$\bar{y}$	125	217
s	44.2	28.7

7.3 Compute the standard error of $(\bar{y}_1 - \bar{y}_2)$ for the following data:

	Sample 1	Sample 2
n	25	29
$\bar{y}$	18	16
s	5	6

7.4 Compute the standard error of $(\bar{y}_1 - \bar{y}_2)$ for the following data:

	Sample 1	Sample 2
n	5	7
$\bar{y}$	44	47
s	6.5	8.4

7.5 Consider the data from Exercise 7.4. Suppose the sample sizes were doubled, but the means and SDs stayed the same, as follows. Compute the standard error of $(\bar{y}_1 - \bar{y}_2)$.

	Sample 1	Sample 2
n	10	14
$\bar{y}$	44	47
s	6.5	8.4

7.6 Data from two samples gave the following results:

	Sample 1	Sample 2
$\bar{y}$	96.2	87.3
SE	3.7	4.6

Compute the standard error of $(\bar{y}_1 - \bar{y}_2)$.

7.7 Data from two samples gave the following results:

	Sample 1	Sample 2
n	22	21
$\bar{y}$	1.7	2.4
SE	0.5	0.7

Compute the standard error of $(\bar{y}_1 - \bar{y}_2)$.

7.8 Two varieties of lettuce were grown for 16 days in a controlled environment. The following table shows the total dry weight (in grams) of the leaves of nine plants of the variety "Salad Bowl" and six plants of the variety "Bibb."[4]

	Salad Bowl	Bibb
	3.06	1.31
	2.78	1.17
	2.87	1.72
	3.52	1.20
	3.81	1.55
	3.60	1.53
	3.30	
	2.77	
	3.62	
$\bar{y}$	3.259	1.413
s	.400	.220

Compute the standard error of $(\bar{y}_1 - \bar{y}_2)$ for these data.

7.9   Some soap manufacturers sell special "antibacterial" soaps. However, one might expect ordinary soap also to kill bacteria. To investigate this, a researcher prepared a solution from ordinary, non-antibiotic soap and a control solution of sterile water. The two solutions were placed onto petri dishes and *E. coli* bacteria were added. The dishes were incubated for 24 hours and the number of bacteria colonies on each dish were counted.[5] The data are given in the following table.

	Control (Group 1)	Soap (Group 2)
	30	76
	36	27
	66	16
	21	30
	63	26
	38	46
	35	6
	45	
n	8	7
$\bar{y}$	41.8	32.4
s	15.6	22.8
SE	5.5	8.6

Compute the standard error of $(\bar{y}_1 - \bar{y}_2)$ for these data.

7.3 CONFIDENCE INTERVAL FOR $(\mu_1 - \mu_2)$

One way to compare two sample means is to construct a confidence interval for the difference in the population means—that is, a confidence interval for the quantity $(\mu_1 - \mu_2)$. Recall from Chapter 6 that a 95% confidence interval for the mean μ of a single population that is normally distributed is constructed as

$$\bar{y} \pm t_{.025} \mathrm{SE}_{\bar{y}}$$

Analogously, a 95% confidence interval for $(\mu_1 - \mu_2)$ is constructed as

$$(\bar{y}_1 - \bar{y}_2) \pm t_{.025} \mathrm{SE}_{(\bar{y}_1 - \bar{y}_2)}$$

The critical value $t_{.025}$ is determined from Student's t distribution using degrees of freedom given as*

$$(7.1) \qquad df = \frac{(\mathrm{SE}_1^2 + \mathrm{SE}_2^2)^2}{\mathrm{SE}_1^4/(n_1 - 1) + \mathrm{SE}_2^4/(n_2 - 1)}$$

where $\mathrm{SE}_1 = s_1/\sqrt{n_1}$ and $\mathrm{SE}_2 = s_2/\sqrt{n_2}$.

---

* Strictly speaking, the distribution needed to construct a confidence interval here depends on the unknown population standard deviations σ_1 and σ_2 and is not a Student's t distribution. However, Student's t distribution with degrees of freedom given by formula (7.1) is a very good approximation. This is sometimes known as Welch's method or Satterthwaite's method.

Of course, calculating the degrees of freedom from formula (7.1) is complicated and time consuming. Most computer software uses formula (7.1), as do some graphing calculators. However, another option, which does not require technology, is to use Student's t distribution with degrees of freedom given by the smaller of $(n_1 - 1)$ and $(n_2 - 1)$. This option gives a confidence interval that is somewhat conservative, in the sense that the true confidence level is a bit larger than 95% when $t_{.025}$ is used. A third approach is to use Student's t distribution with degrees of freedom $n_1 + n_2 - 2$. This approach is somewhat liberal, in the sense that the true confidence level is a bit smaller than 95% when $t_{.025}$ is used.

Intervals with other confidence coefficients are constructed analogously; for example, for a 90% confidence interval one would use $t_{.05}$ instead of $t_{.025}$.

The following example illustrates the construction of a confidence interval for $(\mu_1 - \mu_2)$.

Example 7.7

Fast Plants. The "Wisconsin Fast Plant," *Brassica campestris*, has a very rapid growth cycle that makes it particularly well suited for the study of factors that affect plant growth. In one such study, seven plants were treated with the substance Ancymidol (ancy) and were compared to eight control plants that were given ordinary water. Heights of all of the plants were measured, in cm, after 14 days of growth.[6] The data are given in Table 7.5.

TABLE 7.5 14-Day Height of Control and of Ancy Plants (cm)

	Control (Group 1)	Ancy (Group 2)
	10.0	13.2
	13.2	19.5
	19.8	11.0
	19.3	5.8
	21.2	12.8
	13.9	7.1
	20.3	7.7
	9.6	
n	8	7
$\bar{y}$	15.9	11.0
s	4.8	4.7
SE	1.7	1.8

Parallel dotplots and normal probability plots (Figure 7.5) show that both sample distributions are reasonably symmetric and bell shaped. Moreover, we would expect that a distribution of plant heights might well be normally distributed, since height distributions often follow a normal curve. The dotplots show that the ancy distribution is shifted down a bit from the control distribution; the difference in sample means is $15.9 - 11.0 = 4.9$. The SE for the difference in sample means is

$$\text{SE}_{(\bar{y}_1 - \bar{y}_2)} = \sqrt{\frac{4.8^2}{8} + \frac{4.7^2}{7}} = 2.46$$

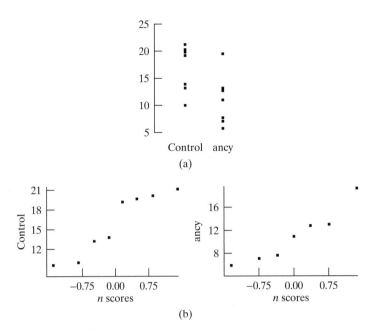

Figure 7.5 Parallel dotplots (a) and normal probability plots (b) of heights of fast plants

Using formula (7.1), we find the degrees of freedom to be 12.8:

$$df = \frac{(1.7^2 + 1.8^2)^2}{1.7^4/7 + 1.8^4/6} = 12.8$$

Using a computer, we can find that for a 95% confidence interval the *t*-multiplier for 12.8 degrees of freedom is $t(12.8)_{.025} = 2.164$. (Without a computer, we could round down the degrees of freedom to 12, in which case the *t*-multiplier is 2.179. This change from 12.8 to 12 degrees of freedom has little effect on the final answer.) The confidence interval formula gives

$$(15.9 - 11.0) \pm (2.164)(2.46)$$

or

$$4.9 \pm 5.32$$

The 95% confidence interval for $(\mu_1 - \mu_2)$ is

$$(-0.42, 10.22).$$

Rounding off, we have

$$(-0.4, 10.2)$$

Thus, we are 95% confident that the population average 14-day height of fast plants when water is used (μ_1) is between 0.4 cm lower and 10.2 cm higher than the average 14-day height of fast plants when ancy is used (μ_2). ∎

Example 7.8

Fast Plants. We said that a conservative method of constructing a confidence interval for a difference in means is to use the smaller of $n_1 - 1$ and $n_2 - 1$. For the data given in Example 7.7, this method would use 6 degrees of freedom and a t-multiplier of 2.447. In this case, the 95% confidence interval for $(\mu_1 - \mu_2)$ is

$$(15.9 - 11.0) \pm (2.447)(2.46)$$

or

$$4.9 \pm 6.02$$

The 95% confidence interval for $(\mu_1 - \mu_2)$ is

$$(-1.1, 10.9)$$

This interval is a bit conservative in the sense that the interval is wider than the interval found in Example 7.7. ∎

Example 7.9

Toluene and the Brain. Abuse of substances containing toluene (for example, glue) can produce various neurological symptoms. In an investigation of the mechanism of these toxic effects, researchers measured the concentrations of various chemicals in the brains of rats who had been exposed to a toluene-laden atmosphere, and also in unexposed control rats. The concentrations of the brain chemical norepinephrine (NE) in the medulla region of the brain, for six toluene-exposed rats and five control rats, are given in Table 7.6 and displayed in Figure 7.6.[7]

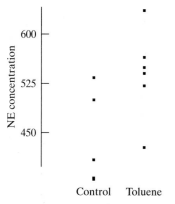

Figure 7.6 Parallel dotplots of NE concentration

TABLE 7.6 NE Concentration (ng/g)

	Toluene (Group 1)	Control (Group 2)
	543	535
	523	385
	431	502
	635	412
	564	387
	549	
n	6	5
$\bar{y}$	540.8	444.2
s	66.1	69.6
SE	27	31

For the data in Table 7.6, the SE for $(\bar{y}_1 - \bar{y}_2)$ is

$$SE_{(\bar{y}_1 - \bar{y}_2)} = \sqrt{\frac{66.1^2}{6} + \frac{69.6^2}{5}} = 41.195$$

Formula (7.1) gives degrees of freedom

$$df = \frac{(27^2 + 31^2)^2}{27^4/5 + 31^4/4} = 8.47$$

For a 95% confidence interval the t-multiplier is $t(8.47)_{.025} = 2.284$*. (We could round the degrees of freedom to 8, in which case the t-multiplier is 2.306. This change from 8.47 to 8 degrees of freedom has only a small effect on the final answer.) The confidence interval formula gives

$$(540.8 - 444.2) \pm (2.284)(41.195)$$

---

* Some software packages may produce slightly different values, but this should not be a cause for concern.

or

$$96.6 \pm 94.17$$

and the 95% confidence interval for $(\mu_1 - \mu_2)$ is

$$(2.43, 190.77)$$

Rounding off, we have

$$(2, 191)$$

According to the confidence interval, we can be 95% confident that the population mean NE concentration for toluene-exposed rats (μ_1) is larger than that for control rats (μ_2) by an amount that might be as small as 2 ng/g or as large as 191 ng/g.

Likewise, for a 90% confidence interval the t-multiplier is $t(8.47)_{.05} = 1.846$. The confidence interval formula gives

$$(540.8 - 444.2) \pm (1.846)(41.195)$$

or

$$96.6 \pm 76.05$$

and the 90% confidence interval for $(\mu_1 - \mu_2)$ is

$$(20.55, 172.65)$$

Rounding off, we have

$$(21, 173)$$

According to the confidence interval, we can be 90% confident that the population mean NE concentration for toluene-exposed rats (μ_1) is larger than that for control rats (μ_2) by an amount that might be as small as 21 ng/g or as large as 173 ng/g. ∎

Conditions for Validity

In Chapter 6 we stated the conditions that make a confidence interval for a mean valid: We require that the data can be thought of as (1) a random sample from (2) a normal population. Likewise, when comparing two means, we require two independent, random samples from normal populations. If the sample sizes are large, then the condition of normality is not crucial (due to the Central Limit Theorem).

Exercises 7.10–7.22

7.10   In Table 7.1, data were presented from a sample of 489 17-year-old males and a sample of 469 17-year-old females. The average hematocrit level of the males was 45.8, with an SD of 2.8. For the females, the average was 40.6 with an SD of 2.9. Use these data to construct a 95% confidence interval for the male-female difference in population averages. *Note*: Formula (7.1) yields 950 degrees of freedom for these data.

7.11   Ferulic acid is a compound that may play a role in disease resistance in corn. A botanist measured the concentration of soluble ferulic acid in corn seedlings grown in the dark or in a light/dark photoperiod. The results (nmol acid per g tissue) were as shown in the table.[8]

	Dark	Photoperiod
n	4	4
$\bar{y}$	92	115
s	13	13

Construct

(a) a 95% confidence interval   (b) a 90% confidence interval
for the difference in ferulic acid concentration under the two lighting conditions. (Assume that the two populations from which the data came are normally distributed.) *Note*: Formula (7.1) yields 6 degrees of freedom for these data.

7.12 A study was conducted to determine whether relaxation training, aided by biofeedback and meditation, could help in reducing high blood pressure. Subjects were randomly allocated to a biofeedback group or a control group. The biofeedback group received training for eight weeks. The table reports the reduction in systolic blood pressure (mm Hg) after eight weeks.[9] *Note*: Formula (7.1) yields 190 degrees of freedom for these data.

(a) Construct a 95% confidence interval for the difference in mean response.
(b) Interpret the confidence interval from part (a) in the context of this setting.

	Biofeedback	Control
n	99	93
$\bar{y}$	13.8	4.0
SE	1.34	1.30

7.13 Consider the data in Exercise 7.12. Suppose we are worried that the blood pressure data do not come from normal distributions. Does this mean that the confidence interval found in Exercise 7.12 is not valid? Why or why not?

7.14 Prothrombin time is a measure of the clotting ability of blood. For ten rats treated with an antibiotic and ten control rats, the prothrombin times (in seconds) were reported as follows:[10]

	Antibiotic	Control
n	10	10
$\bar{y}$	25	23
s	10	8

(a) Construct a 90% confidence interval for the difference in population means. (Assume that the two populations from which the data came are normally distributed.) *Note*: Formula (7.1) yields 17.2 degrees of freedom for these data.
(b) Interpret the confidence interval from part (a) in the context of this setting.

7.15 The accompanying table summarizes the sucrose consumption (mg in 30 minutes) of black blowflies injected with Pargyline or saline (control). (These are the same data shown in Table 7.2.)

	Control	Pargyline
n	900	905
$\bar{y}$	14.9	46.5
s	5.4	11.7

Construct

(a) a 95% confidence interval   (b) a 99% confidence interval
for the difference in population means. *Note*: Formula (7.1) yields 1274 degrees of freedom for these data.

7.16 In a field study of mating behavior in the Mormon cricket *(Anabrus simplex)*, a biologist noted that some females mated successfully while others were rejected by the males before coupling was complete. The question arose whether some aspect of body size might play a role in mating success. The accompanying table summarizes measurements of head width (mm) in the two groups of females.[11]

  (a) Construct a 95% confidence interval for the difference in population means.
     Note: Formula (7.1) yields 35.7 degrees of freedom for these data.
  (b) Interpret the confidence interval from part (a) in the context of this setting.

	Successful	Unsuccessful
n	22	17
$\bar{y}$	8.498	8.440
s	.283	.262

7.17 In an experiment to assess the effect of diet on blood pressure, 154 adults were placed on a diet rich in fruits and vegetables. A second group of 154 adults were placed on a standard diet. The blood pressures of the 308 subjects were recorded at the start of the study. Eight weeks later, the blood pressures of the subjects were measured again and the change in blood pressure was recorded for each person. Subjects on the fruits-and-vegetables diet had an average drop in systolic blood pressure of 2.8 mm Hg more than did subjects on the standard diet. A 97.5% confidence interval for the difference between the two population means is (0.9, 4.7).[12] Interpret this confidence interval. That is, explain what the numbers in the interval mean. (See Examples 7.7 and 7.9.)

7.18 Consider the experiment described in Exercise 7.17. For the same subjects, the change in diastolic blood pressure was 1.1 mm Hg greater, on average, for the subjects on the fruits-and-vegetables diet than for subjects on the standard diet. A 97.5% confidence interval for the difference between the two population means is $(-0.3, 2.4)$. Interpret this confidence interval. That is, explain what the numbers in the interval mean. (See Examples 7.7 and 7.9.)

7.19 Researchers were interested in the short-term effect that caffeine has on heart rate. They enlisted a group of volunteers and measured each person's resting heart rate. Then they had each subject drink six ounces of coffee. Nine of the subjects were given coffee containing caffeine and eleven were given decaffeinated coffee. After ten minutes each person's heart rate was measured again. The data in the following table show the change in heart rate; a positive number means that heart rate went up and a negative number means that heart rate went down.[13]

	Caffeine	Decaf
	28	26
	11	1
	−3	0
	14	−4
	−2	−4
	−4	14
	18	16
	2	8
	2	0
		18
		−10
n	9	11
$\bar{y}$	7.3	5.9
s	11.1	11.2
SE	3.7	3.4

Use these data to construct a 90% confidence interval for the difference in mean affect that caffeinated coffee has on heart rate, in comparison to decaffeinated coffee. *Note*: Formula (7.1) yields 17.3 degrees of freedom for these data.

7.20 Consider the data from Exercise 7.19. Given that there are only a small number of observations in each group, the confidence interval calculated in Exercise 7.19 is only valid if the underlying populations are normally distributed. Is the normality condition reasonable here? Support your answer with appropriate graphs.

7.21 A researcher investigated the effect of green light, in comparison to red light, on the growth rate of bean plants. The following table shows data on the heights of plants (in inches), from the soil to the first branching stem, two weeks after germination.[14]

Red	Green		Red	Green
8.4	8.6		8.4	11.1
8.4	5.9		10.4	5.5
10.0	4.6			8.2
8.8	9.1			8.3
7.1	9.8			10.0
9.4	10.1			8.7
8.8	6.0			9.8
4.3	10.4			9.5
9.0	10.8			11.0
8.4	9.6			8.0
7.1	10.5			
9.6	9.0	n	17	25
9.3	8.6	$\bar{y}$	8.36	8.94
8.6	10.5	s	1.50	1.78
6.1	9.9	SE	0.36	0.36

Use these data to construct a 95% confidence interval for the difference in mean affect that red light has on bean plant growth, in comparison to green light. *Note*: Formula (7.1) yields 38 degrees of freedom for these data.

7.22 The distributions of the data from Exercise 7.21 are somewhat skewed, particularly the Red group. Does this mean that the confidence interval calculated in Exercise 7.21 is not valid? Why or why not?

7.4 HYPOTHESIS TESTING: THE *t* TEST

We have seen that two means can be compared by using a confidence interval for the difference $(\mu_1 - \mu_2)$. In the following sections we will explore another approach to the comparison of means: the procedure known as *testing a hypothesis*. The general idea is to formulate as a hypothesis the statement that μ_1 and μ_2 do not differ, and then to see whether the data are consistent or inconsistent with that hypothesis.

The Null and Alternative Hypotheses

The hypothesis that μ_1 and μ_2 are equal is called a **null hypothesis** and is abbreviated H_0. It can be written as

$$H_0: \mu_1 = \mu_2$$

Its antithesis is the **alternative hypothesis**,

$$H_A: \mu_1 \neq \mu_2$$

which asserts that μ_1 and μ_2 are *not* equal. A researcher would usually express these hypotheses more informally, as in the following example.

Toluene and the Brain.   For the brain NE data of Example 7.9, the observed mean NE in the toluene group ($\bar{y}_1 = 540.8$ ng/g) was substantially higher than the mean in the control group ($\bar{y}_2 = 444.2$ ng/g). We might ask whether this observed difference indicates a real biological phenomenon—the effect of toluene— or whether the truth might be that toluene has no effect and that the observed difference between $\bar{y}_1$ and $\bar{y}_2$ reflects only chance variation. Corresponding hypotheses, informally stated, would be

Example 7.10

H_0^*: Toluene has no effect on NE concentration in rat medulla.

H_A^*: Toluene has some effect on NE concentration in rat medulla.   ∎

We denote the informal statements by different symbols (H_0^* and H_A^* rather than H_0 and H_A) because they make different assertions. In Example 7.10 the informal alternative hypothesis makes a very strong claim—not only that there is a difference, but that the difference is *caused* by toluene.*

A statistical **test of hypothesis** is a procedure for assessing the compatibility of the data with H_0. The data are considered compatible with H_0 if any discrepancies from H_0 could be readily attributed to chance (that is, to sampling error). Data judged to be incompatible with H_0 are taken as evidence in favor of H_A.

The *t* Statistic

We consider the problem of testing the null hypothesis

$$H_0: \mu_1 = \mu_2$$

against the alternative hypothesis

$$H_A: \mu_1 \neq \mu_2$$

Note that the null hypothesis says that the two population means are equal, which is the same as saying that the difference between them is zero:

$$H_0: \mu_1 = \mu_2 \longleftrightarrow H_0: \mu_1 - \mu_2 = 0$$

The alternative hypothesis asserts that the difference is not zero:

$$H_A: \mu_1 \neq \mu_2 \longleftrightarrow H_A: \mu_1 - \mu_2 \neq 0$$

The *t* **test** is a standard method of choosing between the two hypotheses. To carry out the *t* test, the first step is to compute the **test statistic**, which for a t test is defined as

$$t_s = \frac{(\bar{y}_1 - \bar{y}_2) - 0}{SE_{(\bar{y}_1 - \bar{y}_2)}}$$

---

* Of course, our statements of H_0^* and H_A^* are abbreviated. Complete statements would include all relevant conditions of the experiment—adult male rats, toluene 1,000 ppm atmosphere for 8 hours, and so on. Our use of abbreviated statements should not cause any confusion.

Note that we subtract zero from $\bar{y}_1 - \bar{y}_2$ because H_0 states that $\mu_1 - \mu_2$ equals zero; writing $(\bar{y}_1 - \bar{y}_2) - 0$ reminds us of what we are testing. The subscript s on t_s serves as a reminder that this value is calculated from the data (s for "sample"). The quantity t_s is the test statistic for the t test; that is, t_s provides the data summary that is the basis for the test procedure. Notice the structure of t_s: It is a measure of the difference between the sample means ($\bar{y}$'s), expressed in relation to the SE of the difference. We illustrate with an example.

Example 7.11 **Toluene and the Brain.**   For the brain NE data of Example 7.9, the value of t_s is

$$t_s = \frac{(540.8 - 444.2) - 0}{41.195} = 2.34$$

The t statistic shows that $\bar{y}_1$ and $\bar{y}_2$ differ by about 2.3 SEs.   ∎

How shall we judge whether our data are consistent with H_0? *Perfect* agreement with H_0 would be expressed by sample means that were identical and a resulting t statistic equal to zero ($t_s = 0$). But even if the null hypothesis H_0 were true, we do not expect t_s to be exactly zero; we expect the sample means to differ from one another because of sampling variability. Fortunately, we can set limits on this sampling variability; in fact, the chance difference in the $\bar{y}$'s is not likely to exceed a couple of standard errors. To put this more precisely, it can be shown mathematically that

> If H_0 is true, then the sampling distribution of t_s is well approximated by a Student's t distribution with degrees of freedom given by formula (7.1).*

The preceding statement is true if certain conditions are met. Briefly: We require independent random samples from normally distributed populations. These conditions will be considered in detail in Section 7.9.

The essence of the t test procedure is to locate the observed value t_s in the Student's t distribution, as indicated in Figure 7.7. If t_s is near the center, as in Figure 7.7(a), then the data are regarded as compatible with H_0 because the observed difference between $\bar{y}_1$ and $\bar{y}_2$ can be readily attributed to chance variation caused by sampling error. (H_0 predicts that the sample means will be equal, since

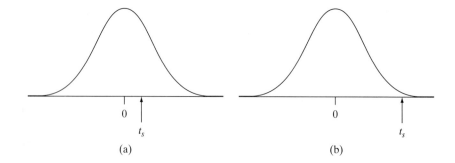

Figure 7.7 Essence of the t test.
(a) Data compatible with H_0;
(b) data incompatible with H_0.

---

* As we stated in Section 7.3, a conservative approximation to formula (7.1) is to use degrees of freedom given by the smaller of $n_1 - 1$ and $n_2 - 1$.

H_0 says that the population means are equal.) If, on the other hand, t_s falls in the far tail of the *t* distribution, as in Figure 7.7(b), then the data are regarded as in-compatible with H_0, because the observed deviation cannot be readily explained as being due to chance variation. To put this another way, if H_0 is true, then it is un-likely that t_s would fall in the far tails of the *t* distribution; consequently, a value of t_s in the far tails is interpreted as evidence against H_0.

The *P*-Value

To judge whether an observed value t_s is "far" in the tail of the *t* distribution, we need a quantitative yardstick for locating t_s within the distribution. This yardstick is provided by the *P*-value, which can be defined (in the present context) as follows:

> The **P-value** of the test is the area under Student's *t* curve in the double tails beyond $-t_s$ and $+t_s$.

Thus, the *P*-value, which is sometimes abbreviated as simply *P*, is the shaded area in Figure 7.8. Note that we have defined the *P*-value as the total area in the *double* tail; this is sometimes called the "two-tailed" *P*-value.

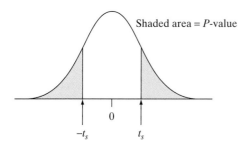

Shaded area = *P*-value

$-t_s$ 0 t_s

Figure 7.8 The two-tailed *P*-value for the *t* test

Toluene and the Brain.   For the brain NE data of Example 7.9, the value of t_s is 2.34. We can ask, "If H_0 were true, so that one would expect $\bar{y}_1 = \bar{y}_2$, on average, what is the probability that $\bar{y}_1$ and $\bar{y}_2$ would differ by as many as 2.34 SEs?" The *P*-value answers this question. Formula (7.1) yields 8.47 degrees of freedom for these data. Thus, the *P*-value is the area under the *t* curve (with 8.47 degrees of freedom) beyond ±2.34. This area, which was found using a computer, is shown in Figure 7.9 to be .0454. ■

Example 7.12

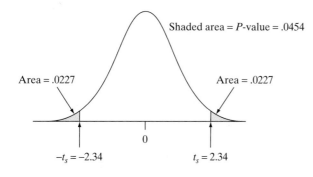

Shaded area = *P*-value = .0454

Area = .0227   Area = .0227

$-t_s = -2.34$ 0 $t_s = 2.34$

Figure 7.9 The two-tailed *P*-value for the Toluene data

> **Definition**
> The **P-value** for a hypothesis test is the probability, computed under the condition that the null hypothesis is true, of the test statistic being at least as extreme as the value of the test statistic that was actually obtained.

From the definition of P-value, it follows that **the P-value is a measure of compatibility between the data and H_0**: A large P-value (close to 1) indicates a value of t_s near the center of the t distribution (compatible with H_0), whereas a small P-value (close to 0) indicates a value of t_s in the far tails of the t distribution (incompatible with H_0).

Drawing Conclusions from a t Test

The P-value is a measure of the compatibility between the data and H_0. But where do we draw the line between compatibility and incompatibility? Most people would agree that P-value = .0001 indicates incompatibility, and that P-value = .80 indicates compatibility, but what about intermediate values? For example, should P-value = .10 be regarded as compatible or incompatible with H_0? The answer is not intuitively obvious.

In much scientific research, it is not necessary to draw a sharp line. However, in many situations a *decision* must be reached. For example, the Food and Drug Administration (FDA) must decide whether the data submitted by a pharmaceutical manufacturer are sufficient to justify approval of a medication. As another example, a fertilizer manufacturer must decide whether the evidence favoring a new fertilizer is sufficient to justify the expense of further research.

Making a decision requires drawing a definite line between compatibility and incompatibility. The threshold value, on the P-value scale, is called the **significance level** of the test, and is denoted by the Greek letter α (alpha). The value of α is chosen by whoever is making the decision. Common choices are $\alpha = .10, .05,$ and $.01$. *If the P-value of the data is less than or equal to α, the data are judged incompatible with H_0; in this case we say that H_0 is* **rejected**, *and that the data provide evidence in favor of H_A. If the P-value of the data is greater than α,* we say that H_0 is **not rejected**, and that the data provide insufficient evidence to claim that H_A is true.

The following example illustrates the use of the t test to make a decision.

Example 7.13

Toluene and the Brain. For the brain NE experiment of Example 7.9, the data are summarized in Table 7.7. Suppose we choose to make a decision at the 5% significance level, $\alpha = .05$. In Example 7.12 we found that the P-value of these data is .0454. This means that one of two things happened: Either (1) H_0 is true and

TABLE 7.7 NE Concentration (ng/g)		
	Toluene	**Control**
n	6	5
$\bar{y}$	540.8	444.2
s	66.1	69.6

we got a strange set of data just by chance, or (2) H_0 is false. If H_0 is true, the kind of discrepancy we observed between $\bar{y}_1$ and $\bar{y}_2$ would only happen about 4.5% of the time. Because the P-value, .0454, is less than .05, we reject H_0 and conclude that the data provide evidence in favor of H_A. The strength of the evidence is expressed by the statement that the P-value is .0454.

Conclusion: The data provide sufficient evidence ($P = .0454$) that at the .05 level of significance we can reject the null hypothesis. We conclude that toluene increases NE concentration.* ■

The next example illustrates a t test in which H_0 is not rejected.

Fast Plants. In Example 7.7 we saw that the mean height of fast plants was smaller when ancy was used than when water (the control) was used. Table 7.8 summarizes the data. The difference between the sample averages is $15.9 - 11.0 = 4.9$. The SE for the difference is

Example 7.14

$$SE_{(\bar{y}_1 - \bar{y}_2)} = \sqrt{\frac{4.8^2}{8} + \frac{4.7^2}{7}} = 2.46$$

TABLE 7.8 14-Day Height of Control and of Ancy Plants

	Control	Ancy
n	8	7
$\bar{y}$	15.9	11.0
s	4.8	4.7

Suppose we choose to use $\alpha = .05$ in testing

$$H_0: \mu_1 = \mu_2 \ (\text{i.e., } \mu_1 - \mu_2 = 0)$$

against the alternative hypothesis

$$H_A: \mu_1 \neq \mu_2 \ (\text{i.e., } \mu_1 - \mu_2 \neq 0)$$

The value of the test statistic is

$$t_s = \frac{(15.9 - 11.0) - 0}{2.46} = 1.99$$

Formula (7.1) gives 12.8 degrees of freedom for the t distribution. The P-value for the test is the probability of getting a t statistic that is at least as far away from zero as 1.99. Figure 7.10 shows that this probability is .0678. (This four-digit P-value was found using a computer.) Because the P-value is greater than α, we do not reject H_0. These data do not provide sufficient evidence to conclude that μ_1 and μ_2 differ; the difference we observed between $\bar{y}_1$ and $\bar{y}_2$ could easily have happened by chance.

---

* Because the alternative hypothesis was $H_A: \mu_1 \neq \mu_2$, some authors would say "We conclude that toluene affects NE concentration," rather than saying that toluene increases NE concentration.

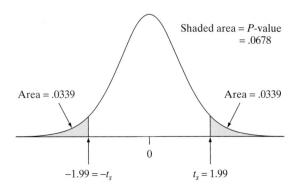

Figure 7.10 The two-sided
P-value for the ancy data

Conclusion: The data do *not* provide sufficient evidence (P-value = .0678) at the .05 level of significance to conclude that ancy and water differ in their effects on fast plant growth (under the conditions of the experiment that was conducted). ■

Note carefully the phrasing of the conclusion in Example 7.14. We do *not* say that there is evidence *for* the null hypothesis, but only that there is insufficient evidence *against* it. When we do not reject H_0, this indicates a lack of evidence that H_0 is false, which is *not* the same thing as evidence that H_0 is true. The astronomer Carl Sagan (in another context) summed up this principle of evidence in this succinct statement:[15]

Absence of evidence is not evidence of absence.

In other words, nonrejection of H_0 is *not* the same as *acceptance* of H_0. (To avoid confusion, it may be best not to use the phrase "accept H_0" at all.)

Nonrejection of H_0 indicates that the data are compatible with H_0, but the data may *also* be quite compatible with H_A. For instance, in Example 7.14 we found that the observed difference between the sample means could be due to sampling variation, but this finding does not rule out the possibility that the observed difference is actually due to a real effect caused by ancy. (Methods for such ruling out of possible alternatives will be discussed in Section 7.7 and optional Section 7.8.)

In testing a hypothesis, the researcher starts out with the assumption that H_0 is true and then asks whether the data contradict that assumption. This logic can make sense even if the researcher regards the null hypothesis as implausible. For instance, in Example 7.14 it could be argued that there is almost certainly *some* difference (perhaps very small) between using ancy and not using ancy. The fact that we did not reject H_0 does not mean that we accept H_0.

Using Tables Versus Using Technology

In analyzing data, how do we determine the P-value of a test? Statistical computer software, and some calculators, will provide exact P-values. If such technology is not available, then we can use formula (7.1) to find the degrees of freedom, but round down to make the value an integer. A conservative alternative to using formula (7.1) is to use the smaller of $n_1 - 1$ and $n_2 - 1$ as the degrees of freedom for the test. A liberal approach is to use $n_1 + n_2 - 2$ as the degrees of freedom.

[Formula (7.1) will always give degrees of freedom between the conservative value of the smaller of $n_1 - 1$ and $n_2 - 1$ and the liberal value of $n_1 + n_2 - 2$.] We can rely on the limited information in Table 4 to *bracket* the P-value, rather than to determine it exactly. The P-value found using the conservative approach will be somewhat larger than the exact P-value; the P-value found using the liberal approach will be somewhat smaller than the exact P-value. The following example illustrates the bracketing process.

Fast Plants.   For the fast plant growth data, the value of the *t* statistic (as determined in Example 7.14) is $t_s = 1.99$. The smaller of $n_1 - 1$ and $n_2 - 1$ is $7 - 1 = 6$, so the conservative degrees of freedom are 6. The liberal degrees of freedom are $8 + 7 - 2 = 13$. Here is a copy of part of Table 4, with key numbers highlighted:

Example 7.15

Upper Tail Probability

df	.05	.04	.03
6	**1.943**	**2.104**	2.313
7	1.895	2.046	2.241
8	1.860	2.004	2.189
9	1.833	1.973	2.150
10	1.812	1.948	2.120
11	1.796	1.928	2.096
12	1.782	1.912	2.076
13	1.771	**1.899**	**2.060**

We begin with the conservative degrees of freedom, 6. From the preceding table (or from Table 4) we find $t(6)_{.05} = 1.943$ and $t(6)_{.04} = 2.104$. The corresponding conservative P-value, based on a *t* distribution with 6 degrees of freedom, is shaded in Figure 7.11. Because t_s is between the .04 and .05 critical values, the upper tail area must be between .04 and .05; thus, the conservative P-value must be between .08 and .10.

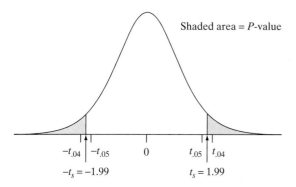

Shaded area = P-value

$-t_{.04}$ $-t_{.05}$ 0 $t_{.05}$ $t_{.04}$

$-t_s = -1.99$ $t_s = 1.99$

Figure 7.11 Conservative
P-value for Example 7.15

The liberal degrees of freedom are $8 + 7 - 2 = 13$. From the preceding table (or from Table 4) we find $t(13)_{.04} = 1.899$ and $t(13)_{.03} = 2.060$. Because t_s is between these .03 and .04 critical values, the upper tail area must be between .03 and .04; thus, the liberal P-value must be between .06 and .08.

Putting these two together, we have

$$0.6 < P\text{-value} < .10$$

■

If the observed t_s is not within the boundaries of Table 4, then the P-value is bracketed on only one side. For example, if t_s is greater than $t_{.0005}$, then the two-sided P-value is bracketed as

$$P\text{-value} < .001$$

Reporting the Results of a t Test

In reporting the results of a t test, a researcher may choose to make a definite decision (to reject H_0 or not to reject H_0) at a specified significance level α, or the researcher may choose simply to describe the results in phrases such as "There is very strong evidence that . . . " or "The evidence suggests that . . . " or "There is virtually no evidence that . . . ." In writing a report for publication, it is very desirable to state the P-value so that the reader can make a decision on his or her own.

The term *significant* is often used in reporting results. For instance, an observed difference is said to be "statistically significant at the 5% level" if it is large enough to justify rejection of H_0 at $\alpha = .05$. In Example 7.13 we saw that the observed difference between the two sample means in the toluene data is statistically significant at the 5% level, since the P-value is .0454, which is less than .05. In contrast, the fast plant data of Example 7.14 do not show a statistically significant difference at the 5% level, since the P-value for the fast plant data is .0678. However, the difference in sample means in the fast plant data *is* statistically significant at the $\alpha = .10$ level, since the P-value is less than .10. When α is not specified, it is usually understood to be .05; we should emphasize, however, that α *is an arbitrarily chosen value and there is nothing "official" about .05*. Unfortunately, the term *significant* is easily misunderstood and should be used with care; we will return to this point in Section 7.7.

Note: In this section we have considered tests of the form $H_0: \mu_1 = \mu_2$ (i.e., $\mu_1 - \mu_2 = 0$) versus $H_A: \mu_1 \neq \mu_2$ (i.e., $\mu_1 - \mu_2 \neq 0$); this is the most common pair of hypotheses. However, it may be that we wish to test that μ_1 is greater than μ_2 by some specific, nonzero amount, say c. To test $H_0: \mu_1 - \mu_2 = c$ versus $H_A: \mu_1 - \mu_2 \neq c$ we use the t test with test statistic given by

$$t_s = \frac{(\bar{y}_1 - \bar{y}_2) - c}{SE_{(\bar{y}_1 - \bar{y}_2)}}$$

From this point on, the test proceeds as before (i.e., as for the case when $c = 0$).

Computer note: The calculations for a two-sample t test or a confidence interval can be carried out with most statistical software. For example, consider the toluene data from Examples 7.9 and 7.12. Suppose the data are entered into MINITAB system in two columns. Then the command

```
MTB > TwoSample 95.0 'Toluene' 'Control';
SUBC > Alternative 0.
```

will produce a 95% confidence interval for the difference in population means together with a t test. The "subcommand" of "Alternative 0" means that we are testing $H_0: \mu_1 = \mu_2$ versus $H_A: \mu_1 \neq \mu_2$. The resulting output of this command is

```
Twosample T for Toluene vs Control
            N    Mean    StDev    SE Mean
Toluene     6    540.8   66.1        27
Control     5    444.2   69.6        31
95% C.I. for mu Toluene - mu Control: (2, 192)
T-Test mu Toluene = mu Control (vs not=):T=2.34
P=0.047 DF=8
```

We had calculated the degrees of freedom for these data, from formula (7.1), to be 8.47. MINITAB rounds this down to 8, which results in slightly different final numbers than those we calculated. Thus, in Example 7.12 we had found the *P*-value to be .0454, whereas MINITAB reports a *P*-value of .047. This discrepancy is minor.

Exercises 7.23–7.38

7.23   For each of the following data sets, use Table 4 to bracket the two-tailed *P*-value of the data as analyzed by the *t* test. Use the degrees of freedom given.

(a)

	Sample 1	**Sample 2**
n	4	3
$\bar{y}$	735	854

$$SE_{(\bar{y}_1 - \bar{y}_2)} = 38 \text{ with df} = 4$$

(b)

	Sample 1	**Sample 2**
n	7	7
$\bar{y}$	5.3	5.0

$$SE_{(\bar{y}_1 - \bar{y}_2)} = .24 \text{ with df} = 12$$

(c)

	Sample 1	**Sample 2**
n	15	20
$\bar{y}$	36	30

$$SE_{(\bar{y}_1 - \bar{y}_2)} = 1.3 \text{ with df} = 30$$

7.24   For each of the following data sets, use Table 4 to bracket the two-tailed *P*-value of the data as analyzed by the *t* test. Use the degrees of freedom given.

(a)

	Sample 1	**Sample 2**
n	8	5
$\bar{y}$	100.2	106.8

$$SE_{(\bar{y}_1 - \bar{y}_2)} = 5.7 \text{ with df} = 10$$

(b)

	Sample 1	**Sample 2**
n	8	8
$\bar{y}$	49.8	44.3

$$SE_{(\bar{y}_1 - \bar{y}_2)} = 1.9 \text{ with df} = 13$$

(c)

	Sample 1	Sample 2
n	10	15
$\bar{y}$	3.58	3.00

$$\text{SE}_{(\bar{y}_1 - \bar{y}_2)} = .12 \text{ with df} = 19$$

7.25 For each of the following situations, suppose $H_0: \mu_1 = \mu_2$ is being tested against $H_A: \mu_1 \neq \mu_2$. State whether or not H_0 would be rejected.

(a) P-value $= .085, \alpha = .10$
(b) P-value $= .065, \alpha = .05$
(c) $t_s = 3.75$ with 19 degrees of freedom, $\alpha = .01$
(d) $t_s = 1.85$ with 12 degrees of freedom, $\alpha = .05$

7.26 For each of the following situations, suppose $H_0: \mu_1 = \mu_2$ is being tested against $H_A: \mu_1 \neq \mu_2$. State whether or not H_0 would be rejected.

(a) P-value $= .046, \alpha = .02$
(b) P-value $= .033, \alpha = .05$
(c) $t_s = 2.26$ with 5 degrees of freedom, $\alpha = .10$
(d) $t_s = 1.94$ with 16 degrees of freedom, $\alpha = .05$

7.27 In a study of the nutritional requirements of cattle, researchers measured the weight gains of cows during a 78-day period. For two breeds of cows, Hereford (HH) and Brown Swiss/Hereford (SH), the results are summarized in the following table.[16] *Note*: Formula (7.1) yields 71.9 df.

	HH	SH
n	33	51
$\bar{y}$	18.3	13.9
s	17.8	19.1

Use a t test to compare the means. Use $\alpha = .10$.

7.28 Backfat thickness is a variable used in evaluating the meat quality of pigs. An animal scientist measured backfat thickness (cm) in pigs raised on two different diets, with the results given in the table.[17]

	Diet 1	Diet 2
$\bar{y}$	3.49	3.05
s	.40	.40

Consider using the t test to compare the diets. Bracket the P-value, assuming that the number of pigs on each diet was

(a) 5   (b) 10   (c) 15

Use $n_1 + n_2 - 2$ as the degrees of freedom.

7.29 Heart disease patients often experience spasms of the coronary arteries. Because biological amines may play a role in these spasms, a research team measured amine levels in coronary arteries that were obtained postmortem from patients who had died of heart disease and also from a control group of patients who had died from other causes. The accompanying table summarizes the concentration of the amine serotonin.[18]

	Serotonin (ng/g)	
	Heart Disease	*Controls*
n	8	12
Mean	3840	5310
SE	850	640

(a) For these data, the SE of $(\bar{y}_1 - \bar{y}_2)$ is 1,064 and df $= 14.3$ (which can be rounded to 14). Use a *t* test to compare the means at the 5% significance level.

(b) State the conclusion of the *t* test in the context of the setting. (See Examples 7.13 and 7.14.)

(c) Verify the value of $SE_{(\bar{y}_1 - \bar{y}_2)}$ given in part (a).

7.30 In a study of the periodical cicada *(Magicicada septendecim)*, researchers measured the hind tibia lengths of the shed skins of 110 individuals. Results for males and females are shown in the accompanying table.[19]

<table>
<tr><td colspan="4" align="center">**Tibia Length (micrometer units)**</td></tr>
<tr><td>**Group**</td><td>*n*</td><td>*Mean*</td><td>*SD*</td></tr>
<tr><td>Males</td><td>60</td><td>78.42</td><td>2.87</td></tr>
<tr><td>Females</td><td>50</td><td>80.44</td><td>3.52</td></tr>
</table>

(a) Use a *t* test to investigate the dependence of tibia length on gender in this species. Use the 5% significance level. *Note*: Formula (7.1) yields 94.3 df.

(b) State the conclusion of the *t* test in the context of the setting. (See Examples 7.13 and 7.14.)

(c) Given the preceding data, if you were told the tibia length of an individual of this species, could you make a fairly confident prediction of its sex? Why or why not?

(d) Repeat the *t* test of part (a), assuming that the means and standard deviations were as given in the table, but that they were based on only one-tenth as many individuals (6 males and 5 females). *Note*: Formula (7.1) yields 7.8 df.

7.31 In a study of the development of the thymus gland, researchers weighed the glands of ten chick embryos. Five of the embryos had been incubated 14 days and five had been incubated 15 days. The thymus weights were as shown in the table.[20] Note: Formula (7.1) yields 7.7 df.

<table>
<tr><td colspan="3" align="center">**Thymus Weight (mg)**</td></tr>
<tr><td></td><td>*14 days*</td><td>*15 days*</td></tr>
<tr><td></td><td>29.6</td><td>32.7</td></tr>
<tr><td></td><td>21.5</td><td>40.3</td></tr>
<tr><td></td><td>28.0</td><td>23.7</td></tr>
<tr><td></td><td>34.6</td><td>25.2</td></tr>
<tr><td></td><td>44.9</td><td>24.2</td></tr>
<tr><td>*n*</td><td>5</td><td>5</td></tr>
<tr><td>$\bar{y}$</td><td>31.72</td><td>29.22</td></tr>
<tr><td>*s*</td><td>8.73</td><td>7.19</td></tr>
</table>

(a) Use a *t* test to compare the means at $\alpha = .10$.

(b) Note that the chicks that were incubated longer had a smaller mean thymus weight. Is this "backward" result surprising, or could it easily be attributed to chance? Explain.

7.32 As part of an experiment on root metabolism, a plant physiologist grew birch tree seedlings in the greenhouse. He flooded four seedlings with water for one day, and kept four others as controls. He then harvested the seedlings and analyzed the roots for ATP content. The results (nmol ATP per mg tissue) are as follows:[21] *Note*: Formula (7.1) yields 5.6 df.

	Flooded	Control
	1.45	1.70
	1.19	2.04
	1.05	1.49
	1.07	1.91
n	4	4
$\bar{y}$	1.190	1.785
s	.184	.241

(a) Use a t test to investigate the effect of flooding. Use $\alpha = .05$.

(b) State the conclusion of the t test in the context of the setting. (See Examples 7.13 and 7.14.)

7.33 After surgery a patient's blood volume is often depleted. In one study, the total circulating volume of blood plasma was measured for each patient immediately after surgery. After infusion of a "plasma expander" into the bloodstream, the plasma volume was measured again and the increase in plasma volume (mL) was calculated. Two of the plasma expanders used were albumin (25 patients) and polygelatin (14 patients). The accompanying table reports the increase in plasma volume.[22] *Note*: Formula (7.1) yields 33.6 df.

(a) Use a t test to compare the mean increase in plasma volume under the two treatments. Let $\alpha = .01$.

(b) State the conclusion of the t test in the context of the setting. (See Examples 7.13 and 7.14.)

	Albumin	**Polygelatin**
n	25	14
Mean increase	490	240
SE	60	30

7.34 Nutritional researchers conducted an investigation of two high-fiber diets intended to reduce serum cholesterol level. Twenty men with high serum cholesterol were randomly allocated to receive an "oat" diet or a "bean" diet for 21 days. The table summarizes the fall (before minus after) in serum cholesterol levels.[23] Use a t test to compare the diets at the 5% significance level. *Note*: Formula (7.1) yields 17.9 df.

Diet	*n*	**Fall in Cholesterol (mg/dL)**	
		Mean	*SD*
Oat	10	53.6	31.1
Bean	10	55.5	29.4

7.35 Suppose we have conducted a t test, with $\alpha = .05$, and the P-value is .03. For each of the following statements, say whether the statement is true or false and explain why.

(a) We reject H_0 with $\alpha = .05$.

(b) We would reject H_0 if α were .10.

(c) If H_0 is true, the probability of getting a test statistic at least as extreme as the value of the t_s that was actually obtained is 3%.

7.36 Suppose we have conducted a t test, with $\alpha = .10$, and the P-value is .07. For each of the following statements, say whether the statement is true or false and explain why.

(a) We reject H_0 with $\alpha = .10$.

(b) We would reject H_0 if α were .05.

(c) The probability that $\bar{y}_1$ is greater than $\bar{y}_2$ is .07.

7.37 The following table shows the number of bacteria colonies present in each of several petri dishes, after *E. coli* bacteria were added to the dishes and they were incubated for 24 hours. The "soap" dishes contained a solution prepared from ordinary soap; the "control" dishes contained a solution of sterile water.5 (These data were seen in Exercise 7.9.)

	Control	Soap
	30	76
	36	27
	66	16
	21	30
	63	26
	38	46
	35	6
	45	
n	8	7
$\bar{y}$	41.8	32.4
s	15.6	22.8
SE	5.5	8.6

(a) Use a *t* test to investigate whether soap affects the number of bacteria colonies that form. Use $\alpha = .10$. *Note*: Formula (7.1) yields 10.4 degrees of freedom for these data.

(b) State the conclusion of the *t* test in the context of the setting. (See Examples 7.13 and 7.14.)

7.38 Researchers studied the effect of a houseplant fertilizer on radish sprout growth. They randomly selected some radish seeds to serve as controls, while others were planted in aluminum planters to which fertilizer sticks were added. Other conditions were held constant between the two groups. The following table shows data on the heights of plants (in cm) two weeks after germination.[24]

Control		Fertilized	
3.4	1.6	2.8	1.9
4.4	2.9	1.9	2.7
3.5	2.3	3.6	2.3
2.9	2.8	1.2	1.8
2.7	2.5	2.4	2.7
2.6	2.3	2.2	2.6
3.7	1.6	3.6	1.3
2.7	1.6	1.2	3.0
2.3	3.0	0.9	1.4
2.0	2.3	1.5	1.2
1.8	3.2	2.4	2.6
2.3	2.0	1.7	1.8
2.4	2.6	1.4	1.7
2.5	2.4	1.8	1.5

	Control	Fertilized
n	28	28
$\bar{y}$	2.58	2.04
s	0.65	0.72

(a) Use a *t* test to investigate whether the fertilizer has an effect on average radish sprout growth. Use $\alpha = .05$. *Note*: Formula (7.1) yields 53.5 degrees of freedom for these data.

(b) State the conclusion of the *t* test in the context of the setting. (See Examples 7.13 and 7.14.)

7.5 FURTHER DISCUSSION OF THE t TEST

In this section we discuss more fully the method and interpretation of the t test.

Relationship Between Test and Confidence Interval

There is a close connection between the confidence interval approach and the hypothesis-testing approach to the comparison of μ_1 and μ_2. Consider, for example, a 95% confidence interval for $(\mu_1 - \mu_2)$ and its relationship to the t test at the 5% significance level. The t test and the confidence interval use the same three quantities—$(\bar{y}_1 - \bar{y}_2)$, $SE_{(\bar{y}_1 - \bar{y}_2)}$, and $t_{.025}$—but manipulate them in different ways.

In the t test, when $\alpha = .05$, we reject H_0 if the P-value is less than or equal to .05. This happens if and only if the test statistic, t_s, is in the tail of the t distribution, at or beyond $\pm t_{.025}$. If the magnitude of t_s (symbolized as $|t_s|$) is greater than or equal to $t_{.025}$, then the P-value is less than or equal to .05 and we reject H_0; if $|t_s|$ is less than $t_{.025}$, then the P-value is greater than .05 and we do *not* reject H_0. Figure 7.12 shows this relationship.

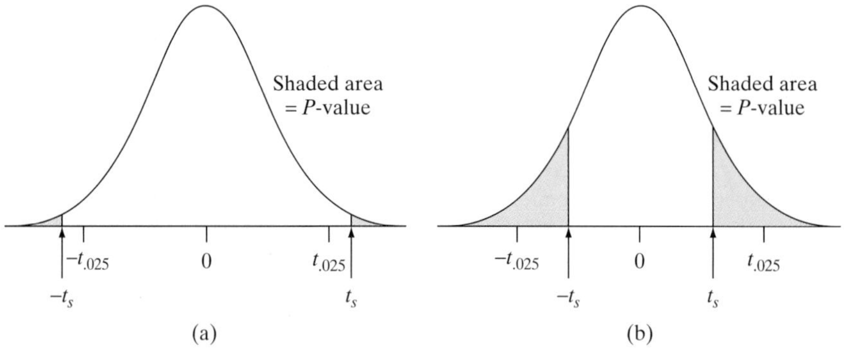

Figure 7.12 Possible outcomes of the t test at $\alpha = .05$. (a) If $|t_s| \geq t_{.025}$, then P-value $\leq .05$ and H_0 is rejected. (b) If $|t_s| < t_{.025}$, then P-value $> .05$ and H_0 is not rejected.

Thus, we fail to reject $H_0: \mu_1 - \mu_2 = 0$ if and only if $|t_s| < t_{.025}$. That is, we fail to reject H_0 when

$$\frac{|\bar{y}_1 - \bar{y}_2|}{SE_{(\bar{y}_1 - \bar{y}_2)}} < t_{.025}$$

This is equivalent to

$$|\bar{y}_1 - \bar{y}_2| < t_{.025}SE_{(\bar{y}_1 - \bar{y}_2)}$$

or

$$-t_{.025}SE_{(\bar{y}_1 - \bar{y}_2)} < (\bar{y}_1 - \bar{y}_2) < t_{.025}SE_{(\bar{y}_1 - \bar{y}_2)}$$

which is equivalent to

$$-(\bar{y}_1 - \bar{y}_2) - t_{.025}SE_{(\bar{y}_1 - \bar{y}_2)} < 0 < -(\bar{y}_1 - \bar{y}_2) + t_{.025}SE_{(\bar{y}_1 - \bar{y}_2)}$$

or

$$(\bar{y}_1 - \bar{y}_2) + t_{.025}SE_{(\bar{y}_1 - \bar{y}_2)} > 0 > (\bar{y}_1 - \bar{y}_2) - t_{.025}SE_{(\bar{y}_1 - \bar{y}_2)}$$

or

$$(\bar{y}_1 - \bar{y}_2) - t_{.025}SE_{(\bar{y}_1 - \bar{y}_2)} < 0 < (\bar{y}_1 - \bar{y}_2) + t_{.025}SE_{(\bar{y}_1 - \bar{y}_2)}$$

Thus, we have shown that we fail to reject $H_0: \mu_1 - \mu_2 = 0$ if and only if the confidence interval for $(\mu_1 - \mu_2)$ includes zero. If the 95% confidence interval for $(\mu_1 - \mu_2)$ does not cover zero, then we reject $H_0: \mu_1 - \mu_2 = 0$ when $\alpha = .05$. (The same relationship holds between the 90% confidence interval and the test at $\alpha = .10$, and so on.) We illustrate with an example.

Wasp Eggs and Parasites.   Many wasps are parasitic in or on caterpillars, which always die as a result of parasitism. Two such wasps are the internal parasite *Copidosomopsis tanytmema* (Ct) and the external parasite *Bracon hebetor* (Bh). Ct is polyembryonic and produces over a hundred offspring for every egg it lays. Bh, on the other hand, stings its host caterpillar to paralyze it prior to parasitizing it, or to feed on its body fluids. A stung caterpillar survives at least until the Ct brood, if present within, emerges as adult wasps. Researchers wanted to determine if Bh-induced paralysis decreased Ct brood size. To do this, Ct-parasitized caterpillars were exposed to Bh stings and then observed until the Ct matured into wasps. The number of wasps per stung caterpillar host—the brood size—was tallied. These data were compared with numbers from nonparalyzed ("unstung"), Ct-parasitized caterpillars.[25] Table 7.9 shows the data from this experiment. Figure 7.13 shows parallel boxplots for the data. The stung distribution is shifted down from the unstung distribution; both distributions are reasonably symmetric.
    For these data the two SEs are $43.5/\sqrt{46} = 6.41$ and $34.6/\sqrt{57} = 4.58$. The degrees of freedom are

$$df = \frac{(6.41^2 + 4.58^2)^2}{6.41^4/45 + 4.58^4/56} = 84.9$$

The quantities needed for a *t* test with $\alpha = .05$ are

$$\bar{y}_1 - \bar{y}_2 = 161.8 - 155.3 = 6.5$$

and

$$SE_{(\bar{y}_1 - \bar{y}_2)} = \sqrt{6.41^2 + 4.58^2} = 7.88$$

The test statistic is

$$t_s = \frac{(161.8 - 155.3) - 0}{7.88} = \frac{6.5}{7.88} = 0.82$$

The *P*-value for this test (found using a computer) is .412, which is greater than .05, so we do not reject H_0. (A quick look at Table 4, using df = 80, shows that the *P*-value is greater than .40.)

Example 7.16

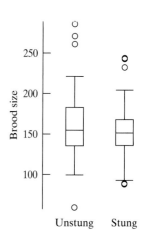

Figure 7.13 Boxplots of the wasp data

TABLE 7.9 Wasp Data: Brood Size for Unstung Larva and for Stung Larva

	Unstung	Stung
n	46	57
$\bar{y}$	161.8	155.3
s	43.5	34.6

If we construct a 95% confidence interval for $(\mu_1 - \mu_2)$, we get

$$6.5 \pm 1.989 \cdot 7.88$$

or $(-9.2, 22.2)$.*

      The confidence interval includes zero, which is consistent with not rejecting H_0: $\mu_1 - \mu_2 = 0$ in the t test. Note that this equivalence between the test and the confidence interval makes common sense; according to the confidence interval, μ_1 may be as much as 9.2 less, or as much as 22.2 more, than μ_2; it is natural, then, to say that we are uncertain as to whether μ_1 is greater than (or less than, or equal to) μ_2. ■

      In the context of the Student's t method, the confidence interval approach and hypothesis-testing approach are different ways of using the same basic information. The confidence interval has the advantage that it indicates the magnitude of the difference between μ_1 and μ_2. The testing approach has the advantage that the P-value describes on a continuous scale the strength of the evidence that μ_1 and μ_2 are really different. In Section 7.7 we will explore further the use of a confidence interval to supplement the interpretation of a t test. In later chapters we will encounter other hypothesis tests that cannot so readily be supplemented by a confidence interval.

Interpretation of α

In analyzing data or making a decision based on data, you will often need to choose a significance level α. How do you know whether to choose $\alpha = .05$ or $\alpha = .01$ or some other value? To make this judgment, it is helpful to have an *operational* interpretation of α. We now give such an interpretation.

      Recall from Section 7.4 that the sampling distribution of t_s, if H_0 is true, is a Student's t distribution. Let us assume for definiteness that df $= 60$ and that α is chosen equal to .05. The critical value (from Table 4) is $t_{.025} = 2.000$. Figure 7.14 shows the Student's t distribution and the values ± 2.000. The total shaded area in the figure is .05; it is split into two equal parts of area .025 each. We can think of Figure 7.14 as a formal guide for deciding whether to reject H_0: If the observed value of t_s falls in the hatched regions of the t_s axis, then H_0 will be rejected. But the chance of this happening is 5%, if H_0 is true. Thus, we can say that

$$\Pr\{\text{reject } H_0\} = .05 \qquad \text{if } H_0 \text{ is true}$$

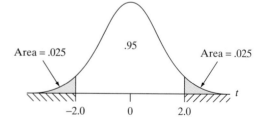

Area = .025   .95   Area = .025

Figure 7.14 A t test at $\alpha = .05$. H_0 is rejected if t_s falls in the hatched region.

−2.0   0   2.0 t

---

* The value of $t_{.025} = 1.989$ is based on 84.9 degrees of freedom. If we were to use 80 degrees of freedom (i.e., if we had to rely on Table 4, rather than a computer) the t-multiplier would be 1.990. This makes almost no difference in the resulting confidence interval.

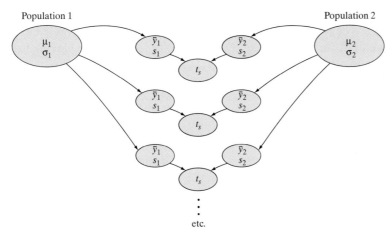

Population 1   Population 2

Figure 7.15 Meta-experiment for the *t* test

This probability has meaning in the context of a meta-experiment (depicted in Figure 7.15) in which we repeatedly sample from two populations and calculate a value of t_s. It is important to realize that the probability refers to a situation in which H_0 is true. In order to concretely picture such a situation, you are invited to suspend disbelief for a moment and come on an imaginary trip in Example 7.17.

Music and Marigolds.   Imagine that the scientific community has developed great interest in the influence of music on the growth of marigolds. One school of investigation centers on whether music written by Bach or Mozart produces taller plants. Plants are randomly allocated to listen to Bach (treatment 1) or Mozart (treatment 2) and, after a suitable period of listening, data are collected on plant height. The null hypothesis is

Example 7.17

> H_0: Marigolds respond equally well to Bach or Mozart.

or

$$H_0: \mu_1 = \mu_2$$

where

> μ_1 = Mean height of marigolds if exposed to Bach
> μ_2 = Mean height of marigolds if exposed to Mozart

Assume for the sake of argument that H_0 is in fact true. Imagine now that many investigators perform the Bach versus Mozart experiment, and that each experiment results in data with 60 degrees of freedom. Suppose each investigator analyzes his or her data with a *t* test at $\alpha = .05$. What conclusions will the investigators reach? In the meta-experiment of Figure 7.15, suppose each pair of samples represents a different investigator. Since we are assuming that μ_1 and μ_2 are actually equal, the values of t_s will deviate from 0 only because of chance sampling error. If all the investigators were to get together and make a frequency distribution of their t_s values, that distribution would follow a Student's *t* curve. The investigators would make their decisions as indicated by Figure 7.14, so we would expect them to have the following experiences:

> 95% of them would not reject H_0;
> 2.5% of them would reject H_0 and conclude that the plants prefer Bach;
> 2.5% of them would reject H_0 and conclude that the plants prefer Mozart.

Thus, a total of 5% of the investigators would reject the true null hypothesis. ∎

Example 7.17 provides an image for interpreting α. Of course, in analyzing data, we are dealing not with a meta-experiment but with a single experiment. When we perform a t test at the 5% significance level, we are playing the role of one of the investigators in Example 7.17, and the others are imaginary. If we reject H_0, there are two possibilities:

1. H_0 is in fact false; or

2. H_0 is in fact true, but we are one of the unlucky 5% who rejected H_0 anyway. In this case, we can think of rejecting H_0 as "setting off a false alarm."

We feel "confident" in our rejection of H_0 because the second possibility is unlikely (assuming that we regard 5% as a small percentage). Of course, we never know (unless someone replicates the experiment) whether or not we are one of the unlucky 5%.

Significance Level Versus P-Value. Students sometimes find it hard to distinguish between significance level (α) and P-value.* For the t test, both α and the P-value are tail areas under Student's t curve. But α is an arbitrary prespecified value; it can be (and should be) chosen before looking at the data. By contrast, the P-value is determined from the data; indeed, giving the P-value is a way of describing the data. You may find it helpful at this point to compare Figure 7.8 with Figure 7.14. The shaded area represents P-value in the former and α in the latter figure.

Type I and Type II Errors

We have seen that α can be interpreted as a probability:

$$\alpha = \Pr\{\text{reject } H_0\} \qquad \text{if } H_0 \text{ is true}$$

The erroneous rejection of H_0 when H_0 is true is called a **Type I error**. In choosing α, we are choosing our level of protection against Type I error. Many researchers regard 5% as an acceptably small risk. If we do not regard 5% as small enough, we might choose to use a more conservative value of α such as $\alpha = .01$; in this case the percentage of true null hypotheses that we reject would be not 5% but 1%.

In practice, the choice of α may depend on the context of the particular experiment. For example, a regulatory agency might demand more exacting proof of efficacy for a toxic drug than for a relatively innocuous one. Also, a person's choice of α may be influenced by his or her prior opinion about the phenomenon under study. For instance, suppose an agronomist is skeptical of claims for a certain soil treatment; in evaluating a new study of the treatment, he might express his skepticism by choosing a very conservative significance level (say, $\alpha = .001$), thus indicating that it would take a lot of evidence to convince him that the treatment is effective. Note that, if a written report of an investigation includes a P-value, then each reader is free to choose his or her own value of α in evaluating the reported results.

---

* Unfortunately, the term *significance level* is not used consistently by all people who write about statistics. A few authors use the terms *significance level* or *significance probability* where we have used P-value.

If H_0 is false (and H_A is true), but we do not reject H_0, then we have made a **Type II error**. Table 7.10 displays the situations in which Type I and Type II errors can occur. For example, if we reject H_0, then we eliminate the possibility of a Type II error, but by rejecting H_0 we may have made a Type I error.

The consequences of Type I and Type II errors can be very different. The following two examples show some of the variety of these consequences.

TABLE 7.10 Possible Outcomes of Testing H_0

		True Situation	
		H_0 *true*	H_0 *false*
Our	Do not reject H_0	Correct	Type II error
Decision	Reject H_0	Type I error	Correct

Marijuana and the Pituitary.   Cannabinoids, which are substances contained in marijuana, can be transmitted from mother to young through the placenta and through the milk. Suppose we conduct the following experiment on pregnant mice: We give one group of mice a dose of cannabinoids and keep another group as controls. We then evaluate the function of the pituitary gland in the offspring. The hypotheses would be

Example 7.18

H_0: Cannabinoids do not affect pituitary of offspring.

H_A: Cannabinoids do affect pituitary of offspring.

If in fact cannabinoids do not affect the pituitary of the offspring, but our data lead us to reject H_0, this would be a Type I error; the consequence might be unnecessary alarm if the conclusion were made public. On the other hand, if cannabinoids do affect the pituitary of the offspring, but our *t* test results in nonrejection of H_0, this would be a Type II error; one consequence might be unjustifiable complacency on the part of marijuana-smoking mothers. ∎

Immunotherapy.   Chemotherapy is standard treatment for a certain cancer. Suppose we conduct a clinical trial to study the efficacy of supplementing the chemotherapy with immunotherapy (stimulation of the immune system). Patients are given either chemotherapy or chemotherapy plus immunotherapy. The hypotheses would be

Example 7.19

H_0: Immunotherapy is not effective in enhancing survival.

H_A: Immunotherapy does effect survival.

If immunotherapy is actually not effective, but our data lead us to reject H_0 and conclude that immunotherapy is effective, then we have made a Type I error. The consequence, if this conclusion is acted on by the medical community, might be the widespread use of unpleasant, dangerous, and worthless immunotherapy. If, on the other hand, immunotherapy is actually effective, but our data do not enable us to detect that fact (perhaps because our sample sizes are too small), then we have made a Type II error, with consequences quite different from those of a Type I error: The standard treatment will continue to be used until someone provides convincing evidence that supplementary immunotherapy is effective. If we still "believe" in immunotherapy, we can conduct another trial (perhaps with larger samples) to try again to establish its effectiveness.   ■

As the foregoing examples illustrate, the consequences of a Type I error are usually quite different from those of a Type II error. The likelihoods of the two types of error may be very different, also. The significance level α is the probability of rejecting H_0 if H_0 is true. Because α is chosen at will, the hypothesis testing procedure "protects" you against Type I error by giving you control over the risk of such an error. This control is independent of the sample size and other factors. The chance of a Type II error, by contrast, depends on many factors, and may be large or small. In particular, an experiment with small sample sizes often has a high risk of Type II error.

We are now in a position to reexamine Carl Sagan's aphorism that "Absence of evidence is not evidence of absence." Because the risk of Type I error is controlled and that of Type II error is not, our state of knowledge is much stronger after rejection of a null hypothesis than after nonrejection. For example, suppose we are testing whether a certain soil additive is effective in increasing the yield of field corn. If we reject H_0, then either (1) we are right; or (2) we have made a Type I error; since the risk of a Type I error is controlled, we can be relatively confident of our conclusion that the additive is effective (although not necessarily very effective). Suppose, on the other hand, that the data are such that we do not reject H_0. Then either (1) we are right (that is, H_0 is true), or (2) we have made a Type II error. Since the risk of a Type II error may be quite high, we cannot say confidently that the additive is ineffective. In order to justify a claim that the additive is ineffective, we would need to supplement our test of hypothesis with further analysis, such as a confidence interval or an analysis of the chance of Type II error. We will consider this in more detail in Sections 7.7 and 7.8.

Power

As we have seen, Type II error is an important concept. The probability of making a Type II error is denoted by β:

$$\beta = \Pr\{\text{do not reject } H_0\} \qquad \text{if } H_0 \text{ is false}$$

The chance of not making a Type II error when H_0 is false—that is, the chance of rejecting H_0 when it is false—is called the **power** of a statistical test:

$$\text{Power} = 1 - \beta = \Pr\{\text{reject } H_0\} \qquad \text{if } H_0 \text{ is false}$$

Thus, the power of a t test is a measure of the sensitivity of the test, or the ability of the test procedure to detect a difference between μ_1 and μ_2 when such a difference really *does* exist. In this way the power is analogous to the resolving power of a microscope.

The power of a statistical test depends on many factors in an investigation, including the sample sizes and the inherent variability of the observations. All other things being equal, using larger samples gives more information and thereby increases power. In addition, we will see that some statistical tests can be more powerful than others, and that some study designs can be more powerful than others.

The planning of a scientific investigation should always take power into consideration. No one wants to emerge from lengthy and perhaps expensive labor in the lab or the field, only to discover upon analyzing the data that the sample sizes were insufficient or the experimental material too variable, so that experimental effects that were considered important were not detected. Two techniques are available to aid the researcher in planning for adequate sample sizes. One technique is to decide how small each standard error ought to be and choose *n* using an analysis such as that of Section 6.4. A second technique is a quantitative analysis of the power of the statistical test. Such an analysis for the *t* test is discussed in Section 7.8.

Exercises 7.39–7.45

7.39   *(Sampling exercise)*   Refer to the collection of 100 ellipses shown with Exercise 3.1, which can be thought of as representing a natural population of the organism *C. ellipticus*. Use random digits (from Table 1 or your calculator) to choose two random samples of five ellipses each. Use a metric ruler to measure the body length of each ellipse; measurements to the nearest millimeter will be adequate.

   (a) Compare the means of your two samples, using a *t* test at $\alpha = .05$.
   (b) Did the analysis of part (a) lead you to a Type I error, a Type II error, or no error?

7.40   *(Sampling exercise)*   Simulate choosing random samples from two different populations, as follows. First, proceed as in Exercise 7.39 to choose two random samples of five ellipses each and measure their lengths. Then add 6 mm to *each* measurement in one of the samples.

   (a) Compare the means of your two samples, using a *t* test at $\alpha = .05$.
   (b) Did the analysis of part (a) lead you to a Type I error, a Type II error, or no error?

7.41   *(Sampling exercise)*   Prepare simulated data as follows. First, proceed as in Exercise 7.39 to choose two random samples of five ellipses each and measure their lengths. Then, toss a coin. If the coin falls heads, add 6 mm to *each* measurement in one of the samples. If the coin falls tails, do not modify either sample.

   (a) Prepare two copies of the simulated data. On the Student Copy, show the data only; on the Instructor Copy, indicate also which sample (if any) was modified.
   (b) Give your Instructor Copy to the instructor and trade your Student Copy with another student when you are told to do so.
   (c) After you have received another student's paper, compare the means of his or her two samples using a two-tailed *t* test at $\alpha = .05$. If you reject H_0, decide which sample was modified.

7.42   Suppose a new drug is being considered for approval by the Food and Drug Administration. The null hypothesis is that the drug is not effective. If the FDA approves the drug, what type of error, Type I or Type II, could not possibly have been made?

7.43   In Example 7.16, the null hypothesis was not rejected. What type of error, Type I or Type II, might have been made in that *t* test?

7.44   Suppose that a 95% confidence interval for $(\mu_1 - \mu_2)$ is calculated to be $(1.4, 6.7)$. If we test $H_0: \mu_1 - \mu_2 = 0$ versus $H_0: \mu_1 - \mu_2 \neq 0$ using $\alpha = .05$, will we reject H_0? Why or why not?

7.45   Suppose that a 95% confidence interval for $(\mu_1 - \mu_2)$ is calculated to be $(-7.4, -2.3)$. If we test $H_0: \mu_1 = \mu_2$ versus $H_0: \mu_1 \neq \mu_2$ using $\alpha = .10$, will we reject H_0? Why or why not?

7.6 ONE-TAILED t TESTS

The t test described in the preceding sections is called a **two-tailed t test** or a **two-sided t test** because the null hypothesis is rejected if t_s falls in either tail of the Student's t distribution and the P-value of the data is a two-tailed area under Student's t curve. A two-tailed t test is used to test the null hypothesis

$$H_0: \mu_1 = \mu_2$$

against the alternative hypothesis

$$H_A: \mu_1 \neq \mu_2$$

This alternative H_A is called a **nondirectional alternative**.

Directional Alternative Hypotheses

In some studies it is apparent from the beginning—*before* the data are collected—that there is only one reasonable direction of deviation from H_0. In such situations it is appropriate to formulate a directional alternative hypothesis. The following is a directional alternative:

$$H_A: \mu_1 < \mu_2$$

Another directional alternative is

$$H_A: \mu_1 > \mu_2$$

The following two examples illustrate situations where directional alternatives are appropriate.

Example 7.20

Niacin Supplementation.   Consider a feeding experiment with lambs. The observation Y will be weight gain in a two-week trial. Ten animals will receive diet 1, and ten animals will receive diet 2, where

$$\text{Diet 1} = \text{Standard ration} + \text{niacin}$$
$$\text{Diet 2} = \text{Standard ration}$$

On biological grounds it is expected that niacin may increase weight gain; there is no reason to suspect that it could possibly decrease weight gain. An appropriate formulation would be

H_0: Niacin is not effective in increasing weight gain ($\mu_1 = \mu_2$).
H_A: Niacin is effective in increasing weight gain ($\mu_1 > \mu_2$). ∎

Example 7.21

Hair Dye and Cancer.   Suppose a certain hair dye is to be tested to determine whether it is carcinogenic (cancer causing). The dye will be painted on the skins of 20 mice (group 1), and an inert substance will be painted on the skins of 20 mice (group 2) who will serve as controls. The observation Y will be the number of tumors appearing on each mouse. An appropriate formulation is

H_0: The dye is not carcinogenic ($\mu_1 = \mu_2$).
H_A: The dye is carcinogenic ($\mu_1 > \mu_2$). ∎

Note: If H_A is directional, then some people would rewrite H_0 to include the "opposite direction." For example, if H_A is $H_A: \mu_1 > \mu_2$, then we could write

H_0 as $H_0: \mu_1 \leq \mu_2$. Thus, the null hypothesis is stating that the mean of population 1 is not greater than the mean of population 2, whereas the alternative hypothesis asserts that the mean of population 1 *is* greater than the mean of population 2. Between these two hypotheses, all possibilities are covered.

The One-Tailed Test Procedure

When the alternative hypothesis is directional, the *t* test procedure must be modified. The modified procedure is called a **one-tailed *t* test** and is carried out in two steps as follows:

Step 1. Check directionality—see if the data deviate from H_0 in the direction specified by H_A:

(a) If not, the P-value is greater than .50.
(b) If so, proceed to step 2.

Step 2. The P-value of the data is the *one-tailed* area beyond t_s.

To conclude the test, we can make a decision at a prespecified significance level α: H_0 is rejected if $P \leq \alpha$.

The rationale of the two-step procedure is that the P-value measures deviation from H_0 in the direction specified by H_A. The one-tailed P-value is illustrated in Figure 7.16 and the two-step testing procedure is illustrated in Example 7.22.

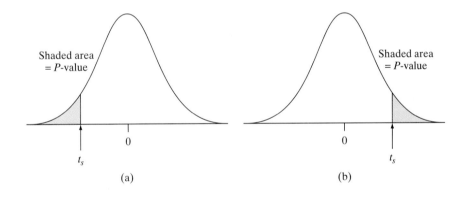

Figure 7.16 One-tailed *P*-value for a *t* test, (a) if the alternative is $H_A: \mu_1 < \mu_2$ and t_s is negative; (b) if the alternative is $H_A: \mu_1 > \mu_2$ and t_s is positive.

Niacin Supplementation. Consider the lamb feeding experiment of Example 7.20. The alternative hypothesis is

$$H_A: \mu_1 > \mu_2$$

Example 7.22

We will reject H_0 if $\bar{y}_1$ is sufficiently greater than $\bar{y}_2$. Suppose formula (7.1) yields df = 18. The critical values from Table 4 are reproduced in Table 7.11.

TABLE 7.11 Critical Values with df = 18

Tail area	.20	.10	.05	.04	.03	.025	.02	.01	.005	.0005
Critical value	0.862	1.330	1.734	1.855	2.007	2.101	2.214	2.552	2.878	3.922

To illustrate the one-tailed test procedure, suppose that we have[26]

$$\text{SE}_{(\bar{y}_1 - \bar{y}_2)} = 2.2 \text{ lb}$$

and that we choose $\alpha = .05$. Let us consider various possibilities for the two sample means.

(a) Suppose the data give $\bar{y}_1 = 10$ lb and $\bar{y}_2 = 13$ lb. This deviation from H_0 is opposite to the assertion of H_A: We have $\bar{y}_1 < \bar{y}_2$, but H_A asserts that $\mu_1 > \mu_2$. Consequently, P-value $> .5$, so we would not reject H_0 at any significance level. (We would never use an α greater than .50.) We conclude that the data provide no evidence that niacin is effective in increasing weight gain.

(b) Suppose the data give $\bar{y}_1 = 14$ lb and $\bar{y}_2 = 10$ lb. This deviation from H_0 is in the direction of H_A (because $\bar{y}_1 > \bar{y}_2$), so we proceed to step 2. The value of t_s is

$$t_s = \frac{(14 - 10) - 0}{2.2} = 1.82$$

The (one-tailed) P-value for the test is the probability of getting a t statistic, with 18 degrees of freedom, that is as large as or larger than 1.82. This upper tail probability (found with a computer) is .043, as shown in Figure 7.17.

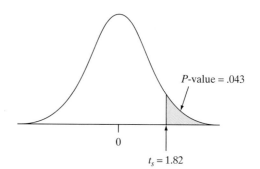

Figure 7.17 One-tailed P-value for the t test in Example 7.22

If we did not have a computer or graphing calculator available, we could use Table 4 to bracket the P-value. From Table 4, we see that the P-value would be bracketed as follows:

$$.04 < \text{one-tailed } P\text{-value} < .05$$

Since $P < \alpha$, we reject H_0 and conclude that there is some evidence that niacin is effective.

(c) Suppose the data give $\bar{y}_1 = 11$ lb and $\bar{y}_2 = 10$ lb. Then, proceeding as in part (b), we compute the test statistic as $t_s = .45$. The P-value is .329.

If we did not have a computer or graphing calculator available, we could use Table 4 to bracket the P-value as

$$P\text{-value} > .20$$

Since $P > \alpha$, we do not reject H_0; we conclude that there is insufficient evidence to claim that niacin is effective. Thus, although these data deviate from H_0 in the direction of H_A, the amount of deviation is not great enough to justify rejection of H_0. ∎

Notice that what distinguishes a one-tailed *t* test from a two-tailed *t* test is the way in which *P*-value is determined, but not the directionality or nondirectionality of the conclusion. If we reject H_0 our conclusion is directional even if our H_A is nondirectional.* (For instance, in Example 7.12 we concluded that toluene increases NE concentration.)

Directional Versus Nondirectional Alternatives

The same data will give a different *P*-value depending on whether the alternative hypothesis is directional or nondirectional. Indeed, if the data deviate from H_0 in the direction specified by H_A, the *P*-value for a directional alternative hypothesis will be 1/2 of the *P*-value for the test that uses a nondirectional alternative. It can happen that the same data will permit rejection of H_0 using the one-tailed procedure but not using the two-tailed procedure, as Example 7.23 shows.

Niacin Supplementation.   Consider part (b) of Example 7.22. In that example we chose $\alpha = .05$ and tested

$$H_0: \mu_1 = \mu_2$$

against the directional alternative hypothesis

$$H_A: \mu_1 > \mu_2$$

With $\bar{y}_1 = 14$ lb and $\bar{y}_2 = 10$ lb, the test statistic was $t_s = 1.82$ and the *P*-value was .043, as indicated in Figure 7.17. Our conclusion was to reject H_0.

However, suppose we had wished to test

$$H_0: \mu_1 = \mu_2$$

against the nondirectional alternative hypothesis

$$H_A: \mu_1 \neq \mu_2$$

With the same data of $\bar{y}_1 = 14$ lb and $\bar{y}_2 = 10$ lb, the test statistic is still $t_s = 1.82$. The *P*-value, however, is .086, as shown in Figure 7.18.

> **Example 7.23**

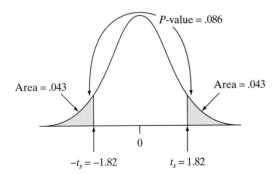

P-value = .086

Area = .043

Area = .043

0

$-t_s = -1.82$ $t_s = 1.82$

Figure 7.18 Two-tailed *P*-value for the *t* test in Example 7.23

Thus, *P*-value $> \alpha$ and we do not reject H_0.

Hence, the one-tailed procedure rejects H_0 but the two-tailed procedure does not. In this sense, it is "easier" to reject H_0 with the one-tailed procedure than with the two-tailed procedure. ∎

------

* Some authors prefer not to draw a directional conclusion if H_A is nondirectional.

Why is the two-tailed *P*-value cut in half when the alternative hypothesis is directional? In Example 7.23, the researcher would conclude by saying, "The data suggest that niacin increases weight gain. But if niacin has no effect, then the kind of data I got in my experiment—having two sample means that differ by 1.82 SEs or more—would happen fairly often (*P*-value .086). Sometimes the niacin diet would come out on top; sometimes the standard diet would come out on top. I cannot reject H_0 on the basis of what I have seen in these data." In Example 7.22(b), the researcher would conclude by saying, "*Before the experiment was run*, I suspected that niacin increases weight gain. The data provide evidence in support of this theory. If niacin has no effect, then the kind of data I got in my experiment—having the niacin diet sample mean exceed the standard diet that differ by 1.82 SEs or more—would rarely happen (*P*-value .043). (Before the experiment was run I dismissed the possibility that the niacin diet mean could be less than the standard diet mean.) Thus, I can reject H_0." The researcher in Example 7.22(b) is using *two* sources of information in rejecting H_0: (1) what the data have to say (as measured by the tail area), and (2) previous expectations (which allow the researcher to ignore the lower tail area—the .043 area under the curve below -1.82 in Figure 7.18).

Note that the modification in procedure, when going from a two-tailed to a one-tailed test, preserves the interpretation of significance level α as given in Section 7.5, that is,

$$\alpha = \Pr\{\text{reject } H_0\} \qquad \text{if } H_0 \text{ is true}$$

For instance, consider the case $\alpha = .05$. Figure 7.19 shows that the total shaded area—the probability of rejecting H_0—is equal to .05 in both a two-tailed test and a one-tailed test. This means that, if a great many investigators were to test a true H_0, then 5% of them would reject H_0 and commit a Type I error; this statement is true whether the alternative H_A is directional or nondirectional.

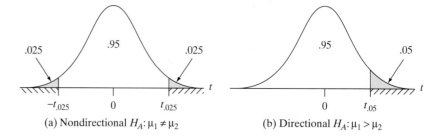

Figure 7.19 Two-tailed and one-tailed *t* test with $\alpha = .05$. H_0 is rejected if t_s falls in the hatched region of the *t*-axis.

The crucial point in justification of the modified procedure for testing against a directional H_A is that *if* the direction of deviation of the data from H_0 is *not* as specified by H_A, then we will not reject H_0. For example, in the carcinogenesis experiment of Example 7.16, if the mice exposed to the hair dye had *fewer* tumors than the control group, we might (1) simply conclude that the data do not indicate a carcinogenic effect, or (2) if the exposed group had *substantially* fewer tumors, so that the test statistic t_s was very far in the wrong tail of the *t* distribution, we might look for methodological errors in the experiment—for example, mistakes in lab technique or in recording the data, nonrandom allocation of the mice to the two groups, and so on—but we would not reject H_0.

A one-tailed t test is especially natural when only one direction of deviation from H_0 is believed to be plausible. However, one-tailed tests are also used in situations where deviation in both directions is possible, but only one direction is of interest. For instance, in the niacin experiment of Example 7.22, it is not necessary that the experimenter believe that it is *impossible* for niacin to reduce weight gain rather than increase it. Deviations in the wrong direction (less weight gain on niacin) would not lead to rejection of H_0, and thus to no claims about the effect of niacin; this is the essential feature that distinguishes a directional from a nondirectional formulation.

Choosing the Form of H_A

When is it legitimate to use a directional H_A, and so to perform a one-tailed test? The answer to this question is linked to the directionality check—step 1 of the two-step test procedure given previously. Clearly such a check makes sense only if H_A was formulated before the data were inspected. (If we were to formulate a directional H_A that was "inspired" by the data, then of course the data would always deviate from H_0 in the "right" direction and the test procedure would always proceed to step 2.) This is the rationale for the following rule.

> **Rule for Directional Alternatives**
> It is legitimate to use a directional alternative H_A only if H_A is formulated before seeing the data.

In research, investigators often get more pleasure from rejecting a null hypothesis than from not rejecting one. In fact, research reports often contain phrases such as "we are unable to reject the null hypothesis" or "the results failed to reach statistical significance." Under these circumstances, we might wonder what the consequences would be if researchers succumbed to the natural temptation to ignore the preceding rule for using directional alternatives. After all, very often we can think of a rationale for an effect ex post facto—that is, after the effect has been observed. A return to the imaginary experiment on plants' musical tastes will illustrate this situation.

Music and Marigolds.   Recall the imaginary experiment of Example 7.17, in which investigators measure the heights of marigolds exposed to Bach or Mozart. Suppose, as before, that the null hypothesis is true, that df $= 60$, and that the investigators all perform t tests at $\alpha = .05$. Now suppose in addition that all of the investigators violate the rule for use of directional alternatives, and that they formulate H_A after seeing the data. Half of the investigators would obtain data for which $\bar{y}_1 > \bar{y}_2$, and they would formulate the alternative

$$H_A: \mu_1 > \mu_2 \quad \text{(plants prefer Bach)}$$

The other half would obtain data for which $\bar{y}_1 < \bar{y}_2$, and they would formulate the alternative

$$H_A: \mu_1 < \mu_2 \quad \text{(plants prefer Mozart)}$$

Now envision what would happen. Since the investigators are using directional alternatives, they will all compute P-values using only one tail of the distribution. We would expect them to have the following experiences:

Example 7.24

90% of them would get a t_s in the middle 90% of the distribution and would not reject H_0;

5% of them would get a t_s in the top 5% of the distribution and would conclude that the plants prefer Bach;

5% of them would get a t_s in the bottom 5% of the distribution and would conclude that the plants prefer Mozart.

Thus, a total of 10% of the investigators would reject the true null hypothesis. Of course, each investigator individually never realizes that the overall percentage of Type I errors is 10% rather than 5%. And the conclusions that plants prefer Bach or Mozart could be supported by ex post facto rationales that would be limited only by the imagination of the investigators. ■

As Example 7.24 illustrates, a researcher who uses a directional alternative when it is not justified pays the price of a doubled risk of Type I error.

Computer note: Switching between a directional alternative and a nondirectional alternative does not affect the calculated value of the test statistic, t_s, but it does change the *P*-value of the test.

When conducting a test using statistical software, we must specify the type of alternative hyposthesis we wish to use. For example, consider again using the MINITAB system to analyze the Toluene data from Examples 7.9 and 7.12. At the end of Section 7.4 we noted that the command

```
MTB  > TwoSample 95.0 'Toluene' 'Control';
SUBC > Alternative 0.
```

will produce a 95% confidence interval for the difference in population means together with a *t* test of $H_0: \mu_1 = \mu_2$ versus $H_A: \mu_1 \neq \mu_2$.

If we want to conduct a directional test, of $H_0: \mu_1 = \mu_2$ versus $H_A: \mu_1 > \mu_2$, we use the command

```
MTB  > TwoSample 95.0 'Toluene' 'Control';
SUBC > Alternative 1.
```

The resulting output of this command is

```
Twosample T for Toluene vs Control
              N      Mean      StDev   SE Mean
Toluene       6      540.8     66.1      27
Control       5      444.2     69.6      31
95% C.I. for mu Toluene - mu Control: (2, 192)
T-Test mu Toluene = mu Control (vs >): T= 2.34 P=0.023
DF= 8
```

Notice that the *P*-value of .023 is half of the *P*-value from the nondirectional test of .047.

It is important to keep track of the order in which the data columns are chosen, so that the direction desired for the alternative hypothesis agrees with the command we give to the computer. In the preceding command, we chose the Toluene column first, then the Control column.

If we wanted to conduct a directional test, of $H_0: \mu_1 = \mu_2$ versus $H_A: \mu_1 < \mu_2$, with data in columns C1 and C2, we would use the command

```
MTB  > TwoSample 95.0 C1 C2;
SUBC > Alternative -1.
```

Exercises 7.46–7.56

7.46 For each of the following data sets, use Table 4 to bracket the one-tailed *P*-value of the data as analyzed by the *t* test, assuming that the alternative hypothesis is $H_A: \mu_1 > \mu_2$.

(a)

	Sample 1	Sample 2
n	10	10
$\bar{y}$	10.8	10.5

$\text{SE}_{(\bar{y}_1 - \bar{y}_2)} = .23$ with df $= 18$

(b)

	Sample 1	Sample 2
n	100	100
$\bar{y}$	750	730

$\text{SE}_{(\bar{y}_1 - \bar{y}_2)} = 11$ with df $= 180$

7.47 For each of the following data sets, use Table 4 to bracket the one-tailed *P*-value of the data as analyzed by the *t* test, assuming that the alternative hypothesis is $H_A: \mu_1 > \mu_2$.

(a)

	Sample 1	Sample 2
n	10	10
$\bar{y}$	3.24	3.00

$\text{SE}_{(\bar{y}_1 - \bar{y}_2)} = 0.61$ with df $= 17$

(b)

	Sample 1	Sample 2
n	6	5
$\bar{y}$	560	500

$\text{SE}_{(\bar{y}_1 - \bar{y}_2)} = 45$ with df $= 8$

(c)

	Sample 1	Sample 2
n	20	20
$\bar{y}$	73	79

$\text{SE}_{(\bar{y}_1 - \bar{y}_2)} = 2.8$ with df $= 35$

7.48 For each of the following situations, suppose $H_0: \mu_1 = \mu_2$ is being tested against $H_A: \mu_1 > \mu_2$. State whether or not H_0 would be rejected.

(a) $t_s = 3.75$ with 19 degrees of freedom, $\alpha = .01$
(b) $t_s = 2.6$ with 5 degrees of freedom, $\alpha = .10$

(c) $t_s = 2.1$ with 7 degrees of freedom, $\alpha = .05$
(d) $t_s = 1.8$ with 7 degrees of freedom, $\alpha = .05$

7.49 For each of the following situations, suppose $H_0: \mu_1 = \mu_2$ is being tested against $H_A: \mu_1 < \mu_2$. State whether or not H_0 would be rejected.

(a) $t_s = -1.6$ with 23 degrees of freedom, $\alpha = .05$
(b) $t_s = -2.3$ with 5 degrees of freedom, $\alpha = .10$
(c) $t_s = 0.4$ with 16 degrees of freedom, $\alpha = .10$
(d) $t_s = -2.8$ with 27 degrees of freedom, $\alpha = .01$

7.50 Ecological researchers measured the concentration of red cells in the blood of 27 field-caught lizards (*Sceloporis occidetitalis*). In addition, they examined each lizard for infection by the malarial parasite *Plasmodium*. The red cell counts ($10^{-3} \cdot$ cells per mm^3) were as reported in the table.[27]

	Infected Animals	Noninfected Animals
n	12	15
$\bar{y}$	972.1	843.4
s	245.1	251.2

We might expect that malaria would reduce the red cell count, and in fact previous research with another lizard species had shown such an effect. Do the data support this expectation? Assume that the data are normally distributed. Test the null hypothesis of no difference against the alternative that the infected population has a lower red cell count. Use a t test at

(a) $\alpha = .05$ (b) $\alpha = .10$

Note: Formula (7.1) yields 24 df.

7.51 A study was undertaken to compare the respiratory responses of hypnotized and nonhypnotized subjects to certain instructions. The 16 male volunteers were allocated at random to an experimental group to be hypnotized and to a control group. Baseline measurements were taken at the start of the experiment. In analyzing the data, the researchers noticed that the baseline breathing patterns of the two groups were different; this was surprising, since all the subjects had been treated the same up to that time. One explanation proposed for this unexpected difference was that the experimental group were more excited in anticipation of the experience of being hypnotized. The accompanying table presents a summary of the baseline measurements of total ventilation (liters of air per minute per square meter of body area). Parallel dotplots of the data are given in the graph that follows.[28] *Note*: Formula (7.1) yields 14 df.

	Experimental	Control
	5.32	4.50
	5.60	4.78
	5.74	4.79
	6.06	4.86
	6.32	5.41
	6.34	5.70
	6.79	6.08
	7.18	6.21
n	8	8
$\bar{y}$	6.169	5.291
s	.621	.652

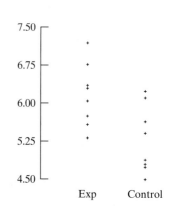

(a) Use a t test to test the hypothesis of no difference against a nondirectional alternative. Let $\alpha = .05$.

(b) Use a t test to test the hypothesis of no difference against the alternative that the experimental conditions produce a larger mean than the control conditions. Let $\alpha = .05$.

(c) Which of the two tests, that of part (a) or part (b), is more appropriate? Explain.

7.52 In a study of lettuce growth, 10 seedlings were randomly allocated to be grown in either standard nutrient solution or in a solution containing extra nitrogen. After 22 days of growth, the plants were harvested and weighed, with the results given in the table.[29] Are the data sufficient to conclude that the extra nitrogen enhances plant growth under these conditions? Use a t test at $\alpha = .10$ against a directional alternative. (Assume that the data are normally distributed.) *Note*: Formula (7.1) yields 7.7 df.

		Leaf Dry Weight (g)	
Nutrient solution	n	*Mean*	*SD*
Standard	5	3.62	.54
Extra nitrogen	5	4.17	.67

7.53 An entomologist conducted an experiment to see if wounding a tomato plant would induce changes that improve its defense against insect attack. She grew larvae of the tobacco hornworm (*Manduca sexta*) on wounded plants or control plants. The accompanying table shows the weights (mg) of the larvae after 7 days of growth.[30] (Assume that the data are normally distributed.) How strongly do the data support the researchers expectation? Use a t test at the 5% significance level. Let H_A be that wounding the plant tends to diminish larval growth. *Note*: Formula (7.1) yields 31.8 df.

	Wounded	**Control**
n	16	18
$\bar{y}$	28.66	37.96
s	9.02	11.14

7.54 A pain-killing drug was tested for efficacy in 50 women who were experiencing uterine cramping pain following childbirth. Twenty-five of the women were randomly allocated to receive the drug, and the remaining 25 received a placebo (inert substance). Capsules of drug or placebo were given before breakfast and again at noon. A pain relief score, based on hourly questioning throughout the day, was computed for each woman. The possible pain relief scores ranged from 0 (no relief) to 56 (complete relief for 8 hours). Summary results are shown in the table.[31] *Note*: Formula (7.1) yields 47.2 df.

		Pain Relief Score	
Treatment	n	*Mean*	*SD*
Drug	25	31.96	12.05
Placebo	25	25.32	13.78

(a) Test for evidence of efficacy using a t test. Use a directional alternative and $\alpha = .05$.

(b) If the alternative hypothesis were nondirectional, how would the answer to part (a) change?

7.55 In Example 7.15 we considered testing $H_0: \mu_1 = \mu_2$ against the nondirectional alternative hypothesis $H_A: \mu_1 \neq \mu_2$ and found that the P-value could be bracketed as $.06 < P\text{-value} < .10$. Recall that the sample mean for the group 1 (the control group) was 15.9, which was less than the sample mean of 11.0 for group 2 (the group treated with Ancymidol). However, Ancymidol is considered to be a growth inhibitor, which means that we would expect the control group to have a larger mean than the treatment group if ancy has any effect on the type of plant being studied (in this case, the Wisconsin Fast Plant). Suppose the researcher had expected ancy to retard growth—before conducting the experiment—and had conducted a test of $H_0: \mu_1 = \mu_2$ against the nondirectional alternative hypothesis $H_A: \mu_1 > \mu_2$, using $\alpha = .05$. What would be the bounds on the P-value? Would H_0 be rejected? Why or why not? What would be the conclusion of the experiment? *Note*: This problem requires almost no calculation.

7.56 *(Computer exercise)* An ecologist studied the habitat of a marine reef fish, the six bar wrasse (*Thalassoma hardwicke*), near an island in French Polynesia that is surrounded by a barrier reef. He examined 48 patch reef settlements at each of two distances from the reef crest: 250 meters from the crest and 800 meters from the crest. For each patch reef, he calculated the "settler density," which is the number of settlers (juvenile fish) per unit of settlement habitat. Before collecting the data, he hypothesized that the settler density might decrease as distance from the reef crest increased, since the way that waves break over the reef crest causes resources (i.e., food) to tend to decrease as distance from the reef crest increases. Here are the data:[32]

250 meters				800 meters	
0.318	0.758	0.318	0.941	0.289	0.399
0.637	0.372	0.524	0.279	0.392	0.955
0.196	0.637	1.404	1.021	0.725	0.531
0.624	1.560	0.000	0.108	1.318	0.252
0.909	0.207	1.061	0.738	0.612	1.179
0.295	0.685	0.590	0.907	0.637	0.442
0.594	0.000	0.363	0.503	0.181	0.291
0.442	1.303	1.567	0.637	0.941	0.579
1.220	0.898	1.577	1.498	0.265	0.252
1.303	1.157	0.312	0.866	0.979	0.373
0.187	0.970	0.758	0.588	0.909	0.000
1.560	0.624	0.505	0.606	0.283	0.463
0.849	1.592	0.909	0.490	0.337	1.248
2.411	1.019	0.362	0.163	0.813	2.010
1.705	0.829	0.329	0.277	0.000	1.213
1.019	0.884	0.909	0.293	0.544	0.808

For 250 meters, the sample mean is 0.818 and the sample SD is 0.514. For 800 meters, the sample mean is 0.628 and the sample SD is 0.413. Do these data provide statistically significant evidence, at the .10 level, to support the ecologist's theory? *Note*: Formula (7.1) yields 89.8 df.

7.7 MORE ON INTERPRETATION OF STATISTICAL SIGNIFICANCE

Ideally, statistical analysis should aid the researcher by helping to clarify whatever message is contained in the data. For this purpose, it is not enough that the statistical calculations be correct; the results must also be correctly interpreted. In this section we explore some principles of interpretation that apply not only to the *t* test, but also to other statistical tests to be discussed later.

Significant Difference Versus Important Difference

The term *significant* is often used in describing the results of a statistical analysis. For example, if an experiment to compare a drug against a placebo gave data with a very small *P*-value, then the conclusion might be stated as "The effect of the drug was highly significant." As another example, if two fertilizers for wheat gave a yield comparison with a large *P*-value, then the conclusion might be stated as "The wheat yields did not differ significantly between the two fertilizers" or "The difference between the fertilizers was not significant." As a third example, suppose a substance is tested for toxic effects by comparing exposed animals and control animals, and that the null hypothesis of no difference is not rejected. Then the conclusion might be stated as "No significant toxicity was found."

Clearly such phraseology using the term *significant* can be seriously misleading. After all, in ordinary English usage, the word *significant* connotes "substantial" or "important." In statistical jargon, however, the statement

"The difference was significant"

means nothing more or less than

"The null hypothesis of no difference was rejected."

This is to say, "We rejected the claim that the difference in sample means was caused by chance error."

By the same token, the statement

"The difference was not significant"

means

"The null hypothesis of no difference was not rejected."

It would perhaps be preferable if a different word were used in place of *significant*, such as *discernible* (meaning that the test discerned a difference). Alas, the specialized usage of the word *significant* has become quite common in scientific writing, and understandably is the source of much confusion.

It is essential to recognize that a statistical test provides information about only one question: Is the difference observed in the data large enough to infer that a difference in the same direction exists in the population? The question of whether a difference is *important*, as opposed to (statistically) significant, cannot be decided on the basis of the *P*-value. The following two examples illustrate this fact.

Serum LD. Lactate dehydrogenase (LD) is an enzyme that may show elevated activity following damage to the heart muscle or other tissues. A large study of serum LD levels in healthy young people yielded the results shown in Table 7.12.[33]

Example 7.25

TABLE 7.12 Serum LD (U/L)		
	Males	**Females**
n	270	264
$\bar{y}$	60	57
s	11	10

The difference between males and females is quite significant; in fact, $t_s = 3.3$, which gives a P-value of $P \approx .001$. However, this does not imply that the difference is large or important. ∎

Example 7.26 | **Body Weight.** Imagine that we are studying the body weight of men and women, and we obtain the fictitious but realistic data shown in Table 7.13.[34]

TABLE 7.13 Body Weight (lb)		
	Males	**Females**
n	2	2
$\bar{y}$	175	143
s	35	34

For these data the t test gives $t_s = .93$ and a P-value of $P \approx .45$. The observed difference between males and females is not small (it is $175 - 143 = 32$ lb), yet it is not statistically significant for any reasonable choice of α. The lack of statistical significance does not imply that the sex difference in body weight is small or unimportant. It means only that the data are inadequate to characterize the difference in the population means. A sample difference of 32 lb could easily happen by chance if the two populations are identical. ∎

Effect Size

The preceding examples show that the statistical significance or nonsignificance of a difference does not indicate whether the difference is important. Nevertheless, the question of "importance" can and should be addressed in most data analyses. To assess importance, we need to consider the *magnitude* of the difference. In Example 7.25 the male versus female difference is "statistically significant," but this is largely due to the sample sizes being quite large. A t test uses the test statistic

$$t_s = \frac{(\bar{y}_1 - \bar{y}_2) - 0}{\text{SE}_{(\bar{y}_1 - \bar{y}_2)}}$$

If n_1 and n_2 are large, then $\text{SE}_{(\bar{y}_1 - \bar{y}_2)}$ will be small and the test statistic will tend to be large. Thus, we might reject H_0 due to the sample size being large, even if μ_1 and μ_2 are nearly equal. The sample size acts like a magnifying glass: The larger the sample size, the smaller the difference that can be detected in a hypothesis test.

The **effect size** in a study is the difference between μ_1 and μ_2, expressed relative to the standard deviation of one of the populations. If the two populations have the same standard deviation, σ, then the effect size is*

$$\text{Effect size} = \frac{|\mu_1 - \mu_2|}{\sigma}$$

Of course, when working with sample data we can only calculate an *estimated* effect size by using sample values in place of the unknown population values.

---

* If the standard deviations are not equal, we can use the larger SD in defining the effect size.

Serum LD.   For the data given in Example 7.25 (Table 7.12) the difference in sample means, $60 - 57 = 3$, is less than one-third of a standard deviation. Using the larger sample SD we can calculate a sample effect size of

$$\text{Effect size} = \frac{(\bar{y}_1 - \bar{y}_2)}{s} = \frac{60 - 57}{11} = 0.27$$

This indicates that there is a lot of overlap between the two groups. Figure 7.20 shows the extent of the overlap that occurs if two normally distributed populations differ on average by .27 SDs. ■

Body Weight.   For the data given in Example 7.26 (Table 7.13) the difference in sample means, $175 - 143 = 32$, is roughly one standard deviation. The sample effect size is

$$\text{Effect size} = \frac{(\bar{y}_1 - \bar{y}_2)}{s} = \frac{175 - 143}{35} = 0.91$$

Figure 7.21 shows the extent of the overlap that occurs if two normally distributed populations differ on average by .91 SDs. ■

The definition of effect size that we are using is probably unfamiliar to the biologically oriented reader. It is more common in biology to "standardize" a difference of two quantities by expressing it as a percentage of one of them. For example, the weight difference given in Table 7.13 between males and females, expressed as a percentage of mean female weight, is

$$\frac{\bar{y}_1 - \bar{y}_2}{\bar{y}_2} = \frac{175 - 143}{143} = .22 \text{ or } 22\%$$

Thus, the males are about 22% heavier than the females. However, from a statistical viewpoint it is often more relevant that the average weights for males and females are .91 SDs apart.

Confidence Intervals to Assess Importance

Calculating the effect size is one way to quantify how far apart two sample means are. Another reasonable approach is to use the observed difference $(\bar{y}_1 - \bar{y}_2)$ to construct a confidence interval for the population difference $(\mu_1 - \mu_2)$. In interpreting the confidence interval, the judgment of what is "important" is made on the basis of experience with the particular practical situation. The following three examples illustrate this use of confidence intervals.

Serum LD.   For the LD data of Example 7.25, a 95% confidence interval for $(\mu_1 - \mu_2)$ is

$$3 \pm 1.8$$

or

$$(1.2, 4.8)$$

Example 7.27

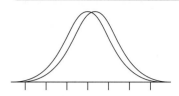

Figure 7.20 Overlap between two normally distributed populations when the effect size is .27

Example 7.28

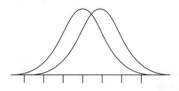

Figure 7.21 Overlap between two normally distributed populations when the effect size is .91

Example 7.29

This interval implies (with 95% confidence) that the population mean difference between the sexes does not exceed 4.8 U/Li. A physician evaluating this information would know that 4.8 U/Li is less than the typical day-to-day fluctuation in a person's LD level, and that therefore the sex difference is negligible from the medical standpoint. For example, the physician might conclude that it is unnecessary to differentiate between the sexes in establishing clinical thresholds for diagnosis of illness. Thus, the sex difference in LD may be said to be statistically significant but medically unimportant. To put this another way, the data suggest that men do in fact tend to have higher levels than women, but not very much higher. ■

Example 7.30

Body Weight.   For the body-weight data of Example 7.26, a 95% confidence interval for $(\mu_1 - \mu_2)$ is

$$32 \pm 149$$

or

$$(-117, 181)$$

From this confidence interval we cannot tell whether the true difference (between the population means) is large favoring females, is small, or is large favoring males. Because the confidence interval contains numbers of small magnitude and numbers of large magnitude, it does not tell us whether the difference between the sexes is important or unimportant. A definite statement about the importance of the difference would require more data. Suppose, for example, that the means and standard deviations were as given in Table 7.13, but that they were based on 2000 rather than 2 people of each sex. Then the 95% confidence interval would be

$$32 \pm 2$$

or

$$(30, 34)$$

This interval would imply (with 95% confidence) that the difference is at least 30 lb, an amount that might reasonably be regarded as important, at least for some purposes. ■

Example 7.31

Yield of Tomatoes.   Suppose a horticulturist is comparing the yields of two varieties of tomatoes; yield is measured as pounds of tomatoes per plant. On the basis of practical considerations, the horticulturist has decided that a difference between the varieties is "important" only if it exceeds 1 pound per plant, on the average. That is, the difference is important if

$$|\mu_1 - \mu_2| > 1.0 \text{ lb}$$

Suppose the horticulturist's data give the following 95% confidence interval:

$$(.2, .3)$$

Because all values in the interval are less than 1.0 lb, the data support (with 95% confidence) the assertion that the difference is *not* important, using the horticulturist's criterion. ■

In many investigations, statistical significance and practical importance are both of interest. The following example shows how the relationship between these two concepts can be visualized using confidence intervals.

Yield of Tomatoes. Let us return to the tomato experiment of Example 7.31. The confidence interval was

Example 7.32

$$(.2, .3)$$

Recall from Section 7.5 that the confidence interval can be interpreted in terms of a t test. Because all values within the confidence interval are positive, a t test (two-tailed) at $\alpha = .05$ would reject H_0. Thus, the difference between the two varieties is statistically significant, although it is not horticulturally important: The data indicate that variety 1 is better than variety 2, but also that it is not much better. The distinction between significance and importance for this example can be seen in Figure 7.22, which shows the confidence interval plotted on the $(\mu_1 - \mu_2)$-axis. Note that the confidence interval lies entirely to one side of zero and also entirely to one side of the "importance" threshold of 1.0.

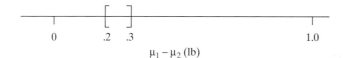

Figure 7.22 Confidence interval for Example 7.32

To further explore the relationship between significance and importance, let us consider other possible outcomes of the tomato experiment. Table 7.14 shows how the horticulturist would interpret various possible confidence intervals, still using the criterion that a difference must exceed 1.0 lb in order to be considered important.

TABLE 7.14 Interpretation of Confidence Intervals

95% Confidence Interval	Is the Difference	
	Significant?	*Important?*
(.2, .3)	Yes	No
(1.2, 1.3)	Yes	Yes
(.2, 1.3)	Yes	Cannot tell
(−.2, .3)	No	No
(−1.2, 1.3)	No	Cannot tell

Table 7.14 shows that a significant difference may or may not be important, and an important difference may or may not be significant. In practice, the assessment of importance using confidence intervals is a simple and extremely useful supplement to a test of hypothesis. ■

Exercises 7.57–7.63

7.57 A field trial was conducted to evaluate a new seed treatment that was supposed to increase soybean yield. When a statistician analyzed the data, the statistician found that the mean yield from the treated seeds was 40 lb/acre greater than that from control plots planted with untreated seeds. However, the statistician declared the difference to be "not (statistically) significant." Proponents of the treatment objected strenuously to the statistician's statement, pointing out that, at current market prices, 40 lb/acre would bring a tidy sum, which would be highly significant to the farmer. How would you answer this objection?[35]

7.58 In a clinical study of treatments for rheumatoid arthritis, patients were randomly allocated to receive either a standard medication or a newly designed medication. After a suitable period of observation, statistical analysis showed that there was no significant difference in the therapeutic response of the two groups, but that the incidence of undesirable side effects was significantly lower in the group receiving the new medication. The researchers concluded that the new medication should be regarded as clearly preferable to the standard medication, because it had been shown to be equally effective therapeutically and to produce fewer side effects. In what respect is the researchers' reasoning faulty? (Assume that the term *significant* refers to rejection of H_0 at $\alpha = .05$.)

7.59 There is an old folk belief that the sex of a baby can be guessed before birth on the basis of its heart rate. In an investigation to test this theory, fetal heart rates were observed for mothers admitted to a maternity ward. The results (in beats per minute) are summarized in the table.[36]

		Heart Rate (bpm)	
	n	*Mean*	*SE*
Males	250	137.21	.62
Females	250	137.18	.53

Construct a 95% confidence interval for the difference in population means. Does the confidence interval support the claim that the population mean sex difference (if any) in fetal heart rates is small and unimportant? (Use your own "expert" knowledge of heart rate to make a judgment of what is "unimportant.")

7.60 Coumaric acid is a compound that may play a role in disease resistance in corn. A botanist measured the concentration of coumaric acid in corn seedlings grown in the dark or in a light/dark photoperiod. The results (nmol acid per g tissue) are given in the accompanying table.[37] *Note*: Formula (7.1) yields 5.7 df.

	Dark	**Photoperiod**
n	4	4
$\bar{y}$	106	102
s	21	27

Suppose the botanist considers the effect of lighting conditions to be "important" if the difference in means is 20%, that is, about 20 nmol/g. Based on a 95% confidence interval, do the preceding data indicate whether the true difference is "important"?

7.61 Repeat Exercise 7.60, assuming that the means and standard deviations are as given in the table, but that the sample sizes are ten times as large (that is, $n = 40$ for "dark" and $n = 40$ for "photoperiod"). *Note*: Formula (7.1) yields 73.5 df.

7.62 As part of a large study of serum chemistry in healthy people, the following data were obtained for the serum concentration of uric acid in men and women aged 18–55 years.[38]

	Serum Uric Acid (mmol/l)	
	Men	*Women*
n	530	420
$\bar{y}$	.354	.263
s	.058	.051

Construct a 95% confidence interval for the true difference in population means. Suppose the investigators feel that the difference in population means is "clinically important" if it exceeds .08 mmol/L. Does the confidence interval indicate whether the difference is "clinically important"? *Note*: Formula (7.1) yields 934 df.

7.63 Repeat Exercise 7.62, assuming that the means and standard deviations are as given in the table, but that the sample sizes are only one-tenth as large (that is, 53 men and 42 women). *Note*: Formula (7.1) yields 92 df.

7.8 PLANNING FOR ADEQUATE POWER (OPTIONAL)

We have defined the power of a statistical test as

$$\text{Power} = \Pr\{\text{reject } H_0\} \quad \text{if } H_0 \text{ is false}$$

To put this another way, the power of a test is the probability that it will yield a statistically significant result when it should (that is, when H_A is true).

Since the power is the probability of *not* making an error (of Type II), high power is desirable: If H_0 is false, a researcher would like to find that out when conducting a study. But power comes at a price. All other things being equal, more observations (larger samples) bring more power, but observations cost time and money. In this section we explain how a researcher can rationally plan an experiment to have adequate power for the purposes of the research project and yet cost as little as possible.

Specifically, we will consider the power of the two-sample t test, conducted at significance level α. We will assume that the populations are normal with equal SDs, and we denote the common value of the SD by σ (that is, $\sigma_1 = \sigma_2 = \sigma$). It can be shown that in this case, for a given total sample size of $2n$, the power is maximized if the sample sizes are equal; thus we will assume that n_1 and n_2 are equal and denote the common value by n (that is, $n_1 = n_2 = n$).

Under the aforementioned conditions, the power of the t test depends on the following factors: (a) α; (b) σ (c) n; (d) $(\mu_1 - \mu_2)$. After briefly discussing each of these factors, we will address the all-important question of choosing the value of n.

Dependence of Power on α

In choosing α, we choose a level of protection against Type I error. However, this protection is traded for vulnerability to Type II error. If, for example, we choose $\alpha = .01$ rather than $\alpha = .05$, then we are choosing to reject H_0 less readily, and so is (perhaps unwittingly) choosing to increase the risk of Type II error and reduce the power. Thus, there is an unavoidable trade-off between the risk of Type I error and the risk of Type II error.

Dependence on σ

The larger σ, the smaller the power (all other things being equal). Recall from Chapter 5 that the reliability of a sample mean is determined by the quantity

$$\sigma_{\bar{Y}} = \frac{\sigma}{\sqrt{n}}$$

The larger σ is, the more variability there is in the sample mean. Thus, having a larger σ implies having samples that produce less reliable information about each population mean, and so less power to discern a difference between them. In order to increase power, then, a researcher usually tries to design the investigation so as to have σ as small as possible. For example, a botanist will try to hold light conditions constant throughout a greenhouse area, a pharmacologist will use genetically identical experimental animals, and so on. Usually, however, σ cannot be reduced to zero; there is still considerable variation in the observations.

Dependence on n

The larger n, the higher the power (all other things being equal). If we increase n, we decrease $\sigma/\sqrt{n}$; this improves the precision of the sample means ($\bar{y}_1$ and $\bar{y}_2$). In addition, larger n gives more information about σ; this is reflected in a reduced critical value for the test (reduced because of more df). Thus, increasing n increases the power of the test in two ways.

Dependence on $(\mu_1 - \mu_2)$

In addition to the factors we have discussed, the power of the t test also depends on the actual difference between the population means, that is, on $(\mu_1 - \mu_2)$. This dependence is very natural, as illustrated by the following example.

Example 7.33   **Heights of People.**   In order to clearly illustrate the concepts, we consider a familiar variable, body height of people. Imagine what would happen if an investigator were to measure the heights of two random samples of eleven people each ($n = 11$), and then conduct a two-tailed t test at $\alpha = .05$.

(a) First, suppose that sample 1 consisted of 17-year-old males and sample 2 consisted of 17-year-old females. The two population means differ substantially; in fact, $(\mu_1 - \mu_2)$ is about 5 inches ($\mu_1 \approx 69.1$ and $\mu_2 \approx 64.1$ inches).[39] It can be shown (as we will see) that in this case the investigator has about a 99% chance of rejecting H_0 and correctly concluding that the males in the population of 17-year-olds are taller (on average) than the females.

(b) By contrast, suppose that sample 1 consisted of 17-year-old females and sample 2 consisted of 14-year-old females. The two population means differ, but by a modest amount; the difference is $(\mu_1 - \mu_2) = .6$ inch ($\mu_1 \approx 64.1$ and $\mu_2 \approx 63.5$ inches). It can be shown that in this case the investigator has less than a 10% chance of rejecting H_0; in other words, there is more than a 90% chance that the investigator will fail to detect the fact that 17-year-old girls are taller than 14-year-old girls. (In fact, it can be shown that there is a 29% chance that $\bar{y}_1$ will be less than $\bar{y}_2$—that is, there is a 29% chance that eleven 17-year-old girls chosen at random will be shorter on the average than eleven 14-year-old girls chosen at random!)

The contrast between cases (a) and (b) is not due to any change in the SDs; in fact, for each of the three populations the value of σ is about 2.5 inches. Rather, the contrast is due to the simple fact that, with a fixed n and σ, it is easier to detect a large difference than a small difference.   ■

Planning a Study

Suppose an investigator is planning a study for which the t test will be appropriate. How shall she take into account all the factors that influence the power of the test?

First consider the choice of significance level α. A simple approach is to begin by determining the cost of an adequately powerful study using a somewhat liberal choice (say, $\alpha = .05$ or .10). If that cost is not high, the investigator can consider reducing α (say, to .01) and see if an adequately powerful study is still affordable.

Suppose, then, that the investigator has chosen a working value of α. Suppose also that the experiment has been designed to reduce σ as far as practicable, and that the investigator has available an estimate or guess of the value of σ.

At this point, the investigator needs to ask herself about the magnitude of the difference she wants to detect. As we saw in Example 7.33, a given sample size may be adequate to detect a large difference in population means, but entirely inadequate to detect a small difference. As a more realistic example, an experiment using five rats in a treatment group and five rats in a control group might be large enough to detect a substantial treatment effect, while detection of a subtle treatment effect would require more rats (perhaps 30) in each group.

The preceding discussion suggests that choosing a sample size for adequate power is somewhat analogous to choosing a microscope: We need high resolving power if we want to see a very tiny structure; for large structures a hand lens will do. In order to proceed with planning the experiment, the investigator needs to decide how large an effect she is looking for.

Recall that in Section 7.7, we defined the effect size in a study as the difference between μ_1 and μ_2, expressed relative to the standard deviation of one of the populations. If, as we are assuming here, the two populations have the same standard deviation, σ, then the effect size is

$$\text{Effect size} = \frac{|\mu_1 - \mu_2|}{\sigma}$$

That is, the effect size is the difference in population means expressed relative to the common population SD. The effect size is a kind of "signal to noise ratio," where $(\mu_1 - \mu_2)$ represents the signal we want to detect and σ represents the background noise that tends to obscure the signal. Figure 7.23(a) shows two normal curves for which the effect size is .5; Figure 7.23(b) shows two normal curves for which the effect size is 4. Clearly, at a fixed sample size it is easier to detect the difference between the curves in graph (b) than it is in graph (a).

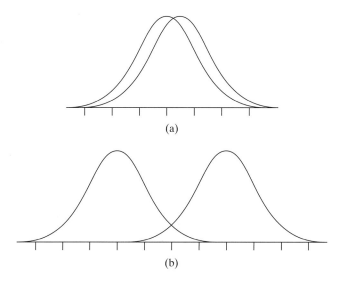

(a)

(b)

Figure 7.23 Normal distributions with an effect size (a) of .5 and (b) of 4

If α and the effect size have been specified, then the power of the t test depends only on the sample sizes (n). Table 5 at the end of the book shows the value of n required in order to achieve a specified power against a specified effect size. Let us see how Table 5 applies to our familiar example of body height.

Example 7.34

Heights of People. In Example 7.33, case (a), we considered samples of 17-year-old males and 17-year-old females. The effect size is

$$\frac{\left|\mu_1 - \mu_2\right|}{\sigma} = \frac{\left|69.1 - 64.1\right|}{2.5} = \frac{5}{2.5} = 2.0$$

For a two-tailed t test at $\alpha = .05$, Table 5 shows that the sample size required for a power of .99 is $n = 11$; this is the basis for the claim in Example 7.33 that the investigator has a 99% chance of detecting the difference between males and females. Figure 7.24 shows the two distributions being considered in Example 7.34.

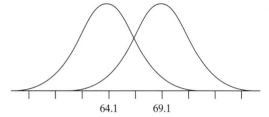

Figure 7.24 Height distributions for Example 7.34

Suppose 100 researchers each conduct the following study: Take a random sample of 11 17-year-old males and a random sample of 11 17-year-old females, find the sample average heights of the two groups, and then conduct a two-tailed t test of $H_0: \mu_1 = \mu_2$ using $\alpha = .05$. We would expect 99 of the 100 researchers to reject H_0 and conclude (correctly) that the average heights of 17-year-old males and females differ. We would expect one of the 100 researchers to conclude that there is not sufficient evidence, at the .05 level of significance, to reject H_0. (So one researcher would make a Type II error.) ∎

We could conduct a computer simulation of the study outlined previously. That is, we could use a computer to (1) get a random sample of size 11 from population 1, with $\mu_1 = 69.1$ and $\sigma = 2.5$; (2) get a random sample of size 11 from population 2, with $\mu_2 = 64.1$ and $\sigma = 2.5$; and (3) conduct the two-tailed t test, using $\alpha = .05$. If we ran this computer program 100 times, we would expect to see 99 cases in which H_0 is rejected and 1 case in which it is not rejected. Of course, we would probably want to run the program more than 100 times. We might run the program 10,000 times—and expect to see H_0 rejected 9900 times not rejected the other 100 times. (For a brief mathematical explanation of the calculations underlying Table 5, see Appendix 7.1.)

As we have seen, in order to choose a sample size the researcher needs to specify not only the size of the effect she wishes to detect, but also how certain she wants to be of detecting it; that is, it is necessary to specify how much power is wanted. Since the power measures the protection against Type II error, the choice of a desired power level depends on the consequences that would result from a Type II error. If the consequences of a Type II error would be very unfortunate (for example, if a promising but risky cancer treatment is being tested on humans and a negative result would discredit the treatment so that it would never be tested again), then the researcher might specify a high power, say .95 or .99. But of course high power is expensive in terms of n. For much research, a Type II error is not a disaster, and a lower power such as .80 is considered adequate.

The following example illustrates a typical use of Table 5 in planning an experiment.

Childhood Asthma.   A group of scientists wished to investigate the claim that chiropractic manipulation of the spine can help children with mild or moderate asthma. One group of children were to be given active treatment by chiropractors, while children in another group (the "control group") were to be given simulated treatment by the chiropractors. That is, the children in the control group were to be seen by chiropractors, but they were to be given a sham treatment, so that they would think that they had been treated, when in fact the chiropractor had not done any manipulation of the spine that was considered to be beneficial. The response variable for the study was the change in "peak expiratory flow," which is a measure of breathing capacity, after four months of therapy. Based on previous experience, it was thought that the SD of peak expiratory flow would be around 15%. The research team wanted to have at least an 80% chance of detecting a difference in mean peak expiratory flow between the two groups of 10%. They planned to conduct a two-tailed t test at the 5% significance level. The team had to decide how many children (n) to put in each group.

The effect size that the team wanted to consider is

$$\frac{|\mu_1 - \mu_2|}{\sigma} = \frac{10}{15} \approx .65$$

| Example 7.35 |

(Note that we are using 10 as the value of $|\mu_1 - \mu_2|$, since the team was interested in detecting a 10% difference between the groups.) For this effect size, and for a power of .80 with a two-tailed test at the 5% significance level, Table 5 yields $n = 39$, which means that 39 children were needed in each group.

At this point, the research team had to consider questions, such as (1) Is it feasible to enroll 78 children with asthma (39 for each group) in the study? If not, then (2) would they perhaps be willing to redefine the size of the difference between the groups that they considered to be important, in order to reduce the required n? With questions such as these, and repeated use of Table 5, they could finally decide on a firm value for n, or possibly decide to abandon the project because an adequate study would be too costly.

Normally the story ends here, but there was an extra wrinkle in the planning of this study: The research team knew from experience that some of the children enrolled in the study would drop out, for one reason or another, before the study ended. There is no formula or table that tells one how many subjects will drop out of a study such as this. Here the only guide is experience. In this case, the research team planned to enroll 100 children, in order to allow for some attrition and still end up with enough data so that they would have the power they wanted.

(*Note*: The research team ended up with 80 subjects and, through random allocation, had $n = 38$ in the active treatment group and $n = 42$ in the control group. The P-value for the resulting t test was .82. The conclusion they stated was "In children with mild or moderate asthma, the addition of chiropractic spinal manipulation to usual medical care provided no benefit." They could say this with the comfort of knowing that they had conducted a study that was large enough to have a good chance (80%) of detecting an important difference if there was one to be found.)[40]   ∎

Exercises 7.64–7.73

7.64 One measure of the meat quality of pigs is backfat thickness. Suppose two researchers, Jones and Smith, are planning to measure backfat thickness in two groups of pigs raised on different diets. They have decided to use the same number (n) of pigs in each group, and to compare the mean backfat thickness using a two-tailed t test at the 5% significance level. Preliminary data indicate that the SD of backfat thickness is about .3 cm.

When the researchers approach a statistician for help in choosing n, she naturally asks how much difference they want to detect. Jones replies, "If the true difference is 1/4 cm or more, I want to be reasonably sure of rejecting H_0." Smith replies, "If the true difference is 1/2 cm or more, I want to be very sure of rejecting H_0."

If the statistician interprets "reasonably sure" as 80% power, and "very sure" as 95% power, what value of n will she recommend

(a) to satisfy Jones's requirement?
(b) to satisfy Smith's requirement?

7.65 Refer to the brain NE data of Example 7.9. Suppose you are planning a similar experiment; you will study the effect of LSD (rather than toluene) on brain NE. You anticipate using a two-tailed t test at $a = .05$. Suppose you have decided that a 10% effect (increase or decrease in mean NE) of LSD would be important, and so you want to have good power (80%) to detect a difference of this magnitude.

(a) Using the data of Example 7.9 as a pilot study, determine how many rats you should have in each group. (The mean NE in the control group in Example 7.9 is 444.2 ng/g and the SD is = 69.6 ng/g.)
(b) If you were planning to use a one-tailed t test, what would be the required number of rats?

7.66 Suppose you are planning a greenhouse experiment on growth of pepper plants. You will grow n individually potted seedlings in standard soil and another n seedlings in specially treated soil. After 21 days, you will measure Y = total stem length (cm) for each plant. If the effect of the soil treatment is to increase the population mean stem length by 2 cm, you would like to have a 90% chance of rejecting H_0 with a one-tailed t test. Data from a pilot study (such as the data in Exercise 2.62) on 15 plants grown in standard soil give $\bar{y} = 12.5$ cm and $s = .8$ cm.

(a) Suppose you plan to test at $\alpha = .05$. Use the pilot information to determine what value of n you should use.
(b) What conditions are necessary for the validity of the calculation in part (a)? Which of these can be checked (roughly) from the data of the pilot study?
(c) Suppose you decide to adopt a more conservative posture and test at $\alpha = .01$. What value of n should you use?

7.67 Diastolic blood pressure measurements on American men aged 18–44 years follow approximately a normal curve with $\mu = 81$ mm Hg and $\sigma = 11$ mm Hg. The distribution for women aged 18–44 is also approximately normal with the same SD but with a lower mean: $\mu = 75$ mm Hg.[41] Suppose we are going to measure the diastolic blood pressure of n randomly selected men and n randomly selected women in the age group 18–44 years. Let E be the event that the difference between men and women will be found statistically significant by a t test. How large must n be in order to have $\Pr\{E\} = .9$

(a) if we use a two-tailed test at $\alpha = .05$?
(b) if we use a two-tailed test at $\alpha = .01$?
(c) if we use a one-tailed test (in the correct direction) at $\alpha = .05$?

7.68 Suppose you are planning an experiment to test the effect of a certain drug treatment on drinking behavior in the rat. You will use a two-tailed t test to compare a treated group of rats against a control group; the observed variable will be Y = one-hour water consumption after 23-hour deprivation. You have decided that, if the effect of the drug is to shift the population mean consumption by 2 mL or more, then you want to have at least an 80% chance of rejecting H_0 at the 5% significance level.

(a) Preliminary data indicate that the SD of Y under control conditions is approximately 2.5 mL. Using this as a guess of σ, determine how many rats you should have in each group.

(b) Suppose that, because the calculation of part (a) indicates a rather large number of rats, you consider modifying the experiment so as to reduce σ. You find that, by switching to a better supplier of rats and by improving lab procedures, you could cut the SD in half; however, the cost of each observation would be doubled. Would these measures be cost effective, that is, would the modified experiment be less costly?

7.69 Data from a large study indicate that the serum concentration of lactate dehydrogenase (LD) is higher in men than in women. (The data are summarized in Example 7.25.) Suppose Dr. Jones proposes to conduct his own study to replicate this finding; however, because of limited resources Jones can enlist only 35 men and 35 women for his study. Supposing that the true difference in population means is 4 U/L and each population SD is 10 U/L, what is the probability that Jones will be successful? Specifically, find the probability that Jones will reject H_0 with a one-tailed t test at the 5% significance level.

7.70 Refer to the painkiller study of Exercise 7.54. In that study, the evidence favoring the drug was marginally significant ($.025 < P < .05$). Suppose Dr. Smith is planning a new study on the same drug in order to try to replicate the original findings, that is, to show the drug to be effective. She will consider this study successful if she rejects H_0 with a one-tailed test at $\alpha = .05$. In the original study, the difference between the treatment means was about half a standard deviation $[(32 - 25)/13 \approx .5]$. Taking this as a provisional value for the effect size, determine how many patients Smith should have in each group in order for her chance of success to be

(a) 80%   (b) 90%

(*Note*: This problem illustrates that surprisingly large sample sizes may be required to make a replication study worthwhile, especially if the original findings were only marginally significant.)

7.71 Consider comparing two normally distributed distributions for which the effect size of the difference is

(a) 3   (b) 1

In each case, draw a sketch that shows how the distributions overlap. (See Figure 7.23.)

7.72 An animal scientist is planning an experiment to evaluate a new dietary supplement for beef cattle. One group of cattle will receive a standard diet and a second group will receive the standard diet plus the supplement. The researcher wants to have 90% power to detect an increase in mean weight gain of 20 kg, using a one-tailed t test at $\alpha = .05$. Based on previous experience, he expects the SD to be 17 kg. How many cattle does he need for each group?

7.73 A researcher is planning to conduct a study that will be analyzed with a two-tailed t test at the 5% significance level. She can afford to collect 20 observations in each of the two groups in her study. What is the smallest effect size for which she has at least 95% power?

7.9 STUDENT'S t: CONDITIONS AND SUMMARY

In the preceding sections we have discussed the comparison of two means using classical methods based on Student's t distribution. In this section we describe the conditions on which these methods are based. In addition, we summarize the methods for convenient reference.

Conditions

The t test and confidence interval procedures we have described are appropriate if the following conditions hold:*

1. **Conditions on the design of the study**

   (a) It must be reasonable to regard the data as random samples from their respective populations. The populations must be large. The observations within each sample must be independent.

   (b) The two samples must be independent of each other.

2. **Conditions on the form of the population distributions**

   (a) If the sample sizes are small, the population distributions must be approximately normal.

   (b) If the sample sizes are large, the population distributions need not be approximately normal. However, we always need to be aware that one or two extreme outliers can have a great effect on the results of any statistical procedure, including the t test.

Condition 2(b) is based on an approximation theorem similar to the Central Limit Theorem. The required "largeness" in condition 2(b) depends on the degree of nonnormality of the populations (as in Section 6.5). In many practical situations, moderate sample sizes (say, $n_1 = 20, n_2 = 20$) are quite "large" enough.

Verification of Conditions

A check of the preceding conditions should be a part of every data analysis.

A check of condition 1(a) would proceed as for a confidence interval (Section 6.5), with the researcher looking for biases in the experimental design and verifying that there is no hierarchical structure within each sample.

Condition 1(b) means that there must be no pairing or dependency between the two samples. The full meaning of this condition will become clear in Chapters 8 and 9.

Sometimes it is known from previous studies whether the populations can be considered to be approximately normal. In the absence of such information, the normality requirement can be checked by making histograms, stem-and-leaf displays, or normal probability plots for each sample separately. Fortunately, the t test is fairly robust against departures from normality.[42] Usually, only a rather conspicuous departure from normality (outliers, or long straggly tails) should be cause for concern. Moderate skewness has very little effect on the t test, even for small samples.

---

* Many authors use the word *assumptions* where we are using the word *conditions*.

Consequences of Inappropriate Use of Student's *t*

Our discussion of the *t* test and confidence interval (in Sections 7.3–7.8) was based on conditions (1) and (2). Violation of the conditions may render the methods inappropriate.

If the conditions are not satisfied, then the *t* test may be inappropriate in two possible ways:

1. It may be invalid in the sense that the actual risk of Type I error is larger than the nominal significance level α. (To put this another way, the *P*-value yielded by the *t* test procedure may be inappropriately small.)
2. The *t* test may be valid but less powerful than a more appropriate test.

If the design includes hierarchical structures that are ignored in the analysis, the *t* test may be seriously invalid. If the samples are not independent of each other, the usual consequence is a loss of power.

One fairly common type of departure from the assumption of normality is for one or both populations to have long straggly tails. The effect of this form of nonnormality is to inflate the SE, and thus to rob the *t* test of power.

Inappropriate use of confidence intervals is analogous to that for *t* tests. If the conditions are violated, then the confidence interval may not be valid, or it may be valid but wider than necessary.

Other Approaches

Because methods based on Student's *t* distribution are not always the most appropriate, statisticians have devised other methods that serve similar purposes. One of these is the Wilcoxon-Mann-Whitney test, which we will describe in Section 7.11. Another approach to the difficulty is to transform the data, for instance to analyze log (Y) instead of Y itself.

Tissue Inflammation.   Researchers took skin samples from 10 patients who had breast implants and from a control group of 6 patients. They recorded the level of interleukin-6 (in pg/mL/10 g of tissue), a measure of tissue inflammation, after each tissue sample was cultured for 24 hours. Table 7.15 shows the data.[43] Parallel dotplots of these data shown in Figure 7.25(a) and normal probability plots shown in Figure 7.26(a) indicate that the distributions are severely skewed, so a transformation is needed before Student's *t* procedure can be used. Taking the natural logarithm of each observation produces the values shown in the right-hand columns of Table 7.15 and in Figure 7.25(b). The normal probability plots in

Example 7.36

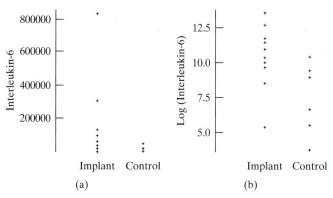

Figure 7.25 Dotplots of tissue inflammation data from Example 7.36 (a) in the original scale; (b) in log scale

TABLE 7.15 Interleukin-6 Levels of Breast Implant Patients and Control Patients

	Original Data		Log$_e$ Scale	
	Breast Implant Patients	Control Patients	Breast Implant Patients	Control Patients
	231	35,324	5.442	10.472
	308,287	12,457	12.639	9.430
	33,291	8,276	10.413	9.021
	124,550	44	11.732	3.784
	17,075	278	9.745	5.628
	22,955	840	10.041	6.733
	95,102		11.463	
	5,649		8.639	
	840,585		13.642	
	58,924		10.984	
$\bar{y}$	150,665	9,537	10.47	7.51
s	259,189	13,613	2.28	2.56

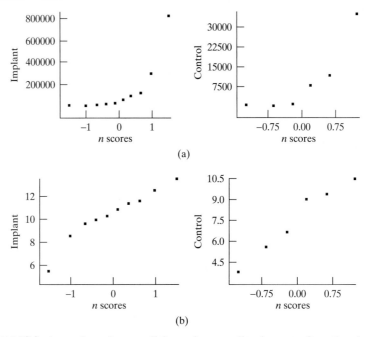

Figure 7.26 Normal probability plots of tissue inflammation data from Example 7.36 (a) in the original scale; (b) in log scale

Figure 7.26(b) show that the condition of normality is met after the data have been transformed to log scale. Thus, we will conduct an analysis of the data in log scale. That is, we will test

$$H_0: \mu_1 = \mu_2$$

against

$$H_A: \mu_1 \neq \mu_2$$

where μ_1 is the population mean of the log of interleukin-6 level for breast implant patients and μ_2 is the population mean of the log of interleukin-6 level for control patients. Suppose we choose $\alpha = .10$. The test statistic is

$$t_s = \frac{(10.47 - 7.51) - 0}{1.27} = 2.33$$

Formula (7.1) yields df $= 9.7$. The P-value for the test is .043. Thus, we have evidence, at the .10 level of significance (and at the .05 level, as well), that the mean log interleukin-6 level is higher in the breast implant population than in the control population. ■

Summary of Formulas

For convenient reference, we summarize in the accompanying boxes the formulas for Student's t method for comparing the means of independent samples.

Standard Error of $\bar{y}_1 - \bar{y}_2$

$$\text{SE}_{(\bar{y}_1 - \bar{y}_2)} = \sqrt{\frac{s_1^2}{n_1} + \frac{s_2^2}{n_2}} = \sqrt{\text{SE}_1^2 + \text{SE}_2^2}$$

Confidence Interval for $\bar{y}_1 - \bar{y}_2$

95% confidence interval:

$$(\bar{y}_1 - \bar{y}_2) \pm t_{.025}\text{SE}_{(\bar{y}_1 - \bar{y}_2)}$$

Critical value $t_{.025}$ from Student's t distribution with

$$\text{df} = \frac{(\text{SE}_1^2 + \text{SE}_2^2)^2}{\text{SE}_1^4/(n_1 - 1) + \text{SE}_2^4/(n_2 - 1)}$$

where $\text{SE}_1 = s_1/\sqrt{n_1}$ and $\text{SE}_2 = s_2/\sqrt{n_2}$
Confidence intervals with other confidence levels (90%, 99%, etc.) are constructed analogously (using $t_{.05}, t_{.005}$, etc.).

---

t Test

$$H_0: \mu_1 = \mu_2$$
$$H_A: \mu_1 \neq \mu_2 \text{ (nondirectional)}$$
$$H_A: \mu_1 < \mu_2 \text{ (directional)}$$
$$H_A: \mu_1 > \mu_2 \text{ (directional)}$$

$$\text{Test statistic: } t_s = \frac{(\bar{y}_1 - \bar{y}_2) - 0}{\text{SE}_{(\bar{y}_1 - \bar{y}_2)}}$$

P-value $=$ tail area under Student's t curve with

$$\text{df} = \frac{(\text{SE}_1^2 + \text{SE}_2^2)^2}{\text{SE}_1^4/(n_1 - 1) + \text{SE}_2^4/(n_2 - 1)}$$

Nondirectional H_A: P-value $=$ two-tailed area beyond t_s and $-t_s$
Directional H_A: Step 1: Check directionality.
Step 2: P-value $=$ single-tail area beyond t_s
Decision: Reject H_0 if P-value $\leq \alpha$

Exercises 7.74–7.75

7.74 Refer to the sucrose consumption data analyzed in Exercise 7.15 and displayed in Figure 7.2.

(a) Does the condition that the populations are normal appear to be reasonable for these data? Explain.

(b) In view of your answer to part (a), on what grounds can you defend the application of Student's t method to these data?

7.75 Refer to the serotonin data of Exercise 7.29. On what grounds might an objection be raised to the use of the t test on these data? (*Hint*: For each sample, calculate the SD and compare it to the sample mean.)

7.10 MORE ON PRINCIPLES OF TESTING HYPOTHESES

Our study of the t test has illustrated some of the general principles of statistical tests of hypotheses. In the remainder of this book we will introduce several other types of tests besides the t test.

A General View of Hypothesis Tests

A typical statistical test involves a null hypothesis H_0, an alternative hypothesis H_A, and a test statistic that measures deviation or discrepancy of the data from H_0. The sampling distribution of the test statistic, under the assumption that H_0 is true, is called the **null distribution** of the test statistic. (For example, the null distribution of the t statistic t_s is—under certain conditions—a Student's t distribution.) The null distribution indicates how much the test statistic can be expected to deviate from H_0 because of chance alone.

In testing a hypothesis, we assess the evidence against H_0 by locating the test statistic within the null distribution; the P-value is a measure of this location that indicates the degree of compatibility between the data and H_0. The dividing line between compatibility and incompatibility is specified by an arbitrarily chosen significance level α. The decision whether to reject the null hypothesis is made according to the following rule:

Reject H_0 if P-value $\leq \alpha$.

In this book, we will sometimes not calculate the P-value exactly, but will bracket it using a table of critical values. If H_A is directional, the bracketing of P-value is a two-step procedure.

Every test of a null hypothesis H_0 has its associated risks of Type I error (rejecting H_0 when H_0 is true) and Type II error (not rejecting H_0 when H_0 is false). The risk of Type I error is always limited by the chosen significance level:

$$\Pr\{\text{reject } H_0\} \leq \alpha \text{ if } H_0 \text{ is true}$$

Thus, the hypothesis testing procedure treats the Type I error as the one to be most stringently guarded against. The risk of Type II error, by contrast, can be quite large if the samples are small.

How is H_0 Chosen?

A common difficulty when first studying hypothesis testing is figuring out what the null hypothesis should be and what the alternative hypothesis should be. In general, the null hypothesis represents the status quo—what one would believe, by default, unless the data showed otherwise.* Often the alternative hypothesis is a statement that the researcher is trying to establish; thus, the alternative hypothesis is sometimes referred to as the *research hypothesis*. For example, if we are testing a new drug against a standard drug, the null hypothesis is that the new drug is no different than the standard—in the absence of evidence, we would expect the two drugs to be equally effective. The typical null hypothesis, $H_0: \mu_1 = \mu_2$, states that the two population means are equal and that any difference between the sample means is simply due to chance error in the sampling process. The alternative hypothesis is that there *is* a difference between the drugs, so that any observed difference in sample means is due to a real effect, rather than being due to chance error. We reject the null hypothesis if the data show a difference in sample means beyond what can reasonably be attributed to chance.

Here are other examples: If we are comparing men and women on some attribute, the usual null hypothesis is that there is no difference, on average, between men and women; if we are studying a measure of biodiversity in two environments, the ususal null hypothesis is that the two environments are equal, on average; if we are studying two diets, the usual null hypothesis is that the diets produce the same average response.

Another Look at *P*-Value

In order to place *P*-value in a general setting, let us consider some verbal interpretations of *P*-value.

First we revisit the *t* test. For a nondirectional H_A, we have defined the *P*-value to be the two-tailed area under the Student's *t* curve beyond the observed value of t_s. Another way of defining the *P*-value is the following:

> The *P*-value of the data is the probability (under H_0) of getting a result as extreme as, or more extreme than, the result that was actually observed.

To put this another way,

> The *P*-value is the probability that, if H_0 were true, a result would be obtained that would deviate from H_0 as much as (or more than) the actual data do.

Actually, this description of *P*-value is a bit too limited. The *P*-value actually depends on the nature of the alternative hypothesis. When we are performing a *t* test against a *directional* alternative, the *P*-value of the data is (if the observed deviation is in the direction of H_A) only a *single-tailed* area beyond the observed value of t_s. The more general definition of *P*-value is the following:

> The *P*-value of the data is the probability (under H_0) of getting a result as deviant as, or more deviant than, the result actually observed—where deviance is measured as discrepancy from H_0 in the direction of H_A.

---

* This general rule is not always true; it is provided only as a guideline.

The P-value measures how easily the observed deviation could be explained as chance variation rather than by the alternative explanation provided by H_A. For example, if the t test yields a P-value of $P = .036$ for our data, then we may say that, if H_0 were true, we would expect data to deviate from H_0 as much as our data did only 3.6% of the time (in the meta-experiment).

Another definition of P-value that is worth thinking about is the following:

The P-value of the data is the value of α for which H_0 would just barely be rejected, using those data.

To interpret this definition, imagine that a research report that includes a P-value is read by a number of interested scientists. The scientists who are quite skeptical of H_A might personally use quite a conservative decision threshold, such as $\alpha = .001$; the scientists who are more favorably disposed toward H_A might use a liberal value such as $\alpha = .10$. The P-value of the data determines the point, within this spectrum of opinion, that separates those who find the data to be convincing in favor of H_A and those who do not. Of course, if the P-value is large—for instance, $P = .40$—then presumably no reasonable person would reject H_0 and be convinced of H_A.

As the preceding discussion shows, the P-value does not describe all facets of the data, but relates only to a test of a particular null hypothesis against a particular alternative. In fact, we will see that the P-value of the data also depends on which statistical test is used to test a given null hypothesis. For this reason, when describing in a scientific report the results of a statistical test, it is best to report the P-value (exact, if possible), the name of the statistical test, and whether the alternative hypothesis was directional or nondirectional.

We repeat here, because it applies to any statistical test, the principle expounded in Section 7.7: The P-value is a measure of the strength of the evidence against H_0, but the P-value does *not* reflect the *magnitude* of the discrepancy between the data and H_0. The data may deviate from H_0 only slightly, yet if the samples are large, the P-value may be quite small. By the same token, data that deviate substantially from H_0 can nevertheless yield a large P-value.

Interpretation of Error Probabilities

A common mistake is to interpret the P-value as the probability that the null hypothesis is true. A related misconception is the belief that, if H_0 has been rejected at (for example) the 5% significance level, then the probability that H_0 is true is 5%. These interpretations are not correct. In fact, the probability that H_0 is true cannot be calculated at all.* This point can be illustrated by an analogy with medical diagnosis.

In applying a diagnostic test for an illness, the null hypothesis is that the person is healthy—this is what we will believe unless the medical test indicates otherwise. Two types of error are possible: A healthy individual may be diagnosed as ill (false positive) or an ill individual may be diagnosed as healthy (false negative). Trying out a diagnostic test on individuals *known* to be healthy or ill will enable us to estimate the proportions of these groups who will be misdiagnosed; yet this information alone will not tell us what proportion of all positive diagnoses are false diagnoses. These ideas are illustrated numerically in the next example.

---

* $\Pr\{H_0$ is true$\}$ *can* be calculated if we use what are known as Bayesian methods, which are beyond the scope of this book.

Medical Testing. Suppose a medical test is conducted to detect an illness. Further, suppose that 1% of the population has the illness in question. If the test indicates that the disease is present, we reject the null hypothesis that the person is healthy. If H_0 is true, then this is a Type I error—a false positive. If the test indicates that the disease is not present, we fail to reject H_0. Suppose that the test has an 80% chance of detecting the disease if the person has it (this is analogous to the power of a hypothesis test being 80%) and a 95% chance of correctly indicating that the disease is absent if the person really does not have the disease (this is analogous to a 5% Type I error rate). Figure 7.27 shows a probability tree for this situation, with bold lines indicating the two ways in which the test result can be positive (i.e., the two ways that H_0 can be rejected).

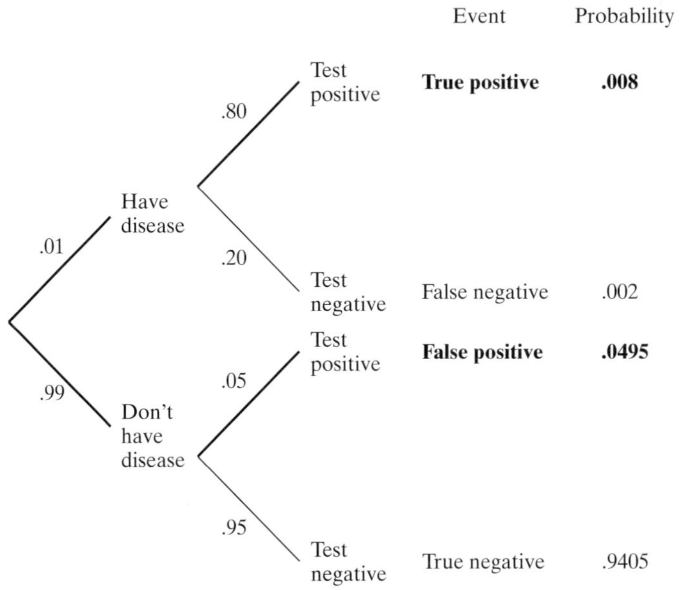

Example 7.37

	Event	Probability
Test positive	**True positive**	**.008**
Test negative	False negative	.002
Test positive	**False positive**	**.0495**
Test negative	True negative	.9405

Probability tree: .01 Have disease → .80 Test positive; .20 Test negative. .99 Don't have disease → .05 Test positive; .95 Test negative.

Figure 7.27 Probability tree for medical testing example

TABLE 7.16 **Hypothetical Results of Medical Test of 100,000 Persons**

		True Situation		
		Healthy (H_0 true)	*Ill (H_0 false)*	*Total*
Test	Negative (do not reject H_0)	94,050	200	94,250
Result	Positive (reject H_0)	4,950	800	5,750
	Total	99,000	1,000	100,000

Now suppose that 100,000 persons are tested and that 1,000 of them (1%) actually have the illness. Then we would expect results like those given in Table 7.16, with 5,750 persons testing positive (which is like rejecting H_0 5,750 times). Of these, 4,950 are false positives. Put another way, the proportion of the time that H_0 is true, given that H_0 was rejected, is $\dfrac{4{,}950}{5{,}750} \approx .86$, which is quite different from .05; this startlingly high proportion of false positives is due to the rarity of the disease. (The proportion of times that H_0 is rejected, given that H_0 is true, is $\dfrac{4{,}950}{99{,}000} = .05$, as expected, but that is a different conditional probability. $\Pr\{A \text{ given } B\} \neq \Pr\{B \text{ given } A\}$: The probability of rainfall, given that there is thunder and lightning, is not the same as the probability of thunder and lightning, given that it is raining.) ∎

The risk of Type I error is a probability computed *under the assumption that H_0 is true*; similarly, the risk of a Type II error is computed assuming that H_A is true. If we have a well-designed study with adequate sample sizes, both of these probabilities will be small. We then have a good test procedure in the same sense that the medical test is a good diagnostic procedure. But this does not in itself guarantee that most of the null hypotheses we reject are in fact false, or that most of those we do not reject are in fact true. The validity or nonvalidity of such guarantees would depend on an unknown and unknowable quantity—namely, the proportion of true null hypotheses among all null hypotheses that are tested (which is analogous to the incidence of the illness in the medical test scenario).

Perspective

We should mention that the philosophy of statistical hypothesis testing that we have explained in this chapter is not shared by all statisticians. The view presented here, which is called the **frequentist view**, is widely used in scientific research. An alternative view, the **Bayesian view**, permits—indeed, requires—the quantitative evaluation of data to depend not only on the observed data but also on the researcher's (or consumer's) prior beliefs about the truth or falsity of H_0.

Exercise 7.76

7.76 Suppose we have conducted a *t* test, with $\alpha = .05$, and the *P*-value is .04. For each of the following statements, say whether the statement is true or false and explain why.

(a) There is a 4% chance that H_0 is true.
(b) We reject H_0 with $\alpha = .05$.
(c) We should reject H_0, and if we repeated the experiment, there is a 4% chance that we would reject H_0 again.
(d) If H_0 is true, the probability of getting a test statistic at least as extreme as the value of the t_s that was actually obtained is 4%.

7.11 THE WILCOXON-MANN-WHITNEY TEST

The **Wilcoxon-Mann-Whitney test** is used to compare two independent samples.* It is a competitor to the *t* test, but unlike the *t* test, the Wilcoxon-Mann-Whitney test is valid even if the population distributions are not normal. The Wilcoxon-Mann-Whitney test is therefore called a **distribution-free** type of test. In addition, the Wilcoxon-Mann-Whitney test does not focus on any particular parameter such as a mean or a median; for this reason it is called a **nonparametric** type of test.

---

* The test presented here is was developed by Wilcoxon in a 1945 article. Mann and Whitney, in a 1947 article, elaborated on the test, which can be conducted in two mathematically equivalent ways. Thus, some books and some computer programs implement the test in a fashion different from the way it is presented here. Also note that some books refer to this as the Wilcoxon test, some as the Mann-Whitney test, and some (including this text) as the Wilcoxon-Mann-Whitney test.

Statement of H_0 and H_A

Let us denote the observations in the two samples by Y_1 and Y_2. A general statement of the null hypothesis of a Wilcoxon-Mann-Whitney test is

H_0: The population distributions of Y_1 and Y_2 are the same.

In practice, it is more natural to state H_0 and H_A in words suitable to the particular application, as illustrated in Example 7.38.

Soil Respiration. Soil respiration is a measure of microbial activity in soil, which affects plant growth. In one study, soil cores were taken from two locations in a forest: (1) under an opening in the forest canopy (the "gap" location) and (2) at a nearby area under heavy tree growth (the "growth" location). The amount of carbon dioxide given off by each soil core was measured (in mol CO_2/g soil/hr). Table 7.17 contains the data.[44]

Example 7.38

TABLE 7.17 Soil Respiration Data (mol CO_2/g soil/hr) from Example 7.38

Growth				Gap			
17	20	170	315	22	29	13	16
22	190	64		15	18	14	6

An appropriate null hypothesis could be stated as

H_0: The populations from which the two samples were drawn have the same distribution of soil respiration

or, more informally, as

H_0: The gap and growth areas do not differ with respect to soil respiration.

A nondirectional alternative could be stated as

H_A: The distribution of soil respiration rates tends to be higher in one of the two populations

or the alternative hypothesis might be directional, for example,

H_A: Soil respiration rates tend to be greater in the growth area than there are in the gap area. ∎

Applicability of the Wilcoxon-Mann-Whitney Test

Figure 7.28 shows dotplots of the soil respiration data from Example 7.38; Figure 7.29 shows normal probability plots of these data. The growth distribution is heavily skewed to the right, whereas the gap distribution is slightly skewed to the left. If both distributions were skewed to the right, we could apply a transformation to the data. However, any attempt to transform the growth distribution, such as taking logarithms of the data, will make the skewness of the gap distribution worse. Hence, the t test is not applicable here. The Wilcoxon-Mann-Whitney test does not require normality of the distributions.

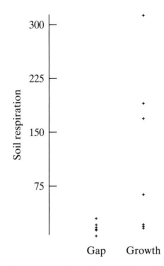

Figure 7.28 Dotplots of the soil respiration data from Example 7.38

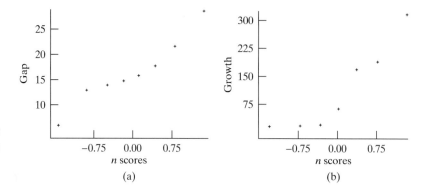

Figure 7.29 Normal probability plots of (a) the gap data and (b) the growth data from Example 7.38

Method

The Wilcoxon-Mann-Whitney test statistic, which is denoted U_s, measures the degree of separation or shift between two samples. A large value of U_s indicates that the two samples are well separated, with relatively little overlap between them. Critical values for the Wilcoxon-Mann-Whitney test are given in Table 6 at the end of this book. The following example illustrates the Wilcoxon-Mann-Whitney test.

Example 7.39

Soil Respiration. Let us carry out a Wilcoxon-Mann-Whitney test on the biodiversity data of Example 7.38.

1. The value of U_s depends on the relative positions of the Y_1's and the Y_2's. The first step in determining U_s is to arrange the observations in increasing order, as is shown in Table 7.18.

TABLE 7.18 Wilcoxon-Mann-Whitney Calculations for Example 7.39

Number of Gap Observations That Are Smaller	Y_1 Growth Data	Y_2 Gap Data	Number of Growth Observations That Are Smaller
5	17	6	0
6	20	13	0
6.5	22	14	0
8	64	15	0
8	170	16	0
8	190	18	1
8	315	22	2.5
		29	3
$K_1 = 49.5$			$K_2 = 6.5$

2. We next determine two counts, K_1 and K_2, as follows:

   (a) *The K_1 count* For each observation in sample 1, we count the number of observations in sample 2 that are smaller in value (that is, to the left). We count 1/2 for each tied observation. In the preceding data, there are five Y_2's less than the first Y_1, there are six Y_2's less than the second Y_1, there are six Y_2's less than the third Y_1 and one equal to it, so we count 6.5. So far we have counts of 5, 6, and 6.5. Continuing in a similar way, we get further counts of 8, 8, 8, and 8. All together there are seven counts, one for each Y_1. The sum of all seven counts is $K_1 = 49.5$.

(b) *The K_2 count* For each observation in sample 2, we count the number of observations in sample 1 that are smaller in value, counting 1/2 for ties. This gives counts of 0, 0, 0, 0, 0, 1, 2.5, and 3. The sum of these counts is $K_2 = 6.5$.

(c) *Check* If the work is correct, the sum of K_1 and K_2 should be equal to the product of the sample sizes:

$$K_1 + K_2 \overset{?}{=} n_1 n_2$$

$$49.5 + 6.5 = 7 \cdot 8$$

3. The test statistic U_s is the larger of K_1 and K_2. In this example, $U_s = 49.5$.

4. To determine critical values, we consult Table 6 with $n =$ the larger sample size, and $n' =$ the smaller sample size. In the present case, $n = 8$ and $n' = 7$. The critical values from Table 6 are reproduced in Table 7.19.

Let us test H_0 against a nondirectional alternative at significance level $\alpha = .05$. From Table 7.19, we note that $U_{.02} = 49$ and $U_{.01} = .50$; since $49 < U_s < 50$, the P-value is between .01 and .02 and H_0 is rejected. There is sufficient evidence to conclude that soil respiration rates are different in the gap and growth areas. ■

TABLE 7.19 Critical Values from Table 6 for $n = 8$, $n' = 7$

Nominal Tail Probability							
One tail	.20	.10	.05	.02	.01	.002	.001
Two tails	.10	.05	.025	.01	.005	.001	.0005
Critical value	40	43	46	49	50	54	55

As Example 7.39 illustrates, Table 6 is used to bracket the P-value for the Wilcoxon-Mann-Whitney test just as Table 4 is used for the *t* test. We simply locate the critical values that bracket the observed U_s; we then bracket the P-value by the corresponding column headings. If U_s is exactly equal to a critical value, then the P-value is less than the column heading.* The following example illustrates the bracketing procedure.

Bracketing the P-Value. Suppose $n = 8$ and $n' = 7$, and H_A is nondirectional. Using the critical values shown in Table 7.19, we would bracket the P-value as follows:

Example 7.40

If $U_s = 46$, then $.02 < P\text{-value} < .05$.

If $U_s = 47$, then $.02 < P\text{-value} < .05$.

If $U_s = 55$, then $P\text{-value} < .001$. ■

---

* In a few cases, the P-value would be exactly equal to (rather than less than) the column heading. To simplify the presentation, we neglect this fine distinction.

Directionality. For the t test, we determine the directionality of the data by seeing whether $\bar{y}_1 > \bar{y}_2$ or $\bar{y}_1 < \bar{y}_2$. Similarly, we can check directionality for the Wilcoxon-Mann-Whitney test by comparing K_1 and K_2: $K_1 > K_2$ indicates a trend for the Y_1's to be larger than the Y_2's, while $K_1 < K_2$ indicates the opposite trend. Often, however, this formal comparison is unnecessary; a glance at the data is enough.

Directional Alternative. If the alternative hypothesis H_A is directional rather than nondirectional, the Wilcoxon-Mann-Whitney procedure must be modified. As with the t test, the modified procedure has two steps and the second step involves halving the P-value.

> *Step 1.* Check directionality—see if the data deviate from H_0 in the direction specified by H_A.
>
> (a) If not, the P-value is greater than .50.
> (b) If so, proceed to step 2.
>
> *Step 2.* The P-value of the data is half as much as it would be if H_A were nondirectional.

To make a decision at a prespecified significance level α, we reject H_0 if P-value $\le \alpha$.

The following example illustrates the two-step procedure.

Example 7.41 | **Directional H_A.** Suppose $n = 8$, $n' = 7$, and H_A is directional. Suppose further that the data do deviate from H_0 in the direction specified by H_A. The critical values shown in Table 7.19 can be used to bracket the P-value as follows:

$$\text{If } U_s = 46, \text{ then } .01 < P\text{-value} < .025.$$
$$\text{If } U_s = 47, \text{ then } .01 < P\text{-value} < .025.$$
$$\text{If } U_s = 55, \text{ then } P\text{-value} < .0005.$$

Note that these P-values are half of those shown in Example 7.40. ∎

Blank Critical Values. In some cases, certain entries in Table 6 are blank. The next example shows how the P-value is bracketed in such a case.

Example 7.42 | **Blank Critical Values.** If $n = 5$ and $n' = 4$, Table 6 reads as follows:

Nominal tail probability							
One tail	.20	.10	.05	.02	.01	.002	.001
Two tails	.10	.05	.025	.01	.005	.001	.0005
Critical value	16	18	19	20			

Suppose H_A is nondirectional. Then the P-value would be bracketed as follows:

$$\text{If } U_s = 19, \text{ then } .02 < P\text{-value} < .05.$$
$$\text{If } U_s = 20, \text{ then } P\text{-value} < .02.$$

For these sample sizes, U_s cannot be larger than 20. The rationale for this bracketing procedure is explained in the next subsection. ∎

Rationale

Let us see why the Wilcoxon-Mann-Whitney test procedure makes sense. To take a specific case, suppose the sample sizes are $n_1 = 5$ and $n_2 = 4$. Then necessarily, regardless of what the data look like, we must have

$$K_1 + K_2 = 5 \cdot 4 = 20$$

The relative magnitudes of K_1 and K_2 indicate the amount of overlap of the Y_1's and the Y_2's. Figure 7.30 shows how this works. For the data of Figure 7.30(a), the two samples do not overlap at all; the data are *least* compatible with H_0, and U_s has its maximum value, $U_s = 20$. Similarly, $U_s = 20$ for Figure 7.30(b). On the other hand, the arrangement *most* compatible with H_0 is the one with maximal overlap, shown in Figure 7.30(c); for this arrangement $K_1 = 10$, $K_2 = 10$, and $U_s = 10$.

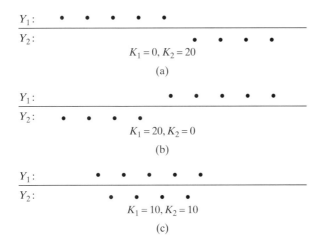

Figure 7.30 Three data arrays for a Wilcoxon-Mann-Whitney test

All other possible arrangements of the data lie somewhere between the three arrangements shown in Figure 7.30; those with much overlap have U_s close to 10, and those with little overlap have U_s closer to 20. Thus, large values of U_s indicate incompatibility of the data with H_0.

We now briefly consider the null distribution of U_s and indicate how the critical values of Table 6 were determined. (Recall from Section 7.10 that, for any statistical test, the reference distribution for critical values is always the null distribution of the test statistic—that is, its sampling distribution under the condition that H_0 is true.) To determine the null distribution of U_s, it is necessary to calculate the probabilities associated with various arrangements of the data, assuming that all the Y's were actually drawn from the same population.* (The method for calculating the probabilities is briefly described in Appendix 7.2.)

Figure 7.31(a) shows the null distribution of K_1 and K_2 for the case $n = 5, n' = 4$. For example, it can be shown that, if H_0 is true, then

$$\Pr\{K_1 = 0, K_2 = 20\} = .008$$

---

* In calculating the probabilities used in this section, it has been assumed that the chance of tied observations is negligible. This will be true for a continuous variable that is measured with high precision. If the number of ties is large, a correction can be made; see Noether (1967).[45]

This is the first probability plotted in Figure 7.31(a). Note that Figure 7.31(a) is roughly analogous to a t distribution; large values of K_1 (right tail) represent evidence that the Y_1's tend to be larger than the Y_2's and large values of K_2 (left tail) represent evidence that the Y_2's tend to be larger than the Y_1's.

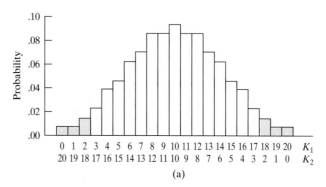

Figure 7.31 Null distributions for the Wilcoxon-Mann-Whitney test when $n = 5, n' = 4$. (a) Null distribution of K_1 and K_2; (b) null distribution of U_s.

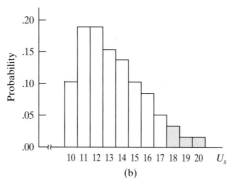

Figure 7.31(b) shows the null distribution of U_s, which is derived directly from the distribution in Figure 7.31(a). For instance, if H_0 is true, then

$$\Pr\{K_1 = 0, K_2 = 20\} = .008$$

and

$$\Pr\{K_1 = 20, K_2 = 0\} = .008$$

so that

$$\Pr\{U_s = 20\} = .008 + .008 = .016$$

which is the rightmost probability plotted in Figure 7.31(b). Thus, both tails of the K distribution have been "folded" into the upper tail of the U distribution; for instance, the one-tailed shaded area in Figure 7.31(b) is equal to the two-tailed shaded area in Figure 7.31(a).

P-values for the Wilcoxon-Mann-Whitney test are upper-tail areas in the U_s distribution. For instance, it can be shown that the shaded area in Figure 7.31(b) is equal to .064; this means that if H_0 is true, then

$$\Pr\{U_s \geq 18\} = .064$$

Thus, a data set that yielded $U_s = 18$ would have an associated P-value .064 (assuming a nondirectional H_A).

The critical values in Table 6 have been determined from the null distribution of U_s. Because the U_s distribution is discrete, the column headings do not correspond exactly to P-values as they do for the t distribution. Rather, the associated P-value is *less than or equal to* the column heading (this is why the columns are labeled *nominal* tail probabilities). For example, the critical value, for a nondirectional test, in the .10 column is 18, while we saw previously that the exact P-value associated with $U_s = 18$ is .064 rather than .10; for these sample sizes (5 and 4) the value .064 is as close as the P-value can get to .10 without exceeding it. Now it should be clear why some of the critical value entries are blank: No value is given for $U_{.01}$, for instance, because the P-value cannot possibly be less than .01; the most extreme case is $U_s = 20$, which has a P-value equal to .016.

If we take a decision-making approach to the Wilcoxon-Mann-Whitney test, then we reject H_0 if P-value $\leq \alpha$. Thus, when H_A is nondirectional, H_0 is rejected if U_s is greater than or equal to the critical value listed in Table 6 under column heading α. If the critical value does not exist, then H_0 can never be rejected. For instance, with sample sizes 5 and 4, a Wilcoxon-Mann-Whitney test against a nondirectional H_A at $\alpha = .01$ can never reject H_0.

Conditions for Use of the Wilcoxon-Mann-Whitney Test

In order for the Wilcoxon-Mann-Whitney test to be applicable, it must be reasonable to regard the data as random samples from their respective populations, with the observations within each sample being independent, and the two samples being independent of each other. Under these assumptions, the Wilcoxon-Mann-Whitney test is valid no matter what the form of the population distributions, provided that the observed variable Y is continuous.[46]

The critical values given in Table 6 have been calculated assuming that ties do not occur. If the data contain only a few ties, then the critical values are approximately correct.*

The Wilcoxon-Mann-Whitney Test Versus the t Test

The Wilcoxon-Mann-Whitney test and the t test are aimed at answering the same question, but they treat the data in very different ways. Unlike the t test, the Wilcoxon-Mann-Whitney test does not use the actual values of the Y's but only their relative positions in a rank ordering. This is both a strength and a weakness of the Wilcoxon-Mann-Whitney test. On the one hand, the test is distribution free because the null distribution of U_s relates only to the various rankings of the Y's, and therefore does not depend on the form of the population distribution. On the other hand, the Wilcoxon-Mann-Whitney test can be inefficient: It can lack power because it does not use all the information in the data. This inefficiency is especially evident for small samples.

---

* Actually, the Wilcoxon-Mann-Whitney test need not be restricted to continuous variables; it can be applied to any ordinal variable. However, if Y is discrete or categorical, then the data may contain many ties, and the test should not be used without appropriate modification of the critical values.

Neither of the competitors—the t test or the Wilcoxon-Mann-Whitney test—is clearly superior to the other. If the population distributions are not approximately normal, the t test may not even be valid. In addition, the Wilcoxon-Mann-Whitney test can be much more powerful than the t test, especially if the population distributions are highly skewed. If the population distributions are approximately normal with equal standard deviations, then the t test is better, but its advantage is not necessarily very great; for moderate sample sizes, the Wilcoxon-Mann-Whitney test can be nearly as powerful as the t test.

There is a confidence interval procedure that is associated with the Wilcoxon-Mann-Whitney test in the same way that the confidence interval for $(\mu_1 - \mu_2)$ is associated with the t test. The procedure is beyond the scope of this book.

Exercises 7.77–7.84

7.77 Consider two samples of sizes $n_1 = 5, n_2 = 7$. Use Table 6 to bracket the P-value, assuming that H_A is nondirectional and that

(a) $U_s = 26$
(b) $U_s = 30$
(c) $U_s = 35$

7.78 Consider two samples of sizes $n_1 = 4, n_2 = 8$. Use Table 6 to bracket the P-value, assuming that H_A is nondirectional and that

(a) $U_s = 25$
(b) $U_s = 31$
(c) $U_s = 32$

7.79 In a pharmacological study, researchers measured the concentration of the brain chemical dopamine in six rats exposed to toluene and six control rats. (This is the same study described in Example 7.9.) The concentrations in the striatum region of the brain were as shown in the table.[47]

Dopamine (ng/g)

Toluene	Control
3,420	1,820
2,314	1,843
1,911	1,397
2,464	1,803
2,781	2,539
2,803	1,990

(a) Use a Wilcoxon-Mann-Whitney test to compare the treatments at $\alpha = .05$. Use a nondirectional alternative.
(b) Proceed as in part (a), but let the alternative hypothesis be that toluene increases dopamine concentration.

7.80 In a study of hypnosis, breathing patterns were observed in an experimental group of subjects and in a control group. The measurements of total ventilation (liters of air per minute per square meter of body area) are shown in the following table.[48] (These are the same data that were summarized in Exercise 7.51.) Use a Wilcoxon-Mann-Whitney test to compare the two groups at $\alpha = .10$. Use a nondirectional alternative.

Experimental	Control
5.32	4.50
5.60	4.78
5.74	4.79
6.06	4.86
6.32	5.41
6.34	5.70
6.79	6.08
7.18	6.21

7.81 In an experiment to compare the effects of two different growing conditions on the heights of greenhouse chrysanthemums, all plants grown under condition 1 were found to be taller than any of those grown under condition 2 (that is, the two height distributions did not overlap). Calculate the value of U_s and bracket the P-value if the number of plants in each group was

(a) 3
(b) 4
(c) 5

(Assume that H_A is nondirectional.)

7.82 In a study of preening behavior in the fruitfly *Drosophila melanogaster*, a single experimental fly was observed for three minutes while in a chamber with ten other flies of the same sex. The observer recorded the timing of each episode ("bout") of preening by the experimental fly. This experiment was replicated 15 times with male flies and 15 times with female flies (different flies each time). One question of interest was whether there is a sex difference in preening behavior. The observed preening times (average time per bout, in seconds) were as follows:[49]

Male: 1.2, 1.2, 1.3, 1.9, 1.9, 2.0, 2.1, 2.2, 2.2, 2.3, 2.3, 2.4, 2.7, 2.9, 3.3

$$\bar{y} = 2.127 \qquad \Sigma(y_i - \bar{y})^2 = 4.969$$

Female: 2.0, 2.2, 2.4, 2.4, 2.4, 2.8, 2.8, 2.8, 2.9, 3.2, 3.7, 4.0, 5.4, 10.7, 11.7

$$\bar{y} = 4.093 \qquad \Sigma(y_i - \bar{y})^2 = 127.2$$

(a) For these data, the value of the Wilcoxon-Mann-Whitney statistic is $U_s = 189.5$. Use a Wilcoxon-Mann-Whitney test to investigate the sex difference in preening behavior. Let H_A be nondirectional and let $\alpha = .01$.
(b) For these data, the standard error of $(\bar{y}_1 - \bar{y}_2)$ is SE $= .7933$ s. Use a t test to investigate the sex difference in preening behavior. Let H_A be nondirectional and let $\alpha = .01$.
(c) What condition is required for the validity of the t test but not for the Wilcoxon-Mann-Whitney test? What feature or features of the data suggest that this condition may not hold in this case?
(d) Verify the value of U_s given in part (a).

7.83 Substances to be tested for cancer-causing potential are often painted on the skin of mice. The question arose whether mice might get an additional dose of the substance by licking or biting their cagemates. To answer this question, the compound benzo(a)pyrene was applied to the backs of ten mice: Five were individually housed and five were group housed in a single cage. After 48 hours, the concentration of the compound in the stomach tissue of each mouse was determined. The results (nmol/g) were as follows:[50]

Singly Housed	Group Housed
3.3	3.9
2.4	4.1
2.5	4.8
3.3	3.9
2.4	3.4

(a) Use a Wilcoxon-Mann-Whitney test to compare the two distributions at $\alpha = .01$. Let the alternative hypothesis be that benzo(a)pyrene concentrations tend to be high in group-housed mice than in singly housed mice.

(b) Why is a directional alternative valid in this case?

7.84 Human beta-endorphin (HBE) is a hormone secreted by the pituitary gland under conditions of stress. An exercise physiologist measured the resting (unstressed) blood concentration of HBE in two groups of men: Group 1 consisted of 11 men who had been jogging regularly for some time, and group 2 consisted of 15 men who had just entered a physical fitness program. The results are given in the following table.[51]

Joggers				Fitness Program Entrants				
39	40	32	60	70	47	54	27	31
19	52	41	32	42	37	41	9	18
13	37	28		33	23	49	41	59

Use a Wilcoxon-Mann-Whitney test to compare the two distributions at $\alpha = .10$. Use a nondirectional alternative.

7.12 PERSPECTIVE

In this chapter we have discussed several techniques—confidence intervals and hypothesis tests—for comparing two independent samples when the observed variable is quantitative. In coming chapters we will introduce confidence interval and hypothesis testing techniques that are applicable in various other situations. Before proceeding, we pause to reconsider the methods of this chapter.

An Implicit Assumption

In discussing the tests of this chapter—the t test and the Wilcoxon-Mann-Whitney test—we have made an unspoken assumption, which we now bring to light. When interpreting the comparison of two distributions, we have assumed that the relationship between the two distributions is relatively simple—that if the distributions differ, then one of the two variables has a consistent tendency to be larger than the other. For instance, suppose we are comparing the effects of two diets on the weight gain of mice, with

$$Y_1 = \text{Weight gain of mice on diet 1}$$
$$Y_2 = \text{Weight gain of mice on diet 2}$$

Our implicit assumption has been that, if the two diets differ at all, then that difference is in a consistent direction for all individual mice. To appreciate the meaning of this assumption, suppose the two distributions are as pictured in Figure 7.32. In this case, even though the mean weight gain is higher on diet 1, it would be an oversimplification to say that mice tend to gain more weight on diet 1 than on diet 2; apparently *some* mice gain *less* on diet 1. Paradoxical situations of this kind do occasionally occur, and then the simple analysis typified by the *t* test and the Wilcoxon-Mann-Whitney test may be inadequate.

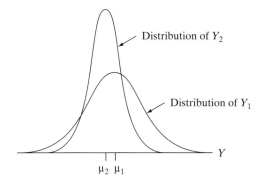

Figure 7.32 Weight gain distributions on two diets

It is relatively easy to compare two distributions that have the same general shape and similar standard deviations. However, if either the shapes or the SDs of two distributions are very different from one another, then making a meaningful comparison of the distributions is difficult. In particular, a comparison of the two means might not be appropriate.

Which Method to Use When

If we are comparing samples from two normally distributed populations, a *t* test can be used to infer whether the population means differ and a confidence interval can be used to estimate how much the two population means might differ, if at all. A confidence interval generally provides more information than does a test, since the test is restricted to a narrow question ("Might the difference between the samples be reasonably attributed to chance?"), whereas the confidence interval addresses a larger question ("How much larger is μ_1 than μ_2?").

Both the confidence interval and the *t* test depend on the condition that the populations are normally distributed. If this condition is not met, then a transformation might be used to make the distributions approximately normal before proceeding. If, despite considering transformations, the normality condition is questionable, then the Wilcoxon-Mann-Whitney test can be used. (Indeed, the Wilcoxon-Mann-Whitney test can be used if the data are normal, although it is less powerful than the *t* test). When in doubt, a good piece of advice is to conduct both a *t* test and a Wilcoxon-Mann-Whitney test. If the two tests give similar, clear, conclusions (i.e., if the *P*-values for the tests are similar and both are considerably larger than α or both are considerably smaller than α), then we can feel comfortable with the conclusion. However, if one test yields a *P*-value somewhat larger than α and the other gives a *P*-value smaller than α, then we might well declare that the tests are inconclusive.

Sometimes an outlier will be present in a data set, calling into question the result of a t test. It is not legitimate to simply ignore the outlier. A sensible procedure is to conduct the analysis with the outlier included and then delete the outlier and repeat the analysis. If the conclusion is unchanged when the outlier is removed, then we can feel confident that no single observation is having undue influence on the inferences we draw from the data. If the conclusion changes when the outlier is removed, then we cannot be confident in the inferences we draw. For example, if the P-value for a test is small with the outlier present but large when the outlier is deleted, then we might state "There is evidence that the populations differ from one another, but this evidence is largely due to a single observation." Such a statement warns the reader that not too much should be read into any differences that were observed between the samples.

Comparison of Variability

It sometimes happens that the variability of Y, rather than its average value, is of primary interest. For instance, in comparing two different lab techniques for measuring the concentration of an enzyme, a researcher might want primarily to know whether one of the techniques is more precise than the other; that is, whether its measurement error distribution has a smaller standard deviation. There are techniques available for testing the hypothesis $H_0: \sigma_1 = \sigma_2$, and for using a confidence interval to compare σ_1 and σ_2. Most of these techniques are very sensitive to the condition that the underlying distributions are normal, which limits their use in practice. The implementation of these techniques is beyond the scope of this book.

Supplementary Exercises 7.85–7.110

Note: Exercises preceded by an asterisk refer to optional sections.

7.85 For each of the following pairs of samples, compute the standard error of $(\bar{y}_1 - \bar{y}_2)$.

(a)

	Sample 1	Sample 2
n	12	13
$\bar{y}$	42	47
s	9.6	10.2

(b)

	Sample 1	Sample 2
n	22	19
$\bar{y}$	112	126
s	2.7	1.9

(c)

	Sample 1	Sample 2
n	5	7
$\bar{y}$	14	16
SE	1.2	1.4

7.86 To investigate the relationship between intracellular calcium and blood pressure, researchers measured the free calcium concentration in the blood platelets of 38 people with normal blood pressure and 45 people with high blood pressure. The results are given in the table and the distributions are shown in the boxplots.[52] Use the t test to compare the means. Let $\alpha = .01$ and let H_A be nondirectional. *Note:* Formula (7.1) yields 67.5 df.

Blood Pressure	Platelet Calcium (nM)		
	n	*Mean*	*SD*
Normal	38	107.9	16.1
High	45	168.2	31.7

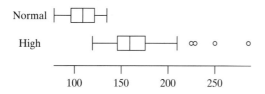

7.87 Refer to Exercise 7.86. Construct a 95% confidence interval for the difference between the population means.

7.88 Refer to Exercise 7.86. The boxplot for the high blood pressure group is skewed to the right and includes outliers. Does this mean that the *t* test is not valid for these data? Why or why not?

7.89 In a study of methods of producing sheep's milk for use in cheese manufacture, ewes were randomly allocated to either a mechanical or a manual milking method. The investigator suspected that the mechanical method might irritate the udder and thus produce a higher concentration of somatic cells in the milk. The accompanying data show the average somatic cell count for each animal.[53]

	Somatic Count ($10^{-3} \cdot$ cells/mL)	
	Mechanical milking	*Manual milking*
	2,966	186
	269	107
	59	65
	1,887	126
	3,452	123
	189	164
	93	408
	618	324
	130	548
	2,493	139
n	10	10
Mean	1,215.6	219.0
SD	1,342.9	156.2

(a) Do the data support the investigator's suspicion? Use a *t* test against a directional alternative at $\alpha = .05$. The standard error of $(\bar{y}_1 - \bar{y}_2)$ is SE = 427.54. and formula (7.1) yields 9.2 df.

(b) Do the data support the investigator's suspicion? Use a Wilcoxon-Mann-Whitney test against a directional alternative at $\alpha = .05$. (The value of the Wilcoxon-Mann-Whitney statistic is $U_s = 69$.) Compare with the result of part (a).

(c) What condition is required for the validity of the *t* test but not for the Wilcoxon-Mann-Whitney test? What features of the data cast doubt on this condition?

(d) Verify the value of U_s given in part (b).

7.90 A plant physiologist conducted an experiment to determine whether mechanical stress can retard the growth of soybean plants. Young plants were randomly allocated to two groups of 13 plants each. Plants in one group were mechanically agitated by shaking for 20 minutes twice daily, while plants in the other group were not agitated. After 16 days of growth, the total stem length (cm) of each plant was measured, with the results given in the accompanying table.[54]

(a) Use a t test to compare the treatments at $\alpha = .01$. Let the alternative hypothesis be that stress tends to retard growth. *Note*: Formula (7.1) yields 23 df.
(b) State the conclusion of the t test in the context of the setting. (See Examples 7.13 and 7.14.)

	Control	Stress
n	13	13
$\bar{y}$	30.59	27.78
s	2.13	1.73

7.91 Refer to Exercise 7.90. Construct a 95% confidence interval for the population mean reduction in stem length. Does the confidence interval indicate whether the effect of stress is "horticulturally important," if "horticulturally important" is defined as a reduction in population mean stem length of at least

(a) 1 cm
(b) 2 cm
(c) 5 cm

7.92 Refer to Exercise 7.90. The raw observations, in increasing order, are shown in the following table. Compare the treatments using a Wilcoxon-Mann-Whitney test at $\alpha = .01$. Let the alternative hypothesis be that stress tends to retard growth.

Control	Stress
25.2	24.7
29.5	25.7
30.1	26.5
30.1	27.0
30.2	27.1
30.2	27.2
30.3	27.3
30.6	27.7
31.1	28.7
31.2	28.9
31.4	29.7
33.5	30.0
34.3	30.6

7.93 One measure of the impact of pollution along a river is the diversity of species in the river floodplain. In one study, two rivers, the Black River and the Vermilion River, were compared. Random 50 m · 20 m plots were sampled along each river and the number of species of trees in each plot was recorded. The following table contains the data.[55]

Vermilion River					Black River			
9	9	16	13		13	10	6	9
12	13	13	13		10	7	6	18
8	11	9	9	10	6			

Conduct a Wilcoxon-Mann-Whitney test, with $\alpha = .10$, of the null hypothesis that the populations from which the two samples were drawn have the same distribution of tree species per plot. Use the directional alternative that biodiversity is greater along the Vermilion River than along the Black River. (The Black River was considered to have been polluted quite a bit more than the Vermilion River, and this was expected to lead to lower biodiversity along the Black River.)

7.94 A developmental biologist removed the oocytes (developing egg cells) from the ovaries of 24 frogs *(Xenopus laevis)*. For each frog the oocyte pH was determined. In addition, each frog was classified according to its response to a certain stimulus with the hormone progesterone. The pH values were as follows:[56]

> *Positive response:* 7.06, 7.18, 7.30, 7.30, 7.31, 7.32, 7.33, 7.34, 7.36, 7.36, 7.40, 7.41, 7.43, 7.48, 7.49, 7.53, 7.55, 7.57
>
> *No response:* 7.55, 7.70, 7.73, 7.75, 7.75, 7.77

Investigate the relationship of oocyte pH to progesterone response using a Wilcoxon-Mann-Whitney test at $\alpha = .05$. Use a nondirectional alternative.

7.95 Refer to Exercise 7.94. Summary statistics for the pH measurements are given in the following table. Investigate the relationship of oocyte pH to progesterone response using a t test at $\alpha = .05$. Use a nondirectional alternative. *Note:* Formula (7.1) yields 14.1 df.

	Positive Response	No Response
n	18	6
$\bar{y}$	7.373	7.708
s	.129	.081

7.96 A proposed new diet for beef cattle is less expensive than the standard diet. The proponents of the new diet have conducted a comparative study in which one group of cattle was fed the new diet and another group was fed the standard. They found that the mean weight gains in the two groups were not statistically significantly different at the 5% significance level, and they stated that this finding supported the claim that the new cheaper diet was as good (for weight gain) as the standard diet. Criticize this statement.

*7.97 Refer to Exercise 7.96. Suppose you discover that the study used 25 animals on each of the two diets, and that the coefficient of variation of weight gain under the conditions of the study was about 20%. Using this additional information, write an expanded criticism of the proponents' claim, indicating how likely such a study would be to detect a 10% deficiency in weight gain on the cheaper diet (using a two-tailed test at the 5% significance level).

7.98 In a study of hearing loss, endolymphatic sac tumors (ELSTs) were discovered in 13 patients. These 13 patients had a total of 15 tumors (i.e., more patients had a single tumor, but two of the patients had two tumors each). Ten of the tumors were associated with the loss of functional hearing in an ear, but for five of the ears with tumors the patient had no hearing loss.[57] A natural question is whether hearing loss is more likely with large tumors than with small tumors. Thus, the sizes of the tumors were measured. Suppose that the sample means and standard deviations were given and that a comparison of average tumor size (hearing loss vs. no hearing loss) were being considered.

(a) Explain why a t test to compare average tumor size is not appropriate here.
(b) If the raw data were given, could a Wilcoxon-Mann-Whitney test be used?

7.99   *(Computer exercise)*   In an investigation of the possible influence of dietary chromium on diabetic symptoms, 14 rats were fed a low-chromium diet and 10 were fed a normal diet. One response variable was activity of the liver enzyme GITH, which was measured using a radioactively labeled molecule. The accompanying table shows the results, expressed as thousands of counts per minute per gram of liver.[58] Use a t test to compare the diets at $\alpha = .05$. Use a nondirectional alternative. *Note:* Formula (7.1) yields 21.9 df.

Low-Chromium Diet		Normal Diet	
42.3	52.8	53.1	53.6
51.5	51.3	50.7	47.8
53.7	58.5	55.8	61.8
48.0	55.4	55.1	52.6
56.0	38.3	47.5	53.7
55.7	54.1		
54.8	52.1		

7.100 *(Computer exercise)*   Refer to Exercise 7.99. Use a Wilcoxon-Mann-Whitney test to compare the diets at $\alpha = .05$. Use a nondirectional alternative.

7.101 *(Computer exercise)*   Refer to Exercise 7.99.

(a) Construct a 95% confidence interval for the difference in population means.
(b) Suppose the investigators believe that the effect of the low-chromium diet is "unimportant" if it shifts mean GITH activity by less than 15%—that is, if the population mean difference is less than about 8 thousand cpm/g. According to the confidence interval of part (a), do the data support the conclusion that the difference is "unimportant"?
(c) How would you answer the question in part (b) if the criterion were 4 thousand rather than 8 thousand cpm/g?

7.102 *(Computer exercise)*   In a study of the lizard *Scelopons occidentalis*, researchers examined field-caught lizards for infection by the malarial parasite *Plasmodium*. To help assess the ecological impact of malarial infection, the researchers tested 15 infected and 15 noninfected lizards for stamina, as indicated by the distance each animal could run in two minutes. The distances (meters) are shown in the table.[59]

Infected	Animals	Uninfected	Animals
16.4	36.7	22.2	18.4
29.4	28.7	34.8	27.5
37.1	30.2	42.1	45.5
23.0	21.8	32.9	34.0
24.1	37.1	26.4	45.5
24.5	20.3	30.6	24.5
16.4	28.3	32.9	28.7
29.1		37.5	

Do the data provide evidence that the infection is associated with decreased stamina? Investigate this question using

(a) a t test

(b) a Wilcoxon-Mann-Whitney test

Let H_A be directional and $\alpha = .05$.

7.103   In a study of the effect of amphetamine on water consumption, a pharmacologist injected four rats with amphetamine and four with saline as controls. She measured the amount of water each rat consumed in 24 hours; the following are the results, expressed as mL water per kg body weight:[60]

Amphetamine	Control
118.4	122.9
124.4	162.1
169.4	184.1
105.3	154.9

(a) Use a t test to compare the treatments at $\alpha = .10$. Let the alternative hypothesis be that amphetamine tends to suppress water consumption.

(b) Use a Wilcoxon-Mann-Whitney test to compare the treatments at $\alpha = .10$, with the directional alternative that amphetamine tends to suppress water consumption.

7.104   Nitric oxide is sometimes given to newborns who experience respiratory failure. In one experiment, nitric oxide was given to 114 infants. This group was compared to a control group of 121 infants. The length of hospitalization (in days) was recorded for each of the 235 infants. The mean in the nitric oxide sample was $\bar{y}_1 = 36.4$; the mean in the control sample was $\bar{y}_2 = 29.5$. A 95% confidence interval for $\mu_1 - \mu_2$ is $(-2.3, 16.1)$, where μ_1 is the population mean length of hospitalization for infants who get nitric oxide and μ_2 is the mean length of hospitalization for infants in the control population.[61] For each of the following, say whether the statement is true or false and say why.

(a) We are 95% confident that μ_1 is greater than μ_2, since most of the confidence interval is greater than zero.

(b) We are 95% confident that the difference between μ_1 and μ_2 is between -2.3 days and 16.1 days.

(c) We are 95% confident that the difference between $\bar{y}_1$ and $\bar{y}_2$ is between -2.3 days and 16.1 days.

(d) 95% of the nitric oxide infants were hospitalized longer than the average control infant.

7.105   Consider the confidence interval for $\mu_1 - \mu_2$ from Exercise 7.104: $(-2.3, 16.1)$. True or false: If we tested $H_0: \mu_1 = \mu_2$ against $H_A: \mu_1 \neq \mu_2$, using $\alpha = .05$, we would reject H_0.

7.106   Researchers studied subjects who had pneumonia and classified them as being in one of two groups: those who were given medical therapy that is consistent with American Thoracic Society (ATS) guidelines and those who were given medical therapy that is inconsistent with ATS guidelines. Subjects in the "consistent" group were generally able to return to work sooner than were subjects in the "inconsistent" group. A Wilcoxon-Mann-Whitney test was applied to the data; the P-value for the test was .04.[62] For each of the following, say whether the statement is true or false and say why.

(a) There is a 4% chance that the "consistent" and "inconsistent" population distributions really are the same.

(b) If the "consistent" and "inconsistent" population distributions really are the same, then a difference between the two samples as large as the difference that these researchers observed would only happen 4% of the time.

(c) If a new study were done that compared the "consistent" and "inconsistent" populations, there is a 4% probability that H_0 would be rejected again.

7.107 A student recorded the number of calories in each of 56 entrees—28 vegetarian and 28 nonvegetarian—served at a college dining hall.[63] The following table summarizes the data. Graphs of the data (not given here) show that both distributions are reasonably symmetric and bell shaped. A 95% confidence interval for $\mu_1 - \mu_2$ is $(-27, 85)$. For each of the following, say whether the statement is true or false and say why.

(a) 95% of the data are between -27 and 85.

(b) We are 95% confident that $\mu_1 - \mu_2$ is between -27 and 85.

(c) 95% of the time $\bar{y}_1 - \bar{y}_2$ will be between -27 and 85.

(d) 95% of the vegetarian entrees have between 27 fewer calories and 85 more calories than the average nonvegetarian entree.

	n	Mean	SD
Vegetarian	28	351	119
Nonvegetarian	28	322	87

7.108 *(Computer exercise)* Lianas are woody vines that grow in tropical forests. Researchers measured liana abundance (stems/ha) in several plots in the central Amazon region of Brazil. The plots were classified into two types: plots that were near the edge of the forest (less than 100 meters from the edge) or plots far from the edge of the forest. The raw data are given below and are summarized in the table.[64]

	n	Mean	SD
Near	34	438	125
Far	34	368	114

Near			Far		
639	601	600	470	339	384
605	581	555	309	395	393
535	531	466	236	252	407
437	423	380	241	215	427
376	362	350	320	228	445
349	346	337	325	267	451
320	317	310	352	294	493
285	271	265	275	356	502
250	450	441	181	418	540
436	432	420	250	425	590
419	407		266	495	
702	676		338	648	

(a) Make normal probability plots of the data to confirm that the distributions are mildly skewed.

(b) Conduct a t test to compare the two types of plots at $\alpha = .05$. Use a nondirectional alternative.

(c) Apply a logarithm transformation to the data and repeat parts (a) and (b).

(d) Compare the *t* tests from parts (b) and (c). What do these results indicate about the effect on a *t* test of mild skewness when the sample sizes are fairly large?

7.109 Androstenedione (andro) is a steroid that is thought by some athletes to increase strength. Researchers investigated this claim by giving andro to one group of men and a placebo to a control group of men. One of the variables measured in the experiment was the increase in "lat pulldown" strength (in pounds) of each subject after 4 weeks. (A lat pulldown is a type of weightlifting exercise.) The raw data are given and are summarized in the table.[65]

	n	**Mean**	**SD**
Andro	10	20.0	12.5
Control	9	14.4	13.3

Andro				**Control**			
30	10	10	30	0	10	0	10
40	20	30	20	10	40	20	10
10	0			30			

(a) Conduct a *t* test to compare the two groups at $\alpha = .10$. Use a nondirectional alternative. *Note*: Formula (7.1) yields 16.5 df.

(b) Prior to the study it was expected that andro would increase strength, which means that a directional alternative might have been used. Redo the analysis in part (a) using the appropriate directional alternative.

7.110 The following is a sample of computer output from a study.[66] Describe the problem and the conclusion, based on the computer output.

```
Y = number of drinks in the previous 7 days
Twosample T for Treatment vs Control:

                    N        Mean        SD
    Treatment      244      13.62      12.39
    Control        238      16.86      13.49
95% C.I. for μ₁ - μ₂:(-5.56, -0.92)
T-Test μ₁ = μ₂(vs <):T=-2.74 P=.0031 DF=474.3
```

CHAPTER 8

Statistical Principles of Design

8.1 INTRODUCTION

In the previous chapters we have given our primary attention to the *analysis* of data, rather than to the *design* of the study that generated the data. Of course, the planning and design of a scientific investigation is a complex process that requires expert knowledge of the phenomenon under study. However, statistical considerations often play a role in the design process. In this chapter we will discuss some of those considerations, including

>   Selection of individuals for study
>
>   Arrangement of experimental material in space and time
>
>   Allocation of experimental units to treatment groups

We have already discussed methods of data analysis in one-group and two-group studies in which the observed variable Y is quantitative. We have also considered confidence intervals for one-group studies in which the observed variable Y is categorical. In the present chapter we broaden our perspective and consider studies with *any* number of groups with any kind of response variable (quantitative or categorical). In this chapter we concentrate our attention on design, not analysis. Some methods of analysis will be presented in later chapters; others are beyond the scope of this book. We begin by considering ways in which information is gathered.

Anecdotal Evidence

Intercessory Prayer.   It has been said that prayer is the most widely used form of medicine. People often offer prayers for others who are ill, asking God to heal them; such prayers are called intercessory prayers. But can intercessory prayer be considered to be a medicine? Does prayer function the way that medicine does? If we pray for some people but not for others, will we see a difference in the health of the two groups? As part of an investigation of the effectiveness of prayer

---

Objectives

In this chapter we study data collection and experimental design. We will

- *learn how experiments differ from observational studies*

- *discuss how randomization affects the scope of a statistical inference*

- *learn about blocking and how to prepare a randomized block design*

- *discuss the concepts of response variable, explanatory variable, placebo effect, blinding, and confounding*

Example 8.1

as a medical treatment, doctors and nurses were asked if they were aware of scientific studies of intercessory prayer. Many people wrote letters stating that they had seen patients recover from serious illness as a result of prayer. However, none of the letter writers could name any *scientific* study of prayer. Rather, they told stories of individual cases.[1] ■

The evidence discussed in Example 8.1 is **anecdotal evidence**. An anecdote is a short story or an example of an interesting event, in this case, of prayer helping someone who is ill. The accumulation of anecdotes often leads to conjecture and to scientific investigation, but it is predictable pattern, not anecdote, that establishes a scientific theory.

Observational Versus Experimental Studies

A major consideration in interpreting the results of a biological study is whether the study was observational or experimental. In an **experiment**, the researcher intervenes in or manipulates the experimental conditions. In an **observational study**, the researcher merely observes an existing situation. The distinction between observational and experimental studies was illustrated in Section 7.1; the following is another example.

Example 8.2   **Cigarette Smoking.**   In studies of the effects of smoking cigarettes, both experimental and observational approaches have been used. Effects in animals can be studied experimentally, because animals (for instance, dogs) can be allocated to treatment groups and the groups can be given various doses of cigarette smoke. Effects in humans are usually studied observationally. In one study, for example, pregnant women were questioned about their smoking habits, dietary habits, and so on.[2] When the babies were born, their physical and mental development was followed. One striking finding related to the babies' birthweights: The smokers tended to have smaller babies than the nonsmokers. The difference was not attributable to chance (the *P*-value was less than 10^{-5}). Nevertheless, it was far from clear that the difference was *caused* by smoking, because the women who smoked differed from the nonsmokers in many other aspects of their lifestyle besides smoking—for instance, they had very different dietary habits. (We will expand this example in Section 8.2.) ■

As Example 8.2 illustrates, it can be difficult to determine the exact nature of a cause-effect relationship in an observational study. In an experiment, on the other hand, a cause-effect relationship may be easy to see, based on the way in which the researcher manipulated the experimental conditions. To help fix the ideas, consider studying cholesterol level. Suppose a group of patients with high cholesterol levels enroll in a clinical trial—that is, in a medical experiment—in which some of the patients are randomly chosen to receive a new drug and others are given a standard drug that has shown only modest effects in the past. If a two-sample *t* test shows that average cholesterol level decreased more for those on the new drug than for those on the standard drug, then the researcher can conclude that the new drug *caused* the superior outcome and is better than the standard drug.

Now consider a two-sample *t* test to compare average cholesterol level in a random sample of 50-year-olds to average cholesterol level in a random sample of 25-year-olds. Suppose a two-sample t test gives a small P-value, with the 50-year-olds having higher cholesterol than the 25-year-olds. We could be fairly confident that cholesterol level tends to increase with age. However, it would be *possible* that some other explanation were at work. For example, maybe diets have changed over time and the 25-year-olds are eating foods that the 50-year-olds don't eat, causing the 25-year-olds to have low cholesterol; perhaps if the 25-year-olds keep the same diet until they are 50, they will still have low cholesterol at age 50.

As a third example, consider comparing a random sample of home-owners to a random sample of renters. Suppose a two-sample t test shows a significantly higher average cholesterol level among the home-owners than among the renters. We should not conclude that buying a home causes one's cholesterol level to rise. Rather, we should consider that people who own homes tend to be older than are renters. It might very well be the case that age is the causal factor, which explains why the home-owners have higher cholesterol than do the renters.

All three of these cases might involve a two-sample t test and the rejection of H_0. Indeed, we might get the same P-value in each test. However, the conclusions we can draw from the three situations are quite different. The scope of the inference we can draw depends on the way in which the data are collected. Experiments allow us to infer cause-effect relationships that can only be guessed at in observational studies. Sometimes an observational study will leave us feeling reasonably confident that we understand the causal mechanism at work. However, we will see in Section 8.2 that drawing such conclusions is fraught with danger.

In Sections 8.3 through 8.5 we consider experimental studies, in which two or more experimental manipulations *(treatments)* are to be compared, in detail. In such studies the allocation of experimental units to treatment groups is under the control of the researcher. We will consider the two simplest schemes for allocation: the *completely randomized design* and the *randomized blocks design*.

We will find it useful to use standard terminology when discussing a study. A **response variable** measures an outcome of interest. An **explanatory variable** is a variable that is used to explain or predict an outcome. **Extraneous variables** are variables that may affect the results but that are not of direct interest. The **observational units** are the persons or things being studied. In Example 8.1, prayer is an explanatory variable, the response variable is the health of the person being prayed for, and the observational units are individual persons. In Example 8.2, the explanatory variable is whether or not a woman smokes, the response variable is birthweight, and the observational units are the babies whose birthweights are measured. Extraneous variables would be things such as the mother's age, income, and education.

8.2 OBSERVATIONAL STUDIES

In this section we explore observational studies in more depth. The difficulties in interpreting observational studies arise from two primary sources:

    Nonrandom selection from populations
    Uncontrolled extraneous variables

The following example illustrates both of these.

Example 8.3

Race and Brain Size. In the nineteenth century, much effort was expended in the attempt to show "scientifically" that certain human races were inferior to others. A leading researcher on this subject was the American physician S. G. Morton, who won widespread admiration for his studies of human brain size. Throughout his life, Morton collected human skulls from various sources, and he carefully measured the cranial capacities of hundreds of these skulls. His data appeared to suggest that (as he suspected) the "inferior" races had smaller cranial capacities. Table 8.1 gives a summary of Morton's data comparing Caucasian skulls to those of Native Americans.[3] According to a t test, the difference between these two samples is "statistically significant" ($P < .001$). But is it *meaningful*?

TABLE 8.1 Cranial Capacity (in.^{3})

	Caucasian	Native American
Mean	87	82
SD	8	10
n	52	144

In the first place, the notion that cranial capacity is a measure of intelligence is no longer taken seriously. Leaving that question aside, we can still ask whether it is true that the mean cranial capacity of Native Americans is less than that of Caucasians. Such an inference beyond the actual data requires that the data be viewed as random samples from their respective populations. Of course in actuality Morton's data are not random samples but "samples of convenience," because Morton measured those skulls that he happened to obtain. But might the data be viewed as if they were generated by random sampling? One way to approach this question is to look for sources of bias. In 1977 the noted biologist Stephen Jay Gould reexamined Morton's data with this goal in mind, and indeed Gould found several sources of bias. For instance, the 144 Native American skulls represent many different groups of Native Americans; as it happens, 25% of the skulls (that is, 36 of them) were from Inca Peruvians, who were a small-boned people with small skulls, while relatively few were from large-skulled tribes such as the Iroquois. Clearly a comparison between Native Americans and Caucasians is meaningless unless somehow adjusted for such imbalances. When Gould made such an adjustment, he found that the difference between Native Americans and Caucasians vanished. ∎

Even though the story of Morton's skulls is more than 100 years old, it can still serve to alert us to the pitfalls of inference. Morton was a conscientious researcher and took great care to make accurate measurements; Gould's reexamination did not reveal any suggestion of conscious fraud on Morton's part. Morton may have overlooked the biases in his data because they were *invisible* biases; that is, they related to aspects of the selection process rather than aspects of the measurements themselves.

When we look at a set of observational data, we can sometimes become so hypnotized by its apparent *solidity* and *objectivity* that we forget to ask how the observational units were selected. The question should always be asked. If the selection was haphazard rather than truly random, the results can be severely distorted.

Confounding

Many observational studies are aimed at discovering some kind of causal relationship. Such discovery can be very difficult because of extraneous variables that enter in an uncontrolled (and perhaps unknown) way. The investigator must be guided by the maxim:

> *Association is not causation.*

For instance, it is known that some populations whose diets are high in fiber enjoy a reduced incidence of colon cancer. But this observation does not in itself show that it is the high-fiber diet, rather than some other factor, that provides the protection against colon cancer.

The following example shows how uncontrolled extraneous variables can cloud an observational study, and what kinds of steps can be taken to clarify the picture.

Smoking and Birthweight.   In a large observational study of pregnant women, it was found that the women who smoked cigarettes tended to have smaller babies than the nonsmokers.[2] (This study was mentioned in Example 8.2.) It is plausible that smoking could cause a reduction in birthweight, for instance by interfering with the flow of oxygen and nutrients across the placenta. But of course plausibility is not proof. In fact, the investigators found that the smokers differed from the nonsmokers with respect to many other variables. For instance, the smokers drank more whiskey than the nonsmokers. Alcohol consumption might plausibly be linked to a deficit in growth. ∎

Example 8.4

In Example 8.4 three variables are presented; let us refer to these as X = smoking, Y = birthweight, and Z = alcohol consumption. There is an association between X and Y, but is there a *causal* link between them? Or is there a causal link between Z and Y? Figure 8.1 gives a schematic representation of the situation. Changes in X are associated with changes in Y. However, changes in Z are also associated with changes in Y. We say that the effect that X has on Y is **confounded** with the effect that Z has on Y. In the context of Example 8.4, we say that the effect that smoking has on birthweight is confounded with the effect that alcohol consumption has on birthweight. In observational studies, confounding of effects is a common problem.

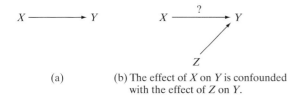

(a)

(b) The effect of X on Y is confounded with the effect of Z on Y.

Figure 8.1 Schematic representation of causation (a) and of confounding (b)

Smoking and Birthweight.   The study presented in Example 8.4 uncovered many confounding variables. For example, the smokers drank more coffee than the nonsmokers. In addition—and this is especially puzzling—it was found that the smokers began to menstruate at younger ages than the nonsmokers. This phenomenon (early onset of menstruation) could not possibly have been *caused* by smoking, because it occurred (in almost all instances) *before* the woman began to

Example 8.5

smoke. One interpretation that has been proposed is that the two populations—women who choose to smoke and those who do not—are different in some biological way; thus, it has been suggested that the reduced birthweight is due "to the *smoker*, not the *smoking*."[4]

A number of more recent studies have attempted to shed some light on the relationship between maternal smoking and infant development. Researchers in one study observed, in addition to smoking habits, about 50 extraneous variables, including the mother's age, weight, height, blood type, upper arm circumference, religion, education, income, and so on.[5] After applying complex statistical methods of adjustment, they concluded that birthweight varies with smoking even when all the extraneous factors are held constant. This says that there is a link between $X = $ smoking and $Y = $ birthweight as shown in Figure 8.1, although several other variables also affect birthweight. The point is that the presence of confounding doesn't mean that a link does not exist between X and Y, only that it is tangled up with other effects, so that we have to be cautious when interpreting the findings of an observational study.

In another study of pregnant women, researchers measured various quantities related to the functioning of the placenta.[6] They found that, compared with non-smokers, women who smoked had more abnormalities of the placenta, and their infants had very much higher blood levels of cotinine, a substance derived from nicotine. They also found evidence that, in the women who smoked, the circulation of blood in the placenta was notably improved by abstaining from smoking for 3 hours.

A third study used a matched design to try to isolate the effect of smoking behavior. The investigators identified 159 women who had smoked during one pregnancy but quit smoking before the next pregnancy.[7] These women were individually matched with 159 women who smoked during two consecutive pregnancies; pairs were matched with respect to the birthweight of the first child, amount of smoking during the first pregnancy, and several other factors. Thus, the members of a pair were believed to have identical "reproductive potential." The researchers then considered the birthweight of the second child; they found that the women who had quit smoking gave birth to infants who weighed more than the infants of their matched controls who continued to smoke. Of course, we cannot rule out the possibility that the women who quit smoking also quit other harmful habits, such as drinking too much alcohol, and that the increased birthweight was not really caused by giving up smoking. ∎

Example 8.5 shows that observational studies can provide information about causality, but must be interpreted cautiously. Researchers generally agree that a causal interpretation of an observed association requires extra support—for instance, that the association be observed consistently in observational studies conducted under various conditions and taking various extraneous factors into account, and also, ideally, that the causal link be supported by experimental evidence. We do not mean to say that an observed association *cannot* be causally interpreted, but only that such interpretation requires particular caution.

Spurious Association

Example 8.6 **Ultrasound.** It is quite common for a physician to use ultrasound examination of the fetus of a pregnant woman. However, when ultrasound technology was first used there were concerns that the procedure might be harmful to the baby. An early study seemed to bear this out: on average, babies exposed to ultrasound in the womb were lighter at birth than were babies not exposed to ultrasound.[8] Later, a study was done in which some women were randomly chosen to have ultrasounds and other were not given ultrasounds. This study found no difference in birth-weight between the two groups.[9] It seems that the reason a difference appeared in the first study was that ultrasound was being used mostly for women who were experiencing problem pregnancies. The complications with the pregnancy were leading to low birthweight, not the use of ultrasound.   ■

Figure 8.2 gives a schematic representation of the situation in Example 8.6. Changes in X (having an ultrasound examination) are associated with changes in Y (lower birthweight). However, X and Y are both dependent on a third variable Z (whether or not there are problems with the pregnancy), which is the variable that is driving the relationship. Changes in X and changes in Y are a common response to the third variable Z. We say that the association between X and Y is **spurious**: When we control for the "lurking variable" Z, the link between X and Y disappears. In the case of Example 8.6, it is not having an ultrasound that influences birthweight; what matters is whether or not there were problems with the pregnancy.

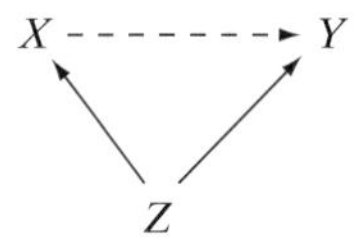

The association between X and Y is spurious; controlling for the lurking variable Z eliminates the X-Y link.

Figure 8.2 Schematic representation of spurious association

Case-Control Studies

Observational investigation of the possible causes of a disease often takes the form of a **case-control study**, in which cases of the disease are compared against controls who do not have the disease. Here is an example.

Diet and Stomach Cancer.   It is suspected that certain foods might increase the risk of stomach cancer. In order to investigate this, suppose that we plan to in-terview some stomach cancer patients ("cases") to learn about their dietary habits, and that we will also interview some control patients who do not have stomach can-cer. Then we might compare the diets of the cases and the controls to see if any dif-ferences emerge. But of course many factors, especially social and economic ones, influence dietary habits, and some of these factors may themselves be linked with cancer. How can controls be chosen who are comparable to the cases with respect to these extraneous factors?

Here is how one group of investigators approached the problem.[10] The cases were 220 Japanese stomach cancer patients from several cooperating hospitals in Hawaii. For each case, two controls were chosen according to an unambiguous rule: The controls were the next older and the next younger Japanese of the same sex in the same section of the hospital at the time of the interview, subject to the condi-tion that the control did not have stomach cancer, other diseases of the stomach, or other cancers of the digestive system. Because of this selection rule, the controls were comparable to the cases in terms of age, sex, and—most important—for many un-measurable extraneous variables (such as socioeconomic status) that were indi-rectly expressed by the location of the hospital. Note that all the cases and controls were Japanese; this factor was not matched, but rather was constant.   ■

Example 8.7

In a case-control study, the cases are usually a sample of convenience—say, all the available patients with the given disease. A primary design consideration is how to choose the controls. Often (as in Example 8.7) the cases are individually matched to the controls with respect to several important extraneous variables. Unfortunately, reports of case-control studies sometimes do not specify how the controls were chosen; this is a serious omission that can undermine the entire report.

Exercises 8.1–8.7

8.1   In 1992, 8.9% of the deaths in the United States were caused by respiratory illness. In Arizona, 10.3% of deaths were due to respiratory illness.[11] Does this mean that living in Arizona exacerbates respiratory problems? If not, how can we explain the Arizona rate being above the national rate?

8.2   It has been hypothesized that silicone breast implants cause illness. In one study it was found that women with implants were more likely to smoke, to be heavy drinkers, to use hair dye, and to have had an abortion than were women in a comparison group who did not have implants.[12] Use the language of statistics to explain why this study casts doubt on the claim that implants cause illness.

8.3   Consider the setting of Exercise 8.2.

(a) What is the explanatory variable?
(b) What is the response variable?
(c) What are the observational units?

8.4   In a study of 1,040 subjects, researchers found that the prevalence of coronary heart disease increased as the number of cups of coffee consumed per day increased.[13]

(a) What is the explanatory variable?
(b) What is the response variable?
(c) What are the observational units?

8.5   For an early study of the relationship between diet and heart disease, the investigator obtained data on heart disease mortality in various countries and on national average dietary compositions in the same countries. The accompanying graph shows, for six countries, the 1948–1949 death rate from degenerative heart disease (among men aged 55–59 years) plotted against the amount of fat in the diet.[14]
In what ways might this graph be misleading? Which extraneous variables might be relevant here? Discuss.

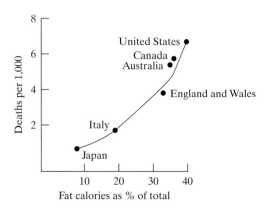

8.6   Shortly before Valentine's Day in 1999 a newspaper article was printed with the headline "Marriage makes for healthier, longer life, studies show." The headline was based on studies that showed that married persons live longer and have lower rates of cancer, heart disease, and stroke than do those who never marry.[15] Use the language of statistics to discuss the headline. Use a schematic diagram similar to Figure 8.1 or Figure 8.2 to support your explanation of the situation.

8.7   In a study of the relationship between birthweight and race, birth records of babies born in Illinois were examined. The researchers found that the percentage of low birthweight babies among babies born to U.S.-born white women was much lower than the percentage of low birthweight babies among babies born to U.S.-born black women. This suggests that race plays an important role in determining the chance that a baby will have a low birthweight. However, the percentage of low birthweight babies among babies born to African-born black women was roughly equal to the percentage among babies born to U.S.-born white women.[16] Use the language of statistics to discuss what these data say about the relationships between low birthweight, race, and mother's birthplace. Use a schematic diagram similar to Figure 8.1 or Figure 8.2 to support your explanation.

8.3 EXPERIMENTS

An **experiment** is a study in which the researcher intervenes and imposes treatment conditions. The following is a simple example.

Headache Pain.   Suppose a researcher gives ibuprofen to some people who have headaches and aspirin to others and then measures how long it takes for each person's headache to disappear. In this case, there are two treatments: ibuprofen and aspirin. By assigning people to treatment groups—ibuprofen and aspirin—the researcher is conducting an experiment.   ■

Example 8.8

   In Section 8.1 we introduced the term *observational unit*. When we are discussing an experiment, we refer to the units to which the treatments are assigned as **experimental units**. In an agricultural experiment, an experimental unit might be a plot of land. In general, an experimental unit is the smallest unit to which a treatment is applied in an experiment. Thus, in Example 8.8 the experimental units are individual people, since treatment is assigned on a person-by-person basis.

   If treatments are assigned at random, for example, by tossing a coin and letting heads mean the person gets ibuprofen, while tails means the person gets aspirin, then the experiment is a *randomized* experiment. Sometimes an experiment is conducted in which one group is given a treatment and a second—the control group—is given nothing. For example, one could investigate the effectiveness of ibuprofen in treating headache pain by giving it to some people, while giving no pain killer to others. In contrast, the experiment in which some people are given ibuprofen and others are given aspirin is said to have an "active" control—the aspirin group.

   A third option, and one that is commonly used, is to give the control group a **placebo**—an inert substance, such as a sugar pill. It is well known that people often exhibit a *placebo response*; that is, they tend to respond favorably to *any* treatment, even if it is only a placebo. This psychological effect can be quite powerful. Research has shown that placebos are effective for roughly one-third of people who are in pain (i.e., one-third of pain sufferers report their pain ending after being giving a "pain killer" that is, in fact, an inert pill). For diseases such as bronchial asthma, angina pectoris (recurrent chest pain caused by decreased blood flow to the heart), and ulcers the use of placebos has been shown to produce clinically beneficial results in over 60% of patients.[17] Of course, if a placebo control is used, then the subjects must not be told which group they are in—the group getting the active treatment or the group getting the placebo.

Example 8.9

Autism.   Autism is a serious condition in which children withdraw from normal social interactions and sometimes engage in aggressive or repetitive behavior. In 1997 an autistic child responded remarkably well to the digestive enzyme secretin. This led an experiment (a "clinical trial") in which secretin was compared to a placebo. In this experiment children who were given secretin improved considerably. However, the children given the placebo also improved considerably. There was no statistically significant difference between the two groups. Thus, the favorable response in the secretin group was considered to be only a "placebo response," meaning that, unfortunately, secretin was not found to be beneficial (beyond inducing a positive response associated simply with taking a substance as part of an experiment).[18] ∎

The word *placebo* means "I shall please." The word *nocebo* ("I shall harm") is sometimes used to describe adverse reactions to perceived, but nonexistent, risks. The following example illustrates the strength that psychological effects can have.

Example 8.10

Bronchial Asthma.   A group of patients suffering from bronchial asthma were given a substance that they were told was a chest-constricting chemical. After being given this substance, several of the patients experienced bronchial spasms. However, during part of the experiment the patients were given a substance that they were told would alleviate their symptoms. In this case, bronchial spasms were prevented. In reality, the second substance was identical to the first substance: Both were distilled water. It appears that it was the power of suggestion that brought on the bronchial spasms; the same power of suggestion prevented spasms.[19] ∎

Similar to placebo treatment is *sham* treatment, which can be used on animals as well as humans. An example of sham treatment is injecting control animals with an inert substance such as saline. In some studies of surgical treatments, control animals (even, occasionally, humans) are given a "mock" surgery.*

Example 8.11

Mammary Artery Ligation.   In the 1950s the surgical technique of internal mammary artery ligation became a popular treatment for patients suffering from angina pectoris. In this operation the surgeon would ligate (tie) the mammary artery, with the goal of increasing collateral blood flow to the heart. Doctors and patients alike enthusiastically endorsed this surgery as an effective treatment. In 1958 studies of internal mammary artery ligation in animals found that it was not effective; this raised doubts about its usefulness on humans. A study was conducted in which patients were randomly assigned to one of two groups. Patients in the treatment group received the standard surgery. Patients in the control group

---

* Example 7.35 in optional Section 7.8, which dealt with manipulation of the spine in children with asthma, provides another example of a sham treatment performed on human subjects.

received a sham operation in which an incision was made, the mammary artery was exposed as in the real operation, but the incision was closed *without* the artery being ligated. These patients had no way of knowing that their operation was a sham. The rates of improvement in the two groups of patients were nearly identical. (Patients who had the sham operation did slightly better than patients who had the real operation, but the difference was small.) A second randomized, controlled study also found that patients who received the sham surgery did as well as those who had the real operation. As a result of these studies, physicians stopped using internal mammary artery ligation.[20]   ◼

Blinding

In experiments on humans, particularly those that involve the use of placebos, **blinding** is often used. This means that the treatment assignment is kept secret from the experimental subject. The purpose of blinding the subject is to minimize the extent to which his or her expectations influence the results of the experiment. If subjects exhibit a psychological reaction to getting a medication, that "placebo response" will tend to balance out between the two groups, so that any difference between the groups can be attributed to the effect of the active treatment.

In many experiments the persons who evaluate the responses of the subjects are kept blind; that is, during the experiment they are kept ignorant of the treatment assignment. Consider, for instance, the following:

> In a study to compare two treatments for lung cancer, a radiologist reads X-rays to evaluate each patient's progress. The X-ray films are coded so that the radiologist cannot tell which treatment each patient received.

> Mice are fed one of three diets; the effects on their liver are assayed by a research assistant who does not know which diet each mouse received.

Of course, *someone* needs to keep track of which subject is in which group, but that person should not be the one who measures the response variable. The most obvious reason for blinding the person making the evaluations is to reduce the possibility of subjective bias influencing the observation process itself: Someone who *expects* or *wants* certain results may unconsciously influence those results. Such bias can enter even apparently "objective" measurements, through subtle variation in dissection techniques, titration procedures, and so on.

In medical studies of human beings, blinding often serves additional purposes. For one thing, a patient must be asked whether he or she consents to participate in a medical study. If the physician who asks the question already knows which treatment the patient would receive, then by discouraging certain patients and encouraging others, the physician can (consciously or unconsciously) create noncomparable treatment groups. The effect of such biased assignment can be surprisingly large, and it has been noted that it generally favors the "new" or "experimental" treatment.[21] Another reason for blinding in medical studies is that a physician may (consciously or unconsciously) provide more psychological encouragement, or even better care, to the patients who are receiving the treatment that the physician regards as superior.

An experiment in which both the subjects and the persons making the evaluations of the response are blinded is called a **double-blind** experiment. The first mammary artery ligation experiment described in Example 8.11 was conducted as a double-blind experiment.

The Need for Control Groups

Example 8.12

Clofibrate. An experiment was conducted in which subjects were given the drug clofibrate, which was intended to lower cholesterol and reduce the chance of death from coronary disease. The researchers noted that many of the subjects did not take all of the medication that the experimental protocol called for them to take. They calculated the percentage of the prescribed capsules that each subject took and divided the subjects into two groups, according to whether or not the subjects took at least 80% of the capsules they were given. Table 8.2 shows that the five-year mortality rate for those who took at least 80% of their capsules was much lower than the corresponding rate for subjects who did not adhere to the proto-col. On the surface, this suggests that taking the medication lowers the chance of death. However, there was a placebo control group in the experiment and many of the placebo subjects took fewer than 80% of their capsules. The mortality rates for the two placebo groups—those who adhered to the protocol and those who did not—are quite similar to the rates for the clofibrate groups.

TABLE 8.2 Mortality Rates for the Clofibrate Experiment

	Clofibrate		Placebo	
Adherence	n	5-Year Mortality	n	5-Year Mortality
≥80%	708	15.0%	1813	15.1%
<80%	357	24.6%	882	28.2%

The clofibrate experiment seems to indicate that there are two kinds of subjects: those who adhere to the protocol and those who do not. The first group had a much lower mortality rate than the second group. This might be due simply to better health habits among people who are willing to follow a scientific proto-col for five years than among people who don't adhere to the protocol. A further conclusion from the experiment is that clofibrate does not appear to be any more effective than placebo in reducing the death rate. Were it not for the presence of the placebo control group, the researchers might well have drawn the wrong con-clusion from the study and attributed the lower death rate among adherers to clofi-brate itself, rather than to other confounded effects that make the adherers different from the nonadherers.[22] ■

Example 8.13

The Common Cold. Many years ago, investigators invited university students who believed themselves to be particularly susceptible to the common cold to be part of an experiment. Volunteers were randomly assigned to either the treat-ment group, in which case they took capsules of an experimental vaccine, or to the control group, in which case they were told that they were taking a vaccine, but in fact were given a placebo—capsules that looked like the vaccine cap-sules, but which contained lactose in place of the vaccine.[23] As shown in Table 8.3, both groups reported having dramatically fewer colds during the study than they had had in the previous year. The average number of colds per person dropped 70% in the treatment group. This would have been startling evidence that the vaccine had an effect, except that the corresponding drop in the con-trol group was 69%. ■

TABLE 8.3

	Vaccine	Placebo
n	201	203
Average number of colds		
Previous year (from memory)	5.6	5.2
Current year	1.7	1.6
% reduction	70%	69%

We can attribute much of the large drop in colds in Example 8.13 to the placebo effect. However, another statistical concern is **panel bias**, which is bias that is attributable to the study having influenced the behavior of the subjects—that is, people who know they are being studied often change their behavior. The students in this study reported from memory the number of colds they had suffered in the previous year. The fact that they were part of a study might have influenced their behavior, so that they were less likely to catch a cold during the study. Being in a study might also have affected the way in which they defined having a cold—during the study, they were "instructed to report to the health service whenever a cold developed"—so that some illness may have gone unreported during the study. (How sick do you have to be before you classify yourself as having a cold?)

Historical Controls

Researchers may be particularly reluctant to use randomized allocation in medical experiments on human beings. Suppose, for instance, that researchers want to evaluate a promising new treatment for a certain illness. It can be argued that it would be unethical to withhold the treatment from any patients, and that therefore all current patients should receive the new treatment. But then who would serve as a control group? One possibility is to use historical controls—that is, previous patients with the same illness who were treated with another therapy. One difficulty with historical controls is that there is often a tendency for later patients to show a better response—even to the same therapy—than earlier patients with the same diagnosis. This tendency has been confirmed, for instance, by comparing experiments conducted at the same medical centers in different years.[24] One major reason for the tendency is that the overall characteristics of the patient population may change with time. For instance, because diagnostic techniques tend to improve, patients with a given diagnosis (say, breast cancer) in 2001 may have a better chance of recovery (even with the same treatment) than those with the same diagnosis in 1991, because they were diagnosed earlier in the course of the disease.

Medical researchers do not agree on the validity and value of historical controls. The following example illustrates the importance of this controversial issue.

Coronary Artery Disease. Disease of the coronary arteries is often treated by surgery (such as bypass surgery), but it can also be treated with drugs only. Many studies have attempted to evaluate the effectiveness of surgical treatment for this common disease. In a review of 29 of these studies, each study was classified as to whether it used randomized controls or historical controls; the conclusions of the 29 studies are summarized in Table 8.4.[25]

It would appear from Table 8.4 that enthusiasm for surgery is much more common among researchers who use historical controls than among those who use randomized controls.

Example 8.14

■

TABLE 8.4 Coronary Artery Disease Studies

Type of Controls	Conclusion about Effectiveness of Surgery		Total Number of Studies
	Effective	*Not Effective*	
Randomized	1	7	8
Historical	16	5	21

Proponents of the use of historical controls argue that statistical adjustment can provide meaningful comparison between a current group of patients and a group of historical controls; for instance, if the current patients are younger than the historical controls, then the data can be analyzed in a way that adjusts, or corrects, for the effect of age. Critics reply that such adjustment may be grossly inadequate.

The concept of historical controls is not limited to medical studies. The issue arises whenever a researcher compares current data with past data. Whether the data are from the lab, the field, or the clinic, the researcher must confront the question: Can the past and current results be meaningfully compared? One should always at least ask whether the experimental material, and/or the environmental conditions, may have changed enough over time to distort the comparison.

How to Randomize

For the reasons given previously, randomized, double-blind experiments are considered the gold standard in research. A straightforward way to allocate experimental units to treatment groups is to use a **completely randomized design**, in which all possible allocations are equally likely. A completely randomized allocation is simple to carry out. The following example illustrates the procedure.

Example 8.15 **Adolescent Pregnancy.** An experiment was conducted to study whether peer support and monetary incentives help prevent repeat adolescent pregnancy. Adolescent mothers whose infants were younger than 5 months were asked if they wanted to be part of a study in which "teams" would be formed of teenaged mothers who did not want to have another child right away. Those who agreed to participate were split into four groups:

T_1: Peer support only

T_2: Monetary incentive only

T_3: Peer support and monetary incentive

T_4: No intervention (control)

The peer-support group met weekly to discuss issues related to pregnancy, such as contraceptive use and the advantages of delaying having a second child. Members of the monetary incentive group were given one dollar for each day that they participated in the experiment, provided they did not become pregnant. Members of the third group, T_3, were given the dollar-a-day incentive and also had weekly support group meetings. The fourth group served as a control and received no intervention.

The subjects were assigned to groups by blindly reaching into a can that contained bracelets of four different colors. The color of the bracelet drawn out determined the group to which the subject was assigned.[26] (We will assume that the bracelet was replaced by another of the same color before the next draw, so that there was always a 1 in 4 chance of being assigned to any of the four treatments.) The process of complete randomization is represented schematically in Figure 8.3.   ■

In the experiment described in Example 8.15 randomization was carried out by reaching into a can and drawing out differently colored bracelets. An alternative to a physical process like this is to use random numbers to make the treatment assignments. For example, generate a random number from 1 to 4—with a computer, calculator, or table of random digits—and assign a subject to treatment group 1 if a 1 is generated, to group 2 if a 2 is generated, and so on.

It is not necessary to have equal sample sizes in a completely randomized design. However, if equal samples sizes are desired, say with n experimental units per group, then the treatment groups can be selected by first choosing a random sample of size n from the available experimental units to comprise the first group, then choosing a random sample of size n from the remaining experimental units to comprise the second treatment group, and so on.

The following two examples illustrate some of the variety of applications of the completely randomized design.

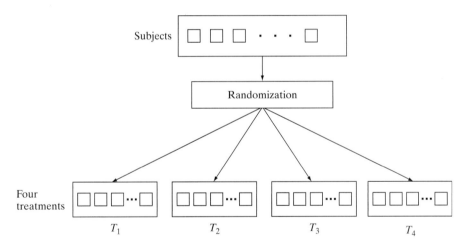

Figure 8.3 Completely randomized allocation

Soybean Growth.   For a study of plant growth, 60 one-week-old soybean seedlings will be set in individual pots in a greenhouse. The 60 pots will be divided into three groups of 20 each, and a different soil additive (A, B, or C) will be applied to each group. After 10 days of growth, 5 plants from each group will be harvested, dried, and weighed. Additional harvests will be made after 20, 30, and 40 days of growth. This experiment has 12 "treatments," as follows:

Example 8.16

T_1: Additive A, harvest at 10 days

T_2: Additive B, harvest at 10 days

$\vdots$

T_{12}: Additive C, harvest at 40 days

A completely randomized allocation could be used, with 5 of the 60 pots assigned to each of the 12 treatment groups.   ■

Example 8.17

White Blood Count.   In order to compare the effects of two drugs, A and B, on white blood count, 20 human volunteers will be divided into two groups, 10 to receive drug A and 10 to receive drug B. Blood specimens will be drawn at 0, 2, 4, 8, 16, and 24 hours after administration of the drug, and white blood cells will be counted in each specimen. For a completely randomized design, the 20 volunteers would be randomly allocated to the two treatment groups (A and B), 10 to each group.   ■

Notice that Examples 8.16 and 8.17 both involve measures over time, but the structure of the two experiments is quite different. In Example 8.17, repeated measurements are made on the same individual; by contrast, in Example 8.16 each plant is measured only once. This is an important distinction, and we will see that the two structures generally require different methods of data analysis.

Why Randomize?

Although the process of randomized allocation is straightforward, it can be tedious and researchers may be tempted to skip it. A haphazard rather than random procedure might appear to be equally good. For instance, suppose we need to allocate rats to four treatment groups: $T_1, T_2, T_3,$ and T_4. Rather than use random digits to allocate the rats, why not just reach in the cage and grab some rats for T_1, then some rats for T_2, and so on? This procedure would have a potentially serious bias: It would tend to assign the more sluggish (easier to catch) rats to T_1, and the most lively rats to T_4. The comparison of the treatments might then be distorted by the systematic differences in the rats. Of course, this particular source of bias is rather obvious. The advantage of randomization is that it guards against *all* biases, even those that are not at all obvious. Note, however, that randomization is not magic; it cannot *guarantee* that the treatment groups are exactly comparable, but it provides a form of insurance that is far better than haphazard allocation. (In some situations statistical adjustment, as mentioned in the previous section, can be used to correct for imbalances remaining after randomization.)

Randomization and the Random Sampling Model

Statistical methods for comparing two groups of observations (such as the *t* test) or several groups of observations are based on the random sampling model—the assumption that the groups to be compared can be regarded as random samples from their respective populations. The use of randomized allocation in an experiment helps to justify the application of the random sampling model in analyzing the data from the experiment. For instance, in Example 8.16 we might define conceptual populations as follows:

Population 1: Weights of soybean plants grown with additive A for 10 days

Population 2: Weights of soybean plants grown with additive B for 10 days

and so on.

In interpreting the experiment, it is crucial that the conceptual populations be identical in all respects except the experimental manipulations (soil additives and time of harvest). The physical act of randomization helps to justify this assumption. Thus, randomization is a kind of insurance against the possibility of hidden differences between the conceptual populations.

Exercises 8.8–8.15

Note: In several of these exercises you are asked to prepare a randomized allocation. For this purpose you can use either Table 1 or random digits from your calculator or a computer.

8.8 Fluoridation of drinking water has long been a controversial issue in the United State. One of the first communities to add fluoride to their water was Newburgh, New York. In March of 1944 a plan was announced to add fluoride to the Newburgh water supply, to begin on April 1 of that year. During the month of April citizens of Newburgh complained of digestive problems, which were attributed to the fluoridation of the water. However, there had been a delay in the installation of the fluoridation equipment, so that fluoridation did not begin until May 2.[27] Explain how the placebo effect/nocebo effect is related to this example.

8.9 Olestra is a no-calorie, no-fat additive that is used in the production of some potato chips. After the Food and Drug Administration approved the use of olestra some consumers complained that olestra caused stomach cramps and diarrhea. A randomized, double-blind experiment was conducted in which some subjects were given bags of potato chip made with olestra and other subjects were given ordinary potato chips. In the olestra group 38% of the subjects reported having gastrointestinal symptoms. However, in the group given regular potato chips the corresponding percentage was 37%. (The two percentages are not statistically significantly different.)[28] Explain how the placebo effect/nocebo effect is related to this example. Also explain why it was important for this experiment to be double-blind.

8.10 *[Hypothetical]* In a study of acupuncture, patients with headaches are randomly divided into two groups. One group is given acupuncture and the other group is given aspirin. The acupuncturist evaluates the effectiveness of the acupuncture and compares it to the results from the aspirin group. Explain how lack of blinding biases the experiment in favor of acupuncture.

8.11 Randomized, controlled experiments have found that vitamin C is not effective in treating terminal cancer patients.[29] However, a 1976 research paper reported that terminal cancer patients given vitamin C survived much longer than did historical controls. The patients treated with vitamin C were selected by surgeons from a group of cancer patients in a hospital.[30] Explain how this experiment was biased in favor of vitamin C.

8.12 For a medical experiment to investigate a certain new therapy, the investigators believe that the results for the new therapy will be of interest in themselves as well as in comparison to the standard therapy, and so they have decided to allocate twice as many patients to the new therapy. Suppose the experiment is to include 30 patients. Prepare a completely randomized allocation, assigning 10 patients to receive the standard therapy and 20 to receive the new therapy.

8.13 For a physiological study, eight monkeys are to be allocated to three treatment groups: Groups 1 and 2 will contain three animals each, and group 3 will contain two animals. Prepare a completely randomized allocation.

8.14 For a greenhouse experiment on lettuce growth, 15 pots, each containing four plants, are to be allocated to five treatment groups, three pots to a group. (All plants in a pot receive the same treatment, so that the experimental units are pots, not plants.) Prepare a completely randomized allocation of pots to treatments.

8.15   Canine parvovirus (CPV) is an intestinal disease that affects dogs. In a study to test a vaccine against CPV, 7-week-old beagle pups will be given either vaccine by injection, vaccine by nose drops, or placebo. The pups will be housed in pens of six pups each and watched for 5 weeks for signs of CPV. Because CPV is highly contagious, the experimental unit will be a pen rather than an individual pup; for instance, each pen will be classified as a success if all its pups remain free of CPV, and as a failure otherwise. All pups in a pen will receive the same treatment. Suppose 18 pens are available; prepare a completely randomized allocation of the 18 pens to the three treatment groups.[31]

8.4 RESTRICTED RANDOMIZATION: BLOCKING AND STRATIFICATION

The completely randomized design makes no distinctions among the experimental units. Often an experiment can be improved by a more refined approach, one that takes advantage of known patterns of variability in the experimental units.

The Randomized Blocks Design

In a **randomized blocks design**, we first group the experimental units into sets, or **blocks**, of relatively similar units and then we randomly allocate treatments within each block. Here is an example.

Example 8.18   **Stacked Cages.**   Methionine is a sulfur-amino acid that is added to the diets of turkeys to enhance growth. In one experiment, three types of methionine, which we will refer to as T_1, T_2, and T_3, were compared. The response variable was weight gain of young turkeys over a three-week period; the experimental units were 12 cages of turkey chicks. The cages were stacked on top of each other, four layers high, in a room, with three cages on the floor, three cages in the second level, three cages in the third level up, and three cages in the top level, near the ceiling. If a completely randomized design had been used, then four cages would have been assigned T_1, four cages would have been assigned T_2, and four cages would have been assigned T_3. However, in such a design it might have turned out that most of the T_1 cages were on the floor, most of the T_2 cages were in the middle levels, and most of the T_3 cages were at the ceiling level. The problem with this arrangement is that cages near the ceiling will tend to be quite a bit warmer than cages near the floor and how much a turkey eats depends in part on the temperature. Thus, if the T_3 turkeys gained less weight than the others, we wouldn't know if this was due to the treatment, T_3, or due to the temperature effect from being near the ceiling.

Because of the concern over temperature, a randomized blocks design was used in this study. The three floor cages were considered to be the first block, the three cages in the second level were the second block, the three cages in the third level were the third block, and the three ceiling cages were the fourth block. In each block one cage was assigned at random to T_1, one to T_2, and one to T_3. This experimental design is indicated schematically in Figure 8.4.[32] By using a randomized blocks design, the effect of temperature was controlled—temperature had an equal affect on all three treatments. Moreover, the variability in turkey weight gain that is caused by temperature was removed from the comparisons of the three treatments, which made it much easier to see which treatment is best.   ■

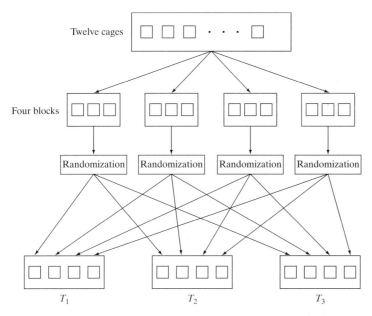

Figure 8.4 Blocking by cage location in turkey growth study

Example 8.18 is an illustration of a randomized blocks design. To carry out a randomized blocks design, the experimenter creates or identifies suitable blocks of experimental units and then randomly assigns treatments within each block in such a way that each treatment appears in each block.* In Example 8.18, the cages at a given level serve as blocks. In general, we create blocks in order to reduce or eliminate variability caused by extraneous variables, so that the precision of the experiment is increased. We want the experimental units within a block to be homogenous; we want the extraneous variability to occur *between* the blocks. Here are more examples of randomized blocks designs in biological experiments.

Blocking by Litter.   How does experience affect the anatomy of the brain? In a typical experiment to study this question, young rats are placed in one of three environments for 80 days:

T_1: *Standard environment.* The rat is housed with a single companion in a standard lab cage.

T_2: *Enriched environment.* The rat is housed with several companions in a large cage, furnished with various playthings.

T_3: *Impoverished environment.* The rat lives alone in a standard lab cage.

At the end of the 80-day experience, various anatomical measurements are made on the rats' brains.

Suppose a researcher plans to conduct the aforementioned experiment using 30 rats. To minimize variation in response, all 30 animals will be male, of the same age and strain. To reduce variation even further, the researcher can take advantage of the similarity of animals from the same litter. In this approach, the researcher would obtain three male rats from each of ten litters. The three littermates from each litter would be assigned at random: one to T_1, one to T_2, and one to T_3.[33]   ■

Example 8.19

---

* Strictly speaking, the design we discuss is termed a *randomized complete blocks design* because every treatment appears in every block. In an *incomplete blocks design*, each block contains some, but not necessarily all, of the treatments.

Another way to visualize the experimental design is in tabular form, as shown in Table 8.5. Each "Y" in the table represents an observation on one rat. Using the layout of Table 8.5, the experimenter can compare the responses of rats that received *different* treatments but are in the *same* litter. Such comparisons are, of course, not affected by any difference (genetic and other) that may exist between one litter and another.

TABLE 8.5 Format for Rat Brain Data

	Treatment		
	T_1	T_2	T_3
Litter 1	Y	Y	Y
Litter 2	Y	Y	Y
Litter 3	Y	Y	Y
⋮	⋮	⋮	⋮
Litter 10	Y	Y	Y

Example 8.20

Within-Subject Blocking. A dermatologist is planning a study to compare two medicated lotions for their effectiveness in treating acne. Twenty patients are to participate in the study. Each patient will use lotion A on one side of his or her face and lotion B on the other; the dermatologist will observe the improvement on each side during a 3-month period. For each patient, the side of the face to receive lotion A is randomly selected; the other side receives lotion B. The bottles of medication have coded labels so that neither the patient nor the physician knows which bottle contains A and which contains B—that is, in addition to blocking, the experiment also makes use of blinding.[34] This example, with blocks of size 2, is an example of **pairing**: The left side of the face is paired with the right side of the face. We will consider the analysis of paired data in Chapter 9. ∎

Example 8.21

Blocking in an Agricultural Field Study. When comparing several varieties of grain, an agronomist will generally plant many field plots of each variety and measure the yield of each plot. Differences in yields may reflect not only genuine differences among the varieties, but also differences among the plots in soil fertility, pH, water-holding capacity, and so on. Consequently, the spatial arrangement of the plots in the field is important. An efficient way to use the available field area is to divide the field into large regions—the blocks—and to subdivide each block into several plots. Within each block the various varieties of grain are then randomly allocated to the plots, with a separate randomization done for each block. For instance, suppose we want to test four varieties of barley. Then each block would contain four plots. The resulting randomized allocation might look like Figure 8.5, which is a schematic map of the field. The "treatments" T_1, T_2, T_3, and T_4 are the four varieties of barley. ∎

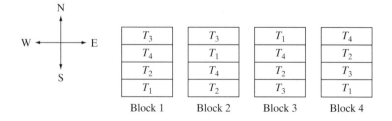

Figure 8.5 Layout of an agricultural randomized blocks design

Creating the Blocks

As the preceding examples show, blocking is a way of *organizing* the inherent variation that exists among experimental units. Ideally, the blocking should be arranged so as to increase the information available from the experiment. To achieve this goal, **the experimenter should try to create blocks that are as homogeneous within themselves as possible, so that the inherent variation between experimental units becomes, as far as possible, variation between blocks rather than *within* blocks**. This principle was illustrated in the preceding examples (e.g., in Example 8.19, where blocking by litter exploits the fact that littermates are more similar to each other than to non-littermates). The following is another illustration.

Agricultural Field Study.   For the barley experiment of Example 8.21, how would agronomists determine the best arrangement or layout of blocks in a field? They would design the blocks to take advantage of any prior knowledge they may have of fertility patterns in the field. For instance, if they know that an east-west fertility gradient exists in the field (perhaps the field slopes from east to west, with the result that the west end has a thicker layer of good soil), then they might choose blocks as in Figure 8.5; the layout maximizes soil differences between the blocks and minimizes differences between plots within each block. (But even if a field appears to be uniform, blocking is usually used in agronomic experiments, because plots closer together in the field are generally more similar than plots farther apart.)

> **Example 8.22**

To add solidity to this example, let us look at a set of data from a randomized blocks experiment on barley. Each entry in Table 8.6 shows the yield (bushels of barley per acre) of a plot 3.5 ft wide by 80 ft long.[35]

TABLE 8.6 Yield (lb) of Barley

	Block 1	Block 2	Block 3	Block 4	Mean
Variety 1	93.5	66.6	50.5	42.4	63.3
Variety 2	102.9	53.2	47.4	43.8	61.8
Variety 3	67.0	54.7	50.0	40.1	53.0
Variety 4	86.3	61.3	50.7	46.4	61.2

It appears from Table 8.6 that the yield potential of the blocks varies greatly; the data indicate a definite fertility gradient from block 1 to block 4. Because of the blocked design, comparison of the varieties is relatively unaffected by the fertility gradient. Of course, there also appears to be substantial variation within blocks. (You might find it an interesting exercise to peruse the data and ask yourself whether the observed differences between varieties are large enough to conclude that, for example, variety 1 is superior [in mean yield] to variety 3; use your intuition rather than a formal statistical analysis. The truth is revealed in Note 35.)   ■

The Randomization Procedure

Once the blocks have been created, the blocked allocation of experimental units is straightforward. Randomization is carried out for each block separately, as illustrated in the following example.

Example 8.23 | **Agricultural Field Study.** Consider the agricultural field experiment of Example 8.21. In block 1, let us label the plots 1, 2, 3, 4, from north to south (see Figure 8.5); we will allocate one plot to each variety. The allocation proceeds as for the completely randomized design, by choosing plots at random from the four, and assigning the first plot chosen to T_1, the second to T_2, and so on. For instance, if we start reading Table 1 at row 06, column 16 and proceed downward, we get the following allocation:

$$\text{Block 1:} \quad \begin{array}{ll} T_1\text{:} & \text{Plot 4} \\ T_2\text{:} & \text{Plot 3} \\ T_3\text{:} & \text{Plot 1} \\ T_4\text{:} & \text{Plot 2} \end{array}$$

This is in fact the assignment shown in Figure 8.5 for block 1. We can then repeat this procedure for block 2. There is no need to find a new starting place in the random digit table. If we continue reading consecutive digits from Table 1, we obtain for block 2:

$$\text{Block 2:} \quad \begin{array}{ll} T_1\text{:} & \text{Plot 2} \\ T_2\text{:} & \text{Plot 4} \\ T_3\text{:} & \text{Plot 1} \\ T_4\text{:} & \text{Plot 3} \end{array}$$

which is the assignment shown in Figure 8.5. We can proceed in this way until all the blocks have been processed. ∎

Stratified Randomization

Stratification is a design strategy similar to blocking, in which discrete groups, or **strata**, of experimental units are identified, and random allocation to treatment groups is carried out separately in each stratum. Usually the number of strata is fairly small, so that each stratum contains a substantial number of experimental units. For example, it is common for a medical trial to be conducted in which subjects are enrolled at each of several centers (e.g., in different cities) and a separate randomization of subjects to treatment groups is conducted at each center. In such a multicenter clinical trial, each center is a stratum. The following is an example of stratification in which the strata are formed on the basis of sex, age, and stage of disease.

Example 8.24 | **Prognostic Strata in Medical Experiments.** Consider a medical experiment to compare several treatments for a certain disease. Suppose the investigators know in advance that the patient's overall outlook, or prognosis, depends on certain factors such as the patient's age, sex, and how far the disease has progressed at the time treatment is started. These prognostic factors can be used to define categories (strata) of patients. For instance, suppose we crudely divide age into two categories (young, old) and stage of disease into two categories (early, advanced). Then the joint criteria of sex, age, and stage of disease determine the following strata:

Stratum 1: Female, young, early

Stratum 2: Male, young, early

Stratum 3: Female, old, early

Stratum 4: Male, old, early
Stratum 5: Female, young, advanced
Stratum 6: Male, young, advanced
Stratum 7: Female, old, advanced
Stratum 8: Male, old, advanced

A stratified randomization would consist of randomly allocating patients to treatment groups, separately for each stratum. Note that, because of the nature of the strata, one would expect the different strata to contain different numbers of patients. ■

Complementarity of Randomization and Blocking

Randomization and blocking (or stratification) are two complementary ways of dealing with the extraneous variables that may threaten an experiment. In biological experimentation, the experimental units may be similar but they are usually not identical; they differ in many respects, both visible and invisible. The advantage of randomization is that it tends to maintain balance among the treatment groups with respect to all extraneous variables—including those the experimenter has not even thought of. Of course, blocking gives a more definite assurance of balance, but blocking can control only a limited number of variables. No matter how carefully the blocks are constructed, the units within the blocks are still different. This is why randomization within blocks should always be used as a complement to blocking.

In many experiments, blocking is not practical at all. For one thing, blocking can be expensive. Another difficulty is that the experimenter may not have a good basis for forming blocks. Blocking on a variable that is not related to the response variable of interest is a waste of time and (as we will see) is inefficient.

Blocking actually has two related purposes. First, it improves the comparability of the various treatment groups by forcing them to be balanced with respect to the variables used in the blocking. Second, and more important, is the effect of blocking on the *precision* of treatment comparisons in an experiment. By minimizing the variability among experimental units within a block, the experimenter enables treatment differences to stand out more clearly. Consequently, an experiment with well-constructed blocks can yield much more information than an unblocked experiment on similar experimental material; the blocked experiment is said to be more **efficient**. In Chapter 9 we will show in detail, for the specific case of comparing two treatments, how blocking can increase information, and how the added information can be extracted from the data by suitable analysis.

In many experiments randomization is used at several stages. For instance, in a greenhouse experiment a botanist may randomly allocate seedlings to pots, treatments to pots, and pots to positions on the greenhouse bench. Often, time as well as space must be considered in the design. For instance, suppose 20 mice are to be sacrificed and their livers assayed for a certain biochemical. If the same researcher is going to do all 20 assays, he or she may not be able to do them all simultaneously. The worst possible plan (although the simplest) would be to assay the animals systematically: all T_1 animals first, then all T_2 animals, and so on. A better plan would be to randomly allocate the 20 animals to the 20 time slots for assay.

Statistical Adjustment for Extraneous Variables

We have noted that it is desirable to minimize or control the influence of extraneous variables in an experiment. Randomization, blocking, and stratification are techniques for exerting this control through the design of the experiment. Control can also be imposed at the analysis stage. A battery of statistical techniques is available for taking account of extraneous variables in the analysis of data. These analytical techniques include poststratification, regression, analysis of variance, and analysis of covariance. We will discuss some of these in subsequent chapters. Many of these analytical methods can be used in conjunction with the design techniques of randomization, blocking, and stratification.

Exercises 8.16–8.24

Note: In several of these exercises you are asked to prepare a randomized allocation. For this purpose you can use either Table 1 or random digits from your calculator or a computer.

8.16   In an experiment to compare six different fertilizers for tomatoes, 36 individually potted seedlings are to be used, six to receive each fertilizer. The tomato plants will be grown in a greenhouse, and the total yield of tomatoes will be observed for each plant. The experimenter has decided to use a randomized blocks design: the pots are to be arranged in six blocks of six plants each on the greenhouse bench. Two possible arrangements of the blocks are shown in the accompanying figure.

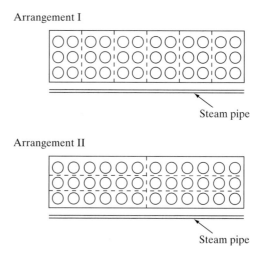

One factor that affects tomato yield is temperature, which cannot be held exactly constant throughout the greenhouse. In fact, a temperature gradient across the bench is likely. Heat for the greenhouse is provided by a steam pipe which runs lengthwise under one edge of the bench, and so the side of the bench near the steam pipe is likely to be warmer.

(a) Which arrangement of blocks (I or II) is better? Why?
(b) Prepare a randomized allocation of treatments to the pots within each block. (Refer to Example 8.23 as a guide; assume that the assignments of seedlings to pots and of pots to positions within the block have already been made.)

8.17 An experiment on vitamin supplements is to be conducted on young piglets, using litters as blocks in a randomized blocks design. There will be five treatments: four types of supplement and a control. Thus, five piglets from each litter will be used. The experiment will include five litters. Prepare a schematic representation of a randomized blocks allocation of piglets to treatments, similar to Figure 8.4.

8.18 Refer to the vitamin experiment of Exercise 8.17. Prepare a randomized blocks allocation of piglets to treatments. (Refer to Example 8.23 as a guide.)

8.19 Refer to the vitamin experiment of Exercise 8.17. Suppose a colleague of the experimenter proposes an alternative design: that all pigs in a given litter receive the same treatment, with the five litters being randomly allocated to the five treatments. He points out that his proposal would save labor and greatly simplify the recordkeeping. If you were the experimenter, how would you reply to this proposal?

8.20 In a pharmacological experiment on eating behavior in rats, 18 rats are to be randomly allocated to three treatment groups: T_1, T_2, and T_3. While under observation, the animals will be kept in individual cages in a rack. The rack has three tiers with six cages per tier. In spite of efforts to keep the lighting uniform, the lighting conditions vary somewhat from one tier to another (the bottom tier is darkest), and the experimenter is concerned about this because lighting is thought to influence eating behavior in rats. The following three plans are proposed for allocating the rats to positions in the rack (to be done after the allocation of rats to treatment groups):

   Plan I. Randomly allocate the 18 rats to the 18 positions in the rack.
   Plan II. Put all T_1 rats on the first tier, all T_2 rats on the second, and all T_3 rats on the third tier.
   Plan III. On each tier, put two T_1 rats, two T_2 rats, and two T_3 rats.

   Put these three plans in order, from best to worst. Explain your reasoning.

8.21 An experimenter is planning an agricultural field experiment to compare the yields of 25 varieties of corn. She will use a randomized blocks design with six blocks; thus, there will be 150 plots, and the yield of each plot must be measured. The experimenter realizes that the time required to harvest and weigh all the plots is so long that rain might interrupt the operation. If rain should intervene, there could be a yield difference between the harvests before and after the rain. The experimenter is considering the following plans.

   Plan I. Harvest all plots of variety 1 first, all of variety 2 next, and so on.
   Plan II. Harvest all plots of block 1 first, all of block 2 next, and so on.

   Which plan is better? Why?

8.22 For an experiment to compare two methods of artificial insemination in cattle, the following cows are available:

   Heifers (14–15 months old): 8 animals
   Young cows (2–3 years old): 8 animals
   Mature cows (4–8 years old): 10 animals

   The animals are to be randomly allocated to the two treatment groups, using the three age groups as strata. Prepare a suitable allocation, randomly dividing each stratum into two equal groups.

8.23   Suppose a drug for treating high blood pressure is to be compared to a standard blood pressure drug in a study of humans.

(a) Describe an experimental design for a study that makes use of blocking. Be careful to note which parts of the design involve randomness and which parts do not.

(b) Can the experiment you described in part (b) involve blinding? If so, explain how blinding could be used.

8.24   True or false (and say why): The primary reason for using a randomized blocks design in an experiment is to reduce bias.

8.5 LEVELS OF REPLICATION

Replication is central to the statistical analysis of experimental data. By considering the variation among experimental units treated alike, the researcher obtains a benchmark for assessing differences between treatments. Increased replication tends to provide increased information (all other things being equal). For instance, we saw in Chapter 6 how the information in an experiment, as measured by the standard error of the mean, depends on the number of replicate observations, through the formula

$$SE_{\bar{y}} = \frac{s}{\sqrt{n}}$$

In Section 6.4 we discussed the important topic of planning an experiment to include sufficient replication (that is, large enough n) to achieve a desired standard error (SE).

In this section we consider an issue that can cause confusion in both the planning and the analysis of biological experiments. Sometimes variation arises at several different hierarchical levels in an experiment, and it can be a challenge to sort these out, and particularly, to correctly identify the quantity n. Example 8.25 illustrates this issue.

Example 8.25   **Germination of Spores.**   In a study of the fungus that causes the anthracnose disease of corn, interest focused on the survival of the fungal spores.[36] Batches of spores, all prepared from a single culture of the fungus, were stored in chambers under various environmental conditions, and then assayed for their ability to germinate, as follows. Each batch of spores was suspended in water, and then plated on agar in a petri dish. Ten "plugs" of 3-mm diameter were cut from each petri dish, and were incubated at 25°C for 12 hours. Each plug was then examined with a microscope for germinated and ungerminated spores. The environmental conditions of storage (the "treatments") included the following:

T$_1$: Storage at 70% relative humidity for one week

T$_2$: Storage at 60% relative humidity for one week

T$_3$: Storage at 60% relative humidity for two weeks

and so on.

All together there were 43 treatments.

The design of the experiment is indicated schematically in Figure 8.6. There were 129 batches of spores, which were randomly allocated to the 43 treatments, three batches to each treatment. Each batch of spores resulted in one petri dish, and each petri dish resulted in ten plugs.

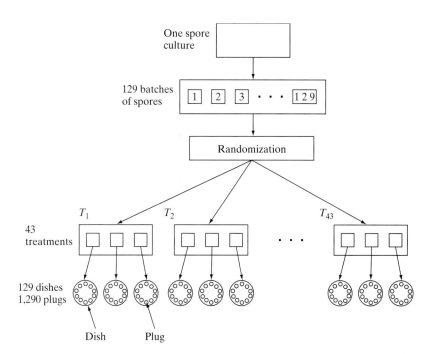

Figure 8.6 Design of spore germination experiment

To get a feeling for the issues raised by this design, let us look at some of the raw data. Table 8.7 shows the percentage of the spores that had germinated for each plug asssayed for treatment 1. Table 8.7 also shows that there is considerable variability both *within* each petri dish and *between* the dishes. The variability within the dishes reflects local variation in the percent germination, perhaps due largely to differences among the spores themselves (some of the spores were more mature

TABLE 8.7 Percentage Germination Under Treatment 1

	Dish I	Dish II	Dish III
	49	66	49
	58	84	60
	48	83	54
	69	69	72
	45	72	57
	43	85	70
	60	59	65
	44	60	68
	44	75	66
	68	68	60
Mean	52.8	72.1	62.1
SD	10.1	9.5	7.4

than others). The variability between dishes is even larger, because it includes not only local variation but also larger-scale variation such as the variability among the original batches of spores, and temperature and relative humidity variations within the storage chambers.

Now consider the problem of comparing treatment 1 to the other treatments. Would it be legitimate to take the point of view that we have 30 observations for each treatment? To focus this question, let us consider the matter of calculating the standard error for the mean of treatment 1. The mean and SD of all 30 observations are

$$\text{Mean} = 62.33$$

$$\text{SD} = 11.88$$

Is it legitimate to calculate the SE of the mean as

$$\text{SE}_{\bar{y}} = \frac{s}{\sqrt{n}} = \frac{11.88}{\sqrt{30}} = 2.2$$

As you may suspect, **this is not legitimate**. There is a hierarchical structure in the data (see Section 6.5), and so we cannot apply the SE formula so naively. An acceptable way to calculate the SE is to consider the mean for each dish as an observation; thus, we obtain the following:*

$$\text{Observations:} \quad 52.8, 72.1, 62.1$$

$$n = 3$$

$$\text{Mean} = 62.33$$

$$\text{SD} = 9.65$$

$$\text{SE}_{\bar{y}} = \frac{s}{\sqrt{n}} = \frac{9.65}{\sqrt{3}} = 5.6$$

Notice that the incorrect analysis gave the same mean (62.33) as this analysis but an inappropriately small SE (2.2 rather than 5.6). If we were comparing several treatments, the same pattern would tend to hold; the incorrect analysis would tend to produce SEs that were (individually and pooled) too small, which might cause us to "overinterpret" the data, in the sense of thinking we see treatment differences where none exists.

We should emphasize that, even though the correct analysis requires combining the measurements on the ten plugs in a dish into a single observation for that dish, nevertheless the experimenter was not wasting effort by measuring ten plugs per dish instead of, say, only one plug per dish. The mean of ten plugs is a much better estimate of the average for the entire dish than is a measurement on one plug; the improved precision for measuring ten plugs is reflected in a smaller between-dish SD. For instance, for treatment 1 the SD was 9.65; if fewer plugs per dish had been measured, this SD would probably have been larger. ■

---

* An alternative way to aggregate the data from the ten plugs in a dish would be to combine the raw counts of germinated and ungerminated spores for the whole dish and express these as an overall percent germination.

The pitfall illustrated by Example 8.25 has trapped many an unwary re-searcher. When hierarchical structures result from repeated measurements on the same individual organism (as illustrated in Section 6.5), they are relatively easy to recognize. But the hierarchical structure in Example 8.25 has a different origin; it is due to the fact that the unit of observation is an individual plug, but individual plugs are not randomly allocated to the treatment groups. Rather, the unit that is randomly allocated to treatment is a batch of spores, which later is plated in a petri dish, which then gives rise to ten plugs. In the language of experimental design, *plugs* are **nested** within petri dishes. *Whenever observational units are nested within the units that are randomly allocated to treatments, a hierarchical structure may potentially exist in the data.* Note that the difficulty is only "potential"; in some cases a nonhierarchical analysis may be acceptable. For instance, if experience had shown that the differences between petri dishes were negligible, then we might ig-nore the hierarchical structure in analyzing the data. The decision can be a diffi-cult one and may require expert statistical advice.

The issue of hierarchical data structures has important implications for the design of an experiment as well as its analysis. The sample size (n) must be appropriately identified in order to determine whether the experiment includes enough replication. As a simple example, suppose it is proposed to do a spore germination experiment such as that of Example 8.25, but with only *one* dish per treatment, rather than three. To see the flaw in this proposal, suppose for simplicity that the proposed experiment is to include only three treatments, with one dish per treatment. How, then, would we distinguish treatment dif-ferences from inherent differences between the dishes? The answer is that we could not. The intertreatment differences and the interdish differences would be mutually entangled, or confounded. You can easily visualize this situation if you look at the data in Table 8.7 and pretend that those data came from the proposed experiment; that is, pretend that dishes I, II, and III had received dif-ferent treatments and that we had no other data. It would be difficult to extract meaningful information about intertreatment differences unless we knew for *certain* that interdish variation was negligible. (This is not to say that a pro-posal to use one dish per treatment is irredeemably flawed. Such a design can succeed if the treatment effects are sufficiently regular; we will see an appli-cation of this idea in Chapter 13.)

Determination of Sample Size

We saw in Section 6.4 how to use a preliminary estimate of the SD to determine the sample size (n) required to attain a desired degree of precision, as expressed by the SE. (A similar development in optional Section 7.8 expressed desired pre-cision in terms of statistical power.) These ideas carry over to experiments in-volving hierarchical data structures. For example, suppose a botanist is planning a spore germination experiment such as that of Example 8.25. If she has already de-cided to use ten plugs per dish, the remaining problem would be to decide on the number of dishes per treatment. This question could be approached as in Section 6.4, considering the dish as the experimental unit, and using a preliminary esti-mate of the SD between dishes (which was 9.65 in Example 8.25). If, however, she wants to choose optimal values for *both* the number of plugs per dish *and* the num-ber of dishes per treatment, she may wish to consult a statistician.

Exercises 8.25-8.26

8.25 Four treatments were compared for their effect on the growth of spinach cells in cell culture flasks. Using a completely randomized design, the experimenter allocated two flasks to each treatment. After a certain time on treatment, he randomly drew three aliquots (1 cc each) from each flask and measured the cell density in each aliquot; thus, he had six cell density measurements for each treatment. In calculating the standard error of a treatment mean, the experimenter calculated the standard deviation of the six measurements and divided by $\sqrt{6}$. On what grounds might an objection be raised to this method of calculating the SE?

8.26 In an experiment on soybean varieties, individually potted soybean plants were grown in a greenhouse, using a completely randomized design with 10 plants of each variety. From the harvest of each plant, five seeds were chosen at random and individually analyzed for their percentage of oil. This gave a total of 50 measurements for each variety. To calculate the standard error of the mean for a variety, the experimenter calculated the standard deviation of the 50 observations and divided by $\sqrt{50}$. Why would this calculation be of doubtful validity?

8.6 SAMPLING CONCERNS (OPTIONAL)

Public health officials often rely on information obtained from surveys of populations. In this section we will consider issues related to sampling, particularly survey sampling.

Sampling Errors

The term **sampling error** refers to any error that arises due to the sampling process. When we construct a confidence interval and discuss the margin of error (the "±part" of the confidence interval), we are talking about sampling error. Another type of sampling error is **selection bias**, which is bias introduced by the way in which individuals are selected. For example, if we use a net to obtain a sample of fish, we are likely to oversample large fish and to undersample very small fish, which can more easily slip through the holes in the net (as discussed in Example 3.2). If we conduct a telephone survey and choose people at random from a telephone book, we exclude people who have unlisted numbers. This kind of error would not arise if we were to take a complete census of the population of interest; thus, it is a sampling error.

Nonsampling Errors

A **nonsampling error** is an error that is not caused by the sampling method; that is, a nonsampling error is one that would have arisen even if the researcher had a census of the population. For example, the way in which questions are worded can greatly influence how people answer them, as Example 8.26 shows.

Example 8.26 **Abortion Funding.**   In 1991 the U.S. Supreme Court made a controversial ruling upholding a ban on abortion counseling in federally financed family-planning clinics. Shortly after the ruling, a sample of 1,000 people were asked, "As you may know, the U.S. Supreme Court recently ruled that the federal government is not required to use taxpayer funds for family planning programs to perform, counsel or refer for abortion as a method of family planning. In general, do you favor or oppose this ruling?" In the sample, 48% favored the ruling, 48% were opposed, and 4% had no opinion.

A separate opinion poll conducted at nearly the same time, but by a different polling organization, asked over 1,200 people, "Do you favor or oppose that Supreme Court decision preventing clinic doctors and medical personnel from discussing abortion in family planning clinics that receive federal funds?" In this sample, 33% favored the decision and 65% opposed it.[37] The difference in the percentages favoring the opinion is too large to be attributed to chance error in the sampling. It seems that the way in which the question was worded had a strong impact on the respondents. ■

Another type of nonsampling error is **nonresponse bias**, which is bias caused by persons not responding to some of the questions in a survey or not returning a written survey. It is common to have only one-third of those receiving a survey in the mail complete the survey and return it to the researchers. (We consider the people receiving the survey to be part of the sample, even if some of them don't complete the entire survey, or even return the survey at all.) If the people who respond are unlike those who choose not to respond—and this is often the case, since people with strong feelings about an issue tend to complete a questionnaire, while others will ignore it—then the data collected will not accurately represent the population.

HIV Testing.   A sample of 949 men were asked if they would submit to an HIV test of their blood. Of the 782 who agreed to be tested, 8 (1.02%) were found to be HIV positive. However, some of the men refused to be tested. The health researchers conducting the study had access to serum specimens that had been taken earlier from these 167 men and found that 9 of them (5.4%) were HIV positive.[38] Thus, those who refused to be tested were much more likely to have HIV than those who agreed to be tested. An estimate of the HIV rate based only on persons who agree to be tested is likely to substantially underestimate the true prevalence. ■

> **Example 8.27**

There are other cases in which an experimenter is faced with the vexing problem of **missing data**—that is, observations that were planned but cannot be made. In addition to nonresponse, this can arise because experimental animals or plants die, because equipment malfunctions, or because human subjects fail to return for a follow-up observation.

A common approach to the problem of missing data is to simply use the remaining data and ignore the fact that some observations are missing. This approach is temptingly simple but must be used with extreme caution, because comparisons based on the remaining data may be seriously biased. For instance, if observations on some experimental mice are missing because the mice died of causes related to the treatment they received, it is obviously not valid to simply compare the mice who survived. As another example, if patients drop out of a medical study because they think their treatment is not working, then analysis of the remaining patients could produce a greatly distorted picture.

Naturally, it is best to make every effort to avoid missing data. But if data are missing, it is crucial that the possible reasons for the omissions be considered in interpreting and reporting the results.

Randomized Response Sampling

Sometimes researchers want to estimate a population proportion, such as the percentage of persons in a given community who are HIV positive or the percentage who have used illegal drugs in the past year, but they cannot simply conduct a survey because the persons selected may refuse to answer the sensitive question asked, or they may be inclined to lie. The **randomized response** sampling method provides a way to obtain an unbiased estimate of a population proportion in such a situation. There are several ways in which a randomized response sample can be conducted. Example 8.28 shows one approach.

Example 8.28 **Use of Illegal Drugs.** What percentage of college students on a given campus use illegal drugs? Let this unknown percentage be denoted p. A sample of 78 students were asked about their use of drugs in the following round-about way.[39] They were told to toss a coin twice, privately. They were then told to answer yes or no to either the "real" question, Q1, or the "decoy" question, Q2, according to whether they got heads or tails on their first coin toss. Those who got heads were to answer Q1; those who got tails were to answer Q2.

Q1: Have you used illegal drugs?

Q2: Did you get heads on your second coin toss?

Before they answered their randomly chosen questions, it was explained to the students that they were protected against self-incrimination in the following way: Only the student knew which question he or she was answering. Thus, students could be honest when answering.

Of the 78 responses, 34 were "yes" and 54 were "no." Thus, the sample proportion "yes" was $34/78 = .436$. Figure 8.7 shows a probability tree for this sampling scheme. Note that there are two ways a respondent can answer "yes": either (1) by getting heads on the first coin toss and having used illegal drugs, which happens with probability $\frac{1}{2}p$, or (2) by getting tails on the first toss and heads on the second toss, which happens with probability $\frac{1}{4}$. Thus, the probability of a "yes" response is $.5p + .25$:

$$\Pr(\text{"yes"}) = .5p + .25$$

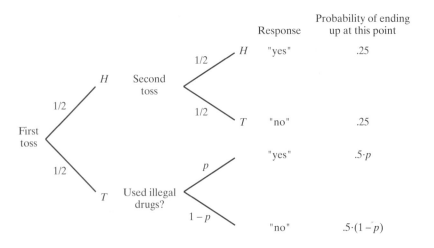

Figure 8.7 Probability tree for randomized response sampling

If we substitute our sample fraction of "yes" responses, .436, we have a way to estimate p.

$$.436 = .5\hat{p} + .25$$

so

$$\hat{p} = \frac{.436 - .25}{.5} = .372$$

Thus, our estimate is that 37.2% of the students have used illegal drugs. ∎

Exercises 8.27–8.31

8.27   Suppose a computer scanner is used to enter data from a study into a database. Suppose the scanner mistakenly reads every "7" as a "1". Is this an example of a sampling error or a nonsampling error? Why?

8.28   Suppose that in a study of birthweights of babies, a weight of 11 lb 4 oz is recorded as 1 lb 4 oz. Is this an example of a sampling error or a nonsampling error? Why?

8.29   Suppose someone conducts a randomized response sample, as described in Example 8.28, and find that 43 out of 104 persons answer "yes." What is the resulting estimate of p, the population percentage for whom the truthful answer to the sensitive question is "yes"?

8.30   Suppose someone conducts a randomized response sample, as described in Example 8.28, and find that 28 out of 74 persons answer "yes." What is the resulting estimate of p, the population percentage for whom the truthful answer to the sensitive question is "yes"?

8.31   The American Heart Association designed a 12-week educational program for women with the goal of improving their nutrition, physical activity, and knowledge of heart disease. A group of 23,171 women were recruited for the program and were sent a manual that contained weekly information regarding risk factors for cardiovascular disease, as well as how to build a support system for lifestyle change. The women were asked to return follow-up evaluations during the program in which they reported on their physical activity, diet, and knowledge of heart disease. However, only 6,389 women sent in evaluations at the end of the second week and only 4,209 women sent in evaluations at the end of the fourth week. Only 3,775 of the women completed the program and sent in evaluations when the program ended after 12 weeks. The participants who completed the program reported a marked increase in exercise (P-value = .001) and a decrease in "excess calories or fat" (P-value = .001) compared with their levels at the beginning of the program.[40] Is it correct to conclude, given the very small P-values, that the program is successful at promoting exercise and improved diets? Discuss.

8.7   PERSPECTIVE

In this chapter we have introduced some of the statistical principles that can guide a scientist in designing an investigation. We have considered several types of observational studies and two basic designs using randomized allocation.

Design of Experiments

There is much more to the subject of randomized allocation than we have space to present. For instance, in many situations it is possible to retrieve much information from an experiment in which replication is absent or incomplete, if the experiment has been designed correctly. The statistical discipline called **design of experiments** is devoted to the investigation of schemes for designing experiments that are informative and make efficient use of the available resources. These include methods of allocation of experimental units to treatments, how and where to incorporate replication, and how to choose which treatments to include (for instance, which doses of a drug, or which combinations of light and temperature).

Only Statistical?

The term *statistical* is sometimes used—or, rather, misused—as an epithet. For instance, some people say that the evidence linking dietary cholesterol and heart disease is "only statistical." What they really mean is "only observational." Statistical evidence can be very strong indeed, if it flows from a randomized experiment rather than an observational study. As we saw in Section 8.2, statistical evidence from an observational study must be interpreted with great care, because of potential distortions caused by extraneous variables. For this reason, serious observational research frequently involves collaboration with a statistician.

Scope of Inference

In statistical inference, randomization can enter in two different, but related, ways. The first is that we might have a random sample from a population. The second is that in an experiment we might assign treatments to the experimental units at random. The combination of these two leads to four possible settings.

1. *Experimental units (subjects) not chosen at random; no random assignment to treatments.* In this setting we cannot make a statistical inference. For example, consider studying the effect of vitamin C on the common cold. Suppose we take the first 50 people who volunteer to be in a study and assign the men to receive vitamin C and the women to receive a placebo. Even if the study shows a difference between vitamin C and placebo in frequency of colds, we are in no position to make a statistical inference since we have confounded the effect of vitamin C with the effect of being male.

2. *Experimental units (subjects) chosen at random; no random assignment to treatments.* In this setting we have an observational study (as in setting 1), so we cannot make a sound inference about cause-effect relationships. However, we can infer that any pattern we see in the study reflects a similar pattern in the population. For example, suppose we take a random sample of people from a population and ask them (a) whether they take vitamin C and (b) how many colds they have had in the past year. If we find that those who take vitamin C have had fewer colds than do others—and if this difference is greater than would be expected by chance alone—then we can conclude that taking vitamin C is associated with having fewer colds, in the population. We cannot, however, conclude that taking vitamin C *causes* a reduction in colds. (Maybe people who take vitamin C differ from others in other ways that are related to cold frequency, such as diet, exercise, stress level, etc.)

3. *Experimental units (subjects) not chosen at random; random assignment to treatments.* Here we can make an inference if we see a statistically significant difference. For example, suppose we have volunteers for a vitamin C study and we randomly assign some of the subjects to vitamin C and some to a placebo. If our experiment shows a difference between vitamin C and placebo in frequency of colds, then we can infer that the difference was caused by vitamin C. However, we can't extend our inference to the general population. (Maybe vitamin C is only effective in people who volunteer to be in experiments.)

4. *Experimental units (subjects) chosen at random; random assignment to treatments.* In this setting we can make a cause-effect inference that extends to the entire population. For example, suppose we take a random sample of people from a population and we randomly assign some of them to vitamin C and some to a placebo. If our experiment shows a difference between vitamin C and placebo in frequency of colds, then we can infer that the difference was caused by vitamin C. Moreover, we can extend our inference to the entire population.

Analysis and Design

Techniques for data analysis must be appropriately tailored to the design of the study that generated the data. In Chapter 7 we described techniques for comparing data from two *independent* samples. In observational studies, the requirement of independence means that the observational units were selected independently from the two populations, with no matching or stratification in the selection process. In randomized studies the requirement of independence means that the design was completely randomized, with no blocking or stratification in the randomization process.

In Chapter 9 we will describe techniques for analyzing a randomized blocks experiment with two treatments. The same techniques can be used to compare two matched observational samples.

In Chapter 10 we will expand the comparison of independent samples to include analysis of categorical data. In Chapter 11 we will consider the comparison of three or more treatments when the response variable is continuous. However, analysis of randomized blocks experiments that have three or more treatments is beyond the scope of this book.

Supplementary Exercises 8.32–8.43

Note: In several of these exercises you are asked to prepare a randomized allocation. For this purpose you can use either Table 1 or random digits from your calculator or a computer.

8.32 It is known that alcohol consumption during pregnancy can harm the fetus. To study this phenomenon, 20 pregnant mice are to be allocated to three treatment groups. Group 1 (ten mice) will receive no alcohol, group 2 (five mice) will receive a low dose of alcohol, and group 3 (five mice) will receive a high dose. Prepare a completely randomized allocation.

8.33 Refer to the alcohol study of Exercise 8.32. When each mouse gives birth, the birthweight of each pup will be measured. Suppose the mice in group 1 give birth to a total of 85 pups, so the experimenter has 85 observations of Y = birthweight. To calculate the standard error of the mean of these 85 observations, the experimenter could calculate the standard deviation of the 85 observations and divide by $\sqrt{85}$. On what grounds might an objection be raised to this method of calculating the SE?

8.34 The following 24 subjects are available for a medical study:

Subject	Sex	Age	Subject	Sex	Age
C.M.	F	19	K.N.	M	43
A.F.	F	23	C.O.	M	45
E.G.	F	32	T.K.	M	48
M.A.	F	39	T.W.	M	50
D.N.	F	46	A.J.	M	52
B.B.	F	51	D.S.	M	56
A.S.	M	22	G.D.	M	56
M.L.	M	27	W.H.	M	59
P.C.	M	34	M.S.	M	61
L.W.	M	38	J.H.	M	63
J.P.	M	40	C.F.	M	66
C.R.	M	40	R.V.	M	72

Suppose the subjects are to be allocated to two treatment groups of 12 subjects each, using a completely randomized design. Prepare a suitable allocation.

8.35 Refer to Exercise 8.34. Suppose a randomized blocks design is to be used to compare the two treatments. Use the given information to form 12 pairs of subjects, matched by sex and (as closely as possible) age. Randomly allocate the members of each pair to the two treatment groups.

8.36 Refer to Exercise 8.34. Suppose a stratified design is to be used to compare the two treatments. Randomly allocate the 24 subjects to two groups of 12 each, but stratify by sex—that is, randomly divide the females into two groups of three each and the males into two groups of nine each.

8.37 Refer to Exercise 8.34. Suppose these 24 subjects are to participate in a study to compare *three* treatments, using a randomized blocks design. Use the given information to form eight triplets of subjects, matched by sex and (as closely as possible) age. Randomly allocate the members of each triplet to the three treatment groups.

8.38 For each of the following cases (a, b, and c),

(I) State whether the study should be observational or experimental.
(II) State whether the study should be run blind, double-blind, or neither. If the study should be run blind or double-blind, who should be blinded?

(a) An investigation of whether taking aspirin reduces one's chance of having a heart attack.

(b) An investigation of whether babies born into poor families (family income below $20,000) are more likely to weigh less than 5.5 pounds at birth than babies born into wealthy families (family income above $50,000).

(c) An investigation of whether the size of the midsagittal plane of the anterior commisssure (a part of the brain) of a man is related to the sexual orientation of the man.

8.39 Researchers wanted to compare two drugs, formoterol and salbutamol, in aerosol solution, to a placebo for the treatment of patients who suffer from exercise-induced asthma. Patients were to take a drug or the placebo, do some exercise, and then have their "forced expiratory volume" measured. There were 30 subjects available.[41]

(a) Should this be an experiment or an observational study? Why?

(b) *Within the context of this setting*, what is the placebo effect?

(c) Briefly explain how to set up a randomized blocks design (RBD) here.

(d) How would an RBD be a helpful? That is, what is the main advantage of using a RBD in a setting like this?

8.40 *[Hypothetical]* In order to assess the effectiveness of a new fertilizer, researchers applied the fertilizer to the tomato plants on the west side of a garden, but did not fertilize the plants on the east side of the garden. They later measured the weights of the tomatoes produced by each plant and found that the fertilized plants grew larger tomatoes than did the nonfertilized plants. They concluded that the fertilizer works.

(a) Was this an experiment or an observational study? Why?

(b) What are the explanatory and response variables in this study?

(c) This study is seriously flawed. Use the language of statistics to explain the flaw and how this affects the validity of the conclusion reached by the researchers.

(d) Could this study have used the concept of "blinding" (i.e., does the word *blind* apply to this study)? If so, how? Could it have been double-blind? If so, how?

8.41 In a controversial study to determine the effectiveness of Azidothymidine (AZT), a group of HIV-positive pregnant women were randomly assigned to get either AZT or a placebo. Some of the babies born to these women were HIV positive, while others were not.[42]

(a) What is the explanatory variable?

(b) What is the response variable?

(c) What are the experimental units?

8.42 Patients suffering from acute respiratory failure were randomly assigned to either be placed in a prone (face down) position or a supine (face up) position. In the prone group, 21 out of 152 patients died. In the supine group, 25 out of 152 patients died.[43]

(a) What is the explanatory variable?

(b) What is the response variable?

(c) What are the experimental units?

8.43 Reseachers studied 1718 persons over age 65 living in North Carolina. They found that those who attended religious services regularly were more likely to have strong immune systems (as determined by the blood levels of the protein interleukin-6) than those who didn't.[44] Does this mean that attending religious services improves one's health? Why or why not?

CHAPTER 9

Comparison of Paired Samples

9.1 INTRODUCTION

In Chapter 7 we considered the comparison of two independent samples when the response variable Y is a quantitative variable. In the present chapter we consider the comparison of two samples that are not independent but are paired. In a **paired design**, the observations (Y_1, Y_2) occur in pairs; the observational units in a pair are linked in some way, so that they have more in common with each other than with members of another pair. The following is an example of a paired design.

Weight Loss. The compound m-chlorophenylpiperazine (mCPP) is thought to affect appetite and food intake in humans. In a study of the effect of mCPP on weight loss, nine moderately obese women were given mCPP in a double-blind, placebo-controled experiment. Some of the women took mCPP for two weeks, then took nothing for two weeks (the "washout period"), and then took a placebo for two weeks. The other women were given the placebo for the first two weeks, then had a two-week washout period, and took mCPP for the final two weeks. The weight loss (in kilograms) for each woman was recorded under each condition. Table 9.1 shows the data.[1] (Note that if a woman gained weight, then her weight loss is negative. For example, subject 2 gained 0.3 kg when on the placebo, so her weight loss is recorded as -0.3.) ■

In Example 9.1 the data arise in pairs; the data in a pair are linked by virtue of being measurements on the same person. A suitable analysis of the data should take advantage of this pairing. That is, we could imagine an experiment in which some women are given mCPP and other women are given a placebo; such an experiment would provide two independent samples of data and could be analyzed using the methods of Chapter 7. But the current experiment used a paired design. Subject 8 lost weight both when on mCPP and when on the placebo; subject 9 gained weight both times. Knowing how a subject did on

Example 9.1

347

TABLE 9.1 Weight Loss (kg) for 9 Women		
	Weight Loss	
Subject	*mCPP*	*Placebo*
1	1.1	0.0
2	1.3	−0.3
3	1.0	0.6
4	1.7	0.3
5	1.4	−0.7
6	0.1	−0.2
7	0.5	0.6
8	1.6	0.9
9	−0.5	−2.0
Mean	.91	−.09
SD	.74	.88

mCPP tells us something about how the subject did on placebo, and vice versa. We want to use this information when we analyze the data.

In Section 9.2 we show how to analyze paired data using methods based on Student's *t* distribution. In Sections 9.4 and 9.5 we describe two nonparametric tests for paired data. Sections 9.3, 9.6, and 9.7 contain more examples and discussion of the paired design.

9.2 THE PAIRED-SAMPLE *t* TEST AND CONFIDENCE INTERVAL

In this section we discuss the use of Student's *t* distribution to obtain tests and confidence intervals for paired data.

Analyzing Differences

In Chapter 7 we considered how to analyze data from two independent samples. When we have paired data we make a simple shift of viewpoint: Instead of considering Y_1 and Y_2 separately, we consider the *difference d*, defined as

$$d = Y_1 - Y_2$$

Note that it is often natural to consider a difference as the response variable of interest in a study. For example, if we were studying the growth rates of plants, we might grow plants under control conditions for a while at the beginning of a study and then apply a treatment for one week. We would measure the growth that takes place during the week after the treatment is introduced as $d = Y_1 - Y_2$, where Y_1 = height one week after applying the treatment and Y_2 = height before the treatment is applied. Sometimes data are paired in a way that is less obvious, but whenever we have paired data, it is the observed differences that we wish to analyze.

Let us denote the mean of the *d*'s as $\bar{d}$. The quantity $\bar{d}$ is related to the individual sample means as follows:

$$\bar{d} = (\bar{y}_1 - \bar{y}_2)$$

The relationship between population means is analogous:

$$\mu_d = \mu_1 - \mu_2$$

Thus, we may say that *the mean of the difference is equal to the difference of the means*. Because of this simple relationship, a comparison of two paired means can be carried out by concentrating entirely on the *d*'s.

The standard error for $\bar{d}$ is easy to calculate. Because $\bar{d}$ is just the mean of a single sample, we can apply the SE formula of Chapter 6 to obtain the following formula:

$$\text{SE}_{\bar{d}} = \frac{s_d}{\sqrt{n_d}}$$

where s_d is the standard deviation of the *d*'s and n_d is the number of *d*'s. The following example illustrates the calculation.

Weight Loss. Table 9.2 shows the weight loss data of Example 9.1 and the differences *d*.

Example 9.2

TABLE 9.2 Weight Loss (kg) for 9 Women

	Weight Change		
	mCPP	*Placebo*	*Difference*
Subject	y_1	y_2	$d = y_1 - y_2$
1	1.1	0.0	1.1
2	1.3	−0.3	1.6
3	1.0	0.6	0.4
4	1.7	0.3	1.4
5	1.4	−0.7	2.1
6	0.1	−0.2	0.3
7	0.5	0.6	−0.1
8	1.6	0.9	0.7
9	−0.5	−2.0	1.5
Mean	.91	−.09	1.00
SD	.74	.88	.72

Note that the mean of the difference is equal to the difference of the means:

$$\bar{d} = 1.00 = .91 - -.09$$

Figure 9.1 shows the distribution of the 9 sample differences.

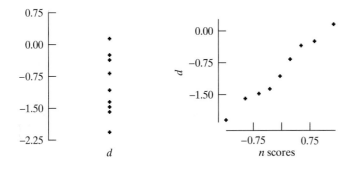

Figure 9.1 Dotplot of differences in weight loss when on mCPP and when on placebo, along with a normal probability plot of the data

We calculate the standard error of the mean difference as follows:

$$s_d = .72$$
$$n_d = 9$$
$$SE_{\bar{d}} = \frac{.72}{\sqrt{9}} = .24 \qquad \blacksquare$$

Confidence Interval and Test of Hypothesis

The standard error described in the preceding subsection is the basis for the **paired-sample t method** of analysis, which can take the form of a confidence interval or a test of hypothesis.

A 95% confidence interval for μ_d is constructed as

$$\bar{d} \pm t_{.025}SE_{\bar{d}}$$

where the constant $t_{.025}$ is determined from Student's t distribution with

$$df = n_d - 1$$

Intervals with other confidence coefficients (such as 90%, 99%, etc.) are constructed analogously (using $t_{.05}, t_{.005}$, etc.). The following example illustrates the confidence interval.

Example 9.3 | **Weight Loss.** For the weight loss data, we have df $= 9 - 1 = 8$. From Table 4 we find that $t(8)_{.025} = 2.306$; thus, the 95% confidence interval for μ_d is

$$1.00 \pm (2.306)(.24)$$

or

$$1.00 \pm .55$$

or

$$(.45, 1.55)$$

Thus, we are 95% confident that the population average weight loss (in a two-week period) is between .45 kg and 1.55 kg greater when taking mCPP than when taking a placebo. $\blacksquare$

We can also conduct a t test. To test the null hypothesis

$$H_0: \mu_d = 0$$

we use the test statistic

$$t_s = \frac{\bar{d} - 0}{SE_{\bar{d}}}$$

Critical values are obtained from Student's t distribution (Table 4) with df $= n_d - 1$. The following example illustrates the t test.

Example 9.4 | **Weight Loss.** For the weight loss data, let us formulate the null hypothesis and nondirectional alternative:

H_0: Mean weight loss is the same when on mCPP and when on placebo.

H_A: Mean weight loss when on mCPP is different than when on placebo.

In symbols,

$H_0: \mu_d = 0$

$H_A: \mu_d \neq 0$

Let us test H_0 against H_A at significance level $\alpha = .05$. The test statistic is

$$t_s = \frac{1.00 - 0}{.24} = 4.17$$

From Table 4, $t(8)_{.005} = 3.355$ and $t(8)_{.0005} = 5.041$, so the upper tail area beyond 4.17 is between .0005 and .005. Thus, the P-value is between .001 and .01. We reject H_0 and find that there is sufficient evidence $(.001 < P < .01)$ to conclude that mean weight loss is greater when on mCPP than when on placebo. (Using a computer gives the P-value as $P = .003$.) ∎

Result of Ignoring Pairing

Suppose that a study is conducted using a paired design, but that the pairing is ignored in the analysis of the data. Such an analysis is not valid because it assumes that the samples are independent when in fact they are not. The incorrect analysis can be misleading, as the following example illustrates.

Hunger Rating.   As part of the weight loss study described in Example 9.1, the subjects were asked to rate how hungry there were at the end of each two week period. The hunger rating data are shown in Table 9.3.[2]

Example 9.5

TABLE 9.3 Hunger Rating for 9 Women

	Hunger Rating		
Subject	mCPP y_1	Placebo y_2	Difference $d = y_1 - y_2$
1	79	78	1
2	48	54	−6
3	52	142	−90
4	15	25	−10
5	61	101	−40
6	107	99	8
7	77	94	−17
8	54	107	−53
9	5	64	−59
Mean	55	85	−30
SD	32	34	33

For the hunger rating data, the SE for the mean difference is

$$SE_{\bar{d}} = \frac{33}{\sqrt{9}} = 11$$

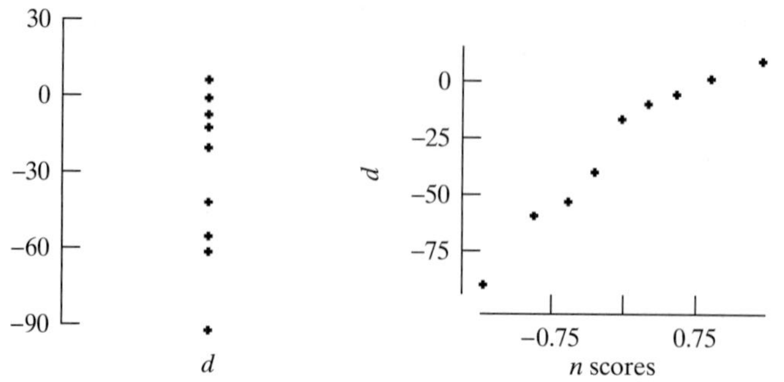

Figure 9.2 Dotplot of differences in hunger rating when on mCPP and when on placebo, along with a normal probability plot of the data

Figure 9.2 shows the distribution of the 9 sample differences.
A test of

$$H_0: \mu_d = 0 \text{ vs. } H_A: \mu_d \neq 0$$

gives a test statistic of

$$t_s = \frac{-30 - 0}{11} = -2.73$$

This test statistic has 8 degrees of freedom. Using a computer gives the *P*-value as $P = .026$.

Looking at the mCPP and placebo data separately, the two sample SDs are $s_1 = 32$ and $s_2 = 34$. If we proceed as if the samples were independent and apply the SE formula of Chapter 7, we obtain

$$SE_{(\bar{y}_1 - \bar{y}_2)} = \sqrt{\frac{s_1^2}{n_1} + \frac{s_2^2}{n_2}}$$

$$= \sqrt{\frac{32^2}{9} + \frac{34^2}{9}} = 15.6$$

This SE is quite a bit larger than the value ($SE_{\bar{d}} = 11$) that we calculated using the pairing. Continuing to proceed as if the samples were independent, the test statistic is

$$t_s = \frac{55 - 85}{15.6} = -1.92$$

The *P*-value for this test is .073, which is much greater than the *P*-value for the correct test, .026.

To compare further the paired and unpaired analysis, let us consider the 95% confidence interval for $(\mu_1 - \mu_2)$. For the unpaired analysis, formula (7.1) yields 15.9 degrees of freedom; this gives a *t*-multiplier of $t(15.9)_{.025} = 2.121$ and yields a confidence interval of

$$(55 - 85) \pm (2.121)(15.6)$$

or

$$-30 \pm 33.1$$

or

$$(-63.1, 3.1)$$

This confidence interval is wider than the correct confidence interval from a paired analysis. A paired analysis yields the narrower interval

$$-30 \pm (2.306)(11)$$

or

$$-30 \pm 25.4$$

or

$$(-55.4, -4.6)$$

The paired-sample interval is narrower because it uses a smaller SE; this effect is slightly offset by a larger value of $t_{.025}$ (2.306 vs. 2.121).

Why is the paired-sample SE smaller than the independent-samples SE calculated from the same data ($SE = 11$ vs. $SE = 15.6$)? Table 9.3 reveals the reason. The data show that there is large variation from one subject to the next. For instance, subject 4 has low hunger ratings (both when on mCPP and when on placebo) and subject 6 has high values. The independent-samples SE formula incorporates all of this variation (expressed through s_1 and s_2); in the paired-sample approach, intersubject variation in hunger rating has no influence on the calculations because only the d's are used. By using each subject as her own control, the experimenter has increased the precision of the experiment. But if the pairing is ignored in the analysis, the extra precision is wasted.   ■

The preceding example illustrates the gain in precision that can result from a paired design coupled with a paired analysis. The choice between a paired and an unpaired design will be discussed in Section 9.3.

Conditions for Validity of Student's *t* Analysis

The conditions for validity of the paired-sample *t* test and confidence interval are as follows:

1. It must be reasonable to regard the *differences* (the *d*'s) as a random sample from some large population.

2. The population distribution of the *d*'s must be normal. The methods are approximately valid if the population distribution is approximately normal or if the sample size (n_d) is large.

The preceding conditions are the same as those given in Chapter 6; in the present case the conditions apply to the *d*'s because the analysis is based on the *d*'s. Verification of the conditions can proceed as described in Chapter 6. First, the design should be checked to assure that the *d*'s are independent of each other, and especially that there is no hierarchical structure within the *d*'s. (Note, however, that the

Y_1's are not independent of the Y_2's because of the pairing.) Second, a histogram, stem-and-leaf display, or dotplot of the d's can provide a rough check for approximate normality. A normal probability plot can also be used to assess normality.

Notice that normality of the Y_1's and Y_2's is not required, because the analysis depends only on the d's. The following example shows a case in which the Y_1's and Y_2's are not normally distributed but the d's are.

Example 9.6

Squirrels. If you walk toward a squirrel that is on the ground, it will eventually run to the nearest tree for safety. A researcher wondered whether he could get closer to the squirrel than the squirrel was to the nearest tree before the squirrel would start to run. He made 11 observations, which are given in Table 9.4. Figure 9.3

TABLE 9.4 **Distances (in Inches) from Person and from Tree When Squirrel Started to Run**

Squirrel	From Person y_1	From Tree y_2	Difference $d = y_1 - y_2$
1	81	137	−56
2	178	34	144
3	202	51	151
4	325	50	275
5	238	54	184
6	134	236	−102
7	240	45	195
8	326	293	33
9	60	277	−217
10	119	83	36
11	189	41	148
Mean	190	118	72
SD	89	101	148

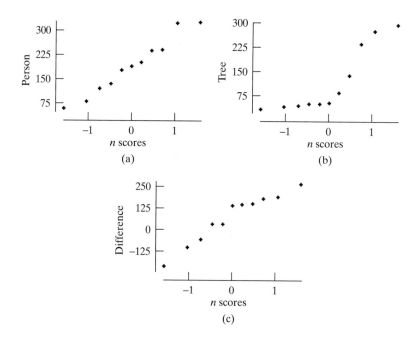

Figure 9.3 Normal probability plots of distance from squirrel to person and from squirrel to tree

shows that the distribution of distances from squirrel to person appear to be reasonably normal, but that the distances from squirrel to tree are far from being normally distributed. However, panel (c) of Figure 9.3 shows that the 11 differences *do* meet the normality condition. Since a paired *t* test analyzes the differences, a *t* test (or confidence interval) is valid here.[3]   ■

Summary of Formulas

For convenient reference, we summarize in the box the formulas for the paired-sample methods based on Student's *t*.

Standard Error of $\bar{d}$

$$\text{SE}_{\bar{d}} = \frac{s_d}{\sqrt{n_d}}$$

t Test

$$H_0: \mu_d = 0$$

$$t_s = \frac{\bar{d} - 0}{\text{SE}_{\bar{d}}}$$

95% Confidence Interval for μ_d

$$\bar{d} \pm t_{.025}\text{SE}_{\bar{d}}$$

Intervals with other confidence levels (e.g., 90%, 99%) are constructed analogously (e.g., using $t_{.05}, t_{.005}$).

Exercises 9.1–9.9

9.1 In an agronomic field experiment, blocks of land were subdivided into two plots of 346 square feet each. The plots were planted with two varieties of wheat, using a randomized blocks design. The plot yields (lb) of wheat are given in the table.[4]

| | Variety | | |
Block	*1*	*2*	**Difference**
1	32.1	34.5	−2.4
2	30.6	32.6	−2.0
3	33.7	34.6	−0.9
4	29.7	31.0	−1.3
Mean	31.53	33.18	−1.65
SD	1.76	1.72	0.68

(a) Calculate the standard error of the mean difference between the varieties.
(b) Test for a difference between the varieties using a paired *t* test at $\alpha = .05$. Use a nondirectional alternative.
(c) Test for a difference between the varieties the wrong way, using an independent-samples test. Compare with the result of part (b).

9.2   In an experiment to compare two diets for fattening beef steers, nine pairs of animals were chosen from the herd; members of each pair were matched as closely as possible with respect to hereditary factors. The members of each pair were randomly allocated, one to each diet. The following table shows the weight gains (lb) of the animals over a 140-day test period on diet 1 (Y_1) and on diet 2 (Y_2).[5]

Pair	Diet 1	Diet 2	Difference
1	596	498	98
2	422	460	−38
3	524	468	56
4	454	458	−4
5	538	530	8
6	552	482	70
7	478	528	−50
8	564	598	−34
9	556	456	100
Mean	520.4	497.6	22.9
SD	57.1	47.3	59.3

(a) Calculate the standard error of the mean difference.
(b) Test for a difference between the diets using a paired t test at $\alpha = .10$. Use a nondirectional alternative.
(c) Construct a 90% confidence interval for μ_d.
(d) Interpret the confidence interval from part (c) in the context of this setting.

9.3   Cyclic adenosine monophosphate (cAMP) is a substance that can mediate cellular response to hormones. In a study of maturation of egg cells in the frog *Xenopts laevis*, oocytes from each of four females were divided into two batches; one batch was exposed to progesterone and the other was not. After two minutes, each batch was assayed for its cAMP content, with the results given in the table.[6] Use a t test to investigate the effect of progesterone on cAMP. Let H_A be nondirectional and let $\alpha = .10$.

	cAMP (pmol/oocyte)		
Frog	*Control*	*Progesterone*	*d*
1	6.01	5.23	0.78
2	2.28	1.21	1.07
3	1.51	1.40	0.11
4	2.12	1.38	0.74
Mean	2.98	2.31	0.68
SD	2.05	1.95	0.40

9.4   Under certain conditions, electrical stimulation of a beef carcass will improve the tenderness of the meat. In one study of this effect, beef carcasses were split in half; one side (half) was subjected to a brief electrical current and the other side was an untreated control. For each side, a steak was cut and tested in various ways for tenderness. In one test, the experimenter obtained a specimen of connective tissue (collagen) from the steak and determined the temperature at which the tissue would shrink; a tender piece of meat tends to yield a low collagen shrinkage temperature. The data are given in the following table.[7]

(a) Construct a 95% confidence interval for the mean difference between the treated side and the control side.
(b) Construct a 95% confidence interval the wrong way, using the independent-samples method. How does this interval differ from the one you obtained in part (a)?

	Collagen Shrinkage Temperature (°C)		
Carcass	*Treated Side*	*Control Side*	*Difference*
1	69.50	70.00	−.50
2	67.00	69.00	−2.00
3	70.75	69.50	1.25
4	68.50	69.25	−.75
5	66.75	67.75	−1.00
6	68.50	66.50	2.00
7	69.50	68.75	.75
8	69.00	70.00	−1.00
9	66.75	66.75	.00
10	69.00	68.50	.50
11	69.50	69.00	.50
12	69.00	69.75	−.75
13	70.50	70.25	.25
14	68.00	66.25	1.75
15	69.00	68.25	.75
Mean	68.750	68.633	.117
SD	1.217	1.302	1.118

9.5 Refer to Exercise 9.4. Use a *t* test to test the null hypothesis of no effect against the alternative hypothesis that the electrical treatment tends to reduce the collagen shrinkage temperature. Let $\alpha = .10$.

9.6 Trichotillomania is a psychiatric illness that causes its victims to have an irresistible compulsion to pull their own hair. Two drugs were compared as treatments for trichotillomania in a study involving 13 women. Each woman took clomipramine during one time period and desipramine during another time period in a double-blind experiment. Scores on a trichotillomania-impairment scale, in which high scores indicate greater impairment, were measured on each woman during each time period. The average of the 13 measurements for clomipramine was 6.2; the average of the 13 measurements for desipramine was 4.2.[8] A paired *t* test gave a value of $t_s = 2.47$ and a two-tailed *P*-value of .03. Interpret the result of the *t* test. That is, what does the test indicate about clomipramine, desipramine, and hair pulling?

9.7 A scientist conducted a study of how often her pet parakeet chirps. She recorded the number of distinct chirps the parakeet made in a 30-minute period, sometimes when the room was silent and sometimes when there was music playing. The data are shown in the following table.[9] Construct a 95% confidence interval for the mean increase in chirps (per 30 minutes) when music is playing over when music is not playing.

	Chirps in 30 Minutes		
Day	*With Music*	*Without Music*	*Difference*
1	12	3	9
2	14	1	13
3	11	2	9
4	13	1	12
5	20	5	15
6	14	3	11
7	10	0	10
8	12	2	10
9	8	6	2
10	13	3	10
11	14	2	12
12	15	4	11
13	12	3	9
14	13	2	11
15	8	0	8
16	18	5	13
17	15	3	12
18	12	2	10
19	17	2	15
20	15	4	11
21	11	3	8
22	22	4	18
23	14	2	12
24	18	4	14
25	15	5	10
26	8	1	7
27	13	2	11
28	16	3	13
Mean	13.7	2.8	10.9
SD	3.4	1.5	3.0

9.8 Consider the data in Exercise 9.7. There are two outliers among the 28 differences: the smallest value, which is 2, and the largest value, which is 18. Delete these two observations and construct a 95% confidence interval for the mean increase, using the remaining 26 observations. Do the outliers have much of an effect on the confidence interval?

9.9 Invent a paired data set, consisting of five pairs of observations, for which $\bar{y}_1$ and $\bar{y}_2$ are not equal, and $SE_{\bar{y}_1} > 0$ and $SE_{\bar{y}_2} > 0$, but $SE_{\bar{d}} = 0$.

9.3 THE PAIRED DESIGN

Ideally, in a paired design the members of a pair are relatively similar to each other—that is, more similar to each other than to members of other pairs—with respect to extraneous variables. The advantage of this arrangement is that, when members of a pair are compared, the comparison is free of the extraneous variation that originates in between-pair differences. We will expand on this theme after giving some examples.

Examples of Paired Designs

Paired designs can arise in a variety of ways, including the following:

Randomized blocks experiments with two experimental units per block
Observational studies with individually matched controls
Repeated measurements on the same individual at two different times
Blocking by time

Randomized Blocks Experiments. A randomized blocks design (Chapter 8) is a paired design if there are only two treatments. Each block would then contain two experimental units, one to receive each treatment. The following is an example.

Fertilizers for Eggplants. In a greenhouse experiment to compare two fertilizer treatments for eggplants, individually potted plants are arranged on the greenhouse bench in blocks of two (that is, pairs). Within each pair, one (randomly chosen) plant will receive treatment 1 and the other will receive treatment 2. ■

Example 9.7

Observational Studies. As noted in Chapter 8, randomized experiments are preferred over observational studies, due to the many confounding variables that can arise within an observational study. If no experiment is possible and an observational study must be carried out, then the researcher can try to use identical twins as the observational units. For example, in a study of the effect of "second-hand smoke" it would be ideal to enroll several sets of nonsmoking twins for which, in each pair, one of the twins lived with a smoker and the other twin did not. Because sets of twins are rarely, if ever, available, **matched-pair designs**, in which two groups are matched with respect to various extraneous variables, are often used.[10] Here is an example.

Smoking and Lung Cancer. In a case-control study of lung cancer, 100 lung cancer patients were identified. For each case, a control was chosen who was individually matched to the case with respect to age, sex, and education level. The smoking habits of the cases and controls were compared. ■

Example 9.8

Repeated Measurements. Many biological investigations involve repeated measurements made on the same individual at different times. These include studies of growth and development, studies of biological processes, and studies in which measurements are made before and after application of a certain treatment. When only two times are involved, the measurements are paired, as in Example 9.1. The following is another example.

Exercise and Serum Triglycerides. Triglycerides are blood constituents that are thought to play a role in coronary artery disease. To see whether regular exercise could reduce triglyceride levels, researchers measured the concentration of triglycerides in the blood serum of seven male volunteers, before and after participation in a 10-week exercise program. The results are shown in Table 9.5.[11] Note that there is considerable variation from one participant to another. For instance, participant 1 had relatively low triglyceride levels both before and after, while participant 3 had relatively high levels. ■

Example 9.9

Blocking by Time. In some situations, blocks or pairs are formed implicitly when replicate measurements are made at different times. The following is an example.

TABLE 9.5 Serum Triglycerides (mmol/Li)		
Participant	**Before**	**After**
1	.87	.57
2	1.13	1.03
3	3.14	1.47
4	2.14	1.43
5	2.98	1.20
6	1.18	1.09
7	1.60	1.51

Example 9.10

Growth of Viruses. In a series of experiments on a certain virus (mengovirus), a microbiologist measured the growth of two strains of the virus—a mutant strain and a nonmutant strain—on mouse cells in petri dishes. Replicate experiments were run on 19 different days. The data are shown in Table 9.6. Each number represents the total growth in 24 hours of the viruses in a single dish.[12]

TABLE 9.6 Virus Growth at 24 Hours					
Run	**Nonmutant Strain**	**Mutant Strain**	**Run**	**Nonmutant Strain**	**Mutant Strain**
1	160	97	11	61	15
2	36	55	12	14	10
3	82	31	13	140	150
4	100	95	14	68	44
5	140	80	15	110	31
6	73	110	16	37	14
7	110	100	17	95	57
8	180	100	18	64	70
9	62	6	19	58	45
10	43	7			

Note that there is considerable variation from one run to another. For instance, run 1 gave relatively large values (160 and 97), whereas run 2 gave relatively small values (36 and 55). This variation between runs arises from unavoidable small variations in the experimental conditions. For instance, both the growth of the viruses and the measurement technique are highly sensitive to environmental conditions such as the temperature and CO_2 concentration in the incubator. Slight fluctuations in the environmental conditions cannot be prevented, and these fluctuations cause the variation that is reflected in the data. In this kind of situation the advantage of running the two strains concurrently (that is, in pairs) is particularly striking. ■

Examples 9.9 and 9.10 both involve measurements at different times. But notice that the pairing structure in the two examples is entirely different. In Example 9.9 the members of a pair are measurements on the same individual at two times, whereas in Example 9.10 the members of a pair are measurements on two petri dishes at the same time. Nevertheless, in both examples the principle of pairing is the same: Members of a pair are similar to each other with respect to extraneous variables. In Example 9.10 time is an extraneous variable, whereas in Example 9.9 the comparison between two times (before and after) is of primary interest and interperson variation is extraneous.

Purposes of Pairing

Pairing in an experimental design can serve to reduce bias, to increase precision, or both. Usually the primary purpose of pairing is to increase precision. We noted in Chapter 8 that blocking or matching can reduce bias by controlling variation due to extraneous variables. The variables used in the matching are necessarily balanced in the two groups to be compared, and therefore cannot distort the comparison. For instance, if two groups are composed of age-matched pairs of people, then a comparison between the two groups is free of any bias due to a difference in age distribution.

In randomized experiments, where bias can be controlled by randomized allocation, a major reason for pairing is to increase precision. Effective pairing increases precision by increasing the information available in an experiment. An appropriate analysis, which extracts this extra information, leads to more powerful tests and narrower confidence intervals. Thus, an effectively paired experiment is more efficient; it yields more information than an unpaired experiment with the same number of observations.

We saw an instance of effective pairing in the hunger rating data of Example 9.5. The pairing was effective because much of the variation in the measurements was due to variation between subjects, which did not enter the comparison between the treatments. As a result, the experiment yielded more precise information about the treatment difference than would a comparable unpaired experiment—that is, an experiment that would compare hunger ratings of 9 women given mCPP to hunger ratings of 9 different control women who were given the placebo.

The effectiveness of a given pairing can be displayed visually in a scatterplot of Y_2 against Y_1; each point in the scatterplot represents a single pair (Y_1, Y_2). Figure 9.4 shows a scatterplot for the virus growth data of Example 9.10, together with a boxplot of the differences; each point in the scatterplot represents a single run. Notice that the points in the scatterplot show a definite upward trend. This upward trend indicates the effectiveness of the pairing: Measurements on the same run (i.e., the same day) have more in common than measurements on different runs, so that a run with a relatively high value of Y_1 tends to have a relatively high value of Y_2, and similarly for low values.

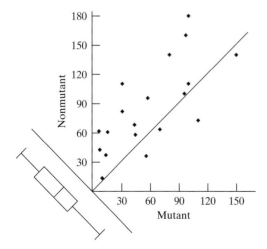

Figure 9.4 Scatterplot for the virus growth data, with a boxplot of the differences

Note that pairing is a strategy of *design*, not of analysis, and is therefore carried out *before* the Y's are observed. It is not correct to use the observations themselves to form pairs. Such a data manipulation could distort the experimental results as severely as outright fakery.

Randomized Pairs Design Versus Completely Randomized Design

In planning a randomized experiment, the experimenter may need to decide between a paired design and a completely randomized design. We have said that effective pairing can greatly enhance the precision of an experiment. On the other hand, pairing in an experiment may not be effective, if the observed variable Y is not related to the factors used in the pairing. For instance, suppose pairs were matched on age only, but in fact Y turned out not to be age related. It can be shown that ineffective pairing actually can yield less precision than no pairing at all. For instance, in relation to a t test, ineffective pairing would not tend to reduce the SE, but it would reduce the degrees of freedom, and the net result would be a loss of power.

The choice of whether to use a paired design depends on practical considerations (pairing may be expensive or unwieldy) and on precision considerations. With respect to precision, the choice depends on how effective the pairing is expected to be. The following example illustrates this issue.

Example 9.11

Fertilizers for Eggplants. A horticulturist is planning a greenhouse experiment with individually potted eggplants. Two fertilizer treatments are to be compared and the observed variable is to be Y = yield of eggplants (pounds). The experimenter knows that Y is influenced by such factors as light and temperature, which vary somewhat from place to place on the greenhouse bench. The allocation of pots to positions on the bench could be carried out according to a completely randomized design, or according to a randomized blocks (paired) design, as in Example 9.7. In deciding between these options, the experimenter must use her knowledge of how effective the pairing would be—that is, whether two pots sitting adjacent on the bench would be very much more similar in yield than pots farther apart. If she judges that the pairing would not be very effective, she may opt for the completely randomized design. ∎

Note that effective pairing is *not* the same as simply holding experimental conditions constant. Pairing is a way of *organizing* the unavoidable variation that still remains after experimental conditions have been made as constant as possible. The ideal pairing organizes the variation in such a way that the variation within each pair is minimal and the variation between pairs is maximal.

Choice of Analysis

The analysis of data should fit the design of the study. If the design is paired, a paired-sample analysis should be used; if the design is unpaired, an independent-samples analysis (as in Chapter 7) should be used.

Note that the extra information made available by an effectively paired design is *entirely wasted* if an unpaired analysis is used. (We saw an illustration of this in Example 9.5.) Thus, the paired design does not increase efficiency unless it is accompanied by a paired analysis.

Exercises 9.10–9.13

9.10 *(Sampling exercise)* This exercise illustrates the application of a matched-pairs design to the population of 100 ellipses (shown with Exercise 3.1). The accompanying table shows a grouping of the 100 ellipses into 50 pairs.

Pair	Ellipse ID Numbers		Pair	Ellipse ID Numbers		Pair	Ellipse ID Numbers	
01	20	45	18	11	46	35	16	66
02	03	49	19	09	29	36	18	58
03	07	27	20	19	39	37	30	50
04	42	82	21	00	10	38	76	86
05	81	91	22	40	55	39	17	83
06	38	72	23	21	56	40	04	52
07	60	70	24	08	62	41	12	64
08	31	61	25	24	78	42	23	57
09	77	89	26	67	93	43	98	99
10	01	41	27	35	80	44	36	96
11	14	48	28	74	88	45	44	84
12	59	87	29	94	97	46	06	51
13	22	68	30	02	28	47	85	90
14	47	79	31	26	71	48	37	63
15	05	95	32	25	65	49	43	69
16	53	73	33	15	75	50	34	54
17	13	33	34	32	92			

To better appreciate this exercise, imagine the following experimental setting. We want to investigate the effect of a certain treatment, T, on the organism *C. ellipticus*. We will observe the variable Y = length. We can measure each individual only once, and so we will compare n treated individuals with n untreated controls. We know that the individuals available for the experiment are of various ages, and we know that age is related to length, so we have formed 50 age-matched pairs, some of which will be used in the experiment. The purpose of the pairing is to increase the power of the experiment by eliminating the random variation due to age. (Of course, the ellipses do not actually have ages, but the pairing shown in the table has been constructed in a way that *simulates* age matching.)

(a) Use random digits (from Table 1 or your calculator) to choose a random sample of five pairs from the list.

(b) For each pair, use random digits (or toss a coin) to allocate one member to treatment (T) and the other to control (C).

(c) Measure the lengths of all ten ellipses. Then, to simulate a treatment effect, add 6 mm to each length in the T group.

(d) Apply a paired-sample t test to the data. Use a nondirectional alternative and let $\alpha = .05$.

(e) Did the analysis of part (d) lead you to a Type II error?

9.11 *(Continuation of Exercise 9.10)* Apply an independent-samples t test to your data. Use a nondirectional alternative and let $\alpha = .05$. Does this analysis lead you to a Type II error?

9.12 *(Sampling exercise)* Refer to Exercise 9.10. Imagine that a matched-pairs experiment is not practical (perhaps because the ages of the individuals cannot be measured), so we decide to use a completely randomized design to evaluate the treatment T.

(a) Use random digits (from Table 1 or your calculator) to choose a random sample of ten individuals from the ellipse population (shown with Exercise 3.1). From these ten, randomly allocate five to T and five to C. (Or, equivalently, just randomly select five from the population to receive T and five to receive C.)

(b) Measure the lengths of all ten ellipses. Then, to simulate a treatment effect, add 6 mm to each length in the T group.

(c) Apply an independent-samples t test to the data. Use a nondirectional alternative and let $\alpha = .05$.

(d) Did the analysis of part (c) lead you to a Type II error?

9.13 Refer to each exercise indicated. Construct a scatterplot of the data. Does the appearance of the scatterplot indicate that the pairing was effective?

(a) Exercise 9.1
(b) Exercise 9.2
(c) Exercise 9.4

9.4 THE SIGN TEST

The **sign test** is a nonparametric test that can be used to compare two paired samples. It is not particularly powerful, but it is very flexible in application and is especially simple to use and understand—a blunt but handy tool.

Method

Like the paired-sample t test, the sign test is based on the differences

$$d = Y_1 - Y_2$$

The only information used by the sign test is the *sign* (positive or negative) of each difference. If the differences are preponderantly of one sign, this is taken as evidence against the null hypothesis. The following example illustrates the sign test.

Example 9.12 **Skin Grafts.** Skin from cadavers can be used to provide temporary skin grafts for severely burned patients. The longer such a graft survives before its inevitable rejection by the immune system, the more the patient benefits. A medical team investigated the usefulness of matching graft to patient with respect to the HL-A (Human Leukocyte Antigen) antigen system. Each patient received two grafts, one with close HL-A compatibility and the other with poor compatibility. The survival times (in days) of the skin grafts are shown in the Table 9.7.[13]

Notice that a t test could not be applied here because two of the observations are incomplete; patient 3 died with a graft still surviving and the observation on patient 10 was incomplete for an unspecified reason. Nonetheless, we can proceed with a sign test, since the sign test depends only on the sign of the difference for each patient and we know that $Y_1 - Y_2$ is positive for both of these patients.

Let us carry out a sign test to compare the survival times of the two sets of skin grafts using $\alpha = .05$. The null hypothesis is

H_0: The survival time distribution is the same for close compatibility as it is for poor compatibility.

TABLE 9.7 Skin Graft Survival Times

| | HL-A Compatibility | | |
| | Close | Poor | Sign of |
Patient	Y_1	Y_2	$d = Y_1 - Y_2$
1	37	29	+
2	19	13	+
3	57+	15	+
4	93	26	+
5	16	11	+
6	23	18	+
7	20	26	−
8	63	43	+
9	29	18	+
10	60+	42	+
11	18	19	−

A directional alternative is appropriate for this experiment:

H_A: Skin grafts tend to last longer when the HL-A compatibility is close.

The first step is to determine the following counts:

N_+ = Number of positive differences

N_- = Number of negative differences

Because H_A is directional and it predicts that most of the differences will be positive, the test statistic B_s is

$$B_s = N_+$$

For the present data, we have

$$N_+ = 9$$
$$N_- = 2$$
$$B_s = 9$$

The next step is to find the *P*-value. We use the letter B in labeling the test statistic B_s because the distribution of B_s is based on the binomial distribution. Let p represent the probability that a difference will be positive. If the null hypothesis is true, then $p = .5$. Thus, the null distribution of B_s is a binomial with $n = 11$ and $p = .5$. That is, the null hypothesis implies that the sign of each difference is like the result of a coin toss, with heads corresponding to a positive difference and tails to a negative difference.

For the skin graft data, the *P*-value for the test is the probability of getting 9 or more positive differences in 11 patients if $p = .5$. This is the probability that a binomial random variable with $n = 11$ and $p = .5$ will be greater than or equal to 9. Using the binomial formula, from Chapter 3, we find that this probability is

$$_{11}C_9(.5)^9(.5)^2 + {}_{11}C_{10}(.5)^{10}(.5)^1 + {}_{11}C_{11}(.5)^{11} = .02686 + .00537 + .00049 = .03272$$

Because the *P*-value is less than α, we reject H_0 and conclude that skin grafts tend to last longer when the HL-A compatibility is close than when it is poor. ■

Example 9.13   **Growth of Viruses.**   Table 9.8 shows the virus growth data of Example 9.10, together with the signs of the differences.

TABLE 9.8 **Virus Growth at 24 Hours**

Run	Nonmutant Strain Y_1	Mutant Strain Y_2	Sign of $d = Y_1 - Y_2$	Run	Nonmutant Strain Y_1	Mutant Strain Y_2	Sign of $d = Y_1 - Y_2$
1	160	97	+	11	61	15	+
2	36	55	−	12	14	10	+
3	82	31	+	13	140	150	−
4	100	95	+	14	68	44	+
5	140	80	+	15	110	31	+
6	73	110	−	16	37	14	+
7	110	100	+	17	95	57	+
8	180	100	+	18	64	70	−
9	62	6	+	19	58	45	+
10	43	7	+				

Let's carry out a sign test to compare the growth of the two strains, using $\alpha = .10$. The null hypothesis and nondirectional alternative are

H_0: The two strains of virus grow equally well.

H_A: One of the strains grows better than the other.

For these data

$$N_+ = 15$$
$$N_- = 4$$

When the alternative is nondirectional, B_s is defined as

$$B_s = \text{Larger of } N_+ \text{ and } N_-$$

so for the virus growth data,

$$B_s = 15$$

The P-value for the test is the probability of getting 15 or more successes, plus the probability of getting 4 or fewer successes, in a binomial experiment with $n = 19$. We could use the binomial formula to calculate the P-value. As an alternative, critical values for the sign test are given in Table 7 (at the end of the book). Using Table 7 with $n_d = 19$, we obtain the critical values shown in Table 9.9.

To bracket the P-value, we find the rightmost column in the table with critical value less than or equal to B_s; then the P-value is bracketed between that column heading and the next one. In the present case the result is

$$.01 < P\text{-value} < .02$$

TABLE 9.9 **Critical Values for the Sign Test When $n_d = 19$**

	Nominal Tail Probability						
Two tails	.20	.10	.05	.02	.01	.002	.001
One tail	.10	.05	.025	.01	.005	.001	.0005
Critical value	13	14	15	15	16	17	17

We reject H_0 and find that the data provide sufficient evidence to conclude that the nonmutant strain grows better (at 24 hours) than the mutant strain of the virus. ■

Bracketing the P-Value. Like the Wilcoxon-Mann-Whitney test, the sign test has a discrete null distribution. The sign test statistic B_s may be exactly equal to an entry in Table 7, and in such a case the P-value is less than the column heading.* Also, certain critical value entries in Table 7 are blank. Both these situations are already familiar from our study of the Wilcoxon-Mann-Whitney test. Table 7 has another peculiarity that is not shared by the Wilcoxon-Mann-Whitney test: Some critical values appear more than once in the same row. This feature is also due to the discreteness of the null distribution and does not cause any particular difficulty; to bracket the P-value, we move to the right in Table 7 as far as possible, stopping when the next critical value in the table is *larger* than the observed value B_s.

Directional Alternative. To use Table 7 if the alternative hypothesis is directional, we proceed with the familiar two-step procedure:

Step 1. Check directionality (see if the data deviate from H_0 in the direction specified by H_A).

(a) If not, the P-value is greater than .50.

(b) If so, proceed to step 2.

Step 2. The P-value, which is half what it would be if H_A were nondirectional, is found by reading the "one tail" column headings.

Caution: Table 7, for the sign test, and Table 4, for the t test, are organized differently: Table 7 is entered with n_d, while Table 4 is entered with $(n_d - 1)$.

Treatment of Zeros. It may happen that some of the differences $(Y_1 - Y_2)$ are equal to zero. Should these be counted as positive or negative in determining B_s? A recommended procedure is to drop the corresponding pairs from the analysis and reduce the sample size n_d accordingly. In other words, each pair whose difference is zero is ignored entirely; such pairs are regarded as providing no evidence against H_0 in either direction. Notice that this procedure has no parallel in the t test; the t test treats differences of zero the same as any other value.

Null Distribution. Consider an experiment with ten pairs, so that $n_d = 10$. If H_0 is true, then the probability distribution of N_+ is a binomial distribution with $n = 10$ and $p = .5$. Figure 9.5(a) shows this binomial distribution, together with the associated values of N_+ and N_-. Figure 9.5(b) shows the null distribution of B_s, which is a "folded" version of Figure 9.5(a). [We saw a similar relationship between parts (a) and (b) of Figure 7.31.]

> **Example 9.14**

   If N_+ is 7 and H_A is directional (and predicts that positive differences are more likely than negative differences), then the P-value is the probability of 7 or more $(+)$ signs in 10 trials. This can be calculated from the binomial formula as

$$_{10}C_7(.5)^7(.5)^3 + {}_{10}C_8(.5)^8(.5)^2 + {}_{10}C_9(.5)^9(.5)^1 + {}_{10}C_{10}(.5)^{10}$$

$$= .11719 + .04395 + .00977 + .00098 = .17189$$

---

* In a few cases the P-value would be exactly equal to (rather than less than) the column heading. To simplify the presentation, we neglect this fine distinction.

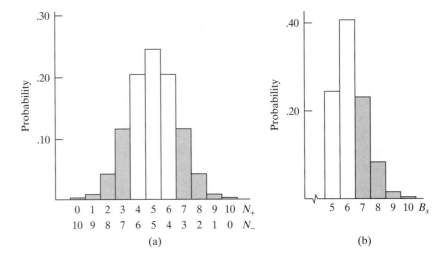

Figure 9.5 Null distributions for the sign test when $n_d = 10$. (a) Distribution of N_+ and N_-. (b) Distribution of B_s.

This value (.17189) is the sum of the shaded bars in the right-hand tail in Figure 9.5(a). If H_A is nondirectional, then the P-value is the sum of the shaded bars in the left-hand tail and of the right-hand tail of Figure 9.5(a). The two shaded areas are both equal to .1719; consequently, the total shaded area, which is the P-value, is

$$P = 2(.17189) = .34378 \approx .34$$

In terms of the null distribution of B_s, the P-value is an upper-tail probability; thus, the sum of the shaded bars in Figure 9.5(b) is equal to .34. ∎

How Table 7 is Calculated. Throughout your study of statistics you are asked to take on faith the critical values given in various tables. Table 7 is an exception; the following example shows how you could (if you wished to) calculate the critical values yourself. Understanding the example will help you to appreciate how the other tables of critical values have been obtained.

Example 9.15 Suppose $n_d = 10$. We saw in Example 9.14 that

If $B_s = 7$ the P-value of the data is .34378.

Similar calculations using the binomial formula show that

If $B_s = 8$, the P-value of the data is .1094.
If $B_s = 9$, the P-value of the data is .0215.
If $B_s = 10$, the P-value of the data is .00195.

For $n_d = 10$, the critical values are given in Table 7 as shown in Table 9.10.

TABLE 9.10 Critical Values for the Sign Test When $n_d = 10$

	Nominal Tail Probability						
Two tails	.20	.10	.05	.02	.01	.002	.001
One tail	.10	.05	.025	.01	.005	.001	.0005
Critical value	8	9	9	10	10	10	

These critical values have been determined from the preceding P-values, using the principle that the P-value corresponding to each entry should be as close as possible to the column heading without exceeding it. Thus, for instance, the critical value in the .05 column is equal to 9 because the P-value for $B_s = 9$ (.0215) is less than .05, but the next larger P-value (.1094) is greater than .05. In fact, .1094 is also greater than .10, and this is why 9 (rather than 8) is also listed in the .10 column. Look now at the .02 column. The only possible P-value less than .02 is .00195, which is also less than .01 and less than .002; this accounts for the three critical values listed as 10. On the other hand, .00195 is the smallest possible P-value and yet is greater than .001; for this reason the .001 column is left blank. ∎

Applicability of the Sign Test

The sign test is valid in any situation where the d's are independent of each other and the null hypothesis can be appropriately translated as

$$H_0: \Pr\{d \text{ is positive}\} = .5$$

Thus, the sign test is distribution free; its validity does not depend on any conditions about the form of the population distribution of the d's. This broad validity is bought at a price: If the population distribution of the d's is indeed normal, then the sign test is much less powerful than the t test.

The sign test is useful because it can be applied quickly and in a wide variety of settings. In fact, sometimes the sign test can be applied to data that do not permit a t test at all, as was shown in Example 9.12. There is another test for paired data, the Wilcoxon signed-ranks test, which is presented in Section 9.5, that is generally more powerful than the sign test and yet is distribution free. However, the Wilcoxon signed-ranks test is more difficult to carry out than the sign test and, like the t test, there are situations in which it cannot be conducted. The following is another example in which only a sign test is possible.

THC and Chemotherapy. Chemotherapy for cancer often produces nausea and vomiting. The effectiveness of THC (tetrahydrocannabinol the active ingredient of marijuana) in preventing these side effects was compared with the standard drug Compazine. Of the 46 patients who tried both drugs (but were not told which was which), 21 expressed no preference, while 20 preferred THC and 5 preferred Compazine. Since "preference" indicates a sign for the difference, but not a magnitude, a t test is impossible in this situation. For a sign test, we have $n_d = 25$ and $B_s = 20$, so that $.002 < P < .01$; even at $\alpha = .01$ we would reject H_0 and find that the data provide sufficient evidence to conclude that THC is preferred to Compazine.[14] ∎

Example 9.16

Exercises 9.14–9.27

9.14 Use Table 7 to bracket the P-value for a sign test (against a nondirectional alternative), assuming that $n_d = 9$ and

(a) $B_s = 6$
(b) $B_s = 7$
(c) $B_s = 8$
(d) $B_s = 9$

9.15 Use Table 7 to bracket the P-value for a sign test (against a nondirectional alternative), assuming that $n_d = 15$ and

(a) $B_s = 10$
(b) $B_s = 11$

(c) $B_s = 12$
(d) $B_s = 13$
(e) $B_s = 14$
(f) $B_s = 15$

9.16   A group of 30 postmenopausal women were given oral conjugated estrogen for one month. Plasma levels of plasminogen-activator inhibitor type 1 (PAI-1) went down for 22 of the women, but went up for 8 women.[15] Use a sign test to test the null hypothesis that oral conjugated estrogen has no effect on PAI-1 level. Use $\alpha = .10$ and use a nondirectional alternative.

9.17   Can mental exercise build "mental muscle"? In one study of this question, twelve littermate pairs of young male rats were used; one member of each pair, chosen at random, was raised in an "enriched" environment with toys and companions, while its littermate was raised alone in an "impoverished" environment. (See Example 8.19.) After 80 days, the animals were sacrificed and their brains were dissected by a researcher who did not know which treatment each rat had received. One variable of interest was the weight of the cerebral cortex, expressed relative to total brain weight. For 10 of the 12 pairs, the relative cortex weight was greater for the "enriched" rat than for his "impoverished" littermate; in the other 2 pairs, the "impoverished" rat had the larger cortex. Use a sign test to compare the environments at $\alpha = .05$; let the alternative hypothesis be that environmental enrichment tends to increase the relative size of the cortex.[16]

9.18   Refer to Exercise 9.17. Calculate the exact P-value of the data as analyzed by the sign test. (Note that H_A is directional.)

9.19   Twenty institutionalized epileptic patients participated in a study of a new anticonvulsant drug, valproate. Ten of the patients (chosen at random) were started on daily valproate and the remaining 10 received an identical placebo pill. During an eight-week observation period, the numbers of major and minor epileptic seizures were counted for each patient. After this, all patients were "crossed over" to the other treatment, and seizure counts were made during a second eight-week observation period. The numbers of minor seizures are given in the accompanying table.[17] Test for efficacy of valproate using the sign test at $\alpha = .05$. Use a directional alternative. (Note that this analysis ignores the possible effect of time—that is, first versus second observation period.)

Patient Number	Placebo Period	Valproate Period	Patient Number	Placebo Period	Valproate Period
1	37	5	11	7	8
2	52	22	12	9	8
3	63	41	13	65	30
4	2	4	14	52	22
5	25	32	15	6	11
6	29	20	16	17	1
7	15	10	17	54	31
8	52	25	18	27	15
9	19	17	19	36	13
10	12	14	20	5	5

*9.20   *(This exercise is based on material from optional Section 5.5.)*   Refer to Exercise 9.19. Use the normal approximation to the binomial distribution (with the continuity correction) to calculate the P-value of the data as analyzed by the sign test. (Note that H_A is directional.)

*9.21 (*This exercise is based on material from optional Section 5.5.*)   Refer to Exercise 9.16. Use the normal approximation to the binomial distribution (with the continuity correction) to calculate the *P*-value of the data as analyzed by the sign test. (Note that H_A is nondirectional.)

9.22 An ecological researcher studied the interaction between birds of two subspecies, the Carolina Junco and the Northern Junco. He placed a Carolina male and a Northern male, matched by size, together in an aviary and observed their behavior for 45 minutes beginning at dawn. This was repeated on different days with different pairs of birds. The table shows counts of the episodes in which one bird displayed dominance over the other—for instance, by chasing it or displacing it from its perch.[18] Use a sign test to compare the subspecies. Use a nondirectional alternative and let $\alpha = .01$.

| | **Number of Episodes in Which** | |
| | **Northern Was** | **Carolina Was** |
Pair	**Dominant**	**Dominant**
1	0	9
2	0	6
3	0	22
4	2	16
5	0	17
6	2	33
7	1	24
8	0	40

9.23 Refer to Exercise 9.22. Calculate the exact *P*-value of the data as analyzed by the sign test. (Note that H_A is nondirectional.)

9.24 (a) Suppose a paired data set has $n_d = 7$ and $B_s = 7$. Calculate the exact *P*-value of the data as analyzed by the sign test (against a nondirectional alternative).
(b) Explain why, in Table 7 with $n_d = 7$, no critical value is given in the .01 column.

9.25 (a) Suppose a paired data set has $n_d = 15$. Calculate the exact *P*-value of the data as analyzed by the sign test (against a nondirectional alternative) if (i) $B_s = 13$; (ii) $B_s = 14$; (iii) $B_s = 15$.
(b) Explain why, in Table 7 with $n_d = 15$, the critical value in the .002 column is $B_s = 14$.
(c) If Table 7 had a .005 column, what would be the entry in that column for $n_d = 15$?

9.26 The study described in Example 9.1, involving the compound mCPP, included a group of men. The men were asked to rate how hungry they were at the end of each two-week period and differences were computed (hunger rating when taking mCPP − hunger rating when taking the placebo). The distribution of the differences was not normal. Nonetheless, a sign test can be conducted using the following information: Out of 8 men who recorded hunger ratings, 3 reported greater hunger on mCPP than on the placebo and 5 reported lower hunger on mCPP than on the placebo.[1] Conduct a sign test at the $\alpha = .10$ level; use a nondirectional alternative.

9.27 Refer to Exercise 9.26. Calculate the exact *P*-value of the data as analyzed by the sign test. (Note that H_A is nondirectional.)

9.5 THE WILCOXON SIGNED-RANK TEST

The **Wilcoxon signed-rank test**, like the sign test, is a nonparametric method that can be used to compare paired samples. Conducting a Wilcoxon signed-rank test is somewhat more complicated than conducting a sign test, but the Wilcoxon test is more powerful than the sign test. Like the sign test, the Wilcoxon signed-rank test does *not* require that the data be a sample from a normally distributed population.

The Wilcoxon signed-rank test is based on the set of differences, $d = Y_1 - Y_2$. It combines the main idea of the sign test—"look at the signs of the differences"—with the main idea of the paired t test—"look at the magnitudes of the differences."

Method

The Wilcoxon signed-rank test proceeds in several steps, which we present here in the context of an example.

Example 9.17

Nerve Cell Density.   For each of nine horses, a veterinary anatomist measured the density of nerve cells at specified sites in the intestine. The results for site I (midregion of jejunum) and site II (mesenteric region of jejunum) are given in the accompanying table.[19] Each density value is the average of counts of nerve cells in five equal sections of tissue. The null hypothesis of interest is that in the population of all horses there is no difference between the two sites.

1. The first step in the Wilcoxon signed-rank test is to calculate the differences, as shown in Table 9.11.

TABLE 9.11 Nerve Cell Density at Each of Two Sites

Animal	Site I	Site II	Difference
1	50.6	38.0	12.6
2	39.2	18.6	20.6
3	35.2	23.2	12.0
4	17.0	19.0	−2.0
5	11.2	6.6	4.6
6	14.2	16.4	−2.2
7	24.2	14.4	9.8
8	37.4	37.6	−.2
9	35.2	24.4	10.8

2. Next we find the absolute value of each difference.
3. We then rank these absolute values, from smallest to largest, as shown in Table 9.12.
4. Next we restore the + and − signs to the ranks of the absolute differences to produce signed ranks, as shown in Table 9.13.
5. We sum the positive signed ranks to get W_+; we sum the absolute values of the negative signed ranks to get W_-. For the nerve cell data, $W_+ = 8 + 9 + 7 + 4 + 5 + 6 = 39$ and $W_- = 2 + 3 + 1 = 6$. The test statistic, W_s, is defined as

$$W_s = \text{Larger of } W_+ \text{ and } W_-$$

For the nerve cell data, $W_s = 39$.

TABLE 9.12

Animal	Difference, d	$|d|$	Rank of $|d|$
1	12.6	12.6	8
2	20.6	20.6	9
3	12.0	12.0	7
4	−2.0	2.0	2
5	4.6	4.6	4
6	−2.2	2.2	3
7	9.8	9.8	5
8	−.2	.2	1
9	10.8	10.8	6

TABLE 9.13

Animal	Difference, d	Rank of $|d|$	Signed Rank
1	12.6	8	8
2	20.6	9	9
3	12.0	7	7
4	−2.0	2	−2
5	4.6	4	4
6	−2.2	3	−3
7	9.8	5	5
8	−.2	1	−1
9	10.8	6	6

6. To bracket the P-value, we consult Table 8 (at the end of the book). Part of Table 8 is reproduced in Table 9.14.

TABLE 9.14 Critical Values for the Wilcoxon Signed-Rank Test When $n_d = 9$

	Nominal Tail Probability						
Two tails	.20	.10	.05	.02	.01	.002	.001
One tail	.10	.05	.025	.01	.005	.001	.0005
Critical value	35	37	40	42	44		

From Table 9.14, we note that our value of W_s is between 37, which is the .10 critical value, and 40, which is the .05 critical value. Thus, for the nerve cell data the P-value is between .10 and .05. We conclude that there is moderately strong evidence $(.05 < P$ value $< .10)$ that there is a difference in nerve cell density in the two regions. (We reject H_0 if α is .10 or larger.) ∎

Bracketing the P-Value. Like the sign test, the Wilcoxon signed-rank test has a discrete null distribution. The test statistic W_s may be exactly equal to an entry in Table 8, and in such a case the P-value is less than the column heading.* For example,

---

* As with the sign test, in a few cases the P-value would be exactly equal to (rather than less than) the column heading. To simplify the presentation, we neglect this fine distinction.

suppose $n_d = 9$ and $W_s = 37$. The entry 37 is in the (two-tailed) .10 column, but the P-value is actually .0977 (to four decimal places). Also, certain critical value entries in Table 8 are blank; this situation is familiar from our study of the Wilcoxon-Mann-Whitney test and the sign test. For example, if $n_d = 9$, then the strongest possible evidence against H_0 occurs when all 9 differences are positive, in which case $W_s = 45$. But the chance that W_s will equal 45 when H_0 is true is $(1/2)^8$, which is approximately .0039. Thus, it is not possible to have a two-tailed P-value smaller than .002, let alone .001. This is why the last two entries are blank in the $n_d = 9$ row of Table 8.

Directional Alternative. To use Table 8 if the alternative hypothesis is directional, we proceed with the familiar two-step procedure:

Step 1. Check directionality (see if the data deviate from H_0 in the direction specified by H_A).

(a) If not, the P-value is greater than .50.

(b) If so, proceed to step 2.

Step 2. The P-value, which is half what it would be if H_A were nondirectional, is found by reading the "one tail" column headings.

Treatment of Zeros. If any of the differences $(Y_1 - Y_2)$ are zero, then those data points are deleted and the sample size is reduced accordingly. For example, if one of the 9 differences in Example 9.17 had been zero, we would have deleted that point when conducting the Wilcoxon test, so that the sample size would have become 8.

Treatment of Ties. If there are ties among the absolute values of the differences (in step 3), we average the ranks of the tied values. If there are ties, then the P-value given by the Wilcoxon signed-rank test is only approximate.

Applicability of the Wilcoxon Signed-Rank Test

The Wilcoxon signed-rank test can be used in any situation in which the d's are independent of each other and come from a symmetric distribution; the distribution need not be normal.* The null hypothesis of "no treatment effect" or "no difference between populations" can be stated as

$$H_0: \mu_d = 0$$

Sometimes the Wilcoxon signed-rank test can be carried out even with incomplete information. For example, a Wilcoxon test is possible for the skin graft data of Example 9.12. It is true that an exact value of d cannot be calculated for two of the patients, but for both of these patients the difference is positive and is larger than either of the negative differences. The data in Table 9.15 show that there only are two negative differences. The smaller of these is -1, for patient 11. This is the smallest difference in absolute value, so it has signed rank -1. The only other negative signed rank is for patient 7; all of the other signed ranks are positive. (The rest of this example is left as an exercise.)

---

* Strictly speaking, the distribution must be continuous, which means that the probability of a tie is zero.

TABLE 9.15 Skin Graft Survival Times

| Patient | HL-A Compatibility | | $d = Y_1 - Y_2$ |
	Close Y_1	Poor Y_2	
1	37	29	8
2	19	13	6
3	57+	15	42+
4	93	26	67
5	16	11	5
6	23	18	5
7	20	26	−6
8	63	43	20
9	29	18	11
10	60+	42	18+
11	18	19	−1

As with the Wilcoxon-Mann-Whitney test for independent samples, there is a procedure associated with the Wilcoxon signed-rank test that can be used to construct a confidence interval for μ_d. The procedure is beyond the scope of this book.

When dealing with paired data we have three inference procedures: the paired t test, the Wilcoxon signed-rank test, and the sign test. The t test requires that the data come from a normally distributed population; if this condition is met, then the t test is recommended, as it is more powerful than the Wilcoxon test or sign test. The Wilcoxon test does not require normality but does require that we can rank the set of differences; it has more power than the sign test. The sign test is the least powerful of the three methods, but the most widely applicable, since it only requires that we determine whether each difference is positive or negative.

Exercises 9.28–9.33

9.28 Use Table 8 to bracket the P-value for a Wilcoxon signed-rank test (against a nondirectional alternative), assuming that $n_d = 7$ and

(a) $W_s = 22$
(b) $W_s = 24$
(c) $W_s = 26$
(d) $W_s = 28$

9.29 Use Table 8 to bracket the P-value for a Wilcoxon signed-rank test (against a nondirectional alternative), assuming that $n_d = 12$ and

(a) $B_s = 55$
(b) $B_s = 63$
(c) $B_s = 71$
(d) $B_s = 73$

9.30 The study described in Example 9.1, involving the compound mCPP, included a group of nine men. The men were asked to rate how hungry they were at the end of each two-week period and differences were computed (hunger rating when taking mCPP − hunger rating when taking the placebo). Data for one of the subjects are not available; the data for the other eight subjects are given in the accompanying table.[1] Analyze these data with a Wilcoxon signed-rank test at the $\alpha = .10$ level; use a nondirectional alternative.

<table>
<tr><th rowspan="3">Subject</th><th colspan="3">Hunger Rating</th></tr>
<tr><th>mCPP</th><th>Placebo</th><th>Difference</th></tr>
<tr><th>y_1</th><th>y_2</th><th>$d = y_1 - y_2$</th></tr>
<tr><td>1</td><td>64</td><td>69</td><td>−5</td></tr>
<tr><td>2</td><td>119</td><td>112</td><td>7</td></tr>
<tr><td>3</td><td>0</td><td>28</td><td>−28</td></tr>
<tr><td>4</td><td>48</td><td>95</td><td>−47</td></tr>
<tr><td>5</td><td>65</td><td>145</td><td>−80</td></tr>
<tr><td>6</td><td>119</td><td>112</td><td>7</td></tr>
<tr><td>7</td><td>149</td><td>141</td><td>8</td></tr>
<tr><td>8</td><td>NA</td><td>NA</td><td>NA</td></tr>
<tr><td>9</td><td>99</td><td>119</td><td>−20</td></tr>
</table>

9.31 As part of the study described in Example 9.1 (and in Exercise 9.30), involving the compound mCPP, weight change was measured for nine men. For each man two measurements were made: weight change when taking mCPP and weight change when taking the placebo. The data are given in the accompanying table.[1] Analyze these data with a Wilcoxon signed-rank test at the $\alpha = .05$ level; use a nondirectional alternative.

<table>
<tr><th rowspan="3">Subject</th><th colspan="3">Weight Change</th></tr>
<tr><th>mCPP</th><th>Placebo</th><th>Difference</th></tr>
<tr><th>y_1</th><th>y_2</th><th>$d = y_1 - y_2$</th></tr>
<tr><td>1</td><td>0.0</td><td>−1.1</td><td>1.1</td></tr>
<tr><td>2</td><td>−1.1</td><td>0.5</td><td>−1.6</td></tr>
<tr><td>3</td><td>−1.6</td><td>0.5</td><td>−2.1</td></tr>
<tr><td>4</td><td>−0.3</td><td>0.0</td><td>−0.3</td></tr>
<tr><td>5</td><td>−1.1</td><td>−0.5</td><td>−0.6</td></tr>
<tr><td>6</td><td>−0.9</td><td>1.3</td><td>−2.2</td></tr>
<tr><td>7</td><td>−0.5</td><td>−1.4</td><td>0.9</td></tr>
<tr><td>8</td><td>0.7</td><td>0.0</td><td>0.7</td></tr>
<tr><td>9</td><td>−1.2</td><td>−0.8</td><td>−0.4</td></tr>
</table>

9.32 Consider the skin graft data of Example 9.12. Table 9.15, at the end of Section 9.5, shows the first steps in conducting a Wilcoxon signed-rank test of the null hypothesis that HL-A compatibility has no effect on graft survival time. Complete this test. Use $\alpha = .05$ and use the directional alternative that survival time tends to be greater when compatibility score is close.

9.33 In an investigation of possible brain damage due to alcoholism, an X-ray procedure known as a computerized tomography (CT) scan was used to measure brain densities in eleven chronic alcoholics. For each alcoholic, a nonalcoholic control was selected who matched the alcoholic on age, sex, education, and other factors. The brain density measurements on the alcoholics and the matched controls are reported in the accompanying table.[20] Use a Wilcoxon signed-rank test to test the null hypothesis of no difference against the alternative that alcoholism reduces brain density. Let $\alpha = .02$.

Pair	Alcoholic	Control	Difference
1	40.1	41.3	−1.2
2	38.5	40.2	−1.7
3	36.9	37.4	−.5
4	41.4	46.1	−4.7
5	40.6	43.9	−3.3
6	42.3	41.9	.4
7	37.2	39.9	−2.7
8	38.6	40.4	−1.8
9	38.5	38.6	−.1
10	38.4	38.1	.3
11	38.1	39.5	−1.4
Mean	39.14	40.66	−1.52
SD	1.72	2.56	1.58

9.6 FURTHER CONSIDERATIONS IN PAIRED EXPERIMENTS

In this section we discuss two additional topics: the interpretation of before–after studies and the reporting of paired data.

Before–After Studies

Many studies in the life sciences compare measurements before and after some experimental intervention. These studies can be difficult to interpret, because the effect of the experimental intervention may be confounded with other changes over time. One way to protect against this difficulty is to use randomized concurrent controls, as in the following example.

Biofeedback and Blood Pressure. A medical research team investigated the effectiveness of a biofeedback training program designed to reduce high blood pressure. Volunteers were randomly allocated to a biofeedback group or a control group. All volunteers received health education literature and a brief lecture. In addition, the biofeedback group received 8 weeks of relaxation training, aided by biofeedback, meditation, and breathing exercises. The results for systolic blood pressure, before and after the 8 weeks, are shown in Table 9.16.[21]

Example 9.18

TABLE 9.16 Results of Biofeedback Experiment

		Systolic Blood Pressure (mm Hg)			
Group	*n*	*Before* Mean	*After* Mean	*Difference* Mean	*SE*
Biofeedback	99	145.2	131.4	13.8	1.34
Control	93	144.2	140.2	4.0	1.30

Let us analyze the before–after changes by paired t tests at $\alpha = .05$. In the biofeedback group, the mean systolic blood pressure fell by 13.8 mm Hg. To evaluate the statistical significance of this drop, the test statistic is

$$t_s = \frac{13.8}{1.34} = 10.3$$

which is highly significant (P-value $\ll .0001$). However, this result alone does not demonstrate the effectiveness of the biofeedback training; the drop in blood pressure might be partly or entirely due to other factors, such as the health education literature or the special attention received by all the participants. Indeed, a paired t test applied to the control group gives

$$t_s = \frac{4.0}{1.30} = 3.08 \qquad .001 < P < .01$$

Thus, the people who received *no* biofeedback training *also* experienced a statistically significant drop in blood pressure.

To isolate the effect of the biofeedback training, we can compare the experience of the two treatment groups, using an independent-samples t test. We again choose $\alpha = .05$. The difference between the mean changes in the two groups is

$$13.8 - 4.0 = 9.8 \text{ mm Hg}$$

and the standard error of this difference is

$$\sqrt{1.34^2 + 1.30^2} = 1.87$$

Thus, the t statistic is

$$t_s = \frac{9.8}{1.87} = 5.24$$

This test provides strong evidence ($P < .0001$) that the biofeedback program is effective. If the experimental design had not included the control group, then this last crucial comparison would not have been possible, and the support for efficacy of biofeedback would have been shaky indeed. ■

Reporting of Data

In communicating experimental results, it is desirable to choose a form of reporting that conveys the extra information provided by pairing. With small samples, a graphical approach can be used, as in the following example.

Example 9.19

Plasma Aldosterone in Dogs. Aldosterone is a hormone involved in maintaining fluid balance in the body. In a veterinary study, six dogs with heart failure were treated with the drug Captopril, and plasma concentrations of aldosterone were measured before and after the treatment. The results are displayed in Figure 9.6.[22] The experience of each dog is represented by two points joined by a line. Note that the lines carry crucial information. For instance, all the lines slope downward, which indicates that all six dogs experienced a fall (rather than a rise) in plasma aldosterone. Also, lines that are parallel represent falls of equal magnitude; in Figure 9.6 four of the lines are approximately parallel. If the lines were omitted from the plot, the reader have difficulty assessing either the statistical or the practical significance of the before–after change. ■

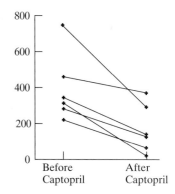

Figure 9.6 Plasma aldosterone in six dogs before and after treatment with Captopril

In published reports of biological research, the crucial information related to pairing is often omitted. For instance, a common practice is to report the means and standard deviations of Y_1 and Y_2 but to omit the standard deviation of the difference, d! This is a serious error. It is best to report some description of d, using either a display like Figure 9.6, or a histogram of the d's, or at least the standard deviation of the d's.

Computer note: Statistical software can be used to check conditions for a paired data analysis and to aid in completing calculations. The inference procedures presented in this chapter—the paired t test and confidence interval, the sign test, and the Wilcoxon signed-rank test—can all be carried out with common statistical software. In the MINITAB system one would first calculate the differences in the paired data and then proceed to a test on the new column of differences. For example, suppose the weight loss data from Example 1 are stored in the columns "mCPP" and "Placebo." Then we can calculate the differences with the command

```
MTB > Let c3 = 'mCPP'-'Placebo'
```

To conduct a paired t test, we use the command

```
MTB > TTest 0.0 C3;
SUBC> Alternative 0.
```

which indicates that the null hypothesis is $H_0: \mu_d = 0$ and the alternative is $H_A: \mu_d \neq 0$. The resulting output is

```
T-Test of the Mean
Test of mu = 0.000 vs mu not = 0.000
Variable     N   Mean   StDev   SE Mean    T    P-Value
C3           9  1.000   0.719    0.240   4.17   0.0032
```

To conduct a Wilcoxon signed-rank test we use the command

```
MTB > WTest 0.0 C3;
SUBC> Alternative 0.
```

which produces

```
Wilcoxon Signed Rank Test
TEST OF MEDIAN = 0.000000 VERSUS MEDIAN N.E. 0.000000
                  N FOR     WILCOXON                ESTIMATED
          N       TEST      STATISTIC    P-VALUE     MEDIAN
C3        9       9          44.0        0.013       1.000
```

Note that MINITAB states the hypotheses in terms of the median, rather than the mean. Since we are assuming that the differences have a symmetric distribution, this is equivalent to stating the hypotheses in terms of the mean.

Finally, for a sign test we use the command

```
MTB > STest 0.0 C3;
SUBC>   Alternative 0.
```

which produces

```
Sign Test for Median
Sign test of median = 0.00000 versus N.E. 0.00000
          N     BELOW    EQUAL    ABOVE    P-VALUE     MEDIAN
C3        9     1        0        8        0.0391      1.100
```

Exercises 9.34–9.35

9.34 Thirty-three men with high serum cholesterol, all regular coffee drinkers, participated in a study to see whether abstaining from coffee would affect their cholesterol level. Twenty-five of the men (chosen at random) drank no coffee for 5 weeks, while the remaining eight men drank coffee as usual. The accompanying table shows the serum cholesterol levels (in mg/dLi) at baseline (at the beginning of the study) and the change from baseline after 5 weeks.[23]

	No Coffee ($n = 25$)		Usual Coffee ($n = 8$)	
	Mean	*SD*	*Mean*	*SD*
Baseline	341	37	331	30
Change from baseline	−35	27	+26	56

For the following *t* tests, use nondirectional alternatives and let $\alpha = .05$.

(a) The no-coffee group experienced a 35 mg/dLi drop in mean cholesterol level. Use a *t* test to assess the statistical significance of this drop.

(b) The usual-coffee group experienced a 26 mg/dLi rise in mean cholesterol level. Use a *t* test to assess the statistical significance of this rise.

(c) Use a *t* test to compare the no-coffee mean change (−35) to the usual-coffee mean change (+26).

(d) State the conclusion of the test from part (c) in the context of this setting.

9.35 Eight young women participated in a study to investigate the relationship between the menstrual cycle and food intake. Dietary information was obtained every day by interview; the study was double blind in the sense that the participants did not know its purpose and the interviewer did not know the timing of their menstrual cycles. The table shows, for each participant, the average caloric intake for the 10 days preceding and the 10 days following the onset of the menstrual period (these data are for one cycle only). For these data, prepare a display like that of Figure 9.6.[24]

	Food Intake (Calories)	
Participant	*Premenstrual*	*Postmenstrual*
1	2,378	1,706
2	1,393	958
3	1,519	1,194
4	2,414	1,682
5	2,008	1,652
6	2,092	1,260
7	1,710	1,239
8	1,967	1,758

9.7 PERSPECTIVE

We have discussed several statistical methods of comparing two samples. In analyzing real data, it is wise to keep in mind that these statistical methods address only limited questions.

The paired *t* test is limited in two ways:

1. It is limited to questions concerning $\bar{d}$.
2. It is limited to questions about *aggregate* differences.

The second limitation is very broad; it applies not only to the methods of this chapter but also to those of Chapter 7 and to many other elementary statistical techniques. We will discuss these two limitations separately.

Limitation of $\bar{d}$

One limitation of the paired *t* test and confidence interval is simple but too often overlooked: When some of the *d*'s are positive and some are negative, the magnitude of $\bar{d}$ does not reflect the "typical" magnitude of the *d*'s. The following example shows how misleading $\bar{d}$ can be.

Measuring Serum Cholesterol. Suppose a clinical chemist wants to compare two methods of measuring serum cholesterol; she is interested in how closely the two methods agree with each other. She takes blood specimens from 400 patients, splits each specimen in half, and assays one half by method A and the other by method B. Table 9.17 shows fictitious data, exaggerated to clarify the issue.

Example 9.20

TABLE 9.17 Serum Cholesterol (mg/dLi)

Specimin	Method A	Method B	$d = A - B$
1	200	234	−34
2	284	272	+12
3	146	153	−7
4	263	250	+13
5	258	232	+26
⋮	⋮	⋮	⋮
400	176	190	−14
Mean	215.2	214.5	.7
SD	45.6	59.8	18.8

In Table 9.17, the sample mean difference is small ($\bar{d} = .7$). Furthermore, the data indicate that the population mean difference is small (a 95% confidence interval is -1.1 mg/dLi $< \mu_d < 2.5$ mg/dLi). But such discussion of $\bar{d}$ or μ_d does not address the central question, which is: How closely do the methods agree? In fact, Table 9.17 indicates that the two methods do not agree well; the individual differences between method A and method B are not small. The mean $\bar{d}$ is small because the positive and negative differences tend to cancel each other. A graph similar to Figure 9.4 (in Section 9.3) would be very helpful in visually determining how well the methods agree. We would examine such a graph to see how closely the points cluster around the $y = x$ line as well as to see the spread in the boxplot of differences. To make a numerical assessment of agreement between the methods we should not focus on the mean, $\bar{d}$. It would be far more relevant to analyze the absolute (unsigned) magnitudes of the d's (that is, 34, 12, 7, 13, 26, and so on). These magnitudes could be analyzed in various ways: We could average them, we could count how many are "large" (say, more than 10 mg/dLi), and so on. ∎

Limitation of the Aggregate Viewpoint

Consider a paired experiment in which two treatments, say A and B, are applied to the same person. If we apply a t test, a sign test, or a Wilcoxon signed-rank test, we are viewing the people as an ensemble rather than individually. This is appropriate if we are willing to assume that the difference (if any) between A and B is in a consistent direction for all people—or, at least, that the important features of the difference are preserved even when the people are viewed *en masse*. The following example illustrates the issue.

Example 9.21 | **Treatment of Acne.** Consider a clinical study to compare two medicated lotions for treating acne. Twenty patients participate. Each patient uses lotion A on one side of his face and lotion B on the other side. After 3 weeks, each side of the face is scored for total improvement.

First, suppose that the A side improves more than the B side in ten patients, while in the other ten the B side improves more. According to a sign test, this result is in perfect agreement with the null hypothesis. And yet, two very different interpretations are logically possible:

Interpretation 1: Treatments A and B are in fact completely equivalent; their action is indistinguishable. The observed differences between A and B sides of the face were entirely due to chance variation.

Interpretation 2: Treatments A and B are in fact completely different. For some people (about 50% of the population), treatment A is more effective than treatment B, whereas in the remaining half of the population treatment B is more effective. The observed differences between A and B sides of the face were biologically meaningful.*

The same ambiguity of interpretation arises if the results favor one treatment over another. For instance, suppose the A side improved more than the B side in 18 of the 20 cases, while B was favored in 2 patients. This result, which is statistically significant ($P < .001$), could again be interpreted in two ways. It could mean that treatment A is in fact superior to B for everybody, but chance variation obscured its superiority in two of the patients; or it could mean that A is superior to B for most people, but for about 10% of the population ($2/10 = .10$) B is superior to A. ■

The difficulty illustrated by Example 9.21 is not confined to randomized blocks experiments. In fact, it is particularly clear in another type of paired experiment—the measurement of change over time. Consider, for instance, the blood pressure data of Example 9.18. Our discussion of that study hinged on an aggregate measure of blood pressure: the mean. If some patients' pressures rose as a result of biofeedback and others fell, these details were ignored in the analysis based on Student's *t*; only the average change was analyzed.

Neither is the difficulty confined to human experiments. Suppose, for instance, that two fertilizers, A and B, are to be compared in an agronomic field experiment using a randomized blocks design, with the data to be analyzed by a paired *t* test. If treatment A is superior to B on acid soils, but B is better than A on alkaline soils, this fact would be obscured in an experiment that included soils of both types.

The issue raised by the preceding examples is a very general one. Simple statistical methods such as the sign test and the *t* test are designed to evaluate treatment effects *in the aggregate*—that is, *collectively*—for a population of people, or of mice, or of plots of ground. The segregation of differential treatment effects in subpopulations requires more delicate handling, both in design and analysis.

This confinement to the aggregate point of view applies to Chapter 7 (independent samples) even more forcefully than to the present chapter. For instance, if treatment A is given to one group of mice and treatment B to another, it is quite impossible to know how a mouse in group A would have responded *if* it had received treatment B; the only possible comparison is an aggregate one. In Section 7.12 we stated that the statistical comparison of independent samples depends on an "implicit assumption"; essentially, the assumption is that the phenomenon under study can be adequately perceived from an aggregate viewpoint.

---

* This may seem farfetched, but phenomena of this kind do occur; as an obvious example, consider the response of patients to blood transfusions of type A or type B blood.

In many, perhaps most, biological investigations the phenomena of interest are reasonably universal, so that this issue of submerging the individual in the aggregate does not cause a serious problem. Nevertheless, we should not lose sight of the fact that aggregation may obscure important individual detail.

Exercises 9.36–9.37

9.36 For each of 29 healthy dogs, a veterinarian measured the glucose concentration in the anterior chamber of the left eye and the right eye, with the results shown in the table.[25]

Animal Number	GLUCOSE (mg/dLi) Right Eye	Left Eye	Animal Number	GLUCOSE (mg/dLi) Right Eye	Left Eye
1	79	79	16	80	80
2	81	82	17	78	78
3	87	91	18	112	110
4	85	86	19	89	91
5	87	92	20	87	91
6	73	74	21	71	69
7	72	74	22	92	93
8	70	66	23	91	87
9	67	67	24	102	101
10	69	69	25	116	113
11	77	78	26	84	80
12	77	77	27	78	80
13	84	83	28	94	95
14	83	82	29	100	102
15	74	75			

Using the paired t method, a 95% confidence interval for the mean difference is -1.1 mg/dLi $< \mu_d < .7$ mg/dLi. Does this result suggest that, for the typical dog in the population, the difference in glucose concentration between the two eyes is less than 1.1 mg/dLi? Explain.

9.37 Tobramycin is a powerful antibiotic. To minimize its toxic side effects, the dose can be individualized for each patient. Thirty patients participated in a study of the accuracy of this individualized dosing. For each patient, the predicted peak concentration of Tobramycin in the blood serum was calculated, based on the patient's age, sex, weight, and other characteristics. Then Tobramycin was administered and the actual peak concentration (μg/mLi) was measured. The results were reported as in the table.[26]

	Predicted	Actual
Mean	4.52	4.40
SD	.90	.85
n	30	30

Does the reported summary give enough information for you to judge whether the individualized dosing is, on the whole, accurate in its prediction of peak concentration? If so, describe how you would make this judgment. If not, describe what additional information you would need and why.

Supplementary Exercises 9.38–9.57

9.38   A volunteer working at an animal shelter conducted a study of the effect of catnip on cats at the shelter. She recorded the number of "negative interactions" each of 15 cats made in 15 minute periods before and after being given a teaspoon of catnip. The paired measurements were collected on the same day within 30 minutes of one another; the data are given in the accompanying table.[27]

Cat	Before (Y_1)	After (Y_2)	Difference
Amelia	0	0	0
Bathsheba	3	6	−3
Boris	3	4	−1
Frank	0	1	−1
Jupiter	0	0	0
Lupine	4	5	−1
Madonna	1	3	−2
Michelangelo	2	1	1
Oregano	3	5	−2
Phantom	5	7	−2
Posh	1	0	1
Sawyer	0	1	−1
Scary	3	5	−2
Slater	0	2	−2
Tucker	2	2	0
Mean	1.8	2.8	−1
SD	1.66	2.37	1.20

(a) Construct a 95% confidence interval for the difference in mean number of negative interactions.

(b) Construct a 95% confidence interval the wrong way, using the independent-samples method. How does this interval differ from the one obtained in part (a)?

9.39   Refer to Exercise 9.38. Compare the before and after populations using a t test at $\alpha = .05$. Use a nondirectional alternative.

9.40   Refer to Exercise 9.38.

(a) Compare the before and after populations using a sign test at $\alpha = .05$. Use a nondirectional alternative.

(b) Calculate the exact P-value for the analysis of part (a).

9.41   Refer to Exercise 9.38. Construct a scatterplot of the data. Does the appearance of the scatterplot indicate that the pairing was effective? Explain.

9.42   As part of a study of the physiology of wheat maturation, an agronomist selected six wheat plants at random from a field plot. For each plant, she measured the moisture content in two batches of seeds: one batch from the "central" portion of the wheat head, and one batch from the "top" portion, with the results shown in the following table.[28] Construct a 90% confidence interval for the mean difference in moisture content of the two regions of the wheat head.

	Percent Moisture	
Plant	*Central*	*Top*
1	62.7	59.7
2	63.6	61.6
3	60.9	58.2
4	63.0	60.5
5	62.7	60.6
6	63.7	60.8

9.43 Biologists noticed that some stream fishes are most often found in pools, which are deep, slow-moving parts of the stream, while others prefer riffles, which are shallow, fast-moving regions. To investigate whether these two habitats support equal levels of diversity (i.e., equal numbers of species), they captured fish at 15 locations along a river. At each location, they recorded the number of species captured in a riffle and the number captured in an adjacent pool. The following table contains the data.[29] Construct a 90% confidence interval for the difference in mean diversity between the types of habitats.

Location	Pool	Riffle	Difference
1	6	3	3
2	6	3	3
3	3	3	0
4	8	4	4
5	5	2	3
6	2	2	0
7	6	2	4
8	7	2	5
9	1	2	−1
10	3	2	1
11	4	3	1
12	5	1	4
13	4	3	1
14	6	2	4
15	4	3	1
Mean	4.7	2.5	2.2
SD	1.91	0.74	1.86

9.44 Refer to Exercise 9.43. What conditions are necessary for the confidence interval to be valid? Are those conditions satisfied? How do you know?

9.45 Refer to Exercise 9.43. Compare the habitats using a t test at $\alpha = .10$. Use a nondirectional alternative.

9.46 Refer to Exercise 9.43.

(a) Compare the habitats using a sign test at $\alpha = .10$. Use a nondirectional alternative.
(b) Calculate the exact P-value for the analysis of part (a).

9.47 Refer to Exercise 9.43. Analyze these data using a Wilcoxon signed-rank test.

9.48 Refer to the Wilcoxon signed-rank test from Exercise 9.47. On what grounds could it be argued that the P-value found in this test might not be accurate? This is, why might it be argued that the Wilcoxon test P-value is not a completely accurate measure of the strength of the evidence against H_0 in this case?

9.49 In a study of the effect of caffeine on muscle metabolism, nine male volunteers underwent arm exercise tests on two separate occasions. On one occasion, the volunteer took a placebo capsule an hour before the test; on the other occasion he received a capsule containing pure caffeine. (The time order of the two occasions was randomly determined.) During each exercise test, the subject's respiratory exchange ratio (RER) was measured. The RER is the ratio of carbon dioxide produced to oxygen consumed, and is an indicator of whether energy is being obtained from carbohydrates or from fats. The results are presented in the accompanying table.[30] Use a *t* test to assess the effect of caffeine. Use a nondirectional alternative and let $\alpha = .05$.

	RER (%)	
Subject	*Placebo*	*Caffeine*
1	105	96
2	119	99
3	92	89
4	97	95
5	96	88
6	101	95
7	94	88
8	95	93
9	98	88

9.50   For the data of Exercise 9.49, construct a display like that of Figure 9.6.

9.51   Refer to Exercise 9.49. Analyze these data using a sign test.

9.52   Certain types of nerve cells have the ability to regenerate a part of the cell that has been amputated. In an early study of this process, measurements were made on the nerves in the spinal cord in rhesus monkeys. Nerves emanating from the left side of the cord were cut, while nerves from the right side were kept intact. During the regeneration process, the content of creatine phosphate (CP) was measured in the left and the right portion of the spinal cord. The following table shows the data for the right (control) side (Y_1), and for the left (regenerating) side (Y_2). The units of measurement are mg CP per 100 g tissue.[31] Use a t test to compare the two sides at $\alpha = .05$. Use a nondirectional alternative.

Animal	Right side (Control)	Left side (Regenerating)	Difference
1	16.3	11.5	4.8
2	4.8	3.6	1.2
3	10.9	12.5	−1.6
4	14.2	6.3	7.9
5	16.3	15.2	1.1
6	9.9	8.1	1.8
7	29.2	16.6	12.6
8	22.4	13.1	9.3
Mean	15.50	10.86	4.64
SD	7.61	4.49	4.89

9.53   *(Computer exercise)*   For an investigation of the mechanism of wound healing, a biologist chose a paired design, using the left and right hindlimbs of the salamander *Notophthalmus viridescens*. After amputating each limb, she made a small wound in the skin and then kept the limb for 4 hours in either a solution containing benzamil or a control solution. She theorized that the benzamil would impair the healing. The accompanying table shows the amount of healing, expressed as the area (mm^2) covered with new skin after 4 hours.[32]

Animal	Control Limb	Benzamil Limb	Animal	Control Limb	Benzamil Limb
1	.55	.14	10	.42	.21
2	.15	.08	11	.49	.11
3	.00	.00	12	.08	.03
4	.13	.13	13	.32	.14
5	.26	.10	14	.18	.37
6	.07	.08	15	.35	.25
7	.20	.11	16	.03	.05
8	.16	.00	17	.24	.16
9	.03	.05			

(a) Assess the effect of benzamil using a t test at $\alpha = .05$. Let the alternative hypothesis be that the researcher's expectation is correct.
(b) Proceed as in part (a) but use a sign test.
(c) Construct a 95% confidence interval for the mean effect of benzamil.
(d) Construct a scatterplot of the data. Does the appearance of the scatterplot indicate that the pairing was effective? Explain.

9.54 *(Computer exercise)* In a study of hypnotic suggestion, 16 male volunteers were randomly allocated to an experimental group and a control group. Each subject participated in a two-phase experimental session. In the first phase, respiration was measured while the subject was awake and at rest. (These measurements were also described in Exercises 7.51 and 7.80.) In the second phase, the subject was told to imagine that he was performing muscular work, and respiration was measured again.

For subjects in the experimental group, hypnosis was induced between the first and second phases; thus, the suggestion to imagine muscular work was "hypnotic suggestion" for experimental subjects and "waking suggestion" for control subjects. The accompanying table shows the measurements of total ventilation (liters of air per minute per square meter of body area) for all 16 subjects.[33]

Experimental Group			Control Group		
Subject	*Rest*	*Work*	*Subject*	*Rest*	*Work*
1	5.74	6.24	9	6.21	5.50
2	6.79	9.07	10	4.50	4.64
3	5.32	7.77	11	4.86	4.61
4	7.18	16.46	12	4.78	3.78
5	5.60	6.95	13	4.79	5.41
6	6.06	8.14	14	5.70	5.32
7	6.32	11.72	15	5.41	4.54
8	6.34	8.06	16	6.08	5.98

(a) Use a t test to compare the mean resting values in the two groups. Use a nondirectional alternative and let $\alpha = .05$. This is the same as Exercise 7.51(a).
(b) Use suitable paired and unpaired t tests to investigate (i) the response of the experimental group to suggestion; (ii) the response of the control group to suggestion; (iii) the difference between the responses of the experimental and control groups. Use directional alternatives (suggestion increases ventilation, and hypnotic suggestion increases it more than waking suggestion) and let $\alpha = .05$ for each test.
(c) Repeat the investigations of part (b) using suitable nonparametric tests (sign and Wilcoxon-Mann-Whitney tests).

(d) Use suitable graphs to investigate the reasonableness of the normality condition underlying the *t* tests of part (b). How does this investigation shed light on the discrepancies between the results of parts (b) and (c)?

9.55 Suppose we want to test whether an experimental drug reduces blood pressure more than does a placebo. We are planning to administer the drug or the placebo to some subjects and record how much their blood pressures are reduced. We have 20 subjects available.

(a) We could form 10 matched pairs, where we form a pair by matching subjects, as best we can, on the basis of age and sex, and then randomly assign one subject in each pair to the drug and the other subject in the pair to the placebo. Explain why using a matched-pairs design might be a good idea.

(b) Briefly explain why a matched-pairs design might *not* be a good idea. That is, how might such a design be inferior to a completely randomized design?

9.56 A group of 20 postmenopausal women were given transdermal estradiol for one month. Plasma levels of plasminogen-activator inhibitor type 1 (PAI-1) went down for 10 of the women and went up for the other 10 women.[34] Use a sign test to test the null hypothesis that transdermal estradiol has no effect on PAI-1 level. Use $\alpha = .05$ and use a nondirectional alternative.

9.57 Six patients with renal disease underwent plasmapheresis. Urinary protein excretion (grams of protein per gram of creatinine) was measured for each patient before and after plasmapheresis. The data are given in the following table.[35] Use these data to investigate whether or not plasmapheresis affects urinary protein excretion in patients with renal disease. (*Hint*: Graph the data and consider whether a *t* test is appropriate in the original scale.)

Patient	Before	After	Difference
1	20.3	.8	19.5
2	9.3	.1	9.2
3	7.6	3.0	4.6
4	6.1	.6	5.5
5	5.8	.9	4.9
6	4.0	.2	3.8
Mean	8.9	0.9	7.9
SD	5.9	1.1	6.0

CHAPTER 10

Analysis of Categorical Data

10.1 INFERENCE FOR PROPORTIONS: THE CHI-SQUARE GOODNESS-OF-FIT TEST

In Chapter 6 we described methods for constructing confidence intervals (1) when the observed variable is quantitative and (2) when the observed variable is categorical. In Chapters 7 and 9 we considered hypothesis testing with quantitative data. In this chapter we present hypothesis testing for categorical, rather than quantitative, data. Recall from Chapter 2 that with a categorical variable each observation is a category, rather than a number; each observed unit belongs to one and only one category. For instance, human blood type is a categorical variable; each person's type is either A, B, AB, or O.

We begin by considering analysis of a single sample of categorical data. We assume that the data can be regarded as a random sample from some population, and we test a null hypothesis, H_0, that specifies the population proportions, or probabilities, of the various categories. Here is an example.

Snapdragon Colors. In the snapdragon (*Antirrhinum majus*), individual plants can be red flowered, pink flowered, or white flowered. According to a certain Mendelian genetic model, self-pollination of pink-flowered plants should produce progeny that are red, pink, and white in the ratio $1:2:1$. This Mendelian prediction can be formulated as the following null hypothesis:

$$H_0: \Pr\{\text{Red}\} = .25, \Pr\{\text{Pink}\} = .50, \Pr\{\text{White}\} = .25$$

This hypothesis asserts that each progeny plant has a 25% chance of being red, a 50% chance of being pink, and a 25% chance of being white. Equivalently, H_0 asserts that, in the conceptual population of all potential progeny, 25% of the individuals are red, 50% are pink, and 25% are white. ∎

Objectives

In this chapter we study categorical data. We will

- *learn how to conduct a chi-square goodness-of-fit test.*

- *discuss independence and association for categorical variables.*

- *learn how to test for independence between two categorical variables.*

- *consider the conditions under which a chi-square test is valid.*

- *learn how to analyze paired categorical data using McNemar's test.*

- *learn how to calculate relative risk and the odds ratio.*

Example 10.1

The Chi-Square Statistic

Given a random sample of n categorical observations, how can one judge whether they agree with a null hypothesis H_0 that specifies the probabilities of the categories? There are two complementary approaches to this question. First, as a way of describing the data, we can calculate the observed relative frequency of each category and graph the data. The observed frequencies serve as estimates of the probabilities of the categories. The following notation for relative frequencies is useful: When a probability $\Pr\{E\}$ is estimated from observed data, the estimate is denoted by a hat ("^"), thus:

$$\hat{\Pr}\{E\}$$

Example 10.2

Snapdragon Colors. A geneticist, investigating the Mendelian prediction of Example 10.1, self-pollinated pink-flowered snapdragon plants and produced 234 progeny with the following colors:[1]

Red: 54 plants

Pink: 122 plants

White: 58 plants

These data are shown in Figure 10.1.

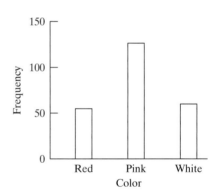

Figure 10.1 Bar chart of snapdragon data

The estimated category probabilities are

$$\hat{\Pr}\{\text{Red}\} = \frac{54}{234} = .231$$

$$\hat{\Pr}\{\text{Pink}\} = \frac{122}{234} = .521$$

$$\hat{\Pr}\{\text{White}\} = \frac{58}{234} = .248$$

These estimated probabilities agree fairly well, but not exactly, with those in the model that are specified by H_0. ∎

The second approach is to use a statistical test, called a **goodness-of-fit test**, to assess the compatibility of the data with H_0. The most widely used goodness-of-fit test is the **chi-square test** or χ^2 test (χ is the Greek letter "chi").

The calculation of the chi-square test statistic is done in terms of the absolute, rather than the relative, frequencies of the categories. For each category, let O represent the **observed frequency** of the category and let E represent the **expected frequency**—that is, the frequency that would be expected according to H_0. The E's are calculated by multiplying each probability specified in H_0 by n, as shown in Example 10.3.

Snapdragon Colors. Consider the null hypothesis specified in Example 10.1 and the data from Example 10.2. If the null hypothesis is true, then we expect 25% of the 234 snapdragons to be red; 25% of 234 is 58.55:

$$\text{Red: } E = (.25)(234) = 58.5$$

The corresponding expected frequencies for pink and white are

$$\text{Pink: } E = (.50)(234) = 117$$
$$\text{White: } E = (.25)(234) = 58.5 \qquad \blacksquare$$

Example 10.3

The test statistic for the chi-square goodness-of-fit test is then calculated from the O's and the E's using the formula given in the accompanying box. Example 10.4 illustrates the calculation of the chi-square statistic.

> **The Chi-Square Statistic**
>
> $$\chi_s^2 = \sum \frac{(O - E)^2}{E}$$

where the summation is over all the categories.

Snapdragon Colors. The observed frequencies of 234 snapdragon colors are as follows:

Example 10.4

Color	Red	Pink	White	Total
Observed	54	122	58	234

The expected frequencies are

Color	Red	Pink	White	Total
Expected	58.5	117	58.5	234

Note that the sum of the expected frequencies is the same as the sum of the observed frequencies (234). The χ^2 statistic is

$$\chi_s^2 = \frac{(54 - 58.5)^2}{58.5} + \frac{(122 - 117)^2}{117} + \frac{(58 - 58.5)^2}{58.5}$$
$$= 0.56 \qquad \blacksquare$$

Computational Notes. The following tips are helpful in calculating a chi-square statistic:

1. The table of observed frequencies must include all categories, so that the sum of the O's is equal to the total number of observations.

2. The O's must be *absolute*, rather than relative, frequencies.

3. It is convenient to add each term $(O - E)^2/E$ to memory after calculating it. If you prefer to write down and reenter the expected frequencies, you may round them to two decimal places.

The χ^2 Distribution

From the way in which χ_s^2 is defined, it is clear that small values of χ_s^2 would indicate that the data agree with H_0, while large values of χ_s^2 would indicate disagreement. In order to base a statistical test on this agreement or disagreement, we need to know how much χ_s^2 may be affected by sampling variation.

We consider the null distribution of χ_s^2—that is, the sampling distribution that χ_s^2 follows if H_0 is true. It can be shown (using the methods of mathematical statistics) that, if the sample size is large enough, then the null distribution of χ_s^2 can be approximated by a distribution known as a χ^2 **distribution**. The form of a χ^2 distribution depends on a parameter called "degrees of freedom" (df). Figure10.2 shows the χ^2 distribution with df = 5.

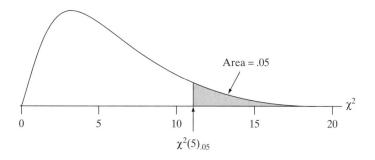

Figure 10.2 The χ^2 distribution with df = 5

Table 9 (at the end of this book) gives critical values for the χ^2 distribution. For instance, for df = 5, the 5% critical value is $\chi^2(5)_{.05} = 11.07$. This critical value corresponds to an area of .05 in the upper tail of the χ^2 distribution, as shown in Figure 10.2.

The Goodness-of-Fit Test

For the chi-square goodness-of-fit test we have presented, the null distribution of χ_s^2 is approximately a χ^2 distribution with*

$$\text{df} = (\text{Number of categories}) - 1$$

For example, for the setting presented in Example 10.1 there are three categories. The null hypothesis specifies the probabilities for each of the three categories. However, once the first two probabilities are specified, the last one is determined, since the three probabilities must sum to 1. There are three categories, but only two of them are "free"; the last one is constrained by the first two.

The test of H_0 is carried out using critical values from Table 9, as illustrated in the following example.

---

* The chi-square test can be extended to more general situations in which parameters are estimated from the data before the expected frequencies are calculated. In general, the degrees of freedom for the test are (number of categories) − (number of parameters estimated) − 1. We are considering only the case in which there are no parameters to be estimated from the data.

Example 10.5

Snapdragon Colors.   For the snapdragon data of Example 10.4, the observed chi-square statistic was $\chi_s^2 = 0.56$. Because there are three color categories, the degrees of freedom for the null distribution are calculated as

$$\text{df} = 3 - 1 = 2$$

From Table 9 with df $= 2$ we find that $\chi^2(2)_{.20} = 3.22$. Since $\chi_s^2 = 0.56$ is less than 3.22, the upper tail area beyond 0.56 is greater than .20. Thus, the P-value is greater than .20 and we would not reject H_0 even at $\alpha = .20$. (Using a computer yields a P-value of .75.) We conclude that the data are reasonably consistent with the Mendelian model. ∎

The chi-square test can be used with any number of categories. In Example 10.6 the test is applied to a variable with six categories.

Example 10.6

Flax Seeds.   Researchers studied a mutant type of flax seed that they hoped would produce oil for use in margarine and shortening. The amount of palmitic acid in the flax seed was an important factor in this research; a related factor was whether the seed was brown or was variegated. The seeds were classified into six combinations of palmitic acid and color, as shown in Table 10.1.[2] According to a hypothesized genetic model, the six combinations should occur in a 3:6:3:1:2:1 ratio.

TABLE 10.1 **Flax Seed Distribution**

Color	Acid level	Observed	Expected
Brown	Low	15	13.5
Brown	Intermediate	26	27
Brown	High	15	13.5
Variegated	Low	0	4.5
Variegated	Intermediate	8	9
Variegated	High	8	4.5
Total		72	72

That is, Brown and Low acid level should occur with probability 3/16, Brown and Intermediate acid level should occur with probability 6/16, and so on. The null hypothesis is that the model is correct; the alternative hypothesis is that the model is incorrect. The χ^2 statistic is

$$\chi_s^2 = \frac{(15 - 13.5)^2}{13.5} + \frac{(26 - 27)^2}{27} + \frac{(15 - 13.5)^2}{13.5}$$
$$+ \frac{(0 - 4.5)^2}{4.5} + \frac{(8 - 9)^2}{9} + \frac{(8 - 4.5)^2}{4.5}$$
$$= 7.71$$

The χ^2 test has $6 - 1 = 5$ degrees of freedom. From Table 9 with df $= 5$ we find that $\chi^2(5)_{.20} = 7.29$ and $\chi^2(5)_{.10} = 9.24$. Thus, the P-value is bracketed as $.10 < P\text{-value} < .20$. If the level of α chosen for the test is .10 or smaller, then the P-value is larger than α and we would not reject H_0. We conclude that the data are reasonably consistent with the Mendelian model. ∎

Note that the critical values for the chi-square test do not depend on the sample size, n. However, the test procedure *is* affected by n, through the value of the chi-square statistic. If we change the size of a sample while keeping its percentage composition fixed, then χ_s^2 varies directly as the sample size, n. For instance, imagine appending a replicate of a sample to the sample itself. Then the expanded sample would have twice as many observations as the original, but they would be in the same relative proportions. The value of each O would be doubled, the value of each E would be doubled, and so the value of χ^2 would be doubled (because in each term of χ_s^2 the numerator $(O - E)^2$ would be multiplied by 4, and the denominator E would be multiplied by 2). That is, the value of χ_s^2 would go up by a factor of 2, despite the fact that the pattern in the data stayed the same! In this way, an increased sample size magnifies any discrepancy between what is observed and what is expected under the null hypothesis.

Compound Hypotheses and Directionality

Let us examine the goodness-of-fit null hypothesis more closely. In a two-sample comparison such as a t test, the null hypothesis contains exactly one assertion—for instance, that two population means are equal. By contrast, a goodness-of-fit null hypothesis can contain more than one assertion. Such a null hypothesis may be called a **compound null hypothesis**. The following is an example.

Example 10.7 | **Snapdragon Colors.** The Mendelian null hypothesis of Example 10.1 is

$$H_0: \Pr\{Red\} = .25, \Pr\{Pink\} = .50, \Pr\{White\} = .25$$

This is a compound hypothesis because it makes two independent assertions, namely

$$\Pr\{Red\} = .25 \quad \text{and} \quad \Pr\{Pink\} = .50$$

Note that the third assertion $(\Pr\{White\} = .25)$ is not an independent assertion because it follows from the other two. ∎

When the null hypothesis is compound, the chi-square test has two special features. First, the alternative hypothesis is necessarily nondirectional. Second, if H_0 is rejected the test does not yield a directional conclusion. The following example illustrates these points.

Example 10.8 | **Snapdragon Colors.** In Example 10.1, the alternative hypothesis (which we did not state explicitly) was as follows:

H_A: At least one of the probabilities specified in H_0 is incorrect or, in other words,

$H_A: \Pr\{Red\} \neq .25$, and/or $\Pr\{Pink\} \neq .50$, and/or $\Pr\{White\} \neq .25$

This alternative hypothesis is nondirectional. (Perhaps "omnidirectional" would be a better term.)

Suppose a geneticist were to obtain the following data on 234 plants:

Red: 34 plants
Pink: 142 plants
White: 58 plants

For these data, $\chi_s^2 = 15.61$; from Table 9 we find that $.0001 < P < .001$. Thus, H_0 would be rejected, even at $\alpha = .001$. The conclusion from this test would be that the

Mendelian prediction does not hold in this situation. However, the test would not yield a directional conclusion such as $\Pr\{Red\} < .25$, $\Pr\{Pink\} > .50$, $\Pr\{White\} < .25$. Indeed, for this particular data set the observed relative frequency of white plants is $\dfrac{58}{234} = .248$, which (you can see intuitively) provides little or no evidence that $\Pr\{White\} < .25$. ∎

When H_0 is compound, the chi-square test is nondirectional in nature because the chi-square statistic measures deviations from H_0 in all directions. Statistical methods are available that do yield directional conclusions and that can handle directional alternatives, but such methods are beyond the scope of this book.

Dichotomous Variables

If the categorical variable analyzed by a goodness-of-fit test is dichotomous, then the null hypothesis is not compound, and directional alternatives and directional conclusions do not pose any particular difficulty.*

Directional Conclusion. The following example illustrates the directional conclusion.

Sexes of Birds. In an ecological study of the Carolina Junco, 53 birds were captured from a certain population; of these, 40 were male.[3] These data are shown in Figure 10.3.

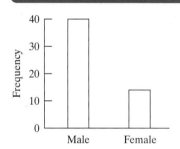

Example 10.9

Is this evidence that males outnumber females in the population? An appropriate null hypothesis is

H_0: Population is 50% male and 50% female.

Equivalently, H_0 can be restated in terms of the probability that a randomly chosen bird will be male or female:

$$H_0: \Pr\{Male\} = .50, \Pr\{Female\} = .50$$

This hypothesis is not compound because it contains only one independent assertion. (Note that the second assertion—$\Pr\{Female\} = .50$—is redundant; it follows from the first.)

Let us test H_0 against the nondirectional alternative

$$H_A: \Pr\{Male\} \neq .50$$

The observed and expected frequencies are shown in Table 10.2.

Figure 10.3 Bar chart of Carolina Junco data

TABLE 10.2 Carolina Junco Data

	Male	Female	Total
Observed	40	13	53
Expected	26.5	26.5	53

* When the data are dichotomous, there is an alternative to the goodness-of-fit test that is known as the Z test for a single proportion. The calculations used in the Z test look quite different from those of the goodness-of-fit test but, in fact, the two tests are mathematically equivalent. However, unlike the goodness-of-fit test, which can handle any number of categories, the Z test can only be used when the data are limited to only two categories. Thus, we do not present it here.

The data yield $\chi_s^2 = 13.8$, and from Table 9 we find that $.0001 < P < .001$. Even at $\alpha = .001$ we would reject H_0 and find that there is sufficient evidence to conclude that the population contains more males than females. ∎

To recapitulate, the directional conclusion in Example 10.9 is legitimate because we know that if H_0 is false, then necessarily either $\Pr\{\text{Male}\} < .5$ or $\Pr\{\text{Male}\} > .5$. By contrast, in Example 10.8 H_0 may be false but $\Pr\{\text{White}\}$ may still be equal to $.25$; the chi-square analysis does not determine which of the probabilities are not as specified by H_0.

Directional Alternative. A chi-square goodness-of-fit test against a directional alternative (when the observed variable is dichotomous) uses the familiar two-step procedure:

Step 1. Check directionality (see if the data deviate from H_0 in the direction specified by H_A).

   (a) If not, the *P*-value is greater than .50.

   (b) If so, proceed to step 2.

Step 2. The *P*-value is half what it would be if H_A were nondirectional.

The following example illustrates the procedure.

Example 10.10

Harvest Moon Festival. Can people who are close to death postpone dying until after a symbolically meaningful occasion? Researchers studied death from natural causes among elderly Chinese women (over age 75) living in California. They chose to study the time around the Harvest Moon Festival because (1) the date of the traditional Chinese festival changes somewhat from year to year, making it less likely that a time-of-year effect would be confounded with the effect they were studying; and (2) it is a festival in which the role of the oldest woman in the family is very important.

Previous research had suggested that there might be a decrease in the mortality rate among elderly Chinese women immediately prior to the festival, with a corresponding increase afterward. The researchers found that over a period of several years there were 33 deaths in the group in the week preceding the Harvest Moon Festival and 70 deaths in the week following the festival.[4] How strongly does this support the interpretation that people can prolong life until a symbolically meaningful event?

We may formulate null and alternative hypotheses as follows:

H_0: Given that an elderly Chinese woman dies within one week of the Harvest Moon Festival, she is equally likely to die before the festival or after the festival.

H_A: Given that an elderly Chinese woman dies within one week of the Harvest Moon Festival, she is more likely to die after the festival than before the festival.

These hypotheses can be translated as

$$H_0: \Pr\{\text{die after festival}\} = \frac{1}{2}$$

$$H_A: \Pr\{\text{die after festival}\} > \frac{1}{2}$$

where it is understood that Pr{die after festival} is the probability of death after the festival, given that the woman dies within one week before or after the festival. The observed and expected frequencies are shown in Table 10.3.

TABLE 10.3 **Harvest Moon Festival Data**			
	Before	**After**	**Total**
Observed	33	70	103
Expected	51.5	51.5	103

From the data on the 103 deaths, we first note that the data do, indeed, deviate from H_0 in the direction specified by H_A, because the observed relative frequency of deaths after the festival is 70/103, which is greater than 1/2. The value of the chi-square statistic is $\chi_s^2 = 13.3$; from Table 9 we see that the P-value would have been bracketed between .0001 and .001 had H_A been nondirectional. However, for the directional alternative hypothesis specified in this test, we bracket the P-value as .00005 < P-value < .0005. We conclude that the evidence is very strong that the death rate among elderly Chinese women goes up after the festival. ∎

Exercises 10.1–10.12

10.1 A cross between white and yellow summer squash gave progeny of the following colors:[5]

Color	White	Yellow	Green
Number of progeny	155	40	10

Are these data consistent with the 12:3:1 ratio predicted by a certain genetic model? Use a chi square test at $\alpha = .10$.

10.2 Refer to Exercise 10 1. Suppose the sample had the same composition but was 10 times as large: 1,550 white, 400 yellow, and 100 green progeny. Would the data be consistent with the 12:3:1 model?

10.3 How do bees recognize flowers? As part of a study of this question, researchers used the following two artificial "flowers":[6]

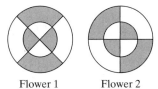

Flower 1   Flower 2

The experiment was conducted as a series of trials on individual bees; each trial consisted of presenting a bee with both flowers and observing which flower it landed on first. (Flower 1 was sometimes on the left, and sometimes on the right.) During the "training" trials, flower 1 contained a sucrose solution and flower 2 did not; thus, the bee was trained to prefer flower 1. During the testing trials, neither flower contained sucrose. In 25 testing trials with a particular bee, the bee chose flower 1 twenty times and flower 2 five times.

(a) Use a goodness-of-fit test to assess the evidence that the bee could remember and distinguish the flower patterns. Use a directional alternative and let $\alpha = .05$.

(b) State your conclusion from part (a) in the context of this setting.

10.4 At a midwestern hospital there were a total of 932 births in 20 consecutive weeks. Of these births, 216 occurred on weekends.[7] Do these data reveal more than chance deviation from random timing of the births? (Test for goodness of fit, with two categories of births: weekday and weekend. Use a nondirectional alternative and let $\alpha = .05$.)

10.5 In a breeding experiment, white chickens with small combs were mated and produced 190 offspring, of the types shown in the accompanying table.[8] Are these data consistent with the Mendelian expected ratios of $9:3:3:1$ for the four types? Use a chi-square test at $\alpha = .10$.

Type	Number of Offspring
White feathers, small comb	111
White feathers, large comb	37
Dark feathers, small comb	34
Dark feathers, large comb	8
Total	190

10.6 Among n babies born in a certain city, 51% were boys.[9] Suppose we want to test the hypothesis that the true probability of a boy is $\frac{1}{2}$. Calculate the value of χ_s^2, and bracket the P-value for testing against a nondirectional alternative, if

(a) $n = 1,000$
(b) $n = 5,000$
(c) $n = 10,000$

10.7 In an agronomy experiment peanuts with shriveled seeds were crossed with normal peanuts. The genetic model that the agronomists were considering predicted that the ratio of normal to shriveled progeny would be $3:1$. They obtained 95 normal and 54 shriveled progeny.[10] Do these data support the hypothesized model?

(a) Conduct a chi-square test with $\alpha = .05$. Use a nondirectional alternative.
(b) State your conclusion from part (a) in the context of this setting.

10.8 An experimental design using litter-matching was employed to test a certain drug for cancer-causing potential. From each of 50 litters of rats, three females were selected; one of these three, chosen at random, received the test drug, and the other two were kept as controls. During a 2-year observation period, the time of occurrence of a tumor, and/or death from various causes, was recorded for each animal. One way to analyze the data is to note simply which rat (in each triplet) developed a tumor first. Some triplets were uninformative on this point because either (a) none of the three littermates developed a tumor, or (b) a rat developed a tumor after its littermate had died from some other cause. The results for the 50 triplets are shown in the table.[11] Use a goodness-of-fit test to evaluate the evidence that the drug causes cancer. Use a directional alternative and let $\alpha = .01$. State your conclusion from part (a) in the context of this setting. (*Hint*: Use only the 20 triplets that provide complete information.)

	Number of Triplets
Tumor first in the treated rat	12
Tumor first in one of the two control rats	8
No tumor	23
Death from another cause	7
Total	50

10.9 A study of color vision in squirrels used an apparatus containing three small translucent panels that could be separately illuminated. The animals were trained to choose, by pressing a lever, the panel that appeared different from the other two. (During these "training" trials, the panels differed in brightness, rather than color.) Then the animals were tested for their ability to discriminate between various colors. In one series of "testing" trials on a single animal, one of the panels was red and the other two were white; the location of the red panel was varied randomly from trial to trial. In 75 trials, the animal chose correctly 45 times and incorrectly 30 times.[12] How strongly does this support the interpretation that the animal can discriminate between the two colors?

(a) Test the null hypothesis that the animal cannot discriminate red from white. Use a directional alternative and let $\alpha = .02$.
(b) State your conclusion from part (a) in the context of this setting.
(c) Why is a directional alternative appropriate is this case?

10.10 Scientists have used Mongolian gerbils when conducting neurological research. A certain breed of these gerbils were crossed and gave progeny of the following colors:[13]

Color	Black	Brown	White
Number of progeny	40	59	42

Are these data consistent with the $1:2:1$ ratio predicted by a certain genetic model? Use a chi-square test at $\alpha = .05$.

10.11 Each of 36 men was asked to touch the foreheads of three women, one of whom was their romantic partner, while blindfolded. The two "decoy" women were the same age, height, and weight as the man's partner. Of the 36 men tested, 18 were able to correctly identify their partner.[14] Of course, we would expect 12 of the 36 to be correct even if the men were guessing each time. Do the data provide sufficient evidence to conclude that men can do better than they would do by merely guessing?

(a) Conduct an appropriate test.
(b) State your conclusion from part (a) in the context of this setting.

10.12 Geneticists studying the inheritance pattern of cowpea plants classified the plants in one experiment according to the nature of their leaves. The data are shown in the following table:[15]

Type	I	II	III
Number	179	44	23

Test the null hypothesis that the three types occur with probabilities 12/16, 3/16, and 1/16. Use a chi-square test with $\alpha = .10$.

10.2 THE CHI-SQUARE TEST FOR THE 2 × 2 CONTINGENCY TABLE

In Section 10.1 we considered the analysis of a single sample of categorical data. The basic techniques were estimation of category probabilities and comparison of category frequencies with frequencies "expected" according to a null hypothesis. In this section these basic techniques will be extended to more complicated situations. To set the stage, here are two examples.

Example 10.11

Treatment of Angina. Angina pectoris is a chronic heart condition in which the sufferer has periodic attacks of chest pain. In a study to evaluate the effectiveness of the drug Timolol in preventing angina attacks, patients were randomly allocated to receive a daily dosage of either Timolol or placebo for 28 weeks. The numbers of patients who became completely free of angina attacks are shown in Table 10.4.[16]

TABLE 10.4 Response to Angina Treatment

	Treatment	
	Timolol	*Placebo*
Angina free	44	19
Not angina free	116	128
Total	160	147

A natural way to express the results is in terms of percentages, as follows:

Of Timolol patients, $\frac{44}{160}$ or 28% were angina free.

Of placebo patients, $\frac{19}{147}$ or 13% were angina free.

In this study, the angina-free response was more common among the Timolol-treated patients than among the placebo-treated patients—28% versus 13%. Figure 10.4 is a bar chart showing the percentages of angina-free patients for the two groups. ∎

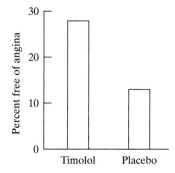

Figure 10.4 Bar chart of angina data

Example 10.12

Mammography. Women who are at high risk for breast cancer are generally advised to have mammograms every year. However, some physicians believe that an annual physical examination is sufficient for early detection of breast cancer. A large, randomized clinical trial was conducted in Canada to determine whether the combination of an annual examination plus mammography is more effective than an annual examination alone in preventing death from breast cancer. Women were enrolled into the study when they were between 50 and 59 years old, were randomly assigned to one of the two treatment groups, and were followed for up to 16 years.[17] The data are shown in Table 10.5.

TABLE 10.5 Mammography Data

		Treatment	
		Exam Plus Mammography	*Examination Only*
Breast cancer death?	**Yes**	107	105
	No	19,604	19,589
	Total	19,711	19,694

Of the women who underwent mammography, the rate of death due to breast cancer was $\dfrac{107}{19711} = .0054$ or 0.54%. Of the women who did not have annual mammography, the rate of death due to breast cancer was $\dfrac{105}{19694} = .0053$ or 0.53%. These two percentages are nearly identical. ∎

Tables such as Tables 10.4 and 10.5 are called **contingency tables**. The focus of interest in a contingency table is the dependence or association between the column variable and the row variable—for instance, between treatment and response in Tables 10.4 and 10.5. (The word *contingent* means "dependent.") In particular, Tables 10.4 and 10.5 are called **2 × 2** ("two-by-two") **contingency tables**, because they consist of two rows (excluding the "total" row) and two columns. Each category in the contingency table is called a **cell**; thus, a 2 × 2 contingency table has four cells.

In Chapter 6 we considered the estimate $\hat{p}$ of a population proportion, or probability, p. In analyzing a 2 × 2 contingency table, it is natural to think in terms of two probabilities, p_1 and p_2, which are to be compared. One form of comparison is through a statistical test of the null hypothesis that the probabilities are equal:

$$H_0: p_1 = p_2$$

The following example illustrates this null hypothesis.

Treatment of Angina.   For the angina study of Example 10.11, the probabilities of interest are

Example 10.13

p_1 = Probability that a patient will become angina free if given Timolol

p_2 = Probability that a patient will become angina free if given a placebo

The estimated probabilities from the data are

$$\hat{p}_1 = \frac{44}{160} = .28$$

$$\hat{p}_2 = \frac{19}{147} = .13$$

The null hypothesis asserts that the corresponding true (population) probabilities are equal. ∎

The Chi-Square Statistic

Clearly, a natural way to test the preceding null hypothesis would be to reject H_0 if $\hat{p}_1$ and $\hat{p}_2$ are different by a sufficient amount. We describe a test procedure that compares $\hat{p}_1$ and $\hat{p}_2$ indirectly, rather than directly. The procedure is a chi-square test, based on the test statistic χ_s^2 that was introduced in Section 10.1:

$$\chi_s^2 = \sum \frac{(O - E)^2}{E}$$

In the formula, the sum is taken over all four cells in the contingency table. Each O represents an observed frequency and each E represents the corresponding expected frequency according to H_0. We now describe how to calculate the E's.

The first step in determining the E's for a contingency table is to calculate the row and column total frequencies (these are called the **marginal frequencies**) and the grand total of all the cell frequencies. The E's then follow from a simple rationale, as illustrated in Example 10.14.

| **Example 10.14** | **Treatment of Angina.** Table 10.6 shows the angina data of Example 10.11, together with the marginal frequencies. |

TABLE 10.6 Observed Frequencies for Angina Study

	Treatment		
	Timolol	*Placebo*	*Total*
Angina free	44	19	63
Not angina free	116	128	244
Total	160	147	307

The E's should agree exactly with the null hypothesis. Because H_0 asserts that the probability of an angina-free response does not depend on the treatment, we can generate an estimate of this probability by pooling the two treatment groups; from Table 10.6, the pooled estimate is $\frac{63}{307}$. That is, if H_0 is true, then the two columns "Timolol" and "Placebo" are equivalent and we can pool them together. Our best estimate of Pr{a patient will become angina free} is then the pooled estimate $\frac{63}{307}$. We can then apply this estimate to each treatment group to yield the number of angina-free patients expected according to H_0, as follows:

$$\text{Timolol group:}\quad \frac{63}{307} \cdot 160 = 32.83 \text{ angina-free patients expected}$$

$$\text{Placebo group:}\quad \frac{63}{307} \cdot 147 = 30.17 \text{ angina-free patients expected}$$

Likewise, the pooled estimate of Pr{a patient will *not* become angina free} is $\frac{244}{307}$. Applying this probability to the two treatment groups gives

$$\text{Timolol group:}\quad \frac{244}{307} \cdot 160 = 127.17 \text{ not-angina-free patients expected}$$

Placebo group: $\dfrac{244}{307} \cdot 147 = 116.83$ not-angina-free patients expected

The expected frequencies are shown in parentheses in Table 10.7. Note that the marginal totals for the E's are the same as for the O's.

TABLE 10.7 Observed and Expected Frequencies for Angina Study

	Treatment		
	Timolol	*Placebo*	*Total*
Angina free	44(32.83)	19(30.17)	63
Not angina free	116(127.17)	128(116.83)	244
Total	160	147	307

In practice, it is not necessary to proceed through a chain of reasoning to obtain the expected frequencies for a contingency table. The procedure for calculating the E's can be condensed into a simple formula. The expected frequency for each cell is calculated from the marginal total frequencies for the same row and column, as follows:

Expected Frequencies in a Contingency Table

$$E = \frac{(\text{Row total}) \cdot (\text{Column total})}{\text{Grand total}}$$

The formula produces the same calculation as does the rationale given in Example 10.14, as the following example shows.

Treatment of Angina. We will apply the above formula to the angina data of Example 10.11. The expected frequency of angina-free Timolol patients is calculated from the marginal totals as

$$E = \frac{(63)(160)}{307} = 32.83$$

Note that this is the same answer obtained in Example 10.14. Proceeding similarly for each cell in the contingency table, we would obtain all the E's shown in Table 10.7. ■

Example 10.15

Note that, although the formula for χ_s^2 for contingency tables is the same as given for goodness-of-fit tests in Section 10.1, the method of calculating the E's is quite different for contingency tables because the null hypothesis is different.

The Test Procedure

The chi-square test for a contingency table is carried out similarly to the chi-square goodness-of-fit test. Large values of χ_s^2 indicate evidence against H_0. Critical values are determined from Table 9; the number of degrees of freedom for a 2 × 2 contingency table is

$$\text{df} = 1$$

The chi-square test for a 2×2 table has 1 degree of freedom because, in a sense, there only is one free cell in the table. Table 10.7 has four cells, but once we have determined that the expected cell frequency for the top-left cell is 32.83, the expected frequency for top-right cell is constrained to be 30.17, since the top row adds across to a total of 63. Likewise, the bottom-left cell is constrained to be 127.17, since the left column adds down to a total of 160. Once these three cells are determined, the remaining cell, on the bottom right, is constrained as well. Thus, there are four cells in the table, but only one of them is "free"; once we have used the null hypothesis to determine the expected frequency for one of the cells, the other cells are constrained.

For a 2×2 contingency table, the alternative hypothesis can be directional or nondirectional. Directional alternatives are handled by the familiar two-step procedure, cutting the P-value in half if the data deviate from H_0 in the direction specified by H_A. Note that χ_s^2 itself does not express directionality; to determine the directionality of the data, we must calculate and compare the estimated probabilities.

The following example illustrates the chi-square test.

Example 10.16 | **Treatment of Angina.** For the angina experiment of Example 10.11, let us apply a chi-square test at $\alpha = .01$. We may state the null hypothesis and a directional alternative informally as follows:

H_0: Timolol is no better than placebo for preventing angina.

H_A: Timolol is better than placebo for preventing angina.

Symbolically, the statements are

$$H_0: p_1 = p_2$$
$$H_A: p_1 > p_2$$

To check the directionality of the data, we calculate the estimated probabilities of response:

$$\hat{p}_1 = \frac{44}{160} = .28$$

$$\hat{p}_2 = \frac{19}{147} = .13$$

and we note that

$$\hat{p}_1 > \hat{p}_2$$

Thus, the data do deviate from H_0 in the direction specified by H_A. We proceed to calculate the chi-square statistic from Table 10.7 as

$$\chi_s^2 = \frac{(44 - 32.83)^2}{32.83} + \frac{(116 - 127.17)^2}{127.17} + \frac{(19 - 30.17)^2}{30.17} + \frac{(128 - 116.83)^2}{116.83}$$

$$= 10.0$$

From Table 9 with df $= 1$, we find that $\chi^2(1)_{.01} = 6.63$ and $\chi^2(1)_{.001} = 10.83$, and so we have $.0005 < P < .005$. Thus, we reject H_0 and find that the data provide sufficient evidence to conclude that Timolol is better than placebo for producing an angina-free response. ■

Note that, even though $\hat{p}_1$ and $\hat{p}_2$ do not enter into the calculation of χ_s^2, *the calculation of $\hat{p}_1$ and $\hat{p}_2$ is an important part of the test procedure*; the information provided by the quantities $\hat{p}_1$ and $\hat{p}_2$ is essential for meaningful interpretation of the results.*

Computational Notes The following tips are helpful in analyzing a 2 × 2 contingency table:

1. The contingency table format is convenient for computations. For presenting the data in a report, however, it is usually better to use a more readable form of display; some examples are shown in the Exercises.

2. For calculating χ_s^2, the observed frequencies (O's) must be *absolute*, rather than relative, frequencies; also, *the table must contain all four cells*, so that the sum of the O's is equal to the total number of observations.

Illustration of the Null Hypothesis

The chi-square statistic measures discrepancy between the data and the null hypothesis in an indirect way; the quantities $\hat{p}_1$ and $\hat{p}_2$ are involved indirectly in the calculation of the expected frequencies. If $\hat{p}_1$ and $\hat{p}_2$ are equal, then the value of χ_s^2 is zero. Here is an example.

Table 10.8 shows fictitious data for an angina study similar to that described in Example 10.11.

Example 10.17

TABLE 10.8 **Fictitious Data for Angina Study**

	Treatment		
	Timolol	*Placebo*	*Total*
Angina free	30	20	50
Not angina free	120	80	200
Total	150	100	250

For the data of Table 10.8, the estimated probabilities of an angina-free response are *equal*:

$$\hat{p}_1 = \frac{30}{150} = .20$$

$$\hat{p}_2 = \frac{20}{100} = .20$$

You can easily verify that, for Table 10.8, the expected frequencies are equal to observed frequencies, so that the value of χ_s^2 is zero. Also notice that the columns of the table are proportional to each other:

$$\frac{30}{120} = \frac{20}{80}$$

■

---

* It is natural to wonder why we do not use a more direct comparison of $\hat{p}_1$ and $\hat{p}_2$. In fact, there is a test procedure based on a *t*-type statistic, calculated by dividing $(\hat{p}_1 - \hat{p}_2)$ by its standard error. This *t*-type procedure is equivalent to the chi-square test. We have chosen to present the chi-square test instead, for two reasons: (1) It can be extended to contingency tables larger than 2 × 2; (2) in certain applications the chi-square statistic is more natural than the *t*-type statistic; some of these applications appear in Section 10.3.

As the preceding example suggests, an "eyeball" analysis of a contingency table is based on checking for proportionality of the columns. If the columns are nearly proportional, then the data agree fairly well with H_0; if they are highly non-proportional, then the data disagree with H_0. The following example shows a case in which the data agree quite well with the expected frequencies under H_0.

Example 10.18

Mammography. The data from Example 10.12 show very similar percentages of breast cancer deaths for women who have annual physical examinations plus mammograms and women who only have annual physical examinations. The natural null hypothesis is that $p_1 = p_2$ and that the sample proportions differ only due to chance error in the sampling process. The expected frequencies are shown in parentheses in Table 10.9. The chi-square test statistic is $\chi_s^2 = 0.017$. From Table 9 with df $= 1$, we find that $\chi^2(1)_{.20} = 1.64$. Thus, the P-value is greater than .20 (using a computer yields $P = .90$) and we do not reject the null hypothesis. Our conclusion is that the data are consistent with the claim that the addition of mammography to annual physical examinations has no effect on breast cancer mortality for women age 50–59.

TABLE 10.9 Observed and Expected Frequencies for Mammography Study

		Treatment		
		Exam Plus Mammography	Examination Only	
Breast cancer death?	Yes	107(106.05)	105(105.95)	212
	No	19,604(19,604.95)	19,589(19,588.05)	39,193
	Total	19,711	19,694	39,405

Note that the actual value of χ_s^2 depends on the sample sizes as well as the degree of nonproportionality; as discussed in Section 10.1, the value of χ_s^2 varies directly with the number of observations if the percentage composition of the data is kept fixed and the number of observations is varied. This reflects the fact that a given percentage deviation from H_0 is less likely to occur by chance with a larger number of observations.

Exercises 10.13–10.26

10.13 The accompanying partially complete contingency table shows the responses to two treatments:

		Treatment	
		1	2
Response	Success	70	
	Failure		
	Total	100	200

(a) Invent a fictitious data set that agrees with the table and for which $\chi_s^2 = 0$.
(b) Calculate the estimated probabilities of success ($\hat{p}_1$ and $\hat{p}_2$) for your data set. Are they equal?

10.14 Proceed as in Exercise 10.13 for the following contingency table:

		Treatment	
		1	*2*
Response	Success	30	
	Failure		
	Total	300	100

10.15 Proceed as in Exercise 10.13 for the following contingency table:

		Treatment	
		1	*2*
Response	Success	5	20
	Failure	10	

10.16 Proceed as in Exercise 10.13 for the following contingency table:

		Treatment	
		1	*2*
Response	Success	20	10
	Failure	80	

10.17 Most salamanders of the species *P. cinereus* are red striped, but some individuals are all red. The all-red form is thought to be a mimic of the salamander *N. viridescens*, which is toxic to birds. In order to test whether the mimic form actually survives more successfully, 163 striped and 41 red individuals of *P. cinereus* were exposed to predation by a natural bird population. After two hours, 65 of the striped and 23 of the red individuals were still alive.[18] Use a chi-square test to assess the evidence that the mimic form survives more successfully. Use a directional alternative and let $\alpha = .05$.

(a) State the null hypothesis in words.
(b) State the null hypothesis in symbols.
(c) Find the value of the test statistic and the *P*-value.
(d) State the conclusion of the test in the context of this setting.

10.18 Can attack of a plant by one organism induce resistance to subsequent attack by a different organism? In a study of this question, individually potted cotton *(Gossypium)* plants were randomly allocated to two groups. Each plant in one group received an infestation of spider mites *(Tetranychus)*; the other group was kept as controls. After two weeks the mites were removed and all plants were inoculated with *Verticillium*, a fungus that causes wilt disease. The accompanying table shows the numbers of plants that developed symptoms of wilt disease.[19] Do the data provide sufficient evidence to conclude that infestation with mites induces resistance to wilt disease? Use a chi-square test against a directional alternative. Let $\alpha = .01$.

		Mites	No mites
Response	Wilt disease	11	17
	No wilt disease	15	4
	Total	26	21

10.19 It has been suspected that prolonged use of a cellular telephone increases the chance of developing brain cancer, due to the microwave-frequency signal that is transmitted by the cell phone. According to this theory, if a cell phone is repeatedly held near one side of the head, then brain tumors are more likely to develop on that side of the head. To investigate this, a group of patients were studied who had used cell phones for a least six months prior to developing brain tumors. The patients were asked whether they routinely held the cell phone to a certain ear and, if so, which ear. The 88 responses (from those who preferred one side over the other) are shown in the following table.[20] Do the data provide sufficient evidence to conclude that use of cellular tellephones leads to an increase in brain tumors on that side of the head? Use a chi-square test against a directional alternative. Let $\alpha = .05$. (*Hint*: Be sure to calculate the two sample proportions.)

		Phone holding side	
		Left	*Right*
Brain tumor side	Left	14	28
	Right	19	27
	Total	33	55

10.20 Phenytoin is a standard anticonvulsant drug that unfortunately has many toxic side effects. A study was undertaken to compare phenytoin with valproate, another drug in the treatment of epilepsy. Patients were randomly allocated to receive either phenytoin or valproate for 12 months. Of 20 patients receiving valproate, 6 were free of seizures for the 12 months while 6 of 17 patients receiving phenytoin were seizure free.[21]

(a) Use a chi-square test to compare the seizure-free response rates for the two drugs. Let H_A be nondirectional and $\alpha = .10$.

(b) Does the test in part (a) provide evidence that valproate and phenytoin are equally effective in preventing seizures? Discuss.

10.21 Estrus synchronization products are used to bring cows into heat at a predictable time so that they can be reliably impregnated by artificial insemination. In a study of two estrus synchronization products, 42 mature cows (aged 4–8 years) were randomly allocated to receive either product A or product B, and then all cows were bred by artificial insemination. The table shows how many of the inseminations resulted in pregnancy.[22] Use a chi-square test to compare the effectiveness of the two products in producing pregnancy. Use a nondirectional alternative and let $\alpha = .05$.

(a) State the null hypothesis in words.
(b) State the null hypothesis in symbols.
(c) Find the value of the test statistic and the *P*-value.
(d) State the conclusion of the test in the context of this setting.

	Treatment	
	Product A	*Product B*
Total number of cows	21	21
Number of cows pregnant	8	15

10.22 Experimental studies of cancer often use strains of animals that have a naturally high incidence of tumors. In one such experiment, tumor-prone mice were kept in a sterile environment; one group of mice were maintained entirely germ free, while another group were exposed to the intestinal bacterium *Eschericbia coli*. The accompanying table shows the incidence of liver tumors.[23]

Treatment	Total number of mice	Mice with liver tumors	
		Number	*Percent*
Germ free	49	19	39%
E. Coli	13	8	62%

(a) How strong is the evidence that tumor incidence is higher in mice exposed to *E. coli?* Use a chi-square test against a directional alternative. Let $\alpha = .05$.
(b) How would the result of part (a) change if the percentages (39% and 62%) of mice with tumors were the same, but the sample sizes were (i) doubled (98 and 26)? (ii) tripled (147 and 39)? [*Hint*: Part (b) requires almost no calculation.]

10.23 In a randomized clinical trial to determine the most effective timing of administration of chemotherapeutic drugs to lung cancer patients, 16 patients were given four drugs simultaneously and 11 patients were given the same drugs sequentially. Objective response to the treatment (defined as shrinkage of the tumor by at least 50%) was observed in 11 of the patients treated simultaneously and in 3 of the patients treated sequentially.[24] Do the data provide evidence as to which timing is superior? Use a chi-square test against a nondirectional alternative. Let $\alpha = .05$.

10.24 Physicians conducted an experiment to investigate the effectiveness of external hip protectors in preventing hip fractures in elderly people. They randomly assigned some people to get hip protectors and others to be the control group. They recorded the number of hip fractures in each group.[25] Do the data in the following table provide sufficient evidence to conclude that hip protectors reduce the likelihood of fracture? Use a chi-square test against a directional alternative. Let $\alpha = .01$.

		Treatment	
		Hip protector	*Control*
Response	Hip fracture	13	67
	No hip fracture	640	1081
	Total	653	1148

10.25 A sample of 276 healthy adult volunteers were asked about the variety of social networks that they were in (e.g., relationships with parents, close neighbors, workmates, etc.). They were then given nasal drops containing a rhinovirus and were quarantined for 5 days. Of the 123 subjects who were in 5 or fewer types of social relationships, 57 (46.3%) developed colds. Of 153 who were in at least 6 types of social relationships, 52 (34.0%) developed colds.[26] Thus, the data suggest that having more types of social relationships helps one develop resistance to the common cold. Determine whether this difference is statistically significant. That is, use a chi-square test to test the null hypothesis that the probability of getting a cold does not depend on the number of social relationships a person is in. Use a nondirectional alternative and let $\alpha = .05$.

10.26 The drug ancrod was tested in a double-blind clinical trial in which subjects who had strokes were randomly assigned to get either ancrod or a placebo. One response variable in the study was whether or not a subject experienced intracranial hemorrhaging.[27] The data are provided in the following table. Use a chi-square test to determine whether the difference in hemorrhaging rates is statistically significant. Use a nondirectional alternative and let $\alpha = .05$.

		Treatment	
		Ancrod	*Placebo*
Hemorrhage?	Yes	13	5
	No	235	247
	Total	248	252

10.3 INDEPENDENCE AND ASSOCIATION IN THE 2 × 2 CONTINGENCY TABLE

The 2 × 2 contingency table is deceptively simple. In this section we explore further the relationships that it can express.

Two Contexts for Contingency Tables

A 2 × 2 contingency table can arise in two contexts, namely

1. Two independent samples with a dichotomous observed variable
2. One sample with two dichotomous observed variables

The first context is illustrated by the angina data of Example 10.11, which can be viewed as two independent samples—the Timolol group and the placebo group—of sizes $n_1 = 160$ and $n_2 = 147$. The observed variable is angina-free status. Any study involving a dichotomous observed variable and completely randomized allocation to two treatments can be viewed this way. The second context is illustrated by the following example.

Example 10.19 **HIV Testing.** A random sample of 120 college students found that 9 of the 61 women in the sample had taken an HIV test, compared to 8 of the 59 men.[28] These data are shown in Table 10.10.

TABLE 10.10 HIV Testing Data		
	Female	**Male**
HIV test	9	8
No HIV test	52	51
Total	61	59

Of the women, $\frac{9}{61}$ or 14.8% had been tested for HIV. Of the men, $\frac{8}{59}$ or 13.6% had been tested for HIV. ∎

Example 10.19 can be viewed as a single sample of $n = 120$ students, observed with respect to two dichotomous variables—sex (male or female) and HIV test status (whether or not the person had been tested for HIV).

The two contexts—two samples with one variable or one sample with two variables—are not always sharply differentiated. For instance, the HIV data of Example 10.19 could have been collected in two samples—61 women and 59 men—observed with respect to one dichotomous variable (HIV test status).

The arithmetic of the chi-square test is the same in both contexts, but the statement and interpretation of hypotheses and conclusions can be very different. To describe relationships in the second context, it is useful to extend the language of probability to include a new concept: conditional probability.*

Conditional Probability

Recall that the probability of an event predicts how often the event will occur. A **conditional probability** predicts how often an event will occur under specified conditions. The notation for a conditional probability is

$$\Pr\{E|C\}$$

which is read "probability of E, given C." When a conditional probability is estimated from observed data, the estimate is denoted by a hat ("^") thus:

$$\hat{\Pr}\{E|C\}$$

The following example illustrates these ideas.

HIV Testing. Suitable conditional probabilities to describe the relation between sex and HIV test status (Example 10.19) would be as follows:

| | **Example 10.20** |

$$\Pr\{\text{HIV test}|\text{F}\} = \text{Probability that a person has been}$$
$$\text{tested for HIV, given that the person is female}$$

$$\Pr\{\text{HIV test}|\text{M}\} = \text{Probability that a person has been}$$
$$\text{tested for HIV, given that the person is male}$$

Here HIV test denotes that the person has been tested for HIV and F and M denote female and male. The estimates of these conditional probabilities from the data of Table 10.10 are

$$\hat{\Pr}\{\text{HIV test}|\text{F}\} = \frac{9}{61} = .148$$

and

$$\hat{\Pr}\{\text{HIV test}|\text{M}\} = \frac{8}{59} = .136 \qquad \blacksquare$$

Note that the conditional probability notation is a substitute for the p notation of Section 10.2. For instance, in Example 10.20 we can make the identification

$$p_1 = \Pr\{\text{HIV test}|\text{F}\}$$
$$p_2 = \Pr\{\text{HIV test}|\text{M}\}$$

---

* Conditional probability is also discussed in optional Section 3.5.

It may seem unnecessary to introduce a new and complicated notation when a simpler notation was already available. However, we will find that we sometimes need the greater flexibility of the conditional probability notation.

Independence and Association

In many contingency tables, the columns of the table play a different role than the rows. For instance, in the angina data of Example 10.11, the columns represent treatments and the rows represent responses. Also, in Example 10.20 it seems more natural to define the columnwise conditional probabilities $\Pr\{\text{HIV test}|\text{F}\}$ and $\Pr\{\text{HIV test}|\text{M}\}$ rather than the rowwise conditional probabilities $\Pr\{\text{F}|\text{HIV test}\}$ and $\Pr\{\text{M}|\text{HIV test}\}$.

On the other hand, in some cases it is natural to think of the rows and the columns of the contingency table as playing interchangeable roles. In such a case, conditional probabilities may be calculated either rowwise or columnwise, and the null hypothesis for the chi-square test may be expressed either rowwise or columnwise. The following is an example.

Example 10.21 **Hair Color and Eye Color.** To study the relationship between hair color and eye color in a German population, an anthropologist observed a sample of 6,800 men, with the results shown in Table 10.11.[29]

TABLE 10.11 Hair Color and Eye Color

		Hair color		
		Dark	Light	Total
Eye	Dark	726	131	857
Color	Light	3,129	2,814	5,943
	Total	3,855	2,945	6,800

The data of Table 10.11 would be naturally viewed as a single sample of size $n = 6,800$ with two dichotomous observed variables—hair color and eye color. To describe the data, let us denote dark and light eyes by DE and LE, and dark and light hair by DH and LH. We may calculate estimated columnwise conditional probabilities as follows:

$$\hat{\Pr}\{\text{DE}|\text{DH}\} = \frac{726}{3855} \approx .19$$

$$\hat{\Pr}\{\text{DE}|\text{LH}\} = \frac{131}{2945} \approx .04$$

A natural way to analyze the data is to compare these values: .19 versus .04. On the other hand, it is just as natural to calculate and compare estimated rowwise conditional probabilities:

$$\hat{\Pr}\{\text{DH}|\text{DE}\} = \frac{726}{857} \approx .85$$

$$\hat{\Pr}\{\text{DH}|\text{LE}\} = \frac{3129}{5943} \approx .53$$

Corresponding to these two views of the contingency table, the null hypothesis for the chi-square test can be stated columnwise as

$$H_0: \Pr\{DE|DH\} = \Pr\{DE|LH\}$$

or rowwise as

$$H_0: \Pr\{DH|DE\} = \Pr\{DH|LE\}$$

As we shall see, these two hypotheses are equivalent—that is, any population that satisfies one of them also satisfies the other. ∎

When a data set is viewed as a single sample with two observed variables, the relationship expressed by H_0 is called **statistical independence** of the row variable and the column variable. Variables that are not independent are called **dependent** or **associated**. Thus, the chi-square test is sometimes called a "test of independence" or a "test for association."

Hair Color and Eye Color.   The null hypothesis of Example 10.21 can be stated verbally as

> H_0: Eye color is independent of hair color

or

> H_0: Hair color is independent of eye color

or, more symmetrically,

> H_0: Hair color and eye color are independent ∎

> **Example 10.22**

The null hypothesis of independence can be stated generically as follows. Two groups, G_1 and G_2, are to be compared with respect to the probability of a characteristic C. The null hypothesis is

$$H_0: \Pr\{C|G_1\} = \Pr\{C|G_2\}$$

Note that each of the two statements of H_0 in Example 10.21 is of this form.

To clarify further the meaning of the null hypothesis of independence, in the following example we examine a data set that agrees *exactly* with H_0.

Plant Height and Disease Resistance.   Consider a (fictitious) species of plant that can be categorized as short (S) or tall (T) and as resistant (R) or nonresistant (NR) to a certain disease. Consider the following null hypothesis:

> H_0: Plant height and disease resistance are independent.

Each of the following is a valid statement of H_0:

1. $H_0: \Pr\{R|S\} = \Pr\{R|T\}$
2. $H_0: \Pr\{NR|S\} = \Pr\{NR|T\}$
3. $H_0: \Pr\{S|R\} = \Pr\{S|NR\}$
4. $H_0: \Pr\{T|R\} = \Pr\{T|NR\}$

The following is not a statement of H_0:

5. $\Pr\{R|S\} = \Pr\{NR|S\}$

> **Example 10.23**

Note the difference between (5) and (1). Statement (1) compares two groups (short and tall plants) with respect to disease resistance, whereas (5) is a statement about the distribution of disease resistance in only *one* group (short plants);

statement (5) merely asserts that half (50%) of short plants are resistant and half are nonresistant.

Suppose, now, that we choose a random sample of 100 plants from the population and we obtain the data in Table 10.12.

TABLE 10.12 Plant Height and Disease Resistance

		Height		
		S	T	Total
Resistance	R	12	18	30
	NR	28	42	70
	Total	40	60	100

The data in Table 10,12 agree exactly with H_0; this agreement can be checked in four different ways, corresponding to the four different symbolic statements of H_0:

1. $\hat{\Pr}\{R|S\} = \hat{\Pr}\{R|T\}$

$$\frac{12}{40} = .30 = \frac{18}{60}$$

2. $\hat{\Pr}\{NR|S\} = \hat{\Pr}\{NR|T\}$

$$\frac{28}{40} = .70 = \frac{42}{60}$$

3. $\hat{\Pr}\{S|R\} = \hat{\Pr}\{S|NR\}$

$$\frac{12}{30} = .40 = \frac{28}{70}$$

4. $\hat{\Pr}\{T|R\} = \hat{\Pr}\{T|NR\}$

$$\frac{18}{30} = .60 = \frac{42}{70}$$

Note that the data in Table 10.11 do *not* agree with statement (5):

$$\hat{\Pr}\{R|S\} = \frac{12}{40} = .30 \quad \text{and} \quad \hat{\Pr}\{NR|S\} = \frac{28}{40} = .70$$

$$.30 \neq .70 \qquad \blacksquare$$

Facts about Rows and Columns

The data in Table 10.12 display independence whether viewed rowwise or columnwise. This is no accident, as the following fact shows.

Fact 10.1. The columns of a 2 × 2 table are proportional if and only if the rows are proportional. Specifically, suppose that $a, b, c,$ and d are any positive numbers, arranged as in Table 10.13.

TABLE 10.13 A General 2 × 2 Contingency Table

			Total
	a	b	$a + b$
	c	d	$c + d$
Total	$a + c$	$b + d$	

Then

$$\frac{a}{c} = \frac{b}{d} \quad \text{if and only if} \quad \frac{a}{b} = \frac{c}{d}$$

Another way to express this is

$$\frac{a}{a + c} = \frac{b}{b + d} \quad \text{if and only if} \quad \frac{a}{a + b} = \frac{c}{c + d}$$

You can easily show that Fact 10.1 is true; just use simple algebra. Because of Fact 10.1, the relationship of independence in a 2 × 2 contingency table is the same whether the table is viewed rowwise or columnwise. Note also that the expected frequencies, and therefore the value of χ_s^2, would remain the same if the rows and columns of the contingency table were interchanged. The following fact shows that the *direction* of dependence is also the same whether viewed rowwise or columnwise.

Fact 10.2. Suppose that $a, b, c,$ and d are any positive numbers, arranged as in Table 10.13. Then

$$\frac{a}{a + c} > \frac{b}{b + d} \quad \text{if and only if} \quad \frac{a}{a + b} > \frac{c}{c + d}$$

Also,

$$\frac{a}{a + c} < \frac{b}{b + d} \quad \text{if and only if} \quad \frac{a}{a + b} < \frac{c}{c + d}.$$

Note: For more discussion of conditional probability and independence, see optional Section 3.5.

Verbal Description of Association

Ideas of logical implication are expressed in everyday English in subtle ways. The following excerpt is from *Alice in Wonderland*, by Lewis Carroll:

"... *you should say what you mean,*" *the March Hare went on.*

"*I do,*" *Alice hastily replied;* "*at least—at least I mean what I say—that's the same thing, you know.*"

"*Not the same thing a bit!*" *said the Hatter.* "*Why, you might just as well say that 'I see what I eat' is the same thing as 'I eat what I see'!*"

... "*You might just as well say,*" *added the Dormouse* ..., "*That 'I breathe when I sleep' is the same thing as 'I sleep when I breathe'!*"

"*It is the same thing with you,*" *said the Hatter* ...

We also use ordinary language to express ideas of probability, conditional probability, and association. For instance, consider the following four statements:

Color-blindness is more common among males than among females.

Maleness is more common among color-blind people than femaleness.

Most color-blind people are male.

Most males are color-blind.

The first three statements are all true; are they actually just different ways of saying the same thing? However, the last statement is false.[30]

In interpreting contingency tables, it is often necessary to describe probabilistic relationships in words. This can be quite a challenge. If you become fluent in such description, then you can always "say what you mean" *and* "mean what you say." The following two examples illustrate some of the issues.

Example 10.24

Plant Height and Disease Resistance. For the plant height and disease resistance study of Example 10.23, we considered the null hypothesis

H_0: Height and resistance are independent.

This hypothesis could also be expressed verbally in various other ways, such as

H_0: Short and tall plants are equally likely to be resistant.

H_0: Resistant and nonresistant plants are equally likely to be tall.

H_0: Resistance is equally common among short and tall plants.   ■

Example 10.25

Hair Color and Eye Color. Let us consider the interpretation of Table 10.11. The chi-square statistic is $\chi_s^2 = 314$; from Table 9 we see that the P-value is tiny, so that the null hypothesis of independence is overwhelmingly rejected. We might state our conclusion in various ways. For instance, suppose we focus on the incidence of dark eyes. From the data we found that

$$\hat{\Pr}\{DE|DH\} > \hat{\Pr}\{DE|LH\}$$

that is,

$$\frac{726}{3855} = .19 > \frac{131}{2945} = .04$$

A natural conclusion from this comparison would be

Conclusion 1: There is sufficient evidence to conclude that dark-haired men have a greater tendency to be dark-eyed than do light-haired men.

This statement is carefully phrased, because the statement

"Dark-haired men have a greater tendency to be dark-eyed"

is ambiguous by itself; it could mean

"Dark-haired men have a greater tendency to be dark-eyed than do light-haired men"

or

"Dark-haired men have a greater tendency to be dark-eyed than to be light-eyed"

The first of these statements says that

$$\hat{\Pr}\{DE|DH\} > \hat{\Pr}\{DE|LH\}$$

whereas the second says that

$$\hat{\Pr}\{DE|DH\} > \hat{\Pr}\{LE|DH\}$$

The second statement asserts that more than half of dark-haired men have dark eyes. Note that the data do not support this assertion; of the 3,855 dark-haired men, only 19% have dark eyes.

Conclusion 1 is only one of several possible wordings of the conclusion from the contingency table analysis. For instance, we might focus on dark hair and find

Conclusion 2: There is sufficient evidence to conclude that dark-eyed men have a greater tendency to be dark-haired than do light-eyed men.

A more symmetrical phrasing would be

Conclusion 3: There is sufficient evidence to conclude that dark hair is associated with dark eyes.

However, the phrasing in conclusion 3 is easily misinterpreted; it may suggest something like

"There is sufficient evidence to conclude that most dark-haired men are dark-eyed"

which is not a correct interpretation.   ■

We emphasize once again the principle that we stated in Section 10.2: *The calculation and comparison of appropriate conditional probabilities or $\hat{p}$'s is an essential part of the chi-square test*. Example 10.25 provides ample illustration of this point.

Exercises 10.27–10.39

10.27 Consider a fictitious population of mice. Each animal's coat is either black (B) or gray (G) in color and is either wavy (W) or smooth (S) in texture. Express each of the following relationships in terms of probabilities or conditional probabilities relating to the population of animals.

(a) Smooth coats are more common among black mice than among gray mice.
(b) Smooth coats are more common among black mice than wavy coats are.
(c) Smooth coats are more often black than are wavy coats.
(d) Smooth coats are more often black than gray.
(e) Smooth coats are more common than wavy coats.

10.28 Consider a fictitious population of mice in which each animal's coat is either black (B) or gray (G) in color, and is either wavy (W) or smooth (S) in texture (as in Exercise 10.27). Suppose a random sample of mice is selected from the population and the coat color and texture are observed; consider the accompanying partially complete contingency table for the data.

		Height	
		B	G
Texture	W		50
	S		
	Total	60	150

(a) Invent fictitious data sets that agree with the table and for which
  (i) $\hat{\Pr}\{W|B\} > \hat{\Pr}\{W|G\}$
  (ii) $\hat{\Pr}\{W|B\} = \hat{\Pr}\{W|G\}$
  In each case, verify your answer by calculating the estimated conditional probabilities.
(b) For each of the two data sets you invented in part (a), calculate $\hat{\Pr}\{B|W\}$ and $\hat{\Pr}\{B|S\}$.

(c) Which of the data sets of part (a) has $\hat{\Pr}\{B|W\} > \hat{\Pr}\{B|S\}$? Can you invent a data set for which

$$\hat{\Pr}\{W|B\} > \hat{\Pr}\{W|G\} \text{ but } \hat{\Pr}\{B|W\} < \hat{\Pr}\{B|S\}$$

If so, do it. If not, explain why not.

10.29 A medical team investigated the relation between immunological factors and survival after a heart attack. Blood specimens from 213 male heart-attack patients were tested for presence of antibody to milk protein. The patients were followed to determine whether they lived for 6 months following their heart attack. The results are given in the table.[31]

		Treatment		
		Positive	*Negative*	*Total*
Survival	Died	29	10	39
	Alive	80	94	174
	Total	109	104	213

(a) Let D and A represent died and alive, respectively, and let P and N represent positive and negative antibody tests. Calculate $\Pr\{D|P\}$, $\Pr\{D|N\}$, $\Pr\{P|D\}$, and $\Pr\{P|A\}$.

(b) The value of the contingency-table chi-square statistic for these data is $\chi_s^2 = 10.27$. Test for a relationship between the antibody and survival. Use a nondirectional alternative and let $\alpha = .05$.

10.30 Refer to Exercise 10.29. Is the antibody test a good predictor of survival? To answer this question, imagine trying to predict survival solely on the basis of the antibody test. Use the data to estimate the probability that such a prediction would be correct (that is, the percentage of heart attack patients for whom the prediction would be correct).

10.31 In a study of behavioral asymmetries, 2,391 women were asked which hand they preferred to use (for instance, to write) and which foot they preferred to use (for instance, to kick a ball). The results are reported in the table.[32]

Preferred Hand	**Preferred Foot**	**Number of Women**
Right	Right	2,012
Right	Left	142
Left	Right	121
Left	Left	116
	Total	2,391

(a) Estimate the conditional probability that a woman is right-footed, given that she is right-handed.

(b) Estimate the conditional probability that a woman is right-footed, given that she is left-handed.

(c) Suppose we want to test the null hypothesis that hand preference and foot preference are independent. Calculate the chi-square statistic for this hypothesis.

(d) Suppose we want to test the null hypothesis that right-handed women are equally likely to be right-footed or left-footed. Calculate the chi-square statistic for this hypothesis.

10.32 Consider a study to investigate a certain suspected disease-causing agent. One thousand people are to be chosen at random from the population; each individual is to be classified as diseased or not diseased and as exposed or not exposed to the agent. The results are to be cast in the following contingency table:

		Exposure	
		Yes	*No*
Disease	Yes		
	No		

Let EY and EN denote exposure and nonexposure and let DY and DN denote presence and absence of the disease. Express each of the following statements in terms of conditional probabilities. (Note that "a majority" means "more than half.")

(a) The disease is more common among exposed than among nonexposed people.
(b) Exposure is more common among diseased people than among nondiseased people.
(c) Exposure is more common among diseased people than is nonexposure.
(d) A majority of diseased people are exposed.
(e) A majority of exposed people are diseased.
(f) Exposed people are more likely to be diseased than are nonexposed people.
(g) Exposed people are more likely to be diseased than to be nondiseased.

10.33 Refer to Exercise 10.32. Which of the statements express the assertion that occurrence of the disease is associated with exposure to the agent? (There may be more than one.)

10.34 Refer to Exercise 10.32. Invent fictitious data sets as specified, and verify your answer by calculating appropriate estimated conditional probabilities. (Your data need not be statistically significant.)

(a) Invent a data set for which

$$\hat{Pr}\{DY|EY\} > \hat{Pr}\{DY|EN\} \text{ but } \hat{Pr}\{EY|DY\} < \hat{Pr}\{EY|DN\}$$

or explain why it is not possible.
(b) Invent a data set that agrees with statement (a) of Exercise 10.27 but with neither (d) nor (e); or, explain why it is not possible.
(c) Invent a data set for which

$$\hat{Pr}\{DY|EY\} > \hat{Pr}\{DY|EN\} \text{ but } \hat{Pr}\{EY|DY\} < \hat{Pr}\{EY|DN\}$$

or explain why it is not possible.

10.35 An ecologist studied the spatial distribution of tree species in a wooded area. From a total area of 21 acres, he randomly selected 144 quadrats (plots), each 38 feet square, and noted the presence or absence of maples and hickories in each quadrat. The results are shown in the table.[33]

		Maples	
		Present	*Absent*
Hickories	Present	26	63
	Absent	29	26

The value of the chi-square statistic for this contingency table is $\chi_s^2 = 7.96$. Test the null hypothesis that the two species are distributed independently of each other. Use a nondirectional alternative and let $\alpha = .01$. In stating your conclusion, indicate whether the data suggest attraction between the species or repulsion. Support your interpretation with estimated conditional probabilities from the data.

10.36 Refer to Exercise 10.35. Suppose the data for fictitious tree species, A and B, were as presented in the accompanying table. The value of the chi-square statistic for this contingency table is $\chi_s^2 = 9.07$. As in Exercise 10.35, test the null hypothesis of independence and interpret your conclusion in terms of attraction or repulsion between the species.

		Species A	
		Present	*Absent*
Species B	Present	30	10
	Absent	49	55

10.37 A randomized experiment was conducted in which patients with coronary artery disease either had angioplasty or bypass surgery. The accompanying table shows the incidence of angina (chest pain) among the patients five years after treatment.[34]

		Treatment		
		Angioplasty	*Bypass*	*Total*
Angina?	Yes	111	74	185
	No	402	441	843
	Total	513	515	1,028

Let A represent angioplasty and B represent bypass.

(a) Calculate $\hat{\Pr}\{\text{Yes}|\text{A}\}$ and $\hat{\Pr}\{\text{Yes}|\text{B}\}$.
(b) Calculate $\hat{\Pr}\{\text{A}|\text{Yes}\}$ and $\hat{\Pr}\{\text{A}|\text{No}\}$.

10.38 Refer to Exercise 10.37. Invent a fictitious data set on coronary treatment and angina for 1,000 patients, for which $\hat{\Pr}\{\text{Yes}|\text{A}\}$ is twice as great as $\hat{\Pr}\{\text{Yes}|\text{B}\}$, but nevertheless the majority of patients who have angina also had bypass surgery (as opposed to angioplasty).

10.39 Suppose pairs of fraternal twins are examined and the handedness of each twin is determined; assume that all the twins are brother-sister pairs. Suppose data are collected for 1,000 twin pairs, with the results shown in the following table.[35] State whether each of the following statements is true or false.

(a) Most of the brothers have the same handedness as their sisters.
(b) Most of the sisters have the same handedness as their brothers.
(c) Most of the twin pairs are either both right-handed or both left-handed.
(d) Handedness of twin sister is independent of handedness of twin brother.
(e) Most left-handed sisters have right-handed brothers.

		Sister		
		Left	*Right*	*Total*
Brother	Left	15	85	100
	Right	135	765	900
	Total	150	850	1,000

10.4 FISHER'S EXACT TEST (OPTIONAL)

In this optional section we consider an alternative to the chi-square test for 2×2 contingency tables. This procedure, known as **Fisher's exact test**, is particularly appropriate when dealing with small samples. Example 10.26 presents a situation in which Fisher's exact test can be used.

Example 10.26

ECMO.   Extracorporeal membrane oxygenation (ECMO) is a potentially life-saving procedure that is used to treat newborn babies who suffer from severe respiratory failure. An experiment was conducted in which 29 babies were treated with ECMO and 10 babies were treated with conventional medical therapy (CMT). The data are shown in Table 10.14.[36]

TABLE 10.14 ECMO Experiment Data

		Treatment		
		CMT	ECMO	Total
Outcome	Die	4	1	5
	Live	6	28	34
	Total	10	29	39

The data in Table 10.14 show that 34 of the 39 babies survived but 5 of them died. The death rate was 40% for those given CMT and was 3.4% for those given ECMO. However, the sample sizes here are quite small. Is it possible that the difference in death rates happened simply by chance?

The null hypothesis of interest is that outcome (live or die) is independent of treatment (CMT or ECMO). If the null hypothesis is true, then we can think of the data in the following way: The two column headings of "CMT" and "ECMO" are arbitrary labels. Five of the babies would have died no matter which treatment group they were in; 4 of these babies ended up in the CMT group by chance.

The alternative hypothesis asserts that probability of death depends on treatment group. This means that there is a real difference between CMT and ECMO survival rates, which accounts for the sample percentages being different.

Thus, a question of interest is this: "If the null hypothesis is true, how likely is it to get a table of data like Table 10.14?" In conducting Fisher's exact test, we find the probability that the observed table, Table 10.14, would arise by chance, given that the marginal totals—5 deaths and 34 survivors, 10 given CMT and 29 given ECMO—are fixed. To make this more concrete, suppose the null hypothesis is true and another experiment is conducted, with 10 babies given CMT and 29 given ECMO. Further, suppose that 5 of these 39 babies are going to die, no matter which group they are in. That is, there are 5 babies who are so seriously ill that neither treatment would be able to save them. What is the probability that 4 of them will be assigned to the CMT group?

To find this probability, we need to determine the following:

1. The number of ways of assigning exactly 4 of the 5 babies who are fated to die to the CMT group

2. The number of ways of assigning exactly 6 of the 34 babies who are going to survive to the CMT group

3. The number of ways of assigning 10 of the 39 babies to the CMT group

The product of (1) and (2), divided by (3), gives the probability in question. ■

Combinations

In Section 3.8 we presented the binomial distribution formula. Part of that formula is the quantity $_nC_j$ (which in Section 3.8 we called a binomial coefficient). The quantity $_nC_j$ is the number of ways in which j objects can be chosen out of a set of

n objects. For instance, the number of ways that a group of 4 babies can be chosen out of 5 babies is $_5C_4$. The numerical value of $_nC_j$ is given by formula (10.1):

$$_nC_j = \frac{n!}{j!(n-j)!} \tag{10.1}$$

where $n!$ ("n factorial") is defined for any positive integer as

$$n! = n(n-1)(n-2)\cdots(2)(1)$$

and $0! = 1$.

For example, if $j = 1$, then we have $_nC_1 = \dfrac{n!}{1!(n-1)!} = n$, which makes sense: There are n ways to choose 1 object from a set of n objects. If $j = n$, then we have $_nC_n = \dfrac{n!}{n!0!} = 1$, since there is only one way to choose all n objects from a set of size n.

Example 10.27

ECMO. We can apply formula (10.1) as follows.

1. The number of way of assigning 4 babies to the CMT group from among the 5 who are fated to die is $_5C_4 = \dfrac{5!}{4!1!} = 5$.

2. The number of way of assigning 6 babies to the CMT group from among the 34 who are going to survive is $_{34}C_6 = \dfrac{34!}{6!29!} = 1{,}344{,}904$.

3. The number of way of assigning 10 babies to the CMT group from among the 19 total babies is $_{39}C_{10} = \dfrac{39!}{10!29!} = 635{,}745{,}396$.*

Thus, the probability of getting the same data as those in Table 10.14, given that the marginal totals are fixed, is $\dfrac{_5C_4 \cdot {}_{34}C_6}{_{39}C_{10}} = \dfrac{5 \cdot 1344904}{635745396} = .01058$. ∎

When conducting Fisher's exact test of a null hypothesis against a directional alternative, we need to find the probabilities of all tables of data (having the same margins as the observed table) that provide evidence as strongly against H_0, in the direction predicted by H_A, as the observed table.

Example 10.28

ECMO. Prior to the experiment described in Example 10.26, there was evidence that suggested that ECMO is better than CMT. Hence, a directional alternative hypothesis is appropriate:

$$H_A: \Pr\{\text{death}|\text{ECMO}\} < \Pr\{\text{death}|\text{CMT}\}$$

---

* It is evident from this example that a computer or a graphing calculator is a very handy tool when conducting Fisher's exact test. This is a statistical procedure that is almost never carried out without the use of technology.

The data in the observed table, Table 10.14, support H_A. There is one other possible table, shown as Table 10.15, that has the same margins as Table 10.14 but is even more extreme in supporting H_A. Given that 5 of 39 babies died and that 10 babies were assigned to CMT, the most extreme possible result supporting the alternative hypothesis (that ECMO is better than CMT) is the table in which none of the ECMO babies die and all 5 deaths occur in the CMT group.

TABLE 10.15 A More Extreme Table That Could Have Resulted from the ECMO Experiment

		Treatment		
		CMT	ECMO	Total
Outcome	Die	5	0	5
	Live	5	29	34
	Total	10	29	39

The probability of Table 10.15 occurring, if H_0 is true, is $\dfrac{{}_5C_5 \cdot {}_{34}C_5}{{}_{39}C_{10}} = \dfrac{1 \cdot 278256}{635745396} = .00044$. The P-value is the probability of obtaining data at least as extreme as those observed, if H_0 is true. In this case, the P-value is the probability of obtaining either the data in Table 10.14 or in Table 10.15, if H_0 is true. Thus, $P = .01058 + .00044 = .01102$. This P-value is quite small, so the experiment provided strong evidence that H_0 is false and that ECMO really is better than CMT.

∎

Comparison to the Chi-Square Test

The chi-square test presented in Section 10.2 is often used for analyzing 2×2 contingency tables. One advantage of the chi-square test is that it can be extended to 2×3 tables and other tables of larger dimension, as will be shown in Section 10.6. The P-value for the chi-square test is based on the chi-square distribution, as the name implies. It can be shown that as the sample size becomes large, this distribution provides a good approximation to the theoretical sampling distribution of the chi-square test statistic χ_s^2. If the sample size is small, however, then the approximation can be poor and the P-value from the chi-square test can be misleading.

Fisher's exact test is called an "exact" test because the P-value is determined exactly, using calculations such as those shown in Example 10.27, rather than being based on an asymptotic approximation. Example 10.29 shows how the exact test and the chi-square test compare for the ECMO data.

ECMO. Conducting a chi-square test on the ECMO experiment data in Table 10.14 gives a test statistic of

$$\chi_s^2 = \frac{(4 - 1.28)^2}{1.28} + \frac{(1 - 3.72)^2}{3.72} + \frac{(6 - 8.72)^2}{8.72} + \frac{(28 - 25.28)^2}{25.28}$$

$$= 8.89$$

Example 10.29

The P-value (using a directional alternative) is .0014. This is quite a bit smaller than the P-value found with the exact test of .01102.

∎

Nondirectional Alternatives and the Exact Test

Typically, the difference between a directional and a nondirectional test is that the P-value for the nondirectional test is twice the P-value for the directional test (assuming that the data deviate from H_0 in the direction specified by H_A). For Fisher's exact test this is not true. The P-value when H_A is nondirectional is not found by simply doubling the P-value from the directional test. Rather, a generally accepted procedure is to find the probabilities of all tables that are as likely or less likely than the observed table. These probabilities are added together to get the P-value for the nondirectional test.* Example 10.30 illustrates this idea.

Example 10.30

Flu Shots. A random sample of college students found that 13 of them had gotten a flu shot at the beginning of the winter and 28 had not. Of the 13 who had a flu shot, 3 got the flu during the winter. Of the 28 who did not get a flu shot, 15 got the flu.[37] These data are shown in Table 10.16. Consider the null hypothesis that the probability of getting the flu is the same whether or not one gets a flu shot. The probability of the data in Table 10.16, given that the margins are fixed, is

$$\frac{_{18}C_3 \cdot {}_{23}C_{10}}{_{41}C_{13}} = .05298.$$

Table		Probability
15	3	
13	10	.05298
16	2	
12	11	.01174
17	1	
11	12	.00138
18	0	
10	13	.00006

Figure 10.5

TABLE 10.16 Flu Shot Data

		No Shot	Flu Shot	Total
Flu?	Yes	15	3	18
	No	13	10	23
	Total	28	13	41

A natural directional alternative would be that getting a flu shot reduces one's chance of getting the flu. Figure 10.5 shows the obtained data (from Table 10.16) along with tables of possible outcomes that more strongly support H_A. The probability of each table is given in Figure 10.5, as well.

The P-value for the directional test is the sum of the probabilities of these tables: $P = .05298 + .01174 + .00138 + .00006 = .06616$.

A nondirectional alternative states that the probability of getting the flu depends on whether or not one gets a flu shot, but does not state whether a flu shot increases or decreases the probability. (Some people might get the flu *because* of the shot, so it is plausible that the overall flu rate is higher among people who get the shot than among those who don't—although public health officials certainly hope otherwise!)

Figure 10.6 shows tables of possible outcomes for which the flu rate is higher among those who got the shot than among those who didn't. The probability of each table is given, as well. The first five tables all have probabilities less than .05298, which is the probability of the observed data in Table 10.16, but the

Table		Probability
5	13	
23	0	.00000
6	12	
22	1	.00002
7	11	
21	2	.00046
8	10	
20	3	.00440
9	9	
19	4	.02443
10	8	
18	5	.08356

Figure 10.6

---

* There is not universal agreement on this process. The P-value can be taken to be the sum of the probabilities of all "extreme" tables, but there are several ways to define "extreme." One alternative to the method presented here is to order tables according to the values of χ_s^2 and to count a table as extreme if it has a value of χ_s^2 that is at least as large as the χ_s^2 from the observed table. Another approach is to order the tables according to $|p_1 - p_2|$. These methods will sometimes lead to a different P-value than the P-value being presented here.

probability of the sixth table is greater than .05298. Thus, the contribution to the *P*-value from this set of tables is the sum of the first five probabilities: .00000 + .00002 + .00046 + .00440 + .02443 = .02931. Adding this to the *P*-value for the directional test of .06616 gives the *P*-value for the nondirectional test: $P = .06616 + .02931 = .09547$.

As this example shows, the calculation of a *P*-value for Fisher's exact test is quite cumbersome, particularly when the alternative is nondirectional. It is highly recommended that statistics software be used to carry out the test. ■

Exercises 10.40–10.46

10.40 Consider conducting Fisher's exact test with the following fictitious table of data. Let the null hypothesis be that treatment and response are independent and let the alternative be the directional hypothesis that treatment B is better than treatment A. List the tables of possible outcomes that more strongly support H_A.

		Treatment		
		A	*B*	*Total*
Outcome	Die	4	2	6
	Live	10	14	24
	Total	14	16	30

10.41 Repeat Exercise 10.40 with the following table of data.

		Treatment		
		A	*B*	*Total*
Outcome	Die	5	3	8
	Live	12	13	25
	Total	17	16	33

10.42 In a randomized, double-blind clinical trial, 156 subjects were given an antidepressant medication to help them stop smoking; a second group of 153 subjects were given a placebo. A significantly higher percentage in the antidepressant group quit smoking than in the placebo group. Moreover, the antidepressant group generally had fewer side effects (such as weight gain) than did the placebo group. However, insomnia was more common in the antidepressant group than in the placebo group; Fisher's exact test of the insomnia data gave a *P*-value of .008.[38] Interpret this *P*-value in the context of the clinical trial.

10.43 *(Computer exercise)* A random sample of 99 students in a Conservatory of Music found that 9 of the 48 women sampled had "perfect pitch" (the ability to identify, without error, the pitch of a musical note), but only 1 of the 51 men sampled had perfect pitch.[39] Conduct Fisher's exact test of the null hypothesis that having perfect pitch is independent of sex. Use a directional alternative and let $\alpha = .05$. Do you reject H_0? Why or why not?

10.44 Consider the data from Exercise 10.43. Conduct a chi-square test and compare the results of the chi-square test to the results of Fisher's exact test.

10.45 *(Computer exercise)* The growth factor pleiotrophin is associated with cancer progression in humans. In an attempt to monitor the growth of tumors, doctors measured serum pleiotrophin levels in patients with pancreatic cancer and in a control group of patients. They found that only 2 of 28 control patients had serum levels more than two standard deviations above the control group mean, whereas 20 of 41 cancer patients had serum levels this high.[40] Use Fisher's exact test to determine whether a discrepancy this large (2 of 28 versus 20 of 41) is likely to happen by chance. Use a directional alternative and let $\alpha = .05$.

10.46 *(Computer exercise)* An experiment involving subjects with schizophrenia compared "personal therapy" to "family therapy." Only 2 out of 23 subjects assigned to the personal therapy group suffered psychotic relapses in the first year of the study, compared to 8 of the 24 subjects assigned to the family therapy group.[41] Is this sufficient evidence to conclude, at the .05 level of significance, that the two types of therapies are not equally effective? Conduct Fisher's exact test using a nondirectional alternative. State your conclusion in the context of the problem.

10.5 THE $r \times k$ CONTINGENCY TABLE

The ideas of Sections 10.2 and 10.3 extend readily to contingency tables that are larger than 2×2. We now consider a contingency table with r rows and k columns, which is termed an *$r \times k$ contingency table*. Here is an example.

Example 10.31

Distribution of Blood Type. Table 10.17 shows the observed distribution of ABO blood type in three samples of African Americans living in different locations.[42]

TABLE 10.17 Frequency Distributions of Blood Type

		Location		
		I (Florida)	*II* (Iowa)	*III* (Missouri)
Blood type	A	122	1,781	353
	B	117	1,351	269
	AB	19	289	60
	O	244	3,301	713
	Total	502	6,722	1,395

To compare the distributions in the three locations, we can calculate the columnwise percentages, as displayed in Table 10.18. (For instance, of the Florida sample, $\frac{122}{502}$ or 24.3% are Type A.) Inspection of Table 10.18 shows that the three percentage distributions (columns) are fairly similar. ■

TABLE 10.18 Percentage Distributions of Blood Type

		Location		
		I (Florida)	*II* (Iowa)	*III* (Missouri)
Blood type	A	24.3	26.5	25.3
	B	23.3	20.1	19.3
	AB	3.8	4.3	4.3
	O	48.6	49.1	51.1
	Total	100.0	100.0	100.0

Figure 10.7 is a bar chart of the data that gives a visual impression of the distributions.

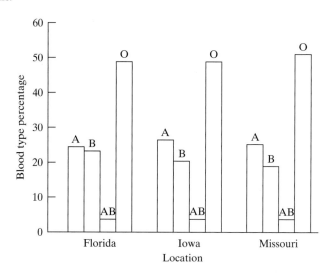

Figure 10.7 Bar chart of blood type data

The Chi-Square Test for the $r \times k$ Table

The goal of statistical analysis of an $r \times k$ contingency table is to investigate the relationship between the row variable and the column variable. Such an investigation can begin with an inspection of the columnwise or rowwise percentages, as in Table 10.18. One route to further analysis is to ask whether the discrepancies in percentages are too large to be explained as sampling error. This question can be answered by a chi-square test. The chi-square statistic is calculated from the familiar formula

$$\chi_s^2 = \sum \frac{(O - E)^2}{E}$$

where the sum is over all cells of the contingency table, and the expected frequencies (E's) are calculated as

$$E = \frac{(\text{Row total}) \cdot (\text{Column total})}{\text{Grand total}}$$

This method of calculating the E's can be justified by a simple extension of the rationale given in Section 10.2. Critical values for the chi-square test are obtained from Table 9 with

$$\text{df} = (r - 1)(k - 1)$$

The following example illustrates the chi-square test.

Distribution of Blood Type. Let us apply the chi-square test to the blood type data of Example 10.31. The null hypothesis is

> H_0: The distribution of blood type is the same in the three populations.

This hypothesis can be stated symbolically in conditional probability notation as follows:

Example 10.32

$$H_0: \begin{cases} \Pr\{A|I\} = \Pr\{A|II\} = \Pr\{A|III\} \\ \Pr\{B|I\} = \Pr\{B|II\} = \Pr\{B|III\} \\ \Pr\{AB|I\} = \Pr\{AB|II\} = \Pr\{AB|III\} \\ \Pr\{O|I\} = \Pr\{O|II\} = \Pr\{O|III\} \end{cases}$$

Note that the percentages in Table 10.18 are the estimated conditional probabilities; that is,

$$\hat{\Pr}\{A|I\} = .243$$

$$\hat{\Pr}\{A|II\} = .265$$

and so on. We test H_0 against the nondirectional alternative hypothesis

H_A: The distribution of blood type is not the same in the three populations.

Table 10.19 shows the observed and expected frequencies.

TABLE 10.19 Observed and Expected Frequencies of Blood Type

| | | Location | | | |
		I	II	III	Total
	A	122 (131.40)	1,781 (1,759.47)	353 (365.14)	2,256
Blood	B	117 (101.17)	1,351 (1,354.69)	269 (281.14)	1,737
type	AB	19 (21.43)	289 (287.00)	60 (59.56)	368
	O	244 (248.00)	3,301 (3,320.83)	713 (689.16)	4,258
	Total	502	6,722	1,395	8,619

From Table 10.19, we can calculate the test statistic as

$$\chi_s^2 = \frac{(122 - 131.40)^2}{131.40} + \frac{(1781 - 1759.47)^2}{1759.47} + \cdots + \frac{(713 - 689.16)^2}{689.16}$$

$$= 5.65$$

For these data, $r = 4$ and $k = 3$, so that

$$df = (4 - 1)(3 - 1) = 6$$

From Table 9 with df = 6, we find that $\chi^2(6)_{.20} = 8.56$, so that $P > .20$. The null hypothesis would not be rejected at any reasonable significance level. Thus, the chi-square test shows that the observed differences among the three blood type distributions are no more than would be expected from sampling variation. ∎

Note that H_0 in Example 10.32 is a compound null hypothesis in the sense defined in Section 10.1—that is, H_0 contains more than one independent assertion. This will always be true for contingency tables larger than 2×2, and consequently for such tables the alternative hypothesis for the chi-square test will always be nondirectional and the conclusion, if H_0 is rejected, will be nondirectional. Thus, the chi-square test will often not represent a complete analysis of an $r \times k$ contingency table.

Two Contexts for $r \times k$ Contingency Tables

We noted in Section 10.3 that a 2×2 contingency table can arise in two different contexts. Similarly, an $r \times k$ contingency table can arise in the following two contexts:

1. k independent samples; a categorical observed variable with r categories
2. One sample; two categorical observed variables—one with k categories and one with r categories

As with the 2×2 table, the calculation of the chi-square statistic is the same for both contexts, but the statement of hypotheses and conclusions can differ. The following example illustrates the second context.

Hair Color and Eye Color. Table 10.20 shows the relationship between hair color and eye color for 6,800 German men.[43] (This is the same study as in Example 10.21.)

Example 10.33

TABLE 10.20 Hair Color and Eye Color

		Hair Color			
		Brown	*Black*	*Fair*	*Red*
Eye	Brown	438	288	115	16
Color	Grey or Green	1,387	746	946	53
	Blue	807	189	1,768	47

Let us use a chi square test to test the hypothesis

H_0: Hair color and eye color are independent.

For the data of Table 10.20, we can calculate $\chi_s^2 = 1,074$. The degrees of freedom for the test are df $= (3 - 1)(4 - 1) = 6$. From Table 9 we find $\chi^2(6)_{.0001} = 27.86$. Thus, H_0 is overwhelmingly rejected and we conclude that there is extremely strong evidence that hair color and eye color are associated. ∎

Computer note: The chi-square test of independence can all be carried out with common statistical software. For example, consider the hair color and eye color data of Example 10.33. In the MINITAB system, suppose the cell counts are entered column 1 for brown hair, column 2 for black hair, column 3 for fair hair, and column 4 for red hair. That is, suppose the columns of data are

C1	C2	C3	C4
438	288	115	16
1387	746	946	53
807	189	1768	47

The command

```
MTB > ChiSquare C1-C4.
```

gives the following output:

```
Chi-Square Test
Expected counts are printed below observed counts
             C1        C2        C3        C4      Total
   1         438       288       115        16       857
          331.71    154.13    356.54     14.62
   2        1387       746       946        53      3132
          1212.27    563.30   1303.00     53.43
   3         807       189      1768        47      2811
          1088.02    505.57   1169.46     47.95
 Total      2632      1223      2829       116      6800
ChiSq=34.059+116.263+163.630+0.130+
      25.185+59.257+97.814+0.003+
      72.584+198.222+306.340+0.019=1073.508
df=6,  p=0.000
```

Exercises 10.47–10.53

10.47 Herpes simplex virus type 2 (HSV-2) is a sexually transmitted disease. As part of the third National Health and Nutrition Examination Survey (NHANES III), prevalence of HSV-2 was determined in four regions of the United States. The data are given in the following table.[44]

	HSV-2 Prevalence		
Region	*Sample Size*	*Number*	*Percent*
Northeast	1488	323	21.7
Midwest	2070	381	18.4
South	5323	1,320	24.8
West	2698	712	26.4

(a) Use a chi-square test to compare the prevalence rates at $\alpha = .01$. (The value of the chi-square statistic is $\chi_s^2 = 49.77$.)
(b) Verify the value of χ_s^2 given in part (a).

10.48 For a study of free-living populations of the fruitfly *Drosophila subobscura*, researchers placed baited traps in two woodland sites and one open-ground area. The numbers of male and female flies trapped in a single day are given in the table.[45]

	Woodland Site I	**Woodland Site II**	**Open Ground**
Males	89	34	74
Females	31	20	136
Total	120	54	210

(a) Use a chi-square test to compare the sex ratios at the three sites. Let $\alpha = .05$.
(b) Construct a table that displays the data in a more readable format, such as the one in Exercise 10.47.

10.49 In a classic study of peptic ulcer, blood types were determined for 1,655 ulcer patients. The accompanying table shows the data for these patients and for an independently chosen group of 10,000 healthy controls from the same city.[46]

Blood Type	Ulcer Patients	Controls
O	911	4,578
A	579	4,219
B	124	890
AB	41	313
Total	1,655	10,000

(a) The value of the chi-square statistic for this contingency table is $\chi_s^2 = 49.0$. Carry out the chi-square test at $\alpha = .01$.

(b) Construct a table showing the percentage distributions of blood type for patients and for controls.

(c) Verify the value of χ_s^2 given in part (a).

10.50 The two claws of the lobster *(Homarus americanus)* are identical in the juvenile stages. By adulthood, however, the two claws normally have differentiated into a stout claw called a "crusher" and a slender claw called a "cutter." In a study of the differentiation process, 26 juvenile animals were reared in smooth plastic trays and 18 were reared in trays containing oyster chips (which they could use to exercise their claws). Another 23 animals were reared in trays containing only one oyster chip. The claw configurations of all the animals as adults are summarized in the table.[47]

		Claw Configuration		
		Right Crusher, Left Cutter	Right Cutter, Left Crusher	Right Cutter, Left Cutter
Treatment	Oyster chips	8	9	1
	Smooth plastic	2	4	20
	One oyster chip	7	9	7

(a) The value of the contingency-table chi-square statistic for these data is $\chi_s^2 = 24.36$. Carry out the chi-square test at $\alpha = .01$.

(b) Verify the value of χ_s^2 given in part (a).

(c) Construct a table showing the percentage distribution of claw configurations for each of the three treatments.

(d) Interpret the table from part (c): In what way is claw configuration related to treatment? (For example, if you wanted a lobster with two cutter claws, which treatment would you choose and why?)

10.51 A randomized, double-blind, placebo-controlled experiment was conducted in which patients with Alzheimer's disease were given either extract of Ginkgo biloba (EGb) or a placebo for one year. The change in each patient's Alzheimer's Disease Assessment Scale–Cognitive subscale (ADAS-Cog) score was measured. The results are given in the table.[48] *Note*: If the ADAS-Cog went down, then the patient improved.

	Change in ADAS-Cog Score				
	−4 or better	−2 to −3	−1 to +1	+2 to +3	+4 or worse
EGb	22	18	12	7	16
Placebo	10	11	19	11	24

(a) Use a chi-square test to compare the prevalence rates at $\alpha = .05$. (The value of the chi-square statistic is $\chi_s^2 = 10.26$.)

(b) Verify the value of χ_s^2 given in part (a).

10.52 Marine biologists have noticed that the color of the outermost growth band on a clam tends to be related to the time of the year in which the clam dies. A biologist conducted a small investigation of whether this is true for the species *Protothaca staminea*. She collected a sample of 78 clam shells from this species and cross-classified them according to (1) month when the clam died and (2) color of the outermost growth band. The data are shown in the following table.[49]

	Color		
	Clear	*Dark*	*Unreadable*
February	9	26	9
March	6	25	3
Total	15	51	12

Use a chi-square test to compare the color distributions for the two months. Let $\alpha = .10$.

10.53 A group of patients with a binge-eating disorder were randomly assigned to take either the experimental drug fluvoxamine or a placebo in a nine-week-long double-blind clinical trial. At the end of the trial the condition of each patient was classified into one of four categories: no response, moderate response, marked response, or remission. The following table shows a cross-classification of the data.[50] Is there statistically significant evidence, at the .10 level, to conclude that there is an association between treatment group (fluvoxamine versus placebo) and condition?

	No Response	Moderate Response	Marked Response	Remission	Total
Fluvoxamine	15	7	3	15	40
Placebo	22	7	3	11	43
Total	37	14	6	26	

10.6 APPLICABILITY OF METHODS

In this section we discuss guidelines for deciding when to use a chi-square test.

Conditions for Validity

A chi-square test is valid under the following conditions:

1. **Design conditions**

   For the chi-square goodness-of-fit test, it must be reasonable to regard the data as a random sample of categorical observations from a large population.

   For the contingency-table chi-square test, it must be appropriate to view the data in one of the following ways:

   (a) As two or more independent random samples, observed with respect to a categorical variable; or

(b) As one random sample, observed with respect to two categorical variables.

For either type of chi-square test, the observations within a sample must be independent of each other.

2. **Sample size conditions**

The sample size must be large enough. The critical values given in Table 9 are only approximately correct for determining the P-value associated with χ_s^2. As a rule of thumb, the approximation is considered adequate if each expected frequency (E) is at least equal to 5.* (If the expected frequencies are small and the data form a 2×2 contingency table, then Fisher's exact test might be appropriate—see optional Section 10.4.)

3. **Form of H_0**

For the chi-square goodness-of-fit test, the null hypothesis must specify numerical values for the category probabilities.[†]
A generic form of the null hypothesis for the contingency-table chi-square test may be stated as follows:

H_0: The row variable and the column variable are independent.

Verification of Design Conditions

To verify the design conditions, we need to identify a population from which the data may be viewed as a random sample. Sometimes the "population" is rather abstract. For instance, in applications to genetics such as Example 10.1, we may think of the progeny of a given mating or cross as a random sample from the conceptual population of all *potential* progeny of that mating or cross.

If the data consist of several samples [situation 1(a) in the preceding list], then the samples are required to be independent of each other. Failure to observe this restriction may result in a loss of power. If the design includes any pairing, blocking, or matching of experimental units, then the samples would not be independent. A method of analysis for dependent samples is described in Section 10.8.

As always, bias in the sampling procedure must be ruled out. Moreover, chi-square methods are not appropriate when complex random sampling schemes such as cluster sampling or stratified random sampling are used. Finally, there must be no dependency or hierarchical structure in the design. Failure to observe this restriction can result in a vastly inflated chance of Type I error (which is usually much more serious than a loss of power). The following examples show the relevance of checking for dependency in the observations.

Food Choice by Insect Larvae. In a behavioral study of the clover root curcuho *Sitona hispidulus*, 20 larvae were released into each of six petri dishes. Each dish contained nodulated and nonnodulated alfalfa roots, arranged in a symmetric pattern. (This experiment was more fully described in Example 1.5.) After 24 hours the location of each larva was noted, with the results shown in Table 10.21.[51]

> **Example 10.34**

---

* For an $r \times k$ table with more than 2 rows and columns, the approximation is adequate if the average expected frequency is at least 5, even if some of the cell counts are smaller.

[†] A slightly modified form of the goodness-of-fit test can be used to test a hypothesis that merely constrains the probabilities rather than specifying them exactly. An example would be testing the fit of a binomial distribution to data (see optional Section 3.9).

TABLE 10.21 Food Choice by *Sitona* Larvae

Dish	Number of larvae		
	Nodulated Roots	Nonnodulated Roots	Other (died, lost, etc.)
1	5	3	12
2	9	1	10
3	6	3	11
4	7	1	12
5	5	1	14
6	14	3	3
Total	46	12	62

Suppose the following analysis is proposed. A total of 58 larvae made a choice; the observed frequencies of choosing nodulated and nonnodulated roots were 46 and 12, and the corresponding expected frequencies (assuming random choice) would be 29 and 29; these data yield $\chi_s^2 = 19.93$, from which (using a directional alternative) we find from Table 9 that $P < .00005$. The validity of this proposed analysis is highly doubtful because it depends on the assumption that all the observations in a given dish are independent of each other; this assumption would certainly be false if (as is biologically plausible) the larvae tend to follow each other in their search for food.

How, then, should the data be analyzed? One approach is to make the reasonable assumption that the observations in one dish are independent of those in another dish. Under this assumption we could use a paired analysis on the six dishes ($n_d = 6$); a paired t test yields $P \approx .005$ and a sign test yields $P \approx .02$. Note that the questionable assumption of independence within dishes led to a P-value that was much too small. ∎

Example 10.35

Pollination of Flowers. A study was conducted to determine the adaptive significance of flower color in the scarlet gilia *(Ipomopsis aggregata)*. Six red-flowered plants and six white-flowered plants were chosen for observation in field conditions; hummingbirds were permitted to visit the flowers, but the other major pollinator, a moth, was excluded by covering the plants at night. Table 10.22 shows, for each plant, the total number of flowers at the end of the season and the number that had set fruit.[52]

TABLE 10.22 Fruit Set in Scarlet Gilia Flowers

	Red-flowered Plants			White-flowered Plants		
	Number of Flowers	Number Setting Fruit	Percent Setting Fruit	Number of Flowers	Number Setting Fruit	Percent Setting Fruit
	140	26	19	125	21	17
	116	11	9	134	17	13
	34	0	0	273	81	30
	79	9	11	146	38	26
	185	28	15	103	17	17
	106	11	10	82	24	29
Sum	660	85		863	198	

The question of interest is whether the percentage of fruit set is different for red-flowered than for white-flowered plants. Suppose this question is approached by regarding the individual flower as the observational unit; then the data could be cast in the contingency table format of Table 10.23.

TABLE 10.23 **Fruit Set in Scarlet Gilia Flowers**

		Flower Color	
		Red	*White*
	Yes	85	198
Fruit set	No	575	665
Total		660	863
	Percent setting fruit	13	23

Table 10.23 yields $\chi_s^2 = 25.0$, for which Table 9 gives $P < .0001$. However, this analysis is not correct, because the observations on flowers on the same plant are not independent of each other; they are dependent because the pollinator (the hummingbird) tends to visit flowers in groups, and perhaps also because the flowers on the same plant are physiologically and genetically related. The chi-square test is invalidated by the hierarchical structure in the data.

A better approach would be to treat the entire plant as the observational unit. For instance, we could take the "Percent Setting Fruit" column of Table 10.22 as the basic observations; applying a t test to the values yields $t_s = 2.88$ (with $.01 < P < .02$), and applying a Wilcoxon-Mann-Whitney test yields $U_s = 32$ (with $.02 < P < .05$). Thus, the P-value from the inappropriate chi-square analysis is much too small. ∎

Power Considerations

In many studies the chi-square test is valid but is not as powerful as a more appropriate test. Specifically, consider a situation in which the rows or the columns (or both) of the contingency table correspond to a *rankable* categorical variable with more than two categories. The following is an example.

Pain Medication.   In a completely randomized study to compare two pain medications, A and B, each patient rated the amount of pain relief on a 4-point subjective scale. The results are shown in Table 10.24.

Example 10.36

TABLE 10.24 Response to Pain Medication

		Treatment	
		Drug A	*Drug B*
	None	3	7
Pain	Some	7	11
relief	Substantial	10	5
	Complete	5	2
	Total	25	25

A contingency-table chi-square test would be valid to compare drugs A and B, but the test would lack power because it does not use the information contained in the *ordering* of the pain relief categories (none, some, substantial, complete). A related weakness of the chi-square test is that, even if H_0 is rejected, the test does not yield a directional conclusion such as "drug A relieves pain better than drug B." ■

Methods are available to analyze contingency tables with rankable row and/or column variables; such methods, however, are beyond the scope of this book.

Exercises 10.54–10.56

10.54 Refer to the chemotherapy data of Exercise 10.23. Are the sample sizes large enough for the approximate validity of the chi-square test?

10.55 In a study of prenatal influences on susceptibility to seizures in mice, pregnant females were randomly allocated to a control group or a "handled" group. Handled mice were given sham injections three times during gestation, while control mice were not touched. The offspring were tested for their susceptibility to seizures induced by a loud noise. The investigators noted that the response varied considerably from litter to litter. The accompanying table summarizes the results.[53]

| | Number of litters | Number of mice | Response to loud noise | | |
| | | | *No response* | *Wild running* | *Seizure* |
Treatment					
Handled	19	104	23	10	71
Control	20	120	47	13	60

If these data are analyzed as a 2×3 contingency table, the chi-square statistic is $\chi_s^2 = 8.45$ and Table 9 gives $.01 < P < .02$. Is this an appropriate analysis for this experiment? Explain. (*Hint*: Does the design meet the conditions for validity of the chi-square test?)

10.56 In control of diabetes it is important to know how blood glucose levels change after eating various foods. Ten volunteers participated in a study to compare the effects of two foods—a sugar and a starch. A blood specimen was drawn before each volunteer consumed a measured amount of food; then additional blood specimens were drawn at eleven times during the next 4 hours. Each volunteer repeated the entire test on another occasion with the other food. Of particular concern were blood glucose levels that dropped below the initial level; the accompanying table shows the number of such values.[54]

Food	No. of Values Less Than Initial Value	Total Number of Observations
Sugar	26	110
Starch	14	110

Suppose we analyze the given data as a contingency table. The test statistic would be

$$\chi_s^2 = \frac{(26 - 20)^2}{20} + \frac{(14 - 20)^2}{20} + \frac{(84 - 90)^2}{90} + \frac{(96 - 90)^2}{90} = 4.40$$

At $\alpha = .05$ we would reject H_0 and find that there is sufficient evidence to conclude that blood glucose values below the initial value occur more often after ingestion of sugar than after ingestion of starch. This analysis contains two flaws. What are they? (*Hint*: Are the conditions for validity of the test satisfied?)

10.7 CONFIDENCE INTERVAL FOR DIFFERENCE BETWEEN PROBABILITIES

The chi-square test for a 2×2 contingency table answers only a limited question: Do the estimated probabilities $\hat{p}_1$ and $\hat{p}_2$ differ enough to conclude that the true probabilities p_1 and p_2 are not equal? A complementary mode of analysis is to use a confidence interval for the magnitude of the difference, $(p_1 - p_2)$.

When we discussed constructing a confidence interval for a single proportion, p, in Section 6.6, we defined an estimate $\tilde{p}$, based on the idea of "adding 2 successes and 2 failures to the data." Making this adjustment to the data resulted in a confidence interval procedure that has good coverage properties. Likewise, when constructing a confidence interval for the difference in two proportions, we will define new estimates that are based on the idea of adding 1 observation to each cell of the 2×2 table (so that a *total* of 2 successes and 2 failures are added to the data).

Consider a 2×2 contingency table that can be viewed as a comparison of two samples, of sizes n_1 and n_2, with respect to a dichotomous response variable. Let the 2×2 table be given as

Sample 1	Sample 2
y_1	y_2
$n_1 - y_1$	$n_2 - y_2$
n_1	n_2

We define

$$\tilde{p}_1 = \frac{y_1 + 1}{n_1 + 2}$$

and

$$\tilde{p}_2 = \frac{y_2 + 1}{n_2 + 2}$$

We will use the difference in the new values, $(\tilde{p}_1 - \tilde{p}_2)$, to construct a confidence interval for $(p_1 - p_2)$. Like all quantities calculated from samples, the quantity $(\tilde{p}_1 - \tilde{p}_2)$ is subject to sampling error. The magnitude of the sampling error can be expressed by the standard error of $(\tilde{p}_1 - \tilde{p}_2)$, which is calculated from the following formula:

$$\mathrm{SE}_{(\tilde{p}_1 - \tilde{p}_2)} = \sqrt{\frac{\tilde{p}_1(1 - \tilde{p}_1)}{n_1 + 2} + \frac{\tilde{p}_2(1 - \tilde{p}_2)}{n_2 + 2}}$$

Note that $\mathrm{SE}_{(\tilde{p}_1 - \tilde{p}_2)}$ is analogous to $\mathrm{SE}_{(\bar{y}_1 - \bar{y}_2)}$ as described in Section 7.2.

An approximate confidence interval can be based on $\mathrm{SE}_{(\tilde{p}_1 - \tilde{p}_2)}$; for instance, a 95% confidence interval is

$$(\tilde{p}_1 - \tilde{p}_2) \pm (1.96)\mathrm{SE}_{(\tilde{p}_1 - \tilde{p}_2)}$$

Confidence intervals constructed this way have good coverage properties (i.e., approximately 95% of all 95% confidence intervals cover the true difference

$p_1 - p_2$) for almost any sample sizes n_1 and n_2.[55] The following example illustrates the construction of the confidence interval.*

Example 10.37

Treatment of Angina.   For the angina data of Example 10.11, the sample sizes are $n_1 = 160$ and $n_2 = 147$, and the estimated probabilities of the angina-free response are

$$\tilde{p}_1 = \frac{45}{162} = .278$$

$$\tilde{p}_2 = \frac{20}{149} = .134$$

The difference between these is

$$\tilde{p}_1 - \tilde{p}_2 = .278 - .134$$
$$= .144$$
$$\approx .14$$

Thus, we estimate that treatment with Timolol increases the probability of the angina-free response by .14, compared to placebo. To set confidence limits on this estimate, we calculate the standard error as

$$SE_{(\tilde{p}_1 - \tilde{p}_2)} = \sqrt{\frac{.278(.722)}{162} + \frac{.134(.866)}{149}}$$
$$= .0449$$

The 95% confidence interval is

$$.144 \pm (1.96)(.0449)$$
$$.144 \pm .088$$
$$.056 < p_1 - p_2 < .232$$

We are 95% confident that Timolol increases the probability of angina-free response by between .056 and .232, compared to placebo. ∎

Relationship to Test.   The chi-square test for a 2×2 contingency table (Section 10.2) is approximately, but not exactly, equivalent to checking whether a confidence interval for $(p_1 - p_2)$ includes zero. [Recall from Section 7.5 that there is an exact equivalence between a *t* test and a confidence interval for $(\mu_1 - \mu_2)$.]

Exercises 10.57–10.62

10.57   Refer to the estrus synchronization data of Exercise 10.21. Let p_1 and p_2 represent the probabilities of pregnancy using products A and B, respectively. Construct a 95% confidence interval for $(p_1 - p_2)$.

10.58   Refer to the liver tumor data of Exercise 10.22. Let p_1 and p_2 represent the probabilities of liver tumors under the germ-free and the *E. coli* conditions, respectively.

---

* In Section 6.6 we presented a general version of the "add 2 successes and 2 failures" idea, in which the formula for $\tilde{p}$ depends on the confidence level (95%, 90%, etc.). When constructing a confidence interval for a difference in two proportions the coverage properties of the interval are best when 1 is added to each cell in the 2×2 table, no matter what confidence level is being used.[56]

(a) Construct a 95% confidence interval for $(p_1 - p_2)$.

(b) Interpret the confidence interval from part (a). That is, explain what the interval tells you about tumor probabilities.

10.59 For women who are pregnant with twins, complete bed rest in late pregnancy is commonly prescribed in order to reduce the risk of premature delivery. To test the value of this practice, 212 women with twin pregnancies were randomly allocated to a bed-rest group or a control group. The accompanying table shows the incidence of preterm delivery (less than 37 weeks of gestation).[57]

	Bed Rest	**Controls**
No. of preterm deliveries	32	20
No. of women	105	107

Let p_1 and p_2 represent the probabilities of preterm delivery in the row conditions. Construct a 95% confidence interval for $(p_1 - p_2)$. Does the confidence interval suggest that bed rest is beneficial?

10.60 Refer to Exercise 10.59. The numbers of infants with low birthweight (2,500 g or less) born to the women are shown in the table.

	Bed Rest	**Controls**
No. of low-birthweight babies	76	92
Total no. of babies	210	214

Let p_1 and p_2 represent the probabilities of a low-birthweight baby in the two-conditions. Explain why the above information is not sufficient to construct a confidence interval for $(p_1 - p_2)$.

10.61 Refer to the blood type data of Exercise 10.49. Let p_1 and p_2 represent the probabilities of Type O blood in the patient population and the control population, respectively.

(a) Construct a 95% confidence interval for $(p_1 - p_2)$.

(b) Interpret the confidence interval from part (a). That is, explain what the interval tells you about the difference in probabilities of Type O blood.

10.62 In an experiment to treat patients with "generalized anxiety disorder," the drug hydroxyzine was given to 71 patients and 30 of them improved. A group of 70 patients were given a placebo and 20 of them improved.[58] Let p_1 and p_2 represent the probabilities of improvement using hydroxyzine and the placebo, respectively. Construct a 95% confidence interval for $(p_1 - p_2)$.

10.8 PAIRED DATA AND 2 × 2 TABLES (OPTIONAL)

In Chapter 9 we considered paired data when the response variable is continuous. In this section we consider the analysis of paired categorical data.

HIV Transmission to Children. A study was conducted to determine a woman's risk of transmitting HIV to her unborn child. A sample of 114 HIV-infected women who gave birth to two children found that HIV infection occurred in 19 of the 114

Example 10.38

older siblings and in 20 of the 114 younger siblings.[59] These data are shown in Table 10.25.

TABLE 10.25 HIV Infection Data		Older Sibling	Younger Sibling
HIV?	Yes	19	20
	No	95	94
	Total	114	114

At first glance, it might appear that a regular chi-square test could be used to test the null hypothesis that the probability of HIV infection is the same for older siblings as for younger siblings. However, as we stated in Section 10.6, for the chi-square test to be valid the two samples—of 114 older siblings and of 114 younger siblings—must be independent of each other. In this case the samples are clearly dependent. Indeed, these are paired data, with a family generating the pair (older sibling, younger sibling).

Table 10.26 presents the data in a different format. This format helps focus attention on the relevant part of the data.

TABLE 10.26 HIV Infection Data Shown by Pairs		Younger Sibling HIV?	
		Yes	*No*
Older sibling HIV?	Yes	2	17
	No	18	77

From Table 10.26 we can see that there are 79 pairs in which both siblings have the same HIV status: 2 are "yes/yes" pairs and 77 are "no/no" pairs. These 79 pairs, which are called **concordant pairs**, do not help us determine whether HIV infection is more likely for younger siblings than for older siblings. The remaining 35 pairs—17 "yes/no" pairs and 18 "no/yes" pairs—do provide information on the relative likelihood of HIV infection for older and younger siblings. These pairs are called **discordant pairs**; we will focus on these 35 pairs in our analysis.

If the chance of HIV infection is the same for older siblings as it is for younger siblings, then the two kinds of pairs—"yes/no" and "no/yes"—are equally likely. Thus, the null hypothesis

H_0: the probability of HIV infection is the same for older siblings as it is for younger siblings

is equivalent to

H_0: among discordant pairs, $\Pr(\text{"yes/no"}) = \Pr(\text{no/yes}) = \dfrac{1}{2}$ ∎

McNemar's Test

The hypothesis that discordant pairs are equally likely to be "yes/no" or "no/yes" can be tested with the chi-square goodness-of-fit test developed in Section 10.2. This application of the goodness-of-fit test is known as **McNemar's test** and has a

TABLE 10.27 A General Table of Paired Proportion Data

	Yes	No
Yes	n_{11}	n_{12}
No	n_{21}	n_{22}

particularly simple form.* Let n_{11} denote the number of "yes/yes" pairs, n_{12} the number of "yes/no" pairs, n_{21} the number of "no/yes" pairs, and n_{22} the number of "no/no" pairs, as shown in Table 10.27. If H_0 is true, the expected number of "yes/no" pairs is $\dfrac{n_{12} + n_{21}}{2}$, as is the expected number of "no/yes" pairs. Thus, the test statistic is

$$\chi_s^2 = \frac{\left(n_{12} - \dfrac{(n_{12} + n_{21})}{2}\right)^2}{\dfrac{(n_{12} + n_{21})}{2}} + \frac{\left(n_{21} - \dfrac{(n_{12} + n_{21})}{2}\right)^2}{\dfrac{(n_{12} + n_{21})}{2}}$$

which simplies to

$$\chi_s^2 = \frac{(n_{12} - n_{21})^2}{n_{12} + n_{21}}$$

The distribution of χ_s^2 under the null hypothesis is approximately a χ^2 distribution with 1 degree of freedom.

HIV Transmission to Children. For the data given in Example 10.38, $n_{12} = 17$ and $n_{21} = 18$. Thus,

Example 10.39

$$\chi_s^2 = \frac{(17 - 18)^2}{17 + 18} = 0.0286$$

From Table 9 we see that the P-value is greater than .20. (Using a computer gives $P = .87$.) The data are very much consistent with the null hypothesis that the probability of HIV infection is the same for older siblings as it is for younger siblings. ∎

Exercises 10.63–10.65

10.63 As part of a study of risk factors for stroke, 155 women who had experienced a hemorrhagic stroke (cases) were interviewed. For each case, a control was chosen who had not experienced a stroke; the control was matched to the case by neighborhood of residence, age, and race. Each woman was asked whether she used oral

---

* The null hypothesis tested by McNemar's test can also be tested by using the binomial distribution. The null hypothesis states that among discordant pairs, Pr("yes/no") = Pr("no/yes") = $\dfrac{1}{2}$. Thus, under the null hypothesis, the number of "yes/no" pairs has a binomial distribution with n = the number of discordant pairs and $p = .5$.

contraceptives. The data for the 155 pairs are displayed in the table. "Yes" and "No" refer to use of oral contraceptives.[60]

		Case	
		No	*Yes*
Control	No	107	30
	Yes	13	5

To test for association between oral contraceptive use and stroke, consider only the 43 discordant pairs (pairs who answered differently) and test the hypothesis that a discordant pair is equally likely to be "yes/no" or "no/yes." Use McNemar's test to test the hypothesis that having a stroke is independent of use of oral contraceptives against a nondirectional alternative at $\alpha = .05$.

10.64 Example 10.38 referred to a sample of HIV-infected women who gave birth to two children. One of the outcomes that was studied was whether the gestational age of the child was less than 38 weeks; this information was recorded for 106 of the families. The data for this variable are shown in the following table. Analyze these data using McNemar's test. Use a nondirectional alternative and let $\alpha = .10$.

(a) State the null hypothesis in words.
(b) Do you reject H_0? Why or why not?
(c) State your conclusion from part (b) in the context of the setting.

		Younger sibling < 38 weeks?	
		Yes	*No*
Older sibling < 38 weeks?	Yes	26	5
	No	21	54

10.65 A study of 85 patients with Hogkin's disease found that 41 had had their tonsils removed. Each patient was matched with a sibling of the same sex. Only 33 of the siblings had undergone tonsillectomy. The data are shown in the following table.[61] Use McNemar's test to the hypothesis that "Yes/No" and "No/Yes" pairs are equally likely. Previous research had suggested that having a tonsillectomy is associated with an increased risk of Hogkin's disease; thus, use a directional alternative. Let $\alpha = .05$.

		Sibling Tonsillectomy?	
		Yes	*No*
Hogkin's Patient Tonsillectomy	Yes	26	15
	No	7	37

10.9 RELATIVE RISK AND THE ODDS RATIO (OPTIONAL)

It is quite common to test the null hypothesis that two population proportions, p_1 and p_2, are equal. A chi-square test, based on a 2×2 table, is often used for this purpose. A confidence interval for $(p_1 - p_2)$ provides information about the magnitude of the difference between p_1 and p_2. In this section we consider two other measures of dependence: the relative risk and the odds ratio.

Relative Risk

Sometimes researchers prefer to compare probabilities in terms of their *ratio*, rather than their difference. When the outcome event is deleterious (such as having a heart attack or getting cancer) the ratio of probabilities is called the **relative risk**, or the risk ratio. The relative risk is defined as p_1/p_2. This measure is widely used in studies of human health. The following is an example.

Smoking and Birthweight. In a study of the effects of smoking, 9,793 pregnant women were asked about their smoking habits. (This study was mentioned in Examples 8.2, 8.4, and 8.5.) Table 10.28 shows the incidence of low birthweight (2500 g or less) among their infants.[62]

Example 10.40

TABLE 10.28 Data on Low Birthweight and Smoking Status

		Smoking Status	
		Smoker	*Nonsmoker*
Birthweight	Low	237	197
	Normal	3489	5870
	Total	3726	6067

The probabilities of primary interest are the columnwise conditional probabilities:

$$p_1 = \text{Pr}\{\text{Low birthweight}|\text{Smoker}\}$$

$$p_2 = \text{Pr}\{\text{Low birthweight}|\text{Nonsmoker}\}$$

The estimates of these from the data are

$$\hat{p}_1 = \frac{237}{3726} = .06361 \approx .064$$

$$\hat{p}_2 = \frac{197}{6067} = .03247 \approx .032$$

The estimated relative risk is

$$\frac{\hat{p}_1}{\hat{p}_2} = \frac{.06361}{.03247} = 1.959 \approx 2$$

Thus, we estimate that the risk (i.e., the conditional probability) of having a low birthweight baby is about twice as great for smokers as for nonsmokers. (Of course, because this is an observational study, we would not be justified in concluding that smoking *causes* the low birthweight.) ∎

The Odds Ratio

Another way to compare two probabilities is in terms of **odds**. The odds of an event E is defined to be the ratio of the probability that E occurs to the probability that E does not occur:

$$\text{odds of } E = \frac{\text{Pr}\{E\}}{1 - \text{Pr}\{E\}}$$

For instance, if the probability of an event is 1/4, then the odds of the event are $\frac{1/4}{3/4} = 1/3$ or $1 : 3$. As another example, if the probability of an event is 1/2, then the odds of the event are $\frac{1/2}{1/2} = 1$ or $1 : 1$.

The **odds ratio** is simply the ratio of odds under two conditions. Specifically, suppose that p_1 and p_2 are the conditional probabilities of an event under two different conditions. Then the odds ratio, which we denote by θ ("theta"), is defined as follows:

$$\theta = \frac{\dfrac{p_1}{1 - p_1}}{\dfrac{p_2}{1 - p_2}}$$

If the estimated probabilities $\hat{p}_1$ and $\hat{p}_2$ are calculated from a 2×2 contingency table, the corresponding estimated odds ratio, denoted $\hat{\theta}$, is calculated as

$$\hat{\theta} = \frac{\dfrac{\hat{p}_1}{1 - \hat{p}_1}}{\dfrac{\hat{p}_2}{1 - \hat{p}_2}}$$

We illustrate with an example. ■

Example 10.41

Smoking and Birthweight.   From the data of Example 10.40, we estimate the odds of a low-birthweight baby as follows:

$$\hat{\text{odds}} = \frac{.06361}{1 - .06361} = .06793 \text{ among smokers}$$

$$\hat{\text{odds}} = \frac{.03247}{1 - .03247} = .03356 \text{ among nonsmokers}$$

The estimated odds ratio is

$$\hat{\theta} = \frac{.06793}{.03356} = 2.024 \approx 2$$

Thus, we estimate that the odds of having a low-birthweight baby are about twice as great for smokers as for nonsmokers. ■

Odds Ratio and Relative Risk

The odds ratio measures association in an unfamiliar way; the relative risk is a more natural measure. Fortunately, in many applications the two measures are approximately equal. In general the relationship between the odds ratio and the relative risk is given by

$$\text{odds ratio} = \text{relative risk} \cdot \frac{1 - p_2}{1 - p_1}$$

Notice that if p_1 and p_2 are small, then the relative risk is approximately equal to the odds ratio. We illustrate with the smoking and birthweight data.

Smoking and Birthweight. For the data in Table 10.28 we found that the estimated relative risk of a low-birthweight baby is

Example 10.42

$$\text{estimated relative risk} = 1.959$$

and the estimated odds ratio is

$$\hat{\theta} = 2.024$$

These are approximately equal because the outcome of interest (low birthweight) is rare, so that $\hat{p}_1$ and $\hat{p}_2$ are small. ∎

Advantage of the Odds Ratio

Both the relative risk p_1/p_2 and the difference $(p_1 - p_2)$ are easier to interpret than the odds ratio. Why, then, is the odds ratio used at all? One important advantage of the odds ratio is that, in certain kinds of studies, the odds ratio can be estimated even though p_1 and p_2 *cannot* be estimated. To explain this property, we must first discuss the question of estimability of conditional probabilities in contingency tables.

In a 2×2 contingency table, the conditional probabilities can be defined by rows or by columns. Whether these probabilities can be estimated from the observed data depends on the study design. The following example illustrates this point.

Smoking and Birthweight. In studying the relationship between smoking and low birthweight, the conditional probabilities of primary interest are

Example 10.43

$$p_1 = \Pr\{\text{Low birthweight}|\text{Smoker}\}$$

and

$$p_2 = \Pr\{\text{Low birthweight}|\text{Nonsmoker}\}$$

These are columnwise probabilities in a table like Table 10.28. We could, however, also consider the following rowwise conditional probabilities:

$$p_1^* = \Pr\{\text{Smoker}|\text{Low birthweight}\}$$

and

$$p_2^* = \Pr\{\text{Smoker}|\text{Normal birthweight}\}$$

(Of course, p_1^* and p_2^* are not particularly meaningful biologically.) From the study described in Example 10.40—that is, a single sample of size $n = 9{,}793$ observed with respect to smoking status and birthweight—we can estimate not only p_1 and p_2 but also p_1^* and p_2^*. However, there are other important study designs that do not provide enough information to estimate all of these conditional probabilities. For example, suppose that a study is conducted by choosing a group of 500 smokers and a group of 500 nonsmokers and then observing the birthweights of their infants. This kind of study is called a prospective study or **cohort study**. Such a study might produce the fictitious but realistic data of Table 10.29.

TABLE 10.29 Fictitious Data for Cohort Study of Smoking and Birthweight

		Smoking Status	
		Smoker	Nonsmoker
Birthweight	**Low**	32	16
	Normal	468	484
	Total	500	500

The data of Table 10.29 can be viewed as two independent samples. From the data we can estimate the conditional probabilities of low birthweight in the two populations (smokers and nonsmokers):

$$\hat{p}_1 = \frac{32}{500} = .064 \qquad \hat{p}_2 = \frac{16}{500} = .032$$

By contrast, the rowwise probabilities p_1^* and p_2^* cannot be estimated from Table 10.29. Because the relative numbers of smokers and nonsmokers were predetermined by the design of the study ($n_1 = 500$ and $n_2 = 500$), the data contain no information about the prevalence of smoking, and therefore no information about the population values of

$$\Pr\{\text{Smoker}|\text{Low birthweight}\} \quad \text{and} \quad \Pr\{\text{Smoker}|\text{Normal birthweight}\}$$

Table 10.29 was generated by fixing the column totals and observing the row variable. Consider now the reverse sort of design. Suppose we choose 500 mothers with low-birthweight babies and 500 mothers with normal-birthweight babies and we then determine the smoking habits of the mothers. This design is called a **case-control design**. Such a design might generate the fictitious but realistic data of Table 10.30.

TABLE 10.30 Fictitious Data for Cohort Study of Smoking and Birthweight

		Smoking Status		
		Smoker	Nonsmoker	Total
Birthweight	Low	273	227	500
	Normal	186	314	500

From Table 10.30 we can estimate the rowwise conditional probabilities

$$\hat{p}_1^* = \frac{273}{500} = .546 \approx .55$$

$$\hat{p}_2^* = \frac{186}{500} = .372 \approx .37$$

However, from the data in Table 10.30 we cannot estimate the columnwise conditional probabilities p_1 and p_2: Because the row totals were predetermined by design, the data contain no information about $\Pr\{\text{Low birthweight}|\text{Smoker}\}$ and $\Pr\{\text{Low birthweight}|\text{Nonsmoker}\}$. ∎

The preceding example shows that, depending on the design, a study may not permit estimation of both columnwise probabilities p_1 and p_2 and rowwise probabilities p_1^* and p_2^*. Fortunately, the odds ratio is the same whether it is determined columnwise or rowwise. Specifically,

$$\theta = \frac{\dfrac{p_1}{1 - p_1}}{\dfrac{p_2}{1 - p_2}} = \frac{\dfrac{p_1^*}{1 - p_1^*}}{\dfrac{p_2^*}{1 - p_2^*}}$$

Because of this relationship, the odds ratio θ can be estimated by estimating p_1 and p_2 or by estimating p_1^* and p_2^*. This fact has important applications, especially for case-control studies, as illustrated by the following example.

Smoking and Birthweight. To characterize the relationship between smoking and birthweight, the columnwise probabilities p_1 and p_2 are more biologically meaningful than the rowwise probabilities p_1^* and p_2^*. If we investigate the relationship using a case-control design, neither p_1 nor p_2 can be estimated from the data. (See Example 10.43.) However, the odds ratio *can* be estimated from the data. For instance, from Table 10.30 we obtain

Example 10.44

$$\hat{\theta} = \frac{\dfrac{\hat{p}_1^*}{1 - \hat{p}_1^*}}{\dfrac{\hat{p}_2^*}{1 - p_2}}$$

$$= \frac{\dfrac{.546}{1 - .546}}{\dfrac{.372}{1 - .372}} = 2.03$$

We can interpret this odds ratio as follows: We know that the outcome event— low birthweight—is rare, and so we know that the odds ratio is approximately equal to the relative risk, p_1/p_2. We therefore estimate that the risk of a low-birthweight baby is about twice as great for smokers as for nonsmokers. ∎

There is an easier way to compute the odds ratio for a 2×2 contingency table. For a general 2×2 table, let n_{11} denote the number of observations in the first row and the first column. Likewise, let n_{12} be the number of observations in the first row and second column, and so on. The general 2×2 table then has the form

n_{11}	n_{12}
n_{21}	n_{22}

The estimated odds ratio from the table is

$$\hat{\theta} = \frac{n_{11} \cdot n_{22}}{n_{12} \cdot n_{21}}$$

Example 10.45

Smoking and Birthweight. From the data in Table 10.28, we can calculate the estimated odds ratio as

$$\hat{\theta} = \frac{237 \cdot 5870}{197 \cdot 3489} = 2.024$$

∎

The case-control design is often the most efficient design for investigating rare outcome events, such as rare diseases. Although Table 10.30 was constructed assuming that the two samples, cases and controls, were chosen independently, a more common design is to incorporate matching of cases and controls with respect to potential confounding factors (for example, age). As we have seen, by taking advantage of the odds ratio, we can estimate the relative risk from a case-control study of a rare event even though we cannot estimate the risks p_1 and p_2 separately.

If the odds ratio (or the relative risk) is equal to 1.0, then the odds (or the risk) are the same for both of the groups being compared. In the smoking and birthweight data of Table 10.28 the calculated odds ratio was *greater* than 1.0, indicating that the odds of a low-birthweight baby are greater for smokers than for nonsmokers. Notice that we could have focused attention on the odds of a normal-birthweight baby. In this case, the odds ratio would be *less* than 1.0, as shown in Example 10.46.

Example 10.46

Smoking and Birthweight. Suppose we rearrange the data in Table 10.28 by putting normal birthweight in the first row and low birthweight in the second row:

		Smoking Status	
		Smoker	*Nonsmoker*
Birthweight	Normal	3489	5870
	Low	237	197
	Total	3726	6067

In this case the odds ratio is the odds of a normal-birthweight baby for a smoker divided by the odds of a normal-birthweight baby for a nonsmoker. We can calculate the estimated odds ratio as

$$\hat{\theta} = \frac{3489 \cdot 197}{5870 \cdot 237} = 0.494$$

This is the reciprocal of the odds ratio calculated in Example 10.45: $\frac{1}{2.024} = 0.494$.

The fact that the odds ratio is less than 1.0 means that the event (a normal-birthweight baby) is less likely for smokers than for nonsmokers. ∎

Confidence Interval for the Odds Ratio

In Chapter 6 we discussed confidence intervals for proportions, which are of the form $\tilde{p} \pm Z_{\alpha/2}\text{SE}_{\tilde{p}}$, where $\tilde{p} = \frac{y+2}{n+4}$. In particular, a 95% confidence interval for p is given by $\tilde{p} \pm Z_{.025}\text{SE}_{\tilde{p}}$. Such confidence intervals are based on the fact that for large samples the sampling distribution of $\tilde{p}$ is approximately normal (according to the central limit theorem).

In a similar way, we can construct a confidence interval for an odds ratio. One problem is that the sampling distribution of $\hat{\theta}$ is not normal. However, if we take the logarithm of $\hat{\theta}$, then we have a distribution that is approximately normal. Hence, we construct a confidence interval for θ by first finding a confidence interval for $\log(\theta)$* and then transforming the endpoints back to the original scale.

In order to construct a confidence interval for $\log(\theta)$, we need the standard error of $\log(\hat{\theta})$. The formula for the standard error of $\log(\hat{\theta})$ is given in the following box.

Standard Error of $\log(\hat{\theta})$

$$SE_{\log(\hat{\theta})} = \sqrt{\frac{1}{n_{11}} + \frac{1}{n_{12}} + \frac{1}{n_{21}} + \frac{1}{n_{22}}}$$

A 95% confidence interval for $\log(\theta)$ is given by $\log(\theta) \pm (1.96)SE_{\log(\hat{\theta})}$. We then exponentiate the two endpoints of the interval to get a 95% confidence interval for θ. Intervals with other confidence coefficients are constructed analogously; for instance, for a 90% confidence interval we would use $Z_{.05}(1.645)$ instead of $Z_{.025}(1.960)$. The process for finding a confidence interval for θ is summarized in the following box.[†]

Confidence Interval for θ
To construct a 95% confidence interval for θ,
1. Calculate $\log(\hat{\theta})$.
2. Construct a confidence interval for $\log(\theta)$ using the formula $\log(\hat{\theta}) \pm (1.96)SE_{\log(\hat{\theta})}$.
3. Exponentiate the endpoints to get a confidence interval for θ.

This process is illustrated in the following examples.

Smoking and Birthweight. From the data in Table 10.28, the estimated odds ratio is

Example 10.47

$$\hat{\theta} = \frac{237 \cdot 5870}{197 \cdot 3489} = 2.024$$

Thus, $\log(\hat{\theta}) = \log(2.024) = .705$. The standard error is given by

$$SE_{\log(\hat{\theta})} = \sqrt{\frac{1}{237} + \frac{1}{197} + \frac{1}{3489} + \frac{1}{5870}} = .0988.$$

A 95% confidence interval for $\log(\theta)$ is $.705 \pm (1.96)(.0988)$ or $.705 \pm .194$. This interval is $(.511, .899)$.

To get a 95% confidence interval for θ, we evaluate $e^{.511} = 1.67$ and $e^{.899} = 2.46$. Thus, we are 95% confident that the population value of the odds ratio is between 1.67 and 2.46. ∎

---

* We will use $\log(\theta)$ to denote the natural log (base e) of θ.

† A confidence interval for the relative risk can be found in a similar manner, for those situations in which the relative risk can be estimated from the data.

Example 10.48

Heart Attacks and Aspirin. During the Physician's Health Study, 11,037 physicians were randomly assigned to take 325 mg of aspirin every other day; 104 of them had heart attacks during the study. Another 11,034 physicians were randomly assigned to take a placebo; 189 of them had heart attacks. These data are shown in Table 10.31.[63] The odds ratio for comparing the heart attack rate on aspirin to the heart attack rate on placebo is

$$\hat{\theta} = \frac{189 \cdot 10933}{104 \cdot 10845} = 1.832$$

Thus, $\log\left(\hat{\theta}\right) = \log(1.832) = .605$.

The standard error is $\mathrm{SE}_{\log(\hat{\theta})} = \sqrt{\dfrac{1}{189} + \dfrac{1}{104} + \dfrac{1}{10845} + \dfrac{1}{10933}} = .123$.

A 95% confidence interval for $\log(\theta)$ is $.605 \pm (1.96)(.123)$ or $.605 \pm .241$. This interval is $(.364, .846)$.

TABLE 10.31	Heart Attacks on Placebo and on Aspirin	
	Placebo	**Aspirin**
Heart attack	189	104
No heart attack	10845	10933
Total	11034	11037

To get a 95% confidence interval for θ, we evaluate $e^{.364} = 1.44$ and $e^{.846} = 2.33$. Thus, we are 95% confident that the population value of the odds ratio is between 1.44 and 2.33. Because heart attacks are relatively rare in this data set, the relative risk is nearly equal to the odds ratio. Thus, we can say that we are 95% confident that the probability of a heart attack is about 1.44 to 2.33 times greater when taking the placebo than when taking aspirin. ■

Exercises 10.66–10.72

10.66 For each of the following tables, calculate (i) the relative risk and (ii) the odds ratio.

(a)
25	23
492	614

(b)
12	8
93	84

10.67 For each of the following tables, calculate (i) the relative risk and (ii) the odds ratio.

(a)
14	16
322	412

(b)
15	7
338	82

10.68 The medical records of heart disease patients who underwent balloon angioplasty at the Mayo Clinic were examined for the period between 1979 and 1995. One outcome that was recorded was whether or not the patient had a myocardial infarction (a heart attack). The data are shown in the following table.[64] Calculate the relative risk of myocardial infarction for smokers compared to nonsmokers.

		Smokers	Nonsmokers
Myocardial	Yes	23	25
infarction?	No	712	1984
	Total	735	2009

10.69 Consider the data from Exercise 10.68.

(a) Calculate the sample value of the odds ratio.
(b) Construct a 95% confidence interval for the population value of the odds ratio.
(c) Interpret the confidence interval from part (b) in the context of this setting.

10.70 As part of the National Health Interview Survey, occupational injury data were collected on thousands of American workers. The following table summarizes part of these data.[65]

		Self-employed	Employed by Others
Injured?	Yes	210	4391
	No	33724	421502
	Total	33934	425893

(a) Calculate the sample value of the odds ratio.
(b) According to the odds ratio, are self-employed workers more likely, or less likely, to be injured than persons who work for others?
(c) Construct a 95% confidence interval for the population value of the odds ratio.
(d) Interpret the confidence interval from part (b) in the context of this setting.

10.71 Many over-the-counter decongestants and appetite suppressants contain the ingredient phenypropanolamine. A study was conducted to investigate whether this ingredient is associated with strokes. The study found that 6 of 702 stroke victims had used an appetite suppressant containing phenypropanolamine, compared to only 1 of 1376 subjects in a control group. The following table summarizes these data.[66]

		Stroke	No Stroke
Appetite Suppressant?	Yes	6	1
	No	696	1375
	Total	702	1376

(a) Calculate the sample value of the odds ratio.
(b) Construct a 95% confidence interval for the population value of the odds ratio.
(c) Upon hearing of these data, some scientists called the study "inconclusive" because the numbers of users of appetite suppressants containing phenypropanolamine (7 total: 6 in one group and 1 in the other) are so small. What is your response to these scientists?

10.72 Two treatments, heparin and enoxaparin, were compared in a double-blind, randomized clinical trial of patients with coronary artery disease. The subjects can be classified as having a positive or negative response to treatment; the data are given in the following table.[67]

Outcome		Heparin	Enoxaparin
	Negative	309	266
	Positive	1255	1341
	Total	1564	1607

(a) Calculate the sample value of the odds ratio.

(b) Construct a 95% confidence interval for the population value of the odds ratio.

(c) Interpret the confidence interval from part (b) in the context of this setting.

10.10 SUMMARY OF CHI-SQUARE TESTS

We have discussed two types of chi-square tests: goodness-of-fit tests and contingency table tests. These tests are similar but are used for different purposes. The following summary should serve as a convenient reference for both tests and as a guide for distinguishing between them.

Summary of Chi-Square Tests

Goodness-of-Fit Test

Null hypothesis:

H_0 specifies the probability of each category.

Calculation of expected frequencies:

$$E = n \cdot \text{Probability specified by } H_0$$

Test statistic:

$$\chi_s^2 = \sum \frac{(O - E)^2}{E}$$

Null distribution (approximate):

χ^2 distribution with df = (Number of categories) − 1

This approximation is adequate if $E \geq 5$ for every category.

Contingency Table

Null hypothesis:

H_0: Row variable and column variable are independent.

Calculation of expected frequencies:

$$E = \frac{(\text{Row total}) \cdot (\text{Column total})}{\text{Grand total}}$$

Test statistic:

$$\chi_s^2 = \sum \frac{(O - E)^2}{E}$$

Null distribution (approximate):

χ^2 distribution with df $= (r - 1)(k - 1)$

where r is the number of rows and k is the number of columns in the contingency table. This approximation is adequate if $E \geq 5$ for every cell. If r and k are large, the condition that $E \geq 5$ is less critical and the χ^2 approximation is adequate if the average expected frequency is at least 5, even if some of the cell counts are smaller.

Supplementary Exercises 10.73–10.99

Note: Exercises preceded by an asterisk refer to optional sections.

10.73 When male mice are grouped, one of them usually becomes dominant over the others. In order to see how a parasitic infection might affect the competition for dominance, male mice were housed in groups, three mice to a cage; two mice in each cage received a mild dose of the parasitic worm *H. polygyrus*. Two weeks later, criteria such as the relative absence of tail wounds were used to identify the dominant mouse in each cage. It was found that the uninfected mouse had become dominant in 15 of 30 cages.[68] Is this evidence that the parasitic infection tends to inhibit the development of dominant behavior? Use a goodness-of-fit test against a directional alternative. Let $\alpha = .05$. (*Hint*: The observational unit in this experiment is not an individual mouse, but a cage of three mice.)

10.74 Are mice right-handed or left-handed? In a study of this question, 320 mice of a highly inbred strain were tested for paw preference by observing which forepaw—right or left—they used to retrieve food from a narrow tube. Each animal was tested 50 times, for a total of $320 \cdot 50 = 16,000$ observations. The results were as follows:[69]

	Right	**Left**
Number of Observations	7,871	8,129

Suppose we assign an expected frequency of 8,000 to each category and perform a goodness-of-fit test; we find that $\chi_s^2 = 4.16$, so that at $\alpha = .05$ we would reject the hypothesis of a 1:1 ratio and find that there is sufficient evidence to conclude that mice of this strain are (slightly) biased toward use of the left paw. This analysis contains a fatal flaw; what is it?

10.75 One explanation for the widespread incidence of the hereditary condition known as sickle-cell trait is that the trait confers some protection against malarial infection. In one investigation, 543 African children were checked for the trait and for malaria. The results are shown in the table.[70] Do the data provide evidence in favor of the explanation? The value of the chi-square statistic for this contingency table is $\chi_s^2 = 5.33$.

(a) Carry out the chi-square test against a directional alternative at $\alpha = .10$.
(b) Interpret the result of the test from part (a) in the context of this setting.

		Malaria		
		Heavy Infection	*Noninfected or Lightly Infected*	
Sickle-cell Trait	Yes	36	100	136
	No	152	255	407
	Total	188	355	543

10.76 As part of a study of environmental influences on sex determination in the fish *Menidia*, eggs from a single mating were divided into two groups and raised in either a warm or a cold environment. It was found that 73 of 141 offspring in the warm environment and 107 of 169 offspring in the cold environment were females.[71] In each of the following chi-square tests, use a nondirectional alternative and let $\alpha = .05$.

(a) Test the hypothesis that the population sex ratio is 1:1 in the warm environment.
(b) Test the hypothesis that the population sex ratio is 1:1 in the cold environment.
(c) Test the hypothesis that the population sex ratio is the same in the warm as in the cold environment.
(d) Define the population to which the conclusions reached in parts (a)–(c) apply. (Is it the entire genus *Menidia*?)

10.77 As part of the study of the inheritance pattern of cowpea plants, geneticists classified the plants in one experiment according to whether the plants had one leaf or three. The data are as follows:[72]

Number of Leaves	1	3
Number of Plants	74	61

Test the null hypothesis that the two types of plants occur with equal probabilities. Use a nondirectional alternative and let $\alpha = .05$.

10.78 People who harvest wild mushrooms sometimes accidentally eat the toxic "death-cap" mushroom, *Amanita phalloides*. In reviewing 205 European cases of death-cap poisoning from 1971 through 1980, researchers found that 45 of the victims had died.[73] Conduct a test to compare this mortality to the 30% mortality that was recorded before 1970. Let the alternative hypothesis be that mortality has decreased with time and let $\alpha = .05$.

10.79 The appearance of leaf pigment glands in the seedling stage of cotton plants is genetically controlled. According to one theory of the control mechanism, the population ratio of glandular to glandless plants resulting from a certain cross should be 11:5; according to another theory it should be 13:3. In one experiment, the cross produced 89 glandular and 36 glandless plants.[74] Use goodness-of-fit tests (at $\alpha = .10$) to determine whether these data are consistent with

(a) the 11:5 theory
(b) the 13:3 theory

10.80 When fleeing a predator, the minnow *Fundulus notti* will often head for shore and jump onto the bank. In a study of spatial orientation in this fish, individuals were caught at various locations and later tested in an artificial pool to see which direction they would choose when released: Would they swim in a direction which, at their place of capture, would have led toward shore? The following are the directional choices (±45°) of 50 fish tested under cloudy skies:[75]

Toward shore	18
Away from shore	12
Along shore to the right	13
Along shore to the left	7

Use chi-square tests at $\alpha = .05$ to test the hypothesis that directional choice under cloudy skies is random,

(a) using the four categories listed.
(b) collapsing to two categories—"toward shore" and "away from or along shore"—and using a directional H_A.
(*Note*: Although the chi-square test is valid in this setting, more powerful tests are available for analysis of orientation data.[76])

10.81 The cilia are hairlike structures that line the nose and help to protect the respiratory tract from dust and foreign particles. A medical team obtained specimens of nasal tissue from nursery school children who had viral upper respiratory infections, and also from healthy children in the same classroom. The tissue was sectioned and the cilia were examined with a microscope for specific defects, with the results shown in the accompanying table.[77] The data show that the percentage of defective cilia was much higher in the tissue from infected children (15.7% versus 3.1%). Would it be valid to apply a chi-square test to compare these percentages? If so, do it. If not, explain why not.

			Cilia with Defects	
	Number of Children	*Total Number of Cilia Counted*	*Number*	*Percent*
Control	7	556	17	3.1
Respiratory Infection	22	1,493	235	15.7

10.82 A group of mountain climbers participated in a trial to investigate the usefulness of the drug acetazolamide in preventing altitude sickness. The climbers were randomly assigned to receive either drug or placebo during an ascent of Mt. Rainier. The experiment was supposed to be double-blind, but the question arose whether some of the climbers might have received clues (perhaps from the presence or absence of side effects or from a perceived therapeutic effect or lack of it) as to which treatment they were receiving. To investigate this possibility, the climbers were asked (after the trial was over) to guess which treatment they had received.[78] The results can be cast in the following contingency table, for which $\chi_s^2 = 5.07$:

		Treatment Received	
		Drug	*Placebo*
Guess	Correct	20	12
	Incorrect	11	21

Alternatively the same results can be rearranged in the following contingency table, for which $\chi_s^2 = .01$:

		Treatment Received	
		Drug	*Placebo*
Guess	Drug	20	21
	Placebo	11	12

Consider the null hypothesis

H_0: The blinding was perfect (the climbers received no clues).

Carry out the chi-square test of H_0 against the alternative that the climbers did receive clues. Let $\alpha = .05$. (You must decide which contingency table is relevant to this question.) (*Hint*: To clarify the issue for yourself, try inventing a fictitious data set in which most of the climbers *have* received strong clues, so that we would expect a large value of χ_s^2; then arrange your fictitious data in each of the two contingency table formats and note which table would yield a larger value of χ_s^2.)

*10.83 Desert lizards (*Dipsosaurus dorsalis*) regulate their body temperature by basking in the sun or moving into the shade, as required. Normally the lizards will maintain a daytime temperature of about 38°C. When they are sick, however, they maintain a temperature about 2° to 4° higher—that is, a "fever." In an experiment to see whether this fever might be beneficial, lizards were given a bacterial infection; then 36 of the animals were prevented from developing a fever by keeping them in a 38° enclosure, while 12 animals were kept at a temperature of 40°. The following table describes the mortality after 24 hours.[79] How strongly do these results support the hypothesis that fever has survival value? Use Fisher's exact test against a directional alternative. Let $\alpha = .05$.

	38°	40°
Died	18	2
Survived	18	10
Total	36	12

10.84 Consider the data from Exercise 10.83. Analyze these data with a chi-square test. Let $\alpha = .05$.

10.85 In a randomized clinical trial, 154 women with breast cancer were assigned to receive chemotherapy. Another 164 women were assigned to receive chemotherapy combined with radiation therapy. Survival data after 15 years are given in the following table.[80] Use these data to conduct a test of the null hypothesis that type of treatment does not affect survival rate. Let $\alpha = .05$.

	Chemotherapy Only	Chemotherapy and Radiation Therapy
Died	78	66
Survived	76	98
Total	154	164

*10.86 Refer to the data in Exercise 10.85.

(a) Calculate the sample odds ratio.
(b) Find a 95% confidence interval for the population value of the odds ratio.

10.87 Two drugs, zidovudine and didanosine, were tested for their effectiveness in preventing progression of HIV disease in children. In a double-blind clinical trial, 276 children with HIV were given zidovudine, 281 were given didanosine, and 274 were given zidovudine plus didanosine. The following table shows the survival data for the three groups.[81] Use these data to conduct a test of the null hypothesis that survival and treatment are independent. Let $\alpha = .10$.

	Zidovudine	Didanosine	Zidovudine and Didanosine
Died	17	7	10
Survived	259	274	264
Total	276	281	274

10.88 A group of inner-city African American adolescents were randomly divided into three groups as part of an experiment to assess the effectiveness of different HIV prevention programs. The first group was given "abstinence HIV intervention," the second group was given "safer-sex HIV intervention," and the third group was given "health promotion intervention," which was to serve as a control. One outcome that was measured was whether subjects who were sexually active during a three-month period reported consistent condom use. The data are shown in the following table.[82] Use these data to conduct a test of the null hypothesis that the response variable is independent of treatment group. Let $\alpha = .05$.

		Abstinence Intervention	Safer-sex Intervention	Control
Consistent	Yes	14	20	21
Condom Use?	No	20	12	20
	Total	34	32	41

10.89 The habitat selection behavior of the fruitfly *Drosophila subobscura* was studied by capturing flies from two different habitat sites. The flies were marked with colored fluorescent dust to indicate the site of capture and then released at a point midway between the original sites. On the following two days, flies were recaptured at the two sites. The results are summarized in the table.[83] The value of the chi-square statistic for this contingency table is $\chi_s^2 = 10.44$. Test the null hypothesis of independence against the alternative that the flies preferentially tend to return to their site of capture. Let $\alpha = .01$.

		Site of Recapture	
		I	II
Site of Original Capture	I	78	56
	II	33	58

10.90 In the garden pea *Pisum sativum*, seed color can be yellow (Y) or green (G), and seed shape can be round (R) or wrinkled (W). Consider the following three hypotheses describing a population of plants:

$$H_0^{(1)}: \Pr\{Y\} = \tfrac{3}{4}$$

$$H_0^{(2)}: \Pr\{R\} = \tfrac{3}{4}$$

$$H_0^{(3)}: \Pr\{R|Y\} = \Pr\{R|G\}$$

The first hypothesis asserts that yellow and green plants occur in a 3:1 ratio; the second hypothesis asserts that round and wrinkled plants occur in a 3:1 ratio, and the third hypothesis asserts that color and shape are independent. (In fact, for a population of plants produced by a certain cross—the dihybrid cross—all three hypotheses are known to be true.)

Suppose a random sample of 1,600 plants is to be observed, with the data to be arranged in the following contingency table:

		Color	
		Y	G
Shape	R		
	W		
		1600	

Invent fictitious data sets as specified, and verify each answer by calculating the estimated conditional probabilities. (*Hint*: In each case, begin with the marginal frequencies.)

(a) A data set that agrees perfectly with $H_0^{(1)}$, $H_0^{(2)}$, and $H_0^{(3)}$
(b) A data set that agrees perfectly with $H_0^{(1)}$ and $H_0^{(2)}$ but not with $H_0^{(3)}$
(c) A data set that agrees perfectly with $H_0^{(3)}$ but not with $H_0^{(1)}$ or $H_0^{(2)}$

*10.91 A study of 36,080 persons who had heart attacks found that men were more likely to survive than were women. The following table shows some of the data collected in the study.[84]

		Men	**Women**
Survived at	Yes	25,339	8,914
Least 24 Hours?	No	1,141	686
	Total	26,480	9,600

(a) Calculate the odds ratio for comparing survival of men to survival of women.
(b) Calculate a 95% confidence interval for the population value of the odds ratio.
(c) Does the odds ratio give a good approximation to the relative risk for these data? Why or why not?

*10.92 In the study described in Exercise 10.71, one of the variables measured was whether the subjects had used *any* products containing phenypropanolamine. The odds ratio was calculated to be 1.49, with stroke victims more likely than the control subjects to have used a product containing phenypropanolamine.[66] A 95% confidence interval for the population value of the odds ratio is (0.84, 2.64). Interpret this confidence interval in the context of this setting.

10.93 Refer to the cortex-weight data of Exercise 9.17.

(a) Use a goodness-of-fit test to test the hypothesis that the environmental manipulation has no effect. As in Exercise 9.17, use a directional alternative and let $\alpha = .05$. (This exercise shows how, by a shift of viewpoint, the sign test can be reinterpreted as a goodness-of-fit test. Of course, the chi-square goodness-of-fit test described in this chapter can be used only if the number of observations is large enough.)
(b) Is the number of observations large enough for the test in part (a) to be valid?

10.94 A biologist wanted to know if the cowpea weevil has a preference for one type of bean over others as a place to lay eggs. She put equal amounts of four types of seeds into a jar and added adult cowpea weevils. After a few days she observed the following data:[85]

Type of Bean	**Number of Eggs**
Pinto	167
Cowpea	176
Navy beans	174
Northern beans	194

Do these data provide evidence of a preference for one type of bean? That is, are the data consistent with the claim that the eggs are distributed randomly among the four types of bean?

10.95 An experiment was conducted in which two types of acorn squash were crossed. According to a genetic model, 1/2 of the resulting plants should have dark stems and dark fruit, 1/4 should have light stems and light fruit, and 1/4 should have light stems and plain fruit. The actual data were 220, 129, and 105 for these three categories.[86] Are these data consistent with the model? Conduct a chi-square test with $\alpha = .10$.

10.96 *(Computer exercises)*   In a study of the effects of smoking cigarettes during pregnancy, researchers examined the placenta from each of 58 women after childbirth. They noted the presence or absence (P or A) of a particular placental abnormality—atrophied villi. In addition, each woman was categorized as a nonsmoker (N), moderate smoker (M), or heavy smoker (H). The following table shows, for each woman, an ID number (#) and the results for smoking (S) and atrophied villi (V).[87]

#	S	V	#	S	V	#	S	V	#	S	V
1	N	A	16	H	P	31	M	A	46	M	A
2	M	A	17	H	P	32	M	A	47	H	P
3	N	A	18	N	A	33	N	A	48	H	P
4	M	A	19	M	P	34	N	A	49	H	A
5	M	A	20	N	P	35	N	A	50	N	P
6	M	P	21	M	A	36	H	P	51	N	A
7	H	P	22	H	A	37	N	A	52	M	P
8	N	A	23	M	P	38	H	P	53	M	A
9	N	A	24	N	A	39	H	P	54	H	P
10	M	P	25	N	P	40	N	A	55	H	A
11	N	A	26	N	A	41	M	A	56	M	P
12	N	P	27	N	A	42	N	A	57	H	P
13	H	P	28	M	P	43	H	A	58	H	P
14	M	A	29	N	A	44	M	A			
15	M	P	30	N	A	45	M	P			

(a) Test for a relationship between smoking status and atrophied villi. Use a chi-square test at $\alpha = .05$.

(b) Prepare a table that shows the total number of women in each smoking category, and the number and percentage in each category who had atrophied villi.

(c) What pattern appears in the table of part (b) that is not used by the test of part (a)?

10.97   Each of 36 men was asked to touch the backs of the hands of three women, one of whom was the man's romantic partner, while blindfolded. The two "decoy" women were the same age, height, and weight as the man's partner. [14] Of the 36 men tested, 16 were able to correctly identify their partner. Are these data consistent with the claim that the men were guessing? Conduct a goodness-of-fit test of the data, using $\alpha = .05$.

10.98   Consider Exercise 10.97. The romantic partners of the 36 men discussed in Exercise 10.97 were also tested, in the same manner as the men (i.e., they were blindfolded and asked to identify their partner by touching the backs of the hands of three men, one of whom was their partner). Among the women, 25 were successful and 11 were not. Are these data consistent with the hypothesis that men and women are equally good at indentifying their partners?

(a) Conduct a test, using $\alpha = .05$; use a nondirectional alternative.

(b) Interpret the result of your test from part (a) in the context of this setting.

*10.99   Researchers studied the cellular telephone records of 699 persons who had automobile accidents. They determined that 170 of the 699 had made a cellular telephone call during the 10-minute period prior to their accident; this period is called the hazard interval. There were 37 persons who had made a call during a corresponding 10-minute period on the day before their accident; this period is called the control interval. Finally, there were 13 who made calls both during the hazard interval and the control interval.[88] Do these data indicate that use of a cellular telephone is associated with an increase in accident rate? Analyze these data using McNemar's test. Use a directional alternative and let $\alpha = .01$.

(a) State the null hypothesis in words.
(b) Do you reject H_0? Why or why not?
(c) State your conclusion from part (b) in the context of the setting.

		Call During Control Interval?	
		Yes	*No*
Call During	Yes	13	157
Hazard Interval?	No	24	505

Comparing the Means of Many Independent Samples

Objectives

In this chapter we study analysis of variance (ANOVA). We will

- *discuss when and why an analysis of variance may be conducted.*

- *learn how ANOVA calculations are carried out.*

- *construct a model for ANOVA and show how that model can be extended.*

- *learn how to verify the conditions under which ANOVA is valid.*

- *learn about randomized blocks ANOVA and factorial ANOVA.*

- *learn how to construct contrasts and other linear combinations of means.*

- *learn how to deal with multiple comparisons.*

11.1 INTRODUCTION

In Chapter 7 we considered the comparison of two independent samples with respect to a quantitative variable Y. The classical techniques for comparing the two sample means $\bar{y}_1$ and $\bar{y}_2$ are the test and the confidence interval based on Student's t distribution. In the present chapter we consider the comparison of the means of I independent samples, where I may be greater than 2. The following example illustrates an experiment with $I = 5$.

Sweet Corn.   When growing sweet corn, can organic methods be used successfully to control harmful insects and limit their effect on the corn? In a study of this question, researchers compared the weights of ears of corn under five conditions in an experiment in which sweet corn was grown using organic methods. In one plot of corn a beneficial soil nematode was introduced. In a second plot a parasitic wasp was used. A third plot was treated with both the nematode and the wasp. In a fourth plot a bacterium was used. Finally, a fifth plot of corn acted as a control; no special treatment was applied here. Thus, the treatments were as follows:

> Treatment 1: Nematodes
> Treatment 2: Wasps
> Treatment 3: Nematodes and wasps
> Treatment 4: Bacteria
> Treatment 5: Control

Example 11.1

Ears of corn were randomly sampled from each plot and weighed. The results are given in Table 11.1 and plotted in Figure 11.1[1] Note that in addition to the differences between the treatment means there is also considerable variation within each treatment group.

TABLE 11.1 Weights (Ounces) of Ears of Sweet Corn				
		Treatment		
1	*2*	*3*	*4*	*5*
16.5	11	8.5	16	13
15	15	13	14.5	10.5
11.5	9	12	15	11
12	9	10	9	10
12.5	11.5	12.5	10.5	14
9	11	8.5	14	12
16	9	9.5	12.5	11
6.5	10	7	9	9.5
8	9	10.5	9	18.5
14.5	8	10.5	9	17
7	8	13	6.5	10
10.5	5	9	8.5	11
Mean 11.5	9.6	10.3	11.1	12.3
SD 3.5	2.4	2.0	3.1	2.9
n 12	12	12	12	12

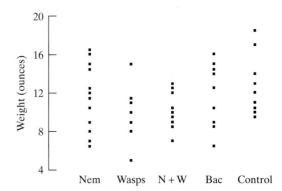

Figure 11.1 Weight of ears of corn receiving five different treatments

We will discuss the classical method of analyzing data from *I* independent samples. The method is called an **analysis of variance**, or **ANOVA**. In applying analysis of variance, the data are regarded as random samples from *I* populations. We will denote the means of these populations as $\mu_1, \mu_2, \ldots, \mu_I$ and the standard deviations as $\sigma_1, \sigma_2, \ldots, \sigma_I$.

Why Not Repeated *t* Tests?

It is natural to wonder why the comparison of the means of *I* samples requires any new methods. For instance, why not just use a two-sample *t* test on each pair of samples? There are three reasons why this is not a good idea.

1. **The problem of multiple comparisons**   The most serious difficulty with a naive "repeated t tests" procedure concerns Type I error: The probability of false rejection of a null hypothesis may be much higher than it appears to be. For instance, suppose $I = 6$. Among six means there are 15 pairs of means, so that 15 hypotheses can be considered; these hypotheses are

$$(1)\ H_0: \mu_1 = \mu_2$$
$$(2)\ H_0: \mu_1 = \mu_3$$
$$(3)\ H_0: \mu_2 = \mu_3$$
$$\vdots$$
$$(15)\ H_0: \mu_5 = \mu_6$$

If t tests are used to test each of these hypotheses at $\alpha = .05$, then there is a 5% risk of a Type I error for *each* of the 15 tests, but the overall risk, of at least one Type I error, is much higher than 5%.

   The consequences of using repeated t tests are indicated by Table 11.2. For tests at $\alpha = .05$, Table 11.2 shows the overall risk of Type I error,* that is,

Overall risk = Probability that at least one of the t tests will reject its

null hypothesis, when in fact $\mu_1 = \mu_2 = \cdots = \mu_I$

If $I = 2$, then the overall risk is .05, as it should be, but with larger I the risk increases rapidly; for $I = 6$ it is .37. It is clear from Table 11.2 that the researcher who uses repeated t tests is highly vulnerable to Type I error unless I is quite small.

TABLE 11.2 Overall Risk of Type I Error in Using Repeated t Tests at $\alpha = .05$

I	Overall Risk
2	.05
3	.12
4	.20
6	.37
8	.51
10	.63

   The difficulties illustrated by Table 11.2 are due to **multiple comparisons**—that is, many comparisons on the same set of data. These difficulties can be reduced when the comparison of several groups is approached through ANOVA.

2. **Estimation of the standard deviation**   The ANOVA technique combines information on variability from all of the samples simultaneously. This global sharing of information can yield improved precision in the analysis.

---

* Table 11.2 was computed assuming that the sample sizes are large and equal and that the population distributions are normal with equal standard deviations.

3. **Structure in the groups**   In many studies the logical structure of the treatments or groups to be compared may inspire questions that cannot be answered by simple pairwise comparisons. For example, we may wish to study the effects of two experimental factors simultaneously. ANOVA can be used to analyze data in such settings (see optional Sections 11.6 and 11.7).

A Graphical Perspective on ANOVA

When data are analyzed by analysis of variance, the usual first step is to test the following global null hypothesis:

$$H_0: \mu_1 = \mu_2 = \cdots = \mu_I$$

which asserts that all of the population means are equal. A statistical test of H_0 will be described in Section 11.4. However, we will first consider analysis of variance from a graphical perspective.

Consider the dotplots shown in Figure 11.2(a). These dotplots were generated in a setting in which H_0 is true. The sample means, which are shown as circles on the graph, differ from one another only as a result of chance error. For the data shown in Figure 11.2(b) H_0 is false. The sample means—again shown as circles—are quite different, which provides evidence that the corresponding population means (μ_1, μ_2, μ_3, and μ_4) are not all equal.

Figure 11.2 (a) H_0 true, (b) H_0 false, with small SDs for the groups.

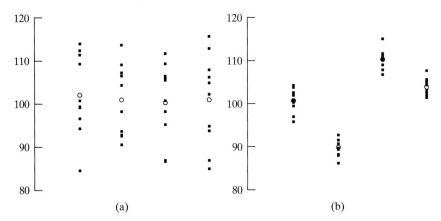

(a)   (b)

Figure 11.3 shows a situation that is less clear. In fact, H_0 is false here—the means in Figure 11.3 are identical to those in Figure 11.2(b). However, the standard deviations in the groups are quite large, which makes it hard to tell that the population means differ.

Figure 11.3 H_0 false, with large SDs for the groups

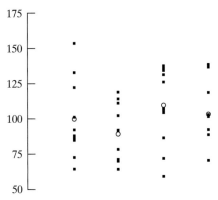

We need to know how much inherent variability there is in the data before we can judge whether a difference in sample means is fairly small and attributable to chance or whether it is too large to be due to chance alone. In order to make an inference about *means*, we compare two kinds of *variability*: (1) variability between sample means and (2) variability within groups. Hence, the procedure is called analysis of variance, although what we are comparing are means.

A Look Ahead

If the global null hypothesis that $\mu_1 = \mu_2 = \cdots = \mu_I$ is rejected, then the data provide sufficient evidence to conclude that at least *some* of the μ's are unequal; the researcher would usually proceed to detailed comparisons to determine the *pattern* of differences among the μ's. If the global null hypothesis is not rejected, then the researcher might choose to construct one or more confidence intervals to characterize the lack of difference among the μ's.

All of the statistical procedures of this chapter—the test of the global null hypothesis and various methods of making detailed comparisons among the means—depend on the same basic calculations. These calculations are presented in Section 11.2.

11.2 THE BASIC ANALYSIS OF VARIANCE

In this section we present the basic ANOVA calculations that are used to describe the data and to facilitate further analysis. The analysis of variance of I samples, or groups, begins with the calculation of quantities that describe the variability of the data *between* the groups and *within* the groups.* (For clarity, in this chapter we will often refer to the samples as "groups" of observations.)

Notation

To describe several groups of quantitative observations, we will use two subscripts: one to keep track of group membership and the other to keep track of observations with the groups. Thus, we will denote observation j in group i as

$$y_{ij} = \text{observation } j \text{ in group } i$$

Thus, the first observation in the first group is y_{11}, the second observation in the first group is y_{12}, the third observation in the second group is y_{23}, and so on.

We will also use the following notation:

$$I = \text{number of groups}$$
$$n_i = \text{number of observations in group } i$$
$$\bar{y}_{i\bullet} = \text{group mean for group } i$$

---

* Grammatically speaking, the word *among* should he used rather than between when referring to three or more groups; however, we will use *between* because it more clearly suggests that the groups are being compared against each other.

A dot subscript indicates that we have averaged over that index. Here the notation $\bar{y}_{i\cdot}$ represents the average, as j goes from 1 to n_i, of the observations in group i. Thus,

$$\bar{y}_{i\cdot} = \frac{(y_{i1} + y_{i2} + \cdots + y_{in_1})}{n_i} = \frac{\sum_{j=1}^{n_i} y_{ij}}{n_i}$$

The total number of observations is

$$n^* = n_1 + n_2 + \cdots + n_I = \sum_{i=1}^{I} n_i$$

Finally, the **grand mean**—the mean of all the observations—is

$$\bar{y}_{\cdot\cdot} = \frac{\sum_{i=1}^{I} \sum_{j=1}^{n_i} y_{ij}}{n^*}$$

The following example illustrates this notation.

Example 11.2   **Weight Gain of Lambs.**   Table 11.3 shows the weight gains (in 2 weeks) of young lambs on three different diets. (These data are fictitious, but are realistic in all respects except for the fact that the group means are whole numbers.[2])

TABLE 11.3 Weight Gains of Lambs (lb)

	Diet 1	Diet 2	Diet 3
	8	9	15
	16	16	10
	9	21	17
		11	6
		18	
n	3	5	4
Sum $= \Sigma y_{ij}$	33	75	48
Mean $= \bar{y}_{i\cdot}$	11	15	12

The total number of observations is

$$n^* = 3 + 5 + 4 = 12$$

and the total of all the observations is

$$\sum_{i=1}^{I} \sum_{j=1}^{n_i} y_{ij} = 33 + 75 + 48 = 156$$

The grand mean is

$$\bar{y}_{\cdot\cdot} = \frac{156}{12} = 13 \text{ lb}$$

∎

If the sample sizes (n_i's) are all equal, then the grand mean $\bar{y}_{\cdot\cdot}$ is just the simple average of the group means (the $\bar{y}_{i\cdot}$'s); but if the sample sizes are unequal, this is not the case. For instance, in Example 11.2 note that

$$\frac{12 + 15 + 11}{3} \neq 13$$

Variation Within Groups

A combined measure of variation within the I groups is the **sum of squares within groups**, or **SS(within)**,* defined as follows:

> ### Sum of Squares Within Groups
>
> $$\text{SS(within)} = \sum_{i=1}^{I} \sum_{j=1}^{n_i} (y_{ij} - \bar{y}_{i\cdot})^2$$

The double sum $\left(\sum_{i=1}^{I} \sum_{j=1}^{n_i} \right)$ is calculated by first adding within each group $\left(\sum_{j=1}^{n_i} \right)$ and then adding across groups $\left(\sum_{i=1}^{I} \right)$. The following example illustrates the calculation of SS(within).

Weight Gain of Lambs. Table 11.4 shows the lamb weight-gain data, together with the group means and sums of squares.

Example 11.3

> **TABLE 11.4 Calculation of SS(Within) for Lamb Weight Gains**
>
> | | Diet 1 | Diet 2 | Diet 3 |
> |---|---|---|---|
> | | 8 | 9 | 15 |
> | | 16 | 16 | 10 |
> | | 9 | 21 | 17 |
> | | | 11 | 6 |
> | | | 18 | |
> | n | 3 | 5 | 4 |
> | Mean $= \bar{y}_{i\cdot}$ | 11 | 15 | 12 |
> | Sum $= \Sigma(y_{ij} - \bar{y}_{i\cdot})^2$ | 38 | 98 | 74 |

To calculate SS(within), we first calculate the quantity $\sum_{j=1}^{n_i} (y_{ij} - \bar{y}_{i\cdot})^2$ for each group, as shown in Table 11.4; for instance,

$$(8 - 11)^2 + (16 - 11)^2 + (9 - 11)^2 = 38$$

Then we add across groups:

$$\text{SS(within)} = 38 + 98 + 74 = 210 \qquad \blacksquare$$

Associated with SS(within) is a quantity called **degrees of freedom within groups**, or **df(within)**, which is defined as follows:

> ### df Within Groups
>
> $$\text{df(within)} = n^* - I$$

---

* Some authors call this the SS(error), rather than SS(within).

Note that df(within) is equal to the sum of the degrees of freedom within each group:

$$df(within) = (n_1 - 1) + (n_2 - 1) + \cdots + (n_I - 1)$$

Finally, we define the **mean square within groups**, or **MS(within)**, as follows:

Mean Square Within Groups

$$MS(within) = \frac{SS(within)}{df(within)}$$

The quantity MS(within) is a measure of variability within the groups. If there were only one group, with n observations, then df(within) would be $n - 1$ and the SS(within) would be $\sum_{j=1}^{n} (y_j - \bar{y})^2$. MS(within) would be $\dfrac{\sum_{j=1}^{n} (y_j - y)^2}{n - 1}$.

Thus, if there were only one group, the MS(within) would be the sample variance, s^2 (i.e., the square of the sample standard deviation for the group).

Analysis of variance deals with several groups simultaneously. Note that $\sum_{j=1}^{n_i} (y_{ij} - \bar{y}_{i\cdot})^2$ is related to the sample variance of group i:

$$\sum_{j=1}^{n_i} (y_{ij} - \bar{y}_{i\cdot})^2 = (n_i - 1) \cdot s_i^2$$

The MS(within) is a combination of the variances of the groups:

$$MS(within) = \frac{(n_1 - 1)s_1^2 + (n_2 - 1)s_2^2 + \cdots + (n_I - 1)s_I^2}{(n_1 - 1) + (n_2 - 1) + \cdots + (n_I - 1)}$$

We can think of the MS(within) calculation as pooling together measurements of variability from the different groups. We will use the notation s_{pooled} to denote the resulting pooled standard deviation:

Pooled Standard Deviation

$$s_{pooled} = \sqrt{MS(within)}$$

The number of degrees of freedom associated with s_{pooled} (that is, the denominator of s_{pooled}^2) is the sum of the df associated with each sample SD:

$$df(within) = n^* - I = (n_1 - 1) + (n_2 - 1) + \cdots + (n_I - 1)$$

These relationships have a simple interpretation. Recall from Section 6.2 that the df associated with a sample SD is the number of independent pieces of information (about variability) upon which the SD is based. If we assume that the population SD is the same in all I populations, then s_{pooled} is an estimate of the population SD and df(within) expresses the total amount of information upon which the estimate is based.

The following example illustrates the calculation and interpretation of MS(within) and s_{pooled}.

Weight Gain of Lambs. For the lamb growth data of Example 11.2, $I = 3$ and $n^* = 12$, so that

$$df(within) = 12 - 3 = 9$$

Example 11.4

We found in Example 11.3 that SS(within) = 210; thus,

$$MS(within) = \frac{210}{9} = 23.333$$

and

$$s_{pooled} = \sqrt{23.333} = 4.83 \, lb$$

If we assume that the population standard deviation of weight gains is the same for all three diets, then we estimate that standard deviation to be 4.83 lb.

Table 11.5 shows the individual sample SDs and their associated df. (Notice that s_{pooled} is within the range of the individual SDs; this will be true for any data set.) The individual SDs are estimates of the corresponding population SDs, and the df reflect the precision of the estimates. The pooled estimate s_{pooled} is based on $2 + 4 + 3 = 9$ df and is related to the individual SDs as follows:

$$s_{pooled}^2 = \frac{2}{9}s_1^2 + \frac{4}{9}s_2^2 + \frac{3}{9}s_3^2$$

TABLE 11.5 Sample SDs and df for Lamb Weight Gains

	Diet 1	Diet 2	Diet 3
SD	4.36	4.95	4.97
df = $n - 1$	2	4	3

The value of MS(within) depends only on the variability within the groups. In Example 11.4 MS(within) = 210. Had the sample means been different, but the SDs of the groups had been the same, the MS(within) would not have changed. Figure 11.4(a) shows the data from Table 11.3. Figure 11.4(b) shows the distributions after adding 7 to each of the observations for Diet 2 and subtracting 5 from each of the observations for Diet 3. MS(within) is 210 in either case.

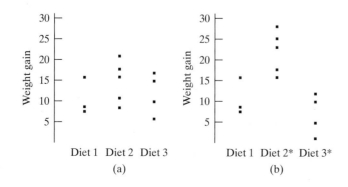

Figure 11.4 (a) Dotplots of the weight-gain data from Table 11.3, with MS(within) = 210; (b) dotplots of modified data, with MS(within) again 210.

Variation Between Groups

For two groups, the difference between the groups is simply described by $(\bar{y}_1 - \bar{y}_2)$. How can we describe between-group variability for more than two groups? It turns out that a convenient measure is the sum of **squares between groups**, or **SS(between)**, defined as follows:

Sum of Squares Between Groups

$$\text{SS(between)} = \sum_{i=1}^{I} n_i(\bar{y}_{i\cdot} - \bar{y}_{\cdot\cdot})^2$$

Each term in SS(between) is the square of a difference between the group mean $\bar{y}_{i\cdot}$ and the grand mean $\bar{y}_{\cdot\cdot}$, multiplied by the group size, n_i; this can be written explicitly as

$$\text{SS(between)} = n_1(\bar{y}_{1\cdot} - \bar{y}_{\cdot\cdot})^2 + n_2(\bar{y}_{2\cdot} - \bar{y}_{\cdot\cdot})^2 + \cdots + n_I(\bar{y}_{I\cdot} - \bar{y}_{\cdot\cdot})^2$$

Associated with SS(between) is the **degrees of freedom between groups**, or **df(between)**, defined as follows:

---

df Between Groups

$$\text{df(between)} = I - 1$$

The **mean square between groups**, or **MS(between)**, is defined as follows:

Mean Square Within Groups

$$\text{MS(between)} = \frac{\text{SS(between)}}{\text{df(between)}}$$

The following example illustrates these definitions.

Example 11.5   **Weight Gain of Lambs.**   For the data of Example 11.2, the quantities that enter SS(between) are shown in Table 11.6.

TABLE 11.6 Calculation of SS(Between) for Lamb Weight Gains

	Diet 1	Diet 2	Diet 3
n	3	5	4
Mean $\bar{y}_{i\cdot}$	11	15	12

Grand mean $\bar{y}_{\cdot\cdot} = 13$

From Table 11.6 we calculate

$$SS(\text{between}) = 3(11 - 13)^2 + 5(15 - 13)^2 + 4(12 - 13)^2 = 36$$

Since $I = 3$, we have

$$df(\text{between}) = 3 - 1 = 2$$

so that

$$MS(\text{between}) = \frac{36}{2} = 18$$

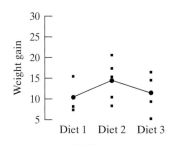

Figure 11.5 Differences in group means for lamb weight gains

The SS(between) and MS(between) measure the variability between the sample means of the groups. This variability is shown graphically in Figure 11.5.

A Fundamental Relationship of ANOVA

The name *analysis of variance* derives from a fundamental relationship involving SS(between) and SS(within). Consider an individual observation y_{ij}. It is obviously true that

$$y_{ij} - \bar{y}_{..} = (y_{ij} - \bar{y}_{i.}) + (\bar{y}_{i.} - \bar{y}_{..})$$

This equation expresses the deviation of an observation from the grand mean as the sum of two parts: a within-group deviation $(y_{ij} - \bar{y}_{i.})$ and a between-group deviation $(\bar{y}_{i.} - \bar{y}_{..})$. It is also true (but not at all obvious) that the analogous relationship holds for the corresponding sums of squares; that is,

$$\sum_{i=1}^{I} \sum_{j=1}^{n_i} (y_{ij} - \bar{y}_{..})^2 = \sum_{i=1}^{I} \sum_{j=1}^{n_i} (y_{ij} - \bar{y}_{i.})^2 + \sum_{i=1}^{I} n_i(\bar{y}_{i.} - \bar{y}_{..})^2 \qquad (11.1)$$

The quantity on the left-hand side of (11.1) is called the total sum of squares, or SS(total):

> **Total Sum of Squares**
>
> $$SS(\text{total}) = \sum_{i=1}^{I} \sum_{j=1}^{n_i} (y_{ij} - \bar{y}_{..})^2$$

Note that SS(total) measures variability among all n^* observations in the I groups. The relationship (11.1) can be written as

> **Relationship Between Sums of Squares**
>
> $$SS(\text{total}) = SS(\text{between}) + SS(\text{within})$$

The preceding fundamental relationship shows how the total variation in the data set can be analyzed, or broken down, into two interpretable components: between-sample variation and within-sample variation. This partition is an analysis of variance.

The total degrees of freedom, or df(total), is defined as follows:

> **Total df**
>
> $$df(\text{total}) = n^* - 1$$

With this definition, the degrees of freedom add, just as the sums of squares do; that is,

$$df(\text{total}) = df(\text{within}) + df(\text{between})$$
$$n^* - 1 = (n^* - I) + (I - 1)$$

Notice that, if we were to consider all n^* observations as a single sample, then the SS for that sample (that is, the numerator of the variance) would be SS(total) and the associated df (that is, the denominator of the variance) would be df(total). The following example illustrates the fundamental relationships between the sums of squares and degrees of freedom.

Example 11.6 **Weight Gain of Lambs.** For the data of Table 11.3, we found $\bar{y}.. = 13$; we calculate SS(total) as

$$SS(\text{total}) = \sum_{i=1}^{I} \sum_{j=1}^{n_1} (y_{ij} - \bar{y}..)^2$$
$$= [(8 - 13)^2 + (16 - 13)^2 + (9 - 13)^2]$$
$$+ [(9 - 13)^2 + (16 - 13)^2 + (21 - 13)^2 + (11 - 13)^2 + (18 - 13)^2]$$
$$+ [(15 - 13)^2 + (10 - 13)^2 + (17 - 13)^2 + (6 - 13)^2]$$
$$= 246$$

For these data, we found that SS(between) = 36 and SS(within) = 210. We verify that

$$246 = 36 + 210$$

Also, we found that df(within) = 9 and df(between) = 2. We verify that

$$df(\text{total}) = 12 - 1 = 11 = 9 + 2 \qquad \blacksquare$$

The ANOVA Table

When working with the ANOVA quantities, it is customary to arrange them in a table. The following example shows a typical format for the ANOVA table.

Example 11.7 **Weight Gain of Lambs.** Table 11.7 shows the ANOVA for the lamb weight gain data. Notice that the ANOVA table clearly shows the additivity of the sums of squares and the degrees of freedom. $\qquad \blacksquare$

TABLE 11.7 ANOVA Table for Lamb Weight Gains

Source	df	SS	MS
Between diets	2	36	18
Within diets	9	210	23.333
Total	11	246	

Summary of Formulas

For convenient reference, we display in the box the definitional formulas for the basic ANOVA quantities.

ANOVA Quantities with Formulas

Source	df	SS(Sum of Squares)	MS(Mean Square)
Between groups	$I - 1$	$\sum_{i=1}^{I} n_i(\bar{y}_{i\cdot} - \bar{y}_{\cdot\cdot})^2$	SS/df
Within groups	$n^* - I$	$\sum_{i=1}^{I} \sum_{i=1}^{n_i} (y_{ij} - \bar{y}_{i\cdot})^2$	
Total	$n^* - 1$	$\sum_{i=1}^{I} \sum_{j=1}^{n_i} (y_{ij} - \bar{y}_{\cdot\cdot})^2$	

Exercises 11.1–11.7

11.1   The accompanying table shows fictitious data for three samples.

	Sample		
	1	*2*	*3*
	48	40	39
	39	48	30
	42	44	32
	43		35
Mean	43	44	34

(a) Compute SS(between) and SS(within).
(b) Compute SS(total), and verify the relationship between SS(between), SS(within), and SS(total).
(c) Compute MS(between), MS(within), and s_{pooled}.

11.2   Proceed as in Exercise 11.1 for the following data:

	Sample		
	1	*2*	*3*
	23	18	20
	29	12	16
	25	15	17
	23		23
			19
Mean	25	15	19

11.3   For the following data, SS(within) = 116 and SS(total) = 338.769.

	Sample		
	1	*2*	*3*
	31	30	39
	34	26	45
	39	35	39
	32	29	37
		30	

(a) Find SS(between).

(b) Compute MS(between), MS(within), and s_{pooled}.

11.4   The following ANOVA table is only partially completed.

Source	df	SS	MS
Between groups	3		45
Within groups	12	337	
Total		472	

(a) Complete the table.

(b) How many groups were there in the study?

(c) How many total observations were there in the study?

11.5   The following ANOVA table is only partially completed.

Source	df	SS	MS
Between groups	4		
Within groups		964	
Total	53	1123	

(a) Complete the table.

(b) How many groups were there in the study?

(c) How many total observations were there in the study?

11.6   The following ANOVA table is only partially completed.

Source	df	SS	MS
Between groups		258	
Within groups	26		
Total	29	898	

(a) Complete the table.

(b) How many groups were there in the study?

(c) How many total observations were there in the study?

11.7   Invent examples of data with

(a) SS(between) = 0 and SS(within) > 0

(b) SS(between) > 0 and SS(within) = 0

For each example, use three samples, each of size 5.

11.3 THE ANALYSIS OF VARIANCE MODEL (OPTIONAL)

In Section 11.2 we introduced the notation y_{ij} for the jth observation in group i. We think of y_{ij} as a random observation from group i, where the population mean of group i is μ_i. We use analysis of variance to investigate the null hypothesis that $\mu_1 = \mu_2 = \cdots = \mu_I$. It can be helpful to think of ANOVA in terms of the following model:

$$y_{ij} = \mu + \tau_i + \varepsilon_{ij}$$

In this model, μ represents the grand population mean—the population mean when all of the groups are combined. If the null hypothesis is true, then μ is the common population mean. If the null hypothesis is false, then at least some of the μ_i's differ from the grand population mean of μ.

The term τ_i represents the effect of group i—that is, the difference between the population mean for group i, μ_i, and the grand population mean, μ (τ is the Greek letter "tau.") Thus,

$$\tau_i = \mu_i - \mu$$

The null hypothesis

$$H_0: \mu_1 = \mu_2 = \cdots = \mu_I$$

is equivalent to

$$H_0: \tau_1 = \tau_2 = \cdots = \tau_I = 0$$

If H_0 is false, then at least some of the groups differ from the others. If τ_i is positive, then observations from group i tend to be greater than the overall average; if τ_i is negative, then data from group i tend to be less than the overall average.

The term ε_{ij} in the model represents random error associated with observation j in group i. Thus, the model

$$y_{ij} = \mu + \tau_i + \varepsilon_{ij}$$

can be stated in words as

observation = overall average + group effect + random error

We estimate the overall average, μ, with the grand mean of the data:

$$\hat{\mu} = \bar{y}_{..}$$

Likewise, we estimate the population average for group i with the sample average for group i:

$$\hat{\mu}_i = \bar{y}_{i.}$$

Since the group effect is

$$\tau_i = \mu_i - \mu$$

we estimate τ_i as

$$\hat{\tau}_i = \bar{y}_{i.} - \bar{y}_{..}$$

Finally, we estimate the random error, ε_{ij}, for observation y_{ij} as

$$\hat{\varepsilon}_{ij} = y_{ij} - \bar{y}_{i.}$$

Putting these estimates together, we have

$$y_{ij} = \bar{y}_{..} + (\bar{y}_{i.} - \bar{y}_{..}) + (y_{ij} - \bar{y}_{i.})$$

or

$$y_{ij} = \hat{\mu} + \hat{\tau}_i + \hat{\varepsilon}_{ij}$$

Note: Some authors use the terminology SS(error) for what we have called SS(within). This is due to the fact that the within-groups component $y_{ij} - \bar{y}_{i.}$ estimates the random error term in the ANOVA model.

Example 11.8

Weight Gain of Lambs.   For the data of Example 11.2, the estimate of the grand population mean is $\hat{\mu} = 13$. The estimated group effects are

$$\hat{\tau}_1 = \bar{y}_{i\cdot} - \bar{y}_{i\cdot\cdot} = 11 - 13 = -2$$
$$\hat{\tau}_2 = 15 - 13 = 2$$

and

$$\hat{\tau}_3 = 12 - 13 = -1$$

Thus, we estimate that Diet 2 increases weight gain by 2 lb on average (when compared to the average of the three diets), Diet 1 decreases weight gain by an average of 2 lb, and Diet 3 decreases weight gain by 1 lb, on average.   ■

When we conduct an analysis of variance, we are comparing the sizes of the sample group effects, the $\hat{\tau}_i$'s, to the sizes of the random errors in the data, the $\hat{\varepsilon}_{ij}$'s. We can see that

$$SS(\text{between}) = \sum_{i=1}^{I} n_i \hat{\tau}_i^2$$

and

$$SS(\text{within}) = \sum_{i=1}^{I} \sum_{j=1}^{n_i} \hat{\varepsilon}_{ij}^2$$

11.4 THE GLOBAL *F* TEST

The global null hypothesis is

$$H_0 : \mu_1 = \mu_2 = \cdots = \mu_I$$

We consider testing H_0 against the nondirectional alternative hypothesis

$$H_A : \text{The } \mu_i\text{'s are not all equal.}$$

Note that H_0 is compound (unless $I = 2$), and so rejection of H_0 does not specify *which* μ_i's are different. If we reject H_0, then we conduct a further analysis to make detailed comparisons among the μ_i's. Testing the global null hypothesis may be likened to looking at a microscope slide through a low-power lens to see if there is anything on it; if we find something, we switch to a greater magnification to examine its fine structure.

The *F* Distributions

The ***F* distributions**, named after the statistician and geneticist R. A. Fisher, are probability distributions that are used in many kinds of statistical analysis. The form of an *F* distribution depends on two parameters, called the **numerator degrees**

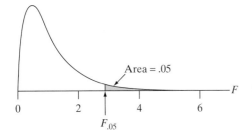

Figure 11.6 The *F* distribution with numerator df = 4 and denominator df = 20

of freedom and the **denominator degrees of freedom**. Figure 11.6 shows an *F* distribution with numerator df = 4 and denominator df = 20. Critical values for the *F* distribution are given in Table 10 at the end of this book. Note that Table 10 occupies ten pages, each page having a different value of the numerator df. As a specific example, for numerator df = 4 and denominator df = 20, we find in Table 10 that $F(4, 20)_{.05} = 2.87$; this value is shown in Figure 11.6.

The *F* Test

The ***F* test** is a classical test of the global null hypothesis. The test statistic, the ***F* statistic**, is calculated as follows:

$$F_s = \frac{\text{MS(between)}}{\text{MS(within)}}$$

From the definitions of the mean squares (Section 11.2), it is clear that F_s will be large if the discrepancies among the group means ($\bar{y}$'s) are large relative to the variability within the groups. Thus, large values of F_s tend to provide evidence against H_0.

To carry out the *F* test of the global null hypothesis, critical values are obtained from an *F* distribution (Table 10) with

$$\text{Numerator df} = \text{df(between)}$$

and

$$\text{Denominator df} = \text{df(within)}$$

It can be shown that (when suitable conditions for validity are met) the null distribution of F_s is an *F* distribution with df as given previously.

The following example illustrates the global *F* test.

Weight Gain of Lambs.   For the lamb feeding experiment of Example 11.2, the global null hypothesis and alternative can be stated verbally as

Example 11.9

H_0: Mean weight gain is the same on all three diets

H_A: Mean weight gain is not the same on all three diets

or symbolically as

H_0: $\mu_1 = \mu_2 = \mu_3$

H_A: The μ_i's are not all equal.

We saw in Figure 11.5 that the three sample means do not differ by much when compared to the variability within the groups, which suggests that H_0 is true. Let us confirm this visual impression by carrying out the F test at $\alpha = .05$. From the ANOVA table (Table 11.7) we find

$$F_s = \frac{18}{23.333} = .77$$

The degrees of freedom can also be read from the ANOVA table as

$$\text{Numerator df} = 2$$
$$\text{Denominator df} = 9$$

From Table 10 we find $F(2, 9)_{.20} = 1.93$, so that $P > .20$. Thus, H_0 is not rejected; there is insufficient evidence to conclude that there is any difference among the diets with respect to population mean weight gain. The observed differences in the mean gains in the samples can readily be attributed to chance variation. ■

Relationship Between F Test and t Test

Suppose only two groups are to be compared ($I = 2$). Then we could test $H_0: \mu_1 = \mu_2$ against $H_A: \mu_1 \neq \mu_2$ using either the F test or the t test. The t test from Chapter 7 can be modified slightly by replacing each sample standard deviation by s_{pooled}, as defined in Section 11.2, before calculating the standard error of $(\bar{y}_1 - \bar{y}_2)$. It can be shown that the F test and this "pooled" t test are actually equivalent procedures. The relationship between the test statistics is $t_s^2 = F_s$; that is, the value of the F statistic for any set of data is necessarily equal to the square of the value of the (pooled) t statistic. The corresponding relationship between the critical values is $t_{.05}^2 = F_{.05}, t_{.01}^2 = F_{.01}$, and so on. For example, suppose $n_1 = 10$ and $n_2 = 7$. Then the appropriate t distribution has df $= n_1 + n_2 - 2 = 15$, and $t(15)_{.05} = 2.131$, whereas the F distribution has numerator df $= I - 1 = 1$ and denominator df $= n^* - I = 15$, so that $F(1, 15)_{.05} = 4.54$; note that $(2.131)^2 = 4.54$. Because of the equivalence of the tests, the application of the F test to compare the means of two samples will always give exactly the same P-value as the pooled t test applied to the same data.

Computer note: It is quite tedious to carry out an analysis of variance with a calculator; statistical software is almost always used. We illustrate ANOVA using a computer for the weight gain data of Example 11.2. In the MINITAB system, suppose the data are entered into columns 1, 2, and 3. That is, suppose the columns of data are

C1	C2	C3
8	9	15
16	16	10
9	21	17
	11	6
	18	

The command

```
MTB < AOVOneway C1 C2 C3.
```

gives the following output:

```
One-Way Analysis of Variance
Analysis of Variance
Source      DF       SS        MS        F        p
Factor       2      36.0      18.0      0.77     0.491
Error        9     210.0      23.3
Total       11     246.0
```

The ANOVA table agrees with our Table 11.7 from Section 11.2. Note that the MINITAB ANOVA table also includes the *F* statistic, given under the *F* heading as 0.77, and the *P*-value, which is given as 0.491. (Recall that in Example 11.9 we stated that $P > .20$.)

Exercises 11.8–11.14

11.8   Monoamine oxidase (MAO) is an enzyme that is thought to play a role in the regulation of behavior. To see whether different categories of schizophrenic patients have different levels of MAO activity, researchers collected blood specimens from 42 patients and measured the MAO activity in the platelets. The results are summarized in the accompanying table. (Values are expressed as nmol benzylaldehyde product/10^8 platelets/hour.)[3] Calculations based on the raw data yielded SS(between) = 136.12 and SS(within) = 418.25.

MAO Activity

Diagnosis	Mean	SD	No. of Patients
Chronic undifferentiated schizophrenic	9.81	3.62	18
Undifferentiated with paranoid features	6.28	2.88	16
Paranoid schizophrenic	5.97	3.19	8

(a) Dotplots of these data are shown below. Based on this graphical display, does it appear that the null hypothesis is true? Why or why not?

(b) Construct the ANOVA table and test the global null hypothesis at $\alpha = .05$.

(c) Calculate the pooled standard deviation, s_{pooled}.

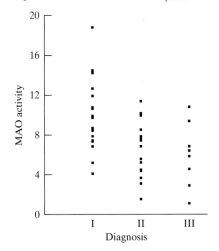

11.9 It is thought that stress may increase susceptibility to illness through suppression of the immune system. In an experiment to investigate this theory, 48 rats were randomly allocated to four treatment groups: no stress, mild stress, moderate stress, and high stress. The stress conditions involved various amounts of restraint and electric shock. The concentration of lymphocytes (cells/mLi $\cdot 10^{-6}$) in the peripheral blood was measured for each rat with the results given in the accompanying table.[4] Calculations based on the raw data yielded SS(between) = 89.036 and SS(within) = 340.24.

	No Stress	**Mild Stress**	**Moderate Stress**	**High Stress**
$\bar{y}$	6.64	4.84	3.98	2.92
s	2.77	2.42	3.91	1.45
n	12	12	12	12

(a) Construct the ANOVA table and test the global null hypothesis at $\alpha = .05$.
(b) Calculate the pooled standard deviation, s_{pooled}.

11.10 Human beta-endorphin (HBE) is a hormone secreted by the pituitary gland under conditions of stress. An exercise physiologist measured the resting (unstressed) blood concentration of HBE in three groups of men: 15 who had just entered a physical fitness program, 11 who had been jogging regularly for some time, and 10 sedentary people. The HBE levels (pg/mLi) are shown in the table.[5] Calculations based on the raw data yielded SS(between) = 240.69 and SS(within) = 6,887.6.

	Fitness Program Entrants	**Joggers**	**Sedentary**
Mean	38.7	35.7	42.5
SD	16.1	13.4	12.8
n	15	11	10

(a) State the null hypothesis in words, in the context of this setting.
(b) State the null hypothesis in symbols.
(c) Construct the ANOVA table and test the null hypothesis. Let $\alpha = .05$.
(d) Calculate the pooled standard deviation, s_{pooled}.

11.11 An experiment was conducted in which the antiviral medication zanamivir was given to patients who had the flu. The length of time until the alleviation of major flu symptoms was measured for three groups: 85 patients who were given inhaled zanamivir, 88 patients who were given inhaled and intranasal zanamivir, and 89 patients who were given a placebo. Summary statistics are given in the table.[6] The ANOVA SS(between) is 53.67 and the SS(within) is 2,034.52.

	Inhaled Zanamivir	**Inhaled and Intranasal Zanamivir**	**Placebo**
Mean	5.4	5.3	6.3
SD	2.7	2.8	2.9
n	85	88	89

(a) State the null hypothesis in words, in the context of this setting.
(b) State the null hypothesis in symbols.
(c) Construct the ANOVA table and test the null hypothesis. Let $\alpha = .05$.
(d) Calculate the pooled standard deviation, s_{pooled}.

11.12 A researcher collected daffodils from four sides of a building and from an open area nearby. She wondered whether the average stem length of a daffodil depends on the side of the building on which it is growing. Summary statistics are given in the table.[7] The ANOVA SS(between) is 871.408 and the SS(within) is 3,588.54.

	North	East	South	West	Open
Mean	41.4	43.8	46.5	43.2	35.5
SD	9.3	6.1	6.6	10.4	4.7
n	13	13	13	13	13

(a) Dotplots of these data are shown below. Based on the dotplots, does it appear that the null hypothesis is true? Why or why not?

(b) State the null hypothesis in symbols.

(c) Construct the ANOVA table and test the null hypothesis. Let $\alpha = .10$.

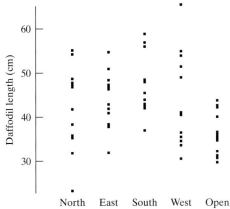

11.13 A researcher studied the flexibility of 10 women in an aerobic exercise class, 10 women in a modern dance class, and a control group of 9 women. One measurement she made on each woman was spinal extension, which was a measure of how far the woman could bend her back. Measurements were made before and after a 16-week training period. The change in spinal extension was recorded for each woman. Summary statistics are given in the table.[8] The ANOVA SS(between) is 7.04 and the SS(within) is 15.08.

	Aerobics	Modern Dance	Control
Mean	−0.18	0.98	0.13
SD	0.80	0.86	0.57
n	10	10	9

(a) Dotplots of these data are shown below. Based on the dotplots, does it appear that the null hypothesis is true? Why or why not?

(b) State the null hypothesis in symbols.

(c) Construct the ANOVA table and test the null hypothesis. Let $\alpha = .01$.

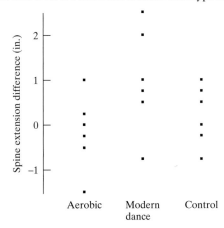

11.14 The computer output below is for an analysis of variance in which yields (bu/acre) of different varieties of oats were compared.[9]

Source	df	Sums of Squares	Mean Square	F Ratio	Prob
Group	2	76.8950	38.4475	0.40245	0.6801
Error	9	859.808	95.5342		
Total	11	936.703			

(a) How many varieties (groups) were in the experiment?
(b) State the conclusion of the ANOVA.
(c) What is the pooled standard deviation, s_{pooled}?

11.5 APPLICABILITY OF METHODS

Like all methods of statistical inference, the calculations and interpretations of ANOVA are based on certain conditions.

Standard Conditions

The ANOVA techniques described in this chapter, including the global F test, are valid if the following conditions hold.

1. **Design conditions**
    (a) It must be reasonable to regard the groups of observations as random samples from their respective populations. The observations within each sample must be independent of each other.
    (b) The I samples must be independent of each other.

2. **Population conditions** The I population distributions must be (approximately) normal with equal standard deviations:

$$\sigma_1 = \sigma_2 = \cdots = \sigma_I$$

These conditions are extensions of the conditions given in Chapter 7 for the independent-samples t test with the added condition that the standard deviations be equal. The condition of normal populations with equal standard deviations is less crucial if the sample sizes (n_i) are large and approximately equal.

Verification of Conditions

The design conditions may be verified as for the independent-samples t test. To check condition 1(a), we look for biases or hierarchical structure in the collection of the data. A completely randomized design assures independence of the samples [condition 1(b)]. If units have been allocated to treatment groups by a randomized blocks design, or if observations on the same experimental unit appear in different samples, then the samples are not independent. (In Chapter 9 we discussed dependence between samples for $I = 2$; analysis of variance for a randomized blocks design is discussed in optional Section 11.6.)

As with the independent-samples t test, the population conditions can be roughly checked from the data. To check normality, a separate histogram, stem-and-leaf display, or normal probability plot can be made for each sample. Another option is to make a single histogram or normal probability plot of the deviations $(y_{ij} - \bar{y}_{i\bullet})$ from all the samples combined.

Equality of the population SDs is checked by comparing the sample SDs; one useful trick is to plot the SDs against the means $(\bar{y}_{i\bullet}\text{'s})$ to check for a trend. Another approach is to make a plot of the deviations $(y_{ij} - \bar{y}_{i\bullet})$ against the means $(\bar{y}_{i\bullet}\text{'s})$. As a rule of thumb, we would like the largest sample SD divided by the smallest sample SD to be less than 2 or so. If this ratio is much larger than 2, then we cannot be confident in the P-value from the ANOVA, particularly if the sample sizes are small and unequal. In particular, if the sample sizes are unequal and the sample SD from a small sample is quite a bit larger than the other SDs, then the P-value can be quite inaccurate.

Weight Gain of Lambs. Consider the lamb feeding experiment of Example 11.2. Figure 11.4 (in Section 11.2) shows that the variability within groups is nearly equal across the three diets: the three sample SDs are 4.36, 4.95, and 4.97. Figure 11.7 is a normal probability plot of the 12 deviations $(y_{ij} - \bar{y}_{i\bullet})$. This plot is close to linear, which supports the normality condition.

Example 11.10

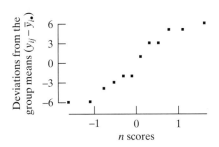

Figure 11.7 Normal probability plot of deviations $(y_{ij} - \bar{y}_{i\bullet})$ in weight gain data

Sweet Corn. Consider the sweet corn data of Example 11.1. Figure 11.8 shows the data for each group plotted above the sample mean for that group. This graph shows that the variability does not change as the mean changes (which is good—if the variability increased as the mean increased, then condition 2 would be violated). Figure 11.9 contains a histogram and a normal probability plot of the 60 deviations $(y_{ij} - \bar{y}_{i\bullet})$. These plots support the normality condition. ■

Example 11.11

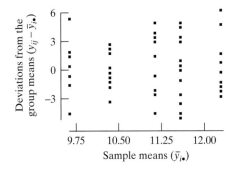

Figure 11.8 Plot of deviations $(y_{ij} - \bar{y}_{i\bullet})$ versus sample mean for the sweet corn data

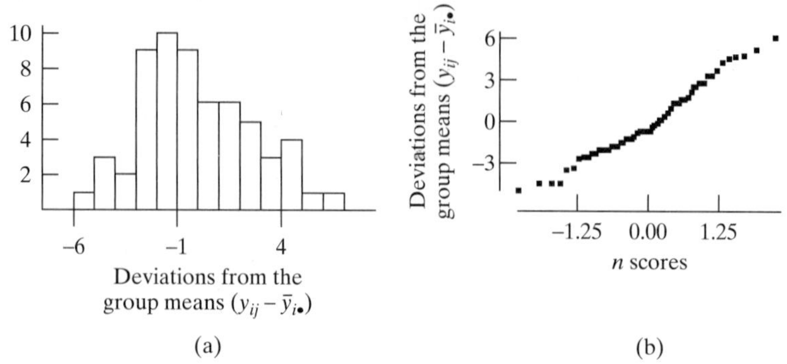

Figure 11.9 Histogram and normal probability plot of deviations $(y_{ij} - \bar{y}_i)$ in sweet corn data

Further Analysis

In addition to their relevance to the F test, the standard conditions underlie many classical methods for further analysis of the data.

If the I populations have the same SD, then a pooled estimate of that SD from the data is

$$s_{pooled} = \sqrt{MS(within)}$$

from the ANOVA. This pooled standard deviation s_{pooled} is a better estimate than any individual sample SD because s_{pooled} is based on more observations.

A simple way to see the advantage of s_{pooled} is to consider the standard error of an individual sample mean, which can be calculated as

$$SE_{\bar{y}} = \frac{s_{pooled}}{\sqrt{n}}$$

where n is the size of the individual sample. The df associated with this standard error is df(within), which is the sum of the degrees of freedom of all the samples. By contrast, if the individual SD were used in calculating $SE_{\bar{y}}$, it would have only $(n-1)$ df. When the SE is used for inference, larger df yield smaller critical values (see Table 4), which in turn lead to improved power and narrower confidence intervals.

In optional Sections 11.7 and 11.8, we will consider methods for detailed analysis of the group means $\bar{y}_1., \bar{y}_2., \ldots \bar{y}_I.$. Like the F test, these methods were designed for independent samples from normal populations with equal standard deviations. The methods use standard errors based on the pooled standard deviation estimate s_{pooled}.

Exercises 11.15–11.16

11.15 Refer to the lymphocyte data of Exercise 11.9. The global F test is based on certain conditions concerning the population distributions.

(a) State the conditions.
(b) Which features of the data suggest that the conditions may be doubtful in this case?

11.16 Patients with advanced cancers of the stomach, bronchus, colon, ovary, or breast were treated with ascorbate. The purpose of the study was to determine if the survival times differ with respect to the organ affected by the cancer. The variable of interest is survival time (in days).[10] Here are parallel dotplots of the raw data. An ANOVA was done after a square root transformation was applied to the raw data. There were two (related) reasons that the data were transformed. What were those two reasons?

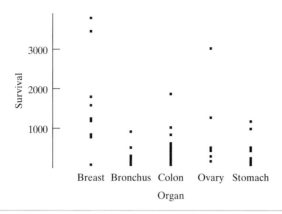

11.6 TWO-WAY ANOVA (OPTIONAL)

When we have several means to compare, there is sometimes additional structure in the data. In this section we take a brief look at analysis of variance for an experiment conducted using a randomized blocks design and analysis of variance when there are two factors of interest. We begin with an example of a randomized blocks design.

Alfalfa and Acid Rain.   Researchers were interested in the effect that acid has on the growth rate of alfalfa plants. They created three treatment groups in an experiment: low acid, high acid, and control. The response variable in their experiment was the average height of the alfalfa plants in a Styrofoam cup after five days of growth. (The observational unit was a cup, rather than individual plants.) They had 5 cups for each of the 3 treatments, for a total of 15 observations. However, the cups were arranged near a window and they wanted to account for the effect of differing amounts of sunlight. Thus, they created 5 blocks and randomly assigned the 3 treatments within each block, as shown in Figure 11.10. The data are given in Table 11.8 and are graphed in Figure 11.11.[11]

Example 11.12

Window	Block 1	Block 2	Block 3	Block 4	Block 5
	high	control	control	control	high
	control	low	high	low	low
	low	high	low	high	control

Organization of blocks for alfalfa experiment

Figure 11.10 Design of the alfalfa experiment

TABLE 11.8 Alfalfa Plant Height after Five Days (cm)

	Low Acid	High Acid	Control	Block Mean
Block 1	1.58	1.10	2.47	1.717
Block 2	1.15	1.05	2.15	1.450
Block 3	1.27	0.50	1.46	1.077
Block 4	1.25	1.00	2.36	1.537
Block 5	1.00	1.50	1.00	1.167
n	5	5	5	
Treatment mean $= \bar{y}_{i\cdot}$	1.250	1.03	1.888	

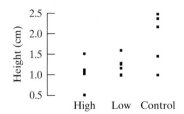

Figure 11.11 Dotplots of the alfalfa data

When we are comparing three treatments, as in Example 11.12, the null hypothesis of interest is that the means of the three treatment populations are equal:

$$H_0: \mu_1 = \mu_2 = \mu_3$$

This hypothesis can be tested with an analysis of variance F test, but first we want to remove the variability in the data that is due to differences between the blocks. To do this, we extend the ANOVA model presented in Section 11.3 to the following model:

$$y_{ijk} = \mu + \tau_i + \beta_j + \varepsilon_{ijk}$$

In this model y_{ijk} is the kth observation when treatment i is applied in block j. (In Example 11.12 there is only one observation for each treatment in each block, but in general there might be more than one.) Here, as before, μ represents the grand population mean and the term τ_i represents the effect of group i (that is, treatment i). The new term in the model is β_j, which represents the effect of the jth block.

In Section 11.2 we discussed how the total sum of squares, SS(total), is broken down into SS(within) and SS(between). For a randomized blocks experiment, SS(within) is subdivided by calculating a part due to differences between the blocks. Thus we have

$$\text{SS(total)} = \text{SS(within)} + \text{SS(treatments)} + \text{SS(blocks)}$$

The sum of squares due to blocks measures the variability between the blocks, just as the SS(treatments) measures variability between treatment means. Usually we are not interested in testing a hypothesis about the blocks, but nonetheless we want to take into consideration the effect that blocking has on the response variable. Refining the ANOVA by calculating SS(blocks) accomplishes this goal.

The sum of squares due to treatments is the same as SS(between) from Section 11.2. The sum of squares due to blocks is calculated by comparing each block average to the grand mean in a way that is analogous to the calculation of SS(between) in Section 11.2. If we define the average of the observations in block j to be $\bar{y}_{\cdot j}$ and we let m_j denote the number of observations in block j, then the sum of squares due to blocks is defined as follows:

> **Sum of Squares Between Blocks**
>
> $$\text{SS(blocks)} = \sum_{j=1}^{b} m_j (\bar{y}_{\cdot j} - \bar{y}_{\cdot \cdot})^2$$

There is a corresponding division of the degrees of freedom. If there are n^* total observations, then there are $n^* - 1$ total degrees of freedom. The I treatments have

$I - 1$ degrees of freedom, and if there are B blocks, then there are $B - 1$ degrees of freedom for the blocks. The degrees of freedom for the error term—the within groups degrees of freedom—can be found by subtraction.

Alfalfa and Acid Rain.   For the alfalfa data in Table 11.8 the total of all the observations is $1.58 + \cdots + 1.0 = 20.84$ and the grand mean is

Example 11.13

$$\bar{y}.. = \frac{20.84}{15} = 1.389$$

We calculate

$$\text{SS(treatments)} = 5(1.25 - 1.389)^2 + 5(1.03 - 1.389)^2$$
$$+ 5(1.888 - 1.389)^2 = 1.986$$

Since $k = 3$, we have

$$\text{df(treatments)} = 3 - 1 = 2$$

so that

$$\text{MS(treatments)} = \frac{1.986}{2} = .993$$

We calculate

$$\text{SS(blocks)} = 3(1.717 - 1.389)^2 + 3(1.450 - 1.389)^2$$
$$+ 3(1.077 - 1.389)^2 + 3(1.537 - 1.389)^2$$
$$+ 3(1.167 - 1.389)^2 = 0.840$$

Since $B = 5$, we have

$$\text{df(blocks)} = 5 - 1 = 4$$

and

$$\text{MS(blocks)} = \frac{0.840}{4} = .210$$

The total sum of squares is found as $(1.58 - 1.389)^2 + \cdots + (1.0 - 1.389)^2 = 4.278$. There are now two ways to obtain SS(within). One way is to note that

$$\text{SS(total)} = \text{SS(within)} + \text{SS(treatments)} + \text{SS(blocks)}$$

which means that

$$4.278 = \text{SS(within)} + 1.986 + 0.840$$

so that SS(within) $= 4.278 - 1.986 - 0.840 = 1.452$. The other approach is to use the following formula:

$$\text{SS(within)} = \sum_{i=1}^{I} \sum_{j=1}^{n_i} (y_{ij} - \bar{y}_{i\cdot} - \bar{y}_{\cdot j} + \bar{y}..)^2$$
$$= (1.58 - 1.25 - 1.717 + 1.389)^2 + \cdots$$
$$+ (1.0 - 1.888 - 1.167 + 1.389)^2 = 1.452$$

We can find the df(within) by subtraction, noting that

$$df(total) = df(within) + df(treatments) + df(blocks)$$

so

$$df(within) = df(total) - df(treatments) - df(blocks)$$

which in this case gives us $14 - 2 - 4 = 8$. Thus, MS(within) $= \dfrac{1.452}{8} = 0.182$. ■

The sums of squares, degrees of freedom, and resulting mean squares are collected in an expanded ANOVA table, which includes a line for the effect of the blocks.

To test the null hypothesis, we calculate

$$F_s = \frac{MS(treatments)}{MS(within)}$$

and reject H_0 if the P-value is too small.

Example 11.14   **Alfalfa and Acid Rain.**   For the alfalfa growth data of Example 11.12, the ANOVA summary is given in Table 11.9. The F statistic is $.993/.1815 = 5.47$, with degrees of freedom 2 for the numerator and 8 for the denominator. From Table 10 we bracket the P-value as $.02 < P\text{-value} < .05$. (Using a computer gives $P = .0318$.) The P-value is small, indicating that the differences between the three sample means are greater than would be expected by chance alone.

TABLE 11.9 ANOVA Table for Alfalfa Experiment

Source	df	SS	MS	F Ratio
Between treatments	2	1.986	0.993	5.47
Between blocks	4	0.840	0.210	
Within groups	8	1.452	.1815	
Total	14	4.278		

■

Factorial ANOVA

In a typical analysis of variance application there is a single explanatory variable or **factor** under study. For example, in the weight gain setting of Example 11.2 the factor is "type of diet," which takes on 3 **levels**: Diet 1, Diet 2, and Diet 3. However, some analysis of variance settings involve the simultaneous study of two or more factors. The following is an example.

Example 11.15   **Growth of Soybeans.**   A plant physiologist investigated the effect of mechanical stress on the growth of soybean plants. Individually potted seedlings were randomly allocated to four treatment groups of 13 seedlings each. Seedlings in two groups were stressed by shaking for 20 minutes twice daily, while two control groups were not stressed. Thus, the first factor in the experiment was presence or absence of stress, with two levels: control or stress. Also, plants were grown in either low or moderate light. Thus, the second factor was amount of light, with two levels: low light or moderate light. This experiment is an example of a $2 \cdot 2$ *factorial experiment*; it includes four treatments:

Treatment 1: Control, low light

Treatment 2: Stress, low light

Treatment 3: Control, moderate light

Treatment 4: Stress, moderate light

After 16 days of growth, the plants were harvested, and the total leaf area (cm^2) of each plant was measured. The results are given in Table 11.10 and plotted in Figure 11.12.[12]

TABLE 11.10 Leaf Area (cm^2) of Soybean Plants

| | Treatment | | |
Control, Low Light	Stress, Low Light	Control, Moderate Light	Stress, Moderate Light
264	235	314	283
200	188	320	312
225	195	310	291
268	205	340	259
215	212	299	216
241	214	268	201
232	182	345	267
256	215	271	326
229	272	285	241
288	163	309	291
253	230	337	269
288	255	282	282
230	202	273	257
Mean 245.3	212.9	304.1	268.8
SD 27.0	29.7	26.9	35.2
n 13	13	13	13

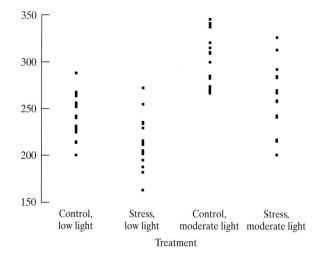

Figure 11.12 Leaf area of soybean plants receiving four different treatments

It is evident in Figure 11.12 that stress reduces leaf area. This is true under low light and under moderate light. Likewise, moderate light increases leaf area, whether or not the seedlings are stressed. ■

A model for this setting is

$$y_{ijk} = \mu + \tau_i + \beta_j + \varepsilon_{ijk}$$

where y_{ijk} is the kth observation of level i of the first factor and level j of the second factor. The term τ_i represents the effect of level i of the first factor (stress condition in Example 11.15) and now the term β_j represents the effect of level j of the second factor (light condition in Example 11.15).

When studying two factors within a single experiment, it helps to organize the sample means in a table that reflects the structure of the experiment and to present the means in a graph that features this structure.

Example 11.16

Growth of Soybeans.   Table 11.11 summarizes the data of Example 11.15. For example, when the first factor is at its first level (control) and the second factor is at its first level (low light), the sample mean is $\bar{y}_{11} = 245.3$. The format of this table permits us easily to consider the two factors—stress condition and light condition—separately and together. The last column shows the effect of light at each stress level. The numbers in this column confirm the visual impression of Figure 11.12: Moderate light increases average leaf area by roughly the same amount when the seedlings are stressed as it does when they are not stressed. Likewise, the last row (-32.4 versus -35.3) shows that the effect of stress is roughly the same at each level of light.

TABLE 11.11 Mean Leaf Areas for Soybean Experiment

		Light Condition		
		Low Light	*Moderate Light*	*Difference*
Shaking	Control	245.3	304.1	58.8
condition	Stress	212.9	268.8	55.9
	Difference	-32.4	-35.3	

If the joint influence of two factors is equal to the sum of their separate influences, the two factors are said to be **additive** in their effects. For instance, consider the soybean experiment of Example 11.15. If stress reduces mean leaf area by the same amount in either light condition, then the effect of stress (a negative effect in this case) is *added* to the effect of light. To visualize this additivity of effects, consider Figure 11.13, which shows the four treatment means. The solid lines connecting treatment means are almost parallel because the data display a pattern of nearly perfect additivity.*

---

*The difference between the mean leaf area for stress under low light (212.9) and the mean leaf area for control under low light of (245.3) is called the **simple effect** of shaking condition under low light. Thus, the simple effect of shaking condition under low light is $212.9 - 245.3 = -32.4$. Likewise, the simple effect of shaking condition under moderate light is $268.8 - 304.1 = -35.3$. A **main effect** is an average of simple effects. For example, the main effect of shaking condition is $(-32.4 + -35.3)/2 = -33.85$. The main effect of light is $(58.8 + 55.9)/2 = 57.35$.

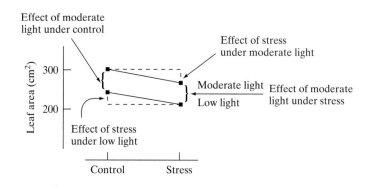

Figure 11.13 Treatment means for soybean experiment

When the effects of factors are additive, we say that there is no **interaction** between the factors. A graph that displays the treatment means is often called an interaction graph. Figure 11.14, which is a simplified version of Figure 11.13, is an interaction graph.

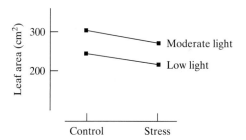

Figure 11.14 Interaction graph for soybean experiment

Sometimes the effect that one factor has on a response variable depends on the level of a second factor. When this happens we say that the two factors interact in their effect on the response. The following is an example.

Carbon Dioxide.   The rate at which trees absorb carbon dioxide (CO_2) depends on the amount of carbon dioxide in the atmosphere, in addition to other factors. Researchers conducted an experiment to learn how two factors affect the rate at which trees in a forested area absorb CO_2. The first factor was CO_2 concentration in the atmosphere, which had the levels "ambient" and "elevated." The second factor was type of soil, which had the levels "unfertilized" and "fertilized." The response variable was annual carbon increment in woody tissue (measured in units of kg C per square meter of ground area). Table 11.12 summarizes the data, which

Example 11.17

TABLE 11.12 Mean Carbon Absorption Values (kg C per Square Meter Ground Area per Year) for CO_2 Experiment.

		Soil Type		
		Unfertilized	*Fertilized*	*Difference*
CO_2	Ambient	.289	.347	.058
concentration	Elevated	.227	.496	.269
	Difference	−.062	.149	

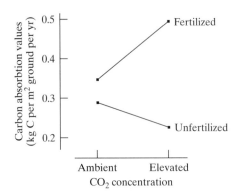

Figure 11.15 Interaction graph for CO_2 experiment

included three observations for each combination of CO_2 concentration and soil type. Figure 11.15 is an interaction plot showing the four means. Note that when the soil is unfertilized, the elevated CO_2 mean is somewhat lower than the ambient CO_2 mean. However, when the soil is fertilized, the elevated CO_2 mean is much higher than the ambient CO_2 mean. Thus, the effect of elevating CO_2 depends on the soil type. We say that CO_2 concentration and soil type interact in their effects on carbon absorption by the trees.[13] ∎

When we suspect that two factors interact in an ANOVA setting, or if we are analyzing data from a randomized blocks design and we suspect that the blocks interact with the treatment, we can extend our model by adding an interaction term:

$$y_{ijk} = \mu + \tau_i + \beta_j + \gamma_{ij} + \varepsilon_{ijk}$$

Here the term γ_{ij} is the effect of the interaction between level i of the first factor and level j of the second factor. As before, if there are n^* total observations, then df(total) $= n^* - 1$. If there are I levels of the first factor, then it has $I - 1$ degrees of freedom. Likewise, if there are J levels of the second factor, then it has $J - 1$ degrees of freedom. There are $(I - 1) \cdot (J - 1)$ interaction degrees of freedom. With I levels of the first factor and J levels of the second factor, there are IJ treatment combinations. Thus, df(within) $= n^* - IJ$.*

A null hypothesis of interest is that all interaction terms are zero:

$$H_0: \gamma_{11} = \gamma_{12} = \cdots = \gamma_{IJ} = 0$$

To test this null hypothesis we calculate

$$F_s = \frac{\text{MS(interaction)}}{\text{MS(within)}}$$

and reject H_0 if the P-value is too small.

---

* This is analogous to the definition of df(within) $= n^* - I$ for one-way ANOVA from Section 11.2. In each setting df(within) = total number of observations − number of treatments.

Example 11.18

Carbon Dioxide. Table 11.13 shows the analysis of variance results for the CO_2 experiment of Example 11.17. This table includes a line for the interaction term.* There were three observations at each combination of CO_2 concentration and soil type; thus $n^* = 12$ and df(total) = 11. In this example $I = J = 2$, so df(CO_2 concentrations) = df(soil types) = df(interaction) = 1. We can find df(within) by subtraction: df(within) = $11 - 1 - 1 - 1 = 8$. (This agrees with the formula df(within) = $n^* - IJ = 12 - (2) \cdot (2)$.)

TABLE 11.13 ANOVA Table for CO_2 Experiment

Source	df	SS	MS	*F* Ratio
Between CO_2 concentrations	1	.005678	.005678	1.19
Between soil types	1	.080197	.080197	16.79
Interaction	1	.033391	.033391	6.99
Within groups	8	.038202	.004775	
Total	11	.157468		

To test whether CO_2 concentration and soil type interact, we use the *F* ratio $.033391/.004775 = 6.99$, which has degrees of freedom 1 for the numerator and 8 for the denominator. From Table 10 we bracket the *P*-value as $.02 < P\text{-value} < .05$. (Using a computer gives $P = .0295$.) The *P*-value is small, indicating that the interaction pattern seen in Figure 11.15 is more pronounced than would be expected by chance alone. Thus, we reject H_0. ∎

The concept of interaction occurs throughout biology. The terms *synergism* and *antagonism* describe interactions between biological agents. The term *epistasis* describes interaction between genes at two loci.

When interactions are present, as in Example 11.17, the main effects of factors don't have their usual interpretations. Regarding Example 11.17, it is difficult to state the effect of soil type because the nature and magnitude of the effect depend on the particular CO_2 concentration. Because of this, we usually test for the presence of interactions first. If interactions are present, as in the CO_2 example, then often we stop the analysis at this stage. If no interaction effect is found (that is, if we do not reject H_0), then we proceed to testing the main effects of the individual factors. The following example illustrates this process.

Example 11.19

Growth of Soybeans. Table 11.14 is an analysis of variance table for the soybean growth data of Example 11.15. The null hypothesis

$$H_0: \gamma_{11} = \gamma_{12} = \gamma_{21} = \gamma_{22} = 0$$

---

* The ANOVA formulas that are used to calculate the sum of squares due to interaction are rather messy and aren't presented here. In particular, it matters whether or not the design is "balanced." The CO_2 experiment is balanced in that there are three observations in each of the four combinations of factor levels shown in Table 11.12. However, unbalanced designs, which lead to complicated calculations and analyses, are possible. We rely here on computer software to calculate the necessary sums of squares.

TABLE 11.14 ANOVA Table for Soybean Growth Experiment

Source	df	SS	MS	*F* Ratio
Between stress levels	1	14858.5	14858.5	16.60
Between light levels	1	42751.6	42751.6	47.75
Interaction	1	26.3	26.3	0.029
Within groups	48	42976.3	895.34	
Total	51	100613		

is tested with the *F* ratio

$$F_s = \frac{\text{MS(interaction)}}{\text{MS(within)}} = \frac{26.3}{895.34} = 0.029$$

Looking in Table 10 with degrees of freedom 1 and 12, we see that the *P*-value is greater than .20; thus we do not reject H_0.

Since there are no interactions, we test the main effect of stress level. Here the *F* ratio is

$$F_s = \frac{\text{MS(between stress levels)}}{\text{MS(within)}} = \frac{14858.5}{895.34} = 16.6$$

This is highly significant (i.e., the *P*-value is very small) and we reject H_0.

Likewise, the test for the main effect of light levels has an *F* ratio of

$$F_s = \frac{\text{MS(between light levels)}}{\text{MS(within)}} = \frac{42751.6}{895.36} = 47.75$$

Again, this is highly significant and we reject H_0.   ∎

Interaction graphs can be used when there are more than two levels for a factor, as in the next example.

Example 11.20

Toads.   Researchers studied the effect that exposure to ultraviolet-B radiation has on the survival of embryos of the western toad *Bufo boreas*. They conducted an experiment in which several *B. borea* embryos were placed at one of three water depths—10 cm, 50 cm, or 100 cm—and one of two radiation settings—exposed to UV-B radiation or shielded. The response variable was the percentage of embryos surviving to hatching. Table 11.15 summarizes the data, which included four observations at each combination of depth and UV-B exposure. Figure 11.16 is an

TABLE 11.15 Percent Embryos Surviving for Toads Experiment

		UV-B		
		Exposed	*Shielded*	*Difference*
Water	10 cm	.425	.759	.334
Depth	50 cm	.729	.748	.019
	100 cm	.785	.766	−.019

interaction graph showing the six means. The presence of interactions here is readily apparent. Table 11.16 summarizes the analysis of variance.[14]

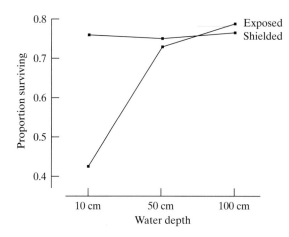

Figure 11.16 Interaction graph for toad experiment

TABLE 11.16 ANOVA Table for Toad Experiment

Source	df	SS	MS	F Ratio
Between water depths	2	.150676	.075338	13.92
Between UV-B levels	1	.074371	.074371	13.74
Interaction	2	.150185	.075093	13.88
Within groups	18	.097401	.005411	
Total	23	.472633		

The topic of interactions is also discussed in Section 11.7

Exercises 11.17–11.22

11.17 A plant physiologist investigated the effect of flooding on root metabolism in two tree species: flood-tolerant river birch and the intolerant European birch. Four seedlings of each species were flooded for one day and four were used as controls. The concentration of adenosine triphosphate (ATP) in the roots of each plant was measured. The data (nmol ATP per mg tissue) are shown in the table.[15]

	River Birch		European Birch	
	Flooded	*Control*	*Flooded*	*Control*
	1.45	1.70	.21	1.34
	1.19	2.04	.58	.99
	1.05	1.49	.11	1.17
	1.07	1.91	.27	1.30
Mean	1.19	1.785	.2925	1.20

Prepare an interaction graph (like Figure 11.14).

11.18 Consider the data from Exercise 11.17. For these data, SS(species of birch) = 2.19781, SS(flooding) = 2.25751, SS(interaction) = 0.097656, and SS(within) = .47438.

(a) Construct the ANOVA table.
(b) Carry out an F test for interactions; use $\alpha = .05$.

(c) Test the null hypothesis that species has no effect on ATP concentration. Use $\alpha = .01$.

(d) Assuming that each of the four populations has the same standard deviation, use the data to calculate an estimate of that standard deviation.

11.19 A completely randomized double-blind clinical trial was conducted to compare two drugs, ticrynafen (T) and hydrochlorothiazide (H), for effectiveness in treatment of high blood pressure. Each drug was given at either a low or a high dosage level for 6 weeks. The accompanying table shows the results for the drop (baseline minus final value) in systolic blood pressure (mm Hg).[16]

	Ticrynafen (T)		**Hydrochlorothiazide (H)**	
	Low Dose	*High Dose*	*Low Dose*	*High Dose*
Mean	13.9	17.1	15.8	17.5
No. of Patients	53	57	55	58

Prepare an interaction graph (like Figure 11.14).

11.20 Consider the data from Exercise 11.19. The difference in response between T and H appears to be larger for the low dose than for the high dose. Carry out an *F* test for interactions to assess whether this pattern can be ascribed to chance variation. Let $\alpha = .10$. For these data SS(interaction) = 31.33 and SS(within) = 30648.81.

11.21 Consider the data from Exercise 11.19. For these data, SS(drug) = 69.22, *SS(dose)* = 330.00, *SS(interaction)* = 31.33, and SS(within) = 30648.81.

(a) Construct the ANOVA table.
(b) Carry out a test of the null hypothesis that the effects of the two drugs (T and H) are equal. Let $\alpha = .05$.

11.22 In a study of lettuce growth, 36 seedlings were randomly allocated to receive either high or low light and to be grown in either a standard nutrient solution or one containing extra nitrogen. After 16 days of growth, the lettuce plants were harvested and the dry weight of the leaves was determined for each plant. The accompanying table shows the mean leaf dry weight (g) of the nine plants in each treatment group.[17]

	Nutrient Solution	
	Standard	*Extra Nitrogen*
Low Light	2.16	3.09
High Light	3.26	4.48

For these data, SS(nutrient solution) = 10.4006, SS(light) = 13.95023, SS(interaction) = 0.18923, and SS(within) = 11.1392.

(a) Construct the ANOVA table.
(b) Carry out an *F* test for interactions; use $\alpha = .05$.
(c) Test the null hypothesis that nutrient solution has no effect on weight. Use $\alpha = .01$.

11.7 LINEAR COMBINATIONS OF MEANS (OPTIONAL)

In many studies, interesting questions can be addressed by considering linear combinations of the group means. A **linear combination** L is a quantity of the form

$$L = m_1 \bar{y}_{1\cdot} + m_2 \bar{y}_{2\cdot} + \cdots + m_I \bar{y}_{I\cdot}.$$

where the m_i's are multipliers of the $\bar{y}_{i\cdot}$'s.

Linear Combinations for Adjustment

One use of linear combinations is to "adjust" for an extraneous variable, as illustrated by the following example.

Forced Vital Capacity.   One measure of lung function is forced vital capacity (FVC), which is the maximal amount of air a person can expire in one breath. In a public health survey, researchers measured FVC in a large sample of people. The results for male ex-smokers, stratified by age, are shown in Table 11.17.[18]

Example 11.21

TABLE 11.17 FVC in Male Ex-Smokers

	FVC (Liters)		
Age (years)	n	Mean	SD
25–34	83	5.29	.76
35–44	102	5.05	.77
45–54	126	4.51	.74
55–64	97	4.24	.80
65–74	73	3.58	.82
25–74	481	4.56	

Suppose it is desired to calculate a summary value for FVC in male ex-smokers. One possibility would be simply to calculate the grand mean of the 481 observed values, which is 4.56 liters. But the grand mean has a serious drawback: It cannot be meaningfully compared with other populations that may have different age distributions. For instance, suppose we were to compare ex-smokers with nonsmokers; the observed difference in FVC would be distorted because ex-smokers as a group are (not surprisingly) older than nonsmokers. A summary measure that does not have this disadvantage is the "age-adjusted" mean, which is an estimate of the mean FVC value in a reference population with a specified age distribution. To illustrate, we will use the reference distribution in Table 11.18, which is (approximately) the distribution for the entire U.S. population.[19]

TABLE 11.18 Age Distribution in Reference Population

Age	Relative Frequency
25–34	.27
35–44	.28
45–54	.21
55–64	.13
65–74	.11

The age-adjusted mean FVC value is the following linear combination:

$$L = .27\bar{y}_{1.} + .28\bar{y}_{2.} + .21\bar{y}_{3.} + .13\bar{y}_{4.} + .11\bar{y}_{5.}$$

Note that the multipliers (m's) are the relative frequencies in the reference population. From Table 11.18, the value of L is

$$L = (.27)(5.29) + (.28)(5.05) + (.21)(4.51) + (.13)(4.24) + (.11)(3.58)$$
$$= 4.73 \text{ liters}$$

This value is an estimate of the mean FVC in an idealized population of people who are biologically like male ex-smokers but whose age distribution is that of the reference population. ■

Contrasts

A linear combination whose multipliers (m's) add to zero is called a **contrast**. The following example shows how contrasts can be used to describe the results of an experiment.

Example 11.22

Growth of Soybeans. Table 11.19 shows the treatment means and sample sizes for the soybean growth experiment of Example 11.15. We can use contrasts to describe the effects of stress in the two temperature conditions.

TABLE 11.19 Soybean Growth Data

Treatment	Mean Leaf Area (cm^2)	n
1. Control, low light	245.3	13
2. Stress, low light	212.9	13
3. Control, moderate light	304.1	13
4. Stress, moderate light	268.8	13

(a) First, note that an ordinary pairwise difference is a contrast. For instance, to measure the effect of stress in low light, we can consider the contrast

$$L = \bar{y}_1 - \bar{y}_2 = 245.3 - 212.9 = 32.4$$

For this contrast, the multipliers are $m_1 = 1, m_2 = -1, m_3 = 0, m_4 = 0$; note that they add to zero.

(b) To measure the effect of stress in moderate light, we can consider the contrast

$$L = \bar{y}_3 - \bar{y}_4 = 304.1 - 268.8 = 35.3$$

For this contrast, the multipliers are $m_1 = 0, m_2 = 0, m_3 = 1, m_4 = -1$.

(c) To measure the overall effect of stress, we can average the contrasts in parts (a) and (b) to obtain the contrast

$$L = \frac{1}{2}(\bar{y}_1 - \bar{y}_2) + \frac{1}{2}(\bar{y}_3 - \bar{y}_4)$$

$$= \frac{1}{2}(32.4) + \frac{1}{2}(35.3) = 33.85$$

For this contrast, the multipliers are $m_1 = \frac{1}{2}, m_2 = -\frac{1}{2}, m_3 = \frac{1}{2}, m_4 = -\frac{1}{2}$. ■

Standard Error of a Linear Combination

Each linear combination L is an estimate, based on the $\bar{y}$'s, of the corresponding linear combination of the population means (μ's). As a basis for statistical inference, we need to consider the standard error of a linear combination, which is calculated as follows.

Standard Error of L
The standard error of the linear combination

$$L = m_1 \bar{y}_1. + m_2 \bar{y}_2. + \cdots + m_I \bar{y}_I.$$

is

$$SE_L = \sqrt{s_{\text{pooled}}^2 \sum_{i=1}^{I} \frac{m_i^2}{n_i}}$$

where $s_{\text{pooled}}^2 = $ MS(within) from the ANOVA.

The SE can be written explicitly as

$$SE_L = \sqrt{s_{\text{pooled}}^2 \left(\frac{m_1^2}{n_1} + \frac{m_2^2}{n_2} + \cdots + \frac{m_I^2}{n_I} \right)}$$

If all the sample sizes (n_i) are equal, the SE can be written as

$$SE_L = \sqrt{\frac{s_{\text{pooled}}^2}{n} (m_1^2 + m_2^2 + \cdots + m_I^2)} = \sqrt{\frac{s_{\text{pooled}}^2}{n} \sum_{i=1}^{I} m_i^2}$$

The following two examples illustrate the application of the standard error formula.

Forced Vital Capacity.   For the linear combination L defined in Example 11.21, we find that

Example 11.23

$$\sum_{i=1}^{I} \frac{m_i^2}{n_i} = \frac{.27^2}{83} + \frac{.28^2}{102} + \frac{.21^2}{126} + \frac{.13^2}{97} + \frac{.11^2}{73}$$

$$= .0023369$$

The ANOVA for these data yields $s_{\text{pooled}}^2 = .59989$. Thus, the standard error of L is

$$SE_L = \sqrt{(.59989)(.0023369)} = .0374 \qquad \blacksquare$$

Growth of Soybeans.   For the linear combination L defined in Example 11.22(a), we find that

Example 11.24

$$\sum_{i=1}^{I} m_i^2 = (1)^2 + (-1)^2 + (0)^2 + (0)^2 = 2$$

so that

$$SE_L = \sqrt{\frac{s_{\text{pooled}}^2}{13}(2)} \qquad \blacksquare$$

Confidence Intervals

Linear combinations of means can be used for testing hypotheses and for constructing confidence intervals. Critical values are obtained from Student's t distribution with

$$df = df(\text{within})$$

from the ANOVA.* Confidence intervals are constructed using the familiar Student's t format. For instance, a 95% confidence interval is

$$L \pm t_{.025} SE_L$$

The following example illustrates the construction of the confidence interval.

Example 11.25   **Growth of Soybeans.**   Consider the contrast defined in Example 11.22(c):

$$L = \frac{1}{2}(\bar{y}_1 - \bar{y}_2) + \frac{1}{2}(\bar{y}_3 - \bar{y}_4)$$

This contrast is an estimate of the quantity

$$L = \frac{1}{2}(\mu_1 - \mu_2) + \frac{1}{2}(\mu_3 - \mu_4)$$

which can be described as the true (population) effect of stress, averaged over the light conditions. Let us construct a 95% confidence interval for this true difference. We found in Example 11.22 that the value of L is

$$L = 33.85$$

To calculate SE_L, we first calculate

$$\sum_{i=1}^{I} \frac{m_i^2}{n_i} = \frac{\left(\frac{1}{2}\right)^2}{13} + \frac{\left(-\frac{1}{2}\right)^2}{13} + \frac{\left(\frac{1}{2}\right)^2}{13} + \frac{\left(-\frac{1}{2}\right)^2}{13} = \frac{1}{13}$$

From the ANOVA, which is shown in Table 11.20, we find that $s^2_{\text{pooled}} = 895.34$; thus,

$$SE_L = \sqrt{s^2_{\text{pooled}} \sum_{i=1}^{I} \frac{m_i^2}{n_i}} = \sqrt{895.34\left(\frac{1}{13}\right)} = 8.299$$

TABLE 11.20 ANOVA Table for Soybean Growth Experiment

Source	df	SS	MS	F Ratio
Between stress depths	1	14858.5	14858.5	16.60
Between light levels	1	42751.6	42751.6	47.75
Interaction	1	26.3	26.3	0.029
Within groups	48	42976.3	895.34	
Total	51	100613		

From Table 4 with df $= 40 \approx 48$, we find $t(40)_{.025} = 2.021$. The confidence interval is

$$33.85 \pm (2.021)(8.299)$$

$$33.85 \pm 16.77$$

$$\text{or } (17.1, 50.6)$$

We are 95% confident that the effect of stress, averaged over the light conditions, is to reduce the leaf area by an amount whose mean value is between 17.1 cm^2 and 50.6 cm^2.   ■

---

* This method of determining critical values does not take account of multiple comparisons. See Section 11.8.

t Tests

To test the null hypothesis that the population value of a contrast is zero, the test statistic is calculated as

$$t_s = \frac{L}{SE_L}$$

and the *t* test is carried out in the usual way. The *t* test will be illustrated in Example 11.26.

Contrasts to Assess Interaction

Sometimes an investigator wishes to study the separate and joint effects of two or more factors on a response variable Y. In Section 11.6 the concept of interaction between two factors was introduced. Linear contrasts provide another way to study such interactions. The following is an example.

Growth of Soybeans. In the soybean growth experiment (Example 11.15 and Example 11.22), the two factors of interest are stress condition and light level. Table 11.21 shows the treatment means, arranged in a new format that permits us easily to consider the factors separately and together.

Example 11.26

TABLE 11.21 Mean Leaf Areas for Soybean Experiment.

		Light Condition		
		Low light	Moderate light	Difference
Shaking	Control	245.3 (1)	304.1 (3)	58.8
Condition	Stress	212.9 (2)	268.8 (4)	55.9
	Difference	−32.4	−35.3	

At each light level, the mean effect of stress can be measured by a contrast:

Effect of stress in low light: $\bar{y}_2 - \bar{y}_1 = 212.9 - 245.3 = -32.4$
Effect of stress in moderate light: $\bar{y}_4 - \bar{y}_3 = 268.8 - 304.1 = -35.3$

Now consider the question, Is the reduction in leaf area due to stress the same in both light conditions? One way to address this question is to compare $(\bar{y}_2 - \bar{y}_1)$ versus $(\bar{y}_4 - \bar{y}_3)$; the difference between these two values is a contrast:

$$L = (\bar{y}_2 - \bar{y}_1) - (\bar{y}_4 - \bar{y}_3)$$
$$= -32.4 - -35.3 = 2.9$$

This contrast L can be used as the basis for a confidence interval or a test of hypothesis. We illustrate the test. The null hypothesis is

$$H_0: (\mu_2 - \mu_1) = (\mu_4 - \mu_3)$$

or, in words,

H_0: The effect of stress is the same in the two light conditions.

For the preceding L, $\sum_{i=1}^{I} m_i^2 = 4$, and the standard error is

$$SE_L = \sqrt{s_{pooled}^2 \sum_{i=1}^{I} \frac{m_i^2}{n_i}} = \sqrt{s_{pooled}^2 \frac{4}{13}} = \sqrt{\frac{(895.34)(4)}{13}} = 16.6$$

The test statistic is

$$t_s = \frac{2.9}{16.6} = .2$$

From Table 4 with df = 40, we find $t(40)_{.20} = 0.851$. The data provide virtually no evidence that the effect of stress is different in the two light conditions. This is consistent with the F test for interactions conducted in Example 11.19 in Section 11.6. ■

The statistical definition of interaction introduced in Section 11.6 and viewed through the lens of contrasts here is rather specialized. It is defined in terms of the observed variable rather than in terms of a biological mechanism. Further, interaction as measured by a contrast is defined by *differences* between means. In some applications the biologist might feel that ratios of means are more meaningful or relevant than differences. The following example shows that the two points of view can lead to different answers.

Example 11.27

Chromosomal Aberrations. A research team investigated the separate and joint effects in mice of exposure to high temperature (35°C) and injection with the cancer drug cyclophosphamide (CTX). A completely randomized design was used, with eight mice in each treatment group. For each animal, the researchers measured the incidence of a certain chromosomal aberration in the bone marrow; the result is expressed as the number of abnormal cells per 1,000 cells. The treatment means are shown in Table 11.22.[20]

TABLE 11.22 Mean Incidence of Chromosomal Aberrations Following Various Treatments

		Injection	
		CTX	*None*
Temperature	Room	23.5	2.7
	High	75.4	20.9

Is the observed effect of CTX greater at room temperature or at high temperature? The answer depends on whether "effect" is measured absolutely or relatively.

Measured as a difference, the effect of CTX is

Room temperature: 23.5 − 2.7 = 20.8

High temperature: 75.4 − 20.9 = 54.5

Thus, the absolute effect of CTX is greater at the high temperature. However, this relationship is reversed if we express the effect of CTX as a ratio rather than as a difference:

Room temperature: $\frac{23.5}{2.7} = 8.70$

High temperature: $\frac{75.4}{20.9} = 3.61$

At room temperature CTX produces almost a ninefold increase in chromosomal aberrations, whereas at high temperature the increase is less than fourfold; thus, in relative terms, the effect of CTX is much greater at room temperature. ■

If the phenomenon under study is thought to be multiplicative rather than additive, so that relative rather than absolute change is of primary interest, then ordinary contrasts should not be used. One simple approach in this situation is to use a logarithmic transformation—that is, to compute $Y' = \log(Y)$, and then analyze Y' using contrasts. The motivation for this approach is that relations of constant *relative* magnitude in the Y scale become relations of constant *absolute* magnitude in the Y' scale.

Exercises 11.23–11.32

11.23 Refer to the FVC data of Example 11.21.

(a) Verify that the grand mean of all 481 FVC values is 4.56.
(b) Taking into account the age distribution among the 481 subjects and the age distribution in the U.S. population, explain intuitively why the grand mean (4.56) is smaller than the age-adjusted mean (4.73).

11.24 To see if there is any relationship between blood pressure and childbearing, researchers examined data from a large health survey. The following table shows the data on systolic blood pressure (mm Hg) for women who had borne no children and women who had borne five or more children. The pooled standard deviation from all eight groups was $s_{pooled} = 18$ mm Hg.[21]

	No Children		Five or More Children	
Age	Mean blood pressure	No. of women	Mean blood pressure	No. of women
18–24	113	230	114	7
25–34	118	110	116	82
35–44	125	105	124	127
45–54	134	123	138	124
18–54	121	568	127	340

Carry out age adjustment, as directed, using the following reference distribution, which is the approximate distribution for U.S. women:[22]

Age	Relative Frequency
18–24	.17
25–34	.29
35–44	.31
45–54	.23

(a) Calculate the age-adjusted mean blood pressure for women with no children.
(b) Calculate the age-adjusted mean blood pressure for women with five or more children.
(c) Calculate the difference between the values obtained in parts (a) and (b). Explain intuitively why the result is smaller than the unadjusted difference of $127 - 121 = 6$ mg Hg.
(d) Calculate the standard error of the value calculated in part (a).
(e) Calculate the standard error of the value calculated in part (c).

11.25 Refer to the ATP data of Exercise 11.17. The sample means and standard deviations are as follows:

| | River Birch | | European Birch | |
	Flooded	Control	Flooded	Control
$\bar{y}$	1.19	1.78	.29	1.20
s	.18	.24	.20	.16

Define linear combinations (that is, specify the multipliers) to measure each of the following:

(a) The effect of flooding in river birch
(b) The effect of flooding in European birch
(c) The difference between river birch and European birch with respect to the effect of flooding (that is, the interaction between flooding and species)

11.26 *(Continuation of Exercise 11.25)*

(a) Use a *t* test to investigate whether flooding has the same effect in river birch and in European birch. Use a nondirectional alternative and let $\alpha = .05$. (The pooled standard deviation is $s_{\text{pooled}} = .199$.)
(b) If the sample sizes were $n = 10$ rather than $n = 4$ for each group, but the means, standard deviations, and s_{pooled} remained the same, how would the result of part (a) change?

11.27 *(Continuation of Exercise 11.25)*   Consider the null hypothesis that flooding has no effect on ATP level in river birch. This hypothesis could be tested in two ways: as a contrast (using the method of Section 11.7), or with a two-sample *t* test (as in Exercise 7.32). Answer the following questions; do not actually carry out the tests.

(a) In what way or ways do the two test procedures differ?
(b) In what way or ways do the conditions for validity of the two procedures differ?
(c) One of the two procedures requires more conditions for its validity, but if the conditions are met, then this procedure has certain advantages over the other one. What are these advantages?

11.28 Consider the data from Exercise 11.19, in which the drugs ticrynafen (T) and hydrochlorothiazide (H) were compared. The data are summarized in the following table. The pooled standard deviation is $s_{\text{pooled}} = 11.83$ mm Hg.

| | Ticrynafen (T) | | Hydrochlorothiazide (H) | |
	Low Dose	High Dose	Low Dose	High Dose
Mean	13.9	17.1	15.8	17.5
No. of Patients	53	57	55	58

If the two drugs have equal effects on blood pressure, then T might be preferable because it has fewer side effects.

(a) Construct a 95% confidence interval for the difference between the drugs (with respect to mean blood pressure reduction), averaged over the two dosage levels.
(b) Interpret the confidence interval from part (a) in the context of this setting.

11.29 Consider the lettuce growth experiment described in Exercise 11.22. The accompanying table shows the mean leaf dry weight (g) of the nine plants in each treatment group. MS(within) from the ANOVA was .3481.

	Nutrient Solution	
	Standard	*Extra Nitrogen*
Low Light	2.16	3.09
High Light	3.26	4.48

Construct a 95% confidence interval for the effect of extra nitrogen, averaged over the two light conditions.

11.30 Refer to the MAO data of Exercise 11.8.

(a) Define a contrast to compare the MAO activity for schizophrenics without paranoid features versus the average of the two types with paranoid features.
(b) Calculate the value of the contrast in part (a) and its standard error.
(c) Apply a t test to the contrast in part (a). Let H_A be nondirectional and $\alpha = .05$.

11.31 Are the brains of left-handed people anatomically different? To investigate this question, a neuroscientist conducted postmortem brain examinations in 42 people. Each person had been evaluated before death for hand preference and categorized as consistently right-handed (CRH) or mixed-handed (MH). The table shows the results on the area of the anterior half of the corpus callosum (the structure that links the left and right hemispheres of the brain).[23] The MS(within) from the ANOVA was 2,498.

	Area (mm²)		
Group	*Mean*	*SD*	*n*
1. Males: MH	423	48	5
2. Males: CRH	367	49	7
3. Females: MH	377	63	10
4. Females: CRH	345	43	20

(a) The difference between MH and CRH is 56 mm² for males and 32 mm² for females. Is this sufficient evidence to conclude that the corresponding population difference is greater for males than for females? Test an appropriate hypothesis. (Use a nondirectional alternative and let $\alpha = .10$.)
(b) As an overall measure of the difference between MH and CRH, we can consider the quantity $.5(\mu_1 - \mu_2) + .5(\mu_3 - \mu_4)$. Construct a 95% confidence interval for this quantity. (This is a sex-adjusted comparison of MH and CRH, where the reference population is 50% male and 50% female.)

11.32 Consider the daffodil data of Exercise 11.12.

(a) Define a contrast to compare the stem length for daffodils from the open area versus the average of the north, south, east, and west sides of the building.
(b) Calculate the value of the contrast in part (a) and its standard error.
(c) Apply a t test to the contrast in part (a). Let H_A be nondirectional and $\alpha = .05$.

11.8 MULTIPLE COMPARISONS (OPTIONAL)

One approach to detailed analysis of the means $\bar{y}_1., \bar{y}_2., \ldots, \bar{y}_I.$ is to make every pairwise comparison among them. Suppose it is desired to test all possible pairwise hypotheses:

$$H_0: \mu_1 = \mu_2$$
$$H_0: \mu_1 = \mu_3$$
$$H_0: \mu_2 = \mu_3$$

and so on. We saw in Section 11.1 that using repeated t tests leads to an increased overall risk of Type I error. There are several methods that can be used to control the overall risk of Type I error. One of these is the **Newman-Keuls procedure**, which we now describe. The procedure is designed for use when the I sample sizes (n) are equal.

The Newman-Keuls Procedure

The Newman-Keuls procedure is a decision-oriented procedure, conducted at a prespecified overall significance level α. For each pair of means (for instance, $\bar{y}_1$. versus $\bar{y}_2$.), the procedure leads to a decision as to whether the corresponding null hypothesis (for instance, H_0: $\mu_1 = \mu_2$) is rejected.

    We give a step-by-step description of the Newman-Keuls procedure and then illustrate with an example. (Although the description is lengthy, the procedure itself is not complicated.)

Step 1.   *Array of means*   Construct an array of the sample means arranged in increasing order.

Step 2.   *Critical values*   Table 11 (at the end of this book) provides constants,* denoted as q_i for the Newman-Keuls procedure at $\alpha = .05$ or $\alpha = .01$. To use Table 11, first determine MS(within) and df(within) from the ANOVA. From the row of Table 11 corresponding to df $=$ df(within), read the values of for $i = 2, 3, \ldots, I$. Next, calculate the following scale factor:

$$\sqrt{\frac{s^2_{\text{pooled}}}{n}}$$

where $s^2_{\text{pooled}} = $ MS(within). (Note that this is SE$_{\bar{y}}$ as given in Section 11.5.) Finally, calculate critical values, denoted R_i, as follows:

$$R_i = q_i \sqrt{\frac{s^2_{\text{pooled}}}{n}}$$

The work for step 2 can be conveniently arranged in a table as follows:

i	2	3	$\cdots$	I
q_i	q_2	q_3	$\cdots$	q_I
R_i	R_2	R_3	$\cdots$	R_I

Step 3.   *The pairwise comparisons*   The R_i's are the critical values with which the differences between sample means will be compared; larger R_i's will be used for the means that are farther apart in the array of means constructed in step 1. As a convenient way of keeping track of the results, nonrejection of a null hypothesis will be indicated by underlining the corresponding pair of means. The procedure is carried out sequentially, as follows:

(a) Compare the difference between the largest and smallest of the I sample means with the critical value R_I. If the difference is smaller than R_I, the corresponding null hypothesis is not rejected;

---

* Technically, the constants q_i are called "percentage points of the Studentized range distribution."

in this case a line is drawn under the entire array of means and the procedure is ended. If the difference is larger than R_I, proceed to step (b).

(b) Ignore the smallest $\bar{y}$ and consider the remaining subarray of $(I - 1)$ means. Compare the difference between the largest and smallest mean in the subarray with R_{I-1}; if the difference is less than R_{I-1}, underline the entire subarray. Now consider the other subarray of $(I - 1)$ means—the means that remain if the largest $\bar{y}$ is ignored. Again, underline this subarray if the difference between its largest and smallest mean is less than R_{I-1}. (When underlining, use a separate line each time; never join a line to one that has already been drawn.)

(c) Continue by looking at all subarrays of $(I - 2)$ means and comparing with R_{I-2}, then subarrays of $(I - 3)$ means and comparing with R_{I-3}, and so on until, finally, each subarray of two means is compared with R_2. During this procedure, however, *never test within any subarray that has already been underlined; all hypotheses in such a subarray are automatically not rejected.*

(d) When the procedure is complete, those pairs of means not connected by an underline correspond to null hypotheses that have been rejected. All other pairwise null hypotheses are not rejected.

Illustration of the Newman-Keuls Procedure

Blood Chemistry in Rats. For an evaluation of diets used for routine maintenance of laboratory rats, researchers used a completely randomized design to allocate weanling male rats to five different diets. After 4 weeks, specimens of blood were collected and various biochemical variables were measured. We consider the results for blood urea concentration (mg/d/Li). The group means were as follows:[24]

Example 11.28

Diet	A	B	C	D	E
$\bar{y}$	40.0	40.7	32.9	29.6	48.8

The ANOVA is shown in Table 11.23.

TABLE 11.23 ANOVA for Blood Urea Data

Source	df	SS	MS
Between diets	4	894.80	223.70
Within diets	15	319.35	21.29
Total	19	1,214.15	

We now apply the Newman-Keuls procedure to compare every pair of diets at $\alpha = .05$.

Step 1. The ordered array is as follows:

Diet	D	C	A	B	E
Mean	29.6	32.9	40.0	40.7	48.8

Step 2. The number of rats in each group was $n = 4$; obtaining MS(within) from Table 11.23, we calculate the scale factor

$$\sqrt{\frac{s^2_{\text{pooled}}}{n}} = \sqrt{\frac{21.29}{4}} = 2.307$$

We read the values from Table 11 with df = 15; we then multiply each q_i by 2.307 to obtain R_i. The results are shown in Table 11.24.

TABLE 11.24 Critical Values for Example 11.28				
i	**2**	**3**	**4**	**5**
q_i	3.01	3.67	4.08	4.37
R_i	6.9	8.5	9.4	10.1

Step 3. We first compare the largest mean against the smallest, using the critical value R_5. We find

$$\bar{y}_E - \bar{y}_D = 48.8 - 29.6 = 19.2$$

$$R_5 = 10.1$$

Because $19.2 > 10.1$, we reject the null hypothesis $H_0: \mu_D = \mu_E$. We then proceed to compare $\bar{y}_E$ against $\bar{y}_C$ using the critical value R_4. The entire sequence of comparisons is shown in Table 11.25.

TABLE 11.25 Newman-Keuls Analysis for Example 11.28		
Value of i	**Comparison**	**Conclusion**
5	$48.8 - 29.6 = 19.2 > 10.1$	Reject
4	$48.8 - 32.9 = 15.9 > 9.4$	Reject
4	$40.7 - 29.6 = 11.1 > 9.4$	Reject
3	$48.8 - 40.0 = 8.8 > 8.5$	Reject
3	$40.7 - 32.9 = 7.8 < 8.5$	Do not reject (line from C to B)
3	$40.0 - 29.6 = 10.4 > 8.5$	Reject
2	$48.8 - 40.7 = 8.1 > 6.9$	Reject
2	$40.7 - 40.0$: Already underlined	Do not reject.
2	$40.0 - 32.9$: Already underlined	Do not reject.
2	$32.9 - 29.6 = 3.3 < 6.9$	Do not reject (line from D to C)

Note that the difference $40.7 - 40.0$ is not compared to 6.9 because the subarray containing that pair is already underlined, and similarly for the difference $40.0 - 32.9$. This is an essential feature of the Newman-Keuls procedure.

The final array is as follows:

Diet	**D**	**C**	**A**	**B**	**E**
Mean	29.6	32.9	40.0	40.7	48.8

Note that D and B are *not* joined by an underline, even though D and C are joined by an underline and C and B are joined by an underline. The overlapping of the underlines reflects the fact that we rejected the null hypothesis $H_0: \mu_D = \mu_B$ but we did not reject $H_0: \mu_D = \mu_C$ or

$H_0: \mu_C = \mu_B$. This may seem contradictory, but it is not if you remember that nonrejection of a null hypothesis is absence of evidence, rather than evidence of absence, of a difference. The data provide enough information to conclude that $\mu_B > \mu_D$, but not enough to conclude that $\mu_B > \mu_C$ nor enough to conclude that $\mu_C > \mu_D$.

We can describe our conclusions verbally as follows. At $\alpha = .05$, there is sufficient evidence to conclude that diet E gives the highest mean blood urea; either diet D or C gives the lowest; μ_A and μ_B are greater than μ_D but not necessarily greater than μ_C; we cannot say which of μ_A or μ_B is greater. Note that the Newman-Keuls procedure takes a strictly decision-oriented approach; there are no P-values associated with the various differences. ∎

Examples of Possible Patterns

In Example 11.28, the Newman-Keuls procedure yielded a moderately complicated pattern of results. Here are some other possible patterns that might emerge for five treatments A, B, C, D, and E.
    Pattern 1:

$$\underline{\text{D} \quad \text{C}} \quad \underline{\text{A} \quad \text{B}} \quad \text{E}$$

In pattern 1 there is no overlapping of underlines, so the treatments fall into distinct groups.
    Pattern 2:

$$\text{D} \quad \text{C} \quad \text{A} \quad \text{B} \quad \text{E}$$

In pattern 2 there are no underlines; all pairwise null hypotheses have been rejected.
    Pattern 3:

$$\text{D} \quad \underline{\underline{\text{C} \quad \text{A}} \quad \underline{\text{B}} \quad \text{E}}$$

Pattern 3 is more subtle to interpret. The overlapping indicates that adjacent treatments are not distinguishable even though more distant ones are (just as black can shade into white through imperceptible stages of gray).

Relation to the *t* Test

The critical value R_2 deserves special comment. The comparison of two means using R_2 is very similar to a two-sample t test; in fact, it differs from a t test only in that the quantity s_{pooled} (which enters the SE calculation) is derived from all I samples rather than just two samples. (This relationship is explained in more detail in Appendix 11.1.)

In spite of the close link between R_2 and the t test, the Newman-Keuls method does not suffer from Type I error risks as high as those given in Table 11.2 for repeated t tests. The Newman-Keuls procedure has less chance of Type I error because many of its comparisons use the larger critical values (R_3, R_4, and so on) rather than R_2, and also because comparisons within an underline are automatically nonsignificant. [In Example 11.28, for instance, notice that the difference $(\bar{y}_B - \bar{y}_C)$ and the difference $(\bar{y}_A - \bar{y}_C)$ are both greater than R_2 and yet the corresponding null hypotheses are not rejected by the Newman-Keuls procedure.]

Relation to the F Test

Although it is customary to precede the Newman-Keuls procedure with an F test of the global null hypothesis, it is not necessary to do so. It is possible for the global F test and the first Newman-Keuls comparison (using R_1) to disagree, but this is rare.

Conditions for Validity

The conditions for validity of the Newman-Keuls procedure consist of the standard conditions given in Section 11.5, together with the requirement that the sample sizes all be equal. In practice, it often happens that the sample sizes are slightly unequal (for instance, an experimental animal may die for reasons unrelated to the experiment); in this case the procedure is approximately valid.

The Bonferroni Method

The **Bonferroni method** is based on a very simple and general relationship: The probability that at least one of several events will occur cannot exceed the sum of the individual probabilities. For instance, suppose we conduct six tests of hypotheses, each at $\alpha = .01$. Then the overall risk of Type I error—that is, the chance of rejecting at least one of the six hypotheses when in fact all of them are true— cannot exceed

$$.01 + .01 + .01 + .01 + .01 + .01 = (6)(.01) = .06$$

Turning this logic around, suppose an investigator plans to conduct six tests of hypotheses and wants the overall risk of Type I error not to exceed .05. A conservative approach is to conduct each of the separate tests at the significance level $\alpha = \dfrac{.05}{6} = .0083$; this is called a **Bonferroni adjustment**.

Note that the Bonferroni technique is very broadly applicable. The separate tests may relate to different response variables, different subsets, and so on; some may be t tests, some chi-square tests, and so on.

The Bonferroni approach can be used by a person reading a research report, if the author has included explicit P-values. For instance, if the report contains six P-values and the reader desires overall 5%-level protection against Type I error, then the reader will not regard a P-value as sufficient evidence of an effect unless it is smaller than .0083.

A Bonferroni adjustment can also be made for confidence intervals. For instance, suppose we wish to construct six confidence intervals, and desire an overall probability of 95% that *all* the intervals contain their respective parameters. Then this can be accomplished by constructing each interval at confidence level 99.17% (because $\dfrac{.05}{6} = .0083$ and $1 - .0083 = .9917$). Note that application of this idea requires unusual critical values, so that standard tables are not sufficient. Table 12 (at the end of this book) provides Bonferroni multipliers for confidence intervals that are based on a t distribution. Example 11.29 illustrates this idea.

Example 11.29 | **Blood Chemistry in Rats.** Suppose we wish to construct 95% confidence intervals for the differences in means of each pair of diets presented in Example 11.28. From the analysis of variance table we have an estimate of the (common) population SD: Table 11.23 shows that MS(within) = 21.29, so $s_{pooled} = \sqrt{21.29} = 4.614$. There were four observations in each of the groups, so the standard error of the difference in any two of the sample means is

$$\sqrt{\frac{21.29}{4} + \frac{21.29}{4}} = 4.614 \cdot \sqrt{\frac{1}{4} + \frac{1}{4}} = 3.263.$$ If we were only comparing two of

the means, say for diets A and B, we would use

$$(\bar{y}_A - \bar{y}_B) \pm t(15)_{.025} \cdot 3.263$$

which is

$$(40.0 - 40.7) \pm 2.131 \cdot 3.263$$

or $-.7 \pm 7.0$.

The Bonferroni method involves adjusting the t-multiplier. There are $_5C_2 = 10$ pairs of means for which a confidence interval could be constructed. Thus, to make a Bonferroni adjustment, we replace $t(15)_{.025}$, which is 2.131, with $t(15)_{.025/10}$ [that is, with $t(15)_{.0025}$], which is found in Table 12 to be 3.286. The Bonferroni-adjusted 95% confidence interval for the population mean difference between diets A and B is

$$(40.0 - 40.7) \pm 3.286 \cdot 3.263$$

or $-.7 \pm 10.7$.

Likewise, the Bonferroni-adjusted 95% confidence interval for the population mean difference between diets A and C is

$$(40.0 - 32.9) \pm 3.286 \cdot 3.263$$

or 7.1 ± 10.7.

Confidence intervals for differences in other pairs of means are constructed in the same way. ∎

A disadvantage of the Bonferroni method is that it is quite conservative. For example, the Bonferroni-adjusted confidence intervals in Example 11.29 are much wider (53% wider, to be precise) than the unadjusted confidence intervals. If there are very many comparisons, the Bonferroni method produces confidence intervals that are very wide. Likewise, the Bonferroni adjustment in a hypothesis test makes it difficult to reject any null hypothesis when there are many tests conducted.

An advantage of the Bonferroni method is that it is widely applicable. In particular, it does not require equal sample sizes. Example 11.30 illustrates the use of the Bonferroni method when the sample sizes differ.

Weight Gain in Lambs. Consider the weight gain data of Example 11.2. From Table 11.7 we have MS(within) = 23.333 and df(within) = 9. There were three diets in this study, so there are $_3C_2 = 3$ pairs of means for which a confidence interval can be constructed. Thus, the Bonferroni adjustment is to replace $t(9)_{.025}$ with $t(9)_{.025/3} = 2.933$. The sample mean for Diet 1 was 11, with a sample size of 3, whereas the sample mean for Diet 2 was 15, with a sample size of 5. A 95% confidence interval for the difference in the corresponding population means is

Example 11.30

$$(11 - 15) \pm 2.933 \cdot \sqrt{23.333} \sqrt{\frac{1}{3} + \frac{1}{5}}$$

or -4 ± 10.3. ∎

In Example 11.30 a Bonferroni adjustment was used in making a 95% confidence interval. This adjustment is based on the idea that three confidence intervals are *possible* (comparing diets 1 and 2, 1 and 3, and 2 and 3); the adjustment made the confidence interval noticeably wider than it would have been had there been no Bonferroni adjustment. We might argue that the adjustment is not needed, since only one confidence interval was actually constructed. Such an attitude is potentially dangerous. It is true that only one confidence interval was constructed, but it is the interval that compares the two most disparate sample means. If there had been prior interest in comparing only diets 1 and 2, then we could justify using an unadjusted confidence interval. (However, we would wonder why three diets were included in the study if there was no interest in the third diet!) If the comparison of diets 1 and 2 was chosen on the basis of having the largest sample difference, then implicitly all three confidence intervals have been considered, which means that a Bonferroni adjustment is called for.

Often a researcher inspects data and chooses, from the multitude of possible analyses, a few that look particularly promising. Thus, the analyses are not preplanned but are "inspired" by the data. Such data-inspired analysis can give rise to serious problems of what is called "hidden multiplicity." This means that there are several simultaneous statistical inferences being made, with some of them not being immediately obvious. The following example shows how the overall risk of Type I error, when testing a data-inspired hypothesis, is inflated just as if many hypotheses had been tested.

Example 11.31

Liver Weight of Mice. Ten treatments were compared for their effect on the liver in mice. There were 12 animals in each treatment group. The mean liver weights are shown in Table 11.26.[25]

TABLE 11.26 Liver Weights in Mice			
Treatment	**Mean Liver Weight (g)**	**Treatment**	**Mean Liver Weight (g)**
1	2.59	6	2.84
2	2.28	7	2.29
3	2.34	8	2.45
4	2.07	9	2.76
5	2.40	10	2.37

Consider two investigators, A and B. Investigator A performs separate t tests (each at $\alpha = .05$) on all possible pairs of the ten treatments. She finds the following differences significant.

$$H_0: \mu_4 = \mu_6 \quad \text{Rejected}$$
$$H_0: \mu_4 = \mu_9 \quad \text{Rejected}$$

Investigator B begins his analysis by inspecting the treatment means. He notices that $\bar{y}_4$ is especially small and that $\bar{y}_6$ and $\bar{y}_9$ are especially large. He then uses t tests to confirm these impressions. His conclusions are as follows:

$$H_0: \mu_4 = \mu_6 \quad \text{Rejected}$$
$$H_0: \mu_4 = \mu_9 \quad \text{Rejected}$$

Among the ten means there are $_{10}C_2 = 45$ possible pairwise tests. Investigator A conducted all 45 tests and investigator B conducted only two of them. But because investigator B's choice of those two hypotheses was inspired by inspection of the data, their conclusions are identical. Investigator B's procedure contains hidden multiplicity. ■

We have presented two approaches that can be used to deal with multiplicity (i.e., the multiple comparisons problem). One is to conduct a global F test of the data first and then only proceed to compare pairs of sample means if the null hypothesis is rejected in the F test. The other approach is to use a multiple comparisons method such as the Bonferroni method. Thus, the investigators in Example 11.31 could make a Bonferroni adjustment to the t tests.

Other Multiple Comparison Procedures

In addition to the Newman-Keuls and Bonferroni methods, statisticians have developed several other methods for multiple comparison of means.* The methods differ from each other in the degree of protection they provide against Type I error. All of the methods have the property that the chance of Type I error is controlled (say, at .05) when the global null hypothesis is true. But Type I errors can occur even when the global null hypothesis is false. For instance, suppose $\mu_1 = 30$, $\mu_2 = 30$, and $\mu_3 = 40$; then rejection of the hypothesis $H_0: \mu_1 = \mu_2$ would be a Type I error. A conservative procedure, which guards stringently against all possible kinds of Type I error, is not as powerful as a less conservative procedure. Compared to the other procedures, the Newman-Keuls and Bonferroni procedures are conservative. (In Appendix 11.1 we give more detail on the different degrees of control of Type I error.)

Certain multiple comparison methods, such as the Bonferroni method, can be used to construct confidence intervals, as was illustrated previously. The Newman-Keuls procedure does not share this advantage.

Exercises 11.33–11.39

11.33 A botanist used a completely randomized design to allocate 45 individually potted eggplant plants to five different soil treatments. The observed variable was the total plant dry weight without roots (g) after 31 days of growth. The treatment means were as shown in the table.[26] The MS(within) was .2246. Use the Newman-Keuls method to compare all pairs of means at $\alpha = .05$.

Treatment	A	B	C	D	E
Mean	4.37	4.76	3.70	5.41	5.38
n	9	9	9	9	9

---

* Two popular methods are the LSD (least significant difference) method and the HSD (honestly significant difference), or Tukey method. The HSD procedure resembles the Newman-Keuls procedure but it uses the largest critical value, R_k, for all comparisons. The LSD procedure uses the smallest critical value, R_2, for all comparisons. For additional protection against Type I error, the LSD procedure begins with the global F test and proceeds to pairwise comparisons only if the global null hypothesis is rejected.

11.34 Proceed as in Exercise 11.33, but let $\alpha = .01$.

11.35 In a study of the dietary treatment of anemia in cattle, researchers randomly divided 144 cows into four treatment groups. Group A was a control group, and groups B, C, and D received different regimens of dietary supplementation with selenium. After a year of treatment, blood samples were drawn and assayed for selenium. The accompanying table shows the mean selenium concentrations $(\mu g/d/Li)$.[27] The MS(within) from the ANOVA was 2.071. Use the Newman-Keuls method to compare all pairs of means at $\alpha = .05$.

Group	Mean	n
A	.8	36
B	5.4	36
C	6.2	36
D	5.0	36

11.36 Proceed as in Exercise 11.35, but let $\alpha = .01$.

11.37 Ten treatments were compared for their effect on the liver in mice. There were 13 animals in each treatment group. The ANOVA gave MS(within) = .5842. The mean liver weights are given in the table.[28]

Treatment	Mean Liver Weight (g)
1	2.59
2	2.28
3	2.34
4	2.07
5	2.40
6	2.84
7	2.29
8	2.45
9	2.76
10	2.37

(a) Use the Newman-Keuls method to compare all pairs of means at $\alpha = .05$.
(b) Suppose the critical value R_2 were used for all comparisons (this would be a form of repeated t tests). Which pairs of means would be declared significantly different?

11.38 Consider the data from Exercise 11.33. Use the Bonferroni method to construct a 95% confidence interval for the difference in population means of treatments E and A.

11.39 Consider the data from Example 11.2 on the weight gain of lambs. The MS(within) from the ANOVA for these data was 23.333. The sample mean of Diet 2 was 15 and of Diet 1 was 11. Use the Bonferroni method to construct a 95% confidence interval for the difference in population means of these two diets.

11.9 PERSPECTIVE

In Chapter 11 we have introduced some statistical issues that arise when analyzing data from more than two samples and we have considered some classical methods of analysis. In this section we review these issues and briefly mention some alternative methods of analysis.

Advantages of Global Approach

Let us recapitulate the advantages of analyzing I independent samples by a global approach rather than by viewing each pairwise comparison separately.

1. **Multiple comparisons**   In Section 11.1 we saw that the use of repeated t tests can greatly inflate the overall risk of Type I error. Some control of Type I error can be gained by the simple device of beginning the data analysis with a global F test. For more stringent control of Type I error, special multiple comparison methods are available. Two of these were described in optional Section 11.8. (Note that the problem of multiple comparisons is not confined to an ANOVA setting.)

2. **Use of structure in the treatments or groups**   Analysis of suitable combinations of group means can be very useful in interpreting data. Many of the relevant techniques are beyond the scope of this book. The discussion in optional Sections 11.6 and 11.7 gave a hint of the possibilities. In Chapter 12 we will discuss some ideas that are applicable when the treatments themselves are quantitative (for instance, doses).

3. **Use of a pooled SD**   We have seen that pooling all of the within-sample variability into a single pooled SD leads to a better estimate of the common population SD and thus to a more precise analysis. This is particularly advantageous if the individual sample sizes (n's) are small, in which case the individual SD estimates are quite imprecise. Of course, using a pooled SD is proper only if the population SDs are equal. It sometimes happens that we cannot take advantage of pooling the SDs because the assumption of equal population SDs is not tenable. One approach that can be helpful in this case is to analyze a transformed variable, such as $\log(Y)$; the SDs may be more nearly equal in the transformed scale.

Other Experimental Designs

The techniques of this chapter are valid only for independent samples. But the basic idea—partitioning variability within and between treatments into interpretable components—can be applied in many experimental designs. For instance, all of the techniques discussed in this chapter can be adapted (by suitable modification of the SE calculation) to analysis of data from a randomized blocks design. (See optional Section 11.6.) These and related techniques belong to the large subject called *analysis of variance*, of which we have discussed only a small part.

Nonparametric Approaches

There are k-sample analogs of the Wilcoxon-Mann-Whitney test and other nonparametric tests. These tests have the advantage of not assuming underlying normal distributions. However, many of the advantages of the parametric techniques—such as the use of linear combinations—do not easily carry over to the nonparametric setting.

Ranking and Selection

In some investigations the primary aim of the investigator is not to answer research questions about the populations but simply to *select* one or several "best" populations. For instance, suppose ten populations (stocks) of laying hens are available and it is desired to select the one population with the highest egg-laying potential. The investigator will select a random sample of n chickens from each stock and will observe for each chicken Y = total number of eggs laid in 500 days.[29] One relevant question is, How large should n be so that the stock that is *actually* best (has the highest μ) is likely to also *appear* best (have the highest $\bar{y}$)? This and similar questions are addressed by a branch of statistics called *ranking and selection theory*.

Supplementary Exercises 11.40–11.56

Note: Exercises preceded by an asterisk refer to optional sections.

11.40 Consider the research described in Exercise 11.13, in which 10 women in an aerobic exercise class, 10 women in a modern dance class, and a control group of 9 women were studied. One measurement made on each woman was change in fat-free mass over the course of the 16-week training period. Summary statistics are given in the table.[8] The ANOVA SS(between) is 2.465 and the SS(within) is 50.133.

	Aerobics	Modern Dance	Control
Mean	0.00	0.44	0.71
SD	1.31	1.17	1.68
n	10	10	9

(a) State in words, in the context of this problem, the null hypothesis that is tested by the analysis of variance.

(b) Construct the ANOVA table and test the null hypothesis. Let α = .05.

11.41 Refer to Exercise 11.40. The F test is based on certain conditions concerning the population distributions.

(a) State the conditions.

(b) The following dotplots show the raw data. Based on these plots and on the information given in Exercise 11.40, does it appear that the F test conditions are met? Why or why not?

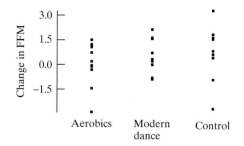

11.42 In a study of the eye disease retinitis pigmentosa (RP), 211 patients were classified into four groups according to the pattern of inheritance of their disease. Visual acuity (spherical refractive error, in diopters) was measured for each eye, and the two values were then averaged to give one observation per person. The accompanying table shows the number of people in each group and the group mean refractive error.[30] The ANOVA of the 211 observations yields SS(between) = 129.49 and SS(within) = 2,506.8. Construct the ANOVA table and carry out the F test at α = .05.

Group	Number of Persons	Mean Refractive Error
Autosomal dominant RP	27	+.07
Autosomal recessive RP	20	−.83
Sex-linked RP	18	−3.30
Isolate RP	146	−.84
Total	211	

11.43 *(Continuation of Exercise 11.42)*   Another approach to the data analysis is to use the eye, rather than the person, as the observational unit. For the 211 persons there were 422 measurements of refractive error; the accompanying table summarizes these measurements. The ANOVA of the 422 observations yields SS(between) = 258.97 and SS(within) = 5,143.9.

Group	Number of Eyes	Mean Refractive Error
Autosomal dominant RP	54	+.07
Autosomal recessive RP	40	−.83
Sex-linked RP	36	−3.30
Isolate RP	292	−.84
Total	422	

(a) Construct the ANOVA table and bracket the *P*-value for the *F* test. Compare with the *P*-value obtained in Exercise 11.37. Which of the two *P*-values is of doubtful validity, and why?

(b) The mean refractive error for the sex-linked RP patients was −3.30. Calculate the standard error of this mean two ways: (i) regarding the person as the observational unit and using s_{pooled} from the ANOVA of Exercise 11.42; (ii) regarding the eye as the observational unit and using s_{pooled} from the ANOVA of this exercise. Which of these standard errors is of doubtful validity, and why?

*11.44   In a study of the mutual effects of the air pollutants ozone and sulfur dioxide, Blue Lake snap beans were grown in open-top field chambers. Some chambers were fumigated repeatedly with sulfur dioxide. The air in some chambers was carbon filtered to remove ambient ozone. There were three chambers per treatment combination, allocated at random. After one month of treatment, total yield (kg) of bean pods was recorded for each chamber, with results shown in the accompanying table.[31] For these data, SS(between) = 1.3538 and SS(within) = .27513. Complete the ANOVA table and carry out the *F* test at $\alpha = .05$.

	Ozone Absent		Ozone Present	
	Sulfur Dioxide		*Sulfur Dioxide*	
	Absent	*Present*	*Absent*	*Present*
	1.52	1.49	1.15	.65
	1.85	1.55	1.30	.76
	1.39	1.21	1.57	.69
Mean	1.587	1.417	1.340	.700
SD	.237	.181	.213	.056

Prepare an interaction graph (like Figure 11.14).

*11.45   Consider the data from Exercise 11.44. For these data, SS(ozone) = 0.696, SS(sulfur) = 0.492, SS(interaction) = 0.166, and SS(within) = 0.275.

(a) Construct the ANOVA table.

(b) Carry out an F test for interactions; use $\alpha = .05$.

(c) Test the null hypothesis that ozone has no effect on yield. Use $\alpha = .05$.

*11.46 Refer to Exercise 11.44. Define contrasts to measure each effect specified, and calculate the value of each contrast.

(a) The effect of sulfur dioxide in the absence of ozone

(b) The effect of sulfur dioxide in the presence of ozone

(c) The interaction between sulfur dioxide and ozone

*11.47 *(Continuation of Exercises 11.45 and 11.46)*   For the snap-bean data, use a t test to test the null hypothesis of no interaction against the alternative that sulfur dioxide is more harmful in the presence of ozone than in its absence. Let $\alpha = .05$. How does this compare with the F test of Exercise 11.45(b) (which has a nondirectional alternative)?

*11.48 *(Computer exercise)*   Refer to the snap-bean data of Exercise 11.44. Apply a reciprocal transformation to the data. That is, for each yield value Y, calculate $Y' = 1/Y$.

(a) Calculate the ANOVA table for Y' and carry out the F test.

(b) It often happens that the SDs are more nearly equal for transformed data than for the original data. Is this true for the snap-bean data when a reciprocal transformation is used?

(c) Make a normal probability plot of the residuals, $(y'_{ij} - \bar{y}'_i)$. Does this plot support the condition that the populations are normal?

*11.49 *(Computer exercise—continuation of Exercises 11.47 and 11.48)*   Repeat the test in Exercise 11.46 using Y' instead of Y, and compare with the results of Exercise 11.46.

11.50 In a study of balloon angioplasty, patients with coronary artery disease were randomly assigned to one of four treatment groups: placebo, probucol (an experimental drug), multivitamins (a combination of beta carotene, vitamin E, and vitamin C), or probucol combined with multivitamins. Balloon angioplasty was performed on each of the patients. Later, "minimal luminal diameter" (a measurement of how well the angioplasty did in dilating the artery) was recorded for each of the patients. Summary statistics are given in the following table.[32]

	Placebo	Probucol	Multivitamins	Probucol and Multivitamins
n	62	58	54	56
Mean	1.43	1.79	1.40	1.54
SD	.58	.45	.55	.61

(a) Complete the following ANOVA table and bracket the P-value for the F test.

Source	df	SS	MS	F
Between treatments		5.4336		
Within treatments				
Total	229	73.9945		

(b) If $\alpha = .01$, do you reject the null hypothesis of equal population means? Why or why not?

*11.51 Refer to Exercise 11.50. Define contrasts to measure each effect specified, and calculate the value of each contrast.

(a) The effect of probucol in the absence of multivitamins

(b) The effect of probucol in the presence of multivitamins

(c) The interaction between probucol and multivitamins

*11.52 Refer to Exercise 11.50. Construct a 95% confidence interval for the effect of probucol in the absence of multivitamins. That is, construct a 95% confidence interval for $\mu_{probucol} - \mu_{placebo}$.

*11.53 Refer to Exercise 11.50. Use the Bonferroni method to construct a 95% confidence interval for the effect of probucol in the absence of multivitamins. That is, construct a Bonferroni-adjusted 95% confidence interval for $\mu_{probucol} - \mu_{placebo}$.

11.54 Three college students collected several pillbugs from a woodpile and used them in an experiment in which they measured the time, in seconds, that it took for a bug to move six inches within an apparatus they had created. There were three groups of bugs: One group was exposed to strong light, for one group the stimulus was moisture, and a third group served as a control. The data are shown in the table.[33]

	Light	Moisture	Control
	23	170	229
	12	182	126
	29	286	140
	12	103	260
	5	330	330
	47	55	310
	18	49	45
	30	31	248
	8	132	280
	45	150	140
	36	165	160
	27	206	192
	29	200	159
	33	270	62
	24	298	180
	17	100	32
	11	162	54
	25	126	149
	6	229	201
	34	140	173
Mean	23.6	169.2	173.5
SD	12.3	83.5	86.0
n	20	20	20

Clearly the SDs show that the variability is not constant between groups, so a transformation is needed. Taking the natural logarithm of each observation results in the following dotplots and summary statistics.

	Light	Moisture	Control
Mean	2.99	4.98	4.99
SD	0.65	0.62	0.66

For the transformed data the ANOVA SS(between) is 53.1103 and the SS(within) is 23.5669.

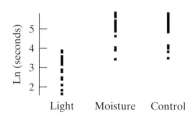

(a) State the null hypothesis in symbols.
(b) Construct the ANOVA table and test the null hypothesis. Let $\alpha = .05$.
(c) Calculate the pooled standard deviation, s_{pooled}.

*11.55 Mountain climbers often experience several symptoms when they reach high altitudes during their climbs. Researchers studied the effects of exposure to high altitude on human skeletal muscle tissue. They set up a $2 \cdot 2$ factorial experiment in which subjects trained for six weeks on a bicycle. The first factor was whether subjects trained under hypoxic conditions (corresponding to an altitude of 3850 m) or normal conditions. The second factor was whether subjects trained at a high level of energy expenditure or at a low level (25% less than the high level). There were either 7 or 8 subjects at each combination of factor levels. The accompanying table shows the results for the response variable "percentage change in vascular endothelial growth factor mRNA."[34]

Energy	Hypoxic		Normal	
	Low Level	High Level	Low Level	High Level
Mean	117.7	173.2	95.1	114.6
No. of Patients	7	7	8	8

Prepare an interaction graph (like Figure 11.14).

*11.56 Consider the data from Exercise 11.55.

(a) Complete the following ANOVA table.

Source	df	SS	MS	F Ratio
Between hypoxic and normal	1	12126.5		
Between energy level	1	10035.7		
Interaction	1			
Within groups	26	56076.0		
Total	29	80738.7		

(b) Conduct a test for interactions. Use $\alpha = .05$.
(c) Test the null hypothesis that energy level has no effect on the response. Use $\alpha = .05$.
(d) Test the null hypothesis that effect on the response of hypoxic training is the same as the effect on the response of normal training. Use $\alpha = .05$.

*11.57 Here are the data from Example 1.7, concerning an experiment in which a new investigational drug was given to 4 male and 4 female dogs, at doses 8 mg/kg and 25 mg/kg. The variable recorded is alkaline phosphatase level (measured in U/Li).

Dose (mg/kg)	Male	Female
8	171	150
	154	127
	104	152
	143	105
Avg	*143*	*133.5*
25	80	101
	149	113
	138	161
	131	197
Avg	*124.5*	*143*

For these data, SS(sex) = 81, SS(dose) = 81, SS(interaction) = 784, and SS(within) = 12604.

(a) Construct the ANOVA table.
(b) Carry out an F test for interactions; use $\alpha = .05$.
(c) Test the null hypothesis that dose has no effect on alkaline phosphatase level. Use $\alpha = .05$.

CHAPTER 12

Linear Regression and Correlation

12.1 INTRODUCTION

In this chapter we discuss some methods for analyzing the relationship between two quantitative variables, X and Y. **Linear regression** and **correlation analysis** are techniques based on fitting a straight line to the data.

Examples

Data for regression and correlation analysis consist of pairs of observations (X, Y). Here are two examples.

Amphetamine and Food Consumption. Amphetamine is a drug that suppresses appetite. In a study of this effect, a pharmacologist randomly allocated 24 rats to three treatment groups to receive an injection of amphetamine at one of two dosage levels, or an injection of saline solution. She measured the amount of food consumed by each animal in the 3-hour period following injection. The results (g of food consumed per kg body weight) are shown in Table 12.1.[1]

Objectives

In this chapter we study correlation and regression. We will

- *study relationships using scatterplots.*

- *learn how least-squares regression models are fit to data.*

- *construct and interpret a regression model.*

- *learn how to test whether a regression relationship is statistically significant.*

- *learn how the correlation coefficient is calculated and interpreted.*

- *learn how regression ideas can be extended to multiple regression, analysis of covariance, and logistic regression.*

Example 12.1

TABLE 12.1 Food Consumption (Y) of Rats (g/kg)

	X = Dose of Amphetamine (mg/kg)		
	0	*2.5*	*5.0*
	112.6	73.3	38.5
	102.1	84.8	81.3
	90.2	67.3	57.1
	81.5	55.3	62.3
	105.6	80.7	51.5
	93.0	90.0	48.3
	106.6	75.5	42.7
	108.3	77.1	57.9
Mean	100.0	75.5	55.0
SD	10.7	10.7	13.3
No. of animals	8	8	8

Figure 12.1 shows a **scatterplot** of

$$Y = \text{Food consumption}$$

against

$$X = \text{Dose of amphetamine}$$

The scatterplot suggests a definite dose-response relationship, with larger values of X tending to be associated with smaller values of Y.* ■

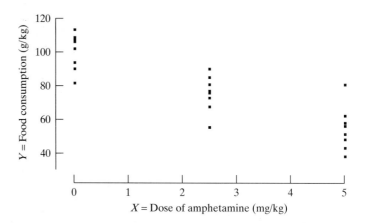

Figure 12.1 Scatterplot of food consumption against dose of amphetamine

| | Example 12.2 | **Fecundity of Crickets.** In a study of reproductive behavior in the Mormon cricket (*Anabrus simplex*), a biologist collected a field sample of 39 females involved in active courtship. For each female, he observed the number of mature eggs (an indicator of fecundity) and the body weight.[2] Figure 12.2 shows a scatterplot of |

$$Y = \text{Number of mature eggs}$$

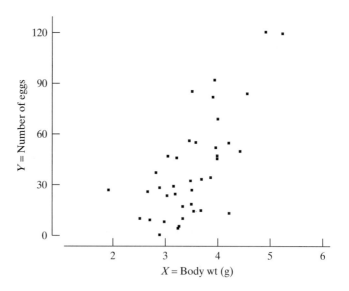

Figure 12.2 Scatterplot of number of eggs against body weight

---

* In many dose-response relationships, the response depends linearly on log(dose) rather than on dose itself. We have chosen a linear portion of the dose-response curve to simplify the exposition.

against

$$X = \text{Body weight}$$

The scatterplot suggests that larger values of X tend to be associated with larger values of Y; in other words, heavier females tend to be more fecund.   ■

In Examples 12.1 and 12.2 the data do not fall on a straight line, even approximately. Nevertheless, the data in these examples are suitable for linear regression analysis because a straight line is a reasonable summary of the *average* trend of Y as related to X.

Two Contexts for Regression and Correlation

Observations of pairs (X, Y) can arise in two different contexts, namely:

1. Y is an observed variable, and the values of X are specified by the experimenter.
2. Both X and Y are observed variables.

The first context is illustrated by Example 12.1, in which amphetamine dose (X) is manipulated by the experimenter, and food consumption (Y) is observed. The second context is illustrated by Example 12.2, in which both body weight (X) and number of eggs (Y) are observed variables.

The distinction between the two contexts is parallel to the distinction for contingency tables that we discussed in Sections 10.3 and 10.5. For regression, as for contingency tables, the distinction between the two contexts is not always sharp. The statistical calculations are the same for the two contexts, but we will see that the emphasis and some of the interpretations can differ.

A Look Ahead

In the following sections we will consider some classical methods for linear analysis of (X, Y) data. Our topics will include

How to fit a straight line to the data

How to describe the closeness of the data points to the fitted line

How to make statistical inferences concerning the fitted line

12.2 THE FITTED REGRESSION LINE

Suppose we have a sample of n pairs (x_i, y_i), where each pair represents the measurements of two variables, X and Y. If a scatterplot of Y versus X shows a general linear trend, then it is natural to try to capture that trend by "fitting" a line to the data. The following example illustrates the kind of situation we wish to consider.

Example 12.3

Length and Weight of Snakes. In a study of a free-living population of the snake *Vipera bertis*, researchers caught and measured nine adult females. Their body lengths (X) and weights (Y) are shown in Table 12.2 and displayed as a scatterplot in Figure 12.3.[3] The number of observations is $n = 9$.

TABLE 12.2 Body Length and Weight of Snakes		
	Length X (cm)	**Weight Y (g)**
	60	136
	69	198
	66	194
	64	140
	54	93
	67	172
	59	116
	65	174
	63	145
Mean	63	152
SD	4.6	35.3

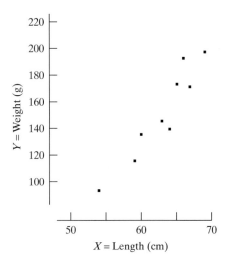

Figure 12.3 Body length and weight of nine snakes

The scatterplot shows a clear upward trend: Greater length is associated with greater weight. Thus, snakes that are longer than the average length of $\bar{x} = 63$ tend to be heavier than the average weight of $\bar{y} = 152$. There are many lines that capture the upward trend and that go through the middle of the data. Figure 12.4 shows three such lines, all of which go through the point $(\bar{x}, \bar{y})$ and all of which do a reasonable job of representing the upward trend in the data. How can we choose one of these as "best"? ∎

We will describe the classical method of fitting a line to the data so that (in a certain sense) the line is as close as possible to the data points. The method of calculation is derived from a criterion called the **least-squares criterion**, and the fitted line is called the **least-squares line** or the **regression line** of Y on X. We will first describe how to determine the regression line and we will then explain the least-squares criterion.

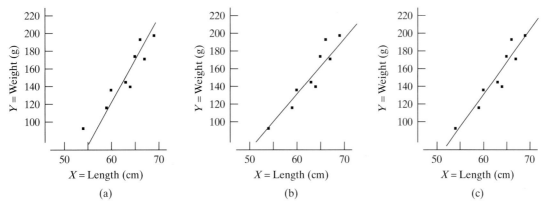

Figure 12.4 Snake data with three lines

Equation of the Regression Line

The equation of a straight line can be written as

$$Y = b_0 + b_1 X$$

where b_0 is the intercept and b_1 is the slope of the line. The slope b_1 is the rate of change of Y with respect to X.

The fitted regression line of Y on X is the line whose slope and intercept are calculated from the data as follows:

Least-Squares Regression Line of Y on X

$$\text{Slope: } b_1 = \frac{\Sigma(x_i - \bar{x})(y_i - \bar{y})}{\Sigma(x_i - \bar{x})^2}$$

$$\text{Intercept: } b_0 = \bar{y} - b_1\bar{x}$$

We illustrate with the snake data from Example 12.3.

Length and Weight of Snakes.   For the data of Example 12.3, we found $\bar{x} = 63$ and $\bar{y} = 152$. Table 12.3 shows that the calculation of $\Sigma(x_i - \bar{x})^2$ gives 172

Example 12.4

TABLE 12.3 Regression Calculations for the Snake Data

x	y	$(x_i - \bar{x})$	$(y_i - \bar{y})$	$(x_i - \bar{x})^2$	$(y_i - \bar{y})^2$	$(x_i - \bar{x})(y_i - \bar{y})$
60	136	−3	−16	9	256	48
69	198	6	46	36	2,116	276
66	194	3	42	9	1,764	126
64	140	1	−12	1	144	−12
54	93	−9	−59	81	3,481	531
67	172	4	20	16	400	80
59	116	−4	−36	16	1,296	144
65	174	2	22	4	484	44
63	145	0	−7	0	49	0
	Sum	0	0	172	9,990	1,237

and the calculation of $\Sigma(x_i - \bar{x})(y_i - \bar{y})$ gives 1,237. Thus, the slope of the fitted regression line is

$$b_1 = \frac{1,237}{172} = 7.19186 \approx 7.19$$

and the intercept is

$$b_0 = 152 - (7.19186)(63) \approx -301$$

(Note that the *unrounded* value of b_1 should be used in calculating b_0.) The equation of the fitted regression line is

$$Y = -301 + 7.19X$$

Figure 12.5 shows the data and the fitted line. ■

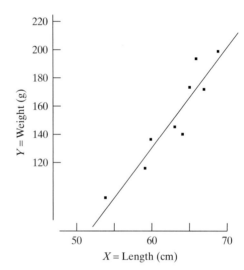

Figure 12.5 Length (X) and weight (Y) of snakes, and the fitted regression line of Y on X

The magnitude of b_1 expresses, in an average sense, the rate of change of Y with respect to X. For instance, for the snake data, $b_1 \approx 7.2$ g/cm; on the average, each centimeter of additional length is associated with an additional 7.2 g of weight.

The formula for the intercept b_0 has a simple interpretation. The formula is

$$b_0 = \bar{y} - b_1\bar{x}$$

but this can be written as

$$\bar{y} = b_0 + b_1\bar{x}$$

This means that *the regression line passes through the joint mean* $(\bar{x}, \bar{y})$ *of the data.*

Plotting Tip: In preparing a graph like Figure 12.5, a convenient way to draw the regression line is to choose two values of X that lie near the extremes of the data, and calculate the corresponding Y's from the regression equation $Y = b_0 + b_1X$. For instance, for the snake data you could choose $X = 54$ and $X = 70$; substituting these in the regression equation yields $Y = 87$ and $Y = 202$, respectively. You would then plot the points $(54, 87)$, and $(70, 202)$ and use a ruler to draw a line between them. As a check, you can verify that the line passes through the joint mean $(\bar{x}, \bar{y})$.

The Residual Sum of Squares

We now consider a statistic that describes the scatter of the points about the fitted regression line. The equation of the fitted line is $Y = b_0 + b_1 X$. For points on the line whose X-values are actual observations x, we use the special notation $\hat{y}$ (read "y-hat"). Thus, for each observed x_i there is a predicted y value of

$$\hat{y}_i = b_0 + b_1 x_i$$

Also associated with each observed pair (x, y) is a quantity called a **residual**, defined as

$$\text{Residual} = y_i - \hat{y}_i$$

Figure 12.6 shows y and the residual for a typical data point (x_i, y_i). It can be shown that the sum of the residuals, taking into account their signs, is always zero, because of "balancing" of data points above and below the fitted regression line. The *magnitude* (disregarding sign) of each residual is the vertical distance of the data point from the fitted line.

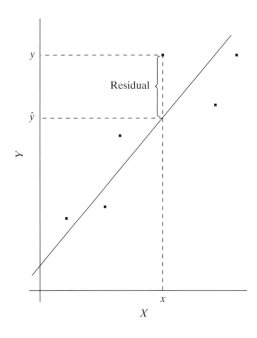

Figure 12.6 $\hat{y}$ and the residual for a typical data point (x, y)

Note that a residual is calculated in terms of *vertical* distance. In using the regression model $Y = b_0 + b_1 X$, we are thinking of the variable X as a predictor and the variable Y as a response that depends on X. We care primarily about how close each observed value, y_i, is to the prediction, $\hat{y}_i$, for it. Thus, we measure vertical distance from each point to the fitted line. A summary measure of the distances of the data points from the regression line is the **residual sum of squares**, or **SS(resid)**, which is defined as follows:

Residual Sum of Squares

$$\text{SS(resid)} = \Sigma(y_i - \hat{y}_i)^2$$

It is clear from the definition that the residual sum of squares will be small if the data points all lie very close to the line.

The following example illustrates SS(resid).

Example 12.5

Length and Weight of Snakes. For the snake data, Table 12.4 indicates how SS(resid) would be calculated from its definition. (The values are abbreviated to improve readability.) ∎

TABLE 12.4 Calculation of SS(Resid)

x	y	$\hat{y}$	$y - \hat{y}$	$(y - \hat{y})^2$
60	136	130.42...	5.57...	31.08...
69	198	195.15...	2.84...	8.11...
66	194	173.57...	20.42...	417.15...
64	140	159.19...	−19.19...	368.32...
54	93	87.27...	5.72...	32.79...
67	172	180.76...	−8.76...	76.86...
59	116	123.23...	−7.23...	52.30...
65	174	166.38...	7.61...	58.00...
63	145	152.00...	−7.00...	49.00...
Sum			0	1,093.66... = SS(resid),

The Least-Squares Criterion

Many different criteria can be proposed to define the straight line that "best" fits a set of data points (x_i, y_i). The classical criterion is the least-squares criterion:

> **Least-Squares Criterion**
>
> The "best" straight line is the one that minimizes the residual sum of squares.

The formulas given for b_0 and b_1 were derived from the least-squares criterion by applying calculus to solve the minimization problem. (The derivation is given in Appendix 12.1.) The fitted regression line is also called the "least-squares line."

The least-squares criterion may seem arbitrary and even unnecessary. Why not fit a straight line by eye with a ruler? Actually, unless the data lie nearly on a straight line, it can be surprisingly difficult to fit a line by eye. The least-squares criterion provides an answer that does not rely on individual judgment, and that (as we shall see in Sections 12.3 and 12.4) can be usefully interpreted in terms of estimating the distribution of Y values for each fixed X. Furthermore, we will see in Section 12.7 that the least-squares criterion is a versatile concept, with applications far beyond the simple fitting of straight lines.

The Residual Standard Deviation

A summary of the results of the linear regression analysis should include a measure of the closeness of the data points to the fitted line. A measure derived from SS(resid), and easier to interpret, is the **residual standard deviation**, denoted $s_{Y|X}$, which is defined as follows:

> **Residual Standard Deviation**
>
> $$s_{Y|X} = \sqrt{\frac{\text{SS(resid)}}{n-2}}$$

The residual standard deviation tells how far above or below the regression line points tend to be. Thus, the residual standard deviation specifies how far off predictions tend to be that are made using the regression model. Notice the factor $(n - 2)$ in the denominator, rather than the usual $(n - 1)$.* The following example illustrates the calculation of $s_{Y|X}$.

Length and Weight of Snakes.   For the snake data, we use SS(resid) from Example 12.5 to calculate

Example 12.6

$$s_{Y|X} = \sqrt{\frac{1093.669}{7}} = \sqrt{156.238} = 12.5 \text{ g}$$

Thus, predictions of snake weight based on the regression model tend to be off by 12.5 g.   ∎

Note that $s_{Y|X}$ is given by

$$s_{Y|X} = \sqrt{\frac{\Sigma(y_i - \hat{y}_i)^2}{n-2}}$$

This formula is closely analogous to the formula for s_Y:

$$s_Y = \sqrt{\frac{\Sigma(y_i - \bar{y})^2}{n-1}}$$

Both of these SDs measure variability in Y, but the residual SD measures variability around the *regression line* and the ordinary SD measures variability around the mean, $\bar{y}$. Roughly speaking, $s_{Y|X}$ is a measure of the typical vertical distance of the data points from the regression line. (Notice that the unit of measurement of $s_{Y|X}$ is the same as that of Y—for instance, grams in the case of the snake data.) Figure 12.7 shows the snake data with the residuals represented as vertical lines and the residual SD indicated as a vertical ruler line. Note that the residual SD roughly indicates the magnitude of a typical residual.

In many cases, $s_{Y|X}$ can be given a more definite quantitative interpretation. Recall from Section 2.6 that for a "nice" data set, we expect roughly 68% of the observations to be within ±1 SD of the mean (and similarly for 95% and ±2 SDs). Recall also that these rules work best if the data follow approximately a normal distribution. Similar interpretations hold for the residual SD: For "nice" data sets that are not too small, we expect roughly 68% of the observed y's to be within ±1 $s_{Y|X}$ of the regression line. In other words, we expect roughly 68% of the data points to be within a vertical distance of $s_{Y|X}$ above and below the regression

---

* The use of $n - 2$ rather than $n - 1$ is discussed in Section 12.4 on page 552.

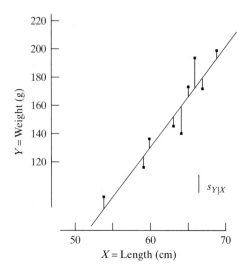

Figure 12.7 Length and weight of snakes, showing the residuals and the residual SD

line (and similarly for 95% and $\pm 2\, s_{Y|X}$). These rules work best if the residuals follow approximately a normal distribution. The following example illustrates the 68% rule.

Example 12.7

Fecundity of Crickets. For the cricket fecundity data provided in Example 12.2, the fitted regression line is $Y = -72 + 31.7X$ and the residual standard deviation is $s_{Y|X} = 22.6$. Figure 12.8 shows the data and the regression line. The dotted lines are a vertical distance $s_{Y|X}$ from the regression line. Of the 39 data points, 27 are within the dotted lines; thus, $\dfrac{27}{39}$ or 69%, of the observed y's are within $\pm 1\, s_{Y|X}$ of the regression line. ∎

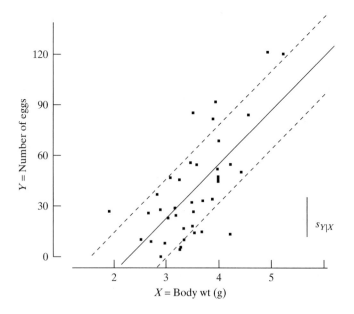

Figure 12.8 Body weight and number of eggs in 39 female crickets. The dotted lines are a vertical distance $s_{Y|X}$ from the regression line.

Computer note: Fitting least-squares regression lines to data sets requires a lot of computation; this is best done with a computer. We illustrate regression using a computer for the snake data of Example 12.3. In the MINITAB system, suppose the data are entered into two columns labeled 'Length' and 'Weight'. That is, suppose the columns of data are

'Length'	'Weight'
60	136
69	198
66	194
64	140
54	93
67	172
59	116
65	174
63	145

The command

```
MTB > Regress 'Weight' 1 'Length';
SUBC > Constant.
```

gives the following output:

```
Regression Analysis
The regression equation is
Weight = - 301 + 7.19 Length
Predictor      Coef       Stdev        t-ratio        p
Constant     -301.09      60.19          -5.00      0.000
Length         7.1919      0.9531          7.55      0.000
s = 12.50   R-sq = 89.1   R-sq(adj) = 87.5%
Analysis of Variance
SOURCE         DF         SS          MS          F          p
Regression     1       8896.3      8896.3      56.94      0.000
Error          7       1093.7       156.2
Total          8       9990.0
```

The output here agrees with the results stated in Example 12.4. Note that MINITAB calls the intercept, b_0, the "Constant." This is common terminology in software packages. MINITAB has also produced many numbers that we haven't yet discussed. Many of these will be considered in later sections of this chapter.

The "Analysis of Variance" table that is part of the output includes an SS(Error) value of 1093.7. This is what we are calling SS(resid), as calculated in Example 12.5. Another part of the output is the value $s = 12.50$. This is the residual standard deviation, which we have labeled $s_{Y|X}$; the calulation of 12.50 agrees with Example 12.6.

Exercises 12.1–12.11

12.1 The table presents a fictitious set of data.

	X	Y
	3	13
	4	15
	1	4
	2	11
	5	22
Mean	3	13

$$\Sigma(x_i - \bar{x})^2 = 10$$
$$\Sigma(x_i - \bar{x})(y_i - \bar{y}) = 40$$

(a) Compute the linear regression of Y on X and compute $\hat{y}$ for each data point.
(b) Plot the data and also the values of $\hat{y}$.
(c) Compute the residual SS.

12.2 Proceed as in Exercise 12.1 for the following data.

	X	Y
	3	10
	7	2
	6	9
	7	4
	2	15
Mean	5	8

$$\Sigma(x_i - \bar{x})^2 = 22$$
$$\Sigma(x_i - \bar{x})(y_i - \bar{y}) = -44$$

12.3 In a study of protein synthesis in the oocyte (developing egg cell) of the frog *Xenopus laevis*, a biologist injected individual oocytes with radioactively labeled leucine. At various times after injection, he made radioactivity measurements and calculated how much of the leucine had been incorporated into protein. The results are given in the accompanying table; each leucine value is the content of labeled leucine in two oocytes. All oocytes were from the same female.[4]

	Time	Leucine
	0	.02
	10	.25
	20	.54
	30	.69
	40	1.07
	50	1.50
	60	1.74
Mean	30	.83

$$\Sigma(x_i - \bar{x})^2 = 2{,}800 \quad \Sigma(y_i - \bar{y})^2 = 2.4308$$
$$\Sigma(x_i - \bar{x})(y_i - \bar{y}) = 81.90$$
$$\text{SS(resid)} = .035225$$

(a) Use linear regression to estimate the rate of incorporation of the labeled leucine.
(b) Plot the data and draw the regression line on your graph.
(c) Calculate the residual standard deviation.

12.4 In an investigation of the physiological effects of alcohol (ethanol), 15 mice were randomly allocated to three treatment groups, each to receive a different oral dose of alcohol. The dosage levels were 1.5, 3.0, and 6.0 g alcohol per kg body weight. The body temperature of each mouse was measured immediately before the alcohol was given, and again 20 minutes afterward. The accompanying table shows the drop (before minus after) in body temperature for each mouse. (The negative value −.1 refers to a mouse whose temperature rose rather than fell.)[5]

Alcohol		Drop in Body Temperature (°C)					
Dose (g/kg)	*Log(dose) X*	*Individual values* (Y)					*Mean*
1.5	.176	.2	1.9	−.1	.5	.8	.66
3.0	.477	4.0	3.2	2.3	2.9	3.8	3.24
6.0	.778	3.3	5.1	5.3	6.7	5.9	5.26

(a) Plot the mean drop in body temperature versus dose. Plot the mean drop in body temperature versus log(dose). Which plot appears more nearly linear?

(b) For the regression of Y on X = log(dose) preliminary calculations yield the following:

$$\bar{x} = .4771 \qquad \bar{y} = 3.053$$
$$\Sigma(x_i - \bar{x})^2 = .906191 \quad \Sigma(y_i - \bar{y})^2 = 63.7773$$
$$\Sigma(x_i - \bar{x})(y_i - \bar{y}) = 6.92369 \quad SS(resid) = 10.8773$$

Calculate the fitted regression line and the residual standard deviation.

(c) Plot the individual (x, y) data points and draw the regression line on your graph.

(d) Draw a ruler line on your graph to show the magnitude of $s_{Y|X}$. (See Figure 12.7.)

12.5 Twenty plots, each 10·4 meters, were randomly chosen in a large field of corn. For each plot, the plant density (number of plants in the plot) and the mean cob weight (g of grain per cob) were observed. The results are given in the table.[6]

Plant Density X	Cob Weight Y	Plant Density X	Cob Weight Y
137	212	173	194
107	241	124	241
132	215	157	196
135	225	184	193
115	250	112	224
103	241	80	257
102	237	165	200
65	282	160	190
149	206	157	208
85	246	119	224

Preliminary calculations yield the following results:

$$\bar{x} = 128.05 \qquad \bar{y} = 224.1$$
$$\Sigma(x_i - \bar{x})^2 = 20,209.0 \quad \Sigma(y_i - \bar{y})^2 = 11,831.8$$
$$\Sigma(x_i - \bar{x})(y_i - \bar{y}) = -14,563.1$$
$$SS(resid) = 1,337.3$$

(a) Calculate the linear regression of Y on X.

(b) Plot the data and draw the regression line on your graph.

(c) Interpret the value of the slope of the regression line, b_1, in the context of this setting.

(d) Calculate s_Y and $s_{Y|X}$ and specify the units of each.

(e) Interpret the value of $s_{Y|X}$ in the context of this setting.

12.6   Laetisaric acid is a compound that holds promise for control of fungus diseases in crop plants. The accompanying data show the results of growing the fungus *Pythium ultimum* in various concentrations of laetisaric acid. Each growth value is the average of four radial measurements of a *P. ultimum* colony grown in a petri dish for 24 hours; there were two petri dishes at each concentration.[7]

Laetisaric Acid Concentration X (μg/mLi)	Fungus Growth Y (mm)
0	33.3
0	31.0
3	29.8
3	27.8
6	28.0
6	29.0
10	25.5
10	23.8
20	18.3
20	15.5
30	11.7
30	10.0

Mean   11.5   23.64

$\Sigma(x_i - \bar{x})^2 = 1{,}303$ $\Sigma(y_i - \bar{y})^2 = 677.349$

$\Sigma(x_i - \bar{x})(y_i - \bar{y}) = -927.75$

SS(resid) = 16.7812

(a) Calculate the linear regression of Y on X.

(b) Plot the data and draw the regression line on your graph.

(c) Calculate $s_{Y|X}$. What are the units of $s_{Y|X}$?

(d) Draw a ruler line on your graph to show the magnitude of $s_{Y|X}$. (See Figure 12.7.)

12.7   To investigate the dependence of energy expenditure on body build, researchers used underwater weighing techniques to determine the fat-free body mass for each of seven men. They also measured the total 24-hour energy expenditure for each man during conditions of quiet sedentary activity. The results are shown in the table.[8] (See also Exercise 12.39.)

Subject	Fat-Free Mass X (kg)	Energy Expenditure Y (kcal)
1	49.3	1,894
2	59.3	2,050
3	68.3	2,353
4	48.1	1,838
5	57.6	1,948
6	78.1	2,528
7	76.1	2,568

Mean   62.40   2,168

$\Sigma(x_i - \bar{x})^2 = 877.74$ $\Sigma(y_i - \bar{y})^2 = 570{,}124$

$\Sigma(x_i - \bar{x})(y_i - \bar{y}) = 21{,}953.7$

SS(resid) = 21,026.1

(a) Calculate the linear regression of Y on X.

(b) Plot the data and draw the regression line on your graph.

(c) Interpret the value of the slope of the regression line, b_1, in the context of this setting.

(d) Calculate $s_{Y|X}$ and specify the units.

12.8   The rowan (*Sorbus aucuparia*) is a tree that grows in a wide range of altitudes. To study how the tree adapts to its varying habitats, researchers collected twigs with attached buds from 12 trees growing at various altitudes in North Angus, Scotland. The buds were brought back to the laboratory and measurements were made of the dark respiration rate. The accompanying table shows the altitude of origin (in meters) of each batch of buds and the dark respiration rate (expressed as μL of oxygen per hour per mg dry weight of tissue).[9]

Altitude of Origin X (m)	Respiration Rate Y (μL/hr/mg)
90	.11
230	.20
240	.13
260	.15
330	.18
400	.16
410	.23
550	.18
590	.23
610	.26
700	.32
790	.37

Mean	433.3	.210

$\Sigma(x_i - \bar{x})^2 = 506{,}667$ $\Sigma(y_i - \bar{y})^2 = .0654$
$\Sigma(x_i - \bar{x})(y_i - \bar{y}) = 161.40$
SS(resid) = .013986

(a) Calculate the linear regression of Y on X.

(b) Plot the data and draw the regression line on your graph.

(c) Interpret the value of the slope of the regression line, b_1, in the context of this setting.

(d) Calculate the residual standard deviation.

12.9   Scientists studied the relationship between the length of the body of a bullfrog and how far it can jump. Eleven bullfrogs were included in the study. The results are given in the table.[10]

Bullfrog	Length X (mm)	Maximum Jump Y (cm)
1	155	71
2	127	70
3	136	100
4	135	120
5	158	103.3
6	145	116
7	136	109.2
8	172	105
9	158	112.5
10	162	114
11	162	122.9
Mean	149.64	103.99

Preliminary calculations yield the following results:

$$\Sigma(x_i - \bar{x})^2 = 2{,}094.55 \quad \Sigma(y_i - \bar{y})^2 = 3{,}218.99$$

$$\Sigma(x_i - \bar{x})(y_i - \bar{y}) = 731.36$$

$$SS(\text{resid}) = 2{,}963.61$$

(a) Calculate the linear regression of Y on X.
(b) Interpret the value of the slope of the regression line, b_1, in the context of this setting.
(c) Calculate s_Y and $s_{Y|X}$ and specify the units of each.
(d) Interpret the value of $s_{Y|X}$ in the context of this setting.

12.10 The peak flow rate of a person is the fastest rate at which the person can expel air after taking a deep breadth. Peak flow rate is measured in units of liters per minute and gives an indication of the person's respiratory health. Researchers measured peak flow rate and height for each of a sample of 17 men. The results are given in the table.[11]

Subject	Height X (cm)	Peak Flow Rate Y (Li/min)
1	174	733
2	183	572
3	176	500
4	169	738
5	183	616
6	186	787
7	178	866
8	175	670
9	172	550
10	179	660
11	171	575
12	184	577
13	200	783
14	195	625
15	176	470
16	176	642
17	190	856
Mean	180.4	660

Preliminary calculations yield the following results:

$$\Sigma(x_i - \bar{x})^2 = 1{,}172 \quad \Sigma(y_i - \bar{y})^2 = 222{,}766$$

$$\Sigma(x_i - \bar{x})(y_i - \bar{y}) = 5{,}288$$

$$SS(\text{resid}) = 198{,}909$$

(a) Calculate the linear regression of Y on X.
(b) For each subject, calculate the predicted peak flow rate, using the regression equation from part (a).
(c) For each subject, calculate the residual, using the results from part (b).
(d) Calculate $s_{Y|X}$ and specify the units.
(e) What percentage of the data points are within $\pm s_{Y|X}$ of the regression line? That is, what percentage of the 17 residuals are in the interval $(-s_{Y|X}, s_{Y|X})$?

12.11 For each data set indicated below, prepare a plot like Figure 12.8, showing the data, the fitted regression line, and two lines whose vertical distance above and below the regression line is $s_{Y|X}$. What percentage of the data points are within $\pm s_{Y|X}$ of the regression line?

(a) The body temperature data of Exercise 12.4
(b) The corn yield data of Exercise 12.5

12.3 PARAMETRIC INTERPRETATION OF REGRESSION: THE LINEAR MODEL

One use of regression analysis is simply to provide a concise description of the data. The quantities b_0 and b_1 locate the regression line, and $s_{Y|X}$ describes the scatter of the points about the line.

For many purposes, however, data description is not enough. In this section we consider inference from the data to a larger population. In previous chapters we have spoken of one or several populations of Y values. Now, to encompass the X variable as well, we need to expand the notion of a population.

Conditional Populations and Conditional Distributions

A **conditional population** of Y values is a population of Y values associated with a fixed, or given, value of X. Within a conditional population we may speak of the **conditional distribution** of Y. The mean and standard deviation of a conditional population distribution are denoted as

$$\mu_{Y|X} = \text{Population mean } Y \text{ value for a given } X$$

$$\sigma_{Y|X} = \text{Population SD of } Y \text{ values for a given } X$$

(Note that the "given" symbol "|" is the same one used for conditional probability in Chapters 3 and 10.) The following example illustrates this notation.

Amphetamine and Food Consumption. In the rat experiment of Example 12.1, the response variable Y was food consumption and the three values of X (dose) were $X = 0$, $X = 2.5$, and $X = 5$. If we were to view the food consumption data as three independent samples (as for an ANOVA), then we would denote the three population means as μ_1, μ_2, and μ_3. In regression notation these means would be denoted as

$$\mu_{Y|X=0} \qquad \mu_{Y|X=2.5} \qquad \mu_{Y|X=5}$$

respectively. Similarly, the three population standard deviations, which would be denoted as σ_1, σ_2, and σ_3 in an ANOVA context, would be denoted as

$$\sigma_{Y|X=0} \qquad \sigma_{Y|X=2.5} \qquad \sigma_{Y|X=5}$$

respectively. In other words, the symbols

$$\mu_{Y|X} \quad \text{and} \quad \sigma_{Y|X}$$

Example 12.8

represent the mean and standard deviation of food consumption values for rats given dose X of amphetamine. ∎

Sometimes conditional distributions pertain to actual subpopulations, as in the following example.

Example 12.9 **Height and Weight of Young Men.** Consider the variables

$$X = \text{Height}$$

and

$$Y = \text{Weight}$$

for a population of young men. The conditional means and standard deviations are

$$\mu_{Y|X} = \text{Mean weight of men who are } X \text{ inches tall}$$
$$\sigma_{Y|X} = \text{SD of weights of men who are } X \text{ inches tall}$$

Thus, $\mu_{Y|X}$ and $\sigma_{Y|X}$ are the mean and standard deviation of weight in the *subpopulation* of men whose height is X. Of course, there is a different subpopulation for each value of X. ∎

The Linear Model

When we conduct a linear regression analysis, we think of Y as having a distribution that depends on X. The analysis can be given a parametric interpretation if two conditions are met. These conditions, which constitute the **linear model**, are given in the box.

The Linear Model
1. **Linearity.** $Y = \mu_{Y|X} + \varepsilon$, where $\mu_{Y|X}$ is a linear function of X; that is,

$$\mu_{Y|X} = \beta_0 + \beta_1 X$$
$$\text{Thus, } Y = \beta_0 + \beta_1 X + \varepsilon$$

2. **Constancy of standard deviation.** $\sigma_{Y|X}$ does not depend on X.

In the linear model $Y = \beta_0 + \beta_1 X + \varepsilon$, the ε term represents **random error**. We include this term in the model to reflect the fact that Y varies, even when X is fixed. The following two examples show the meaning of the linear model.

Example 12.10 **Amphetamine and Food Consumption.** For the rat food consumption experiment, the linear model asserts that (1) the population mean food consumption is a linear function of dose, and that (2) the population standard deviation of food consumption values is the same for all doses. Notice that the second condition is closely analogous to the condition in ANOVA that the population SDs are equal: $\sigma_1 = \sigma_2 = \sigma_3$. The linear model also allows for the fact that there is variability in Y when X is fixed. For example, there were 8 observations for which $X = 5$. The 8 y values averaged 55.0, but none of the observations was equal to 55.0; there was substantial variability within the 8 y values. This variability is quantified by the SD of 13.3. ∎

Height and Weight of Young Men. We consider an idealized fictitious population of young men whose joint height and weight distribution fits the linear model exactly. For our fictitious population we will assume that the conditional means and SDs of weight given height are as follows:

$$\mu_{Y|X} = -145 + 4.25X$$
$$\sigma_{Y|X} = 20$$

Thus, the regression parameters of the population are $\beta_0 = -145$ and $\beta_1 = 4.25$. (This fictitious population resembles that of U.S. 17-year-olds.[12]) Thus, the model is $Y = -145 + 4.25X + \varepsilon$.

Table 12.5 shows the conditional means and SDs of Y = weight for a few selected values of X = height. Figure 12.9 shows the conditional distributions of Y given X for these selected subpopulations.

TABLE 12.5 Conditional Means and SDs of Weight Given Height in a Population of Young Men

Height (in.) X	Mean Weight (lb) $\mu_{Y|X}$	Standard Deviation of Weights (lb) $\sigma_{Y|X}$
64	127	20
68	144	20
72	161	20
76	178	20

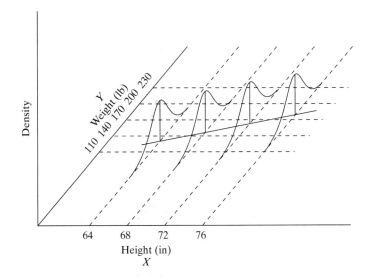

Figure 12.9 Conditional distributions of weight given height in a population of young men

Note, for example, that if height = 68 (in.), then the mean weight is 144 (lb) and the SD of the weights is 20 (lb). For this subpopulation, $Y = 144 + \varepsilon$. If a particular young man who is 68 inches tall weighs 145 pounds, then $\varepsilon = 1$ for him. If another 68-inch-tall young man weighs 140 pounds, then $\varepsilon = -4$ in his case. Of course, β_0, β_1, and ε are generally not observable. This example is fictitious. ■

Example 12.11

Remark. Actually, the term *regression* is not confined to linear regression. In general, the relationship between $\mu_{Y|X}$ and X is called the *regression of Y on X*. The linearity assumption asserts that the regression of Y on X is linear rather than, for instance, a curvilinear function.

Estimation in the Linear Model

Consider now the analysis of a set of (X, Y) data. Suppose we assume that the linear model is an adequate description of the true relationship of Y and X. Suppose further that we are willing to adopt the following **random subsampling model**:

> **Random Subsampling Model**
> For each observed pair (x, y), we regard the value y as having been sampled at random from the conditional population of Y values associated with the X value x.

(We will discuss the definition of the random subsampling model more fully in Section 12.6.)

Within the framework of the linear model and the random subsampling model, the quantities b_0, b_1, and $s_{Y|X}$ calculated from a regression analysis can be interpreted as estimates of population parameters:

b_0 is an estimate of β_0,

b_1 is an estimate of β_1,

$s_{Y|X}$ is an estimate of $\sigma_{Y|X}$.

Example 12.12

Length and Weight of Snakes. For the snake data of Example 12.3, we found that $b_0 = -301$, $b_1 = 7.19$, and $s_{Y|X} = 12.5$. Thus,

-301 is our estimate of β_0,

7.19 is our estimate of β_1,

12.5 is our estimate of $\sigma_{Y|X}$. ∎

The application of the linear model to the snake data has yielded two benefits. First, the slope of the regression line, 7.19 g/cm, is an estimate of a morphological parameter ("weight per unit length"), which is of potential biological interest in characterizing the population of snakes. Second, we have obtained an estimate (12.5 g) of the variability of weight among snakes of fixed length, even though no direct estimate of this variability was possible because no two of the observed snakes were the same length.

The Graph of Averages

If we have several observations of Y at a given level of X, we can estimate $\mu_{Y|X}$ by simply using the sample average of Y, $\bar{y}$, for that given value of X; we can denote this sample average as $\bar{y}|X$. Sometimes we are able to calculate a sample average, $\bar{y}$, for each of several X values. A graph of $\bar{y}|X$ is known as a **graph of averages**, since it shows the (observed) average of Y for different values of X.

Amphetamine and Food Consumption.   Figure 12.10 is a graph of averages for the food consumption data in Table 12.1, showing the average y value for each of the 3 levels of X. Note that the 3 $\bar{y}$'s almost lie on a line. This supports the use of the linear model with these data.   ■

Example 12.13

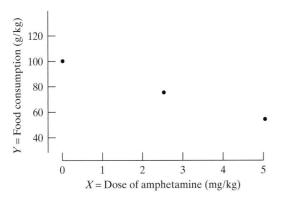

Figure 12.10 Graph of averages for food consumption data from Example 12.1

If the $\bar{y}$'s in a graph of averages fall exactly on a line, then that line is the regression line and $\mu_{Y|X}$ is estimated with $\bar{y}|X$. Usually, however, the $\bar{y}$'s are not perfectly collinear. In this case, the regression line is a *smoothed* version of the graph of averages, resulting in a fitted model in which all of the estimates of $\mu_{Y|X}$ fall on a line. By smoothing the graph of averages into a line, we use information from *all* of the observations to estimate $\mu_{Y|X}$ at any level of X.

Amphetamine and Food Consumption.   If we apply the formulas of Section 12.2 to the food consumption data in Table 12.1, we obtain $b_0 = 99.3$ and $b_1 = -9.01$. Thus, the estimate of $\mu_{Y|X=0}$ is 99.3. This estimate differs slightly from $\bar{y}|X = 0$, which is 100.0. The estimate 99.3 makes use of (1) the 8 y values when $X = 0$ (which averaged to 100.0) and (2) the linear trend established by the other 16 data points, which showed higher food consumption associated with lower doses. Likewise, $\mu_{Y|X=2.5}$ is $99.3 - 9.01 \cdot 2.5 = 76.775$, which differs slightly from $\bar{y}|X = 2.5$, which is 75.5, and $\mu_{Y|X=5}$ is $99.3 - 9.01 \cdot 5 = 54.25$, which differs slightly from $\bar{y}|X = 5$, which is 55.0.   ■

Example 12.14

Interpolation in the Linear Model

The idea of smoothing the graph of averages into a straight line carries over to the setting in which we have only a single observation at each level of X. When we draw a line through a set of (X, Y) data, we are expressing a belief that the underlying dependence of Y on X is smooth, even though the data may show the relationship only roughly. Linear regression is one formal way of providing a smooth description of the data.

Taking advantage of this assumption of smoothness, a fitted regression is sometimes used to estimate the distribution of Y for an X for which there are no data. The following is an example.

Amphetamine and Food Consumption.   Figure 12.11 shows the data and the fitted regression line for the food consumption data from Table 12.1; for these data $s_{Y|X} = 11.4$. Let us predict the response of rats given amphetamine at a dose

Example 12.15

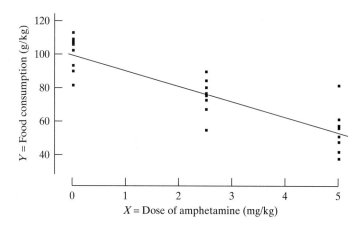

Figure 12.11 Rat food consumption data and fitted regression line

of $X = 3.5\,\text{mg/kg}$. The fitted regression equation is $Y = 99.3 - 9.01X$; substituting $X = 3.5$ yields $Y = 67.8$. Thus, we estimate that rats given 3.5 mg/kg of amphetamine would show a mean food consumption of 67.8 g/kg and an SD of 11.4 g/kg. ∎

Note that estimation of the mean uses the linearity assumption of the linear model, while estimation of the standard deviation uses the assumption of constant standard deviation. In some situations only the linearity assumption may be plausible, and then only the mean would be estimated.

Example 12.15 is an example of **interpolation**, because the X we chose ($X = 3.5$) was within the range of observed values of X. By contrast, **extrapolation** is the use of a regression line (or other curve) to predict Y for values of X that are outside the range of the data. Extrapolation should be avoided whenever possible, because there is usually no assurance that the relationship between $\mu_{Y|X}$ and X remains linear for X values outside the range of those observed. Many biological relationships are linear for only part of the possible range of X values. The following is an example.

Example 12.16

Amphetamine and Food Consumption. The dose-response relationship for the rat food consumption experiment looks approximately like Figure 12.12.[13] The data of Example 12.1 cover only the linear portion of the relationship. Clearly it would be unwise to extrapolate the fitted line out to $X = 10$ or $X = 15$. ∎

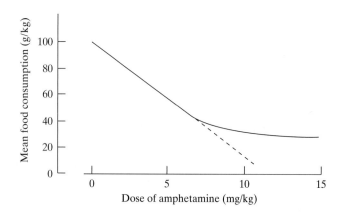

Figure 12.12 Dose-response curve (mean response vs. dose) for rat food consumption experiment

Prediction and the Linear Model

Consider the setting of using height, X, to predict weight, Y, for a large group of young men for whom the average weight is 150 pounds. Suppose a young man is chosen at random and we must predict his weight.

1. If we don't know anything about the height of the man, then the best estimate we can give of his weight is the overall average weight, $\bar{y} = 150$.

2. Suppose we learn that the man's height is 76 inches. If we know that the average weight of all 76-inch-tall men in the group is 180 pounds, then we can use this conditional average, $\bar{y}|x = 76$, as our prediction of the man's weight. We expect this prediction, which essentially is using the graph of averages (but without smoothing), to be more accurate than the one given in part (1).

3. Suppose we learn that the man's height is 76 inches and we also know that the least-squares regression equation is $Y = -140 + 4.3X$. Then we can use the value $x = 76$ to get a prediction, which would be $-140 + 4.3 \cdot 76 = 186.8$.

Is the prediction in (3) better than the prediction made in (2)? Since using the regression equation amounts to smoothing the graph of averages, we expect prediction (3) to be better than prediction (2) *to the extent that we believe that there is a linear relationship between height and weight*. Prediction (3) has the advantage of using information from all of the data points, not just those for which $x = 76$. Method (3) also has the advantage of allowing for predictions when the x value (the height) is not one that is in the original data set (as discussed in the preceding subsection "Interpolation in the Linear Model"), so that $\bar{y}|x$ is not known. However, method (3) will give poor predictions if the linear relationship does not hold. Thus it is very important to think about such relationships, and to explore them graphically, before using a regression model.

Exercises 12.12–12.18

12.12 For the data in Exercise 12.6 there were two observations for which $X = 0$. The average response (Y value) for these points is $\dfrac{33.3 + 31.0}{2} = 32.15$. However, the intercept of the regression line, b_0, is not 32.15. Why not? Why is b_0 a better estimate of the average fungus growth when laetisaric acid concentration is zero than 32.15?

12.13 Refer to the body temperature data of Exercise 12.4. Assuming that the linear model is applicable, estimate the mean and the standard deviation of the drop in body temperature that would be observed in mice given alcohol at a dose of 2 g/kg.

12.14 Refer to the cob weight data of Exercise 12.5. Assume that the linear model holds.

(a) Estimate the mean cob weight to be expected in a plot containing (i) 100 plants; (ii) 120 plants.

(b) Assume that each plant produces one cob. How much grain would we expect to get from a plot containing (i) 100 plants? (ii) 120 plants?

12.15 Refer to the fungus growth data of Exercise 12.6. Assuming that the linear model is applicable, find estimates of the mean and standard deviation of fungus growth at a laetisaric acid concentration of 15 μg/mLi.

12.16 Refer to the energy expenditure data of Exercise 12.7. Assuming that the linear model is applicable, estimate the mean 24-hour energy expenditure of a man whose fat-free mass is 55 kg.

12.17 Refer to the bullfrog data of Exercise 12.9. Assuming that the linear model is applicable, estimate the maximum jump length of a bullfrog whose body length is 150 mm.

12.18 Refer to the peak flow data of Exercise 12.10. Assuming that the linear model is applicable, find estimates of the mean and standard deviation of peak flow for men 180 cm tall.

12.4 STATISTICAL INFERENCE CONCERNING β_1

The linear model provides interpretations of b_0, b_1, and $s_{Y|X}$ that take them beyond data description into the domain of statistical inference. In this section we consider inference about the true slope β_1 of the regression line. The methods are based on the linear model and the random subsampling model. In addition, the methods are based on the condition that the conditional population distribution of Y for each value of X is a normal distribution. This is equivalent to stating that in the linear model of $Y = \beta_0 + \beta_1 X + \varepsilon$, the ε values come from a normal distribution.

The Standard Error of b_1

Within the context of the linear model, b_1 is an estimate of β_1. Like all estimates calculated from data, b_1 is subject to sampling error. The standard error of b_1 is calculated as follows:

> **Standard Error of b_1**
>
> $$\text{SE}_{b_1} = \frac{s_{Y|X}}{\sqrt{\Sigma(x_i - \bar{x})^2}}$$

The following example illustrates the calculation of SE_{b_1}.

Example 12.17 **Length and Weight of Snakes.** For the snake data, we found in Example 12.4 that $\Sigma(x_i - \bar{x})^2 = 172$, and in Example 12.6 that $s_{Y|X} = 12.5$. The standard error of b_1 is

$$\text{SE}_{b_1} = \frac{12.5}{\sqrt{172}} = .9531$$

To summarize, the slope of the fitted regression line (from Example 12.4) is

$$b_1 = 7.19 \text{ g/cm}$$

and the standard error of this slope is

$$\text{SE}_{b_1} = .95 \text{ g/cm}$$

■

Structure of the SE.   Let us see how the standard error of b_1 depends on various aspects of the data. First, note that SE_{b_1} depends, through $s_{Y|X}$, on the scatter of the data points about the fitted regression line; naturally, smaller scatter gives more precise information about β_1. Second, note that SE_{b_1} depends on $\Sigma(x_i - \bar{x})^2$; larger $\Sigma(x_i - \bar{x})^2$ gives a smaller SE. $\Sigma(x_i - \bar{x})^2$ can be made larger in two ways: (a) by increasing n (so that there are more terms in the sum), and (b) by increasing the dispersion or spread in the X values. The dependence on the spread in the X values is illustrated in Figure 12.13, which shows two data sets with the same value of $s_{Y|X}$ and the same value of n, but different values of $\Sigma(x_i - \bar{x})^2$. Imagine using a ruler to fit a straight line by eye; it is intuitively clear that the data set in case (b)—with the larger $\Sigma(x_i - \bar{x})^2$—would determine the slope of the line more precisely.

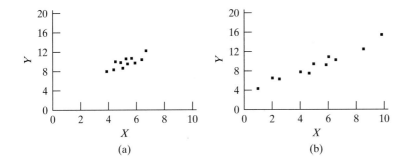

Figure 12.13 Two data sets with the same value of n and of $s_{Y|X}$ but different $\Sigma(x_i - \bar{x})^2$: (a) Smaller $\Sigma(x_i - \bar{x})^2$; (b) larger $\Sigma(x_i - \bar{x})^2$.

As another way of thinking about this, imagine holding your arms out in front of you, extending the index finger on each hand, and balancing a meter stick on your two fingers. If you move your hands far apart from each other, balancing the meter stick is easy—this is like case (b). However, if you move your hands close together, balancing the meter stick becomes more difficult—this is like case (a). Having the base of support spread out increases stability. Likewise, having the x values spread out decreases the standard error of the slope.

Implications for Design.   The preceding discussion implies that, for the purpose of gaining precise information about β_1, it is best to have the values of X as widely dispersed as possible. This fact can guide the experimenter when the design of the experiment includes choosing values of X. Other factors also play a role, however. For instance, if X is the dose of a drug, the criterion of widely dispersed X's would lead to using only two dosages, one very low and one very high. But in practice an experimenter would want to have at least a few observations at intermediate doses, to verify that the relation is actually linear within the range of the data.

Confidence Interval for β_1

In many studies the quantity β_1 is a biologically meaningful parameter, and a primary aim of the data analysis is to estimate β_1. A confidence interval for β_1 can be constructed by the familiar method based on the SE and Student's t distribution. For instance, a 95% confidence interval is constructed as

$$b_1 \pm t_{.025}\, SE_{b_1}$$

where the critical value $t_{.025}$ is determined from Student's t distribution with

$$\text{df} = n - 2$$

Intervals with other confidence coefficients are constructed analogously; for instance, for a 90% confidence interval one would use $t_{.05}$.

Example 12.18

Length and Weight of Snakes. Let us use the snake data to construct a 95% confidence interval for β_1. We found that $b_1 = 7.19186$ and $\text{SE}_{b_1} = .9531$. There are $n = 9$ observations; we refer to Table 4 with df $= 9 - 2 = 7$, and obtain

$$t(7)_{.025} = 2.365$$

The confidence interval is

$$7.19186 \pm (2.365)(.9531)$$

or

$$4.9 \, \text{g/cm} < \beta_1 < 9.4 \, \text{g/cm}$$

We are 95% confident that the true slope of the regression of weight on length for this snake population is between 4.9 g/cm and 9.4 g/cm; this is a rather wide interval because the sample size is not very large. ■

Testing the Hypothesis H_0: $\beta_1 = 0$

In some investigations it is not a foregone conclusion that there is any relationship between X and Y. It then may be relevant to consider the possibility that any apparent trend in the data is illusory and reflects only sampling variability. In this situation it is natural to formulate the null hypothesis

H_0: $\mu_{Y|X}$ does not depend on X.

Within the linear model, this hypothesis can be translated as

$$H_0: \beta_1 = 0$$

A t test of H_0 is based on the test statistic

$$t_s = \frac{b_1}{\text{SE}_{b_1}}$$

Critical values are obtained from Student's t distribution with

$$\text{df} = n - 2$$

The following example illustrates the application of this t test.

Example 12.19

Blood Pressure and Platelet Calcium. It is suspected that calcium in the cells may be related to blood pressure. As part of a study of this relationship, researchers recruited 38 subjects whose blood pressure was normal (that is, not abnormally elevated). For each subject, two measurements were made: X = blood pressure (average of systolic and diastolic measurements) and Y = free calcium concentration in the blood platelets. The data are shown in Figure 12.14.[14] Calculations from the data yield $\bar{x} = 94.5$, $\bar{y} = 107.868$, $\Sigma(x_i - \bar{x})^2 = 2,397.50$, and $\Sigma(x_i - \bar{x})(y_i - \bar{y}) = 2,792.50$, from which we can calculate

$$b_0 = -2.2009 \quad \text{and} \quad b_1 = 1.16475$$

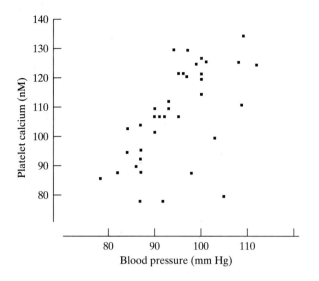

Figure 12.14 Blood pressure and platelet calcium for 38 persons with normal blood pressure

The residual sum of squares is 6,311.7618. Thus,

$$s_{Y|X} = \sqrt{\frac{6,311.76}{38 - 2}} = 13.24 \quad \text{and} \quad SE_{b_1} = \frac{13.24}{\sqrt{2,397.5}} = .2704$$

The values of b_0, b_1, SS(resid), and SE_{b_1} are generally found using computer software. The following computer output is typical:

```
Dependent variable is: Platelet calcium
No Selector
R squared = 34.0%    R squared (adjusted) = 32.2%
s = 13.24  with  38 - 2 = 36  degrees of freedom
Source      Sum of Squares    df    Mean Square  F-ratio
Regression      3252.58        1       3252.58    18.6
Residual        6311.76        36      175.327
Variable     Coefficient  s.e.of Coeff  t-ratio     prob
Constant      -2.20092      25.65       -0.086     0.9321
Blood pressure 1.16475      0.2704       4.31      0.0001
```

We will test the null hypothesis

$$H_0: \beta_1 = 0$$

against the nondirectional alternative

$$H_A: \beta_1 \neq 0$$

These hypotheses are translations, within the linear model, of the verbal hypotheses

H_0: Mean platelet calcium does not depend on blood pressure.

H_A: Mean platelet calcium does depend on blood pressure.

(Note, however, that *depend* does not necessarily refer to causal dependence. We will return to this point in Section 12.7.)

Let us choose $\alpha = .05$. The test statistic is

$$t_s = \frac{1.16475}{.2704} = 4.308$$

From Table 4 with df $= n - 2 = 36 \approx 40$, we find $t(40)_{.0005} = 3.551$. Thus, we find $P < .0005$ and we reject H_0. The data provide sufficient evidence to conclude that the true slope of the regression of platelet calcium on blood pressure in this population is positive (that is, $\beta_1 > 0$). ∎

Note that the test on β_1 does not ask *whether* the relationship between $\mu_{Y|X}$ and X is linear. Rather, the test asks whether, *assuming* that the linear model holds, we can conclude that the slope is nonzero. It is therefore necessary to be careful in phrasing the conclusion from this test. For instance, the statement "There is a significant linear trend" could be easily misunderstood.*

As is the case with other hypothesis tests, if we wish to use a directional alternative hypothesis, we follow the two-step procedure of (1) checking that the specified direction is correct (which in a regression setting means checking that the slope of the regression line has the correct $+$ or $-$ sign) and (2) cutting the P-value in half if this condition is met.

Why $(n - 2)$? The confidence interval and test based on b_1 have associated df $= n - 2$. Also, $(n - 2)$ is the denominator of $s_{Y|X}^2$. The origin of the $(n - 2)$ is easy to explain. It takes two points to determine a straight line, and so (under the linear model) the data provide $(n - 2)$ independent pieces of information concerning $\sigma_{Y|X}$. (Note that if $n = 2$, the regression line will fit the data exactly, but $s_{Y|X}$ cannot be calculated.) Thus, as in earlier contexts related to t distributions and F distributions (Chapters 6, 7, 9, and 11), the number of df is the number of pieces of information provided by the data about the "noise" from which the investigator wants to extract the "signal."

Exercises 12.19–12.26

12.19 Refer to the leucine data given in Exercise 12.3. For these data, $SE_{b_1} = .00159$.

(a) Construct a 95% confidence interval for β_1.

(b) Interpret the confidence interval from part (a) in the context of this setting.

12.20 Refer to the body temperature data of Exercise 12.4. Construct a 95% confidence interval for β_1.

12.21 Refer to the cob weight data of Exercise 12.5.

(a) Construct a 95% confidence interval for β_1.
(b) Interpret the confidence interval from part (a) in the context of this setting.

12.22 Refer to the fungus growth data of Exercise 12.6.

(a) Calculate the standard error of the slope, b_1.

---

* There are tests that can (in some circumstances) test whether the true relationship is linear. Furthermore, there are tests that can test for a linear component of trend without assuming that the relationship is linear. These tests are beyond the scope of this book.

(b) Consider the null hypothesis that laetisaric acid has no effect on growth of the fungus. Assuming that the linear model is applicable, formulate this as a hypothesis about the true regression line, and test the hypothesis against the alternative that laetisaric acid inhibits growth of the fungus. Let $\alpha = .05$.

12.23 Refer to the energy expenditure data of Exercise 12.7.

(a) Construct a 95% confidence interval for β_1.
(b) Construct a 90% confidence interval for β_1.

12.24 Refer to the respiration data of Exercise 12.8. Assuming that the linear model is applicable, test the null hypothesis of no relationship against the alternative that trees from higher altitudes tend to have higher respiration rates. Let $\alpha = .05$.

12.25 The following is MINITAB output from fitting a regression model to the snake length data of Example 12.3. Use this output to construct a 95% confidence interval for β_1.

```
Regression Analysis
The regression equation is
Weight = -301 + 7.19 Length

Predictor        Coef        Stdev       t-ratio        p
Constant       -301.09       60.19        -5.00      0.000
Length           7.1919       0.9531       7.55      0.000

s = 12.50       R-sq = 89.1%       R-sq(adj) = 87.5%
Analysis of Variance

SOURCE          DF        SS         MS         F         p
Regression       1      8896.3     8896.3     56.94    0.000
Error            7      1093.7      156.2
Total            8      9990.0
```

12.26 Refer to the peak flow data of Exercise 12.10. Assume that the linear model is applicable.

(a) Test the null hypothesis of no relationship against the alternative that peak flow is related to height. Use a nondirectional alternative with $\alpha = .10$.
(b) Repeat the test from part (a), but this time use the directional alternative that peak flow tends to increase with height. Again let $\alpha = .10$.

12.5 THE CORRELATION COEFFICIENT

Consider collecting data on the variable Y = weight for a group of persons. Weight varies from person to person, with $\bar{y}$ representing the average weight. The quantity **SS(total)** $= \Sigma(y_i - \bar{y})^2$ measures the total variability in weight for the sample. Suppose that we also know the height, X, of each person. If there is a linear relationship in the data between height, X, and weight, Y, then we can fit a regression model and use height to predict weight; the predictions are given by $\hat{y} = b_0 + b_1 x$. Some people are heavier than others, and this is partly explained by the fact that they are taller than others. To the extent that the predicted weights from the regression model, the $\hat{y}$'s, agree with the actual weights, the y's, we can say that variation in height "explains" variation in weight (through the regression model).

The residuals, $y_i - \hat{y}_i$, represent variation in Y that is *not* explained by X through the regression model. The quantity SS(resid) $= \Sigma(y_i - \hat{y}_i)^2$ measures this unexplained variability in Y.

The difference between SS(total) and SS(resid) is the quantity **SS(reg)** $= \Sigma(\hat{y}_i - \bar{y})^2$, which measures variability that is due to the regression model, through the predictions, the $\hat{y}$'s. Thus, the three sums of squares are related as follows:

$$SS(total) = SS(reg) + SS(resid)$$

That is, the *total* variability in Y equals the variability *explained* by the regression model plus the *unexplained* residual variability:

Total variability = explained variability + unexplained variability

Although we have been talking about height and weight of persons, the ideas carry over to any regression setting.

Example 12.20 **Length and Weight of Snakes.** For the snake data of Example 12.3, the three sums of squares are

$$SS(total) = 9,990$$
$$SS(reg) = 8,896.33$$
$$SS(resid) = 1,093.67$$

Note that $9,990 = 8,896.33 + 1,093.67$ ∎

The Coefficient of Determination

The **coefficient of determination** is defined as the ratio of SS(reg) to SS(total) and is denoted by r^2:

$$\text{coefficient of determination} = r^2 = \frac{SS(reg)}{SS(total)}$$

The coefficient of determination can be interpreted as the proportion of the variation in Y that is "accounted for" or "explained" by the linear regression of Y on X. Likewise, the fraction $\frac{SS(resid)}{SS(total)}$ can be interpreted as the proportion of the variation in Y that is *not* "accounted for" or "explained" by the regression. Note that

$$r^2 = 1 - \frac{SS(resid)}{SS(total)}$$

The coefficient of determination is often expressed as a percentage, as in the following example.

Example 12.21 **Length and Weight of Snakes.** For the snake data the coefficient of determination is

$$r^2 = \frac{8896.33}{9990} = .8905 \approx .89$$

Thus, one might say that 89% of the variation in weight among these snakes is explained, or accounted for, by variation in length. A complementary interpretation is that 11% of the variation in weight is "residual," or not accounted for by variation in length. (In interpreting these phrases, however, it should be remembered that *accounted for* means "accounted for by linear regression.") ■

Note that the value of r^2 is always between 0 and 1:

$$0 \le r^2 \le 1$$

If the data points fall exactly on a line, then $\hat{y}_i = y_i$, so that $y_i - \hat{y}_i = 0$ and SS(resid) = 0. In this case, SS(reg) = SS(total) and $r^2 = 1$. In such a case we could say that 100% of the variation in Y is explained by variation in X.

At the other extreme, it might happen that there is no linear relationship between X and Y. In this case using X to predict Y is worthless; SS(resid) = SS(total), SS(reg) = 0, and $r^2 = 0$. We would say that none (0%) of the variation in Y is explained by variation in X. Most regression settings fall between these two extremes of perfect linear association and no linear association.

The Correlation Coefficient

Related to the coefficient of determination, r^2, is the **correlation coefficient**, r.* The correlation coefficient, r, is the square root of r^2 multiplied by the sign of the slope of the regression line. The correlation coefficient is related to the slope of the regression line through the following formula:

$$b_1 = r \frac{s_Y}{s_X}$$

That is, the slope of the regression line equals the correlation coefficient multiplied by the ratio of the standard deviations of Y and of X. If the regression line has a positive slope, then r is positive; if the regression line has a negative slope, then r is negative. Unlike b_1, however, r is dimensionless—that is, it is not measured in units such as g or g/cm.

Since $0 \le r^2 \le 1$, it follows that $-1 \le r \le 1$. If there is no linear relationship between X and Y, then $r = 0$. If there is a perfect linear trend between X and Y, with a positive slope, then $r = 1$. If there is a perfect linear trend between X and Y, with a negative slope, then $r = -1$. It is the *magnitude* of r that tells how strong the linear relationship is between X and Y; the sign of r only tells whether the slope is positive or negative.

The correlation coefficient can also be found as follows:

Formula for the Correlation Coefficient

$$r = \frac{\Sigma(x_i - \bar{x})(y_i - \bar{y})}{\sqrt{\Sigma(x_i - \bar{x})^2 \Sigma(y_i - \bar{y})^2}}$$

The following example illustrates the calculation of r.

---

* A more complete name for this statistic is *Pearson's product-moment correlation coefficient*.

Example 12.22

Length and Weight of Snakes. We found in Table 12.3 that for the snake data $\Sigma(x_i - \bar{x})^2 = 172$, $\Sigma(y_i - \bar{y})^2 = 9{,}990$, and $\Sigma(x_i - \bar{x})(y_i - \bar{y}) = 1{,}237$. Thus, the correlation coefficient is

$$r = \frac{1{,}237}{\sqrt{(172)(9{,}990)}} = .9437 \approx .94$$

We could also find r for the snake data by noting that the slope between length and weight is positive, so r is the positive square root of r^2. In Example 12.21 we found that $r^2 = .8905$. Thus, $r = \sqrt{.8905} = .9437$. ■

    The magnitude of r is a rough indication of the shape of the scatterplot. A value of r close to $+1$ or -1 suggests that the cloud of data points is long and narrow, with the points clustered close to a line. A small magnitude of r suggests that the data cloud is diffuse, with the points loosely scattered. (In Section 12.6 we will discuss exceptions to these interpretations.) The following example illustrates the general idea.

Example 12.23

Examples of Correlations. Figure 12.15 shows fictitious data sets with various values of r. Figure 12.15(a) shows 30 observations with $r = .98$; the visual impression is a long narrow cloud of points, indicating a tight linear association between X and Y. Figure 12.15(b) shows 30 observations with $r = .65$; the visual impression is a loosely scattered cloud of points that shows an overall upward trend. Figure 12.15(c) shows 30 observations with $r = .35$; the visual impression is a very loosely scattered cloud of points that nevertheless seem to show an overall upward trend. The data in Figures 12.15(d), (e), and (f) have correlations of $r = -.98$, $r = -.65$, and $r = -.35$.

    Note that the value of r does *not* reflect the steepness or shallowness of the slope relating Y and X. In fact, the fitted regression lines for the data sets in Figures 12.15(a), (b), and (c) are identical. [To see this intuitively, imagine superimposing (a) on (b) or on (c).] Similarly, the regression lines in (d), (e) and (f) are identical. ■

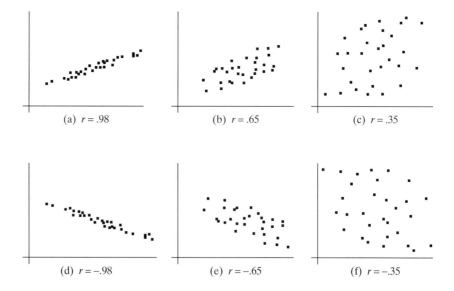

(a) $r = .98$ (b) $r = .65$ (c) $r = .35$

(d) $r = -.98$ (e) $r = -.65$ (f) $r = -.35$

Figure 12.15 Data sets to illustrate the correlation coefficient

How r Describes the Regression

We have said that the magnitude of r describes the tightness of the linear relationship between X and Y. This interpretation is made more specific by the following fact. (This fact is proved in Appendix 12.2.)

> **Fact 12.1: Approximate Relationship of r to $s_{Y|X}$ and s_Y**
> The correlation coefficient r obeys the following approximate relationship:
> $$\frac{s_{Y|X}}{s_Y} \approx \sqrt{1 - r^2}$$

(The approximation in Fact 12.1 is best for large n, but it holds reasonably well even for n as small as 5.) If we know r, then by using Fact 12.1 we can deduce the approximate ratio of the residual SD to the ordinary (marginal) SD of Y. The following two examples illustrate this idea.

Length and Weight of Snakes. For the snake data, we found in Example 12.22 that the correlation coefficient is $r = .9437$. From Fact 12.1 we conclude that

$$\frac{s_{Y|X}}{s_Y} \approx \sqrt{1 - (.9437)^2} = .33$$

Example 12.24

That is, from the value of r we can deduce that the residual SD of weight, after regression on length, is only about 33% of the overall SD of weight; this in turn means that the linear relationship is fairly tight. The two SDs are shown as ruler lines in Figure 12.16. ∎

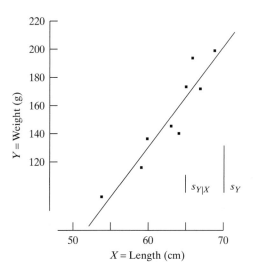

Figure 12.16 Relationship between s_Y and $s_{Y|X}$ for snake data

Fecundity of Crickets. For the cricket data of Example 12.2, $r = .6873$ and $\sqrt{1 - r^2} = .73$. Thus, the value of r tells us that $s_{Y|X}$ is relatively large; it is about 73% of s_Y. This indicates that points are rather widely scattered about the regression line. The two SDs are shown as ruler lines in Figure 12.17. ∎

Example 12.25

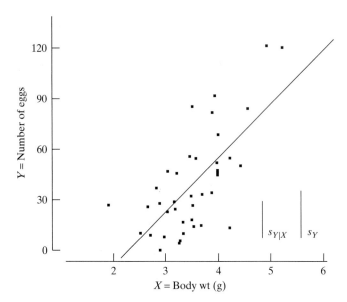

Figure 12.17 Relationship between s_Y and $s_{Y|X}$ for cricket data

The Symmetry of r

Recall that the formula for r is

$$r = \frac{\Sigma(x_i - \bar{x})(y_i - \bar{y})}{\sqrt{\Sigma(x_i - \bar{x})^2 \Sigma(y_i - \bar{y})^2}}$$

From this formula it is clear that *X and Y enter r symmetrically.* Therefore, if we were to interchange the labels X and Y of our variables, r would remain unchanged. In fact, this is one of the advantages of the correlation coefficient as a summary statistic: In interpreting r, it is not necessary to know (or to decide) which variable is labeled X and which is labeled Y.

A Paradox and Its Resolution. The symmetry of r may appear puzzling. We have interpreted r^2, and thus r, in terms of the variation in one of the variables—namely, Y—as it relates to regression on the other variable—namely, X. This asymmetric interpretation of r appears to contradict the symmetric formula for r.

We can resolve this apparent paradox by considering reverse regression. Suppose we keep our variable labels X and Y fixed, but we regress in the reverse direction—that is, we regress X on Y.* The reverse regression minimizes the sum of squares of the *horizontal* distances $x_i - \hat{x}_i$ and the residual sum of squares is $\Sigma(x_i - \hat{x}_i)^2$.

Because the least-squares criterion is applied to vertical distances in one case and horizontal distances in the other, there are actually two regression lines associated with any set of data—the regression of Y on X and the regression of X on Y. Remarkably, however, *the closeness of the data points to the lines, as measured by r, is the same for both regression lines.* Specifically, we can say that, for the regression of Y on X,

$$r^2 = 1 - \frac{\Sigma(y_i - \hat{y}_i)^2}{\Sigma(y_i - \bar{y}_i)^2}$$

_____

* The equations for the reverse regression are not given here because they are of no practical importance in data analysis. They are given in Appendix 12.3.

and for the regression of X on Y,

$$r^2 = 1 - \frac{\Sigma(x_i - \hat{x}_i)^2}{\Sigma(x_i - \bar{x}_i)^2}$$

Similarly, either regression yields an approximate relation between r and the standard deviations. For the regression of Y on X,

$$\frac{s_{Y|X}}{s_Y} \approx \sqrt{1 - r^2}$$

and for the regression of X on Y,

$$\frac{s_{X|Y}}{s_X} \approx \sqrt{1 - r^2}$$

The following example illustrates these relationships.

Fecundity of Crickets.   Consider the cricket data of Example 12.2 on $X =$ body weight and $Y =$ number of eggs. The marginal means are $\bar{x} = 3.5\,\mathrm{g}$ and $\bar{y} = 40$ eggs. For the regression of Y on X, the fitted line is $Y = -71.7 + 31.7X$, the residual SD is $s_{Y|X} = 22.6$ eggs, and the marginal SD of Y is $s_Y = 30.7$ eggs. For the regression of X on Y, the fitted line is $X = 2.93 + .0149Y$, the residual SD is $s_{X|Y} = .490\,\mathrm{g}$, and the marginal SD of X is $s_X = .665\,\mathrm{g}$. Figure 12.18 shows the data and the two regression lines. The marginal and residual SDs are shown as ruler lines. Note that both regression lines pass through the joint mean $(\bar{x}, \bar{y})$.

| Example 12.26 |

For these data, $r = .6873$, and $\sqrt{1 - r^2} = .73$. We verify that

$$\frac{s_{Y|X}}{s_Y} = \frac{22.6}{30.7} = .74 \approx .73 = \sqrt{1 - r^2}$$

$$\frac{s_{X|Y}}{s_X} = \frac{.490}{.665} = .74 \approx .73 = \sqrt{1 - r^2}$$ ∎

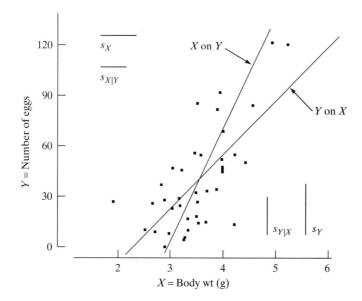

Figure 12.18 Body weight (X) and number of eggs (Y) for 39 female crickets, showing the regression lines of Y on X and of X on Y

Statistical Inference Concerning Correlation

We have described various ways in which the correlation coefficient describes a data set. Now we consider statistical inference based on r.

We saw in Section 12.3 that statistical inference about the regression line is based on the random subsampling model—that is, the view that the Y values were selected by random sampling from the conditional populations defined by the X values. But the correlation approach, unlike the regression approach, treats X and Y symmetrically. In order to regard the sample correlation coefficient r as an estimate of a population parameter, it must be reasonable to assume that both the X and the Y values were selected at random, as in the following **bivariate random sampling model**:

> **Bivariate Random Sampling Model:**
> We regard each pair (x_i, y_i) as having been sampled at random from a population of (x, y) pairs.

In the bivariate random sampling model, the sample correlation coefficient r is an estimate of the population correlation coefficient, which is denoted ρ (the Greek letter rho). Also, in the bivariate random sampling model, the observed x's are regarded as a random sample and the observed y's are also regarded as a random sample, so that the marginal statistics $\bar{x}, \bar{y}, s_X$, and s_Y are estimates of corresponding population values μ_X, μ_Y, σ_X, and σ_Y.

The following example illustrates a case where the bivariate random sampling model is reasonable.

Example 12.27

Blood Pressure and Platelet Calcium. For the data of Example 12.19, the correlation coefficient between blood pressure and platelet calcium is $r = .58$. The data were obtained from 38 adult volunteers who did not suffer from high blood pressure. One might regard the observed pairs (x_i, y_i) as a random sample from potential measurements on a corresponding population (adults not suffering from high blood pressure). In this setting, the observed value of r, .58, is an estimate of the population correlation coefficient ρ. Similarly, the observed mean and SD of blood pressure, and the observed mean and SD of platelet calcium, would be estimates of corresponding quantities in the population. ■

For many investigations the random subsampling model is reasonable, but the additional assumption of a bivariate random sampling model is not. This is generally the case when the values of X are specified by the experimenter. The following is an example.

Example 12.28

Amphetamine and Food Consumption. In the food consumption experiment of Example 12.1, X represents the dose of amphetamine. The observed x's are not a random sample, but were specified by the experimenter. We can calculate $\bar{x}$ and s_X from the data, but these statistics are not estimates of any population parameters. Similarly, $\bar{y}, s_Y$, and r are not estimates of population parameters. ■

We now consider statistical inference concerning the population correlation coefficient ρ. Suppose we wish to test the hypothesis $H_0: \rho = 0$, which asserts that

X and Y are uncorrelated in the population. It turns out that no new technique is required; we can simply reinterpret the t test described in Section 12.4. Specifically, it can be shown that, if the linear model holds, then ρ is related to the population regression slope β_1 as follows:

$$\beta_1 = \rho \frac{\sigma_Y}{\sigma_X}$$

Because of this relationship, the two hypotheses

$$H_0: \beta_1 = 0 \quad \text{and} \quad H_0: \rho = 0$$

are equivalent. Recall from Section 12.4 that the t test for the hypothesis $H_0: \beta_1 = 0$ is based on the test statistic

$$t_s = \frac{b_1}{SE_{b_1}}$$

and uses critical values from Student's t distribution with df $= n - 2$. Under the bivariate random sampling model, this test can be reinterpreted as a test of the hypothesis $H_0: \rho = 0$. It can also be shown that the t statistic can be rewritten in terms of r, as follows:

The t Statistic in Terms of r

$$t_s = \frac{b_1}{SE_{b_1}} = r\sqrt{\frac{n - 2}{1 - r^2}}$$

The following example illustrates the equivalence of the two tests.

Blood Pressure and Platelet Calcium.   For the platelet calcium data of Example 12.19, the observed correlation is $r = .5832$. Let us test the hypothesis

Example 12.29

$\qquad H_0$: Platelet calcium and blood pressure are uncorrelated in the population

that is,

$$H_0: \rho = 0$$

The test statistic can be calculated from r as

$$t_s = r\sqrt{\frac{n - 2}{1 - r^2}}$$

$$= .5832\sqrt{\frac{36}{1 - (.5832)^2}} = 4.308$$

which is the same as the value obtained by regression of platelet calcium on blood pressure in Example 12.19. The reverse regression—of blood pressure on platelet calcium—would also yield the same value of t_s. At $\alpha = .05$ there is sufficient evidence to conclude that blood pressure and platelet calcium are positively correlated in the population. (This is equivalent to asserting that β_1 is positive.) ∎

Cautionary Notes

1.   To describe the results of testing a correlation coefficient, investigators often use the term *significant*, which can be misleading. For instance, a

statement such as "A highly significant correlation was noted . . ." is easily misunderstood. It is important to remember that statistical significance simply indicates rejection of a null hypothesis; it does not necessarily indicate a large or important effect. A "significant" correlation may in fact be quite a weak one; its significance" means only that it cannot easily be explained away as a chance pattern. From the formula $t_s = r\sqrt{\dfrac{n-2}{1-r^2}}$ we can see that for a fixed value of r, t_s increases as n increases. Thus, if the sample size is large enough, t_s will be large enough for the correlation to be "significant" no matter how small r is.

2. The correlation coefficient is highly sensitive to extreme points. For example, Figure 12.19(a) shows a scatterplot of 25 points with a correlation of $r = .2$; one of the points has been plotted as an ◆. Figure 12.19(b) shows the same points, except that the point plotted as an ◆ has been changed. The change of that single point causes the correlation coefficient to climb from .2 to .6. Figure 12.19(c) shows a third version of the data. In this case $r = -.3$. These three graphs illustrate how a single point can greatly influence the size of the correlation coefficient. It is important to always plot the data before using r (or any other statistic) to summarize the data.

Confidence Interval for ρ (Optional)

If the sample size is large, it is possible to construct a confidence interval for ρ. The sampling distribution of the sample correlation coefficient, r, is skewed, so in order to construct the confidence interval we apply what is known as the Fisher transformation of r:

$$z_r = \frac{1}{2}\ln\left[\frac{1+r}{1-r}\right]$$

where ln is the natural logarithm (base e). We can then construct a 95% confidence interval for $\frac{1}{2}\ln\left[\dfrac{1+\rho}{1-\rho}\right]$ as

$$z_r \pm 1.96\frac{1}{\sqrt{n-3}}$$

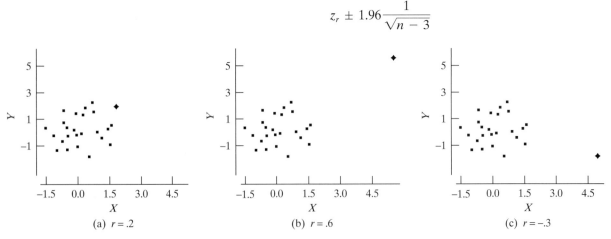

(a) $r = .2$ (b) $r = .6$ (c) $r = -.3$

Figure 12.19 Data sets to illustrate the effect of extreme points on the correlation coefficient. (a) $r = .2$; (b) $r = .6$; (c) $r = -.3$

Finally, we can convert the limits of the confidence interval for $\frac{1}{2} \ln\left[\frac{1 + \rho}{1 - \rho}\right]$ into a confidence interval for ρ by solving for ρ in the equations given by

$$\frac{1}{2} \ln\left[\frac{1 + \rho}{1 - \rho}\right] = z_r \pm 1.96 \frac{1}{\sqrt{n - 3}}$$

Intervals with other confidence levels are constructed analogously. For example, to construct a 90% confidence interval, replace 1.96 with 1.645. The construction of a confidence interval for a correlation coefficient is illustrated in Example 12.30.

Blood Pressure and Platelet Calcium.   For the platelet calcium data of Example 12.19, the sample size is $n = 38$ and the sample correlation is $r = .5832$. The Fisher transformation of r gives

| | **Example 12.30** |

$$z_r = \frac{1}{2} \ln\left[\frac{1 + .5832}{1 - .5832}\right] = \frac{1}{2} \ln\left[\frac{1.5832}{.4168}\right] = .6673$$

A 95% confidence interval for $\frac{1}{2} \ln\left[\frac{1 + \rho}{1 - \rho}\right]$ is

$$.6673 \pm 1.96 \frac{1}{\sqrt{38 - 3}}$$

or $.6673 \pm .3313$, which is $(.3360, .9986)$.

Setting

$$\frac{1}{2} \ln\left[\frac{1 + \rho}{1 - \rho}\right] = .3360 \text{ gives } \rho = \frac{e^{2(.3360)} - 1}{e^{2(.3360)} + 1} = .32$$

Setting

$$\frac{1}{2} \ln\left[\frac{1 + \rho}{1 - \rho}\right] = .9986 \text{ gives } \rho = \frac{e^{2(.9986)} - 1}{e^{2(.9986)} + 1} = .76$$

We are 95% confident that the correlation between blood pressure and platelet calcium in the population is between .32 and .76. Thus, a 95% confidence interval for ρ is $(.32, .76)$.   ∎

Exercises 12.27–12.36

12.27   A plant physiologist grew 13 individually potted soybean seedlings in a greenhouse. The table gives measurements of the total leaf area (cm^2) and total plant dry weight (g) for each plant after 16 days of growth.[15]

Plant	Leaf Area X	Dry weight Y
1	411	2.00
2	550	2.46
3	471	2.11
4	393	1.89
5	427	2.05
6	431	2.30
7	492	2.46
8	371	2.06
9	470	2.25
10	419	2.07
11	407	2.17
12	489	2.32
13	439	2.12
Mean	443.8	2.174

$$\Sigma(x_i - \bar{x})^2 = 28{,}465.7 \quad \Sigma(y_i - \bar{y})^2 = .363708$$
$$\Sigma(x_i - \bar{x})(y_i - \bar{y}) = 82.8977$$
$$\text{SS(resid)} = .1223$$

(a) Calculate the correlation coefficient.

(b) Calculate s_Y and $s_{Y|X}$; specify the units for each. Verify the approximate relationship between s_Y and $s_{Y|X}$, and r.

(c) Calculate the regression line of Y on X.

(d) Construct a scatterplot of the data and draw the regression line on your graph. Place ruler lines on the scatterplot to show the magnitudes of s_Y and $s_{Y|X}$. (See Figure 12.16.)

12.28 Proceed as in Exercise 12.27, but use the cob weight data of Exercise 12.5.

12.29 Proceed as in Exercise 12.27, but use the energy expenditure data of Exercise 12.7.

12.30 In a study of 2,669 adult men, the correlation between age and systolic blood pressure was found to be $r = .43$. The SD of systolic blood pressures among all 2,669 men was 19.5 mm Hg. Assuming that the linear model is applicable, estimate the SD of systolic blood pressures among men 50 years old.[16]

12.31 Consider the data from Exercise 12.30. The correlation coefficient, r, is .43.

(a) Find the value of r^2.

(b) Interpret the value of r^2 found in part (a) in the context of this problem.

12.32 A veterinary anatomist measured the density of nerve cells at specified sites in the intestine of nine horses. Each density value is the average of counts of nerve cells in five equal sections of tissue. The results are given in the accompanying table for site I (midregion of jejunum) and site II (mesenteric region of jejunum).[17] (These data were also given in Example 9.17.)

Animal	Site I	Site II
1	50.6	38.0
2	39.2	18.6
3	35.2	23.2
4	17.0	19.0
5	11.2	6.6
6	14.2	16.4
7	24.2	14.4
8	37.4	37.6
9	35.2	24.4
Mean	29.36	22.02

$$\Sigma(x_i - \bar{x})^2 = 1{,}419.82 \quad \Sigma(y_i - \bar{y})^2 = 853.396$$
$$\Sigma(x_i - \bar{x})(y_i - \bar{y}) = 893.689$$

(a) Calculate the correlation coefficient between the densities at the two sites.

(b) Construct a scatterplot of the data.

(c) Four potential sources of variation in these data are (i) errors in counting the nerve cells, (ii) sampling error due to choosing certain slices for counting, (iii) variation from one horse to another, and (iv) variation from site to site within a horse. Which of these sources of variation would tend to produce positive correlation between the sites? (*Hint:* Ask yourself how each source contributes to the appearance of the scatterplot.)

12.33 Refer to Exercise 12.32. Test the hypothesis that the true (population) correlation coefficient is zero against the alternative that it is positive. Let $\alpha = .05$.

12.34 In a study of natural variation in blood chemistry, blood specimens were obtained from 284 healthy people. The concentrations of urea and of uric acid were measured for each specimen, and the correlation between these two concentrations was found to be $r = .2291$. Test the hypothesis that the population correlation coefficient is zero against the alternative that it is positive.[18] Let $\alpha = .05$.

12.35 Researchers measured the number of neurons in the CA1 region of the hippocampus in the brains of eight persons who had died of causes unrelated to brain function. They found that these data were negatively correlated with age. The sample value of r was $-.63$.[19] Is this correlation coefficient significantly different from zero? Conduct a test using $\alpha = .10$.

12.36 Consider the data from Exercise 12.35. The correlation coefficient, r, is $-.63$.

(a) Find the value of r^2.

(b) Interpret the value of r^2 found in part (a) in the context of this problem.

12.6 GUIDELINES FOR INTERPRETING REGRESSION AND CORRELATION

Any set of (x, y) data can be submitted to a regression analysis and values of $b_0, b_1, s_{Y|X}$, and r can be calculated. But these quantities require care in interpretation. In this section we discuss guidelines and cautions for interpretation of linear regression and correlation. We first consider the use of regression and correlation for purely descriptive purposes, and then turn to inferential uses.

When Is Linear Regression Descriptively Inadequate?

Linear regression and correlation may provide inadequate description of a data set if any of the following features is present:

    curvilinearity

    outliers

    influential points

We briefly discuss each of these.

    If the dependence of Y on X is actually curvilinear rather than linear, the application of linear regression and correlation can be very misleading. The following example shows how this can happen.

Example 12.31

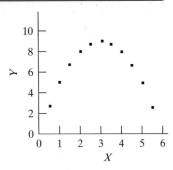

Figure 12.20 Data for which X and Y are uncorrelated but have a strong curvilinear relationship

Figure 12.20 shows a set of fictitious data that obey an exact relationship: $Y = 6X - X^2$. Nevertheless, X and Y are uncorrelated: $r = 0$ and $b_1 = 0$. The best straight line through the data would be a horizontal one, but of course the line would be a poor summary of the curvilinear relationship between X and Y. The residual SD is $s_{Y|X} = 2.43$; this value does not measure random variation but rather measures deviation from linearity. ∎

Generally, the consequences of curvilinearity are that (1) the fitted line does not adequately represent the data; (2) the correlation is misleadingly small; (3) $s_{Y|X}$ is inflated. Of course, Example 12.31 is an extreme case of this distortion. A data set with mild, but still noticeable, curvilinearity is shown in Figure 12.21.

Outliers in a regression setting are data points that are unusually far from the linear trend formed by the data. Outliers can distort regression analysis in two ways: (1) by inflating $s_{Y|X}$ and reducing correlation; (2) by unduly influencing the regression line. The following example illustrates both of these.

Example 12.32

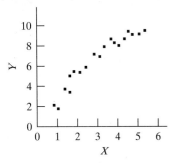

Figure 12.21 Data displaying mild curvilinearity

Figure 12.22(a) shows a data set with a single outlier. A straight line would fit quite well through all the data points except the outlier. Figure 12.22(b) shows the regression line fitted to all the data. Notice how poorly the line fits the points. ∎

Note that a point can be an outlier in a scatterplot without being an outlier in either the distribution of x values or the distribution of y values. Indeed, this is the case in Example 12.32. Figure 12.23 shows boxplots of the x values and of the y values; the outlier in the scatterplot is not an outlier in either of these distributions.

Influential points are points that have a great deal of influence on the fitted regression model. If a point is far removed from the majority of the data in the

Figure 12.22 The effect of an outlier on the regression line. (a) A data set with an outlier; (b) the regression line fitted to all the data.

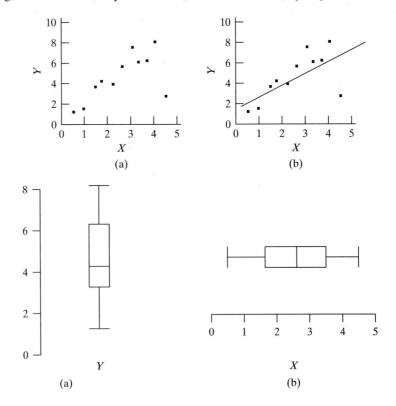

Figure 12.23 Data from Example 12.32. (a) Boxplot of the y values; (b) boxplot of the x values.

x direction, then it will tend to have a large effect when a regression model is fit to the data. Moreover, such a point can greatly affect the size of the correlation coefficient, as is demonstrated in Example 12.33.

Figure 12.24(a) shows a data set and a regression line. Figure 12.24(b) shows the same data set, but with an influential point added. Including the influential point in the data set changes the regression line noticeably. The influential point is not an outlier, in the usual sense, since the residual for this point is not very large.

    The correlation coefficient for the data in Figure 12.24(a) is .35. Adding the influential point to the data set changes the correlation to .86 for the data in Figure 12.24(b). ∎

> **Example 12.33**

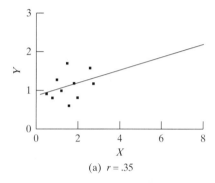

(a) $r = .35$

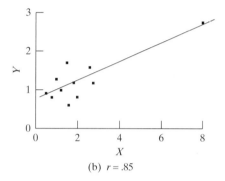

(b) $r = .85$

Figure 12.24 The effect of an influential point on the regression line. (a) A data set; (b) the same data with an influential point added.

Conditions for Inference

The quantities $b_0, b_1, s_{Y|X}$, and r can be used to describe a scatterplot that shows a linear trend. However, statistical inference based on these quantities depends on certain conditions concerning the design of the study, the parameters, and the conditional population distributions. We summarize these conditions and then discuss guidelines and cautions concerning them.

1. **Design conditions.** We have discussed two sampling models for regression and correlation:

   (a) Random subsampling model: For each observed x, the corresponding observed y is viewed as randomly chosen from the conditional population distribution of Y values for that X.*

   (b) Bivariate random sampling model: Each observed pair (x, y) is viewed as randomly chosen from the joint population distribution of bivariate pairs (X, Y).

   In either sampling model, each observed pair (x, y) must be independent of the others. This means that the experimental design must not include any pairing, blocking, or hierarchical structure.

2. **Conditions concerning parameters.** The linear model states that

   (a) $\mu_{Y|X} = \beta_0 + \beta_1 X$.

   (b) $s_{Y|X}$ does not depend on X.

---

* If the X variable includes measurement error, then X in the linear model must be interpreted as the measured value of X rather than some underlying "true" value of X. A linear model involving the "true" value of X leads to a different kind of regression analysis.

3.   **Condition concerning population distributions.**   The confidence interval and t test are based on the condition that the conditional population distribution of Y for each fixed X is a normal distribution.

Other than the linearity condition, that $\mu_{Y|X} = \beta_0 + \beta_1 X$, these conditions can be summarized mnemonically with the letters SINR:

Same SD, $s_{Y|X}$, for all levels of X
Independent observations
Normal distribution of Y for each fixed X
Random sample

The random subsampling model is required if b_0, b_1, and $s_{Y|X}$ are to be viewed as estimates of the parameters β_0, β_1, and $\sigma_{Y|X}$ mentioned in the linear model. The bivariate random sampling model is required if r is to be viewed as an estimate of a population parameter ρ. It can be shown that, if the bivariate random sampling model is applicable, then the random subsampling model is also applicable. Thus, regression parameters can always be estimated if correlation can be estimated, but not vice versa.

Guidelines Concerning The Sampling Conditions

Departures from the sampling conditions not only affect the validity of formal techniques such as the confidence interval for β_1, but also can lead to faulty interpretation of the data even if no formal statistical analysis is performed. Two errors of interpretation that sometimes occur in practice are (1) failure to take into account dependency in the observations, and (2) insufficient caution in interpreting r when the x's do not represent a random sample.

The following two examples illustrate studies with dependent observations.

Example 12.34   **Serum Cholesterol and Serum Glucose.**   A data set consists of 20 pairs of measurements on serum cholesterol (X) and serum glucose (Y) in humans. However, the experiment included only two subjects; each subject was measured on ten different occasions. Because of the dependency in the data, it is not correct to naively treat all 20 data points alike. Figure 12.25 illustrates the difficulty; the figure shows that there is no evidence of any correlation between X and Y, except for the modest fact that the subject who has larger X values happens also to have larger Y values. Clearly it would be impossible to properly interpret the scatterplot if all 20 points were plotted with the same symbol. By the same token, application of regression or correlation formulas to the 20 observations would be seriously misleading.[20]   ∎

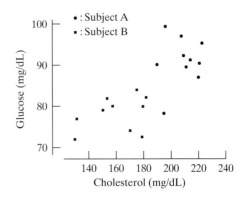

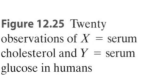

Figure 12.25 Twenty observations of X = serum cholesterol and Y = serum glucose in humans

Growth of Beef Steers. Figure 12.26 shows 20 pairs of measurements on the weight (Y) of beef steers at various times (X) during a feeding trial. The data represent four animals, each weighed at five different times; observations on the same animal are joined by lines in the figure. An ordinary regression analysis on the 20 data points would ignore the information carried in the lines and would yield inflated SEs and weak tests. Similarly, an ordinary scatterplot (without the lines) would be an inadequate representation of the data.[21] ■

Example 12.35

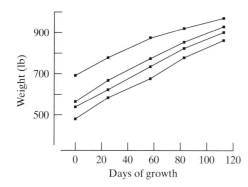

Figure 12.26 Twenty observations of X = Days and Y = Weight in steers. Data for individual animals are joined by lines.

In Example 12.34, ignoring the dependency in the observations would lead to *overinterpretation* of the data—that is, concluding that a relationship exists when there is actually very little evidence for it. By contrast, ignoring the dependency in Example 12.35 would lead to *underinterpretation* of the data—that is, insufficiently extracting the "signal" from the "noise."

In interpreting the correlation coefficient r, one should recognize that r is influenced by the degree of spread in the values of X. If the regression quantities b_0, b_1, and $s_{Y|X}$ are unchanged, *more spread in the X values leads to a stronger correlation (larger magnitude of r)*. The following example shows how this happens.

Figure 12.27 shows fictitious data that illustrate how r can be affected by the distribution of X. The data points in parts (a) and (b) have been plotted together in part (c). The regression line is the same in all three scatterplots, but notice that X and Y appear more highly correlated in (c) than in either (a) or (b). The residuals associated with the data points in (a) and (b) appear relatively smaller when viewed from the perspective of (c), with its expanded range of X. The contrasting appearance of the scatterplots is reflected in the correlation coefficients; in fact, $r = .5$ for (a), $r = .5$ for (b), but $r = .87$ for (c). ■

Example 12.36

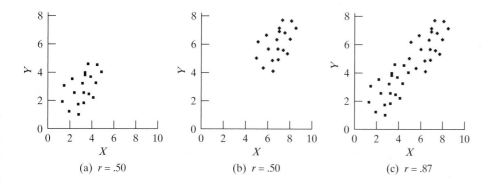

(a) $r = .50$ (b) $r = .50$ (c) $r = .87$

Figure 12.27 Dependence of r on the distribution of X. The data of (a) and (b) are plotted together in (c).

The fact that r depends on the distribution of X does not mean that r is invalid as a descriptive statistic. But it does mean that, when the values of X cannot be viewed as a random sample, r must be interpreted cautiously. For instance, suppose two experimenters conduct separate studies of response (Y) to various doses (X) of a drug. Each of them could calculate r as a description of her or his own data, but they should *not* expect to obtain *similar* values of r unless they both use the same choice of doses (X values). By contrast, they might reasonably expect to obtain similar regression lines and similar residual standard deviations, regardless of their choice of X values, as long as the dose-response relationship remains the same throughout the range of doses used.

Labeling X and Y. If the bivariate random sampling model is applicable, then the investigator is free to decide which variable to label X and which to label Y. Of course, for calculation of r the labeling does not matter. For regression calculations, the decision depends on the purpose of the analysis. The regression of Y on X yields (within the linear model) estimates of $\mu_{Y|X}$—that is, the population mean Y value for fixed X. Similarly, the regression of X on Y is aimed at estimating $\mu_{X|Y}$ —that is, the mean X value for fixed Y. These approaches do not lead to the same regression line because they are directed at answering different questions. The following is an intuitive example.

Example 12.37 | **Height and Weight of Young Men.** For the population of young men described in Example 12.11, the mean weight of young men $76''(6'4'')$ tall is 178 lb. Now consider this question: What would be the mean height of young men who weigh 178 lb? There is no reason that the answer should be $76''$. Intuition suggests that the answer should be less than $76''$—and in fact it is about $71''$. ∎

Guidelines Concerning the Linear Model and Normality Condition

The test and confidence interval for β_1 are based on the linear model and the condition of normality. The interpretation of these inferences can be seriously degraded if the linearity condition is not met; after all, we have seen earlier in this section that even the descriptive usefulness of regression is reduced if curvilinearity or outliers are present.

In addition to linearity, the linear model specifies that $\sigma_{Y|X}$ is the same for all the observations. A common pattern of departure from this assumption is a trend for larger means to be associated with larger SDs. Mild nonconstancy of the SDs does not seriously affect the interpretation of b_0, b_1, SE_{b_1}, and r (although it does invalidate the interpretation of $s_{Y|X}$ as a pooled estimate of a common SD).

Because of the Central Limit Theorem, the condition of normality is less important if n is large. The most troublesome form of nonnormality is the presence of long tails or, especially, outliers.

Residual Plots

Formal statistical tests for curvilinearity, unequal standard deviations, nonnormality, and outliers are beyond the scope of this book. However, the single most useful instrument for detecting these features is the human eye, aided by scatterplots. For instance, notice how easily the eye detects the mild curvilinearity in

Figure 12.21 and the outlier in Figure 12.22. Notice also in Figure 12.22 that examination of the marginal distributions of X and Y separately would not have revealed the outlier.

In addition to scatterplots of Y versus X, it is often useful to look at various displays of the residuals. A scatterplot of each residual $(y_i - \hat{y}_i)$ against $\hat{y}_i$ is called a **residual plot**. Residual plots are very useful for detecting curvature; they can also reveal trends in the conditional standard deviation. Figure 12.28 shows the data from Figure 12.21 together with a residual plot of those data.

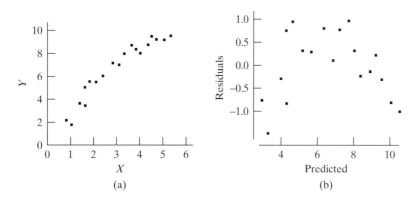

(a)

(b)

Figure 12.28 (a) Data displaying mild curvilinearity; (b) a residual plot of the data

A residual plot shows the data after the linear trend has been removed, which makes it easier to see nonlinear patterns in the data. The curvature in Figure 12.28(a) is apparent, but it is much more visible in the residual plot of Figure 12.28(b).

If the linear model holds, with no outliers, then the fitted regression line captures the trend in the data, leaving a random pattern in the residual plot. Thus, *we hope to see no striking pattern in a residual plot*. For example, Figure 12.29 shows a residual plot of the snake data of Example 12.3. The lack of unusual features in this plot supports the use of a regression model for these data.

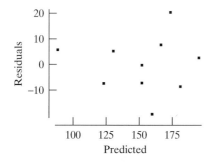

Figure 12.29 Residual plot of the snake data

If the condition of normality is met, then the distribution of the residuals should look roughly like a normal distribution.* A normal probability plot of the residuals provides a useful check of the normality condition. The normal probability plot of the snake data in Figure 12.30 is fairly linear, which supports the use of the t test and the confidence interval presented in Section 12.4.

---

* This is the basis for the 68% and 95% interpretations of $s_{Y|X}$ given in Section 12.2.

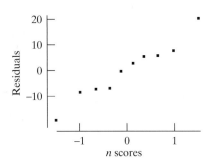

Figure 12.30 Normal probability plot of the snake data

The Use of Transformations

If the conditions of linearity, constancy of standard deviation, and normality are not met, a remedy that is sometimes useful is to transform the scale of measurement of either Y, or X, or both. The following example illustrates the use of a logarithmic transformation.

Example 12.38

Growth of Soybeans. A botanist placed 60 one-week-old soybean seedlings in individual pots. After 12 days of growth, she harvested, dried, and weighed 12 of the young soybean plants. She weighed another 12 plants after 23 days of growth, and groups of 12 plants each after 27 days, 31 days, and 34 days. Figure 12.31 shows the 60 plant weights plotted against days of growth; the group means are connected by solid lines. It is easy to see from Figure 12.31 that the relationship between mean plant weight and time is curvilinear rather than linear and that the conditional standard deviation is not constant but is strongly increasing.[22]

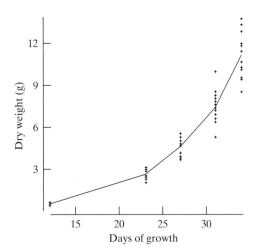

Figure 12.31 Weight of soybean plants plotted against days of growth

Figure 12.32 shows the logarithms (base 10) of the plant weights, plotted against days of growth; the means of the logarithms are connected by solid lines. Notice that the logarithmic transformation has simultaneously straightened the curve and more nearly equalized the standard deviations. It would not be unreasonable to assume that the linear model is valid for the variables $Y = \log(\text{dry weight})$ and $X = \text{days of growth}$. Table 12.6 shows the means and standard deviations before and after the logarithmic transformation. Note especially the effect of the transformation on the equality of the SDs. ∎

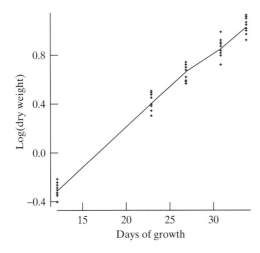

Figure 12.32 Log(weight) of soybean plants plotted against days of growth

TABLE 12.6 Summary of Soybean Growth Data in Original Scale and After Log Transformation

		Dry Weight (g)		Log(Dry Weight)	
Days of Growth	**Number of Plants**	*Mean*	*SD*	*Mean*	*SD*
12	12	.50	.06	−.31	.055
23	12	2.63	.37	.42	.062
27	12	4.67	.70	.67	.066
31	12	7.57	1.19	.87	.069
34	12	11.20	1.62	1.04	.064

Correlation and Causation

We noted in Chapter 8 that an observed association between two variables does not necessarily indicate any causal connection between them. It is important to remember this caution when interpreting correlation.

The following example shows that even strongly correlated variables may be causally unrelated.

Reproduction of an Alga.   Akinetes are sporelike reproductive structures produced by the green alga *Pithophora oedogonia*. In a study of the life cycle of the alga, researchers counted akinetes in specimens of alga obtained from an Indiana lake on 26 occasions over a 17-month period. Low counts indicated germination of the akinetes. The researchers also recorded the water temperature and the photoperiod (hours of daylight) on each of the 26 occasions.

The data showed a rather strong negative correlation between akinete counts and photoperiod; the correlation coefficient was $r = -.72$. But the researchers recognized that this observed correlation might not reflect a causal relationship. Longer days (increasing photoperiod) also tend to bring higher temperatures, and the akinetes might actually be responding to temperature rather than photoperiod. To resolve the question, the researchers conducted laboratory experiments in which temperature and photoperiod were varied independently; these experiments showed that temperature, not photoperiod, was the causal agent.[23]   ∎

Example 12.39

As Example 12.39 shows, one way to establish causality is to conduct a controlled experiment in which the putative causal factor is varied and all other factors are either held constant or controlled by randomization. When such an experiment is not possible, indirect approaches using statistical analysis can shed some light on causal relationships. (One such approach will be illustrated in Example 12.42.)

Exercises 12.37–12.43

12.37 In a metabolic study, four male swine were tested three times: when they weighed 30 kg, again when they weighed 60 kg, and again when they weighed 90 kg. During each test, the experimenter analyzed feed intake and fecal and urinary output for 15 days, and from these data calculated the nitrogen balance, which is defined as the amount of nitrogen incorporated into body tissue per day. The results are shown in the accompanying table.[24]

		NITROGEN BALANCE (g/day)		
Animal Number	**Body Weight**	*30 kg*	*60 kg*	*90 kg*
1		15.8	21.3	16.5
2		16.4	20.8	18.2
3		17.3	23.8	17.8
4		16.4	22.1	17.5
Mean		16.48	22.00	17.50

Suppose these data are analyzed by linear regression. With X = body weight and Y = nitrogen balance, preliminary calculations yield $\bar{x} = 60$ and $\bar{y} = 18.7$. The slope is $b_1 = .017$, with standard error $SE_{b_1} = .032$. The t statistic is $t_s = .53$, which is not significant at any reasonable significance level. According to this analysis, there is insufficient evidence to conclude that nitrogen balance depends on body weight under the conditions of this study.

The preceding analysis is flawed in two ways. What are they? (*Hint*: Look for ways in which the conditions for inference are not met. There may be several minor departures from the conditions, but you are asked to find two major ones. No calculation is required.)

12.38 For measuring the digestibility of forage plants, two methods can be used: The plant material can be fermented with digestive fluids in a glass container, or it can be fed to an animal. In either case, digestibility is expressed as the percentage of total dry matter that is digested. Two investigators conducted separate studies to compare the methods by submitting various types of forage to both methods and comparing the results. Investigator A reported a correlation of $r = .8$ between the digestibility values obtained by the two methods, and investigator B reported $r = .3$. The apparent discrepancy between these results was resolved when it was noted that one of the investigators had tested only varieties of canary grass (whose digestibilities ranged from 56% to 65%), whereas the other investigator had used a much wider spectrum of plants, with digestibilities ranging from 35% for corn stalks to 72% for timothy hay.[25]

Which investigator (A or B) used only canary grass? How does the different choice of test material explain the discrepancy between the correlation coefficients?

12.39 Refer to the energy expenditure data of Exercise 12.7. Each expenditure value (Y) is the average of two measurements made on different occasions. (See Example 1.8.) It might be proposed that it would be better to use the two measurements as separate data points, thus yielding 14 observations rather than 7. If this proposed approach were used, one of the assumptions for inference would be highly doubtful. Which one, and why?

12.40 Refer to the fungus growth data of Exercise 12.6.

(a) Calculate the correlation coefficient.
(b) Suppose a second investigator were to replicate the experiment, using concentrations of 0, 2, 4, 6, 8, and 10 mg, with two petri dishes at each concentration. Would you predict that the value of r calculated by this second investigator would be about the same as that calculated in part (a), smaller in magnitude, or larger in magnitude? Explain.

12.41 The three residual plots, (i), (ii), and (iii), were generated after fitting regression lines to the three scatterplots, (a), (b), and (c). Which residual plot goes with which scatterplot? How do you know?

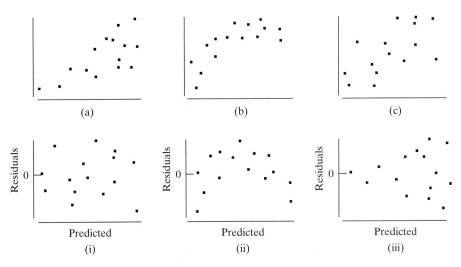

12.42 Sketch the residual plot that would be produced by fitting a regression line to the following scatterplot. One of the points is plotted with an "x." Indicate this point on the residual plot.

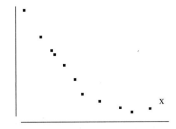

12.43 *(Computer exercise)* Researchers measured the diameters of 20 trees in a central Amazon rain forest and used ^{14}C-dating to determine the ages of these trees. The data are given in the following table.[26] Consider the use of diameter, X, as a predictor of age, Y.

(a) Make a scatterplot of Y = age versus X = diameter and fit a regression line to the data.
(b) Make a residual plot from the regression in part (a). Then make a normal probability plot of the residuals. How do these plots call into question the use of a linear model and regression inference procedures?

(c) Take the logarithm of each value of age. Make a scatterplot of $Y = \log(\text{age})$ versus $X = $ diameter and fit a regression line to the data.

(d) Make a residual plot from the regression in part (c). Then make a normal probability plot of the residuals. Based on these plots, does a regression model in log scale, from part (c), seem appropriate?

Diameter (cm)	Age (yr)	Diameter (cm)	Age (yr)
180	1372	115	512
120	1167	140	512
100	895	180	455
225	842	112	352
140	722	100	352
142	657	118	249
139	582	82	249
150	562	130	227
110	562	97	227
150	552	110	172

12.7 PERSPECTIVE

To put the methods of Chapter 12 in perspective, we will discuss their relationship to methods described in earlier chapters and to methods that might be included in a second statistics course. We begin by relating regression to the methods of Chapters 7 and 11.

Regression and the t Test

When there are several Y values for each of two values of X, we could analyze the data with a two-sample t test or with a regression analysis. Each approach uses the data to estimate the conditional mean of Y for each fixed X; these parameters are estimated by the fitted line $b_0 + b_1 x$ in the regression approach and by the individual sample means $\bar{y}$ in the t test approach. To test the null hypothesis of no dependence of Y on X, each approach translates the null hypothesis into its own terms. The following example illustrates the approaches.

Example 12.40

Toluene and the Brain. In Chapter 7 we analyzed data on norepinephrine (NE) concentrations in the brains of six rats exposed to toluene and of five control rats. The data are reproduced in Table 12.7.

TABLE 12.7 NE Concentrations (ng/g)		
	Toluene	**Control**
	543	535
	523	385
	431	502
	635	412
	564	387
	549	
n	6	5
$\bar{y}$	540.83	444.20
s	66.12	69.64

In Chapter 7 the null hypothesis

$$H_0: \mu_1 - \mu_2 = 0$$

was tested using the (unpooled) two-sample t test. The test statistic was

$$t_s = \frac{(540.83 - 444.20) - 0}{41.195} = 2.346$$

These data could be analyzed using a pooled t test (or, equivalently, with analysis of variance). The pooled variance is

$$s_{\text{pooled}}^2 = \frac{(6 - 1)66.12^2 + (5 - 1)69.64^2}{(6 + 5 - 2)} = 4584.24 = 67.71^2$$

and the pooled SE is

$$SE_{\text{pooled}} = 67.71\sqrt{\left(\frac{1}{6} + \frac{1}{5}\right)} = 41.00$$

This leads to a test statistic of

$$t_s = \frac{(540.83 - 444.20) - 0}{41.00} = 2.357$$

which is not much different than the unpooled t test result.

These data can also be analyzed with a regression model. To use regression, we define an **indicator variable**—a variable that indicates group membership—as follows. Let $X = 0$ for observations in the control group and let $X = 1$ for observations in the toluene group. Then we can present the data graphically with a scatterplot, as in Figure 12.33.

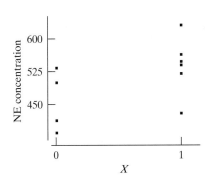

Figure 12.33 NE concentration data. $X = 0$ represents the control group; $X = 1$ represents the toluene group.

We can analyze the data in the scatterplot with the linear model

$$Y = \beta_0 + \beta_1 X + \varepsilon$$

which states that $\mu_{Y|X} = \beta_0 + \beta_1 X$.

The linear model states that for rats in the control group, the (population) mean NE concentration is given by

$$\mu_{Y|X=0} = \beta_0 + \beta_1(0) = \beta_0$$

For rats in the toluene group, NE concentration is given by

$$\mu_{Y|X=1} = \beta_0 + \beta_1(1) = \beta_0 + \beta_1$$

The difference between the two group means is β_1. The null hypothesis

$$H_0: \mu_{Y|X=0} - \mu_{Y|X=1} = 0$$

is equivalent to the null hypothesis

$$H_0: \beta_1 = 0$$

The fitted regression line is $Y = 444.2 + 96.63X$. Note that when $X = 0$, the fitted regression line gives a value of $Y = 444.2$, which is the sample mean of the control group. When $X = 1$, the fitted regression line gives a value of $Y = 444.2 + 96.63 = 540.83$, which is the sample mean of the toluene group. That is, the sample value of the slope is equal to the change in the sample means when going from the control group $(X = 0)$ to the toluene group $(X = 1)$, as shown in Figure 12.34.

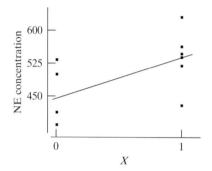

Figure 12.34 NE concentration data with regression line added

The test statistic for testing the hypothesis $H_0: \beta_1 = 0$ is

$$t_s = \frac{96.63}{41.0} = 2.36$$

This is identical to the pooled two-sample t test statistic found previously. (Note that the regression analysis assumes that $\sigma_{Y|X}$ is constant. Thus, regression is similar to the pooled t test, rather than the unpooled t test.) The following computer output shows the coefficients for the fitted regression line as well as the t statistic.

```
Dependent variable is: NE concentration
No Selector
R squared = 38.2%        R squared (adjusted) = 31.3%
s = 67.70 with   11 - 2 = 9 degrees of freedom
```

Source	Sum of Squares	df	Mean Square	F-ratio
Regression	25467.3	1	25467.3	5.56
Residual	41255.6	9	4583.96	

Variable	Coefficient	s.e. of Coeff	t-ratio	prob
Constant	444.200	30.28	14.7	≤0.0001
X	96.6333	41.00	2.36	0.0428

The following example compares the regression approach and the two-sample approach to a data set for which (unlike Example 12.40) X varies within as well as between the samples.

Blood Pressure and Platelet Calcium.   In Example 12.19 we described blood pressure (X) and platelet calcium (Y) measurements on 38 subjects. Actually, the study included two groups of subjects: 38 volunteers with normal blood pressure, selected from hospital lab personnel and other nonpatients, and 45 patients with a diagnosis of high blood pressure. Table 12.8 summarizes the platelet calcium measurements in the two groups and Figure 12.35 shows the blood pressure and calcium measurements for all 83 subjects.[12]

Example 12.41

TABLE 12.8 Platelet Calcium (nM) in Two Groups of Subjects

	Normal Blood Pressure	High Blood Pressure
$\bar{y}$	107.9	168.2
s	16.1	31.7
n	38	45

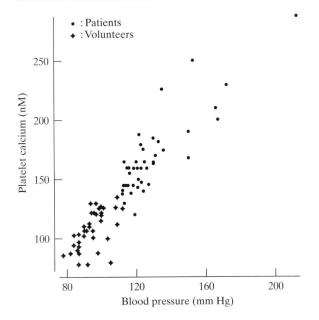

Figure 12.35 Blood pressure and platelet calcium for 83 subjects

Two ways to analyze the data are (1) as two independent samples; (2) by regression analysis. To test for a relationship between blood pressure and platelet calcium, (1) a two-sample t test of $H_0: \mu_1 = \mu_2$ can be applied to Table 12.8; (2) a regression t test of $H_0: \beta_1 = 0$ can be applied to the data in Figure 12.35. The two-sample t statistic (unpooled) is $t_s = 11.2$ and the regression t statistic is $t_s = 20.8$. Both of these are highly significant, but the latter is more so because the regression analysis extracts more information from the data.

For these data, the regression approach is more enlightening and convincing than the two-sample approach. Figure 12.35 suggests that platelet calcium is correlated with blood pressure, not only between but also within the two groups. Relevant regression analyses would include (1) testing for a correlation within each group separately (as in Examples 12.19 and 12.27); (2) testing for an overall correlation (as in the previous paragraph); (3) testing whether the regression lines in the two groups are identical (using methods not described in this book).

Formal testing aside, notice the advantage of the scatterplot as a tool for understanding the data and for communicating the results. Figure 12.35 provides eloquent testimony to the reality of the relationship between blood pressure and platelet calcium. (We emphasize once again, however, that a "real" relationship is not necessarily a causal relationship. Further, even if the relationship is causal, the data do not indicate the direction of causality—that is, whether high calcium causes high blood pressure or vice versa.*) ∎

Example 12.41 illustrates a general principle: If quantitative information on a variable X is available, it is usually better to use that information than to ignore it.

Extensions of Least Squares

We have seen that the classical method of fitting a straight line to data is based on the least-squares criterion. This versatile criterion can be applied to many other statistical problems. For instance, in **curvilinear regression**, the least-squares criterion is used to fit curvilinear relationships such as

$$Y = b_0 + b_1 X + b_2 X^2$$

Another application is **multiple regression and correlation**, in which the least-squares criterion is used to fit an equation relating Y to several X variables—X_1, X_2, and so on; for instance,

$$Y = b_0 + b_1 X_1 + b_2 X_2$$

The following example illustrates both curvilinear and multiple regression.

Example 12.42

Serum Cholesterol and Blood Pressure.   As part of a large health study, various measurements of blood pressure, blood chemistry, and physique were made on 2,599 men.[27] The researchers found a positive correlation between blood pressure and serum cholesterol ($r = .23$ for systolic blood pressure). But blood pressure and serum cholesterol also are related to age and physique. To untangle the relationships, the researchers used the method of least squares to fit the following equation:

$$Y = b_0 + b_1 X_1 + c_1 X_1^2 + b_2 X_2 + b_3 X_3 + b_4 X_4$$

where

Y = Systolic blood pressure
X_1 = Age
X_2 = Serum cholesterol
X_3 = Blood glucose
X_4 = Ponderal index (height divided by the cube root of weight)

Note that the regression is curvilinear with respect to age (X_1) and linear in the other X variables.

---

* In fact, the authors of the study remark that "It remains possible ... that an increased intracellular calcium concentration is a consequence rather than a cause of elevated blood pressure."

By applying multiple regression and correlation analysis, the investigators determined that there is little or no correlation between blood pressure and serum cholesterol, if age and ponderal index are held constant. They concluded that the observed correlation between serum cholesterol and blood pressure was an indirect consequence of the correlation of each of these with age and physique. ◼

Nonparametric and Robust Regression and Correlation

We have discussed the classical least-squares methods for regression and correlation analysis. There are also many excellent modern methods that are not based on the least-squares criterion. Some of these methods are *robust*—that is, they work well even if the conditional distributions of Y given X have long straggly tails or outliers. The nonparametric methods assume little or nothing about the form of dependence—linear or curvilinear—of Y on X, or about the form of the conditional distributions.

Analysis of Covariance

Sometimes regression ideas can add greatly to the power of a data analysis, even if the relationship between X and Y is not of primary interest. The following is an example.

Caterpillar Head Size. Can diet affect the size of a caterpillar's head? Such an effect is plausible, because a caterpillar's chewing muscles occupy a large part of the head. To study the effect of diet, a biologist raised caterpillars (*Pseudaletia unipuncta*) on three different diets: diet 1, an artificial soft diet; diet 2, soft grasses; and diet 3, hard grasses. He measured the weight of the head and of the entire body in the final stage of larval development. The results are shown in Figure 12.36,

Example 12.43

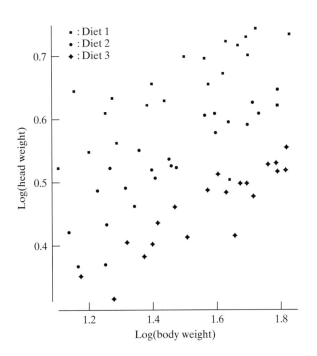

Figure 12.36 Head weight versus body weight (on logarithmic scales) for caterpillars on three different diets

where $Y = $ log(head weight) is plotted against $X = $ log(body weight), with different symbols for the three diets.[28] Note that the effect of diet is striking; there is very little overlap between the three groups of points. But if we were to ignore X and consider Y only, then the effect of diet would be much less clear; to see this, imagine projecting all the data points onto the Y axis. ∎

Example 12.43 shows how comparison of several groups with respect to a variable Y can be strengthened by using information on an auxiliary variable X that is correlated with Y. A classical method of statistical analysis for such data is **analysis of covariance**, which proceeds by fitting regression lines to the (x, y) data. But even without this formal technique, an investigator can often clarify the interpretation of data simply by constructing a scatterplot like Figure 12.35. Plotting the data against X has the effect of removing that part of the variability in Y which is accounted for by X, causing the treatment effect to stand out more clearly against the residual background variation.

Logistic Regression

Regression and correlation are used to analyze the relationship between two quantitative variables, X and Y. Sometimes data arise in which a quantitative variable X is used to predict the response of a categorical variable Y. For example, we might wish to use $X = $ cholesterol level as a predictor of whether or not a person has heart disease. Here we could define a variable Y as 1 if a person has heart disease and 0 otherwise. We could then study how Y depends on X. When the response variable is dichotomous, as in this case, a technique known as **logistic regression** can be used to model the relationship. For example, logistic regression could be used to model how the probability of heart disease depends on blood pressure.

Example 12.44 provides a more detailed look at the use of logistic regression.

Example 12.44

Esophageal Cancer. Esophageal cancer is a serious and very aggressive disease. Scientists conducted a study of 31 patients with esophageal cancer in which they studied the relationship between the size of the tumor that a patient had and whether or not the cancer had spread (metastasized) to the lymph nodes of the patient. In this study the response variable is dichotomous: $Y = 1$ if the cancer had spread to the lymph nodes and $Y = 0$ if not. The predictor variable is the size (recorded as the maximum dimension, in cm) of the tumor found in the esophagus. The data are given in Table 12.9 and plotted in Figure 12.37.[29]

The idea of logistic regression is to model the relationship between X and Y by fitting a response curve that is always between 0 and 1. Thus, unlike linear regression, in which we model Y as a linear function of X (which does not remain between 0 and 1), with logistic regression we model the relationship between X and Y as having an "S" shape, as shown in Figure 12.38.

One way to begin understanding the data is to form groups on the basis of size, X, and calculate for each group the proportion of the y values that are 1's. (This is somewhat analogous to finding the graph of averages described in Section 12.3, except that here we group together data points with differing x values.)

TABLE 12.9 Esophageal Cancer Data

Patient Number	Tumor Size (cm), X	Lymph Node Metastasis, Y	Patient Number	Tumor Size (cm), X	Lymph Node Metastasis, Y
1	6.5	1	17	6.2	1
2	6.3	0	18	2.0	0
3	3.8	1	19	9.0	1
4	7.5	1	20	4.0	0
5	4.5	1	21	3.0	1
6	3.5	1	22	6.0	1
7	4.0	0	23	4.0	0
8	3.7	0	24	4.0	0
9	6.3	1	25	4.0	0
10	4.2	1	26	5.0	1
11	8.0	0	27	9.0	1
12	5.2	1	28	4.5	1
13	5.0	1	29	3.0	0
14	2.5	0	30	3.0	1
15	7.0	1	31	1.7	0
16	5.3	0			

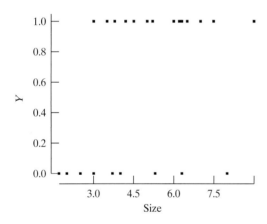

Figure 12.37 Lymph node metastasis, Y, as a function of tumor size, X

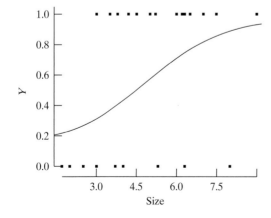

Figure 12.38 Lymph node metastasis, Y, as a function of tumor size, X, with smooth curve added

Table 12.10 provides such a summary, which is shown graphically in Figure 12.39. Note that the proportion of 1's (that is, the proportion of patients for whom the cancer has metastasized) increases as tumor size increases (except for the last category of 7.6–9.0, which only has three cases).

TABLE 12.10 Esophageal Cancer Data in Groups

Size Range	Points with $Y = 1$	Points with $Y = 0$	Fractio $Y = 1$	% $Y = 1$
1.5–3.0	2	4	2/6	.33
3.1–4.5	5	6	5/11	.45
4.6–6.0	4	1	4/5	.80
6.1–7.5	5	1	5/6	.83
7.6–9.0	2	1	2/3	.67

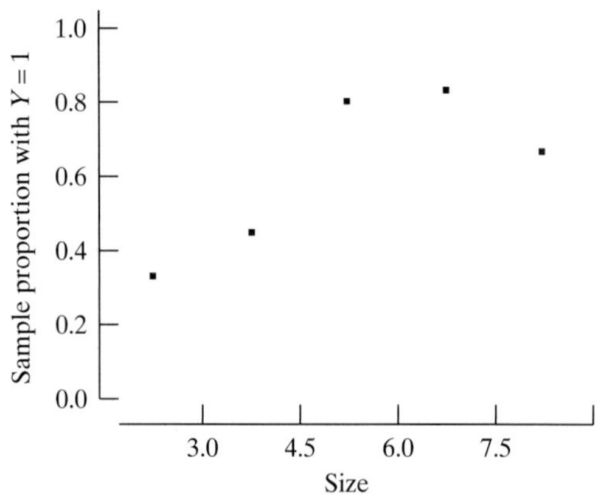

Figure 12.39 Sample proportion of patients with lymph node metastasis $(Y = 1)$ for patients grouped by tumor size, X

We can fit a smooth, continuous function to the data, to smooth out the percentages in the last column of Table 12.10. We can also impose the condition that the function be monotonically increasing, meaning that the probability of metastatis $(Y = 1)$ strictly increases as tumor size increases. To do this, we use a computer to fit a **logistic response function**.* The fitted logistic response function for the esophageal cancer data is

$$\Pr\{Y = 1\} = \frac{e^{-2.086 + .5117(\text{size})}}{1 + e^{-2.086 + .5117(\text{size})}}$$

For example, suppose the size of a tumor is 4.0 cm. Then the predicted probability that the cancer has metastasized is

$$\frac{e^{-2.086 + .5117(4)}}{1 + e^{-2.086 + .5117(4)}} = \frac{e^{-.0392}}{1 + e^{-.0392}} = \frac{.96156}{1 + .96156} = .49$$

---

* Fitting a logistic model is quite a bit more complicated than is fitting a linear regression model. A technique known as maximum likelihood estimation is commonly used, with the help of a computer.

On the other hand, suppose the size of a tumor is 8.0 cm. Then the predicted probability that the cancer has metastasized is

$$\frac{e^{-2.086+.5117(8)}}{1 + e^{-2.086+.5117(8)}} = \frac{e^{2.0076}}{1 + e^{2.0076}} = \frac{7.4454}{1 + 7.4454} = .88$$

We can calculate a predicted probability that $Y = 1$ for each value of X. Figure 12.40 shows a graph of such predictions, which have, generally speaking, an S shape. ∎

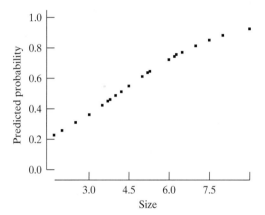

Figure 12.40 Predicted probability that $Y = 1$ as a function of tumor size, X

The S shape of the logistic curve is easier to see if we extend the range of X, as shown in Figure 12.41. As X grows, the logistic curve approaches, but never exceeds, 1. Likewise, if we were to extend the curve into the region where X is less than zero, we would see that as X gets smaller and smaller, the logistic curve approaches, but never drops below, 0. (Of course, in the setting of Example 12.44 it does not make sense to talk about tumor sizes that are negative. Thus, we only show the logistic curve for positive values of X.)

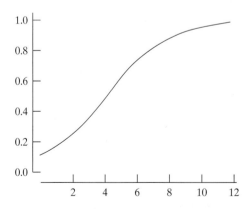

Figure 12.41 Logistic response function for the cancer data, shown over a larger range

In general, if we have a logistic response function

$$\Pr\{Y = 1\} = \frac{e^{b_0 + b_1(x)}}{1 + e^{b_0 + b_1(x)}}$$

with b_1 positive, then as X grows, $\Pr\{Y = 1\}$ approaches one and as X gets smaller, $\Pr\{Y = 1\}$ approaches zero. Thus, unlike a linear regression model, a logistic curve stays between zero and one, which makes it appropriate for modeling a response probability.

12.8 SUMMARY OF FORMULAS

For convenient reference, we summarize the formulas presented in Chapter 12.

Fitted Regression Line

$$Y = b_0 + b_1 X$$

where

$$b_1 = \frac{\Sigma(x_i - \bar{x})(y_i - \bar{y})}{\Sigma(x_i - \bar{x})^2}$$

$$b_0 = \bar{y} - b_1 \bar{x}$$

Residuals: $y_i - \hat{y}_i$ where $\hat{y}_i = b_0 + b_1 x_i$

Residual Sum of Squares:

$$\text{SS(resid)} = \Sigma(y_i - \hat{y}_i)^2$$

Residual Standard Deviation:

$$s_{Y|X} = \sqrt{\frac{\text{SS(resid)}}{n - 2}}$$

Correlation Coefficient

$$r = \frac{\Sigma(x_i - \bar{x})(y_i - \bar{y})}{\sqrt{\Sigma(x_i - \bar{x})^2 \Sigma(y_i - \bar{y})^2}}$$

Fact 12.1: $\dfrac{s_{Y|X}}{s_Y} \approx \sqrt{1 - r^2}$

Inference

Standard Error of b_1:

$$\text{SE}_{b_1} = \frac{s_{Y|X}}{\sqrt{\Sigma(x_i - \bar{x})^2}}$$

95% confidence interval for β_1:

$$b_1 \pm t_{.025}\, \text{SE}_{b_1}$$

Test of $H_0: \beta_1 = 0$ or $H_0: \rho = 0$:

$$t_s = \frac{b_1}{\text{SE}_{b_1}} = r\sqrt{\frac{n - 2}{1 - r^2}}$$

Critical values for the test and confidence interval are determined from Student's t distribution with df $= n - 2$.

Supplementary Exercises 12.44–12.62

12.44 In a study of the Mormon cricket (*Anabrus simplex*), the correlation between female body weight and ovary weight was found to be $r = .836$. The standard deviation of the ovary weights of the crickets was .429 g. Assuming that the linear model is applicable, estimate the standard deviation of ovary weights of crickets whose body weight is 4 g.[30]

12.45 In a study of crop losses due to air pollution, plots of Blue Lake snap beans were grown in open-top field chambers, which were fumigated with various concentrations of sulfur dioxide. After a month of fumigation, the plants were harvested and the total yield of bean pods was recorded for each chamber. The results are shown in the table.[31]

	X = **Sulfur Dioxide Concentration (ppm)**			
	0	*.06*	*.12*	*.30*
	1.15	1.19	1.21	.65
Y = yield (kg)	1.30	1.64	1.00	.76
	1.57	1.13	1.11	.69
Mean	1.34	1.32	1.11	.70

Preliminary calculations yield the following results:

$$\bar{x} = .12 \qquad\qquad \bar{y} = 1.117$$

$$\Sigma(x_i - \bar{x})^2 = .1512 \qquad \Sigma(y_i - \bar{y})^2 = 1.069067$$

$$\Sigma(x_i - \bar{x})(y_i - \bar{y}) = -.342$$

$$\text{SS(resid)} = .2955$$

(a) Calculate the linear regression of Y on X.
(b) Plot the data and draw the regression line on your graph.
(c) Calculate $s_{Y|X}$. What are the units of $s_{Y|X}$?

12.46 Refer to Exercise 12.45.

(a) Assuming that the linear model is applicable, find estimates of the mean and the standard deviation of yields of beans exposed to .24 ppm of sulfur dioxide.
(b) Which condition of the linear model appears doubtful for the snap bean data?

12.47 Refer to Exercise 12.45. Consider the null hypothesis that sulfur dioxide concentration has no effect on yield. Assuming that the linear model holds, formulate this as a hypothesis about the true regression line. Use the data to test the hypothesis against a directional alternative. Let $\alpha = .05$.

12.48 Another way to analyze the data of Exercise 12.45 is to take each treatment mean as the observation Y; then the data would be summarized as in the accompanying table.

Sulfur Dioxide X (ppm)	Mean Yield Y (kg)
0	1.34
.06	1.32
.12	1.11
.30	.70
Mean .12	1.117

$$\Sigma(x_i - \bar{x})^2 = .0504 \quad \Sigma(y_i - \bar{y})^2 = .264875$$
$$\Sigma(x_i - \bar{x})(y_i - \bar{y}) = -.114 \quad SS(resid) = .00702$$

(a) For the regression of mean yield on X, calculate the regression line and the residual standard deviation, and compare with the results of Exercise 12.36. Explain why the discrepancy is not surprising.

(b) Calculate the correlation coefficient between mean yield and X. Also, calculate the correlation coefficient between individual chamber yield and X. Explain why the discrepancy is not surprising.

12.49 In a study of the tufted titmouse (*Parus bicolor*), an ecologist captured seven male birds, measured their wing lengths and other characteristics, and then marked and released them. During the ensuing winter, he repeatedly observed the marked birds as they foraged for insects and seeds on tree branches. He noted the branch diameter on each occasion, and calculated (from 50 observations) the average branch diameter for each bird. The results are shown in the table.[32]

Bird	Wing Length X (mm)	Branch Diameter Y (cm)
1	79.0	1.02
2	80.0	1.04
3	81.5	1.20
4	84.0	1.51
5	79.5	1.21
6	82.5	1.56
7	83.5	1.29
Mean	81.4	1.26

$$\Sigma(x_i - \bar{x})^2 = 23.7143 \quad \Sigma(y_i - \bar{y})^2 = .265486$$
$$\Sigma(x_i - \bar{x})(y_i - \bar{y}) = 2.01571$$
$$SS(resid) = .09415$$

(a) Calculate the correlation coefficient between wing length and branch diameter.

(b) Calculate s_Y and $s_{Y|X}$. Specify the units for each. Verify the approximate relationship between s_Y and $s_{Y|X}$, and r.

(c) Construct a scatterplot of the data.

12.50 Refer to Exercise 12.49.

(a) Do the data provide sufficient evidence to conclude that the diameter of the forage branches chosen by male titmice is correlated with their wing length? Test an appropriate hypothesis against a nondirectional alternative. Let $\alpha = .05$.

(b) The test in part (a) was based on seven observations, but each branch diameter value was the mean of 50 observations. If we were to test the hypothesis of part (a) using the raw numbers, we would have 350 observations rather than only 7. Why would this approach not be valid?

12.51 Exericise 12.9 deals with data on the relationship between body length and jumping distance of bullfrogs. A third variable that was measured in that study was the mass of each bullfrog. The following table shows these data.[10]

Bullfrog	Mass X (g)	Maximum Jump Y (cm)
1	404	71
2	240	70
3	296	100
4	303	120
5	422	103.3
6	308	116
7	252	109.2
8	533.8	105
9	470	112.5
10	522.9	114
11	356	122.9
Mean	373.43	103.99

Preliminary calculations yield the following results:

$$\Sigma(x_i - \bar{x})^2 = 108{,}768.21 \qquad \Sigma(y_i - \bar{y})^2 = 3{,}218.99$$

$$\Sigma(x_i - \bar{x})(y_i - \bar{y}) = 3{,}406.54$$

$$\text{SS(resid)} = 3{,}112.3$$

(a) Calculate the linear regression of Y on X.
(b) Interpret the value of the slope of the regression line, b_1, in the context of this setting.
(c) Calculate and interpret the value of $s_{Y|X}$ in the context of this setting.

12.52 Consider the data from Exercise 12.51.

(a) Find the value of r^2.
(b) Interpret the value of r^2 found in part (a) in the context of this problem.

12.53 An exercise physiologist used skinfold measurements to estimate the total body fat, expressed as a percentage of body weight, for 19 participants in a physical fitness program. The body fat percentages and the body weights are shown in the table.[33]

Participant	Weight X (kg)	Fat Y (%)	Participant	Weight X (kg)	Fat Y (%)
1	89	28	11	57	29
2	88	27	12	68	32
3	66	24	13	69	35
4	59	23	14	59	31
5	93	29	15	62	29
6	73	25	16	59	26
7	82	29	17	56	28
8	77	25	18	66	33
9	100	30	19	72	33
10	67	23			

Actually, participants 1–10 are men, and participants 11–19 are women. Preliminary calculations yield the following results:

Men:

$$\bar{x} = 79.4 \qquad\qquad \bar{y} = 26.3$$

$$\Sigma(x_i - \bar{x})^2 = 1{,}578.40 \quad \Sigma(y_i - \bar{y})^2 = 62.100$$

$$\Sigma(x_i - \bar{x})(y_i - \bar{y}) = 292.80$$

Women:

$$\bar{x} = 63.1 \qquad\qquad \bar{y} = 30.7$$

$$\Sigma(x_i - \bar{x})^2 = 268.89 \quad \Sigma(y_i - \bar{y})^2 = 66.000$$

$$\Sigma(x_i - \bar{x})(y_i - \bar{y}) = 108.33$$

Both sexes:

$$\bar{x} = 71.7 \qquad\qquad \bar{y} = 28.4$$

$$\Sigma(x_i - \bar{x})^2 = 3{,}104.1 \quad \Sigma(y_i - \bar{y})^2 = 218.42$$

$$\Sigma(x_i - \bar{x})(y_i - \bar{y}) = 64.211$$

(a) Calculate the correlation coefficient between X and Y (i) for men; (ii) for women; and (iii) for all participants. The answers may surprise you.

(b) Draw a scatterplot with the points for men and women denoted by different symbols. After studying the scatterplot, try to sketch by eye the regression line of Y on X (pooling the sexes); it will be helpful to visually estimate the mean Y given X for a small X, an intermediate X, and a large X.

(c) Compute the regression line that you estimated in part (b) and draw it on the scatterplot.

(d) Using the insight gained from parts (b) and (c), can you explain the discrepancy between the correlation coefficients computed in part (a)? Discuss.

12.54 Refer to the respiration rate data of Exercise 12.8. Construct a 95% confidence interval for β_1.

12.55 The following plot is a residual plot from fitting a regression model to some data. Make a sketch of the scatterplot of the data that led to this residual plot. (*Note:* There are two possible scatterplots—one in which b_1 is positive and one in which b_1 is negative.)

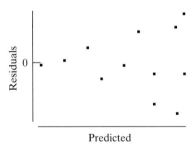

12.56 Biologists studied the relationship between embryonic heart rate and egg mass for 20 species of birds. They found that heart rate, Y, has a linear relationship with the logarithm of egg mass, X. The data are given in the following table.[34]

Species	Egg Mass (g)	Log(Egg Mass) X	Heart Rate Y (beats/min)
Zebra finch	.96	−.018	335
Bengalese finch	1.10	.041	404
Marsh tit	1.39	.143	363
Bank swallow	1.42	.152	298
Great tit	1.59	.201	348
Varied tit	1.69	.228	356
Tree sparrow	2.09	.320	335
Budgerigar	2.19	.340	314
House martin	2.25	.352	357
Japenese bunting	2.56	.408	370
Red-cheeked starling	4.14	.617	358
Cockatiel	5.08	.706	300
Brown-eared bulbul	6.4	.806	333
Domestic pigeon	17.1	1.233	247
Fantail pigeon	19.7	1.294	267
Homing pigeon	19.8	1.297	230
Barn owl	20.1	1.303	219
Crow	20.5	1.312	297
Cattle egret	27.5	1.439	251
Lanner falcon	41.2	1.615	242
Mean	9.94	.690	311

For these data the fitted regression equation is

$$Y = 368.06 - 82.452X$$

and SS(resid) = 15748.6.

(a) Interpret the value of the intercept of the regression line, b_0, in the context of this setting.

(b) Interpret the value of the slope of the regression line, b_1, in the context of this setting.

(c) Calculate $s_{Y|X}$ and specify the units.

(d) Interpret the value of $s_{Y|X}$ in the context of this setting.

12.57 *(Computer exercise)* The accompanying table gives two data sets: (a) and (b). The values of X are the same for both data sets and are only given once.

X	(a) Y	(b) Y	X	(a) Y	(b) Y
.61	.88	.96	2.56	1.97	1.20
.93	1.02	.97	2.74	2.02	3.59
1.02	1.12	.07	3.04	2.26	3.09
1.27	1.10	2.54	3.13	2.27	1.55
1.47	1.44	1.41	3.45	2.43	.71
1.71	1.45	.84	3.48	2.57	3.05
1.91	1.41	.32	3.79	2.53	2.54
2.00	1.59	1.46	3.96	2.73	3.33
2.27	1.58	2.29	4.12	2.92	2.38
2.33	1.66	2.51	4.21	2.96	3.08

(a) Generate scatterplots of the two data sets.

(b) For each data set (i) estimate r visually and (ii) calculate r.

(c) For data set (a), multiply the values of X by 10, and multiply the values of Y by 3 and add 5. Recalculate r and compare with the value before the transformation. How is r affected by the linear transformation?

(d) Find the equations of the regression lines and verify that the regression lines for the two data sets are virtually identical (even though the correlation coefficients are very different).

(e) Draw the regression line on each scatterplot.

(f) Construct a scatterplot in which the two data sets are superimposed, using different plotting symbols for each data set.

12.58 *(Computer exercise)*   This exercise shows the power of scatterplots to reveal features of the data that may not be apparent from the ordinary linear regression calculations. The accompanying table gives three fictitious data sets, A, B, and C. The values of X are the same for each data set, but the values of Y are different.[35]

Data set:	A	B	C
X	Y	Y	Y
10	8.04	9.14	7.46
8	6.95	8.14	6.77
13	7.58	8.74	12.74
9	8.81	8.77	7.11
11	8.33	9.26	7.81
14	9.96	8.10	8.84
6	7.24	6.13	6.08
4	4.26	3.10	5.39
12	10.84	9.13	8.15
7	4.82	7.26	6.42
5	5.68	4.74	5.73

(a) Verify that the fitted regression line is almost exactly the same for all three data sets. Are the residual standard deviations the same? Are the values of r the same?

(b) Construct a scatterplot for each of the data sets. What does each plot tell you about the appropriateness of linear regression for the data set?

(c) Plot the fitted regression line on each of the scatterplots.

12.59 *(Computer exercise)*   In a pharmacological study, 12 rats were randomly allocated to receive an injection of amphetamine at one of two dosage levels or an injection of saline. Shown in the table is the water consumption of each animal (mLi water per kg body weight) during the 24 hours following injection.[36]

Dose of Amphetamine (mLi/kg)		
0	*1.25*	*2.5*
122.9	118.4	134.5
162.1	124.4	65.1
184.1	169.4	99.6
154.9	105.3	89.0

(a) Calculate the regression line of water consumption on dose of amphetamine, and calculate the residual standard deviation.

(b) Construct a scatterplot of water consumption against dose.

(c) Draw the regression line on the scatterplot.

(d) Use linear regression to test the hypothesis that amphetamine has no effect on water consumption against the alternative that amphetamine tends to reduce water consumption. (Use $\alpha = .05$.)

(e) Use analysis of variance to test the hypothesis that amphetamine has no effect on water consumption. (Use $\alpha = .05$.) Compare with the result of part (d).

(f) What conditions are necessary for the validity of the test in part (d) but not for the test in part (e)?

(g) Calculate the pooled standard deviation from the ANOVA, and compare it with the residual standard deviation calculated in part (a).

12.60 *(Computer exercise)* Consider the Amazon tree data from Exercise 12.43. The researchers in this study were interested in how age, Y, is related to $X =$ "growth rate," where growth rate is defined as diameter/age (i.e., cm of growth per year).

(a) Create the variable "growth rate" by dividing each diameter by the corresponding tree age.

(b) Make a scatterplot of $Y =$ age versus $X =$ growth rate and fit a regression line to the data.

(c) Make a residual plot from the regression in part (b). Then make a normal probability plot of the residuals. How do these plots call into question the use of a linear model and regression inference procedures?

(d) Take the logarithm of each value of age and of each value of growth rate. Make a scatterplot of $Y =$ log (age) versus $X =$ log(growth rate) and fit a regression line to the data.

(e) Make a residual plot from the regression in part (d). Then make a normal probability plot of the residuals. Based on these plots, does a regression model in log scale, from part (d), seem appropriate?

12.61 *(Computer exercise)* Researchers measured the blood pressures of 22 students in two situations: when the students were relaxed and when the students were taking an important examination. The following table lists the systolic and diastolic pressures for each student in each situation.[37]

(a) Compute the change in systolic pressure by subtracting systolic pressure when relaxed from systolic pressure during the exam; call this variable X.

(b) Repeat part (a) for diastolic pressure. Call the resulting variable Y.

(c) Make a scatterplot of Y versus X and fit a regression line to the data.

(d) Make a residual plot from the regression in part (c).

(e) Note the outlier in the residual plot (and on the scatterplot from part (c)). Delete the outlier from the data set. Then repeat parts (c) and (d).

(f) What is the fitted regression model (after the outlier has been removed)?

During Exam		Relaxed	
Systolic Pressure (mm Hg)	*Diastolic Pressure (mm Hg)*	*Systolic Pressure (mm Hg)*	*Diastolic Pressure (mm Hg)*
132	75	110	70
124	170	90	75
110	65	90	65
110	65	110	80
125	65	100	55
105	70	90	60
120	70	120	80
125	80	110	60
135	80	110	70
105	80	110	70
110	70	85	65
110	70	100	60
110	70	120	80
130	75	105	75
130	70	110	70
130	70	120	80
120	75	95	60
130	70	110	65
120	70	100	65
120	80	95	65
120	70	90	60
130	80	120	70

12.62 *(Continuation of Exercise 12.61)*   Consider the data from Exercise 12.61, part (f).

(a) Construct a 95% confidence interval for β_1.

(b) Interpret the confidence interval from part (a) in the context of this setting.

CHAPTER 13

A Summary of Inference Methods

13.1 INTRODUCTION

In Chapters 6, 7, 9, 10, 11, and 12 we introduced many statistical methods for analyzing data and for making inferences. Statistics students are often overwhelmed by the number and variety of procedures that have been presented. What a statistician sees as a clearly arranged set of tools for analyzing data can appear as a blur to the novice. In this chapter we will review the methods presented in earlier chapters and provide some guidelines that are useful in deciding how to make an inference from a given set of data.

When presented with a set of data, it is useful to ask a series of questions:

1. *What question were the researchers attempting to answer when they collected these data?* Data analysis is done for a purpose: to extract information and to aid decision making. When looking at data, it helps to bear in mind the purpose for which the data were collected. For example, were the researchers trying to compare groups, perhaps patients given a new drug and patients given a placebo? Were they trying to see how two quantitative variables are related, so that they can use one variable to make predictions of the other? Were they checking whether a hypothesized model gives accurate predictions of the probabilities associated with a categorical variable? A good understanding of why the data were collected often clarifies the next question:

2. *What is the response variable in the study?* For example, if the researchers were concerned with the effect of a medication on blood pressure, then the likely response variable is $Y =$ change in blood pressure of an individual. If they were concerned with whether or not a medication cures an illness, then the response variable is categorical: yes if a person is cured, no if a person is not cured.

Objectives

In this chapter we summarize inference methods presented throughout the text. We will

- *learn how to choose an appropriate inference technique from among those presented in earlier chapters.*

- *consider several examples of choosing an inference method.*

3. *What predictor variables, if any, were involved?* For example, if a new drug is being compared to a placebo, then the predictor variable is group membership: A patient is either in the group that gets the new drug or else the patient is in the placebo group. If height is used to predict weight, then height is the predictor (and weight is the response variable). Sometimes there is no predictor variable. For example, a researcher might be interested in the distribution of cholesterol levels in adults. In this case, the response variable is cholesterol level, but there is no predictor variable. (One might argue that there is a predictor: whether or not someone is an adult. If we wished to compare cholesterol levels of adults to those of children, then whether or not someone is an adult would be a predictor. But if there is no comparison to be made, so that everyone in the study is part of the same group [adults], then it is not accurate to speak of a predictor *variable*, since group membership does not vary from person to person.)

The answers to these questions help frame the analysis to be conducted. Sometimes the analysis will be entirely descriptive and will not include any statistical inference, such as when the data are not collected by way of a random sample. Even when a statistical inference is called for, there is generally more than one way to proceed. Two statisticians analyzing the same set of data will often use somewhat different methods and may draw different conclusions. However, there are commonly used statistical procedures in various situations. The flowchart given in Figure 13.1 helps to organize the inference methods that have been presented in this book.

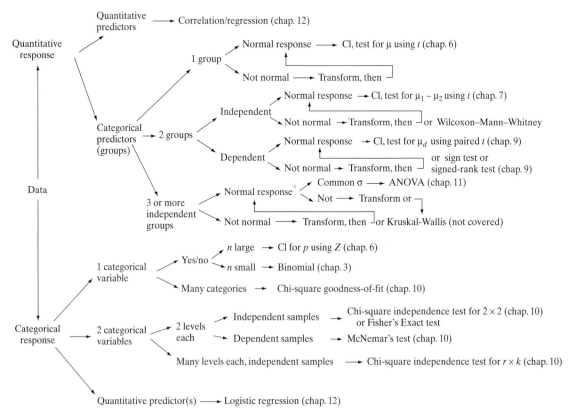

Figure 13.1 A flowchart of inference methods

To use this flowchart, we start by asking whether the response variable is quantitative or categorical. We then consider the type of predictor variables in the study and whether the samples collected are independent of one another or are dependent (e.g., matched pairs). Many of the methods, such as the confidence interval for an average presented in Chapter 6, depend on the data being from a population that has a normal distribution. (This condition is less important for large samples than it is for small samples, due to the Central Limit Theorem.) Nonnormal data can often be transformed to approximate normality and normal-based methods then applied. If such transformation fails to achieve approximate normality, then nonparametric methods, such as the Wilcoxon-Mann-Whitney test or the Wilcoxon signed-rank test, can be used.

Note that the flowchart only directs attention to the collection of inference methods presented in the previous chapters; this is not an exhaustive list. Beware of the Mark Twain fallacy: "When your only tool is a hammer, every problem looks like a nail." Not every statistical inference problem can be addressed with the methods presented here. In particular, these methods center on consideration of parameters, such as a population average, μ, or proportion, p. Sometimes researchers are interested in other aspects of distributions, such as the 75th percentile. When in doubt about how to proceed in an analysis, consult a statistician.

No matter what type of analysis is being considered, it is always a good idea to start by making one or more graphs of the data. The choice of graphics depends on the type of data being analyzed. For example, when comparing two samples of quantitative data, side-by-side dotplots or boxplots are informative—both as a visual comparison of the two samples and for assessing whether or not the data satisfy the normality condition. When analyzing categorical data, bar charts are useful. When dealing with two quantitative variables, scatterplots are helpful.

Bear in mind that a statistical analysis is intended to help us understand the scientific problem at hand. Thus, conclusions should be stated in the context of the scientific study. In Section 13.2 we present some examples of data sets and the kinds of analyses that might be performed on them.

13.2 DATA ANALYSIS EXAMPLES

In this section we consider several data sets and the kinds of analyses that are appropriate for each. The three questions stated in Section 13.1 and the flowchart given in Figure 13.1 provide a framework for the discussion of the following examples.

Gibberellic Acid.   Gibberellic acid (GA) is thought to elongate the stems of plants. Researchers conducted an experiment to investigate the effect of GA on a mutant strain of the genus *Brassica* called *ros*. They applied GA to 17 plants and applied water to 15 control plants. After 14 days they measured the growth of each of the 32 plants. For the 15 control plants the average growth was 26.7 mm, with an SD of 37.5 mm. For the 17 plants treated with GA the average growth was 92.6 mm, with an SD of 41.7 mm. The data are given in Table 13.1 and are graphed in Figure 13.2.[1]

Example 13.1

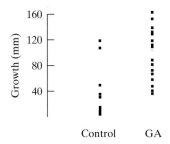

Figure 13.2 Dotplots of growth of *ros* plants (mm) after 14 days

TABLE 13.1 Growth of *Ros* Plants (mm) After 14 Days	
Control	**GA**
3	71
2	87
34	117
13	80
6	112
118	66
14	128
107	153
30	131
9	45
3	38
3	137
49	57
4	163
6	47
	108
	35
Mean 26.7	92.6
SD 37.5	41.7

Let us turn to the three questions stated in Section 13.1: (1) In this experiment, the researchers were trying to establish whether GA affects the growth rate of *ros*; (2) the response variable is 14-day growth of *ros*, which is quantitative; (3) the predictor variable is group membership (GA group or control group) and is categorical; the two groups are independent of one another.

The flowchart in Figure 13.1 directs us to consider a two-sample *t* test, if the data are normal or can be transformed to normality, or a Wilcoxon-Mann-Whitney test. Figure 13.3 shows that the distribution of the control sample of data is markedly nonnormal; thus, a transformation is called for.

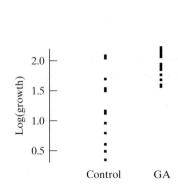

Figure 13.4 Dotplots of Log(growth) of *ros* plants (mm) after 14 days

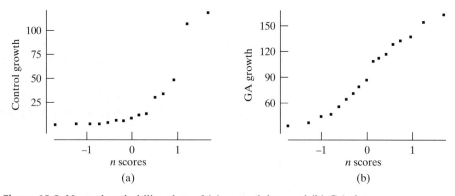

Figure 13.3 Normal probability plots of (a) control data and (b) GA data

Taking logarithms of each of the observations produces the dotplots and normal probability plots in Figures 13.4 and 13.5.

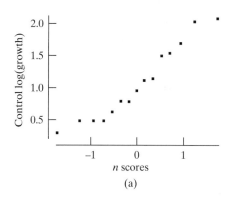
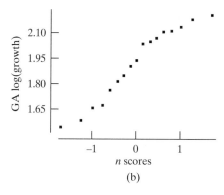

Figure 13.5 Normal probability plots of (a) control data and (b) GA data in log scale

In log scale, the normality condition is satisfied, so we can proceed with a two-sample t test. The standard deviations of the two samples are clearly quite different, as can be seem from Figure 13.4. However, an unpooled t test is still appropriate. The following computer output shows that $t_s = -5.392$ and the P-value is very small. Thus, we have strong evidence that GA increases growth of *ros*. ■

```
Test Ho:μ(Log(control))-μ(Log(GA)) = 0 vs
Ha:μ(Log(control))-μ(Log(GA)) ≠ 0
Difference Between Means = -0.8589   t-Statistic =
-5.392 w/17 df
Reject Ho at Alpha = 0.05
p ≤ 0.0001
```

Example 13.2

Whale Swimming Speed.   A biologist was interested in the relationship between the velocity at which a beluga whale swims and the tail-beat frequency of the whale. A sample of 19 whales were studied and measurements were made on swimming velocity, measured in units of body-lengths of the whale per second (so that a value of 1.0 means that the whale is moving forward by one body length, L, per second) and tail-beat frequency, measured in units of hertz (so that a value of 1.0 means one tail-beat cycle per second).[2] Here are the data:

Whale	Velocity (L/sec)	Frequency (Hz)	Whale	Velocity (L/sec)	Frequency (Hz)
1	0.37	0.62	11	0.68	1.20
2	0.50	0.675	12	0.86	1.38
3	0.35	0.68	13	0.68	1.41
4	0.34	0.71	14	0.73	1.44
5	0.46	0.80	15	0.95	1.49
6	0.44	0.88	16	0.79	1.50
7	0.51	0.88	17	0.84	1.50
8	0.68	0.92	18	1.06	1.56
9	0.51	1.08	19	1.04	1.67
10	0.67	1.14			

We could look at these data in two ways, either by asking "Does velocity depend on frequency?" or by asking "Does frequency depend on velocity?" The biologist conducting the study focused on the second question, for which the response variable is frequency, which is quantitative. The predictor is velocity, which is also quantitative. Thus, we can consider using regression analysis to study the relationship between velocity and frequency. Figure 13.6 is a scatterplot of the data, which shows an increasing trend in frequency as velocity increases.

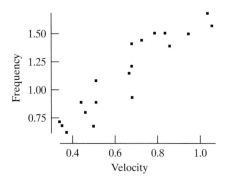

Figure 13.6 Scatterplot of frequency versus velocity

A regression model for these data is $Y = \beta_0 + \beta_1 X + \varepsilon$. Fitting the model to the data gives the equation $Y = 0.19 + 1.439X$, or Frequency = $0.19 + 1.439 \cdot$ Velocity, as shown in the following computer output. Figure 13.7 shows the residual plot for this fit. The fact that this plot does not have any patterns in it supports the use of the regression model.

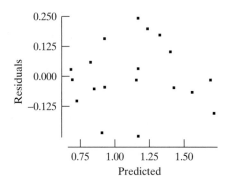

Figure 13.7 Residual plot for frequency regression fit

```
Dependent variable is: Frequency
No Selector
R squared = 85.3%          R squared (adjusted) = 84.4%
s = 0.1396  with   19 - 2 = 17   degrees of freedom
Source       Sum of Squares    df    Mean Square    F-ratio
Regression      1.91688         1      1.91688        98.4
Residual        0.331320        17     0.019489
Variable   Coefficient    s.e. of Coeff    t-ratio     prob
Constant     0.189513       0.1004          1.89      0.0763
Velocity     1.43935        0.1451          9.92      ≤0.0001
```

The null hypothesis

$$H_0: \beta_1 = 0$$

is tested with a t test, as shown in the regression output. A normal probability plot of the residuals, given in Figure 13.8, supports the use of the t test here, since it indicates that the distribution of the 19 residuals is consistent with what we would expect to see if the random errors came from a normal distribution. The t statistic has 17 degrees of freedom and a P-value of less than .0001. Thus, the evidence that frequency is related to velocity is quite strong; we reject the claim that the linear trend in the data arose by chance.

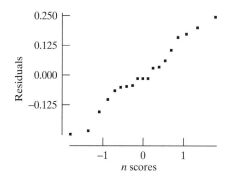

Figure 13.8 Normal probability plot of residuals for frequency regression fit

Continuing the analysis, the computer output shows that r^2 is 85.3%. Thus, in the sample 85.3% of the variability in frequency is accounted for by variability in velocity. (This is significantly different from zero, as indicated with the t test for $H_0: \beta_1 = 0$.) ∎

Tamoxifen. In a randomized, double-blind, experiment the drug tamoxifen was given to 6,681 women and a placebo was given to 6,707 other women. After four years there were 89 cases of breast cancer in the tamoxifen group, compared with 175 in the placebo group.[3]

The purpose of this experiment was to determine whether tamoxifen is effective in preventing cancer. The response variable is whether or not a woman developed cancer. The predictor variable is group membership (i.e., whether or not a woman was given tamoxifen). Figure 13.9 is a bar chart of the data, showing that cancer was much more common in the placebo group.

These data can be organized into a 2×2 contingency table, such as Table 13.2. A chi-square test of independence yields $\chi_S^2 = 28.2$. With 1 degree of freedom, the P-value for this test is nearly zero. There is very strong evidence that tamoxifen reduces the probability of breast cancer.

Example 13.3

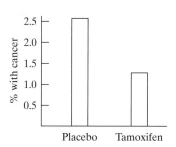

Figure 13.9
Bar chart of the tamoxifen data

TABLE 13.2 **Tamoxifen Data**			
	Treatment		
	Placebo	*Tamoxifen*	
Cancer	175	89	264
No Cancer	6532	6592	13124
Total	6707	6681	13388

We can also construct a confidence interval with these data. Of placebo patients, $\frac{175}{6707}$ or 2.61% developed cancer; thus $\hat{p}_1 = .0261$. Of tamoxifen patients, $\frac{89}{6681}$ or 1.33% developed cancer; thus $\hat{p}_2 = .0133$. The standard error of the difference is

$$\text{SE}_{(\hat{p}_1 - \hat{p}_2)} = \sqrt{\frac{.0261(.9739)}{6707} + \frac{.0133(.9867)}{6681}}$$
$$= .0024$$

A 95% confidence interval for $p_1 - p_2$ is $(.0261 - .0133) \pm 1.96(.0024)$ or $(.0081, .0175)$. Thus, we are 95% confident that tamoxifen reduces the probability of breast cancer by between .81% and 1.75%.

We can also calculate the relative risk of cancer. The estimated relative risk is

$$\frac{\hat{p}_1}{\hat{p}_2} = \frac{.0261}{.0133} = 1.96$$

Thus, we estimate that breast cancer is 1.96 times as likely when taking placebo as when taking tamoxifen. ∎

Example 13.4

Chromosome Puffs. Heat shock proteins (HSPs) are a type of protein produced by some organisms as protection against damage from exposure to high temperature. In the fruit fly *Drosophila melanogaster* the genes that encode HSPs are found on chromosomes that uncoil and appear to puff out. This chromosome puffing can be seen under a microscope. A biologist counted the number of puffs per chromosomal arm from the salivary glands of 40 Drosophila larvae that had been heat shocked at 37°C for 30 minutes, 40 larvae that had been heat shocked for 60 minutes, and 40 control larvae.

The purpose of this experiment was to determine the effect, if any, of heat shock on the HSPs. The response variable is the number of puffs on a chromosome arm, which is quantitative. The predictor variable is group membership (control, 30 minutes, or 60 minutes). Dotplots of the data are given in Figure 13.10; the data are summarized in Table 13.3.[4]

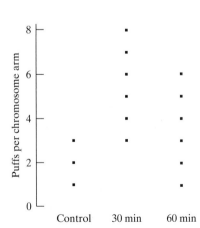

Figure 13.10 Dotplots of puffs per chromosome arm for Drosophila heat shock experiment

Damselflies. A res
them to one of three ;
the wing were artificia
wing spots were enlar
damselflies were ther
each of the three grou
in each of the three gr
enlarged with red ink'
and 57 survivors in the

The response v
variable. These data c
lyzed with a chi-squar

Tobacco Use Preven
school districts in the
of size, location, and p
the study. In each pair
group and the other v
the intervention grou
curriculum on preven
special training to hel]
later with the next ne
then followed for seve
whether or not studen

The experimen
as the response varia
The predictor is categ
groups, which are pai
pairs, there were 13 pa
trict and 7 pairs in whi
A sign test could be u

Exercises 13.1–13

13.1 Researchers co
patients with sc
haloperidol. Aft
ically importan
haloperidol gro
these data, but

13.2 Consider the da

13.3 A biologist col
(PEF—a measu
10 women.[14] He

Subject
1
2
3
4
5

TABLE 13.3 Puffs per Chromosome Arm for Drosophila Heat Shock Experiment			
Group	***n***	**Mean**	**SD**
Control	40	1.88	.76
30 min	40	5.20	1.54
60 min	40	3.45	1.18

The dotplots show an effect due to heat shock. This visual impression can be confirmed with an analysis of variance. Figure 13.11 contains histograms for the three groups. These plots show that the distributions take on only a few values each, so that the normality condition for ANOVA is not met. Nonetheless, the histograms show that the distributions are reasonably symmetric. Moreover, the sample sizes are moderately large and are equal. Under these conditions we can have confidence in the ANOVA P-value. The following ANOVA computer output confirms that there is strong evidence against $H_0: \mu_1 = \mu_2 = \mu_3$. We conclude that heat shock does, indeed, increase the number of puffs per chromosome arm.

```
Analysis of Variance For      Puffs
Source  df  Sums of Squares  Mean Square  F-ratio  Prob
Grp      2       221.317       110.658     76.757  ≤0.0001
Error  117       168.675         1.44167
Total  119       389.992
```

As an extension of the ANOVA, we could consider a contrast that compares the control mean to the average of the two heat shock means.

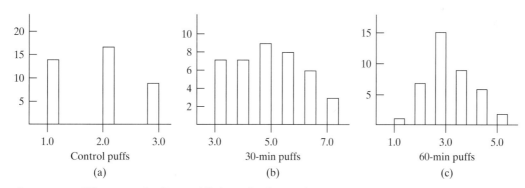

Figure 13.11 Histograms for Drosophila heat shock experiment

Therapeutic Touch. Therapeutic touch (TT) is a form of alternative medicine in which a practitioner manipulates the human energy field of the patient. However, many persons have questioned the ability of TT practitioners to detect the human energy field—and whether the human energy field even exists. An experimenter tested the abilities of 28 TT practitioners as follows. A screen was set up between the experimenter and the practitioner, who sat on opposite sides of a table. The practitioner extended his or her hands under the screen and rested them, palms up, on the table. The researcher tossed a coin to choose one of the practitioner's hands.

Example 13.5

Identify the type of statistical method that is appropriate for these data, but do not actually conduct the analysis.

13.4 A geneticist self-pollinated pink-flowered snapdragon plants and produced 97 progeny with the following colors: 22 red plants, 52 pink plants, and 23 white plants.[15] The purpose of this experiment was to investigate a genetic model that states that the probabilities of red, pink, and white are .25, .50, and .25. Identify the type of statistical method that is appropriate for these data, but do not actually conduct the analysis.

13.5 Consider the data of Exercise 13.4. Conduct an appropriate analysis of the data.

13.6 The effect of diet on heart disease has been widely studied. As part of this general area of investigation, researchers were interested in the short-term effect of diet on endothelial function, such as the effect on triglyceride level. To study this, they designed an experiment in which twenty healthy subjects were given, in random order, a high-fat breakfast and a low-fat breakfast at 8 A.M., following a 12-hour fast, on days one week apart from each other. Serum triglyceride levels were measured on each subject before each breakfast and again four hours after each breakfast.[16] If you had access to all of the measurements collected in this experiment, how would you analyze the data?

13.7 Biologists were interested in the distribution of trees in a wooded area. They intended to use the number of trees per 100-square-meter plot as their unit of measurement. However, they were concerned that the shapes of the plots might affect the data collection. To investigate the possibility, they counted the numbers of trees in square plots, round plots, and rectangular plots. The data are shown in the following table.[17] What type of analysis is appropriate for these data?

	Plot Shape	
Square	*Round*	*Rectangular*
5	5	10
5	7	2
5	5	3
8	2	12
8	4	9
7	4	5
4	4	3
9	7	6
9	7	5
7	10	3
5	9	8
2	2	9
8	7	3
Mean	6.3	5.6	6.0
SD	2.14	2.47	3.27

13.8 Consider the data of Exercise 13.7. Conduct an appropriate analysis of the data.

13.9 A sample of 15 patients were randomly split into two groups as part of a double-blind experiment to compare two pain relievers.[18] The 7 patients in the first group were given Demerol and reported the following numbers of hours of pain relief:

2, 6, 4, 13, 5, 8, 4

The 8 patients in the second group were given an experimental drug and reported the following numbers of hours of pain relief.

$$0, 8, 1, 4, 2, 2, 1, 3$$

How might these data be analyzed?

13.10 Consider the data of Exercise 13.9. Conduct an appropriate analysis of the data.

13.11 A researcher was interested in the relationship between forearm length and height. He measured the forearm lengths and heights of a sample of 16 women and obtained the data shown.[19] How might these data be analyzed?

Height (cm)	Forearm Length (cm)	Height (cm)	Forearm Length (cm)
163	25.5	157	26
161	26	178	27
151	25	163	24.5
163	25	161	26
166	27.2	173	28
168	26	160	24.5
170	26	158	25
163	26	170	26

13.12 A randomized, double-blind, clinical trial was conducted on patients who had coronary angioplasty to compare the drug lovastatin to a placebo. The percentage of stenosis (narrowing of the blood vessels) following angioplasty was measured on 160 patients given lovastatin and on 161 patients given the placebo. For the lovastatin group the average was 46%, with an SD of 20%. For the placebo group the average was 44%, with an SD of 21%.[20] What type of analysis is appropriate for these data?

13.13 Consider the data of Exercise 13.12. Conduct an appropriate analysis of the data.

13.14 Researchers studied persons who had received intravenous immune globulin (IGIV) to see if they had developed infections of hepatitis C virus (HCV). In part of their analysis, they considered doses of Gammagard (an IGIV product) received by 210 patients. They divided the patients into 4 groups according to the number of doses of "Gammagard made from unscreened or first-generation anti-HCV-screened plasma." Among 48 persons who received 0 to 3 doses, there were 4 cases of HCV infection. There were 2 cases of HCV infection among 45 persons who received 4 to 20 doses, there were 7 cases of HCV infection in the 57 persons who received between 21 and 65 doses, and there were 10 cases of HCV infection among the 51 persons who received more than 65 doses.[21] What type of analysis is appropriate for these data?

13.15 Consider the data of Exercise 13.14. Conduct an appropriate analysis of the data.

13.16 An experiment was conducted to study the effect of tamoxifen on patients with cervical cancer. One of the measurements made, both before and again after tamoxifen was given, was microvessel density (MVD). MVD, which is measured as number of vessels per mm^2, is a measurement that relates to the formation of blood vessels that feed a tumor and allow it to grow and spread. Thus, small values of MVD are better than are large values. Data for 18 patients are shown.[22] How might these data be analyzed?

Patient	MVD Before	MVD After	Patient	MVD Before	MVD After
1	98	75	10	70	60
2	100	60	11	60	65
3	82	25	12	88	45
4	100	55	13	45	36
5	93	78	14	159	144
6	119	102	15	65	27
7	70	58	16	98	90
8	78	70	17	66	16
9	104	90	18	67	53

13.17 Consider the data of Exercise 13.16. Conduct an appropriate analysis of the data.

13.18 As part of a large experiment, researchers planted 2,400 sweetgum, 2,400 sycamore, and 1,200 green ash seedlings. After 18 years the survival rates were 93% for the sweetgum trees, 88% for the sycamore trees, and 95% for the green ash trees.[23] What type of analysis is appropriate for these data?

13.19 Consider the data of Exercise 13.18. Conduct an appropriate analysis of the data.

13.20 A group of female college students were divided into three groups according to upper body strength. Their leg strength was tested by measuring how many consecutive times they could leg press 246 pounds before exhaustion. (The subjects were allowed only one second of rest between consecutive lifts.) The data are shown in the following table.[24] What type of analysis is appropriate for these data?

	Upper Body Strength Group		
	Low	*Middle*	*High*
	55	40	181
	70	200	85
	45	250	416
	246	192	228
	240	117	257
	96	215	316
	225		134
Mean	140	169	231
SD	93	77	112

13.21 Consider the data of Exercise 13.20. Conduct an appropriate analysis of the data.

Chapter Appendices

APPENDIX 3.1

Generating Pseudorandom Numbers

The following is a simple method for calculating a sequence of pseudorandom numbers.

(a) Arbitrarily choose any number between 0 and 1; this number is called the **seed** of the sequence. Let the seed be denoted by u_0.

(b) Calculate u_1 according to the formula

$$u_1 = \text{Fractional part of } \left[(\pi + u_0)^5 \right]$$

where $\pi = 3.1415927 \ldots$. (The fractional part of a number is the part to the right of the decimal point; thus, the fractional part of 27.403911 is .403911.)

(c) Continuing in the same way, calculate $u_2, u_3, \ldots$, as follows:

$$u_2 = \text{Fractional part of } \left[(\pi + u_1)^5 \right]$$
$$u_3 = \text{Fractional part of } \left[(\pi + u_2)^5 \right]$$

and so on.

(d) Each u (except u_0) is a pseudorandom number between 0 and 1. If you plotted the values of a long sequence of u's you would find that they would be more or less uniformly distributed between 0 and 1.

(e) The digits of $u_1, u_2, u_3, \ldots$, are pseudorandom digits. They can be read singly or in pairs or in triplets, and so on, as explained in Chapter 3.

Remarks:

(1) We do not give an example of a sequence of u's generated by this method, because the sequence may vary from one type of calculator to the next, even using the same seed. This occurs because the u's are highly sensitive to how many significant figures are used in the computations. To see this great sensitivity, try using $\pi = 3.141593$ instead of 3.1415927 and notice how much the sequence of u's changes. This instability of the u's is not a drawback of the method; on the contrary, the instability enhances the pseudorandom properties of the sequence.

(2) There are many other methods of generating pseudorandom numbers. If computer simulation is used to study a complex system, then it is important that the pseudorandom numbers used for the simulation have good properties (i.e., that they are uniformly distributed and are independent of one another). The method presented here is intended only as an example of random number generation, in order to convey the general concept.

More on the Binomial Distribution Formula

In this appendix we explain more about the reasoning behind the binomial distribution formula.

The Binomial Distribution Formula

We begin by deriving the binomial distribution formula for $n = 3$. Suppose that we conduct three independent trials and that each trial results in success (S) or failure (F). Further, suppose that on each total the probabilities of success and failure are

$$\Pr\{S\} = p$$
$$\Pr\{F\} = 1 - p$$

There are eight possible outcomes of the three trials. Reasoning as in Example 3.28 shows that the probabilities of these outcomes are as follows:

Outcome	Probability
FFF	$(1 - p)^3$
FFS	$p(1 - p)^2$
FSF	$p(1 - p)^2$
SFF	$p(1 - p)^2$
FSS	$p^2(1 - p)$
SFS	$p^2(1 - p)$
SSF	$p^2(1 - p)$
SSS	p^3

Again by reasoning parallel to Example 3.28, these probabilities can be combined to obtain the binomial distribution formula for $n = 3$, as shown in the table:

Number of Successes, j	Failures, $n - j$	Probability
0	3	$1p^0(1 - p)^3$
1	2	$3p^1(1 - p)^2$
2	1	$3p^2(1 - p)^1$
3	0	$1p^3(1 - p)^0$

This distribution illustrates the origin of the binomial coefficients. The coefficient $_3C_1(\, = 3)$ is the number of ways in which 2 S's and 1 F can be arranged; the coefficient $_3C_2(\; = 3)$ is the number of ways in which 1 S and 2 F's can be arranged.

An argument similar to the preceding shows that the general formula (for any n) is

$$\Pr\{j \text{ successes and } n - j \text{ failures}\} = {}_nC_j\, p^j(1 - p)^{n-j}$$

where

$_nC_j =$ the number of ways in which j S's and $(n - j)$ F's can be arranged.

The Binomial Coefficients: Connections

The binomial coefficients are related to other ideas that may be familiar.

The Binomial Expansion. The binomial coefficients appear in the algebraic identity known as the **binomial expansion**. If a and b are any numbers, then the binomial expansion for the quantity $(a + b)^n$ is

$$(a + b)^n = {}_nC_0\, a^n + {}_nC_1\, a^{n-1}b + {}_nC_2\, a^{n-2}b^2 + \cdots + {}_nC_n\, b^n$$

The most familiar special case is the binomial expansion for $n = 2$:

$$(a + b)^2 = a^2 + 2ab + b^2$$

Combinations. The binomial coefficient $_nC_j$ is also known as the number of combinations of n items taken j at a time; it is equal to the number of different subsets of size j that can be formed from a set of n items.

Pascal's Triangle. Pascal's triangle is a triangular array of numbers in which the borders are 1's, and each interior entry is the sum of the two entries above it. The first seven rows of Pascal's triangle are shown here:

$$
\begin{array}{ccccccccccccc}
& & & & & & 1 & & 1 & & & & \\
& & & & & 1 & & 2 & & 1 & & & \\
& & & & 1 & & 3 & & 3 & & 1 & & \\
& & & 1 & & 4 & & 6 & & 4 & & 1 & \\
& & 1 & & 5 & & 10 & & 10 & & 5 & & 1 \\
& 1 & & 6 & & 15 & & 20 & & 15 & & 6 & & 1 \\
1 & & 7 & & 21 & & 35 & & 35 & & 21 & & 7 & & 1
\end{array}
$$

If you compare this array with Table 2, you will see that the numbers in Pascal's triangle are the binomial coefficients. Indeed, Pascal used the triangle to help solve a probability problem that involved the binomial distribution.

The Binomial Coefficients: A Formula

Binomial coefficients can be calculated from the formula

$$_nC_j = \frac{n!}{j!(n - j)!}$$

where $x!$ ("x-factorial") is defined for any positive integer x by

$$x! = x(x - 1)(x - 2)\cdots(2)(1)$$

and $0! = 1$.

For example, for $n = 7$ and $j = 4$ the formula gives

$$_7C_4 = \frac{7!}{4!3!} = \frac{7 \cdot 6 \cdot 5 \cdot 4 \cdot 3 \cdot 2 \cdot 1}{(4 \cdot 3 \cdot 2 \cdot 1)(3 \cdot 2 \cdot 1)}$$

$$= 35$$

To see why this is correct, let us consider in detail why the number of ways of rearranging 4 S's and 3 F's should be equal to

$$\frac{7!}{4!3!}$$

Suppose 4 S's and 3 F's were written on cards, like this:

$$\boxed{S_1} \quad \boxed{S_2} \quad \boxed{S_3} \quad \boxed{S_4} \quad \boxed{F_1} \quad \boxed{F_2} \quad \boxed{F_3}$$

Temporarily we put subscripts on the S's and F's to distinguish them. First, let us see how many ways there are to arrange the 7 cards in a row:

There are 7 choices for which card goes first;
for each of these, there are 6 choices for which card goes second;
for each of these, there are 5 choices for which card goes third;
for each of these, there are 4 choices for which card goes fourth;
for each of these, there are 3 choices for which card goes fifth;
for each of these, there are 2 choices for which card goes sixth;
for each of these, there is 1 choice for which card goes last.

It follows that there are 7! ways of arranging the 7 cards. Now consider the locations of the 4 S's. There are 4! ways in which the S's can be rearranged among themselves. Likewise, there are 3! ways in which the F's can be rearranged among themselves. If we were to ignore the subscripts on the S's and F's, then some of the 7! ways of arranging the 7 cards would be indistinguishable. Indeed, any rearrangement of the S's *among themselves* leaves the 7 card arrangement looking the same. Similarly, any rearrangement of the F's *among themselves* leaves the 7 card arrangement looking the same. Thus, the number of *distinguishable* arrangements is

$$\frac{7!}{4!3!}$$

APPENDIX 3.3

Mean and Standard Deviation of the Binomial Distribution

Suppose that Y is a binomial random variable with n trials and p as the probability of success on each trial. Then we can think of Y as the sum of n variables X_1, $X_2, \ldots, X_n$, where each X_i is equal to either 0 or 1—0 for a failure or 1 for a success. That is, $Y = \Sigma X_i$, with $\Pr\{X_i = 0\} = 1 - p$ and $\Pr\{X_i = 1\} = p$. The n X_i's are a random sample from a hypothetical population of X's that has average $\mu_X = p$ (since $0 \cdot (1 - p) + 1 \cdot (p) = p$).

Now consider the population standard deviation, σ_X, for the population of X's. Recall, from Section 2.8 that for a variable X the definition of σ is

$$\sigma = \sqrt{\text{Population average value of } (X - \mu)^2}$$

For the population of X's, the mean is $\mu_X = p$. Thus, for this population,

$$\sigma_X = \sqrt{\text{Population average value of } (X - p)^2}$$

In the population of X's, the quantity $(X - p)^2$ takes on only two possible values:

$$(X - p)^2 = \begin{cases} (0 - p)^2 & \text{if } X = 0 \\ (1 - p)^2 & \text{if } X = 1 \end{cases}$$

Furthermore, these values occur in the proportions $(1 - p)$ and p, respectively, so that the population average value of $(X - p)^2$ is equal to

$$(0 - p)^2 \cdot (1 - p) + (1 - p)^2 \cdot p$$

This can be simplified to

$$p^2 \cdot (1 - p) + (1 - p)^2 \cdot p = p \cdot (1 - p) \cdot \{p + (1 - p)\} = p \cdot (1 - p) \cdot \{1\}$$
$$= p(1 - p).$$

Hence, the population average value of $(X - p)^2$ is $p(1 - p)$, so $\sigma_X = \sqrt{p(1 - p)}$.

The binomial random variable Y is ΣX_i. To find the mean and standard deviation of Y, we need two facts:

Fact 1. For any collection of random variables $X_1, X_2, \ldots, X_n$ the mean of $\Sigma X_i = \Sigma$ (mean of X_i).

Fact 2. For a collection of independent random variables $X_1, X_2, \ldots, X_n$ the variance of $\Sigma X_i = \Sigma$ (variance of X_i).

(Recall that the variance, σ^2, is the square of the standard deviation, σ.)

Using Fact 1, we see that the mean of Y is the mean of ΣX_i, which is Σp. Thus, the mean of Y is $\mu_Y = np$.

Using Fact 2, the variance of Y is the variance of ΣX_i, which equals Σ(Variance of X_i) or $np(1 - p)$. Thus, the standard deviation of Y is $\sigma_Y = \sqrt{np(1 - p)}$.

Areas of Indefinitely Extended Regions

Consider the region bounded between a normal curve and the horizontal axis. Because the curve never touches the axis, the region extends indefinitely far to the left and to the right. Yet the area of the region is exactly equal to 1.0. How is it possible for an indefinitely extended region to have a finite area?

To gain insight into this paradoxical situation, consider Figure A.1, which shows a region that is simpler than that bounded by a normal curve. In this region, the width of each bar is 1.0; the height of the first bar is $\frac{1}{2}$, the second bar is half as high as the first, the third is half as high as the second, and so on. The bars form a region that is indefinitely extended. Nevertheless, we shall see that it makes sense to say that the area of the region is equal to 1.0.

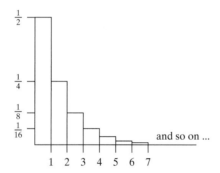

Figure A.1

Let us first consider the areas of the individual bars. The area of the first bar is $\frac{1}{2}$, the area of the second bar is $\frac{1}{4}$, the third $\frac{1}{8}$, and so on. Now suppose that we choose a number, say k, and add up the areas of the first k bars, as follows:

Bar	Height of Bar	Cumulative Total Area
1	$\frac{1}{2}$	$\frac{1}{2}$
2	$\frac{1}{4}$	$\frac{3}{4}$
3	$\frac{1}{8}$	$\frac{7}{8}$
4	$\frac{1}{16}$	$\frac{15}{16}$
$\vdots$	$\vdots$	$\vdots$
k	$\dfrac{1}{2^k}$	$\dfrac{2^k - 1}{2^k}$

The total area of the first two bars is $\frac{3}{4}$, the total area of the first three bars is $\frac{7}{8}$, and so on. In fact, the total area of the first k bars is equal to

$$\frac{2^k - 1}{2^k} = 1 - \frac{1}{2^k}$$

If k is very large, this area is very close to 1.0. In fact, we can make the area as close to 1.0 as we wish, simply by choosing k large enough. In these circumstances it is reasonable to say that the total area of the entire, indefinitely extended region is equal to exactly 1.0.

The preceding example shows that an indefinitely extended region can have a finite area. Likewise, the total area under the normal curve is 1.0 (but the proof of this fact requires fairly advanced calculus).

Relationship Between Central Limit Theorem and Normal Approximation to Binomial Distribution

Consider sampling from a dichotomous population. Theorem 5.2 (Section 5.5) states that the sampling distribution $\hat{p}$, and the equivalent binomial distribution, can be approximated by normal distributions. In this appendix we show how these approximations are related to Theorem 5.1 and the Central Limit Theorem (Section 5.3).

As shown in Appendix 3.3, if Y is a binomial random variable with n trials and p as the probability of success on each trial, then we can think of Y as the sum of n variables $X_1, X_2, \ldots, X_n$, where each X_i is equal to either 0 or 1—0 for a failure or 1 for a success. For a population of 0's and 1's, where the proportion of 1's is given by p, the mean is p and the standard deviation is $\sigma = \sqrt{p(1-p)}$. The sample mean of $X_1 X_2, \ldots, X_n$ is $\overline{X}$, which is the same as the proportion of 1's in the sample (that is, $\hat{p}$). Thus, the sample proportion $\hat{p}$ can be regarded as a sample mean, and so its sampling distribution is described by Theorem 5.1.

From part 3 of Theorem 5.1 (the Central Limit Theorem), the sampling distribution of $\hat{p}$ is approximately normal if n is large. From part 1 of Theorem 5.1, the mean of the sampling distribution of $\hat{p}$ is equal to the population mean—that is, p; this value is given in Theorem 5.2(b). From part 2 of Theorem 5.1, the standard deviation of the sampling distribution of $\hat{p}$ is equal to

$$\frac{\sigma}{\sqrt{n}}$$

where σ represents the standard deviation of the population of 0's and 1's, which is $\sqrt{p(1-p)}$. Thus, the standard deviation of the sampling distribution of $\hat{p}$ is equal to

$$\frac{\sqrt{p(1-p)}}{\sqrt{n}} = \sqrt{\frac{p(1-p)}{n}}$$

which is the value given in Theorem 5.2(b).

Note that the binomial distribution is just a rescaled version of the sampling distribution of $\hat{p}$: $\hat{p} = Y/n$, so $Y = n\hat{p}$. It follows that the binomial distribution also can be approximated by a normal curve with suitably rescaled mean and standard deviation. The mean of $\hat{p}$ is p and the SD of $\hat{p}$ is $\sqrt{p(1-p)/n}$. The rescaled mean is np and the rescaled standard deviation is

$$n\sqrt{\frac{p(1-p)}{n}} = \sqrt{np(1-p)}$$

which are as given in Theorem 5.2(a).

APPENDIX 6.1

Significant Digits

In this appendix we review the concept of significant digits. Let us begin with an example.

Suppose a university president reports that there are 23,000 students at the university. How many significant digits are in the number

23,000?

When the number is expressed this way—in ordinary rather than scientific notation—it is not really possible to tell for sure how many significant digits it has. Does the president *really* mean

23,000 rather than 23,001 or 22,999?

If she does, then all five of the digits are significant. If (as is probable) she really means

23,000 rather than 22,000 or 24,000

then only the 2 and the 3 are significant digits, since only those digits are known with certainty; the three 0's in 23,000 are place-holders. Scientific notation removes the ambiguity:

$2.3 \cdot 10^4$ has 2 significant digits
$2.3000 \cdot 10^4$ has 5 significant digits

As the preceding example illustrates, you can clarify how many significant digits are in a number by expressing the number in scientific notation. Here are some examples:

Ordinary Notation	Scientific Notation	Number of Significant Digits
60,700	$6.07 \cdot 10^4$	3
60,700	$6.0700 \cdot 10^4$	5
60.7	$6.07 \cdot 10^1$	3
60.70	$6.070 \cdot 10^1$	4
.0607	$6.07 \cdot 10^{-2}$	3
.06070	$6.070 \cdot 10^{-2}$	4

In the preceding numbers, note that the interior zero (between 6 and 7) is always a significant digit; the leading zeros (before the 6) are not significant; the terminal zeros (after the 7) are significant in scientific notation and ambiguous in ordinary notation. Digits other than zero are always significant.

Here are some examples of rounding a number to two significant digits:

Number	Rounded to Two Significant Digits
60,700	61,000 (that is, 6.1×10^4)
60.7	61
.0607	.061
.0592	.059
.0596	.060

More on Confidence Intervals for a Proportion

In this appendix we present some of the technical details behind the confidence interval for a proportion introduced in Section 6.6. For a more complete discussion of these ideas, see the paper by Agresti and Coull that is given as reference 32 for Chapter 6.

Suppose we want to develop a $100(1 - \alpha)\%$ confidence interval for a proportion p; for example, a 95% confidence level corresponds to $\alpha = .05$. A common procedure is to use

$$\hat{p} \pm Z_{\alpha/2}\sqrt{\frac{\hat{p}(1 - \hat{p})}{n}}$$

This is called the **Wald confidence interval**.

Another way to construct a confidence interval is to find all values of p such that

$$-Z_{\alpha/2} \leq \frac{\hat{p} - p}{\sqrt{p(1 - p)/n}} \leq Z_{\alpha/2} \tag{1}$$

This is known as "inverting a hypothesis test"; hypothesis testing for proportions is discussed in Chapter 10. The basic idea here is that the sampling distribution of $\hat{p}$ can be approximated by a normal distribution and that we should take values that correspond to the middle $100(1 - \alpha)\%$ (e.g., the middle 95%) of the normal distribution.

In the following presentation we will let Z denote $Z_{\alpha/2}$. Solving inequality (1) for p gives an interval of the form

$$\frac{\hat{p} + \frac{Z^2}{2n} \pm Z\sqrt{\frac{\hat{p}(1 - \hat{p}) + \frac{Z^2}{4n}}{n}}}{1 + \frac{Z^2}{n}}$$

This is called the **score confidence interval**.

Most books present the Wald confidence interval (without giving it that name), since it is much more simple in form than the score confidence interval. However, the Wald confidence interval has poor coverage properties: A nominal 95% Wald confidence interval might actually cover p only 80% of the time, rather than 95% of the time. The score confidence interval has excellent coverage properties, but is quite complex.

The formulation of the **Wilson confidence interval**, presented in Section 6.6, is based on approximating the score interval. The midpoint of the score interval is

$$\hat{p}\left(\frac{n}{n + Z^2}\right) + \frac{1}{2}\left(\frac{Z^2}{n + Z^2}\right) \tag{2}$$

which is a weighted average of $\hat{p}$ and $\frac{1}{2}$, with weights $\dfrac{n}{n + Z^2}$ and $\dfrac{Z^2}{n + Z^2}$. Note that as n increases, more weight is given to $\hat{p}$; for small n, more weight is given to $\frac{1}{2}$. Because $\hat{p} = y/n$, the midpoint given by (2) is

$$\frac{y}{n + Z^2} + \frac{\frac{1}{2}Z^2}{n + Z^2}$$

For a 95% confidence interval, $Z = 1.96 \approx 2$, so that the midpoint is approximated by

$$\frac{y}{n + 4} + \frac{2}{n + 4} = \frac{y + 2}{n + 4}$$

which we called $\tilde{p}$ in Section 6.6.

The standard error that is used in the Wald confidence interval is $\sqrt{\dfrac{\hat{p}(1 - \hat{p})}{n}}$. This is based on the fact that the variance of Y is equal to $np(1 - p)$, as derived in Appendix 3.3, so that the variance of $\hat{p}$ is equal to $\dfrac{np(1 - p)}{n^2}$ or $\dfrac{p(1 - p)}{n}$. Hence the standard deviation of $\hat{p}$ is equal to $\sqrt{\dfrac{p(1 - p)}{n}}$, as discussed in Appendix 5. 1; the standard error uses the sample proportion $\hat{p}$ in place of the unknown value of p. Likewise, the variance of $\tilde{p}$ is equal to $\dfrac{np(1 - p)}{(n + Z^2)^2}$, which is approximately $\dfrac{p(1 - p)}{n + Z^2}$. Hence the standard error for the Wilson confidence interval is given as $\sqrt{\dfrac{\tilde{p}(1 - \tilde{p})}{n + Z^2}}$; for a 95% confidence interval, $Z \approx 2$, giving $\sqrt{\dfrac{\tilde{p}(1 - \tilde{p})}{n + 4}}$.

As a closing note, we mention that some authors advocate the use of an "exact" confidence interval, based on the binomial distribution. However, as Agresti and Coull show, the exact confidence interval is quite conservative, which is one reason that it is not widely used. (It is also quite complex.)

How Power Is Calculated

The required sample sizes given in Table 5 were determined by calculating the power of the t test. For large samples, an appropriate power calculation can be based on the normal curve (Table 3). In this appendix we indicate how such an approximate calculation is done.

Recall that the power is the probability of rejecting H_0 when H_A is true. In order to calculate power, therefore, we need to know the sampling distribution of t_s when H_A is true. For large samples, the sampling distribution can be approximated by a normal curve, as shown by the following theorem.

Theorem A.1. Suppose we choose independent random samples, each of size n, from normal populations with means μ_1 and μ_2 and a common standard deviation σ. If n is large, the sampling distribution of t_s can be approximated by a normal distribution with

$$\text{Mean} = \sqrt{\frac{n}{2}}\left(\frac{\mu_1 - \mu_2}{\sigma}\right)$$

and

$$\text{Standard deviation} = 1$$

To illustrate the use of Theorem A.1 for power calculations, suppose we are considering a one-tailed t test with $\alpha = .025$. The hypotheses are

$$H_0: \quad \mu_1 = \mu_2$$
$$H_A: \quad \mu_1 > \mu_2$$

If we want a power of .80 for an effect size of .4, then Table 5 recommends samples of size of $n = 100$. Let us confirm this recommendation using Theorem A.1.

If H_0 is true, so that $\mu_1 = \mu_2$, then the sampling distribution of t_s is approximately a normal distribution with mean equal to 0 and SD equal to 1. This is the null distribution of t_s; it is shown as the dashed curve in Figure A.2.

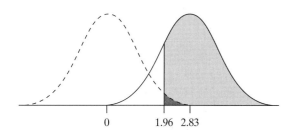

0   1.96 2.83   **Figure A.2**

Suppose that in fact H_A is true, that the effect size is

$$\frac{\mu_1 - \mu_2}{\sigma} = .4$$

and that we are using samples of size $n = 100$. Then, according to Theorem A.1, the sampling distribution of t_s will be approximately a normal distribution with SD equal to 1 and mean equal to

$$\sqrt{\frac{n}{2}}\left(\frac{\mu_1 - \mu_2}{\sigma}\right) = \sqrt{\frac{100}{2}}(.4) = 2.83$$

This distribution is the solid curve in the figure.

For $n_1 = n_2 = 100$, we have df $\approx \infty$, so from Table 4 the critical value is equal to 1.96. Thus, the P-value would be less than .025 and we would reject H_0 if

$$t_s > 1.96$$

Using the dashed curve, the probability of this event is equal to .025; this is shown in the figure as a darker area. Using the solid curve, the probability that $t_s > 1.96$ is the shaded area in the figure. The shaded area can be determined from Table 3 using

$$Z = 1.96 - 2.38 = -.87$$

From Table 3, the area is .8078 $\approx$ .81. Thus, we have shown that, for $n_1 = n_2 = 100$,

$$\text{if} \quad \frac{\mu_1 - \mu_2}{\sigma} = .4, \quad \text{then} \quad \Pr\{\text{reject } H_0\} \approx .81$$

We have found that the power against the specified alternative is approximately equal to .81; this agrees well with Table 5, which claims that the power is equal to .80.

If we were concerned with a two-tailed test at $\alpha = .05$, the critical value would again be 1.96, and so the power would again be approximately equal to .81, because the area under the solid curve corresponding to the left-hand tail of the dashed curve is negligible.

Of course, in constructing Table 5, we begin with the specified power (.80) and determine n, rather than the other way around. This "inverse" problem can be solved using an approach similar to the foregoing. In the figure, the shaded area (.80) would be given; this would determine the Z value and in turn determine n, once the effect size is specified.

More on the Wilcoxon-Mann-Whitney Test

In Section 7.11 we saw how critical values for the Wilcoxon-Mann-Whitney test are related to the null distribution of K_1, K_2, and U_s. In this appendix we indicate how these null distributions can be determined by simple counting methods.

Let us consider the sample sizes of $n = 5$ and $n' = 4$. In Figure 7.30 (page 293), the Y_1's and Y_2's are plotted as dots. To save space, let us now represent the data in a more compact way: We will represent each Y_1 by a "1" and each Y_2 by a "2." Thus, the arrangement in Figure 7.30(a) (where the Y_1's are entirely to the left of the Y_2's) would be represented as

$$1 \quad 1 \quad 1 \quad 1 \quad 1 \quad 2 \quad 2 \quad 2 \quad 2$$

For sample sizes $n = 5$ and $n' = 4$, there are 126 possible arrangements of the Y_1's and Y_2's. Here is a partial list of those arrangements and the associated values of K_1 and K_2:

Number	Arrangement	K_1	K_2
1	1 1 1 1 1 2 2 2 2	0	20
2	1 1 1 1 2 1 2 2 2	1	19
3	1 1 1 1 2 2 1 2 2	2	18
4	1 1 1 2 1 1 2 2 2	2	18
5	1 1 2 1 1 1 2 2 2	3	17
6	1 1 1 2 1 2 1 2 2	3	17
7	1 1 1 1 2 2 2 1 2	3	17
8	1 2 1 1 1 1 2 2 2	4	16
9	1 1 2 1 1 2 1 2 2	4	16
10	1 1 1 2 1 2 2 1 2	4	16
11	1 1 1 2 2 1 1 2 2	4	16
12	1 1 1 1 2 2 2 2 1	4	16
	...and so on...		
126	2 2 2 2 1 1 1 1 1	20	0

To determine the null distributions from this list, we need to know the likelihood of the various arrangements, assuming that H_0 is true. According to H_0, all 9 observations (Y's) were drawn at random from the same population. Under this assumption, it can be shown that the 126 arrangements are all *equally likely*. Because of this simple and elegant fact, the null distribution of K_1 and K_2 (and therefore U_s) can be determined by straightforward counting. Working from the preceding list, we find the following probabilities:

K_1	K_2	Probability
0	20	$\frac{1}{126}$
1	19	$\frac{1}{126}$
2	18	$\frac{2}{126}$
3	17	$\frac{3}{126}$
4	16	$\frac{5}{126}$
	...and so on...	
20	0	$\frac{1}{126}$
		Total 1

These probabilities constitute the null distribution of K_1 and K_2—plotted in Figure 7.31(a). For instance, the first probability in the null distribution is

$$\Pr\{K_1 = 0, K_2 = 20\} = \tfrac{1}{126} \approx .008$$

as stated in Section 7.11.

Why is the Wilcoxon-Mann-Whitney test distribution free? The reason should be clear from the preceding discussion. If the two population distributions are the same, then all possible arrangements of the Y's are equally likely and the specific shape of the population distributions does not matter (except, of course, that we have assumed that there would be no ties; the null distributions are altered if ties are possible.)

The Wilcoxon-Mann-Whitney null distribution can always be determined by straightforward counting such as illustrated above (although for larger sample sizes the counting is very tedious and approximate methods are used instead). The number of possible arrangements for samples of size n and n' is equal to

$$\frac{(n + n')!}{n!n'!}$$

For example, for sample sizes 5 and 4 (as before), we find

$$\frac{9!}{5!4!}$$

(To see why this formula works, refer to the discussion of the formula for binomial coefficients at the end of Appendix 3.2; the reasoning is exactly parallel.)

APPENDIX 11.1

More on the Newman-Keuls Procedure

The Newman-Keuls Procedure and the t Test

In Section 11.7 we mentioned the link between the Newman-Keuls procedure and the t test. We will now show the link more explicitly.

Suppose $\bar{y}_1$ and $\bar{y}_2$ are to be compared using a pooled two-sample t test at $\alpha = .05$. The null hypothesis is $H_0: \mu_1 = \mu_2$. H_0 would be rejected if

$$\frac{|\bar{y}_1 - \bar{y}_2|}{\mathrm{SE}_{(\bar{y}_1 - \bar{y}_2)}} > t_{.025}$$

If $n_1 = n_2 = n$, then

$$\mathrm{SE}_{(\bar{y}_1 - \bar{y}_2)} = \sqrt{s^2_{\text{pooled}}\left(\frac{1}{n} + \frac{1}{n}\right)} = \sqrt{2}\sqrt{\frac{s^2_{\text{pooled}}}{n}}$$

and consequently the t test rejects H_0 if

$$|\bar{y}_1 - \bar{y}_2| > (\sqrt{2}\,t_{.025})\sqrt{\frac{s^2_{\text{pooled}}}{n}} \tag{1}$$

On the other hand, suppose $\bar{y}_1$ and $\bar{y}_2$ are to be compared using the critical value R_2 from the Newman-Keuls procedure. H_0 would be rejected if

$$|\bar{y}_1 - \bar{y}_2| > R_2 \quad \text{that is, if} \quad |\bar{y}_1 - \bar{y}_2| > q_2\sqrt{\frac{s^2_{\text{pooled}}}{n}} \tag{2}$$

Let us compare conditions (1) and (2). The second factor on the right-hand side is the same in both expressions—although of course s_{pooled} is based on two samples for the t test and on all k samples for the Newman-Keuls procedure. To complete the correspondence, it can be shown that the first factors on the right-hand sides of (1) and (2) are equal; that is,

$$\sqrt{2}\,t_{.025} = q_2$$

where q_2 is determined from Table 10 with $\alpha = .05$ and with the same df as $t_{.025}$. For instance, suppose df $= 15$. Then Table 10 gives $q_2 = 3.01$ and Table 4 gives $t_{.025}(15) = 2.131$; and indeed it is true that

$$(\sqrt{2})(2.131) = 3.01$$

Thus, except for the difference in computing s_{pooled}, a comparison of two means using R_2 is equivalent to a pooled two-sample t test against a nondirectional alternative. (Of course, the Newman-Keuls procedure does not use R_2 for all its comparisons.)

Varieties of Type I Error

In Section 11.7 we mentioned that different multiple comparison procedures differ with respect to the degree of control of Type I error. To be more specific about this, let us consider various ways in which the risk of Type I error might be computed. Suppose the number of groups to be compared is $k = 4$. Let H_{ij} represent the null hypothesis $H_0: \mu_i = \mu_j$. The following are a few of the probability calculations that express various aspects of the risk of Type I error:

1. Assume $\mu_1 = \mu_2 = \mu_3 = \mu_4$ and calculate
   $\Pr\{\text{at least one of the } H_{ij} \text{ is rejected}\}$

2. Assume $\mu_1 = \mu_2 = \mu_3$ and calculate
   $\Pr\{H_{12} \text{ or } H_{13} \text{ or } H_{23} \text{ is rejected}\}$

3. Assume $\mu_1 = \mu_2$ and calculate
   $\Pr\{H_{12} \text{ is rejected}\}$

4. Assume $\mu_1 = \mu_2$ and $\mu_3 = \mu_4$, and calculate
   $\Pr\{H_{12} \text{ or } H_{34} \text{ is rejected}\}$

If the Newman-Keuls procedure is performed at $\alpha = .05$, then probabilities such as those defined in (1), (2), and (3) are held at .05, but probabilities such as that in (4) are not controlled. Some multiple comparison procedures control only the probability in (1), whereas some procedures control all the probabilities in (1), (2), (3), and (4).

Researchers do not all agree on how stringently a multiple comparison procedure should control the risk of Type I error. Consequently, there is no multiple comparison procedure that is unambiguously the "best" one.

Least-Squares Formulas

In this appendix we show that the least-squares criterion leads to the formulas

$$b_1 = \frac{\Sigma(x_i - \bar{x})(y_i - \bar{y})}{\Sigma(x_i - \bar{x})^2} \quad \text{and} \quad b_0 = \bar{y} - b_1\bar{x}$$

We will make use of the fact that the minimum of a quadratic function

$$Q(x) = Ax^2 + Bx + C$$

occurs at $x = -B/2A$ (assuming $A > 0$).

Preliminary Result. Given a set of data $y_1, y_2, \ldots, y_n$, the number c that minimizes the quantity $\Sigma(y_i - c)^2$ is the mean $\bar{y}$.

To see this, first expand $(y_i - c)^2$ to get $y_i^2 - 2y_ic + c^2$, then distribute the summation to get

$$\Sigma(y_i - c)^2 = \Sigma y_i^2 - 2(\Sigma y_i)c + nc^2$$

The last expression is a quadratic function $Q(c) = Ac^2 + Bc + C$ where $A = n$ and $B = -2(\Sigma y_i)$.

Therefore, the minimum occurs at

$$c = -\frac{B}{2A} = -\frac{-2(\Sigma y_i)}{2n} = \bar{y}$$

Recall that if $\hat{y}_i = b_0 + b_1x_i$ is the least-squares regression line, then b_0 and b_1 are the values that minimize the residual sum of squares, given by

$$\text{SS(resid)} = \Sigma(y_i - \hat{y}_i)^2 = \Sigma[y_i - (b_0 + b_1x_i)]^2$$

We can write $y_i - (b_0 + b_1x_i)$ as $(y_i - b_1x_i) - b_0$. Next, we apply the preliminary result with $(y_i - b_1x_i)$ in the place of y_i and b_0 in the place of c. Thus, the minimum of the residual sum of squares occurs when

$$b_0 = \text{the mean of } (y_i - b_1x_i) = \bar{y} - b_1\bar{x} \tag{1}$$

Thus, the least-squares line goes through the point of averages, $(\bar{x}, \bar{y})$. Substituting this value of b_0 into the residual sum of squares gives

$$\text{SS(resid)} = \Sigma(y_i - (\bar{y} - b_1\bar{x}) - b_1x_i)^2 = \Sigma(y_i - \bar{y} + b_1\bar{x} - b_1x_i)^2$$

$$= \Sigma[(y_i - \bar{y}) - b_1(x_i - \bar{x})]^2$$

This can be expanded as

$$\text{SS(resid)} = \Sigma[(y_i - \bar{y})^2 - 2b_1(x_i - \bar{x})(y_i - \bar{y}) + b_1^2(x_i - \bar{x})^2]$$

$$= \Sigma(y_i - \bar{y})^2 - 2b_1\Sigma(x_i - \bar{x})(y_i - \bar{y}) + b_1^2\Sigma(x_i - \bar{x})^2$$

Thus, the residual sum of squares is a quadratic function in b_1 with

$$A = \Sigma(x_i - \bar{x})^2 \quad \text{and} \quad B = -2\Sigma(x_i - \bar{x})(y_i - \bar{y})$$

Hence, the minimum of SS(resid) occurs at $-B/2A$, or

$$b_1 = -\frac{-2\Sigma(x_i - \bar{x})(y_i - \bar{y})}{2\Sigma(x_i - \bar{x})^2} = \frac{\Sigma(x_i - \bar{x})(y_i - \bar{y})}{\Sigma(x_i - \bar{x})^2} \tag{2}$$

Together equations (1) and (2) give the formulas for the coefficients of the least-squares regression line.

Derivation of Fact 12.1

By definition

$$r^2 = \frac{SS(reg)}{SS(total)}$$

and $SS(total) = SS(reg) + SS(resid)$. Thus,

$$1 - r^2 = \frac{SS(total) - SS(reg)}{SS(total)} = \frac{SS(resid)}{SS(total)}$$

Also, by definition

$$s_{Y|X}^2 = \frac{SS(resid)}{n-2} \quad \text{and} \quad s_Y^2 = \frac{SS(total)}{n-1}$$

Therefore,

$$1 - r^2 = \frac{(n-2)s_{Y|X}^2}{(n-1)s_Y^2} = f^2\left(\frac{s_{Y|X}^2}{s_Y^2}\right) \quad \text{where} \quad f^2 = \frac{n-2}{n-1}$$

Thus,

$$\frac{s_{Y|X}}{s_Y} = f\sqrt{1 - r^2} \quad \text{where} \quad f = \sqrt{\frac{n-2}{n-1}}$$

The factor f is close to 1 unless n is quite small. Here are some values of f:

n	f
5	.87
10	.94
15	.96

Thus, we have shown that

$$\frac{s_{Y|X}}{s_Y} \approx \sqrt{1 - r^2}$$

The approximation is reasonably good if $n \geq 5$.

APPENDIX 12.3

Calculations for Example 12.26

In this appendix we describe the calculations for the regression analysis presented in Example 12.26, including the regression of X on Y.

We first recall the formulas for regression of Y on X. The regression line is

$$Y = b_0 + b_1 X \quad \text{where} \quad b_1 = \frac{\Sigma(x_i - \bar{x})(y_i - \bar{y})}{\Sigma(x_i - \bar{x})^2} \quad \text{and} \quad b_0 = \bar{y} - b_1 \bar{x}$$

The residual sum of squares is

$$\Sigma(y_i - \hat{y}_i)^2$$

and the residual standard deviation is

$$s_{Y|X} = \sqrt{\frac{\text{SS(resid)}}{n - 2}} = \sqrt{\frac{\Sigma(y_i - \hat{y}_i)^2}{n - 2}}$$

The formulas for the reverse regression—of X on Y—are exactly analygous. The regression line can be written as

$$X = b'_0 + b'_1 Y \quad \text{here} \quad b'_1 = \frac{\Sigma(x_i - \bar{x})(y_i - \bar{y})}{\Sigma(y_i - \bar{y})^2} \quad \text{and} \quad b'_0 = \bar{x} - b'_1 \bar{y}$$

The residual sum of squares for the regression of X on Y is

$$\Sigma(x_i - \hat{x}_i)^2$$

and the residual standard deviation is

$$s_{X|Y} = \sqrt{\frac{\text{SS(resid)}}{n - 2}} = \sqrt{\frac{\Sigma(x_i - \hat{x}_i)^2}{n - 2}}$$

The regression analysis displayed in Figure 12.18 is based on the preceding formulas. The basic statistics for the cricket data are

$$n = 39, \qquad \bar{x} = 3.5186, \qquad \bar{y} = 39.846$$

$$\Sigma(x_i - \bar{x})^2 = 16.8040, \quad \Sigma(y_i - \bar{y})^2 = 35{,}769.1, \quad \Sigma(x_i - \bar{x})(y_i - \bar{y}) = 532.858$$

Application of the preceding formulas to these basic statistics gives the regression lines and residual SDs as described in Example 12.26.

Chapter Notes

Chapter 1

1. Nicolle, J. (1961). *Louis Pasteur: The Story of His Major Discoveries.* New York: Basic Books. p. 170. © 1961 by Jacques Nicolle. © 1961 English translation Hutchinson & Co. (Publishers) Ltd. Reprinted by permission of Basic Books.
2. Mizutani, T. and Mitsuoka, T. (1979). Effect of intestinal bacteria on incidence of liver tumors in gnotobiotic C3H/He male mice. *Journal of the National Cancer Institute* **63**, 1365–1370.
3. Tripepi, R. R. and Mitchell, C. A. (1984) Metabolic response of river birch and European Birch roots to hypoxia. *Plant Physiology* **76**, 31–35. Raw data courtesy of the authors.
4. Adapted from Potkin, S. G., Cannon, H. E., Murphy, D. L., and Wyatt, R. J. (1978). Are paranoid schizophrenics biologically different from other schizophrenics? *New England Journal of Medicine* **298**, 61–66. The data are approximate, having been reconstructed from the histograms and summary information given by Potkin, et al. Reprinted by permission of the *New England Journal of Medicine*.
5. Wolfson, J. L. (1987). Impact of *Rhizobium* nodules on *Sitona hispidulus*, the clover root curculio. *Entomologia Experimentalis et Applicata* **43**, 237–243. Data courtesy of the author. The experiment actually included 11 dishes.
6. Allen, L. S. and Gorski, R. A. (1992). Sexual orientation and the size of the anterior commissure in the human brain. *Proceedings of the National Academy of Science* **89**, 7199–7202. The data are approximate, having been reconstructed from the dotplots and summary information given by Allen and Gorski. Regarding the first concern mentioned in Example 1.6, the authors were mindful of the effect that the two largest observations could have on their conclusions and calculated the average for the homosexual men a second time, after deleting these two values. As for the second concern, the authors calculated the averages for those who had AIDS and those who did not, in each group of men. They found that AIDS is associated with smaller, not larger, AC areas, so that when only persons without AIDS are compared, the difference between homosexual and heterosexual men is even larger than the difference found in the full data set.
7. Bradstreet, T. E. (1992). Favorite data sets from early phases of drug research–part 2. *Proceedings of the Section on Statistical Education of the American Statistical Association.* 219–223.
8. Webb, P. (1981). Energy expenditure and fat-free mass in men and women. *American Journal of Clinical Nutrition* **34**, 1816–1826.

Chapter 2

1. Stewart, R. N. and Arisumi, T. (1966). Genetic and histogenic determination of pink bract color in poinsettia. *Journal of Heredity* **57**, 217–220.
2. Data of Wiener, A. S., Moor-Jankowski, J., and Gordon, E. B. (1966). Reproduced in Erskine, A. G. and Socha, W. W. (1978), *The Principles and Practices of Blood Grouping,* 2nd edition. St. Louis: Mosby, p. 64.

3. *The Time Almanac 2000*. Boston, MA: Time, Inc.

4. Unpublished data courtesy of C. M. Cox and K. J. Drewry.

5. Unpublished data courtesy of W. F. Jacobson.

6. Unpublished data courtesy of M. Kimmel.

7. Knoll, A. E. and Barghoorn, E. 5. (1977). Archean microfossils showing cell division from the Swaziland system of South Africa. *Science* **198**, 396–398.

8. Nurse, C. A. (1981). Interactions between dissociated rat sympathetic neurons and skeletal muscle cells developing in cell culture. II. Synaptic mechanisms. *Developmental Biology* **88**, 71–79.

9. Topinard, P. (1888). Le poids de l'encephale d'apres les registres de Paul Broca. *Memoires Societe d'Anthropologie Paris*, 2nd series, vol. 3, p. 1–41. The data shown are a subset of the data published by Topinard.

10. Johannsen, W. (1903). *Ueber Erblichkeit in Populationen und in reinen Linien*. Jena: G. Fischer. Data reproduced in Strickberger, M. W. (1976). *Genetics*, New York: Macmillan, p. 277; and Peters, J. A. (ed.) (1959). *Classic Papers in Genetics*, Englewood Cliffs, New Jersey: Prentice-Hall, p. 23.

11. Unpublished data courtesy of W. F. Jacobson.

12. Simpson, G. G., Roe, A., and Lewontin, R. C. (1960). *Quantitative Zoology*. New York: Harcourt Brace. p. 51.

13. Adapted from Potkin, S. G., Cannon, H. F., Murphy, D. L., and Wyatt, R. J. (1978). Are paranoid schizophrenics biologically different from other schizophrenics? *New England Journal of Medicine* **298**, 61–66. The data given are approximate, having been reconstructed from the histogram and summary information given by Potkin et al. Reprinted by permission of the *New England Journal of Medicine*.

14. Peters, H. G. and Bademan, H. (1963). The form and growth of stellate cells in the cortex of the guinea-pig. *Journal of Anatomy* (London) **97**, 111–117.

15. Data courtesy of R. F. Jones, Indiana State Dairy Association, Inc.

16. Unpublished data courtesy of D. J. Honor and W. A. Vestre.

17. Connolly, K. (1968). The social facilitation of preening behaviour in Drosophila melanogaster. *Animal Behaviour* **16**, 385–391.

18. Bruce, D., Harvey, D., Hamerton, A. E., and Bruce, L. (1913). Morphology of various strains of the trypanosome causing disease in man in Nyasaland. I. The human strain. *Proceedings of the Royal Society of London, Series B* **86**, 285–302. See also Pearson, K. (1914). On the probability that two independent distributions of frequency are really samples of the same population, with reference to recent work on the identity of trypanosome strains. *Biometnka* **10**, 85–143.

19. Shields, D. R. (1981). The influence of niacin supplementation on growing ruminants and *in vivo* and *in vitro* rumen parameters. Ph.D. thesis, Purdue University. Raw data courtesy of the author and D. K. Colby.

20. Gwynne, D. T. (1981). Sexual difference theory: Mormon crickets show role reversal in mate choice. *Science* **213**, 779–780. Copyright 1981 by the AAAS. Raw data courtesy of the author.

21. Unpublished data courtesy of M. A. Morse and G. P. Carlson.

22. Adapted from Anderson, J. W., Story, L., Sieling, B., Chen, W. L., Petro, M. S., and Story, J. (1984). Hypocholesterolemic effects of oat-bran or bean intake for hypercholesterolemic men. *American Journal of Clinical Nutrition* **40**, 1146–1155. There were actually 20 men in the study.

23. Unpublished data courtesy of C. H. Noller.

24. Luria, S. F. and Delbruck, M. (1943). Mutations of bacteria from virus sensitivity to virus resistance. *Genetics* **28**, 491–511.

25. Fictitious but realistic data. See Roberts, J. (1975). Blood pressure of persons 18–74 years, United States, 1971–72. *U.S. National Center for Health Statistics, Vital and Health Statistics Series 11, No. 150.* Washington, D.C.: U.S. Department of Health, Education and Welfare.

26. Unpublished data collected from a sample of Oberlin College students.

27. Unpublished data courtesy of F. Delgado.

28. Unpublished data collected from a sample of Oberlin College students by Christie Schroth and Scott Houghtaling.

29. Fictitious but realistic data. Based on Beyl, C. A. and Mitchell, C. A. (1977). Characterization of mechanical stress dwarfing in chrysanthemum. *Journal of the American Society for Horticultural Science* **102**, 591–594.

30. Based on a subset of the data in Tuddenham, R. D. and Snyder, M. M. (1954). Physical growth of California boys and girls from birth to age 18. *Calif. Publ. Child Develop.* **1**, 183–364. Data as reported in Weisberg, S. (1985). *Applied Linear Regression*, 2nd ed. New York: Wiley, p. 57.

31. Nelson, L. A. (1980). *Report of the Indiana Beef Evaluation Program, Inc.* Purdue University, West Lafayette, Indiana.

32. Data collected by J. Witmer at a statistics workshop at Johns Hopkins University, July 1995.

33. Day, K. M., Patterson, F. L., Luetkemeier, O.W., Ohm, H. W., Polizotto, K., Roberts, J. J., Shaner, G. E., Huber, D. M., Finney, R. F., Foster, J. F., and Gallun, R. L. (1980). Performance and adaptation of small grains in Indiana. Station Bulletin No. 290. West Lafayette, Indiana: Agricultural Experiment Station of Purdue University. Raw data provided courtesy of W. F. Nyquist.

34. Tripepi, R. R. and Mitchell, C. A. (1984). Metabolic response of river birch and European birch roots to hypoxia. *Plant Physiology* **76**, 31–35. Raw data courtesy of the authors.

35. Ogilvie, R. I., Macleod, S., Fernandez, P., and McCullough, W. (1974). Timolol in essential hypertension. In *Beta-Adrenergic Blocking Agents in the Management of Hypertension and Angina Pectoris,* B. Magnani (ed.). New York: Raven Press. pp. 31–43.

36. Unpublished data courtesy of J. F. Nash and J. E. Zabik.

37. Schall, J.J., Bennett, A. F., and Putnam, R. W. (1982). Lizards infected with malaria: Physiological and behavioral consequences. *Science* **217**, 1057–1059. Copyright 1982 by the AAAS. Raw data courtesy of J. J. Schall.

38. Fictitious but realistic data. Each observation is the average of several measurements made on the same woman at different times. See Royston, J. P. and Abrams, R. M. (1980). An objective method for detecting the shift in basal body temperature in women. *Biometrics* **36**, 217–224.

39. Adapted from data in Cicirelli, M. F., Robinson, K. R., and Smith, L. D. (1983). Internal pH of *Xenopus* oocytes: A study of the mechanism and role of pH changes during meiotic maturation. *Developmental* Biology **100**, 133–146.

40. Adapted from Royston, J. P. and Abrams, R. M. (1980). An objective method for detecting the shift in basal body temperature in women. *Biometrics* **36**, 217–224.

41. Adapted from data provided courtesy of L. A. Nelson.

42. Ikin, E. W., Prior, A. M., Race, R. R., and Taylor, G. L. (1939). The distribution of the A_1A_2BO blood groups in England. *Annals of Eugenics,* London **9**, 409–411. Reprinted with permission of Cambridge University Press.

43. Borg, S., Kvande, H., and Sedvall, G. (1981). Central norepinephrine metabolism during alcohol intoxication in addicts and healthy volunteers. *Science* **213**, 1135–1137. Copyright 1981 by the AAAS. Raw data courtesy of S. Borg.

44. Long, T. F. and Murdock, L. L. (1983). Stimulation of blowfly feeding behavior by octopaminergic drugs. *Proceedings of the National Academy of Sciences* **80**, 4159–4163. Raw data courtesy of the authors and L. C. Sudlow.

45. Fictitious but realistic population. Adapted from LeClerg, E. L., Leonard, W. H., and Clark, A. G. (1962). *Field Plot Technique.* Minneapolis: Burgess.

46. Selawry, O. S. (1974). The role of chemotherapy in the treatment of lung cancer. *Seminars in Oncology* **1**, No. 3, 259–272.

47. Hayes, H. K., East, E. M., and Bernhart, E. G. (1913). *Connecticut Agricultural Experiment Station Bulletin 176.* Data reproduced in Strickberger, M. W. (1976). *Genetics,* New York: Macmillan, p. 288.

48. Unpublished data courtesy of J. Y. Latimer and C. A. Mitchell.

49. The results of similar assays are reported in Pascholati, S. F. and Nicholson, R. L. (1983). *Helminthosporum maydis* suppresses expression of resistance to *Helminthosporum carbonum* in corn. *Phytopathologische Zeitschrift* **107**, 97–105. Unpublished data courtesy of the investigators.

50. Richens, A. and Ahmad, S. (1975). Controlled trial of valproate in severe epilepsy. *British Medical Journal* **4**, 255–256.

51. Fleming, W. E. and Baker, F. E. (1936). A method for estimating populations of larvae of the Japanese beetle in the field. *Journal of Agricultural Research* **53**, 319–331. Data reproduced in *Statistical Ecology, Volume 1* (1971). University Park: The Pennsylvania State University Press, p. 327.

52. Chiarotti, R. M. (1972). An investigation of the energy expenditure of women squash players. Master's thesis, The Pennsylvania State University. Raw data courtesy of R. M. Lyle (nee Chiarotti).

53. Masty, J. (1983). Innervation of the equine small intestine. Master's thesis, Purdue University. Raw data courtesy of the author.

54. Fictitious but realistic data. Adapted from data presented in Falconer, D. S. (1981). *Introduction to Quantitative Genetics,* 2nd edition. New York: Longman, Inc., p. 97.

55. Dow, T. G. B., Rooney, P. J., and Spence, M. (1975). Does anaemia increase the risks to the fetus caused by smoking in pregnancy? *British Medical Journal* **4**, 253–254.

56. Christophers, S. R. (1924). The mechanism of immunity against malaria in communities living under hyper-endemic conditions. *Indian Journal of Medical Research* **12**, 273–294. Data reproduced in Williams, C. B. (1964). *Patterns in the Balance of Nature.* London: Academic Press, p. 243.

57. Data taken from *Climatological Data, Ohio*, and *Local Climatological Data, Cleveland, Ohio*; National Oceanic and Atmospheric Administration, U.S. Dept. of Commerce.

58. These data were published on p. 8-A of the *Cleveland Plain Dealer*, 6 February 1997, from information compiled by the United Network for Organ Sharing. The mortality rate and volume variables are averages over a four-year period beginning in October 1987. There are 31 hospitals in the low volume group and 76 in the high volume group.

59. Erne, P., Bolli, P., Buergisser, E., and Buehler, F. R. (1984). Correlation of platelet calcium with blood pressure. *New England Journal of Medicine* **310**, 1084–1088. Reprinted by permission. Raw data courtesy of F. R. Buehler. The original data set had 47 subjects; we have omitted nine patients with "borderline" high blood pressure.

Chapter 3

1. Parks, N. J., Krohn, K. A., Mathis, C. A., Chasko, J. H., Geiger, K. R., Gregor, M. E., and Peek, N. F. (1981). Nitrogen-13-labelled nitrite and nitrate: Distribution and metabolism after intratracheal administration. *Science*, **212**, 58–61.

2. Fictitious but realistic population. Adapted from Hubbs, C. L. and Schultz, L. P. (1932). *Cottus tubulatus,* a new sculpin from Idaho. *Occasional Papers of the Museum of Zoology, University of Michigan*, **242**, 1–9. Data reproduced in Simpson, G. G., Roe, A., and Lewontin, R. C. (1960). *Quantitative Zoology.* New York: Harcourt, Brace, p. 81.

3. Based on an article by the Neonatal Inhaled Nitric Oxide Study Group (1997). See Inhaled nitic oxide in full-term and nearly full-term infants with hypoxic respiratory failure. *New England Journal of Medicine*, **336**, 597–604.

4. This table is a modified version of data adapted from Ammon, O. (1899). *Zur Anthropologie der Badener.* Jena: G. Fischer. Ammon's data appear in Goodman, L. A. and Kruskal, W. H. (1954). Measures of association for cross classifications. *Journal of the American Statistical Association*, **49**, 732–764. The numbers in the table have been rounded off to aid the exposition.

5. Unpublished data courtesy of Diana Zumas and Lisa Yasuhara, Oberlin College.

6. Adapted from Taira, D. A., Safran, D. G., Seto, T. B., Rogers, W. H., and Tarlov, A. R. (1997). The Relationship Between Patient Income and Physician Discussion of Health Risk Behaviors. *Journal of the American Medical Association*, **278**, 1412–1417.

7. The population is fictitious, but resembles the population of American women aged 18–24, excluding known or suspected diabetics, as reported in Gordon, T. (1964). Glucose tolerance of adults, United States 1960–62. *U.S. National Center for Health Statistics, Vital and Health Statistics Series 11, No. 2.* Washington, D.C.: U.S. Department of Health, Education and Welfare.

8. Meyer, W. H. (1930). Diameter distribution series in even-aged forest stands. *Yale University School of Forestry Bulletin*, **28**. The curve is fitted in Bliss, C. I., and Reinker, K. A. (1964). A lognormal approach to diameter distributions in even-aged stands. *Forest Science*, **10**, 350–360.

9. Pearson, K. (1914). On the probability that two independent distributions of frequency are really samples of the same population, with reference to recent work on the identity of trypanosome strains. *Biometrika*, **10**, 85–143. Reprinted by permission of the Biometrika Trustees.

10. Adapted from unpublished data courtesy of Gloria Zender, Oberlin College.

11. Fictitious but realistic situation. Based on data given by Lack, D. (1948). Natural selection and family size in the starling. *Evolution*, **2**, 95–110. Data reproduced by Riclefs, R. E. (1973). *Ecology.* Newton, Massachusetts: Chiron Press. p. 37.

12. Adapted from unpublished data courtesy of Marni Hansill, Oberlin College.

13. This is one of the crosses performed by Gregor Mendel in his classic studies of heredity; heterozygous plants (which are yellow-seeded because yellow is dominant) are crossed with each other.

14. Fictitious but realistic value. See Hutchison, J. G. P., Johnston, N. M., Plevey, M. V. P., Thangkhiew, I., and Aidney, C. (1975). Clinical trial of Mebendazole, a broad-spectrum anthelminthic. *British Medical Journal*, **2**, 309–310.

15. Fictitious but realistic population. Adapted from Owen, D. F. (1963). Polymorphism and population density in the African land snail, *Limicolaria martensiana. Science,* **140**, 666–667.

16. *The World Almanac and Book of Facts, 1982.* New York: Newspaper Enterprise Association.

17. Adapted from discussion in Galen, R. S. and Gambino, S. R. (1980). *Beyond Normality: The Predictive Value and Efficiency of Medical Diagnoses.* New York: Wiley, pp. 71–74.

18. This would be true for some central-city populations. See Annest, J. L., Mahaffey, K. R., Cox, D. H., and Roberts, J. (1982). Blood lead levels for persons 6 months-74 years of age: United States, 1976–80. *U.S. National Center for Health Statistics, Advance Data from Vital and Health Statistics*, No. 79. Hyattsville, Maryland: U.S. Department of Health and Human Services.

19. Geissler, A. (1889). Beitrage zur Frage des Geschlechtsverhaltnisses der Geborenen. *Zeitschrft des K. Sachsischen Statistischen Bureaus* **35**, 1–24. Data reproduced by Edwards, A. W. F. (1958). An analysis of Geissler's data on the human sex ratio. *Annals of Human Genetics,* **23**, 6–15. The data are also discussed by Stern, C. (1960). *Human Genetics.* San Francisco: Freeman.

20. Haseman, J. K. and Soares, E. R. (1976). The distribution of fetal death in control mice and its implications on statistical tests for dominant lethal effects. *Mutation Research,* **41**, 277–288.

21. Data courtesy of S. N. Postlethwaite.

22. Adapted from Looker, A., et al. (1997). Prevalence of iron deficiency in the United States. *Journal of the American Medical Association,* **277**, 973–976.

23. Fictitious but realistic situation. See Krebs, C. J. (1972). *Ecology: The Experimental Analysis of Distribution and Abundrince.* New York: Harper & Row, p. 142.

24. See Mather, K. (1943). *Statistical Analysis in Biology.* London: Methuen. p. 38.

25. The technique is described in Waid, W. M., Orne, E. C., Cook, M. R., and Orne, M. T. (1981). Meprobamate reduces accuracy of physiological detection of deception. *Science,* **212**, 71–73.

26. Fictitious but realistic population, closely resembling the population of males aged 45–59 years as described in Roberts, J. (1975). Blood pressure of persons 18–74 years, United States, 1971–72. *U.S. National Center for Health Statistics, Vital and Health Statistics Series 11, No. 150.* Washington, D.C.: U.S. Department of Health, Education and Welfare.

Chapter 4

1. Levy, P. S., Hamill, P. V. V., Heald, F., and Rowland, M. (1976). Total serum cholesterol values of youths 12–17 years, United States. *U.S. National Center for Health Statistics, Vital and Health Statistics Series 11, No. 156.* Washington, D.C.: U.S. Department of Health, Education and Welfare.

2. Ikeme, A. I., Roberts, C., Adams, R. L., Hester, P. Y., and Stadelman, W. J. (1983). Effects of supplementary water-administered vitamin D_3 on egg shell thickness. *Poultry Science* **62**, 1120–1122. The normal curve was fitted to raw data provided courtesy of W. J. Stadelman and A. I. Ikeme.

3. Hengstenberg, R. (1971). Das Augenmuskelsystem der Stubenfliege Musca domestica. 1. Analyse der "clock-spikes" und ihrer Quellen. *Kybernetik* **2**, 56–57.

4. Adapted from Magath, T. B. and Betkson, J. (1960). Electronic blood-cell counting. *American Journal of Clinical Pathology* **34**, 203–213. Actually, the percentage error is somewhat less for high counts and somewhat more for low counts. Described in Coulter Electronics (1982). *Performance Characteristics and Specifications for Coulter Counter Model S-560*. Hialeah, Florida: Coulter Electronics.

5. Fictitious but realistic population. Adapted from data given by Hildebrand, S. F. and Schroeder, W. C. (1927). Fishes of Chesapeake Bay. *Bulletin of the United States Bureau of Fisheries* **43**, Part 1, p. 88. The fish are young of the year, observed in October; they are quite small. (The distribution of lengths in older populations is not approximately normal.)

6. Adapted from Pearl, R. (1905). Biometrical studies on man. I. Variation and correlation in brain weight. *Biometrika* **4**, 13–104.

7. Adapted from Swearingen, M. L. and Halt, D. A. (1976). Using a "blank" trial as a teaching tool. *Journal of Agronomic Education* **5**, 3–8. The standard deviation given in this problem is realistic for an idealized "uniform" field, in which yield differences between plots are due to local random variation rather than large-scale and perhaps systematic variation.

8. Adapted from Coulter Electronics (1982). *Performance Characteristics and Specifications for the Coulter Counter Model S-560*. Hialeah, Florida: Coulter Electronics.

9. Unpublished data courtesy of Susan Whitehead, Oberlin College.

10. Unpublished data courtesy of Kaelyn Stiles, Oberlin College.

11. Some software programs create normal probability plots with the normal scores on the vertical axis and the observed data on the horizontal axis.

12. Unpublished data courtesy of Paul Harnik and Lydia Ries, Oberlin College.

13. Data taken from *Climatological Data, Ohio*, and *Local Climatological Data, Cleveland, Ohio*; National Oceanic and Atmospheric Administration, U.S. Dept. of Commerce.

14. Fictitious but realistic population. Adapted from data given by Falconer, D. S. (1981). *Introduction to Quantitative Genetics*, 2nd edition. New York: Longman, p. 97.

15. Fictitious but realistic population. Adapted from data given in Falconer, D. S. (1981). *Introduction to Quantitative Genetics*, 2nd edition. New York: Longman, p. 97.

16. Fictitious but realistic population. Based on unpublished data provided by W. F. Jacobson.

17. Long, E. C. (1976). *Liquid Scintillation Counting Theory and Techniques*. Irvine, California: Beckman Instruments. The distribution is actually a discrete distribution called a Poisson distribution; however, a Poisson distribution with large mean is approximately normal.

18. Fictitious but realistic population, based on data of Emerson, R. A. and East, E. M. (1913). Inheritance of quantitative characters in maize. *Nebraska Experimental Station Research Bulletin* **2**. Data reproduced by Mather, K. (1943). Statistical Analysis in Biology. London: Methuen, pp. 29, 34. Modern hybrid corn is taller and less variable than this population.

19. These percentiles are based on data in the National Health and Nutrition Examination Survey (NHANES), conducted by the National Center for Health Statistics Centers for Disease Control and Prevention. The following web URL provides a link to the data table:
http://www.cdc.gov/nchs/about/major/nhanes/hgtfem.pdf.

20. This is the standard reference distribution for Stanford-Binet scores. See Sattler, J. M. (1982). *Assessment of Children's Intelligence and Special Abilities*, 2nd edition. Boston: Allyn and Bacon, p. 19 and back cover.

21. Unpublished data courtesy of Forrest Crawford and Yvonne Piper, Oberlin College.

Chapter 5

1. This value is approximately correct for American adults. See Roberts, J. (1964). Binocular visual acuity of adults, United States 1960–62. *U.S. National Center for Health Statistics, Vital and Health Statistics Series* **11**, *No. 3*. Washington, D.C.: U.S. Department of Health, Education and Welfare.

2. Fictitious but realistic population. See Example 2.13.

3. The mean and standard deviation are realistic for American women aged 25–34. See O'Brien, R. J. and Drizd, T. A. (1981). Basic data on spirometry in adults 25–74 years of age: *United States, 1971–75. U.S. National Center for Health Statistics, Vital and Health Statistics Series* **11**, *No. 222*. Washington, D.C.: U.S. Department of Health and Human Services. The normality assumption may or may not be realistic.

4. Adapted from data given in Sebens, K. P. (1981). Recruitment in a sea anemone population; juvenile substrate becomes adult prey. *Science* **213**, 785–787.

5. Fictitious but realistic data. Adapted from distribution given for men aged 45–59 in Roberts, J. (1975). Blood pressure of persons 18–74 years, United States, 1971–72. *U.S. National Center for Health Statistics, Vital and Health Statistics Series* **11**, *No. 150*. Washington, D.C.: U.S. Department of Health, Education and Welfare.

6. Based on data in Roberts, J. D., et al., (1997) Inhaled Nitric Oxide and Persistent Pulmonary Hypertension of the Newborn. *New England Journal of Medicine* **336**, 605–10.

7. The distribution in Figure 5.13 is based on data given in Zeleny, C. (1922). The effect of selection for eye facet number in the white bar-eye race of *Drosophila melanogaster. Genetics* **7**, 1–115. The data are displayed in Falconer, D. S. (1981). *Introduction to Quantitative Genetics*, 2nd edition. New York: Longman, p. 97.

8. The distribution in Figure 5.15 is adapted from data described by Bradley, J. V. (1980). Nonrobustness in one-sample Z and *t* tests: A large-scale sampling study. *Bulletin of the Psychonomic Society 15 (1)*, 29–32, used by permission of the Psychonomic Society, Inc.; and Bradley, J. V. (1977). A common situation conducive to bizarre distribution shapes. *American Statistician* **31**, 147–150. Bradley's distribution included additional peaks, because sometimes the subject fumbled the button more than once on a single trial.

9. Fictitious but realistic situation, adapted from data given in Bradley, D. D., Krauss, R. M., Petitte, D. B., Ramcharin, S., and Wingird, I. (1978). Serum high-density lipoprotein cholesterol in women using oral contraceptives, estrogens, and progestins. *New England Journal of Medicine* **299**, 17–20.

10. Kahneman, D. and Tversky, A. (1972). Subjective probability: A judgment of representativeness. *Cognitive Psychology* **3**, 430–454.

11. Strickberger, M. W. (1976). *Genetics*, 2nd edition. New York: Macmillan, p. 206.

12. Fictitious but realistic situation. See Waugh, G.D. (1954). The occurrence of Mytilicola intestinalis (Steuer) on the east coast of England. *Journal of Animal Ecology* **23**, 364–367.

13. Mosteller, F. and Tukey, J. W. (1977). Data *Analysis and Regression*. Reading, Massachusetts: Addison-Wesley, p. 25.

14. This is typical for U.S. populations. See, for example, Maccready, R. A. and Mannin, M. C. (1951). A typing study of one hundred and fifty thousand bloods. *Journal of Laboratory and Clinical Medicine* **37**, 634–636.

15. Fictitious but realistic population, resembling the population of young American men aged 18–24, as described in Abraham, S., Johnson, C. L., and Najjar, M. F. (1979). Weight and height of adults 18–74 years of age: United States 1971–1974. *U.S. National Center for Health Statistics Series* **11**, No. 211. Washington, D.C.: U.S. Department of Health, Education and Welfare.

16. The mean and standard deviation are realistic, based on unpublished data provided courtesy of J. Y. Ustimer and C. A. Mitchell. The normality assumption may or may not be realistic.

17. Fictitious but realistic situation. See Krebs, C. J. (1972). *Ecology: The Experimental Analysis of Distribution and Abundance*. New York: Harper and Row.

18. The mean and standard deviation are realistic, based on unpublished data provided courtesy of S. Newman and D. L. Harris. The normality assumption may or may not be realistic.

Chapter 6

1. Pappas, T. and Mitchell, C. A. (1984). Effects of seismic stress on the vegetative growth of *Glycine max* (L.) Merr. cv. Wells II. *Plant, Cell and Environment* **8**, 143–148. Reprinted with permission of Blackwell Scientific Publications Limited. Raw data courtesy of the authors. The actual experiment included several groups of plants grown under different environmental conditions.

2. Newman, S., Everson, D. O., Gunsett, F. C., and Christian, R. E. (1984). Analysis of two- and three-way crosses among Ramhouillet, Targhee, Columbia and Suffolk sheep for three preweaning traits. Unpublished manuscript. Raw data courtesy of S. Newman.

3. Adapted from the following two papers. Potkin, S. G., Cannon, H. E., Murphy, D. L., and Wyatt, R. J. (1978). Are paranoid schizophrenics biologically different from other schizophrenics? *New England Journal of Medicine* **298**, 61–66. Murphy, D. L., Wright, C., Buchsbaum, M., Nichols, A., Costa, J. L., and Wyatt, R. J. (1976). Platelet and plasma amine oxidase activity in 680 normals: Sex and age differences and stability over time. *Biochemical Medicine* **16**, 254–265. The data displayed are fictitious but realistic, having been reconstructed from the histograms and summary information given by Potkin et al. and Murphy et al.

4. Based on data reported in Rea, T. M., Nash, J. F., Zabik, J. E., Born, G. S., and Kessler, W. V. (1984). Effects of toluene inhalation on brain biogenic amines in the rat. *Toxicology* **31**, 143–150.

5. Based on an experiment by M. Morales.

6. Adapted from Cherney, J. H., Volenec, J. J., and Nyquist, W. E. (1985). Sequential fiber analysis of forage as influenced by sample weight. *Crop Science* **25**, 6, Nov/Dec. 1985, 1113–1115 (Table 1). By permission of the Crop Science Society of America, Inc. Raw data courtesy of W. E. Nyquist.

7. Dice, L. R. (1932). Variation in the geographic race of the deermouse, *Peromyscus maniculatus bairdii*. *Occasional Papers of the Museum of Zoology, University of Michigan*, No. 239. Data reproduced in Simpson, G. G., Roe, A.,

and Lewontin, R. C. (1960). *Quantitative Zoology.* New York: Harcourt, Brace, p. 79.

8. Bodor, N. and Simpkins, J. W. (1983). Redox delivery system for brain-specific, sustained release of dopamine. *Science* **221**, 65–67.

9. Student (W. S. Gosset) (1908). The probable error of a mean. *Biometrika* **6**, 1–25.

10. The Writing Group for the PEPI Trial (1996). Effects of Hormone Therapy on Bone Mineral Density. *Journal of the American Medical Association* **276**, 1389–1396. This study compared change in bone mineral density over 36 months for four medications and a placebo. (Hip bone mineral density was measured at the beginning of the experiment and again 36 months later.) Only the data for those women who adhered to the experimental protocol are used in the example. Standard deviations are calculated based on the standard errors reported in the article.

11. Data collected by Denise D'Abundo, Oberlin College, April 1991.

12. Bockman, D. E. and Kirby, M. L. (1984). Dependence of thymus development on derivatives of the neural crest. *Science* **223**, 498–500. Copyright 1984 by the AAAS.

13. Brown, S. A., Riviere, J. E., Coppoc, G. L., Hinsman, E. J., Carlton, W. W., and Steckel, R. R. (1985). Single intravenous and multiple intramuscular dose pharmacokinetics and tissue residue profile of gentamicin in sheep. *American Journal of Veterinary Research* **46**, 69–74. Raw data courtesy of S. A. Brown and G. L. Coppoc.

14. Lobstein, D. D. (1983). A multivariate study of exercise training effects on beta-endorphin and emotionality in psychologically normal, medically healthy men. Ph.D. thesis, Purdue University. Raw data courtesy of the author.

15. Nicholson, R. L. and Moraes, W. B. C. (1980). Survival of *Colletotrichum graminicola:* Importance of the spore matrix. *Phytopathology* **70**, 255–261.

16. Adapted from Morris, J. G., Gripe, W. S., Chapman, H. L., Jr., Walker, D. F., Armstrong, J. B., Alexander, J. D., Jr., Miranda, R., Sanchez, A., Jr., Sanchez, B., Blair-West, J. R., and Denton, D. A. (1984). Selenium deficiency in cattle associated with Heinz bodies and anemia. *Science* **223**, 491–492. Copyright 1984 by the AAAS.

17. Shaffer, P. L. and Rock, G. C. (1983). Tufted apple budmoth (Lepidoptera: Tortricidae): Effects of constant daylengths and temperatures on larval growth rate and determination of larval-pupal ecdysis. *Environmental Entomology* **12**, 76–80.

18. Bishop, N. J., Morley, R., Day, J. P., and Lucas, A. L. (1997). Aluminum neurotoxicity in preterm infants receiving intravenous-feeding solutions. *New England Journal of Medicine* **336**, 1557–1561.

19. Kaufman, J. S., Reda, D. J., Fye, C. L., Goldfarb, D. S., Henderson, W. G., Kleinman, J. G., and Vaamonde, C. A., (1998). Subcutaneous compared with intravenous epoetin in patients receiving hemodialysis. *New England Journal of Medicine* **339**, 578–583.

20. Based on data provided by C. H. Noller.

21. This is roughly the SD for the U.S. population of middle-aged men. See Moore, F. E. and Gordon, T. (1973). Serum cholesterol levels in adults, United States 1960–62. *U.S. National Center for Health Statistics, Vital and Health Statistics Series* **11**, *No. 22*. Washington, D.C.: U.S. Department of Health, Education and Welfare.

22. Noll, S. L., Waibel, P. E., Cook, R. D., and Witmer, J. A. (1984). Biopotency of methionine sources for young turkeys. *Poultry Science* **63**, 2458–2470.

23. Schaeffer, J., Andrysiak, T., and Ungerleider, J. T. (1981). Cognition and long-term use of ganja (cannabis). *Science* **213**, 465–466.

24. Desai, R. (1982). An anatomical study of the canine male and female pelvic diaphragm and the effect of testosterone on the status of the levator ani of male dogs. *Journal of the American Animal Hospital Association* **18**, 195–202.

25. The probabilities in Table 6.4 were estimated by computer simulation carried out by M. Samuels and R. P. Becker. The standard error of each probability estimate is less than .0015. The sources of the parent distributions are given in Notes 7 and 8 to Chapter 5.

26. Burnett, A. and Haywood, A. (1997). A statistical analysis of differences in sediment yield over time on the West Branch of the Black River. Unpublished manuscript, Oberlin College.

27. Hessell, E. A., Johnson, D. D., Ivey, T. D., and Miller, D. W. (1980). Membrane vs bubble oxygenator for cardiac operations. *Journal of Thoracic and Cardiovascular Surgery* **80**, 111–122.

28. Peters, H. G. and Bademan, H. (1963). The form and growth of stellate cells in the cortex of the guinea-pig. *Journal of Anatomy (London)* **97**, 111–117.

29. Kaneto, A., Kosaka, K., and Nakao, K. (1967). Effects of stimulation of the vagus nerve on insulin secretion. *Endocrinology* **80**, 530–536. Copyright © 1967 by The Endocrine Society.

30. Simmons, F. J. (1943). Occurrence of superparasitism in *Nemeritis canescens*. *Revue Canadienne de Biologie* **2**, 15–40. Data reproduced in Williams, C. B. (1964). *Patterns in the Balance of Nature.* London: Academic Press, p. 223.

31. These data are diameters at breast height of American Sycamore trees in the floodplain of the Vermilion River. Data collected Emily Norland, Oberlin College, March 1995.

32. Looker, A. C., Dallman, P. R., Carroll, M. D., Gunter, E. W., and Johnson, C. L. (1997). Prevalence of iron deficiency in the United States. *Journal of the American Medical Association* **277**, 973–976. The figure of 71 out of 786 is calculated from the reported sample proportion of 9 percent, but is not reported directly in the article.

33. Agresti, A., and Coull, B. A. (1998). Approximate is better than "exact" for interval estimation of binomial proportions. *The American Statistician* **52**, 119–126. The authors show that 95% confidence intervals based on $\tilde{p}$ are superior to other commonly used confidence intervals. They also note that if one uses $\tilde{p}$, then it is not necessary to construct tables or rules for how large the sample size needs to be in order for the confidence interval to have good coverage properties.

34. Couch, F. J., et al. (1997). *BRCA1* mutations in women attending clinics that evaluate the risk of breast cancer. *New England Journal of Medicine* **336**, 1409–1415.

35. Ware, J. H. (1989). Investigating therapies of potentially great benefit: ECMO. *Statistical Science* **4**, 298–306. The ECMO data are discussed in greater detail in Section 10.4.

36. Oldfield, R. C. (1971). The assessment and analysis of handedness: The Edinburgh inventory. *Neuropsychologia* **9**, 97–113.

37. Adapted from McCloskey, R. V., Goren, R., Bissett, D., Bentley, J., and Tutlane, V. (1982). Cefotaxime in the treatment of infections of the skin and skin structure. *Reviews of Infectious Diseases* **4**, Supp., S444–S447.

38. Adapted from Petras, M. L. (1967). Studies of natural populations of *Mus*. III. Coat color polymorphisms. *Canadian Journal of Genetic Cytology* **9**, 287–296.

39. Miller, C. L., Pollock, T. M., and Clewer, A. D. F. (1974). Whooping-cough vaccination: An assessment. *The Lancet*, **ii**, 510–513.

40. Hayes, D. L., et al. (1997). Interference with cardiac pacemakers by cellular telephones. *New England Journal of Medicine* **336**, 1473–1479. The data cited are for CDMA telephones. Although interference was recorded in 15.7% of the tests, a much smaller percentage of the tests caused symptoms that were clinically significant.

41. Erskine, A. G. and Socha, W. W. (1978). *The Principles and Practices of Blood Grouping.* St. Louis: Mosby. p. 209.

42. Curtis, H. (1983). *Biology*, 4th edition. New York: Worth, p. 908.

43. Mourant, A. E., Kopec, A. C., and Domaniewska-Sobczak, K. (1976). *The Distribution of Human Blood Groups and Other Polymorphisms*, 2nd edition. London: Oxford University Press, p. 44.

44. Lamb, M. L., Fishbein, M., Douglas, J. M., Rhodes, F., Rogers, J., Bolan, G., Zenilman, J., Hoxworth, T., Malotte, K., Iatesta, M., Kent, C., Lentz, A., Graziano, S., Byers, R. H., and Peterman, T. A. (1998). Efficacy of risk-reduction counseling to prevent human immunodeficiency virus and sexually transmitted diseases. *Journal of the American Medical Association* **280**, 1161–1167.

45. Based on an experiment described in Oellerman, C. M., Patterson, F. L., and Gallun, R. L. (1983). Inheritance of resistance in "Luso" wheat to Hessian fly. *Crop Science* **23**, 221–224.

46. Parks, N.J., Krohn, K. A., Mathis, C. A., Chasko, J. H., Geiger, K. R., Gregor, M. E., and Peek, N. F. (1981). Nitrogen-13-labelled nitrite and nitrate: Distribution and metabolism after intratracheal administration. *Science* **212**, 58–61. Copyright 1981 by the AAAS. Raw data courtesy of N. J. Parks.

47. Krick, J. A. (1982). Effects of seeding rate on culm diameter and the inheritance of culm diameter in soft red winter wheat (*Triticum aestivum* L. em Thell). M. S. thesis, Department of Agronomy, Purdue University. Raw data courtesy of J. A. Krick and H. W. Ohm. Each diameter is the mean of measurements taken at six prescribed locations on the stem.

48. Data collected by Deborah Ignatoff, Oberlin College, spring 1997.

49. Bailey, J. and Marshall, J. (1970). The relationship of the post-ovulatory phase of the menstrual cycle to total cycle length. *Journal of Biosocial Science* **2**, 123–132.

50. Nansen, C., Tchabi, A., and Meikle, W. G. (2001). Successional sequence of forest types in a disturbed dry forest reserve in southern Benin, West Africa. *Journal of Tropical Ecology* **17**, 525–539.

51. Unpublished data courtesy of W. F. Jacobson.

52. Dale, E. M. and Housley, T. L. (1986). Sucrose synthase activity in developing wheat endosperms differing in maximum weight. *Plant Physiology* **82**, 7–10. Raw data courtesy of the authors.

53. Cheatum, F. L. and Severinghaus, C. W. (1950). Variations in fertility of white-tailed deer related to range conditions. *Transactions of the North American Wildlfe Conference* **15**, 170–189.

54. Duggan, D. J., Gorospe, J. R., Fanin, M., Hoffman, E. P., and Angelini, C. (1997). Mutations in the sarcoglycan genes in patients with myopathy. *New England Journal of Medicine* **336**, 618–624.

55. Looker, A. C., et al. op cit.

56. Roberts, J. (1964). Binocular visual acuity of adults, United States 1960–62. *U.S. National Center for Health Statistics Series* **11**, *No. 3*. Washington, D.C.: U.S. Department of Health, Education and Welfare.

57. Erne, P., Bolli, P., Buergisser, E., and Buehler, F. R. (1984). Correlation of platelet calcium with blood pressure. *New England Journal of Medicine* **310**, 1084–1088. Reprinted by permission. Raw data courtesy of F. R. Buehler.

Chapter 7

1. Heald, F. (1974). Hematocrit values of youths 12–17 years, United States. *U.S. National Center for Health Statistics, Vital and Health Statistics Series* **11**, *No. 146*. Washington, D.C.: U.S. Department of Health, Education and Welfare. Actually, the data were obtained by a sampling scheme more complicated than simple random sampling.

2. Long, T. F. and Murdock, L. L. (1983). Stimulation of blowfly feeding behavior by octopaminergic drugs. *Proceedings of the National Academy of Sciences*, **80**, 4159–4163. Raw data courtesy of the authors and L. C. Sudlow.

3. Hunter, A. and Terasaki, T. (1993). Statistical analysis comparing vital capacities of brass majors in the Conservatory and a normal population. Unpublished manuscript, Oberlin College. All subjects were men, age 18–21, with heights between 175 and 183 cm. Because vital capacity is related to height, the raw data were adjusted slightly, using linear regression, to control for the effect of height.

4. Knight, S. L. and Mitchell, C. A. (1983). Enhancement of lettuce yield by manipulation of light and nitrogen nutrition. *Journal of the American Society for Horticultural Science*, **108**, 750–754. Raw data courtesy of the authors. (The actual sample sizes were equal; some observations have been omitted from the exercise.)

5. O'Marra, S. (1996). Antibacterial soaps: myth or reality. Unpublished manuscript, Oberlin College. The primary purpose of this study was to assess the effectiveness of antibacterial soaps. A solution made from antibacteria soap killed all *E. coli*, in contrast to the non-antibacterial soap and the control. The soap solution was a 1 : 4 solution of soap and water.

6. Ahern, T. (1998). Statistical analysis of EIN plants treated with ancymidol and H_2O. Unpublished manuscript, Oberlin College. The mutant strain EIN (elongated in-ternode) of Brassica was used in this experiment. The data presented here are a randomly selected subset of the full dataset.

7. Rea, T. M., Nash, J. F., Zabik, J. E., Born, G. S., and Kessler, W. V. (1984). Effects of toluene inhalation on brain biogenic amines in the rat. *Toxicology*, **31**, 143–150. Raw data courtesy of J. F. Nash and J. E. Zabik.

8. Hagerman, A. E. and Nicholson, R. L. (1982). High-performance liquid chromatographic determination of hydroxycinnamic acids in the maize mesocotyl. *Journal of Agricultural and Food Chemistry*, **30**, 1098–1102. Reprinted with permission. Copyright 1982 American Chemical Society.

9. Patel, C., Marmot, M. M., and Terry, D. J. (1981). Controlled trial of biofeedback-aided behavioral methods in reducing mild hypertension. *British Medical Journal*, **282**, 2005–2008.

10. Lipsky, J. J., Lewis, J. C., and Novick, W. J., Jr. (1984). Production of hypoprothrombinemia by Moxalactam and 1-methyl-5-thiotetrazole in rats. *Antimicrobial Agents and Chemotherapy*, **25**, 380–381.

11. Gwynne, D. T. (1981). Sexual difference theory: Mormon crickets show role reversal in mate choice. *Science*, **213**, 779–780. Copyright 1981 by the AAAS. Data provided courtesy of the author.

12. Appel, L. J., et al. (1997). A clinical trial of the effects of dietary patterns on blood pressure. *New England Journal of Medicine*, **336**, 1117–1124.

13. Crawford, F. and Piper, Y. (1999). How does caffeine influence heart rate. Unpublished manuscript, Oberlin College. There were 10 subjects in the caffeine group, but an outlier was deleted from the data.

14. Gent, A. (1999). Unpublished data collected at Oberlin College. The colors of light were created using gels: thin pieces of colored plastic used in theater lighting.

15. Sagan, C. (1977). *The Dragons of Eden*. New York: Ballantine, p. 7.

16. Lemenager, R. P., Nelson, L. A., and Hendrix, K. S. (1980). Influence of cow size and breed type on energy requirements. *Journal of Animal Science*, **51**, 566–576. Some of the animals *lost* weight during the 78 days, so that the mean weight gains are based on both positive and negative values.

17. Adapted from Miyada, V. S. (1978). Uso da levedura seca de distilarias de alcool de cana de acucar na alimentacao de suinos em crescimento e acabamento. Master's thesis, University of Sao Paulo, Brazil.

18. Kalsner, S. and Richards, R. (1984). Coronary arteries of cardiac patients are hyperreactive and contain stores of amines: A mechanism for coronary spasm. *Science*, **223**, 1435–1437. Copyright 1984 by the American Association for the Advancement of Science (AAAS).

19. Adapted from Dybas, H. S. and Lloyd, M. (1962). Isolation by habitat in two synchronized species of periodical cicadas (Homoptera, Cicadidae, *Magicicada*). *Ecology*, **43**, 444–459.

20. Bockman, D. E. and Kirby, M. L. (1984). Dependence of thymus development on derivatives of the neural crest. *Science*, **223**, 498–500. Copyright 1984 by the AAAS.

21. Tripepi, R. R. and Mitchell, C. A. (1984). Metabolic response of river birch and European birch roots to hypoxia. *Plant Physiology*, **76**, 31–35. Raw data courtesy of the authors.

22. Lamke, L. O. and Liljedahl, S. O. (1976). Plasma volume changes after infusion of various plasma expanders. *Resuscitation*, **5**, 93–102.

23. Anderson, J. W., Story, L., Sieling, B., Chen, W. J. L., Petro, M. S., and Story, J. (1984). Hypocholesterolemic effects of oat-bran or bean intake for hypercholesterolemic men. *The American Journal of Clinical Nutrition*, **40**, 1146–1155.

24. Ahne, A and Myers, S. (1999). The effect of Miracle Grow on radish growth. Unpublished manuscript, Oberlin College. The data presented here are a subset of the full dataset. (The means and standard deviations for the full dataset are similar to those for the subset presented here. In particular, the sample mean for the control group is greater than for the fertilizer group.)

25. Katzner, T. (1991). Interspecific competition between an ecto- and an endoparasite of *Anagasta kuehniella*. Unpublished manuscript, Oberlin College. This research found that total brood size does not differ between the two groups. However, the mortality rates of the two groups do differ: the average size of the adult population is smaller for the stung larva than it is for the unstung larva.

26. Adapted from data provided courtesy of D. R. Shields and D. K. Colby. See Shields, D. R. (1981). The influence of niacin supplementation on growing ruminants and *in vivo* and *in vitro* rumen parameters. Ph.D. thesis, Purdue University.

27. Schall, J. J., Bennett, A. F., and Putnam, R. W. (1982). Lizards infected with malaria: Physiological and behavioral consequences. *Science*, **217**, 1057–1059. Copyright 1982 by the AAAS.

28. Agosti, E. and Camerota, G. (1965). Some effects of hypnotic suggestion on respiratory function. *International Journal of Clinical and Experimental Hypnosis*, **13**, 149–156.

29. Adapted from Knight, S. L. and Mitchell, C. A. (1983). Enhancement of lettuce yield by manipulation of light and nitrogen nutrition. *Journal of the American Society for Horticultural Science*, **108**, 750–754.

30. Unpublished data courtesy of J. L. Wolfson.

31. Fictitious but realistic data.

32. Shima, J. S. (2001). Recruitment of a coral reef fish: roles of settlement, habitat, and postsettlement losses. *Ecology*, **82**, 2190–2199. Raw data courtesy of the author.

33. Williams, G. Z., Widdowson, G. M., and Penton, J. (1978). Individual character of variation in time-series studies of healthy people. II. Differences in values for clinical chemical analytes in serum among demographic groups, by age and sex. *Clinical Chemistry*, **24**, 313–320. Copyright American Association for Clinical Chemistry, Inc. Reprinted with permission from AACC.

34. Fictitious but realistic data. See Abraham, S., Johnson, C. L., and Najjar, M. F. (1979). Weight and height of adults 18–74 years of age, United States 1971–74. *U.S. National Center for Health Statistics, Vital and Health Statistics Series* **11**, *No. 211*. Washington, D.C.: U.S. Department of Health, Education and Welfare.

35. Example communicated by D. A. Holt.

36. Petrie, B. and Segalowitz, S. J. (1980). Use of fetal heart rate, other perinatal and maternal factors as predictors of sex. *Perceptual and Motor Skills*, **50**, 871–874. Reprinted by permission of the authors and publisher.

37. Hagerman, A. E. and Nicholson, R. L. (1982). High-performance liquid chromatographic determination of hydroxycinnamic acids in the maize mesocotyl. *Journal of Agricultural and Food Chemistry*, **30**, 1098–1102. Copyright 1982 American Chemical Society. Reprinted with permission.

38. Adapted from Williams, G. Z., Widdowson, G. M., and Penton, J. (1978). Individual character of variation in time-series studies of healthy people. II. Difference in values for clinical chemical analytes in serum among demographic groups, by age and sex. *Clinical Chemistry*, **24**, 313–320. Copyright American Association for Clinical Chemistry, Inc. Reprinted with permission from AACC.

39. Hamill, P. V. V., Johnston, F. E., and Lemeshow, S. (1973). Height and weight of youths 12–17 years, United States. *U.S. National Center for Health Statistics, Vital and Health Statistics Series* **11**, *No. 124*. Washington, D.C.: U.S. Department of Health, Education and Welfare.

40. Balon, J., Aker, P. D., Erne, P., Crowther, E. R., Danielson, C., Cox, P. G., O'Shaughnessy, D., Walker, C., Goldsmith, C. H., Duku, E., and Sears, M. R. (1998). A comparison of active treatment and simulated chiropractic manipulation as adjunctive treatment for childhood asthma. *New England Journal of Medicine*, **339**, 1013–1020.

41. Roberts, J. (1975). Blood pressure of persons 18–74 years, United States, 1971–72. *U.S. National Center for Health Statistics, Vital and Health Statistics Series* **11**, *No. 150*. Washington, D.C.: U.S. Department of Health, Education and Welfare. However, the distribution of systolic blood pressure is more skewed (see Exercise 5.27).

42. Pearson, E. S. and Please, N. W. (1975). Relation between the shape of population distribution and the robustness of four simple test statistics. *Biometrika*, **62**, 223–241.

43. Mena, E. A., Kossovsky, N., Chu, C., and Hu, C. (1995). Inflammatory intermediates produced by tissues encasing silicone breast implants. *Journal of Investigative Surgery*, **8**, 31–42. *Note*: There were two control groups in this study. The control group included in this analysis is "patients undergoing reverse augmentation mammaplasty" (the "scar" group discussed in the article). Also, the authors neglected to transform the data before conducting a *t* test. Thus, they got a large *P*-value, although they noted that the two groups looked quite different.

44. Fierer, N. (1994). Statistical analysis of soil respiration rates in a light gap and surrounding old-growth forest. Unpublished manuscript, Oberlin College.

45. Noether, G. E. (1967). *Elements of nonparametric statistics*. New York: Wiley.

46. It is sometimes stated that the validity of the Mann-Whitney test requires that the two population distributions have the same shape, and differ only by a shift. This is not correct. The computations underlying Table 6 require only that the common population distribution (under the null hypothesis) be continuous. A further property, technically called *consistency* of the test, requires that the two distributions be *stochastically ordered*, which is the technical way of saying that one of the variables has a consistent tendency to be larger than the other. In fact, the title of Mann and Whitney's original paper is "On a test of whether one of two random variables is stochastically larger than the other" (*Annals of Mathematical Statistics*, **18**, 1947). In Section 7.12 we discuss the requirement of stochastic ordering, calling it an "implicit assumption." (The confidence interval procedure mentioned at the end of Section 7.11 does require the stronger assumption that the distributions have the same shape.)

47. Rea, T. M., Nash, J. F., Zabik, J. E., Born, G. S., and Kessler, W. V. (1984). Effects of toluene inhalation on brain biogenic amines in the rat. *Toxicology*, **31**, 143–150. Raw data courtesy of J. F. Nash.

48. Agosti, E. and Camerota, G. (1965). Some effects of hypnotic suggestion on respiratory function. *International Journal of Clinical and Experimental Hypnosis*, **13**, 149–156.

49. Connolly, K. (1968). The social facilitation of preening behaviour in *Drosophila melanogaster. Animal Behaviour*, **16**, 385–391.

50. Unpublished data courtesy of G. P. Carlson and M. A. Morse.

51. Lobstein, D. D. (1983). A multivariate study of exercise training effects on beta-endorphin and emotionality in psychologically normal, medically healthy men. Ph.D. thesis, Purdue University. Raw data courtesy of the author.

52. Erne, P., Bolli, P., Buergisser, E., and Buehler, F. R. (1984). Correlation of platelet calcium with blood pressure. *New England Journal of Medicine*, **310**, 1084–1088. Reprinted by permission of the *New England Journal of Medicine*. Summary statistics calculated from raw data provided courtesy of F. R. Buehler.

53. Adapted from unpublished data provided by F. Delgado. The extremely high somatic cell counts probably represent cases of mastitis.

54. Pappas, T. and Mitchell, C. A. (1985). Effects of seismic stress on the vegetative growth of *Glycine max* (L.) Merr. cv. Wells II. *Plant, Cell and Environment*, **8**, 143–148. Reprinted with permission of Blackwell Scientific Publications Limited. Raw data courtesy of the authors. The original experiment included more than two treatment groups.

55. Wee, K. (1995). Species diversity in floodplain forests. Unpublished manuscript, Oberlin College.

56. Cicirelli, M. F., Robinson, K. R., and Smith, L. D. (1983). Internal pH of *Xenopus* oocytes: A study of the mechanism and role of pH changes during meintic maturation. *Developmental Biology*, **100**, 133–146. Raw data courtesy of M. F. Cicirelli.

57. Manski, T. J., et al. (1997). Endolymphatic sac tumors: a source of morbid hearing loss in von Hippel-Lindau disease. *Journal of the American Medical Association*, **277**, 1461–1466.

58. Unpublished data courtesy of J. A. Henricks and V. J. K. Liu.

59. Schall, J. J., Bennett, A. F., and Putnam, R. W. (1982). Lizards infected with malaria: Physiological and behavioral consequences. *Science*, **217**, 1057–1059. Copyright 1982 by the AAAS. Raw data courtesy of J. J. Schall.

60. Unpublished data courtesy of M. B. Nichols and R. P. Maickel.

61. The neonatal inhaled nitric oxide study group (1997). Inhaled nitric oxide in full-term and nearly full-term infants with hypoxic respiratory failure. *New England Journal of Medicine*, **336**, 597–604.

62. Gleason, P. P., et al. (1997). Medical outcomes and antimicrobial costs with the use of American Thoracic Society guidelines for outpatients with community-acquired pneumonia. *Journal of the American Medical Association*, **278**, 32–39.

63. Hodapp, M. (1998). A Study of CDS Nutrition. Unpublished manuscript, Oberlin College.

64. Laurance, W. F., Perez-Salicrup, D., Delamonica, P., Fearside, P. M., D'Angelo, S., Jerozolinski, A., Pohl, L. and Lovejoy, T. E. (2001). Rain forest fragmentation and the structure of Amazonian liana communities. *Ecology*, **82**, 105–116. The data presented were read by J. Witmer from Figure 2 in the paper and may not be completely accurate.

65. King, D. S., Sharp, R. L., Vukovich, M. D., Brown, G. A., Reifenrath, T. A., Uhl, N. L., and Parsons, K. A. Effect of oral androstenedione on serum testosterone and adaptations to resistance training in young men. *Journal of the American Medical Association*, **281**, 2020–2028. Raw data courtesy of the authors. The response variable shown here is change in "maximum muscle strength," which is the greatest weight the subject could lift. There were several other measurements taken in the experiment; generally they showed the same results seen in the lat pulldown data. A primary purpose of the experiment was to study the effect of andro on testosterone level. The researchers found that andro had no effect on serum testosterone level.

66. Fleming, M. F., Barry, K. L., Manwell, L. B., Johnson, K., and London, R. (1997). Brief physician advice for problem alcohol drinkers. *Journal of the American Medical Association*, **277**, 1039–1045.

Chapter 8

1. Witmer, J., and Zimmerman, M. (1991). Intercessory prayer as medical treatment? An inquiry. *The Skeptical Inquirer*, **15**, 177–180. The authors placed notices in medical journals asking readers to contact them if they were aware of any controlled studies, published or unpublished, assessing the efficacy of intercessory prayer. The primary response to this inquiry was a series of letters and telephone calls detailing anecdotes, rather than studies. The only scientific studies of intercessory prayer that were uncovered gave mixed results regarding the efficacy of prayer.

2. Yerushalmy, J. (1971). The relationship of parents' cigarette smoking to outcome of pregnancy—implications as to the problem of inferring causation from observed associations. *American Journal of Epidemiology*, **93**, 443–456.

3. Gould, S. J. (1981). *The Mismeasure of Man*. New York: Norton, pp. 50ff. The SDs were estimated from the ranges reported by Gould.

4. Yerushalmy, J. (1972). Infants with low birth weight born before their mothers started to smoke cigarettes. *American Journal of Obstetrics and Gynecology*, **112**, 277–284.

5. Anderson, G. D., Blidner, I. N., McClemont, S., and Sinclair, J. C. (1984). Determinants of size at birth in a Canadian population. *American Journal of Obstetrics and Gynecology*, **150**, 236–244.

6. Mochizuki, M., Marno, T., Masuko, K., and Ohtsu, T. (1984). Effects of smoking on fetoplacental-maternal system during pregnancy. *American Journal of Obstetrics and Gynecology*, **149**, 413–420.

7. Wainright, R. L. (1983). Change in observed birth weight associated with a change in maternal cigarette smoking. *American Journal of Epidemiology*, **117**, 668–675.

8. Moore, R. M., Diamond, E. L., and Cavalieri, R. L. (1988). The relationship of birthweight and intrauterine diagnostic ultrasound exposure. *Obstetrics and Gynecology*, **71**, 513–517.

9. Waldenstrom, U., Nilsson, S., Fall, O., Axelsson, O., Eklund, G., Lindeberg, S., and Sjodin, Y. (1988). Effects of routine one-stage ultrasound screening in pregnancy: a randomized clinical trial. *Lancet* (10 Sept.), 585–588.

10. Haenszel, W., Kurihara, M., Segi, M., and Lee, R. K. C. (1972). Stomach cancer among Japanese in Hawaii. *Journal of the National Cancer Institute*, **49**, 969–988.

11. *Vital Statistics of the United States*, Vol. II, Part A. Data are taken from Section 1, Table VIII.

12. Cook, L. S., et al. (1997). Characteristics of women with and without breast augmentation. *Journal of the American Medical Association*, **277**, 1612–1617.

13. LaCroix, A. Z., Mead, L. A., Liang, K., Thomas, C. B., and Pearson, T. A. (1986). Coffee consumption and the incidence of coronary heart disease. *New England Journal of Medicine*, **315**, 977–982.

14. Yerushalmy, J., and Hilleboe, H. E. (1957). Fat in the diet and mortality from heart disease. *New York State Journal of Medicine*, **57**, 2343–2354. Reprinted by permission. Copyright by the Medical Society of the State of New York.

15. *Cleveland Plain Dealer* February 10, 1999, page 17-A.

16. David, R. J., and Collins, J. W. (1997). Differing birth weight among infants of U.S.-born blacks, African-born blacks, and U.S.-born whites. *New England Journal of Medicine*, **337**, 1209–1214. Low birthweight means a weight of less than 1500 g, which the authors referred to as "very low birth weight" in the article.

17. Benson, H., and Friedman, R. (1996). Harnessing the power of the placebo effect and renaming it "remembered wellness." In *Annual Review of Medicine*, **47**, 193–199. Annual Reviews, Inc., Palo Alto, CA.

18. Sandler, A. D., Sutton, K. A., DeWeese, J., Girardi, M. A., Sheppard, V., and Bodfish, J. W. (1999). Lack of benefit of a single dose of synthetic human secretin in the treatment of autism and pervasive developmental disorder. *New England Journal of Medicine*, **341**, 1801–1806. The improvement in the placebo group was somewhat better than the improvement in the secretin group for

the response variable of change in total Autism Behavior Checklist score, but the *P*-value for the difference was .11.

19. Butler, C., and Steptoe, A. (1986). Placebo response: an experimental study of asthmatic volunteers. *British Journal of Clinical Psychology*, **25**, 173–183.

20. Barsamian, E. M. (1977). The rise and fall on internal mammary artery ligation in the treatment of angina pectoris and the lessons learned; in Bunker, J. P., Barnes, B. A., and Mosteller, F. (eds.), *Costs, Risks, and Benefits of Surgery*, New York: Oxford Univ. Press, pp. 212–220.

21. Chalmers, T. C., Celano, P., Sacks, H. S., and Smith, H. (1983). Bias in treatment assignment in controlled clinical trials. *New England Journal of Medicine*, **309**, 1358–1361.

22. The Coronary Drug Project Research Group (1980). Influence of adherence to treatment and response of cholesterol on mortality in the coronary drug project. *New England Journal of Medicine*, **303**, 1038–1041. Several variables were measured on each subject at the start of the experiment. Adjusting for the effects of these covariates within the placebo group only slightly reduces the difference in mortality rates between adherers and nonadherers. Thus, differences in overall health only explain a small part of the "adherer vs. nonadherer" mortality rate difference.

23. Diehl, H. S., Baker, A. B., and Cowan, D. W. (1938). Cold vaccines: an evaluation based on a controlled study. *Journal of the American Medical Association*, **111**, 1168–1173.

24. Peto, R., Pike, M. C., Armitage, P., Breslow, N. E., Cox, D. R., Howard, S. V., Mantel, N., McPherson, K., Peto, J., and Smith, P. G. (1976). Design and analysis of randomized clinical trials requiring prolonged observation of each patient. I. Introduction and design. *British Journal of Cancer*, **34**, 585–612.

25. Sacks, H., Chalmers, T. C., and Smith, H. (1982). Randomized versus historical controls for clinical trials. *American Journal of Medicine*, **72**, 233–240.

26. Stevens-Simon, C., Dolgan, J. I., Kelly, L., and Singer, D. (1997). The effect of monetary incentives and peer support groups on repeat adolescent pregnancies: a randomized trial of the dollar-a-day program. *Journal of the American Medical Association*, **277**, 977–982. Partway through the experiment the researchers determined that treatment 1, peer-group support without monetary incentive, was not effective, so they stopped assigning subjects to that group. At the end of the experiment, which lasted for two years, the researchers concluded that the use of monetary incentives increases attendance at weekly support-group meetings, but that these meetings do not prevent repeat pregnancies.

27. Dublin, L. I. (1957). *Water fluoridation: facts, not myths*. New York: Public Affairs Committee, Inc.

28. Sandler, R. S., Zorich, N. L., Filloon, T. G., Wiseman, H. B., Lietz, D. J., Brock, M. H., Royer, M. G., and Miday, R. K. (1999). Gastrointestinal Symptoms in 3181 volunteers ingesting snack foods containing olestra or triglycerides. A 6-week randomized, placebo-controlled trial. *Annals of Internal Medicine*, **130**, 253–261.

29. Moertel, C. G., Fleming, T. R., Creagan, E. T., Rubin, J., O'Connell, M. J., and Ames, M. M. (1985). High-dose vitamin C versus placebo in the treatment of patients with advanced cancer who have had no prior chemotherapy. *New England Journal of Medicine*, **312**, 137–141.

30. Pauling, L., and Cameron, E. (1976). Supplemental ascorbate in the supportive treatment of cancer: prolongation of survival times in terminal human cancer. *Proceedings of the National Academy of Sciences*, **73**, 3685–3789.

31. Based on a study described in Glickman, L. T., and Appel, M. J. G. (1982). A controlled field trial of an attenuated canine origin parvovirus vaccine. *The Compendium of Continuing Education for the Practicing Veterinarian*, **4**, 888–892.

32. This example is based on a study conducted by S. Noll at the University of Minnesota. The actual experiment was slightly different from the description in Example 8.18; some details have been changed for pedagogical purposes.

33. This is the design described in the following papers: Rosenzweig, M. R., Bennett, E. L., and Diamond, M. C. (1972). Brain changes in response to experience. *Scientific American*, **226**, No. 2, 22–29. Bennett, E. L., Diamond, M. C., Krech, D., and Rosenzweig, M. R. (1964). Chemical and anatomical plasticity of brain. *Science*, **146**, 610–619.

34. Based on an experiment by Resh, W., and Stoughton, R. B. (1976). Topically applied antibiotics in acne vulgaris. *Archives of Dermatology*, **112**, 182–184.

35. Swearingen, M. L., and Holt, D. A. (1976). Using a "blank" trial as a teaching tool. *Journal of Agronomic Education*, **5**, 3–8. Reprinted by permission of the American Society of Agronomy, Inc. In fact, in order to demonstrate the variability of plot yields, the experimenters planted the *same* variety of barley in all 16 plots.

36. Nicholson, R. L., and Moraes, W. B. C. (1980). Survival of *Colletotrichum graminicola*: Importance of the spore matrix. *Phytopathology*, **70**, 255–261. Raw data courtesy of R. L. Nicholson.

37. *Cleveland Plain Dealer*, 25 June 1991, page 3-A.

38. Hull, H. F., Bettinger, C. J., Gallaher, M. M., Keller, N. M., Wilson, J., and Mertz, G. J. (1988). Comparison of HIV-antibody prevalence in patients consenting to and declining HIV-antibody testing in an STD clinic. *Journal of the American Medical Association*, **260**, 935–938.

39. These data were collected at Oberlin College. A reference for the randomized response technique is Warner, S. L. (1966). Randomized response: a survey technique for eliminating evasive answer bias. *Journal of the American Statistical Association*, **60**, 63–69.

40. Koffman, D. M., Bazzarre, T., Mosca, L., Redberg, R., Schmid, T., and Wattigney, W. A. (2001). An evaluation of Choose to Move 1999. *Archives of Internal Medicine*, **161**, 2193–2199.

41. Tsoy, A. N., Cheltzov, O. V., Zaseyeva, V., Shilinsh, L. A., and Yashina, L. A. (1990). *European Respiratory Journal*, **3**, 235.

42. Conner, E. M., Sperling, R. S., Gerber, R., Kisalev, P., Scott, G., O'Sullivan, M. J., Van Dyke, R., Bey, M., Shearer, W. Jacobsen, R. L., Jimenez, E., O'Neill, E., Bazin, B., Delfraissy, J.-F., Culname, M., Coombs, R., Elkins, M., More, J., Stratton, P., and Balsley, J. (1994). Reduction of maternal-infant transmission of Human Immunodeficiency Virus Type I with zidovudine treatment. *New England Journal of Medicine*, **331**, 1173–1180. Some people feel that this study should not have been conducted as a randomized experiment, since there was reason to believe that AZT would be helpful in preventing the transfer of HIV to the babies and since HIV is such a serious disease.

43. Gattinoni, L., Tognoni, G., Pesenti, A., Taccone, P., Mascheroni, D., Labarta, V., Malacrida, R., Di Giulio, P., Fumagalli, R., Pelosi, P., Brazzi, L., and Latini, R. (2001). Effect of prone positioning on the survival of patients with acute respiratory failure. *New England Journal of Medicine*, **345**, 568–573.

44. *Cleveland Plain Dealer*, October 23, 1997, p. 15-A.

Chapter 9

1. Sargent, P. A., Sharpley, A. L., Williams, C., Goodall, E. M., and Cowen, P. J. (1997). 5-HT$_{2C}$ receptor activation decreases appetite and body weight in obese subjects. *Psychopharmacology*, **133**, 309–312.

2. Data from Sargent et al. Hunger ratings were recorded "on 10 cm visual analogue scales."

3. Unpublished data courtesy of R. Buchman. The data were collected in Oberlin, Ohio, during the spring of 2001.

4. Day, K. M., Patterson, F. L., Luetkemeier, O. W., Ohm, H. W., Polizotto, K., Roberts, J. J., Shaner, G. E., Huber, D. M., Finney, R. E., Foster, J. E., and Gallun, R. L. (1980). Performance and adaptation of small grains in Indiana. *Station Bulletin*, No. 290. West Lafayette, Indiana: Agricultural Experiment Station of Purdue University. Raw data provided courtesy of W. E. Nyquist. The actual trial included more than two varieties.

5. Unpublished data courtesy of C. H. Noller.

6. Cicirelli, M. F., and Smith, L. D. (1985). Cyclic AMP levels during the maturation of *Xeno pus* oocytes. *Developmental Biology*, **108**, 254–258. Raw data courtesy of M. F. Cicirelli.

7. Judge, M. D., Aberle, E. D., Cross, H. R., and Schanbacher, B. D. (1984). Thermal shrinkage temperature of intramuscular collagen of bulls and steers. *Journal of Animal Science*, **59**, 706–709. Raw data courtesy of the authors and E. W. Mills.

8. Swedo, S. E., Leonard, H. L., Rapoport, J. L., Lenane, M. C., Goldberger, E. L., and Cheslow, B. S. (1989). A double-blind comparison of clomipramine and desipramine in the treatment of trichotillomania (hair pulling). *New England Journal of Medicine*, **321**, 497–501.

9. Unpublished data courtesy of A. Ladavac. The data were collected in Oberlin, Ohio, in November of 1996.

10. In a study in which there is no natural pairing (for example, if identical twins are not available), one may wish to take two equal size groups and create pairs by using covariates such as age and weight. If an experiment is conducted in which members of a pair are randomly assigned to opposite treatment groups, then a paired data analysis has good properties. However, if an observational study is conducted (so that there is no random assignment within pairs), then a paired analysis, such as a paired *t* test, will tend to understate the true variability of the difference being studied and the true type I error rate of a *t* test will be greater than the nominal level of the test. For discussion, see David, H. A., and Gunnink, J. L. (1997). The paired *t* test under artificial pairing. *The American Statistician*, **51**, 9–12.

11. Schriewer, H., Guennewig, V., and Assmann, G. (1983). Effect of 10 weeks endurance training on the concentration of lipids and lipoproteins as well as on the composition of high-density lipoproteins in blood serum. *International Journal of Sports Medicine*, **4**, 109–115. Reprinted with permission of Georg Thieme Verlag Stuttgart.

12. Data from experiments reported in several papers, for example, Fout, G. S., and Simon, E. H. (1983). Antiviral activities directed against wild-type and interferon-sensitive mengovirus. *Journal of General Virology*, **64**, 1543–1555. Raw data courtesy of E. H. Simon. The unit of measurement is proportional to the number of plaques formed by the virus on a monolayer of mouse cells.

Because they are obtained by a serial dilution technique, the measurements have varying numbers of significant digits; the final zeroes of the three-digit numbers are not significant digits.

13. Adapted from Batchelor, J. R., and Hackett, M. (1970). HL-A matching in treatment of burned patients with skin allografts. *Lancet*, **2**, 581–583.

14. Sallan, S. E., Cronin, C., Zelen, M., and Zinberg, N. E. (1980). Antiemetics in patients receiving chemotherapy for cancer. *New England Journal of Medicine*, **302**, 135–138. Reprinted by permission.

15. Koh, K. K., Mincemoyer, R., Bui, M. N., Csako, G., Pucino, F., Guetta, V., Waclawiw, M., Cannon, R. O. (1997). Effects of hormone replacement therapy on fibrinolysis in postmenopausal women. *New England Journal of Medicine*, **336**, 683–690. Raw data courtesy of K. K. Koh.

16. Rosenzweig, M. R., Bennett, E. L., and Diamond, M. C. (1972). Brain changes in response to experience. *Scientific American*, **226**, No. 2, 22–29. Also Bennett, E. L., Diamond, M. C., Krech, D., and Rosenzweig, M. R. (1964). Chemical and anatomical plasticity of brain. *Science*, **146**, 610–619. Copyright 1964 by the American Association for the Advancement of Science (AAAS).

17. Richens, A., and Ahmad, S. (1975). Controlled trial of valproate in severe epilepsy. *British Medical Journal*, **4**, 255–256.

18. Wiedenmann, R. N., and Rabenold, K. N. (1987). The effects of social dominance between two subspecies of dark-eyed juncos, *Junco hyemalis*. *Animal Behavior*, **35**, 856–864. Raw data courtesy of the authors.

19. Masty, J. (1983). Innervation of the equine small intestine. Master's thesis, Purdue University. Raw data courtesy of the author.

20. Golden, C. J., Graber, B., Blose, I., Berg, R., Coffman, J., and Block, S. (1981). Difference in brain densities between chronic alcoholic and normal control patients. *Science*, **211**, 508–510. Raw data courtesy of C. J. Golden. Copyright 1981 by the AAAS.

21. Patel, C., Marmot, M. G., and Terry, D. J. (1981). Controlled trial of biofeedback-aided behavioural methods in reducing mild hypertension. *British Medical Journal*, **282**, 2005–2008.

22. Knowlen, G. G., Kittleson, M. D., Nachreiner, R. F., and Eyster, G. E. (1983). Comparison of plasma aldosterone concentration among clinical status groups of dogs with chronic heart failure. *Journal of the American Veterinary Medical Association*, **183**, 991–996.

23. Forde, O. H., Knutsen, S. F., Arnesen, E., and Thelle, D. S. (1985). The Tromso heart study: Coffee consumption and serum lipid concentrations in men with hypercholesterolaemia: A randomised intervention study. *British Medical Journal*, **290**, 893–895. (The sample sizes are unequal because the 25 no-coffee men actually represent three different treatment groups, which followed the same regimen for the first 5 weeks of the study and different regimens thereafter.)

24. Dalvit, S. P. (1981). The effect of the menstrual cycle on patterns of food intake. *American Journal of Clinical Nutrition*, **34**, 1811–1815.

25. Unpublished data courtesy of D. J. Honor and W. A. Vestre.

26. Sesin, G. P. (1984). Pharmacokinetic dosing of Tobramycin sulfate. *American Pharmacy NS24*, 778. Vakoutis, J., Stein, G. E., Miller, P. B., and Clayman, A. E. (1981). Aminoglycoside monitoring program. *American Journal of Hospital Pharmacy*, **38**, 1477–1480. Copyright 1981, American Society of Hospital Pharmacists, Inc. All rights reserved. Reprinted with permission.

27. Jovan, S. (2000). Catnip bonanza. *Stats*, No. 27, 25–27.

28. Dale, E. M., and Housley, T. L. (1986). Sucrose synthase activity in developing wheat endosperms differing in maximum weight. *Plant Physiology*, **82**, 7–10. Raw data courtesy of the authors.

29. Unpublished data courtesy of M. Heithaus and D. Rogers. The samples were taken from the Vermilion River in northern Ohio during the spring of 1995.

30. Salib, N. M. (1985). The effect of caffeine on the respiratory exchange ratio of separate submaximal arms and legs exercise of middle distance runners. Master's thesis, Purdue University.

31. Adapted from Bodian, D. (1947). Nucleic acid in nerve-cell regeneration. *Symposia of the Society for Experimental Biology, No. 1. Nucleic Acid*, 163–178. Reprinted with permission of Cambridge University Press.

32. Robinson, L. R. (1985). The effects of electrical fields on wound healing in *Notophthalmus viridescens*. Master's thesis, Purdue University. Raw data courtesy of the author and J. W. Vanable, Jr.

33. Agosti, E., and Camerota, G. (1965). Some effects of hypnotic suggestion on respiratory function. *International Journal of Clinical and Experimental Hypnosis*, **13**, 149–156. The experiment actually included a third phase.

34. Koh, K. K., Mincemoyer, R., Bui, M. N., Csako, G., Pucino, F., Guetta, V., Waclawiw, M., Cannon, R. O. (1997). Effects of hormone replacement therapy on fibrinolysis in postmenopausal women. *New England Journal of Medicine*, **336**, 683–690. Raw data courtesy of K. K. Koh.

35. Savin, V. J., Sharma, R., Sharma, M., McCarthy, E. T., Swan, S. K., Ellis, E., Lovell, H., Warady, B., Gunwar, S., Chonko, A. M., Artero, M., and Vincenti, F. (1996). Circulating factor associated with increased glomerular permeability to albumin in recurrent focal segmental glomerulosclerosis. *New England Journal of Medicine*, **334**, 878–883. Raw data courtesy of V. J. Savin.

Chapter 10

1. Baur, E., Fischer, E., and Lenz, F. (1931). *Human Heredity*, 3rd edition. New York: Macmillan, p. 52.

2. Saeidi, G., and Rowland, G. G. (1997) The inheritance of variegated seed color and palmitic acid in flax. *Journal of Heredity*, **88**, 466–468.

3. Rabenold, K. R., and Rabenold, P. P. (1985). Variation in altitudinal migration, winter segregation, and site tenacity in two subspecies of dark-eyed juncos in the Southern Appalachians. *The Auk*, **102**, 805–819.

4. Phillips, D. P., and Smith, D. G. (1990). Postponement of death until symbolically meaningful occasions. *Journal of the American Medical Association*, **263**, 1947–1951. For comparison purposes, the authors also examined deaths among elderly Jewish women during the same time period; they did not find any excess of deaths after the festival for this comparison group.

5. Sinnott, E. W., and Durham, G. B. (1922). Inheritance in the summer squash. *Journal of Heredity*, **13**, 177–186.

6. Adapted from Gould, J. L. (1985). How bees remember flower shapes. *Science*, **227**, 1492–1494. Figure copyright 1985 by the American Association for the Advancement of Science; used by permission.

7. Adapted from 1983 birth data for West Lafayette, Indiana.

8. Bateson, W., and Saunders, E. R. (1902). *Reports to the Evolution Committee of the Royal Society*, **1**, 1–160. Feather color and comb shape are controlled

independently; white feather is dominant and small comb is dominant. The parents in the experiment were first-generation hybrids (F_1) and thus were necessarily double heterozygotes.

9. This is a realistic value. See Exercise 3.31.

10. Jakkula, L. R., Knault, D. A., and Gorbet, D. W. (1997). Inheritance of a shriveled seed trait in peanut. *Journal of Heredity*, **88**, 47–51. The data are taken from Table 5 of the paper.

11. Adapted from Mantel, N., Bohidar, N. R., and Ciminera, J. L. (1977). Mantel-Haenszel analyses of litter-matched time-to-response data, with modifications for recovery of inter-litter information. *Cancer Research*, **37**, 3863–3868. (A more powerful analysis, which uses the partially informative triplets, is described in the paper.)

12. Adapted from Jacobs, G. H. (1978). Spectral sensitivity and colour vision in the ground-dwelling sciurids: Results from golden mantled ground squirrels and comparisons for five species. *Animal Behaviour*, **26**, 409–421. See also Jacobs, G. H. (1981). *Comparative Color Vision*. New York: Academic Press.

13. Petrij, F., van Veen, K., Mettler, M., and Bruckmann, V. (2001). A second acromelanistic allelomorph at the albino locus of the Mongolian gerbil (*Meriones unguiculatus*). *Journal of Heredity*, **92**, 74–78. The gerbils we call "brown" are referred to as "Siamese" by the authors.

14. Kaitz, M. (1992). Recognition of familiar individuals by touch. *Physiology and Behavior*, **52**, 565–567.

15. Fawole, I. (2001). Genetic analysis of mutations at loci controlling leaf form in cowpea (*Vigna unguiculata* [L.] Walp). *Journal of Heredity*, **92**, 43–50. These data come from the 1993a generation listed in Table 8 of the article. The types we call I, II, and III are identified by the authors as trifoliolate, trifoliolate orbicular, and unifoliolate orbicular.

16. Brailovsky, D. (1974). Timolol maleate (MK-950): A new beta-blocking agent for the prophylactic management of angina pectoris. A multicenter, multinational, cooperative trial. In *Beta-Adrenergic Blocking Agents in the Management of Hypertension and Angina Pectoris*, B. Magnani (ed.). New York: Raven Press.

17. Miller, A. B., To, T., Baines, C. J., and Wall, C. (2000). Canadian national breast screening study-2: 13-year results of a randomized trial in women aged 50–59 years. *Journal of the National Cancer Institute*, **92**, 1490–1499.

18. Brodie, E. D., Jr., and Brodie, E. D. III. (1980). Differential avoidance of mimetic salamanders by free-ranging birds. *Science*, **208**, 181–182. Copyright 1980 by the AAAS.

19. Karban, R., Adamchak, R., and Schnathorst, W. C. (1987). Induced resistance and interspecific competition between spider mites and a vascular wilt fungus. *Science*, **235**, 678–680. Copyright 1987 by the AAAS.

20. Inskip, P. D., Targone, R. E., Hatch, E. E., Wilcosky, T. C., Shapiro, W. R., Selker, R. G., Fine, H. A., Black, P. M., Loeffler, J. S., and Linet, M. S. (2001). Cellular-telephone use and brain tumors. *New England Journal of Medicine*, **344**, 79–86. The data are taken from Table 4 of the paper.

21. Turnbull, D. M., Rawlins, M. D., Weightman, D., and Chadwick, D. W. (1982). A comparison of phenytoin and valproate in previously untreated adult epileptic patients. *Journal of Neurology, Neurosurgery, and Psychiatry*, **45**, 55–59.

22. Unpublished data courtesy of W. Singleton and K. Hendrix.

23. Mizutani, T., and Mitsuoka, T. (1979). Effect of intestinal bacteria on incidence of liver tumors in gnotobiotic C3H/He male mice. *Journal of the National Cancer Institute*, **63**, 1365–1370.

24. Selawry, O. S. (1974). The role of chemotherapy in the treatment of lung cancer. *Seminars in Oncology*, **1**, 259–272.

25. Kannus, P., Parkkari, J., Niemi, S., Pasanen, M., Palvanen, M., Jarvinen, M., and Vuori, I. (2000). Prevention of hip fracture in elderly people with use of a hip protector. *New England Journal of Medicine*, **343**, 1506–1513.

26. Cohen, S., Doyle, W. J., Skoner, D. P., Rabin, B. S., and Gwaltney, J. M. (1997). Social ties and susceptibility to the common cold. *Journal of the American Medical Association*, **277**, 1940–1944.

27. Sherman, D. G., Atkinson, R. P., Chippendale, T., Levin, K. A., Ng, K., Futrell, N., Hsu, C. Y., and Levy, D. E. (2000). Intravenous ancrod for treatment of acute ischemic stroke. *Journal of the American Statistical Association*, **283**, 2395–2403.

28. Unpublished data courtesy of D. Wallace, collected at Oberlin College in the fall of 1995.

29. Adapted from Ammon, O. (1899). *Zur Anthropologie der Badener*. Jena: G. Fischer. Ammon's data appear in Goodman, L. A., and Kruskal, W. H. (1954). Measures of association for cross classifications. *Journal of the American Statistical Association*, **49**, 732–764. Light hair was blonde or red; dark hair was brown or black. Light eyes were blue, grey, or green; dark eyes were brown.

30. Cruz-Coke, R. (1970). *Color blindness; an evolutionary approach*. Springfield, IL: Thomas.

31. Davies, D. F., Johnson, A. P., Rees, B. W. G., Elwood, P. C., and Abernethy, M. (1974). Food antibodies and myocardial infarction. *The Lancet*, **i**, 1012–1014. See also Galen, R. S. (1974). Food antibodies and myocardial infarction. *The Lancet*, **ii**, 832.

32. Adapted from Porac, C., and Coren, S. (1981). *Lateral Preferences and Human Behavior*. New York: Springer-Verlag. The frequencies given are approximate, having been deduced from percentages on pages 36 and 45. People with neutral preference were counted as Left.

33. Upton, G., and Fingleton, B. (1985). *Spatial Data Analysis by Example: Point Pattern and Quantitative Data, Vol. 1*. New York: Wiley, p. 230. Adapted from Diggle, P. J. (1979). Statistical methods for spatial point patterns in ecology, pp. 95–150 in *Spatial and Temporal Analysis in Ecology*, R. M. Cormack and J. K. Ord (eds.). Fairland, MD: International Cooperative Publishing House.

34. Based on an article by the Writing Group for Bypass Angioplasty Revascularization Investigation (BARI) Investigators (1997). See Five-year clinical and functional outcome comparing bypass surgery and angioplasty in patients with multivessel coronary disease. *Journal of the American Medical Association*, **277**, 715–721.

35. These data are fictitious, but the proportions of left-handed males and females are realistic and the independence of the twins is in agreement with published data. See Porac, C., and Coren, S. (1981). *Lateral Preferences and Human Behavior*. New York: Springer-Verlag, p. 36; and Morgan, M. C., and Corballis, M. J. (1978). On the biological basis of human laterality: I. Evidence for a maturational left-right gradient. *The Behavioral and Brain Sciences*, **2**, p. 274.

36. Ware, J. H. (1989). Investigating therapies of potentially great benefit: ECMO. *Statistical Science*, **4**, 298–306. There is controversy surrounding this experiment. An earlier experiment using a nonstandard randomization scheme had shown ECMO to be highly effective. Thus, some statisticians question whether this second experiment was necessary. For a discussion of these issues, see the articles on pages 306–340 that follow that the Ware article in *Statistical Science*, **4**.

37. Remus, J. K., and Zahren, L. (1995). An investigation of the influenza virus at Oberlin College. Unpublished manuscript, Oberlin College. This study actually involved more students than are reported here. For simplicity, we restrict attention to the 41 students who had at least two colds during the 1994–95 school year.

38. Hurt, R. D., Sachs, D. P. L., Glover, E. D., Offord, K. P., Johnston, J. A., Dale, L. C., Khayrallah, M. A., Schroeder, D. R., Glover, P. N., Sullivan, C. R., Croghan, I. T., and Sullivan, P. M. (1997). A comparison of sustained-release bupropion and placebo for smoking cessation. *New England Journal of Medicine*, **337**, 1195–1202.

39. Unpublished data courtesy of B. Rogers, collected at the Oberlin College Conservatory of Music in the spring of 1991.

40. Souttou, B., Juhl, H., Hackenbruck, J., Rockseisen, M., Klomp, H.-J., Raulais, D., Vigny, M., and Wellstein, A. (1998). Relationship between serum concentrations of the growth factor pleiotrophin and pleiotrophin-positive tumors. *Journal of the National Cancer Institute*, **90**, 1468–1473.

41. Hogarty, G. E., Kornblith, S. J., Greenwald, D., DiBarry, A. L., Cooley, S., Ulrich, R. F., Carter, M., and Flesher, S. (1997). Three-year trials of personal therapy among schizophrenic patients with or independent of family, I: Description of study and effects on relapse rates. *American Journal of Psychiatry*, **154**, 1504–1513.

42. The three data sets, collected in the 1950s by three different investigators, are reproduced in Mourant, A. E., Kopec, A. C., and Domaniewska-Sobczak, K. (1976). *The Distribution of the Human Blood Groups and Other Polymorphisms*. London: Oxford University Press, p. 204.

43. Adapted from Ammon, O. (1899). *Zur Anthropologie der Badener*. Jena: G. Fischer. Ammon's data appear in Goodman, L. A., and Kruskal, W. H. (1954). Measures of association for cross classifications. *Journal of the American Statistical Association*, **49**, 732–764.

44. Fleming, D. T., McQuillan, G. M., Johnson, R. E., Nahmias, A. J., Aral, S. O., Lee, F. K., and St. Louis, M. E. (1997). Herpes simplex virus type 2 in the United States, 1976 to 1994. *New England Journal of Medicine*, **337**, 1105–1111. The frequencies given are approximate, having been deduced from the regional percentages, which are given in the article.

45. Inglesfield, C., and Begon, M. (1981). Open-ground individuals and population structure in *Drosophila subobscura* Collin. *Biological Journal of the Linnean Society*, **15**, 259–278.

46. Aird, I., Bentall, H. H., Mehigan, J. A., and Roberts, J. A. F. (1954). The blood groups in relation to peptic ulceration and carcinoma of colon, rectum, breast, and bronchus: An association between the ABO blood groups and peptic ulceration. *British Medical Journal*, **ii**, 315–321.

47. Govind, C. K., and Pearce, J. (1986). Differential reflex activity determines claw and closer muscle asymmetry in developing lobsters. *Science*, **233**, 354–356. Copyright 1986 by the AAAS.

48. LeBars, P. L., Katz, M. M., Berman, N., Itil, T. M., Freedman, A. M. and, Schatzberg, A. F. (1997). A placebo-controlled, double-blind, randomized trial of an extract of Gingko biloba for dementia. *Journal of the American Medical Association*, **278**, 1327–1332.

49. Unpublished data courtesy of L. Solimine.

50. Hudson, J. I., McElroy, S. L., Raymond, N. C., Crow, S., Keck, P. E., Carter, W. P., Mitchell, J. E., Strakowski, S. M., Pope, H. G., Coleman, B. S., and Jonas, J. M. (1998). Fluvoxamine in the treatment of binge-eating disorder: a multicenter placebo-controlled, double-blind trial. *American Journal of Psychiatry*, **155**, 1756–1762. The response variable has ordered categories, so there are more powerful methods, beyond the scope of this text, that can be used to analyze the data.

51. Wolfson, J. L. (1987). Impact of *Rhizobium* nodules on *Sitona hispidulus*, the clover root curculio. *Entomologia Experimentalis et Applicata*. **43**, 237–243. Data courtesy of the author. The experiment actually included 11 dishes.

52. Adapted from Paige, K. N., and Whitham, T. G. (1985). Individual and population shifts in flower color by scarlet gilia: A mechanism for pollinator tracking. *Science*, **227**, 315–317. The raw data given are fictitious but have been constructed to agree with the summary statistics given by Paige and Whitham.

53. Beck, S. L., and Gavin, D. L. (1976). Susceptibility of mice to audiogenic seizures is increased by handling their dams during gestation. *Science*, **193**, 427–428. Copyright 1976 by the AAAS.

54. Pittet, P. G., Acheson, K. J., Wuersch, P., Maeder, E., and Jequier, E. (1981). Effects of an oral load of partially hydrolyzed wheatflour on blood parameters and substrate utilization in man. *The American Journal of Clinical Nutrition*, **34**, 2438–2445.

55. Agresti, A., and Caffo, B. (2000). Simple and effective confidence intervals for proportions and differences of proportions result from adding two successes and two failures. *The American Statistician*, **54**, 280–288. Agresti and Caffo conduct a series of simulations that show that adding 1 to each cell results in good coverage properties when the sample sizes, n_1 and n_2, are as small as 10. Unpublished calculations done by the J. Witmer show that these good properties are also obtained when n_1 and n_2 are as small as 5, provided p_1 and p_2 are not both close to 0 or both close to 1, in which case the interval becomes quite conservative (i.e., the coverage rate approaches 100% for a nominal 95% confidence interval).

56. Agresti, A. Personal communication.

57. Saunders, M. C., Dick, J. S., Brown, I. M., McPherson, K., and Chalmers, I. (1985). The effects of hospital admission for bed rest on the duration of twin pregnancy: A randomised trial. *The Lancet*, **ii**, 793–795.

58. Lader, M., and Scotto, J.-C. (1998). A multicentre double-blind comparison of hydroxyzine, buspirone and placebo in patients with generalized anxiety disorder. *Psychopharmacology*, **139**, 402–406. Improvement is taken to be a 50% or greater reduction in Hamilton Anxiety (HAM-A) score. There was a third treatment group in this study, which we are ignoring here.

59. Nesheim, S. R., Shaffer, N., Vink, P., Thea, D. M., Palumbo, P., Greenberg, B., Weedon, J, and Simmons, R. J. (1996). Lack of increased risk for perinatal human immunodeficiency virus transmission to subsequent children born to infected women. *Pediatric Infectious Disease Journal*, **15**, 886–890.

60. Collaborative Group for the Study of Stroke in Young Women (1973). Oral contraception and increased risk of cerebral ischemia or thrombosis. *New England Journal of Medicine*, **288**, 871–878. Reprinted by permission.

61. Johnson, S. K., and Johnson, R. E. (1972). Tonsillectomy history in Hodgkin's disease. *New England Journal of Medicine*, **287**, 1122–1125.

62. Yerushalmy, J. (1971). The relationship of parents' cigarette smoking to outcome of pregnancy—implications as to the problem of inferring causation from observations. *American Journal of Epidemiology*, **93**, 443–356.

63. The steering committee of the physicians' health study research group. (1988). Preliminary report: Findings from the aspirin component of the ongoing physicians' health study. *New England Journal of Medicine*, **318**, 262–264.

64. Hasdai, D., Garratt, K. N., Grill, D. E., Lerman, A., and Holmes, D. R. (1997). Effect of smoking status on the long-term outcome after successful percutaneous coronary revascularization. *New England Journal of Medicine*, **336**, 755–761. Actually, some of the patients underwent coronary revascularization procedures other than balloon angioplasty. However, we include all patients in one group here for simplicity.

65. Zwerling, C., Whitten, P. S., Davis, C. S., and Sprince, N. L. (1997). Occupational injuries among workers with disabilities. *Journal of the American Medical Association*, **278**, 2163–2166. In this study an injury means a occupational injury in the year preceding when the person was interviewed.

66. Kernan, W. N., Viscoli, C. M., Brass, L. M., Broderick, J. P., Brott, T., Feldmann, E., Morgenstern, L. B., Wilterdink, J. L., and Horwitz, R. I. (2000). Phenypropanolamine and the risk of hemorrhagic stroke. *New England Journal of Medicine*, **343**, 1826–1832.

67. Cohen, M., Demers, C., Gurfinkel, E. P., Turpie, A. G. G., Fromell, G. J., Goodman, S., Langer, A., Califf, R. M., Fox, K. A. A., Premmereur, J., and Bigonzi, F. (1997). A comparison of low-molecular-weight heparin with unfractioned heparin for unstable coronary artery disease. *New England Journal of Medicine*, **337**, 447–452. A negative outcome here is taken to be death, myocardial infarction, or recurrent angina during the first 14 days after treatment.

68. Freeland, W. J. (1981). Parasitism and behavioral dominance among male mice. *Science*, **213**, 461–462. Copyright 1981 by the AAAS.

69. Collins, R. L. (1970). The sound of one paw clapping: An inquiry into the origin of left-handedness. In Lindzey, G., and Thiessen, D. D. (eds.). *Contributions to Behavior-Genetic Analysis: The Mouse as Prototype*. New York: Appleton-Century-Crofts.

70. Allison, A. C., and Clyde, D. F. (1961). Malaria in African children with deficient erythrocyte dehydrogenase. *British Medical Journal*, **1**, 1346–1349.

71. Conover, D. O., and Kynard, B. E. (1981). Environmental sex determination: Interaction of temperature and genotype in a fish. *Science*, **213**, 577–579. Copyright 1981 by the AAAS.

72. Fawole, I. op cit. These data are from Table 3 of the paper.

73. Floersheim, G. L., Weber, O., Tschumi, P., and Ulbrich, M. (1983). Research cited in *Scientific American*, April 1983, **248**, No. 4, p. 75.

74. Fuchs, J. A., Smith, J. D., and Bird, L. S. (1972). Genetic basis for an 11:5 dihybrid ratio observed in *Gossypium hirsutum*. *Journal of Heredity*, **63**, 300–303. The genetic basis for the 13:3 and 11:5 ratios is explained in Strickberger, M. W. (1976). *Genetics*, 2nd edition. New York: Macmillan, pp. 206–208.

75. Adapted from Goodyear, C. P. (1970). Terrestrial and aquatic orientation in the starhead top-minnow, *Fundulus noti. Science*, **168**, 603–605. Copyright 1970 by the AAAS.

76. See Batschelet, E. (1981). *Circular Statistics in Biology*. New York: Academic Press.

77. Carson, J. L., Collier, A. M., and Hu, S. S. (1985). Acquired ciliary defects in nasal epithelium of children with acute viral upper respiratory infections. *New England Journal of Medicine*, **312**, 463–468. Reprinted by permission.

78. Larson, E. B., Roach, R. C., Schoene, R. B., and Hombein, T. F. (1982). Acute mountain sickness and acetazolamide. *Journal of the American Medical Association*, **248**, 328–332. Copyright 1982 American Medical Association.

79. Kluger, M. J., Ringler, D. H., and Anver, M. R. (1975). Fever and survival. *Science*, **188**, 166–168. Copyright 1975 by the AAAS. The original article contains a misprint, but Dr. Kluger has kindly provided the correct mortality at 40°C.

80. Ragaz, J., Jackson, S. M., Le, N., Plenderleith, I. H., Spinelli, J. J., Basco, V. E., Wilson, K. S., Knowling, M. A., Coppin, C. M. L., Paradis, M., Coldman, A. J., and Olivotto, I. A. (1997). Adjuvant radiotherapy and chemotherapy in node-positive premenopausal women with breast cancer. *New England Journal of Medicine*, **337**, 956–962.

81. Englund, J. A., Baker, C. J., Raskino, C., McKinney, R. E., Petrie, B., Fowler, M. G., Pearson, D., Gershon, A., McSherry, G. D., Abrams, E. J., Schliozberg, J., and Sullivan, J. L. (1997). Zidovudine, didanosine, or both as the initial treatment for symptomatic HIV-infected children. *New England Journal of Medicine*, **336**, 1704–1712. The data presented here are for an interim analysis that was conducted approximately two years into the study. As a result of the interim analysis of death rates and of rates of disease progression, the use of zidovudine alone was stopped before the end of the trial.

82. Jemmott, J. B. III, Jemmott, L. S., and Fong, G. T. (1998). Abstinence and safer sex HIV risk-reduction interventions for African American adolescents. *Journal of the American Medical Association*, **279**, 1529–1536.

83. Shorrocks, B., and Nigro, L. (1981). Microdistribution and habitat selection in *Drosophila subobscura* collin. *Biological Journal of the Linnean Society*, **16**, 293–301.

84. Malacrida, R., Genoni, M., Maggioni, A. P., Spatato, V., Parish, S., Palmer, A., Collins, R., and Moccetti, T. (1998). A comparison of the early outcome of acute myocardial infarction in women and men. *New England Journal of Medicine*, **338**, 8–14. Although the odds ratio for these data shows that men are more likely to survive than are women, the authors discuss the effect that age has on this finding. They calculate a new odds ratio after adjusting for age and other covariates (using methods that are beyond the scope of this text) and conclude that much of the difference in survival probability is due to these covariates.

85. Unpublished data courtesy J. L. Wolfson, collected at Bard College in 1997.

86. Paris, H. S. (1997). Genes for developmental fruit coloration of acorn squash. *Journal of Heredity*, **88**, 52–56. The experiment included crossing Table Queen squash (TQE) with Vegetable Spaghetti (VSP). The data presented in the exercise are from a back-cross of VSP with a TQE × VSP cross.

87. Mochizuki, M., Marno, T., Masuko, K., and Ohtsu, T. (1984). Effects of smoking on fetoplacental-maternal system during pregnancy. *American Journal of Obstetrics and Gynecology*, **149**, 413–420.

88. Redelmeier, D. A., and Tibshirani, R. J. (1997). Association between cellular-telephone calls and motor vehicle collisions. *New England Journal of Medicine*, **336**, 453–458. Also see Redelmeier, D. A., and Tibshirani, R. J. (1997). Is using a car phone like driving drunk? *Chance*, **10**, No. 2, 5–9.

Chapter 11

1. Martinez, J. (1998). Organic practices for the cultivation of sweet corn. Unpublished manuscript, Oberlin College. For pedagogical purposes, the data presented here are a random sample from a larger study. The nematode used was *Steinernema carpocapsae*, the bacterium was *Bacillus thuringiensis*, and the wasp was *Trichogramma pretiosum*.

2. Shields, D. R. (1981). The influence of niacin supplementation on growing ruminants and *in vivo* and *in vitro* rumen parameters. Ph.D. thesis, Purdue University. Adapted from raw data provided courtesy of the author and D. K. Colby.

3. Adapted from Potkin, S. G., Cannon, H. E., Murphy, D. L., and Wyatt, R. J. (1978). Are paranoid schizophrenics biologically different from other schizophrenics? *New England Journal of Medicine*, **298**, 61–66. Reprinted by permission. The calculations are based on the data in Example 1.4 in this book, which are an approximate reconstruction from the histograms and summary information given by Potkin et al.

4. Adapted from Keller, S. E., Weiss, J. M., Schleifer, S. J., Miller, N. E., and Stein, M. (1981). Suppression of immunity by stress: Effect of a graded series of stressors on lymphocyte stimulation in the rat. *Science*, **213**, 1397–1400. Copyright 1981 by the AAAS. The SDs and SSs were estimated from the SE's given by Keller et al.

5. Lobstein, D. D. (1983). A multivariate study of exercise training effects on beta-endorphin and emotionality in psychologically normal, medically healthy men. Ph.D. thesis, Purdue University. Raw data courtesy of the author.

6. Hayden, F. G., Osterhaus, A. D., Treanor, J. J., Fleming, D. M., Aoki, F. Y., Nicholson, K. G., Bohnen, A. M., Hirst, H. M., Keene, O., and Wightman, K. (1997). Efficacy and safety of the neuraminidase inhibitor zanamivir in the treatment of influenzavirus infections. *New England Journal of Medicine*, **337**, 874–880. The sums of squares have been calculated from the means and SDs given in the article.

7. Person, A. (1999). Daffodil stem lengths. Unpublished manuscript, Oberlin College. The full data set is somewhat larger than that presented here.

8. Kotler, D. (2000). A comparison of aerobics and modern dance training on health-related fitness in college women. Unpublished manuscript, Oberlin College.

9. Unpublished data courtesy of H. W. Ohm.

10. Cameron, E., and Pauling, L. (1978). Supplemental ascorbate in the supportive treatment of cancer: re-evaluation of prolongation of survival times in terminal human cancer. *Proceedings of the National Academy of Science USA*, **75**, 4538–4542.

11. Neumann, A., Richards, A.-L., and Randa, J. (2001). Effects of acid rain on alfalfa plants. Unpublished manuscript, Oberlin College. The low acid group was given three drops of 1.5 M HCl as well as two droppers full of water each day. For the high acid group 3.0 M HCl was used. The control group was only given water.

12. Pappas, T., and Mitchell, C. A. (1985). Effects of seismic stress on the vegetative growth of *Glycine max* (L.) Merr. cv. Wells II. *Plant, Cell and Environment*, **8**, 143–148. Reprinted with permission of Blackwell Scientific Publications Limited. Raw data courtesy of the authors. The original experiment included more than four treatments.

13. Oren, R., Ellsworth, D. S., Johnsen, K. H., Phillips, N., Ewers, B. E., Maier, C., Schafer, K. V. R., McCarthy, H., Hendrey, G., McNulty, S. G., and Katul, G. G. (2001). Soil fertility limits carbon sequestration by forest ecosystems in a CO_2–enriched atmosphere. *Nature*, **411**, 469–471. Sample means and standard deviation were read off of Figure 2b in the article.

14. Kiesecker, J. M., Blaustein, A. R., and Belden, L. K. (2001). Complex causes of amphibian population declines. *Nature*, **410**, 681–684. Sample means and standard deviation were read off of Figure 2a in the article.

15. Tripepi, R. R., and Mitchell, C. A. (1984). Metabolic response of river birch and European birch roots to hypoxia. *Plant Physiology*, **76**, 31–35. Raw data courtesy of the authors.

16. Adapted from Veterans Administration Cooperative Study Group on Antihypertensive Agents (1979). Comparative effects of ticrynafen and hydrochlorothiazide in the treatment of hypertension. *New England Journal of Medicine*, **301**, 293–297. Reprinted by permission. The value of s_{pooled} was calculated from the SE's given in the article.

17. Knight, S. L., and Mitchell, C. A. (1983). Enhancement of lettuce yield by manipulation of light and nitrogen nutrition. *Journal of the American Society for Horticultural Science*, **108**, 750–754. Calculations based on raw data provided by the authors.

18. Fictitious but realistic data, adapted from O'Brien, R. J., and Drizd, T. A. (1981). Basic data on spirometry in adults 25–74 years of age: United States, 1971–75. *U.S. National Center for Health Statistics, Vital and Health Statistics Series* **11**, *No. 222*. Washington, D.C.: U.S. Department of Health and Human Services.

19. U.S. Bureau of the Census (1997). *Statistical Abstract of the United States*, 1997, 117th edition. Washington, D.C: U.S. Government Printing Office. The age distribution varies from year to year; we have given the 1996 distribution.

20. Chrisman, C. L., and Baumgartner, A. P. (1980). Micronuclei in bone-marrow cells of mice subjected to hyperthermia. *Mutation Research*, **77**, 95–97. The original experiment included six treatments.

21. Baird, J. T., and Quinlivan, L. G. (1973). Parity and hypertension. *U.S. National Center for Health Statistics, Vital and Health Statistics Series 11, No. 38*. Washington, D.C.: U.S. Department of Health, Education and Welfare.

22. U.S. Bureau of the Census (1997). *Statistical Abstract of the United States*, 1997, 117th edition. Washington, D.C: U.S. Government Printing Office.

23. Adapted from Witelson, S. F. (1985). The brain connection: The corpus callosum is larger in left-handers. *Science*, **229**, 665–668. Copyright 1985 by the AAAS. Reprinted with permission of *Science* and S. F. Witelson. The SDs and MS(within) have been calculated from the standard errors given by Witelson.

24. Adapted from Clapp, M. J. L. (1980). The effect of diet on some parameters measured in toxicological studies in the rat. *Laboratory Animals*, **14**, 253–261.

25. Fictitious but realistic data. Adapted from Mizutani, T., and Mitsuoka, T. (1979). Effect of intestinal bacteria on incidence of liver tumors in gnotobiotic C3H/He male mice. *Journal of the National Cancer Institute*, **63**, 1365–1370.

26. Latimer, J. (1985). Adapted from unpublished data provided by the investigator.

27. Adapted from Morris, J. G., Cripe, W. S., Chapman, H. L., Jr., Walker, D. F., Armstrong, J. B., Alexander, J. D., Jr., Miranda, R., Sanchez, A., Jr., Sanchez, B., Blair-West, J. R., and Denton, D. A. (1984). Selenium deficiency in cattle associated with Heinz bodies and anemia. *Science*, **223**, 491–492. Copyright 1984 by the AAAS. The MS(within) is fictitious but agrees with the standard errors given by Morris et al.

28. Fictitious but realistic data. Adapted from Mizutani, T. and Mitsuoka, T. (1979). Effect of intestinal bacteria on incidence of liver tumors in gnotobiotic C3H/He male mice. *Journal of the National Cancer Institute*, **63**, 1365–1370.

29. Becker, W. A. (1961). Comparing entries in random sample tests. *Poultry Science*, **40**, 1507–1514.

30. Adapted from Rosner, B. (1982). Statistical methods in ophthalmology: An adjustment for the intraclass correlation between eyes. *Biometrics*, **38**, 105–114. Reprinted with permission from the Biometric Society. The medical study is reported in Berson, E. L., Rosner, B., and Simonoff, E. (1980). An outpatient population of retinitis pigmentosa and their normal relatives: Risk factors for genetic typing and detection derived from their ocular examinations. *American Journal of Ophthalmology*, **89**, 763–775. The means and sums of squares have been estimated from data given by Rosner, after estimating missing values for two patients for whom only one eye was measured.

31. Heggestad, H. E., and Bennett, J. H. (1981). Photochemical oxidants potentiate yield losses in snap beans attributable to sulfur dioxide. *Science*, **213**, 1008–1010. Copyright 1981 by the AAAS. Raw data courtesy of H. E. Heggestad.

32. Tardif, J., Cote, G., Lesperance, J., Bourassa, M., Lambert, J., Doucet, S., Bilodeau, L., Nattel, S., and DeGuise, P. (1997). Probucol and multivitamins in the prevention of restenosis after coronary angioplasty. *New England Journal of Medicine*, **337**, 365–372.

33. Walker, P., Osredkar, M., and Bilancini, S. (1999). The effect of stimuli on pillbug movement. Unpublished manuscript, Oberlin College.

34. Hoppeler, H., and Vogt, M. (2001). Muscle tissue adaptations to hypoxia. *The Journal of Experimental Biology*, **204**, 3133–3139.

Chapter 12

1. Unpublished data courtesy of M. B. Nichols and R. P. Maickel. The original experiment contained more than three treatment groups.

2. Gwynne, D. T. (1981). Sexual difference theory: Mormon crickets show role reversal in mate choice. *Science*, **213**, 779–780. Copyright 1981 by the AAAS. Raw data courtesy of the author.

3. Adapted from Andren, C., and Nilson, G. (1981). Reproductive success and risk of predation in normal and melanistic colour morphs of the adder, *Vipera berus. Biological Journal of the Linnean Society*, **15**, 235–246. (The data are for the melanistic females; the values have been manipulated slightly to simplify the exposition.)

4. Cicirelli, M. F., Robinson, K. R., and Smith, L. D. (1983). Internal pH of *Xenopus* oocytes: A study of the mechanism and role of pH changes during meiotic maturation. *Developmental Biology*, **100**, 133–146. Raw data courtesy of M. F. Cicirelli.

5. Maickel, R. P., and Nash, J. F., Jr. (1985). Differing effects of short-chain alcohols on body temperature and coordinated muscular activity in mice. *Neuropharmacology*, **24**, 83–89. Reprinted with permission. Copyright 1985, Pergamon Journals, Ltd. Raw data courtesy of J. F. Nash, Jr.

6. Smith, R. D. (1978–1979). *Institute of Agricultural Engineering Annual Report*. Salisbury, Zimbabwe: Department of Research and Specialist Services, Ministry of Agriculture. Raw data courtesy of R. D. Smith.

7. Bowers, W. S., Hoch, H. C., Evans, P. H., and Katayama, M. (1986). Thallophytic allelopathy: Isolation and identification of laetisaric acid. *Science*, **232**, 105–106. Copyright 1986 by the AAAS. Raw data courtesy of the authors.

8. Webb, P. (1981). Energy expenditure and fat-free mass in men and women. *American Journal of Clinical Nutrition*, **34**, 1816–1826.

9. Adapted from Barclay, A. M., and Crawford, R. M. M. (1984). Seedling emergence in the rowan *(Sorbus aucuparia)* from an altitudinal gradient. *Journal of Ecology*, **72**, 627–636. Reprinted with permission of Blackwell Scientific Publications Limited.

10. Olson, J. M., and Mardh, R. L. (1998). Activation patterns and length changes in hindlimb muscles of the bullfrog *Rana catesbeiana* during jumping. *The Journal of Experimental Biology*, **201**, 2763–2777.

11. Sulcove, J. A., and Lacuesta, N. N. (1998). The effect of gender and height on peak flow rate. Unpublished manuscript, Oberlin College.

12. Hamill, P. V. V., Johnston, F. E., and Lemeshow, S. (1973). Height and weight of youths 12–17 years, United States. *U.S. National Center for Health Statistics, Vital and Health Statistics Series* **11**, *No. 124.* Washington, D.C.: U.S. Department of Health, Education and Welfare. The conditional distributions of weight given height are plotted in Figure 12.9 as normal distributions. The fictitious population agrees well with the real population (as described by Hamill et al.) in the central portion of each conditional distribution, but the real conditional distributions have shorter left tails and longer right tails than the fictitious normal conditional distributions.

13. Maickel, R. P. Personal communication.

14. Erne, P., Bolli, P., Buergisser, E., and Buehler, F. R. (1984). Correlation of platelet calcium with blood pressure. *New England Journal of Medicine*, **310**, 1084–1088. Reprinted by permission. Raw data courtesy of F. R. Buehler. To simplify the discussion, we have omitted nine patients with "borderline" high blood pressure.

15. Pappas, T., and Mitchell, C. A. (1985). Effects of seismic stress on the vegetative growth of *Glycine max* (L.) Merr. cv. Wells II. *Plant, Cell and Environment*, **8**, 143–148. Reprinted with permission of Blackwell Scientific Publications Limited. Raw data courtesy of the authors.

16. Florey, C. du V., and Acheson, R. M. (1969). Blood pressure as it relates to physique, blood glucose, and serum cholesterol. *U.S. National Center for Health Statistics, Series* **11**, *No. 34.* Washington, D.C.: U.S. Department of Health, Education and Welfare.

17. Masty, J. (1983). Innervation of the equine small intestine. Master's thesis, Purdue University. Raw data courtesy of the author.

18. Albert, A. (1981). Atypicality indices as reference values for laboratory data. *American Journal of Clinical Pathology*, **76**, 421–425.

19. Harding, A. J., Wong, A., Svoboda, M., Kril, J. J., and Halliday, G. M. (1997). Chronic alcohol consumption does not cause hippocampal neuron loss in humans. *Hippocampus*, **7**, 78–87. The value $r = -.63$ was calculated from data displayed graphically by Harding et al.

20. Fictitious but realistic data, based on inter- and intra-individual variation as described in Williams, G. Z., Widdowson, G. M., and Penton, J. (1978). Individual character of variation in time-series studies of healthy people. II. Differences in values for clinical chemical analytes in serum among demographic groups, by age and sex. *Clinical Chemistry*, **24**, 313–320.

21. Stewart, T. S., Nelson, L. A., Perry, T. W., and Martin, T. G. (1985). Unpublished data provided courtesy of T. S. Stewart.

22. Pappas, T., and Mitchell, C. A. (1985). Effects of seismic stress on the vegetative growth of *Glycine max* (L.) Merr. cv. Wells II. *Plant, Cell and Environment*, **8**, 143–149. Reprinted with permission of Blackwell Scientific Publications Limited. Raw data courtesy of the authors.

23. Adapted from Spencer, D. F., Volpp, T. R., and Lembi, C. A. (1980). Environmental control of *Pithophora oedogonia* (Chlorophyceae) akinete germination. *Journal of Phycology*, **16**, 424–427. The value $r = -.72$ was calculated from data displayed graphically by Spencer et al.

24. Fialho, E. T., Ferreira, A. S., Freitas, A. R., and Albino, L. F. T. (1982). Energy and nitrogen balance of ration (corn-soybean meal) for male castrated and noncastrated swine of different weights and breeds (in Portuguese). *Revista Sociedade Brasileira de Zootecnia*, **11**, 405–419. Raw data courtesy of E. T. Fialho.

25. Example communicated by D. A. Holt.

26. Chambers, J. Q., Higuhi, N., and Schimel, J. (1998). Ancient trees in Amazonia. *Nature*, **391**, 135–136. Raw data courtesy of J. Chambers.

27. Florey, C. du V., and Acheson, R. M. (1969). Blood pressure as it relates to physique, blood glucose, and serum cholesterol. *U.S. National Center for Health Statistics, Series* **11**, *No. 34*. Washington, D.C.: U.S. Department of Health, Education and Welfare.

28. Bernays, E. A. (1986). Diet-induced head allometry among foliage-chewing insects and its importance for graminovores. *Science*, **231**, 495–497. Copyright 1986 by the AAAS. Raw data courtesy of the author.

29. Hibi, K., Taguchi, M., Nakayama, H., Takase, T., Kasai, Y., Ito, K., Akiyama, S., and Nakao, A. (2001). Molecular detection of p16 promoter methylation in the serum of patients with esophageal squamous cell carcinoma. *Clinical Cancer Research*, **7**, 3135–3138. There were 38 patients in the study; only the 31 patients for whom "tumor DNA was methylated" are included in this analysis.

30. Gwynne, D. T. (1981). Sexual difference theory: Mormon crickets show role reversal in mate choice. *Science*, **213**, 779–780. Copyright 1981 by the AAAS. Calculations based on raw data provided courtesy of the author.

31. Heggestad, H. E., and Bennett, J. H. (1981). Photochemical oxidants potentiate yield losses in snap beans attributable to sulfur dioxide. *Science*, **213**, 1008–1010. Copyright 1981 by the AAAS. Raw data courtesy of H. E. Heggestad.

32. Thirakhupt, K. (1985). Foraging ecology of sympatric parids: Individual and population responses to winter food scarcity. Ph.D. thesis, Purdue University. Raw data courtesy of the author and K. N. Rabenold.

33. Unpublished data courtesy of A. H. Ismail and L. S. Verity.

34. Tazawa, H., Pearson, J. T., Komoro, T., and Ar, A. (2001). Allometric relationships between embryonic heart rate and fresh egg mass in birds. *The Journal of Experimental Biology*, **204**, 165–174.

35. These data sets were invented by F. J. Anscombe. See Anscombe, F. J. (1973). Graphs in statistical analysis. *The American Statistician*, **27**, 17–21.

36. Unpublished data courtesy of M. B. Nichols and R. P Maickel. The experiment actually contained more than three groups. The data in Example 12.1 are from another part of the same study, using a different chemical form of amphetamine.

37. Marazziti, D., DiMuro, A., and Castrogiovanni, P. (1992). Psychological stress and body temperature changes in humans. *Psychology & Behavior*, **52**, 393–395.

Chapter 13

1. Unpublished data courtesy of M. A. Johnson and F. Bretos. Data collected at Oberlin College in the fall of 1997.
2. Fish, F. E. (1998). Comparative kinematics and hydrodynamics of odontocete cetaceans: morphological and ecological correlates with swimming performance. *Journal of Experimental Biology*, **201**, 2867–2877. The data presented here were read off of Figure 1 in the article.
3. Fisher, B., Costantino, J. P., Wickerman, L., Redmond, C. K., Kavanah, M., Cronin, W. M., Vogel, V., Robidoux, A., Dimitrov, N., Atkins, J., Daly, M., Wieand, S., Tan-Chiu, E., Ford, L., Wolmark, N., et al. (1998). Tamoxifen for prevention of breast cancer: report of the national surgical adjuvant breast and bowel project P-1 study. *Journal of the National Cancer Institute*, **30**, 1371–1388.
4. Unpublished data courtesy of K. Pretl. Data collected at Oberlin College in the spring of 1997.
5. Rosa, L., Rosa, E., Sarner, L., and Barrett, S. (1998). A close look at therapeutic touch. *Journal of the American Medical Association*, **279**, 1005–1010.
6. Morgan, M. B., and Cowles, D. L. (1996). The effects of temperature on the behaviour and physiology of *Phataria unifascialis* (Gray) (Echinodermata, Asteroidea); implications for the species' distribution in the Gulf of California, Mexico. *Journal of Experimental Marine Biology and Ecology*, **208**, 13–27.
7. Kujala, U. M., Kaprio, J., Sarna, S., and Koshenvuo, M. (1998). Relationship of leisure-time physical activity and mortality: the Finnish cohort study. *Journal of the American Medical Association*, **279**, 440–444. The "exerciser" category actually had two sub-categories of "occasional exerciser" and "conditioning exerciser," which we have combined here.
8. Espigares, T., and Peco, B. (1995). Mediterranean annual pasture dynamics: impact of autumn drought. *Journal of Ecology*, **83**, 135–142.
9. Izurieta, H. S., Strebel, P. M., and Blake, P. A. (1997). Postlicensure effectiveness of varicella vaccine during an outbreak in a child care center. *Journal of the American Medical Association*, **278**, 1495–1499.
10. Rose, D. P., Fern, M., Liskowski, L., and Milbrath, J. R. (1977). Effect of treatment with estrogen conjugates on endogenous plasma steroids. *Obstetrics and Gynecology*, **49**, 80–82.
11. Grether, G. (1997). Survival cost of an intrasexually selected ornament in a damselfly. *Proceedings of the Royal Society of London*, **B 264**, 207–210.
12. Peterson, A. V., Kealey, K. A., Mann, S. L., Marek, P. M., and Sarason, I. G. (2000). Hutchinson smoking prevention project: long-term randomized trial in school-based tobacco use prevention—results on smoking. *Journal of the National Cancer Institute*, **92**, 1979–1991. The authors analyzed these data not with a sign test, but with a permutation test. The basic idea of a permutation test is to consider permutations of the observed data within pairings and to ask, for each possible permutation, how large or small is the calculated sample value of the overall difference in smoking rates. One then checks to see where the observed difference in smoking rates fits within the permutation distribution. Like the sign test, the permutation test does not require normally distributed data.

13. Rosenheck, R., Cramer, J., Xu, W., Thomas, J., Henderson, W., Frisman, L., Fye, C., and Charney, D., for the Department of Veterans Affairs Cooperative Study Group on Clozapine in Refractory Schizophrenia (1997). A comparison of clozapine and haloperidol in hospitalized patients with refractory schizophrenia. *New England Journal of Medicine*, **337**, 809–815.

14. Unpublished data courtesy of K. Roberts. Data collected at Oberlin College in the spring of 1997.

15. Baur, E., Fischer, E., and Lenz, F. (1931). *Human Heredity*, 3rd edition. New York: Macmillan, p. 45.

16. Plotnick, G. D., Corretti, M. C., and Vogel, R. A. (1997). Effect of antioxidant vitamins on the transient impairment of endothelium-dependent brachial artery vasoactivity following a single high-fat meal. *Journal of the American Medical Association*, **278**, 1682–1686.

17. Unpublished data collected by E. Lohan and M. Josephy, Oberlin College, in the fall of 1997.

18. Meier, P., Free, S. M., and Jackson, G. L. (1958). Reconsideration of methodology in studies of pain relief. *Biometrics*, **14**, 330–342.

19. Unpublished data collected by J. Amundson, Oberlin College, in the fall of 1997.

20. Weintraub, W. S., Boccuzzi, S. J., Klein, J. L., Kosinski, A. S., King, S. B., Ivanhoe, R., Cedarholm, J. C., Stillabower, M. E., Talley, J. D., DeMaio, S. J., O'Neill, W. W., Frazier, J. E., Cohen-Bernstein, C. L., Robbins, D. C., Brown, C. L., Alexander, R. W., and the Lovastatin Restenosis Trial Study Group. (1994). Lack of effect of lovastatin on restenosis after coronary angioplasty. *New England Journal of Medicine*, **331**, 1331–1337.

21. Bresee, J., Mast, E. E., Coleman, P. J., Baron, M. J., Schonberger, L. B., Alter, M. J., Jonas, M. M., Yu, M. W., Renzi, P. M., and Schneider, L. C. (1996). Hepatitis C virus infection associated with administration of intravenous immune globulin: a cohort study. *Journal of the American Medical Association*, **276**, 1563–1567.

22. Ferrandina, G., Ranelletti, F. O., Larocca, L. M., Maggiano, N., Fruscella, E., Legge, F., Santeusanio, G., Bombonati, A., Mancuso, S., and Scambia, G. (2001). Tamoxifen modulates the expression of Ki67, apoptosis, and microvessel density in cervical cancer. *Cancer Clinical Research*, **7**, 2656–2661.

23. Devine, W. D., Houston, A. E., and Tyler, D. D. (2000). Growth of three hardword species through 18 years on a former agricultural bottomland. *Southern Journal of Applied Forestry*, **24**, 159–165.

24. Unpublished data collected by S. Haaz, Oberlin College, in the spring of 1999. The division into groups is based on bench press strength, adjusted for height, weight, and whether or not the study was on an athletic team.

Statistical Tables

TABLE 1 Random Digits

	01	06	11	16	21	26	31	36	41	46
01	06048	96063	22049	86532	75170	65711	29969	06826	39208	80631
02	25636	73908	85512	78073	19089	66458	06597	93985	14193	69366
03	61378	45410	43511	54364	97334	01267	28304	35047	38789	84896
04	15919	71559	12310	00727	54473	51547	09816	83641	72973	75367
05	47328	20405	88019	82276	33679	10328	25116	59176	64675	95141
06	72548	80667	53893	64400	81955	15163	06146	58549	75530	19582
07	87154	04130	55985	44508	37515	71689	80765	46598	45539	12792
08	68379	96636	32154	94718	22845	80265	92747	66238	58474	23783
09	89391	54041	70806	36012	30833	83132	39338	54753	00722	44568
10	15816	60231	28365	61924	66934	21243	09896	92428	51611	46756
11	29618	55219	18394	11625	27673	08117	89314	42581	36897	03738
12	30723	42988	30002	95364	45473	46107	34222	00739	84847	49096
13	54028	04975	92323	53836	76128	84762	32050	59516	40831	59687
14	40376	02036	48087	05216	26684	97959	85601	86622	70750	15603
15	64439	37357	90935	57330	79738	65361	85944	23619	30504	61564
16	83037	30144	29166	20915	53462	42573	75204	50064	08847	07082
17	71071	01636	31085	71638	77357	14256	89174	15184	81701	21592
18	67891	43187	58159	24144	29683	04276	02987	04571	18334	04291
19	52487	39499	97330	40045	47304	98528	00422	82693	87547	73525
20	67550	82107	27302	79145	73213	27217	19211	59784	63929	04609
21	86472	80165	70773	90519	49710	31921	36102	45042	04203	01439
22	08699	38051	60404	06609	98435	91560	22634	98014	43316	61099
23	59596	13000	07655	74837	81211	71530	28341	83110	72289	25180
24	31810	54868	92799	09893	97499	96509	71548	06462	40498	22628
25	71753	90756	21382	84209	95900	11119	34507	61241	17641	83147

TABLE 1 Random Digits *(continued)*

	51	56	61	66	71	76	81	86	91	96
01	64825	74126	86159	26710	49256	04655	06001	73192	67463	16746
02	46184	63916	89160	87844	53352	43318	70766	23625	09906	65847
03	79976	48891	69431	86571	25979	58755	08884	36704	01107	12308
04	10656	47210	48512	06805	42114	98741	51440	06070	49071	02700
05	18058	84528	56753	02623	81077	60045	06678	53748	10386	37895
06	58979	98046	88467	27762	24781	12559	98384	40926	79570	34746
07	12705	41974	14473	49872	29368	80556	95833	20766	76643	35656
08	39660	83664	18592	82388	27899	24223	36462	61582	95173	36155
09	00360	42077	84161	04464	45042	29560	37916	29889	00342	82533
10	09873	64084	34685	53542	09254	23257	14713	44295	94139	00403
11	12957	84063	79808	23633	77133	41422	26559	29131	74402	82213
12	06090	71584	48965	60201	02786	88929	19861	99361	27535	38297
13	66812	57167	28185	19708	74672	25615	61640	18955	40854	50749
14	91701	36216	66249	04256	31694	33127	67529	73254	72065	74294
15	02775	78899	36471	37098	50270	58933	91765	95157	01384	75388
16	75892	53340	92363	58300	77300	08059	63743	12159	05640	87014
17	18581	70057	82031	68349	55759	46851	33632	28855	74633	08598
18	69698	18177	52824	61742	58119	04168	57843	37870	50988	80316
19	30023	30731	00803	09336	87709	39307	09732	66031	04904	91929
20	94334	05698	97910	37850	77074	56152	67521	48973	29448	84115
21	64133	14640	28418	45405	86974	06666	07879	54026	92264	23418
22	93895	83557	17326	28030	09113	56793	79703	18804	75807	20144
23	54438	83097	52533	86245	02182	11746	58164	90520	99255	44830
24	90565	76710	42456	22612	00232	18919	24019	32254	30703	00678
25	90848	81871	24382	16218	98216	42323	75061	68261	09071	68776

TABLE 1 Random Digits *(continued)*

	01	06	11	16	21	26	31	36	41	46
26	17155	07370	65655	04824	53417	20737	70510	92615	89967	50216
27	36211	24724	94769	16940	43138	25260	75318	69037	95982	28631
28	94777	66946	16120	56382	58416	92391	81457	28101	69766	32436
29	52994	58881	81841	51844	75566	48567	18552	66829	91230	39141
30	84643	32635	51440	96854	35739	66440	82806	82841	56302	31640
31	95690	34873	11297	60518	72717	47616	55751	37187	31413	31132
32	64093	92948	21565	51686	40368	66151	82877	99951	85069	54503
33	89484	50055	67586	16439	96385	67868	66597	51433	44764	66573
34	70184	38164	74646	90244	83169	85276	07598	69242	90088	32308
35	75601	91867	80848	94484	98532	36183	28549	17704	28653	80027
36	99044	78699	34681	31049	40790	50445	79897	68203	11486	93676
37	10272	18347	89369	02355	76671	34097	03791	93817	43142	24974
38	69738	85488	34453	80876	43018	59967	84458	71906	54019	70023
39	93441	58902	17871	45425	29066	04553	42644	54624	34498	27319
40	25814	74497	75642	58350	64118	87400	82870	26143	46624	21404
41	29757	84506	48617	48844	35139	97855	43435	74581	35678	69793
42	56666	86113	06805	09470	07992	54079	00517	19313	53741	25306
43	26401	71007	12500	27815	86490	01370	47826	36009	10447	25953
44	40747	59584	83453	30875	39509	82829	42878	13844	84131	48524
45	99434	51563	73915	03867	24785	19324	21254	11641	25940	92026
46	50734	88330	39128	14261	00584	94266	99677	19852	49673	18680
47	89728	32743	19102	83279	68308	41160	32365	25774	39699	50743
48	71395	61945	41082	93648	99874	82577	26507	07054	29381	16995
49	50945	68182	23108	95765	81136	06792	13322	41631	37118	35881
50	36525	26551	28457	75699	74537	68623	50099	91909	23508	35751

TABLE 1 Random Digits *(continued)*

	51	56	61	66	71	76	81	86	91	96
26	41169	08175	69938	61958	72578	31791	74952	71055	40369	00429
27	84627	70347	41566	00019	24481	15677	54506	54545	89563	50049
28	67460	49111	54004	61428	61034	47197	90084	88113	39145	94757
29	99231	60774	52238	05102	71690	72215	61323	13326	01674	81510
30	95775	73679	04900	27666	18424	59793	14965	22220	30682	35488
31	42179	98675	69593	17901	48741	59902	98034	12976	60921	73047
32	91196	05878	92346	45886	31080	21714	19168	94070	77375	10444
33	18794	03741	17612	65467	27698	20456	91737	36008	88225	58013
34	88311	93622	34501	70402	12272	65995	66086	04938	52966	71909
35	17904	33710	42812	72105	91848	39724	26361	09634	50552	98769
36	05905	28509	69631	69177	39081	58818	01998	53949	47884	91326
37	23432	22211	65648	71866	49532	45529	00189	80025	68956	26445
38	29684	43229	54771	90604	48938	13663	24736	83199	41512	43364
39	26506	65067	64252	49765	87650	72082	48997	04845	00136	98941
40	08807	43756	01579	34508	94082	68736	67149	00209	76138	95467
41	50636	70304	73556	32872	07809	20787	85921	41748	10553	97988
42	32437	41588	46991	36667	98127	05072	63700	51803	77262	31970
43	32571	97567	78420	04633	96574	88830	01314	04811	10904	85923
44	28773	22496	11743	23294	78070	20910	86722	50551	37356	92698
45	65768	76188	07781	05314	26017	07741	22268	31374	53559	46971
46	68601	06488	73776	45361	89059	59775	59149	64095	10352	11107
47	98364	17663	85972	72263	93178	04284	79236	04567	31813	82283
48	95308	70577	96712	85697	55685	19023	98112	96915	50791	31107
49	68681	24419	15362	60771	09962	45891	03130	09937	15775	51935
50	30721	22371	65174	57363	37851	71554	19708	23880	86638	05880

TABLE 2 Binomial Coefficients $_nC_j$

n	0	1	2	3	4	5	j 6	7	8	9	10
1	1	1									
2	1	2	1								
3	1	3	3	1							
4	1	4	6	4	1						
5	1	5	10	10	5	1					
6	1	6	15	20	15	6	1				
7	1	7	21	35	35	21	7	1			
8	1	8	28	56	70	56	28	8	1		
9	1	9	36	84	126	126	84	36	9	1	
10	1	10	45	120	210	252	210	120	45	10	1
11	1	11	55	165	330	462	462	330	165	55	11
12	1	12	66	220	495	792	924	792	495	220	66
13	1	13	78	286	715	1,287	1,716	1,716	1,287	715	286
14	1	14	91	364	1,001	2,002	3,003	3,432	3,003	2,002	1,001
15	1	15	105	455	1,365	3,003	5,005	6,435	6,435	5,005	3,003
16	1	16	120	560	1,820	4,368	8,008	11,440	12,870	11,440	8,008
17	1	17	136	680	2,380	6,188	12,376	19,448	24,310	24,310	19,448
18	1	18	153	816	3,060	8,568	18,564	31,824	43,758	48,620	43,758
19	1	19	171	969	3,876	11,628	27,132	50,388	75,582	92,378	92,378
20	1	20	190	1,140	4,845	15,504	38,760	77,520	125,970	167,960	184,756

TABLE 3 Areas Under the Normal Curve

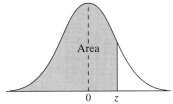

z	.00	.01	.02	.03	.04	.05	.06	.07	.08	.09
−3.4	0.0003	0.0003	0.0003	0.0003	0.0003	0.0003	0.0003	0.0003	0.0003	0.0002
−3.3	0.0005	0.0005	0.0005	0.0004	0.0004	0.0004	0.0004	0.0004	0.0004	0.0003
−3.2	0.0007	0.0007	0.0006	0.0006	0.0006	0.0006	0.0006	0.0005	0.0005	0.0005
−3.1	0.0010	0.0009	0.0009	0.0009	0.0008	0.0008	0.0008	0.0008	0.0007	0.0007
−3.0	0.0013	0.0013	0.0013	0.0012	0.0012	0.0011	0.0011	0.0011	0.0010	0.0010
−2.9	0.0019	0.0018	0.0017	0.0017	0.0016	0.0016	0.0015	0.0015	0.0014	0.0014
−2.8	0.0026	0.0025	0.0024	0.0023	0.0023	0.0022	0.0021	0.0021	0.0020	0.0019
−2.7	0.0035	0.0034	0.0033	0.0032	0.0031	0.0030	0.0029	0.0028	0.0027	0.0026
−2.6	0.0047	0.0045	0.0044	0.0043	0.0041	0.0040	0.0039	0.0038	0.0037	0.0036
−2.5	0.0062	0.0060	0.0059	0.0057	0.0055	0.0054	0.0052	0.0051	0.0049	0.0048
−2.4	0.0082	0.0080	0.0078	0.0075	0.0073	0.0071	0.0069	0.0068	0.0066	0.0064
−2.3	0.0107	0.0104	0.0102	0.0099	0.0096	0.0094	0.0091	0.0089	0.0087	0.0084
−2.2	0.0139	0.0136	0.0132	0.0129	0.0125	0.0122	0.0119	0.0116	0.0113	0.0110
−2.1	0.0179	0.0174	0.0170	0.0166	0.0162	0.0158	0.0154	0.0150	0.0146	0.0143
−2.0	0.0228	0.0222	0.0217	0.0212	0.0207	0.0202	0.0197	0.0192	0.0188	0.0183
−1.9	0.0287	0.0281	0.0274	0.0268	0.0262	0.0256	0.0250	0.0244	0.0239	0.0233
−1.8	0.0359	0.0352	0.0344	0.0336	0.0329	0.0322	0.0314	0.0307	0.0301	0.0294
−1.7	0.0446	0.0436	0.0427	0.0418	0.0409	0.0401	0.0392	0.0384	0.0375	0.0367
−1.6	0.0548	0.0537	0.0526	0.0516	0.0505	0.0495	0.0485	0.0475	0.0465	0.0455
−1.5	0.0668	0.0655	0.0643	0.0630	0.0618	0.0606	0.0594	0.0582	0.0571	0.0559
−1.4	0.0808	0.0793	0.0778	0.0764	0.0749	0.0735	0.0722	0.0708	0.0694	0.0681
−1.3	0.0968	0.0951	0.0934	0.0918	0.0901	0.0885	0.0869	0.0853	0.0838	0.0823
−1.2	0.1151	0.1131	0.1112	0.1093	0.1075	0.1056	0.1038	0.1020	0.1003	0.0985
−1.1	0.1357	0.1335	0.1314	0.1292	0.1271	0.1251	0.1230	0.1210	0.1190	0.1170
−1.0	0.1587	0.1562	0.1539	0.1515	0.1492	0.1469	0.1446	0.1423	0.1401	0.1379
−0.9	0.1841	0.1814	0.1788	0.1762	0.1736	0.1711	0.1685	0.1660	0.1635	0.1611
−0.8	0.2119	0.2090	0.2061	0.2033	0.2005	0.1977	0.1949	0.1922	0.1894	0.1867
−0.7	0.2420	0.2389	0.2358	0.2327	0.2296	0.2266	0.2236	0.2206	0.2177	0.2148
−0.6	0.2743	0.2709	0.2676	0.2643	0.2611	0.2578	0.2546	0.2514	0.2483	0.2451
−0.5	0.3085	0.3050	0.3015	0.2981	0.2946	0.2912	0.2877	0.2843	0.2810	0.2776
−0.4	0.3446	0.3409	0.3372	0.3336	0.3300	0.3264	0.3228	0.3192	0.3156	0.3121
−0.3	0.3821	0.3783	0.3745	0.3707	0.3669	0.3632	0.3594	0.3557	0.3520	0.3483
−0.2	0.4207	0.4168	0.4129	0.4090	0.4052	0.4013	0.3974	0.3936	0.3897	0.3859
−0.1	0.4602	0.4562	0.4522	0.4483	0.4443	0.4404	0.4364	0.4325	0.4286	0.4247
−0.0	0.5000	0.4960	0.4920	0.4880	0.4840	0.4801	0.4761	0.4721	0.4681	0.4641

Continued

TABLE 3 Areas Under the Normal Curve *(continued)*

z	.00	.01	.02	.03	.04	.05	.06	.07	.08	.09
0.0	0.5000	0.5040	0.5080	0.5120	0.5160	0.5199	0.5239	0.5279	0.5319	0.5359
0.1	0.5398	0.5438	0.5478	0.5517	0.5557	0.5596	0.5636	0.5675	0.5714	0.5753
0.2	0.5793	0.5832	0.5871	0.5910	0.5948	0.5987	0.6026	0.6064	0.6103	0.6141
0.3	0.6179	0.6217	0.6255	0.6293	0.6331	0.6368	0.6406	0.6443	0.6480	0.6517
0.4	0.6554	0.6591	0.6628	0.6664	0.6700	0.6736	0.6772	0.6808	0.6844	0.6879
0.5	0.6915	0.6950	0.6985	0.7019	0.7054	0.7088	0.7123	0.7157	0.7190	0.7224
0.6	0.7257	0.7291	0.7324	0.7357	0.7389	0.7422	0.7454	0.7486	0.7517	0.7549
0.7	0.7580	0.7611	0.7642	0.7673	0.7704	0.7734	0.7764	0.7794	0.7823	0.7852
0.8	0.7881	0.7910	0.7939	0.7967	0.7995	0.8023	0.8051	0.8078	0.8106	0.8133
0.9	0.8159	0.8186	0.8212	0.8328	0.8264	0.8289	0.8315	0.8340	0.8365	0.8389
1.0	0.8413	0.8438	0.8461	0.8485	0.8508	0.8531	0.8554	0.8577	0.8599	0.8621
1.1	0.8643	0.8665	0.8686	0.8708	0.8729	0.8749	0.8770	0.8790	0.8810	0.8830
1.2	0.8849	0.8869	0.8888	0.8907	0.8925	0.8944	0.8962	0.8980	0.8997	0.9015
1.3	0.9032	0.9049	0.9066	0.9082	0.9099	0.9115	0.9131	0.9147	0.9162	0.9177
1.4	0.9192	0.9207	0.9222	0.9236	0.9251	0.9265	0.9278	0.9292	0.9306	0.9319
1.5	0.9332	0.9345	0.9357	0.9370	0.9382	0.9394	0.9406	0.9418	0.9429	0.9441
1.6	0.9452	0.9463	0.9474	0.9484	0.9495	0.9505	0.9515	0.9525	0.9535	0.9545
1.7	0.9554	0.9564	0.9573	0.9582	0.9591	0.9599	0.9608	0.9616	0.9625	0.9633
1.8	0.9641	0.9649	0.9656	0.9664	0.9671	0.9678	0.9686	0.9693	0.9699	0.9706
1.9	0.9713	0.9719	0.9726	0.9732	0.9738	0.9744	0.9750	0.9756	0.9761	0.9767
2.0	0.9772	0.9778	0.9783	0.9788	0.9793	0.9798	0.9803	0.9808	0.9812	0.9817
2.1	0.9821	0.9826	0.9830	0.9834	0.9838	0.9842	0.9846	0.9850	0.9854	0.9857
2.2	0.9861	0.9864	0.9868	0.9871	0.9875	0.9878	0.9881	0.9884	0.9887	0.9890
2.3	0.9893	0.9896	0.9898	0.9901	0.9904	0.9906	0.9909	0.9911	0.9913	0.9916
2.4	0.9918	0.9920	0.9922	0.9925	0.9927	0.9929	0.9931	0.9932	0.9934	0.9936
2.5	0.9938	0.9940	0.9941	0.9943	0.9945	0.9946	0.9948	0.9949	0.9951	0.9952
2.6	0.9953	0.9955	0.9956	0.9957	0.9959	0.9960	0.9961	0.9962	0.9963	0.9964
2.7	0.9965	0.9966	0.9967	0.9968	0.9969	0.9970	0.9971	0.9972	0.9973	0.9974
2.8	0.9974	0.9975	0.9976	0.9977	0.9977	0.9978	0.9979	0.9979	0.9980	0.9981
2.9	0.9981	0.9982	0.9982	0.9983	0.9984	0.9984	0.9985	0.9985	0.9986	0.9986
3.0	0.9987	0.9987	0.9987	0.9988	0.9988	0.9989	0.9989	0.9989	0.9990	0.9990
3.1	0.9990	0.9991	0.9991	0.9991	0.9992	0.9992	0.9992	0.9992	0.9993	0.9993
3.2	0.9993	0.9993	0.9994	0.9994	0.9994	0.9994	0.9994	0.9995	0.9995	0.9995
3.3	0.9995	0.9995	0.9995	0.9996	0.9996	0.9996	0.9996	0.9996	0.9996	0.9997
3.4	0.9997	0.9997	0.9997	0.9997	0.9997	0.9997	0.9997	0.9997	0.9997	0.9998

TABLE 4 Critical Values of Student's *t* Distribution

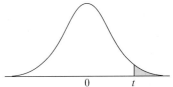

df	UPPER TAIL PROBABILITY									
	0.20	0.10	0.05	0.04	0.03	0.025	0.02	0.01	0.005	0.0005
1	1.376	3.078	6.314	7.916	10.579	12.706	15.895	31.821	63.657	636.619
2	1.061	1.886	2.920	3.320	3.896	4.303	4.849	6.965	9.925	31.599
3	0.978	1.638	2.353	2.605	2.951	3.182	3.482	4.541	5.841	12.924
4	0.941	1.533	2.132	2.333	2.601	2.776	2.999	3.747	4.604	8.610
5	0.920	1.476	2.015	2.191	2.422	2.571	2.757	3.365	4.032	6.869
6	0.906	1.440	1.943	2.104	2.313	2.447	2.612	3.143	3.707	5.959
7	0.896	1.415	1.895	2.046	2.241	2.365	2.517	2.998	3.499	5.408
8	0.889	1.397	1.860	2.004	2.189	2.306	2.449	2.896	3.355	5.041
9	0.883	1.383	1.833	1.973	2.150	2.262	2.398	2.821	3.250	4.781
10	0.879	1.372	1.812	1.948	2.120	2.228	2.359	2.764	3.169	4.587
11	0.876	1.363	1.796	1.928	2.096	2.201	2.328	2.718	3.106	4.437
12	0.873	1.356	1.782	1.912	2.076	2.179	2.303	2.681	3.055	4.318
13	0.870	1.350	1.771	1.899	2.060	2.160	2.282	2.650	3.012	4.221
14	0.868	1.345	1.761	1.888	2.046	2.145	2.264	2.624	2.977	4.140
15	0.866	1.341	1.753	1.878	2.034	2.131	2.249	2.602	2.947	4.073
16	0.865	1.337	1.746	1.869	2.024	2.120	2.235	2.583	2.921	4.015
17	0.863	1.333	1.740	1.862	2.015	2.110	2.224	2.567	2.898	3.965
18	0.862	1.330	1.734	1.855	2.007	2.101	2.214	2.552	2.878	3.922
19	0.861	1.328	1.729	1.850	2.000	2.093	2.205	2.539	2.861	3.883
20	0.860	1.325	1.725	1.844	1.994	2.086	2.197	2.528	2.845	3.850
21	0.859	1.323	1.721	1.840	1.988	2.080	2.189	2.518	2.831	3.819
22	0.858	1.321	1.717	1.835	1.983	2.074	2.183	2.508	2.819	3.792
23	0.858	1.319	1.714	1.832	1.978	2.069	2.177	2.500	2.807	3.768
24	0.857	1.318	1.711	1.828	1.974	2.064	2.172	2.492	2.797	3.745
25	0.856	1.316	1.708	1.825	1.970	2.060	2.167	2.485	2.787	3.725
26	0.856	1.315	1.706	1.822	1.967	2.056	2.162	2.479	2.779	3.707
27	0.855	1.314	1.703	1.819	1.963	2.052	2.158	2.473	2.771	3.690
28	0.855	1.313	1.701	1.817	1.960	2.048	2.154	2.467	2.763	3.674
29	0.854	1.311	1.699	1.814	1.957	2.045	2.150	2.462	2.756	3.659
30	0.854	1.310	1.697	1.812	1.955	2.042	2.147	2.457	2.750	3.646
40	0.851	1.303	1.684	1.796	1.936	2.021	2.123	2.423	2.704	3.551
50	0.849	1.299	1.676	1.787	1.924	2.009	2.109	2.403	2.678	3.496
60	0.848	1.296	1.671	1.781	1.917	2.000	2.099	2.390	2.660	3.460
70	0.847	1.294	1.667	1.776	1.912	1.994	2.093	2.381	2.648	3.435
80	0.846	1.292	1.664	1.773	1.908	1.990	2.088	2.374	2.639	3.416
100	0.845	1.290	1.660	1.769	1.902	1.984	2.081	2.364	2.626	3.390
140	0.844	1.288	1.656	1.763	1.896	1.977	2.073	2.353	2.611	3.361
1000	0.842	1.282	1.646	1.752	1.883	1.962	2.056	2.330	2.581	3.300
∞	0.842	1.282	1.645	1.751	1.881	1.960	2.054	2.326	2.576	3.291
	60%	80%	90%	92%	94%	95%	96%	98%	99%	99.9%
	CONFIDENCE LEVEL									

TABLE 5 Number of Observations for Independent-Samples *t* Test

	SIGNIFICANCE LEVEL (TWO-TAILED TEST)																			
	α = .01					α = .02					α = .05					α = .10				
POWER→	.99	.95	.90	.80	.50	.99	.95	.90	.80	.50	.99	.95	.90	.80	.50	.99	.95	.90	.80	.50
$\frac{\|\mu_1 - \mu_2\|}{\sigma}$																				
.20																				137
.25																				88
.30										123					87					61
.35					110					90					64				102	45
.40					85					70				100	50			108	78	35
.45				118	68				101	55			105	79	39		108	86	62	28
.50				96	55			106	82	45		106	86	64	32		88	70	51	23
.55			101	79	46		106	88	68	38		87	71	53	27	112	73	58	42	19
.60		101	85	67	39		90	74	58	32	104	74	60	45	23	89	61	49	36	16
.65		87	73	57	34	104	77	64	49	27	88	63	51	39	20	76	52	42	30	14
.70	100	75	63	50	29	90	66	55	43	24	76	55	44	34	17	66	45	36	26	12
.75	88	66	55	44	26	79	58	48	38	21	67	48	39	29	15	57	40	32	23	11
.80	77	58	49	39	23	70	51	43	33	19	59	42	34	26	14	50	35	28	21	10
.85	69	51	43	35	21	62	46	38	30	17	52	37	31	23	12	45	31	25	18	9
.90	62	46	39	31	19	55	41	34	27	15	47	34	27	21	11	40	28	22	16	8
.95	55	42	35	28	17	50	37	31	24	14	42	30	25	19	10	36	25	20	15	7
1.00	50	38	32	26	15	45	33	28	22	13	38	27	23	17	9	33	23	18	14	7
	α = .005					α = .01					α = .025					α = .05				
	SIGNIFICANCE LEVEL (ONE-TAILED TEST)																			

TABLE 5 Number of Observations for Independent-Samples t Test (continued)

SIGNIFICANCE LEVEL (TWO-TAILED TEST)

$\frac{\|\mu_1 - \mu_2\|}{\sigma}$	α = .01					α = .02					α = .05					α = .10				
POWER →	.99	.95	.90	.80	.50	.99	.95	.90	.80	.50	.99	.95	.90	.80	.50	.99	.95	.90	.80	.50
1.1	42	32	27	22	13	38	28	23	19	11	32	23	19	14	8	27	19	15	12	6
1.2	36	27	23	18	11	32	24	20	16	9	27	20	16	12	7	23	16	13	10	5
1.3	31	23	20	16	10	28	21	17	14	8	23	17	14	11	6	20	14	11	9	5
1.4	27	20	17	14	9	24	18	15	12	8	20	15	12	10	6	17	12	10	8	4
1.5	24	18	15	13	8	21	16	14	11	7	18	13	11	9	5	15	11	9	7	4
1.6	21	16	14	11	7	19	14	12	10	6	16	12	10	8	5	14	10	8	6	4
1.7	19	15	13	10	7	17	13	11	9	6	14	11	9	7	4	12	9	7	6	3
1.8	17	13	11	10	6	15	12	10	8	5	13	10	8	6	4	11	8	7	5	
1.9	16	12	11	9	6	14	11	9	8	5	12	9	7	6	4	10	7	6	5	
2.0	14	11	10	8	6	13	10	9	7	5	11	8	7	6	4	9	7	6	4	
2.1	13	10	9	8	5	12	9	8	7	5	10	8	6	5	3	8	6	5	4	
2.2	12	10	8	7	5	11	9	7	6	4	10	7	6	5		8	6	5	4	
2.3	11	9	8	7	5	10	8	7	6	4	9	7	6	5		7	5	5	4	
2.4	11	9	8	6	5	10	8	7	6	4	8	6	5	4		7	5	4	4	
2.5	10	8	7	6	4	9	7	6	5	4	8	6	5	4		6	5	4	3	
3.0	8	6	6	5	4	7	6	5	4	3	6	5	4	4		5	4	3		
3.5	6	5	5	4	3	6	5	4	4		5	4	4	3		4	3			
4.0	6	5	4	4		5	4	4	3		4	4	3			4				
	α = .005					α = .01					α = .025					α = .05				

SIGNIFICANCE LEVEL (ONE-TAILED TEST)

Source: "Number of observations for *t*-test of difference between two means." *Research*, Volume 1 (1948), pp. 520–525. Used with permission of the Longman Group UK Ltd. and Butterworth Scientific Publications.

TABLE 6 Critical Values of *U*, the Wilcoxon–Mann–Whitney Statistic

Note: Because the Wilcoxon-Mann-Whitney null distribution is discrete, the actual tail probability corresponding to a given critical value is typically somewhat *less* than the column heading.

			NOMINAL TAIL PROBABILITY						
		Two tails:	.20	.10	.05	.02	.01	.002	.001
n	*n′*	One tail:	.10	.05	.025	.01	.005	.001	.0005
3	2		6						
	3		8	9					
4	2		8						
	3		11	12					
	4		13	15	16				
5	2		9	10					
	3		13	14	15				
	4		16	18	19	20			
	5		20	21	23	24	25		
6	2		11	12					
	3		15	16	17				
	4		19	21	22	23	24		
	5		23	25	27	28	29		
	6		27	29	31	33	34		
7	2		13	14					
	3		17	19	20	21			
	4		22	24	25	27	28		
	5		27	29	30	32	34		
	6		31	34	36	38	39	42	
	7		36	38	41	43	45	48	49
8	2		14	15	16				
	3		19	21	22	24			
	4		25	27	28	30	31		
	5		30	32	34	36	38	40	
	6		35	38	40	42	44	47	48
	7		40	43	46	49	50	54	55
	8		45	49	51	55	57	60	62
9	1		9						
	2		16	17	18				
	3		22	23	25	26	27		
	4		27	30	32	33	35		
	5		33	36	38	40	42	44	45
	6		39	42	44	47	49	52	53
	7		45	48	51	54	56	60	61
	8		50	54	57	61	63	67	68
	9		56	60	64	67	70	74	76
10	1		10						
	2		17	19	20				
	3		24	26	27	29	30		
	4		30	33	35	37	38	40	
	5		37	39	42	44	46	49	50
	6		43	46	49	52	54	57	58
	7		49	53	56	59	61	65	67
	8		56	60	63	67	69	74	75
	9		62	66	70	74	77	82	83
	10		68	73	77	81	84	90	92

TABLE 6 Critical Values of *U*, the Wilcoxon–Mann–Whitney Statistic *(continued)*

		NOMINAL TAIL PROBABILITY						
		Two tails:	.20	.10	.05	.02	.01	.002	.001
n	*n'*	One tail:	.10	.05	.025	.01	.005	.001	.0005
11	1		11						
	2		19	21	22				
	3		26	28	30	32	33		
	4		33	36	38	40	42	44	
	5		40	43	46	48	50	53	54
	6		47	50	53	57	59	62	64
	7		54	58	61	65	67	71	73
	8		61	65	69	73	75	80	82
	9		68	72	76	81	83	89	91
	10		74	79	84	88	92	98	100
	11		81	87	91	96	100	106	109
12	1		12						
	2		20	22	23				
	3		28	31	32	34	35		
	4		36	39	41	42	45	48	
	5		43	47	49	52	54	58	59
	6		51	55	58	61	63	68	69
	7		58	63	66	70	72	77	79
	8		66	70	74	79	81	87	89
	9		73	78	82	87	90	96	98
	10		81	86	91	96	99	106	108
	11		88	94	99	104	108	115	117
	12		95	102	107	113	117	124	127
13	1		13						
	2		22	24	25	26			
	3		30	33	35	37	38		
	4		39	42	44	47	49	51	52
	5		47	50	53	56	58	62	63
	6		55	59	62	66	68	73	74
	7		63	67	71	75	78	83	85
	8		71	76	80	84	87	93	95
	9		79	84	89	94	97	103	106
	10		87	93	97	103	106	113	116
	11		95	101	106	112	116	123	126
	12		103	109	115	121	125	133	136
	13		111	118	124	130	135	143	146
14	1		14						
	2		24	25	27	28			
	3		32	35	37	40	41		
	4		41	45	47	50	52	55	56
	5		50	54	57	60	63	67	68
	6		59	63	67	71	73	78	79
	7		67	72	76	81	83	89	91
	8		76	81	86	90	94	100	102
	9		85	90	95	100	104	111	113
	10		93	99	104	110	114	121	124
	11		102	108	114	120	124	132	135
	12		110	117	123	130	134	143	146
	13		119	126	132	139	144	153	157
	14		127	135	141	149	154	164	167

TABLE 6 Critical Values of U, the Wilcoxon–Mann–Whitney Statistic (continued)

		NOMINAL TAIL PROBABILITY						
		Two tails: .20	.10	.05	.02	.01	.002	.001
n	n'	One tail: .10	.05	.025	.01	.005	.001	.0005
15	1	15						
	2	25	27	29	30			
	3	35	38	40	42	43		
	4	44	48	50	53	55	59	60
	5	53	57	61	64	67	71	72
	6	63	67	71	75	78	83	85
	7	72	77	81	86	89	95	97
	8	81	87	91	96	100	106	109
	9	90	96	101	107	111	118	120
	10	99	106	111	117	121	129	132
	11	108	115	121	128	132	141	144
	12	117	125	131	138	143	152	155
	13	127	134	141	148	153	163	167
	14	136	144	151	159	164	174	178
	15	145	153	161	169	174	185	189
16	1	16						
	2	27	29	31	32			
	3	37	40	42	45	46		
	4	47	50	53	57	59	62	63
	5	57	61	65	68	71	75	77
	6	67	71	75	80	83	88	90
	7	76	82	86	91	94	101	103
	8	86	92	97	102	106	113	115
	9	96	102	107	113	117	125	128
	10	106	112	118	124	129	137	140
	11	115	122	129	135	140	149	152
	12	125	132	139	146	151	161	165
	13	134	143	149	157	163	173	177
	14	144	153	160	168	174	185	189
	15	154	163	170	179	185	197	201
	16	163	173	181	190	196	208	213
17	1	17						
	2	28	31	32	34			
	3	39	42	45	47	49	51	
	4	50	53	57	60	62	66	67
	5	60	65	68	72	75	80	81
	6	71	76	80	84	87	93	95
	7	81	86	91	96	100	106	109
	8	91	97	102	108	112	119	122
	9	101	108	114	120	124	132	135
	10	112	119	125	132	136	145	148
	11	122	130	136	143	148	158	161
	12	132	140	147	155	160	170	174
	13	142	151	158	166	172	183	187
	14	153	161	169	178	184	195	199
	15	163	172	180	189	195	208	212
	16	173	183	191	201	207	220	225
	17	183	193	202	212	219	232	238

TABLE 6 Critical Values of *U*, the Wilcoxon–Mann–Whitney Statistic *(continued)*

			NOMINAL TAIL PROBABILITY					
		Two tails:	.20	.10	.05	.02	.01	.002	.001
n	*n'*	One tail:	.10	.05	.025	.01	.005	.001	.0005
18	1		18						
	2		30	32	34	36			
	3		41	45	47	50	52	54	
	4		52	56	60	63	66	69	71
	5		63	68	72	76	79	84	86
	6		74	80	84	89	92	98	100
	7		85	91	96	102	105	112	115
	8		96	103	108	114	118	126	129
	9		107	114	120	126	131	139	142
	10		118	125	132	139	143	153	156
	11		129	137	143	151	156	166	170
	12		139	148	155	163	169	179	183
	13		150	159	167	175	181	192	197
	14		161	170	178	187	194	206	210
	15		172	182	190	200	206	219	224
	16		182	193	202	212	218	232	237
	17		193	204	213	224	231	245	250
	18		204	215	225	236	243	258	263
19	1		18	19					
	2		31	34	36	37	38		
	3		43	47	50	53	54	57	
	4		55	59	63	67	69	73	74
	5		67	72	76	80	83	88	90
	6		78	84	89	94	97	103	106
	7		90	96	101	107	111	118	120
	8		101	108	114	120	124	132	135
	9		113	120	126	133	138	146	150
	10		124	132	138	146	151	161	164
	11		136	144	151	159	164	175	178
	12		147	156	163	172	177	188	193
	13		158	167	175	184	190	202	207
	14		169	179	188	197	203	216	221
	15		181	191	200	210	216	230	235
	16		192	203	212	222	230	244	249
	17		203	214	224	235	242	257	263
	18		214	226	236	248	255	271	277
	19		226	238	248	260	268	284	291
20	1		19	20					
	2		33	36	38	39	40		
	3		45	49	52	55	57	60	
	4		58	62	66	70	72	77	78
	5		70	75	80	84	87	93	95
	6		82	88	93	98	102	108	111
	7		94	101	106	112	116	124	126
	8		106	113	119	126	130	139	142
	9		118	126	132	140	144	154	157
	10		130	138	145	153	158	168	172
	11		142	151	158	167	172	183	187
	12		154	163	171	180	186	198	202
	13		166	176	184	193	200	212	217
	14		178	188	197	207	213	226	231
	15		190	200	210	220	227	241	246
	16		201	213	222	233	241	255	261
	17		213	225	235	247	254	270	275
	18		225	237	248	260	268	284	287
	19		237	250	261	273	281	298	304
	20		249	262	273	286	295	312	319

TABLE 7 Critical Values of *B* for the Sign Test

Note: Because the sign-test null distribution is discrete, the actual tail probability corresponding to a given critical value is typically somewhat *less* than the column heading.

		NOMINAL TAIL PROBABILITY						
	Two tails:	.20	.10	.05	.02	.01	.002	.001
n_d	One tail:	.10	.05	.025	.01	.005	.001	.0005
1								
2								
3								
4								
5		5	5					
6		6	6	6				
7		6	7	7	7			
8		7	7	8	8	8		
9		7	8	8	9	9		
10		8	9	9	10	10	10	
11		9	9	10	10	11	11	11
12		9	10	10	11	11	12	12
13		10	10	11	12	12	13	13
14		10	11	12	12	13	13	14
15		11	12	12	13	13	14	14
16		12	12	13	14	14	15	15
17		12	13	13	14	15	16	16
18		13	13	14	15	15	16	17
19		13	14	15	15	16	17	17
20		14	15	15	16	17	18	18
21		14	15	16	17	17	18	19
22		15	16	17	17	18	19	19
23		16	16	17	18	19	20	20
24		16	17	18	19	19	20	21
25		17	18	18	19	20	21	21
26		17	18	19	20	20	22	22
27		18	19	20	20	21	22	23
28		18	19	20	21	22	23	23
29		19	20	21	22	22	24	24
30		20	20	21	22	23	24	25

TABLE 8 Critical Values of *W* for the Wilcoxon Signed-rank Test

Note: Because the Wilcoxon signed-rank test null distribution is discrete, the actual tail probability corresponding to a given critical value is typically somewhat *less* than the column heading.

		NOMINAL TAIL PROBABILITY						
	Two tails:	.20	.10	.05	.02	.01	.002	.001
n_d	One tails:	.10	.05	.025	.01	.005	.001	.0005
1								
2								
3								
4		10						
5		13	15					
6		18	19	21				
7		23	25	26	28			
8		28	31	33	35	36		
9		35	37	40	42	44		
10		41	45	47	50	52	55	
11		49	53	56	59	61	65	66
12		57	61	65	69	71	76	77
13		65	70	74	79	82	87	89
14		75	80	84	90	93	99	101
15		84	90	95	101	105	112	114

If $n_d \geq 16$, then W has a distribution that is approximately normal. Thus, for $n_d \geq 16$ compute W_s and reject H_0 if $W_s > \dfrac{n_d(n_d+1)}{4} + Z_{\alpha/2}\sqrt{\dfrac{n_d(n_d+1)(2n_d+1)}{24}}$ for a two-tailed test.

For a one-tailed test, reject H_0 if $W_s > \dfrac{n_d(n_d+1)}{4} + Z_{\alpha}\sqrt{\dfrac{n_d(n_d+1)(2n_d+1)}{24}}$

TABLE 9 Critical Values of the Chi-Square Distribution

Note: If H_A is directional (for df = 1), column headings should be multiplied by 1/2 when bracketing the *P*-value.

df	TAIL PROBABILITY						
	.20	.10	.05	.02	.01	.001	.0001
1	1.64	2.71	3.84	5.41	6.63	10.83	15.14
2	3.22	4.61	5.99	7.82	9.21	13.82	18.42
3	4.64	6.25	7.81	9.84	11.34	16.27	21.11
4	5.99	7.78	9.49	11.67	13.28	18.47	23.51
5	7.29	9.24	11.07	13.39	15.09	20.51	25.74
6	8.56	10.64	12.59	15.03	16.81	22.46	27.86
7	9.80	12.02	14.07	16.62	18.48	24.32	29.88
8	11.03	13.36	15.51	18.17	20.09	26.12	31.83
9	12.24	14.68	16.92	19.68	21.67	27.88	33.72
10	13.44	15.99	18.31	21.16	23.21	29.59	35.56
11	14.63	17.28	19.68	22.62	24.72	31.26	37.37
12	15.81	18.55	21.03	24.05	26.22	32.91	39.13
13	16.98	19.81	22.36	25.47	27.69	34.53	40.87
14	18.15	21.06	23.68	26.87	29.14	36.12	42.58
15	19.31	22.31	25.00	28.26	30.58	37.70	44.26
16	20.47	23.54	26.30	29.63	32.00	39.25	45.92
17	21.61	24.77	27.59	31.00	33.41	40.79	47.57
18	22.76	25.99	28.87	32.35	34.81	42.31	49.19
19	23.90	27.20	30.14	33.69	36.19	43.82	50.80
20	25.04	28.41	31.41	35.02	37.57	45.31	52.39
21	26.17	29.62	32.67	36.34	38.93	46.80	53.96
22	27.30	30.81	33.92	37.66	40.29	48.27	55.52
23	28.43	32.01	35.17	38.97	41.64	49.73	57.08
24	29.55	33.20	36.42	40.27	42.98	51.18	58.61
25	30.68	34.38	37.65	41.57	44.31	52.62	60.14
26	31.79	35.56	38.89	42.86	45.64	54.05	61.66
27	32.91	36.74	40.11	44.14	46.96	55.48	63.16
28	34.03	37.92	41.34	45.42	48.28	56.89	64.66
29	35.14	39.09	42.56	46.69	49.59	58.30	66.15
30	36.25	40.26	43.77	47.96	50.89	59.70	67.63

TABLE 10 Critical Values of the *F* Distribution

Numerator df = 1

Denom. df	TAIL PROBABILITY						
	.20	.10	.05	.02	.01	.001	.0001
1	9.47	39.86	161	101^1	405^1	406^3	405^5
2	3.56	8.53	18.51	48.51	98.50	998	100^2
3	2.68	5.54	10.13	20.62	34.12	167	784
4	2.35	4.54	7.71	14.04	21.20	74.14	242
5	2.18	4.06	6.61	11.32	16.26	47.18	125
6	2.07	3.78	5.99	9.88	13.75	35.51	82.49
7	2.00	3.59	5.59	8.99	12.25	29.25	62.17
8	1.95	3.46	5.32	8.39	11.26	25.41	50.69
9	1.91	3.36	5.12	7.96	10.56	22.86	43.48
10	1.88	3.29	4.96	7.64	10.04	21.04	38.58
11	1.86	3.23	4.84	7.39	9.65	19.69	35.06
12	1.84	3.18	4.75	7.19	9.33	18.64	32.43
13	1.82	3.14	4.67	7.02	9.07	17.82	30.39
14	1.81	3.10	4.60	6.89	8.86	17.14	28.77
15	1.80	3.07	4.54	6.77	8.68	16.59	27.45
16	1.79	3.05	4.49	6.67	8.53	16.12	26.36
17	1.78	3.03	4.45	6.59	8.40	15.72	25.44
18	1.77	3.01	4.41	6.51	8.29	15.38	24.66
19	1.76	2.99	4.38	6.45	8.18	15.08	23.99
20	1.76	2.97	4.35	6.39	8.10	14.82	23.40
21	1.75	2.96	4.32	6.34	8.02	14.59	22.89
22	1.75	2.95	4.30	6.29	7.95	14.38	22.43
23	1.74	2.94	4.28	6.25	7.88	14.20	22.03
24	1.74	2.93	4.26	6.21	7.82	14.03	21.66
25	1.73	2.92	4.24	6.18	7.77	13.88	21.34
26	1.73	2.91	4.23	6.14	7.72	13.74	21.04
27	1.73	2.90	4.21	6.11	7.68	13.61	20.77
28	1.72	2.89	4.20	6.09	7.64	13.50	20.53
29	1.72	2.89	4.18	6.06	7.60	13.39	20.30
30	1.72	2.88	4.17	6.04	7.56	13.29	20.09
40	1.70	2.84	4.08	5.87	7.31	12.61	18.67
60	1.68	2.79	4.00	5.71	7.08	11.97	17.38
100	1.66	2.76	3.94	5.59	6.90	11.50	16.43
140	1.66	2.74	3.91	5.54	6.82	11.30	16.05
∞	1.64	2.71	3.84	5.41	6.63	10.83	15.14

Notation: 406^3 means 406×10^3

Continued

TABLE 10 Critical Values of the _F_ Distribution
(continued)

Numerator df = 2

Denom. df	TAIL PROBABILITY						
	.20	.10	.05	.02	.01	.001	.0001
1	12.00	49.50	200	125[1]	500[1]	500[3]	500[5]
2	4.00	9.00	19.00	49.00	99.00	999	100[2]
3	2.89	5.46	9.55	18.86	30.82	149	695
4	2.47	4.32	6.94	12.14	18.00	61.25	198
5	2.26	3.78	5.79	9.45	13.27	37.12	97.03
6	2.13	3.46	5.14	8.05	10.92	27.00	61.63
7	2.04	3.26	4.74	7.20	9.55	21.69	45.13
8	1.98	3.11	4.46	6.64	8.65	18.49	36.00
9	1.93	3.01	4.26	6.23	8.02	16.39	30.34
10	1.90	2.92	4.10	5.93	7.56	14.91	26.55
11	1.87	2.86	3.98	5.70	7.21	13.81	23.85
12	1.85	2.81	3.89	5.52	6.93	12.97	21.85
13	1.83	2.76	3.81	5.37	6.70	12.31	20.31
14	1.81	2.73	3.74	5.24	6.51	11.78	19.09
15	1.80	2.70	3.68	5.14	6.36	11.34	18.11
16	1.78	2.67	3.63	5.05	6.23	10.97	17.30
17	1.77	2.64	3.59	4.97	6.11	10.66	16.62
18	1.76	2.62	3.55	4.90	6.01	10.39	16.04
19	1.75	2.61	3.52	4.84	5.93	10.16	15.55
20	1.75	2.59	3.49	4.79	5.85	9.95	15.12
21	1.74	2.57	3.47	4.74	5.78	9.77	14.74
22	1.73	2.56	3.44	4.70	5.72	9.61	14.41
23	1.73	2.55	3.42	4.66	5.66	9.47	14.12
24	1.72	2.54	3.40	4.63	5.61	9.34	13.85
25	1.72	2.53	3.39	4.59	5.57	9.22	13.62
26	1.71	2.52	3.37	4.56	5.53	9.12	13.40
27	1.71	2.51	3.35	4.54	5.49	9.02	13.21
28	1.71	2.50	3.34	4.51	5.45	8.93	13.03
29	1.70	2.50	3.33	4.49	5.42	8.85	12.87
30	1.70	2.49	3.32	4.47	5.39	8.77	12.72
40	1.68	2.44	3.23	4.32	5.18	8.25	11.70
60	1.65	2.39	3.15	4.18	4.98	7.77	10.78
100	1.64	2.36	3.09	4.07	4.82	7.41	10.11
140	1.63	2.34	3.06	4.02	4.76	7.26	9.84
∞	1.61	2.30	3.00	3.91	4.61	6.91	9.21

TABLE 10 Critical Values of the *F* Distribution
(continued)

Numerator df = 3

Denom. df	TAIL PROBABILITY						
	.20	.10	.05	.02	.01	.001	.0001
1	13.06	53.59	216	135[1]	540[1]	540[3]	540[5]
2	4.16	9.16	19.16	49.17	99.17	999	100[2]
3	2.94	5.39	9.28	18.11	29.46	141	659
4	2.48	4.19	6.59	11.34	16.69	56.18	181
5	2.25	3.62	5.41	8.67	12.06	33.20	86.29
6	2.11	3.29	4.76	7.29	9.78	23.70	53.68
7	2.02	3.07	4.35	6.45	8.45	18.77	38.68
8	1.95	2.92	4.07	5.90	7.59	15.83	30.46
9	1.90	2.81	3.86	5.51	6.99	13.90	25.40
10	1.86	2.73	3.71	5.22	6.55	12.55	22.04
11	1.83	2.66	3.59	4.99	6.22	11.56	19.66
12	1.80	2.61	3.49	4.81	5.95	10.80	17.90
13	1.78	2.56	3.41	4.67	5.74	10.21	16.55
14	1.76	2.52	3.34	4.55	5.56	9.73	15.49
15	1.75	2.49	3.29	4.45	5.42	9.34	14.64
16	1.74	2.46	3.24	4.36	5.29	9.01	13.93
17	1.72	2.44	3.20	4.29	5.18	8.73	13.34
18	1.71	2.42	3.16	4.22	5.09	8.49	12.85
19	1.70	2.40	3.13	4.16	5.01	8.28	12.42
20	1.70	2.38	3.10	4.11	4.94	8.10	12.05
21	1.69	2.36	3.07	4.07	4.87	7.94	11.73
22	1.68	2.35	3.05	4.03	4.82	7.80	11.44
23	1.68	2.34	3.03	3.99	4.76	7.67	11.19
24	1.67	2.33	3.01	3.96	4.72	7.55	10.96
25	1.66	2.32	2.99	3.93	4.68	7.45	10.76
26	1.66	2.31	2.98	3.90	4.64	7.36	10.58
27	1.65	2.30	2.96	3.87	4.60	7.27	10.41
28	1.65	2.29	2.95	3.85	4.57	7.19	10.26
29	1.65	2.28	2.93	3.83	4.54	7.12	10.12
30	1.64	2.28	2.92	3.81	4.51	7.05	9.99
40	1.62	2.23	2.84	3.67	4.31	6.59	9.13
60	1.60	2.18	2.76	3.53	4.13	6.17	8.35
100	1.58	2.14	2.70	3.43	3.98	5.86	7.79
140	1.57	2.12	2.67	3.38	3.92	5.73	7.57
∞	1.55	2.08	2.60	3.28	3.78	5.42	7.04

Continued

TABLE 10 Critical Values of the *F* Distribution
(continued)

Numerator df = 4

Denom. df	TAIL PROBABILITY						
	.20	.10	.05	.02	.01	.001	.0001
1	13.64	55.83	225	141[1]	562[1]	562[3]	562[5]
2	4.24	9.24	19.25	49.25	99.25	999	100[2]
3	2.96	5.34	9.12	17.69	28.71	137	640
4	2.48	4.11	6.39	10.90	15.98	53.44	172
5	2.24	3.52	5.19	8.23	11.39	31.09	80.53
6	2.09	3.18	4.53	6.86	9.15	21.92	49.42
7	1.99	2.96	4.12	6.03	7.85	17.20	35.22
8	1.92	2.81	3.84	5.49	7.01	14.39	27.49
9	1.87	2.69	3.63	5.10	6.42	12.56	22.77
10	1.83	2.61	3.48	4.82	5.99	11.28	19.63
11	1.80	2.54	3.36	4.59	5.67	10.35	17.42
12	1.77	2.48	3.26	4.42	5.41	9.63	15.79
13	1.75	2.43	3.18	4.28	5.21	9.07	14.55
14	1.73	2.39	3.11	4.16	5.04	8.62	13.57
15	1.71	2.36	3.06	4.06	4.89	8.25	12.78
16	1.70	2.33	3.01	3.97	4.77	7.94	12.14
17	1.68	2.31	2.96	3.90	4.67	7.68	11.60
18	1.67	2.29	2.93	3.84	4.58	7.46	11.14
19	1.66	2.27	2.90	3.78	4.50	7.27	10.75
20	1.65	2.25	2.87	3.73	4.43	7.10	10.41
21	1.65	2.23	2.84	3.69	4.37	6.95	10.12
22	1.64	2.22	2.82	3.65	4.31	6.81	9.86
23	1.63	2.21	2.80	3.61	4.26	6.70	9.63
24	1.63	2.19	2.78	3.58	4.22	6.59	9.42
25	1.62	2.18	2.76	3.55	4.18	6.49	9.24
26	1.62	2.17	2.74	3.52	4.14	6.41	9.07
27	1.61	2.17	2.73	3.50	4.11	6.33	8.92
28	1.61	2.16	2.71	3.47	4.07	6.25	8.79
29	1.60	2.15	2.70	3.45	4.04	6.19	8.66
30	1.60	2.14	2.69	3.43	4.02	6.12	8.54
40	1.57	2.09	2.61	3.30	3.83	5.70	7.76
60	1.55	2.04	2.53	3.16	3.65	5.31	7.06
100	1.53	2.00	2.46	3.06	3.51	5.02	6.55
140	1.52	1.99	2.44	3.02	3.46	4.90	6.35
∞	1.50	1.94	2.37	2.92	3.32	4.62	5.88

TABLE 10 Critical Values of the *F* Distribution (*continued*)

Numerator df = 5

Denom. df	TAIL PROBABILITY						
	.20	.10	.05	.02	.01	.001	.0001
1	14.01	57.24	230	144[1]	576[1]	576[3]	576[5]
2	4.28	9.29	19.30	49.30	99.30	999	100[2]
3	2.97	5.31	9.01	17.43	28.24	135	628
4	2.48	4.05	6.26	10.62	15.52	51.71	166
5	2.23	3.45	5.05	7.95	10.97	29.75	76.91
6	2.08	3.11	4.39	6.58	8.75	20.80	46.75
7	1.97	2.88	3.97	5.76	7.46	16.21	33.06
8	1.90	2.73	3.69	5.22	6.63	13.48	25.63
9	1.85	2.61	3.48	4.84	6.06	11.71	21.11
10	1.80	2.52	3.33	4.55	5.64	10.48	18.12
11	1.77	2.45	3.20	4.34	5.32	9.58	16.02
12	1.74	2.39	3.11	4.16	5.06	8.89	14.47
13	1.72	2.35	3.03	4.02	4.86	8.35	13.29
14	1.70	2.31	2.96	3.90	4.69	7.92	12.37
15	1.68	2.27	2.90	3.81	4.56	7.57	11.62
16	1.67	2.24	2.85	3.72	4.44	7.27	11.01
17	1.65	2.22	2.81	3.65	4.34	7.02	10.50
18	1.64	2.20	2.77	3.59	4.25	6.81	10.07
19	1.63	2.18	2.74	3.53	4.17	6.62	9.71
20	1.62	2.16	2.71	3.48	4.10	6.46	9.39
21	1.61	2.14	2.68	3.44	4.04	6.32	9.11
22	1.61	2.13	2.66	3.40	3.99	6.19	8.87
23	1.60	2.11	2.64	3.36	3.94	6.08	8.65
24	1.59	2.10	2.62	3.33	3.90	5.98	8.46
25	1.59	2.09	2.60	3.30	3.85	5.89	8.28
26	1.58	2.08	2.59	3.28	3.82	5.80	8.13
27	1.58	2.07	2.57	3.25	3.78	5.73	7.99
28	1.57	2.06	2.56	3.23	3.75	5.66	7.86
29	1.57	2.06	2.55	3.21	3.73	5.59	7.74
30	1.57	2.05	2.53	3.19	3.70	5.53	7.63
40	1.54	2.00	2.45	3.05	3.51	5.13	6.90
60	1.51	1.95	2.37	2.92	3.34	4.76	6.25
100	1.49	1.91	2.31	2.82	3.21	4.48	5.78
140	1.48	1.89	2.28	2.78	3.15	4.37	5.59
∞	1.46	1.85	2.21	2.68	3.02	4.10	5.15

TABLE 10 Critical Values of the _F_ Distribution
(continued)

Numerator df = 6

Denom. df	TAIL PROBABILITY						
	.20	.10	.05	.02	.01	.001	.0001
1	14.26	58.20	234	146[1]	586[1]	586[3]	586[5]
2	4.32	9.33	19.33	49.33	99.33	999	100[2]
3	2.97	5.28	8.94	17.25	27.91	133	620
4	2.47	4.01	6.16	10.42	15.21	50.53	162
5	2.22	3.40	4.95	7.76	10.67	28.83	74.43
6	2.06	3.05	4.28	6.39	8.47	20.03	44.91
7	1.96	2.83	3.87	5.58	7.19	15.52	31.57
8	1.88	2.67	3.58	5.04	6.37	12.86	24.36
9	1.83	2.55	3.37	4.65	5.80	11.13	19.97
10	1.78	2.46	3.22	4.37	5.39	9.93	17.08
11	1.75	2.39	3.09	4.15	5.07	9.05	15.05
12	1.72	2.33	3.00	3.98	4.82	8.38	13.56
13	1.69	2.28	2.92	3.84	4.62	7.86	12.42
14	1.67	2.24	2.85	3.72	4.46	7.44	11.53
15	1.66	2.21	2.79	3.63	4.32	7.09	10.82
16	1.64	2.18	2.74	3.54	4.20	6.80	10.23
17	1.63	2.15	2.70	3.47	4.10	6.56	9.75
18	1.62	2.13	2.66	3.41	4.01	6.35	9.33
19	1.61	2.11	2.63	3.35	3.94	6.18	8.98
20	1.60	2.09	2.60	3.30	3.87	6.02	8.68
21	1.59	2.08	2.57	3.26	3.81	5.88	8.41
22	1.58	2.06	2.55	3.22	3.76	5.76	8.18
23	1.57	2.05	2.53	3.19	3.71	5.65	7.97
24	1.57	2.04	2.51	3.15	3.67	5.55	7.79
25	1.56	2.02	2.49	3.13	3.63	5.46	7.62
26	1.56	2.01	2.47	3.10	3.59	5.38	7.48
27	1.55	2.00	2.46	3.07	3.56	5.31	7.34
28	1.55	2.00	2.45	3.05	3.53	5.24	7.22
29	1.54	1.99	2.43	3.03	3.50	5.18	7.10
30	1.54	1.98	2.42	3.01	3.47	5.12	7.00
40	1.51	1.93	2.34	2.88	3.29	4.73	6.30
60	1.48	1.87	2.25	2.75	3.12	4.37	5.68
100	1.46	1.83	2.19	2.65	2.99	4.11	5.24
140	1.45	1.82	2.16	2.61	2.93	4.00	5.06
∞	1.43	1.77	2.10	2.51	2.80	3.74	4.64

TABLE 10 Critical Values of the *F* Distribution (continued)

Numerator df = 7

Denom. df	TAIL PROBABILITY						
	.20	.10	.05	.02	.01	.001	.0001
1	14.44	58.91	237	148[1]	593[1]	593[3]	593[5]
2	4.34	9.35	19.35	49.36	99.36	999	100[2]
3	2.97	5.27	8.89	17.11	27.67	132	614
4	2.47	3.98	6.09	10.27	14.98	49.66	159
5	2.21	3.37	4.88	7.61	10.46	28.16	72.61
6	2.05	3.01	4.21	6.25	8.26	19.46	43.57
7	1.94	2.78	3.79	5.44	6.99	15.02	30.48
8	1.87	2.62	3.50	4.90	6.18	12.40	23.42
9	1.81	2.51	3.29	4.52	5.61	10.70	19.14
10	1.77	2.41	3.14	4.23	5.20	9.52	16.32
11	1.73	2.34	3.01	4.02	4.89	8.66	14.34
12	1.70	2.28	2.91	3.85	4.64	8.00	12.89
13	1.68	2.23	2.83	3.71	4.44	7.49	11.79
14	1.65	2.19	2.76	3.59	4.28	7.08	10.92
15	1.64	2.16	2.71	3.49	4.14	6.74	10.23
16	1.62	2.13	2.66	3.41	4.03	6.46	9.66
17	1.61	2.10	2.61	3.34	3.93	6.22	9.19
18	1.60	2.08	2.58	3.27	3.84	6.02	8.79
19	1.58	2.06	2.54	3.22	3.77	5.85	8.45
20	1.58	2.04	2.51	3.17	3.70	5.69	8.16
21	1.57	2.02	2.49	3.13	3.64	5.56	7.90
22	1.56	2.01	2.46	3.09	3.59	5.44	7.68
23	1.55	1.99	2.44	3.05	3.54	5.33	7.48
24	1.55	1.98	2.42	3.02	3.50	5.23	7.30
25	1.54	1.97	2.40	2.99	3.46	5.15	7.14
26	1.53	1.96	2.39	2.97	3.42	5.07	6.99
27	1.53	1.95	2.37	2.94	3.39	5.00	6.86
28	1.52	1.94	2.36	2.92	3.36	4.93	6.75
29	1.52	1.93	2.35	2.90	3.33	4.87	6.64
30	1.52	1.93	2.33	2.88	3.30	4.82	6.54
40	1.49	1.87	2.25	2.74	3.12	4.44	5.86
60	1.46	1.82	2.17	2.62	2.95	4.09	5.27
100	1.43	1.78	2.10	2.52	2.82	3.83	4.84
140	1.42	1.76	2.08	2.48	2.77	3.72	4.67
∞	1.40	1.72	2.01	2.37	2.64	3.47	4.27

TABLE 10 Critical Values of the *F* Distribution (*continued*)

Numerator df = 8

Denom. df	TAIL PROBABILITY						
	.20	.10	.05	.02	.01	.001	.0001
1	14.58	59.44	239	149[1]	598[1]	598[3]	598[5]
2	4.36	9.37	19.37	49.37	99.37	999	100[2]
3	2.98	5.25	8.85	17.01	27.49	131	609
4	2.47	3.95	6.04	10.16	14.80	49.00	157
5	2.20	3.34	4.82	7.50	10.29	27.65	71.23
6	2.04	2.98	4.15	6.14	8.10	19.03	42.54
7	1.93	2.75	3.73	5.33	6.84	14.63	29.64
8	1.86	2.59	3.44	4.79	6.03	12.05	22.71
9	1.80	2.47	3.23	4.41	5.47	10.37	18.50
10	1.75	2.38	3.07	4.13	5.06	9.20	15.74
11	1.72	2.30	2.95	3.91	4.74	8.35	13.80
12	1.69	2.24	2.85	3.74	4.50	7.71	12.38
13	1.66	2.20	2.77	3.60	4.30	7.21	11.30
14	1.64	2.15	2.70	3.48	4.14	6.80	10.46
15	1.62	2.12	2.64	3.39	4.00	6.47	9.78
16	1.61	2.09	2.59	3.30	3.89	6.19	9.23
17	1.59	2.06	2.55	3.23	3.79	5.96	8.76
18	1.58	2.04	2.51	3.17	3.71	5.76	8.38
19	1.57	2.02	2.48	3.12	3.63	5.59	8.04
20	1.56	2.00	2.45	3.07	3.56	5.44	7.76
21	1.55	1.98	2.42	3.02	3.51	5.31	7.51
22	1.54	1.97	2.40	2.99	3.45	5.19	7.29
23	1.53	1.95	2.37	2.95	3.41	5.09	7.09
24	1.53	1.94	2.36	2.92	3.36	4.99	6.92
25	1.52	1.93	2.34	2.89	3.32	4.91	6.76
26	1.52	1.92	2.32	2.86	3.29	4.83	6.62
27	1.51	1.91	2.31	2.84	3.26	4.76	6.50
28	1.51	1.90	2.29	2.82	3.23	4.69	6.38
29	1.50	1.89	2.28	2.80	3.20	4.64	6.28
30	1.50	1.88	2.27	2.78	3.17	4.58	6.18
40	1.47	1.83	2.18	2.64	2.99	4.21	5.53
60	1.44	1.77	2.10	2.51	2.82	3.86	4.95
100	1.41	1.73	2.03	2.41	2.69	3.61	4.53
140	1.40	1.71	2.01	2.37	2.64	3.51	4.36
∞	1.38	1.67	1.94	2.27	2.51	3.27	3.98

TABLE 10 Critical Values of the *F* Distribution *(continued)*

Numerator df = 9

Denom.	TAIL PROBABILITY						
df	.20	.10	.05	.02	.01	.001	.0001
1	14.68	59.86	241	151[1]	602[1]	602[3]	602[5]
2	4.37	9.38	19.38	49.39	99.39	999	100[2]
3	2.98	5.24	8.81	16.93	27.35	130	606
4	2.46	3.94	6.00	10.07	14.66	48.47	155
5	2.20	3.32	4.77	7.42	10.16	27.24	70.13
6	2.03	2.96	4.10	6.05	7.98	18.69	41.73
7	1.93	2.72	3.68	5.24	6.72	14.33	28.99
8	1.85	2.56	3.39	4.70	5.91	11.77	22.14
9	1.79	2.44	3.18	4.33	5.35	10.11	18.00
10	1.74	2.35	3.02	4.04	4.94	8.96	15.27
11	1.70	2.27	2.90	3.83	4.63	8.12	13.37
12	1.67	2.21	2.80	3.66	4.39	7.48	11.98
13	1.65	2.16	2.71	3.52	4.19	6.98	10.92
14	1.63	2.12	2.65	3.40	4.03	6.58	10.09
15	1.61	2.09	2.59	3.30	3.89	6.26	9.42
16	1.59	2.06	2.54	3.22	3.78	5.98	8.88
17	1.58	2.03	2.49	3.15	3.68	5.75	8.43
18	1.56	2.00	2.46	3.09	3.60	5.56	8.05
19	1.55	1.98	2.42	3.03	3.52	5.39	7.72
20	1.54	1.96	2.39	2.98	3.46	5.24	7.44
21	1.53	1.95	2.37	2.94	3.40	5.11	7.19
22	1.53	1.93	2.34	2.90	3.35	4.99	6.98
23	1.52	1.92	2.32	2.87	3.30	4.89	6.79
24	1.51	1.91	2.30	2.83	3.26	4.80	6.62
25	1.51	1.89	2.28	2.81	3.22	4.71	6.47
26	1.50	1.88	2.27	2.78	3.18	4.64	6.33
27	1.49	1.87	2.25	2.76	3.15	4.57	6.21
28	1.49	1.87	2.24	2.73	3.12	4.50	6.09
29	1.49	1.86	2.22	2.71	3.09	4.45	5.99
30	1.48	1.85	2.21	2.69	3.07	4.39	5.90
40	1.45	1.79	2.12	2.56	2.89	4.02	5.26
60	1.42	1.74	2.04	2.43	2.72	3.69	4.69
100	1.40	1.69	1.97	2.33	2.59	3.44	4.29
140	1.39	1.68	1.95	2.29	2.54	3.34	4.12
∞	1.36	1.63	1.88	2.19	2.41	3.10	3.75

TABLE 10 Critical Values of the *F* Distribution
(continued)

Numerator df = 10

Denom. df	TAIL PROBABILITY						
	.20	.10	.05	.02	.01	.001	.0001
1	14.77	60.19	242	151[1]	606[1]	606[3]	606[5]
2	4.38	9.39	19.40	49.40	99.40	999	100[2]
3	2.98	5.23	8.79	16.86	27.23	129	603
4	2.46	3.92	5.96	10.00	14.55	48.05	154
5	2.19	3.30	4.74	7.34	10.05	26.92	69.25
6	2.03	2.94	4.06	5.98	7.87	18.41	41.08
7	1.92	2.70	3.64	5.17	6.62	14.08	28.45
8	1.84	2.54	3.35	4.63	5.81	11.54	21.68
9	1.78	2.42	3.14	4.26	5.26	9.89	17.59
10	1.73	2.32	2.98	3.97	4.85	8.75	14.90
11	1.69	2.25	2.85	3.76	4.54	7.92	13.02
12	1.66	2.19	2.75	3.59	4.30	7.29	11.65
13	1.64	2.14	2.67	3.45	4.10	6.80	10.60
14	1.62	2.10	2.60	3.33	3.94	6.40	9.79
15	1.60	2.06	2.54	3.23	3.80	6.08	9.13
16	1.58	2.03	2.49	3.15	3.69	5.81	8.60
17	1.57	2.00	2.45	3.08	3.59	5.58	8.15
18	1.55	1.98	2.41	3.02	3.51	5.39	7.78
19	1.54	1.96	2.38	2.96	3.43	5.22	7.46
20	1.53	1.94	2.35	2.91	3.37	5.08	7.18
21	1.52	1.92	2.32	2.87	3.31	4.95	6.94
22	1.51	1.90	2.30	2.83	3.26	4.83	6.73
23	1.51	1.89	2.27	2.80	3.21	4.73	6.54
24	1.50	1.88	2.25	2.77	3.17	4.64	6.37
25	1.49	1.87	2.24	2.74	3.13	4.56	6.23
26	1.49	1.86	2.22	2.71	3.09	4.48	6.09
27	1.48	1.85	2.20	2.69	3.06	4.41	5.97
28	1.48	1.84	2.19	2.66	3.03	4.35	5.86
29	1.47	1.83	2.18	2.64	3.00	4.29	5.76
30	1.47	1.82	2.16	2.62	2.98	4.24	5.66
40	1.44	1.76	2.08	2.49	2.80	3.87	5.04
60	1.41	1.71	1.99	2.36	2.63	3.54	4.48
100	1.38	1.66	1.93	2.26	2.50	3.30	4.08
140	1.37	1.64	1.90	2.22	2.45	3.20	3.93
∞	1.34	1.60	1.83	2.12	2.32	2.96	3.56

TABLE 11 Critical Constants for the Newman–Keuls Procedure

$\alpha = .05$

df \ j	2	3	4	5	6	7	8	9	10
1	18.0	27.0	32.8	37.1	40.4	43.1	45.4	47.4	49.1
2	6.08	8.33	9.80	10.9	11.7	12.4	13.0	13.5	14.0
3	4.50	5.91	6.82	7.50	8.04	8.48	8.85	9.18	9.46
4	3.93	5.04	5.76	6.29	6.71	7.05	7.35	7.60	7.83
5	3.64	4.60	5.22	5.67	6.03	6.33	6.58	6.80	6.99
6	3.46	4.34	4.90	5.30	5.63	5.90	6.12	6.32	6.49
7	3.34	4.16	4.68	5.06	5.36	5.61	5.82	6.00	6.16
8	3.26	4.04	4.53	4.89	5.17	5.40	5.60	5.77	5.92
9	3.20	3.95	4.41	4.76	5.02	5.24	5.43	5.59	5.74
10	3.15	3.88	4.33	4.65	4.91	5.12	5.30	5.46	5.60
11	3.11	3.82	4.26	4.57	4.82	5.03	5.20	5.35	5.49
12	3.08	3.77	4.20	4.51	4.75	4.95	5.12	5.27	5.39
13	3.06	3.73	4.15	4.45	4.69	4.88	5.05	5.19	5.32
14	3.03	3.70	3.11	4.41	4.64	4.83	4.99	5.13	5.25
15	3.01	3.67	4.08	4.37	4.59	4.78	4.94	5.08	5.20
16	3.00	3.65	4.05	4.33	4.56	4.74	4.90	5.03	5.15
17	2.98	3.63	4.02	4.30	4.52	4.70	4.86	4.99	5.11
18	2.97	3.61	4.00	4.28	4.49	4.67	4.82	4.96	5.07
19	2.96	3.59	3.98	4.25	4.47	4.65	4.79	4.92	5.04
20	2.95	3.58	3.96	4.23	4.45	4.62	4.77	4.90	5.01
24	2.92	3.53	3.90	4.17	4.37	4.54	4.68	4.81	4.92
30	2.89	3.49	3.85	4.10	4.30	4.46	4.60	4.72	4.82
40	2.86	3.44	3.79	4.04	4.23	4.39	4.52	4.63	4.73
60	2.83	3.40	3.74	3.98	4.16	4.31	4.44	4.55	4.65
120	2.80	3.36	3.68	3.92	4.10	4.24	4.36	4.47	4.56
∞	2.77	3.31	3.63	3.86	4.03	4.17	4.29	4.39	4.47

Continued

TABLE 11 Critical Constants for the Newman–Keuls Procedure *(continued)*

$\alpha = .01$

df \ j	2	3	4	5	6	7	8	9	10
1	90.0	135	164	186	202	216	227	237	246
2	14.0	19.0	22.3	24.7	26.6	28.2	29.5	30.7	31.7
3	8.26	10.6	12.2	13.3	14.2	15.0	15.6	16.2	16.7
4	6.51	8.12	9.17	9.96	10.6	11.1	11.5	11.9	12.3
5	5.70	6.97	7.80	8.42	8.91	9.32	9.67	9.97	10.2
6	5.24	6.33	7.03	7.56	7.97	8.32	8.61	8.87	9.10
7	4.95	5.92	6.54	7.01	7.37	7.68	7.94	8.17	8.37
8	4.74	5.63	6.20	6.63	6.96	7.24	7.47	7.68	7.87
9	4.60	5.43	5.96	6.35	6.66	6.91	7.13	7.32	7.49
10	4.48	5.27	5.77	6.14	6.43	6.67	6.87	7.05	7.21
11	4.39	5.14	5.62	5.97	6.25	6.48	6.67	6.84	6.99
12	4.32	5.04	5.50	5.84	6.10	6.32	6.51	6.67	6.81
13	4.26	4.96	5.40	5.73	5.98	6.19	6.37	6.53	6.67
14	4.21	4.89	5.32	5.63	5.88	6.08	6.26	6.41	6.54
15	4.17	4.83	5.25	5.56	5.80	5.99	6.16	6.31	6.44
16	4.13	4.78	5.19	5.49	5.72	5.92	6.08	6.22	6.35
17	4.10	4.74	5.14	5.43	5.66	5.85	6.01	6.15	6.27
18	4.07	4.70	5.09	5.38	5.60	5.79	5.94	6.08	6.20
19	4.05	4.67	5.05	5.33	5.55	5.73	5.89	6.02	6.14
20	4.02	4.64	5.02	5.29	5.51	5.69	5.84	5.97	6.09
24	3.96	4.54	4.91	5.17	5.37	5.54	5.69	5.81	5.92
30	3.89	4.45	4.80	5.05	5.24	5.40	5.54	5.65	5.76
40	3.82	4.37	4.70	4.93	5.11	5.27	5.39	5.50	5.60
60	3.76	4.28	4.60	4.82	4.99	5.13	5.25	5.36	5.45
120	3.70	4.20	4.50	4.71	4.87	5.01	5.12	5.21	5.30
∞	3.64	4.12	4.40	4.60	4.76	4.88	4.99	5.08	5.16

Source: Harter, H. L. "Tables of range and Studentized range." *Annals of Mathematical Statistics*, Volume 31 (1960), pp. 1122–1147.

TABLE 12 Bonferroni Multipliers for 95% Confidence Intervals

The values given in the table are t (df)$_{.025/k}$ where k is the number of tests.

df	\multicolumn{10}{c}{NUMBER OF TESTS}
	1	2	3	4	5	6	8	10	15	20
1	12.706	25.452	38.185	50.923	63.657	76.384	101.856	127.321	190.946	254.647
2	4.303	6.205	7.648	8.860	9.925	10.885	12.590	14.089	17.275	19.963
3	3.182	4.177	4.857	5.392	5.841	6.231	6.895	7.453	8.575	9.465
4	2.776	3.495	3.961	4.315	4.604	4.851	5.261	5.598	6.254	6.758
5	2.571	3.163	3.534	3.810	4.032	4.219	4.526	4.773	5.247	5.604
6	2.447	2.969	3.287	3.521	3.707	3.863	4.115	4.317	4.698	4.981
7	2.365	2.841	3.128	3.335	3.499	3.636	3.855	4.029	4.355	4.595
8	2.306	2.752	3.016	3.206	3.355	3.479	3.677	3.833	4.122	4.334
9	2.262	2.685	2.933	3.111	3.250	3.364	3.547	3.690	3.954	4.146
10	2.228	2.634	2.870	3.038	3.169	3.277	3.448	3.581	3.827	4.005
11	2.201	2.593	2.820	2.981	3.106	3.208	3.370	3.497	3.728	3.895
12	2.179	2.560	2.779	2.934	3.055	3.153	3.308	3.428	3.649	3.807
13	2.160	2.533	2.746	2.896	3.012	3.107	3.256	3.372	3.584	3.735
14	2.145	2.510	2.718	2.864	2.977	3.069	3.214	3.326	3.529	3.675
15	2.131	2.490	2.694	2.837	2.947	3.036	3.177	3.286	3.484	3.624
16	2.120	2.473	2.673	2.813	2.921	3.008	3.146	3.252	3.444	3.581
17	2.110	2.458	2.655	2.793	2.898	2.984	3.119	3.222	3.410	3.543
18	2.101	2.445	2.639	2.775	2.878	2.963	3.095	3.197	3.380	3.510
19	2.093	2.433	2.625		2.861	2.944	3.074	3.174	3.354	3.481
20	2.086	2.423	2.613	2.744	2.845	2.927	3.055	3.153	3.331	3.455
25	2.060	2.385	2.566	2.692	2.787	2.865	2.986	3.078	3.244	3.361
30	2.042	2.360	2.536	2.657	2.750	2.825	2.941	3.030	3.189	3.300
40	2.021	2.329	2.499	2.616	2.704	2.776	2.887	2.971	3.122	3.227
50	2.009	2.311	2.477	2.591	2.678	2.747	2.855	2.937	3.083	3.184
60	2.000	2.299	2.463	2.575	2.660	2.729	2.834	2.915	3.057	3.156
70	1.994	2.291	2.453	2.564	2.648	2.715	2.820	2.899	3.039	3.137
80	1.990	2.284	2.445	2.555	2.639	2.705	2.809	2.887	3.026	3.122
100	1.984	2.276	2.435	2.544	2.626	2.692	2.793	2.871	3.007	3.102
140	1.977	2.266	2.423	2.530	2.611	2.676	2.776	2.852	2.986	3.079
1000	1.962	2.245	2.398	2.502	2.581	2.643	2.740	2.813	2.942	3.031
∞	1.960	2.241	2.394	2.498	2.576	2.638	2.734	2.807	2.935	3.023

Answers to Selected Exercises

2.2 (a) (i)   Height and weight
        (ii) Continuous variables
        (iii) A child
        (iv) 37
    (b) (i)   Blood type and cholesterol level
        (ii) Blood type is categorical, cholesterol level is continuous
        (iii) A person
        (iv) 129

2.4 (a) There is no single correct answer. One possibility is

Molar width	Frequency (no. specimens)
5.4–5.5	1
5.6–5.7	5
5.8–5.9	7
6.0–6.1	12
6.2–6.3	8
6.4–6.5	2
6.6–6.7	1
Total	36

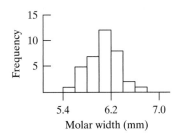

    (b) The distribution is fairly symmetric.

2.10 There is no single correct answer. One possibility is

Glucose (%)	Frequency (no. of dogs)
70–74	3
75–79	5
80–84	10
85–89	5
90–94	2
95–99	2
100–104	1
105–109	1
110–114	0
115–119	1
120–124	0
125–129	0
130–134	1
Total	31

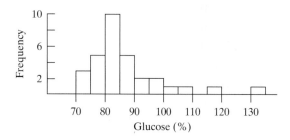

2.11

```
 7 | 8 4 0 6 0 9 5 5
 8 | 1 8 5 4 1 4 2 6 9 9 0 2 1 4 2
 9 | 3 3 9 6
10 | 2 6
11 | 5
12 |
13 | 1
```
Key 7|8 = 78%

2.16 $\bar{y} = 6.40$ nmol/g; median $= 6.3$ nmol/g

2.18 $\bar{y} = 293.8$ mg/dLi; median $= 283$ mg/dLi

2.19 $\bar{y} = 309$ mg/dLi; median $= 292$ mg/dLi

2.24 Median $= 10.5$ piglets

2.26 Mean $\approx$ median ≈ 50

2.27 25%

2.30 (a) Median $= 15, Q_1 = 14, Q_3 = 20$

  (b) IQR $= 6$

  (c) Upper fence $= 29$

2.31 (a) Median $= 9.2, Q_1 = 7.4, Q_3 = 11.9$

(b) IQR $= 4.5$

(c)

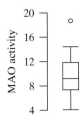

2.40 (a) $s = 2.45$

(b) $s = 3.32$

2.43 (a) $\bar{y} = 33.10\,\text{lb}; s = 3.444\,\text{lb}$

(b) Coeff. of var. $= 10.4\%$

2.45 $\bar{y} = -12.4\,\text{mm Hg}; s = 17.6\,\text{mm Hg}$

2.53 4%

2.55 $\bar{y} = 100; s = 21$

2.56 Mean $= 37.3; \text{SD} = 12.9$

2.56 (a)

9	8 8 9
10	0 0 6 7 7 7 8
11	0 0 1 4 5 5 6 6 6 9
12	0 1 2 3 3 4
13	0

Key $9|8 = .098$

2.73 (a) Median $= 38$

(b) $Q_1 = 36, Q_3 = 41$

(c) 66.4%

Chapter 3

3.5 (a) .51

(b) .94

(c) .46

(d) .54

3.9 (a) .107

3.12 (a) .185

(b) .117

(c) No; $\text{Pr}\{\text{Smoke}\} \neq \text{Pr}\{\text{Smoke}|\text{High income}\}$

3.16 (a) .62

3.22 .9

3.23 .794

3.31 (a) .3746

(b) .0688

(c) .1254

3.34   (a)   .1181
            (b)   .2699
            (c)   .2891
            (d)   .3229
3.35   Expected frequencies: 939.5; 5,982.5; 15,873.1; 22,461.8;
            17,879.3; 7590.2; 1,342.6
3.40   .3369
3.45   .0376
3.47   (a)   .0209

Chapter 4

4.3   (a)   84.13%
           (b)   61.47%
           (c)   77.34%
           (d)   22.66%
           (e)   20.38%
           (f)   20.38%
4.4   (a)   22.66%
           (b)   20.38%
4.8   (a)   90.7 lb
           (b)   85.3 lb
4.12 (a)   98.76%
           (b)   98.76%
           (c)   1.24%
4.20 (b)

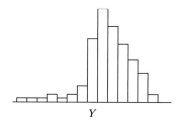

4.24 (a)   .2843
           (b)   .1256
           (c)   .4980
4.29 (a)   97.98%
           (b)   12.71%
           (c)   46.39%
           (d)   10.69%
           (e)   35.51%
           (f)   5.59%
           (g)   59.10%
4.30 .122
4.31 173.2 cm
4.33 1.96
4.40 .1056

Chapter 5

5.2 (a) .227
       (b) .435
5.4 (a) .2501
       (b) .0352
5.5 (a) (i)   .3164
            (ii)   .4219
            (iii) .2109
            (iv)   .0469
            (v)   .0039
5.9 .5053
5.15 (a) 25.86%
        (b) 68.26%
        (c) .6826
5.16 (a) .6680
5.20 (a) .1861
        (b) .9044
5.23 (a) .1056
        (b) .0150
5.29 (a) .66
        (b) .29
5.32 (a) .1762
        (b) .1742
5.34 (a) .7198
5.37 (a) .6102
5.42 .2611
5.47 (a) .2182
        (b) .5981
5.48 (a) .2206
        (b) .5990
5.51 9.68%

Chapter 6

6.1 (a) 51.3 ng/g
       (b) 26.5 ng/g
6.10 (a) SE $= 3.9$ mg
        (b) $(23.4, 40.0)$
6.16 (a) $4.1 < \mu < 21.9$ pg/mLi
        (b) We are 95% confident that the average drop in HBE levels from January to May in the population of all participants in physical fitness programs like the one in the study is between 4.1 and 21.9 pg/mLi.
6.20 $1.17 < \mu < 1.23$ mm
6.24 2.81
6.28 178 men
6.31 The SD is larger than the mean, but negative values are not possible. Thus, the distribution must be skewed to the right.

6.37 (a) .040

(b) .020

6.38 (a) (.134, .290)

(b) (.164, .242)

6.40 (a) (.164, .250)

(b) We are 95% confident that the probability of adverse reaction in infants who receive their first injection of vaccine is between .164 and .250.

6.43 $n \geq 146$

6.52 (a) $\bar{y} = 2.275$; $s = .238$; SE $= .084$

(b) (2.08, 2.47)

(c) $\mu =$ population mean stem diameter of plants of Tetrastichon wheat three weeks after flowering

6.54 63 plants

6.59 (a) We must be able to view the data as a random sample of independent observations from a large population that is approximately normal.

(b) Normality of the population.

(c) Independence of the observations would be questionable, because birthweights of the members of a twin pair might be dependent.

6.65 (.707, .853)

Chapter 7

Remark Concerning Tests of Hypotheses The answer to a hypothesis testing exercise includes verbal statements of the hypotheses and a verbal statement of the conclusion from the test. In phrasing these statements, we have tried to capture the essence of the biological question being addressed; nevertheless the statements are necessarily oversimplified and they gloss over many issues that in reality might be quite important. For instance, the hypotheses and conclusion may refer to a causal connection between treatment and response; in reality the validity of such a causal interpretation usually depends on a number of factors related to the design of the investigation (such as unbiased allocation of animals to treatment groups) and to the specific experimental procedures (such as the accuracy of assays or measurement techniques). In short, the student should be aware that the verbal statements are intended to clarify the *statistical* concepts; their *biological* content may be open to question.

7.1 2.41

7.7 .86

7.10 $4.84 < \mu_1 - \mu_2 < 5.56$

7.14 (a) $-5 < \mu_1 - \mu_2 < 9 \sec$

(b) We are 90% confident that the population mean prothrombin time for rats treated with an antibiotic (μ_1) is smaller than that for control rats (μ_2) by an amount that might be as much as 5 seconds or is larger than that for control rats (μ_2) by an amount that might be as large as 9 seconds.

7.23 (a) $t_s = -3.13$ so $.02 < P < .04$
(b) $t_s = 1.25$ so $.20 < P < .40$
(c) $t_s = 4.62$ so $P < .001$

7.25 (a) yes
(b) no
(c) yes
(d) no

7.29 (a) H_0: Mean serotonin concentration is the same in heart patients and in controls ($\mu_1 = \mu_2$); H_A: Mean serotonin concentration is not the same in heart patients and in controls ($\mu_1 \neq \mu_2$). $t_s = -1.38$. H_0 is not rejected.
(b) There is insufficient evidence ($.10 < P < .20$) to conclude that serotonin levels are different in heart patients than in controls.

7.32 (a) H_0: Flooding has no effect on ATP ($\mu_1 = \mu_2$); H_A: Flooding has some effect on ATP ($\mu_1 \neq \mu_2$). $t_s = -3.92$. H_0 is rejected.
(b) There is sufficient evidence ($.001 < P < .01$) to conclude that flooding tends to lower ATP in birch seedlings.

7.42 Type II

7.44 Yes; because zero is outside of the confidence interval, we know that the P-value is less than $.0$, so we reject the hypothesis that $\mu_1 - \mu_2 = 0$.

7.46 (a) $.10 < P < .20$
(b) $.03 < P < .04$

7.48 (a) yes
(b) yes
(c) yes
(d) no

7.53 H_0: Wounding the plant has no effect on larval growth ($\mu_1 = \mu_2$); H_A: Wounding the plant tends to diminish larval growth ($\mu_1 < \mu_2$), where 1 denotes wounded and 2 denotes control. $t_s = -2.69$. H_0 is rejected. There is sufficient evidence ($.005 < P < .01$) to conclude that wounding the plant tends to diminish larval growth.

7.54 (a) H_0: The drug has no effect on pain ($\mu_1 = \mu_2$); H_A: The drug increases pain relief ($\mu_1 > \mu_2$). $t_s = 1.81$. H_0 is rejected. There is sufficient evidence ($.03 < P < .04$) to conclude that the drug is effective.
(b) The P-value would be between $.06$ and $.08$. At $\alpha = .05$ we would not reject H_0.

7.60 No, according to the confidence interval the data do not indicate whether the true difference is "important."

7.62 Yes, according to the confidence interval the data indicate that the true difference is "clinically important."

7.64 (a) 23
(b) 11

7.67 (a) 71
(b) 101
(c) 58

7.69 .5

7.77   (a) $P > .20$
(b) $.02 < P < .05$
(c) $.002 < P < .01$

7.79   (a) H_0: Toluene has no effect on dopamine in rat striatum; H_A: Toluene has some effect on dopamine in rat striatum. $U_s = 32$. H_0 is rejected. There is sufficient evidence ($.02 < P < .05$) to conclude that toluene increases dopamine in rat striatum.

7.86 H_0: Mean platelet calcium is the same in people with high blood pressure as in people with normal blood pressure ($\mu_1 = \mu_2$); H_A: Mean platelet calcium is different in people with high blood pressure than in people with normal blood pressure ($\mu_1 \neq \mu_2$). $t_s = 11.2$. H_0 is rejected. There is sufficient evidence ($P < .0001$) to conclude that platelet calcium is higher in people with high blood pressure.

7.87 $49.5 < \mu_1 - \mu_2 < 71.1$

7.92 H_0: Stress has no effect on growth; H_A: Stress tends to retard growth. $U_s = 148.5$. H_0 is rejected. There is sufficient evidence ($P < .0005$) to conclude that stress tends to retard growth.

7.105   False: Zero is in the confidence interval.

Chapter 8

8.1   People with respiratory problems move to Arizona (because the dry air is good for them).

8.4   (a) Coffee consumption rate
(b) Coronary heart disease (present or absent)
(c) Subjects (i.e., the 1,040 persons)

8.13   There is no single correct answer. One possibility is as follows:
Group 1:   Animals 2, 5, 6

Group 2:   Animals 1, 3, 7

Group 3:   Animals 4, 8

8.17   There is no single correct answer. One possibility is as follows:

	Piglet				
Treatment	*Litter 1*	*Litter 2*	*Litter 3*	*Litter 4*	*Litter 5*
1	2	5	2	4	5
2	1	4	1	1	2
3	4	2	5	2	4
4	5	3	3	3	3
5	3	1	4	5	1

8.21   Plan II is better. We want units within a block to be similar to each other; plan II achieves this. Under plan I the effect of rain could be confounded with the effect of a variety.

8.29   .327

8.41   (a) Treatment group membership (AZT or placebo)
(b) HIV status of a baby
(c) The babies

Chapter 9

9.1 (a) .34

9.3 H_0: Progesterone has no effect on cAMP ($\mu_1 = \mu_2$); H_A: Progesterone has some effect on cAMP ($\mu_1 \neq \mu_2$). $t_s = 3.4$. H_0 is rejected. There is sufficient evidence ($.04 < P < .05$) to conclude that progesterone decreases cAMP under these conditions.

9.4 (a) $-.50 < \mu_1 - \mu_2 < .74°C$, where 1 denotes treated and 2 denotes control

9.14 (a) $P > .20$

(b) $.10 < P < .20$

(c) $.02 < P < .05$

(d) $.002 < P < .01$

9.17 H_0: Weight of the cerebral cortex is not affected by environment ($p = .5$); H_A: Environmental enrichment increases cortex weight ($p > .5$). $B_s = 310$. H_0 is rejected. There is sufficient evidence ($.01 < P < .025$) to conclude that environmental enrichment increases cortex weight.

9.18 .0193

9.24 (a) .0156

(b) With $n_d = 7$, the smallest possible P-value is .0156; thus P cannot be less than .01

9.28 (a) $P > .20$

(b) $.10 < P < .20$

(c) $.02 < P < .05$

(d) $.01 < P < .02$

9.30 H_0: Hunger rating is not affected by treatment (mCPP versus placebo); H_A: Treatment does affect hunger rating. $W_s = 27$ and $n_d = 8$. H_0 is not rejected. There is insufficient evidence ($P > .20$) to conclude that treatment has an effect.

9.45 H_0: The average number of species is the same in pools as in riffles ($\mu_1 = \mu_2$); H_A: The average numbers of species in pools and in riffles differ ($\mu_1 \neq \mu_2$). $t_s = 4.58$. H_0 is rejected. There is sufficient evidence ($P < .001$) to conclude that the average number of species in pools is greater than in riffles.

9.49 H_0: Caffeine has no effect on RER ($\mu_1 = \mu_2$); H_A: Caffeine has some effect on RER ($\mu_1 \neq \mu_2$). $t_s = 3.94$. H_0 is rejected. There is sufficient evidence ($.001 < P < .01$) to conclude that caffeine tends to decrease RER under these conditions.

Chapter 10

10.1 H_0: The population ratio is 12:3:1 ($\Pr\{\text{white}\} = .75, \Pr\{\text{yellow}\} = .1875, \Pr\{\text{green}\} = .0625$); H_A: The ratio is not 12:3:1. $\chi_s^2 = .69$. H_0 is not rejected. There is little or no evidence ($P > .20$) that the model is not correct; the data are consistent with the model.

10.2 H_0 and H_A as in Exercise 10.1. $\chi_s^2 = 6.9$. H_0 is rejected. There is sufficient evidence ($.02 < P < .05$) to conclude that the model is incorrect; the data are not consistent with the model.

10.8 H_0: The drug does not cause tumors $(\Pr\{T\} = \frac{1}{3})$; H_A: The drug causes tumors $(\Pr\{T\} > \frac{1}{3}$, where T denotes the event that a tumor occurs first in the treated rat). $\chi_s^2 = 6.4$. H_0 is rejected. There is sufficient evidence $(.005 < P < .01)$ to conclude that the drug does
cause tumors.

10.15 (a)

5	20
10	40

(b) $\hat{p}_1 = \dfrac{1}{3}, \hat{p}_2 = \dfrac{1}{3}$; yes

10.18 H_0: Mites do not induce resistance to wilt $(p_1 = p_2)$; H_A: Mites do induce resistance to wilt $(p_1 < p_2)$, where p denotes the probability
of wilt and 1 denotes mites and 2 denotes no mites. $\chi_s^2 = 7.21$. H_0 is
rejected. There is sufficient evidence $(.0005 < P < .005)$ to conclude that mites do induce resistance to wilt.

10.23 H_0: The two timings are equally effective $(p_1 = p_2)$; H_A: The two
timings are not equally effective $(p_1 \neq p_2)$. $\chi_s^2 = 4.48$. H_0 is rejected. There is sufficient evidence $(.02 < P < .05)$ to conclude that the
simultaneous timing is superior to the sequential timing.

10.29 (a) $\hat{\Pr}\{D|P\} = .266, \hat{\Pr}\{D|N\} = .096, \hat{\Pr}\{P|D\} = .744,$
$\hat{\Pr}\{P|A\} = .460.$

(b) H_0: There is no association between antibody and survival
$(\Pr\{D|P\} = \Pr\{D|N\})$; H_A: There is some association be-
tween antibody and survival $(\Pr\{D|P\} \neq \Pr\{D|N\})$. H_0 is re-
jected. There is sufficient evidence $(.001 < P < .01)$ to
conclude that men with antibody are less likely to survive 6
months than men without antibody $(\Pr\{D|P\} > \Pr\{D|N\})$.

10.30 $\hat{\Pr}\{\text{correct prediction}\} = .577$

10.31 (a) $\hat{\Pr}\{\text{RF}|\text{RH}\} = .934$

(b) $\hat{\Pr}\{\text{RF}|\text{LH}\} = .511$

(c) $\chi_s^2 = 398$

(d) $\chi_s^2 = 1{,}623$

10.40

5	1
9	15
6	0
8	16

10.49 H_0: The blood type distributions are the same for ulcer patients
and controls $(\Pr\{O|UP\} = \Pr\{O|C\}, \Pr\{A|UP\} = \Pr\{A|C\},$
$\Pr\{B|UP\} = \Pr\{B|C\}, \Pr\{AB|UP\} = \Pr\{AB|C\})$; H_A: The blood
type distributions are not the same. H_0 is rejected. There is sufficient
evidence $(P < .0001)$ to conclude that the blood type distribution of
ulcer patients is different from that of controls.

10.59 $.003 < p_1 - p_2 < .233$. No; the confidence interval suggests that bed
rest may actually be harmful.

10.61 (a) $.067 < p_1 - p_2 < .119$

(b) We are 95% confident that the proportion of persons with type O blood among ulcer patients is higher than the proportion of persons with type O blood among healthy individuals by between .067 and .119. That is, we are 95% confident that p_1 exceeds p_2 by between .067 and .119.

10.63 H_0: There is no association between oral contraceptive use and stroke ($p = .5$); H_A: There is an association between oral contraceptive use and stroke ($p \neq .5$), where p denotes the probability that a discordant pair will be Yes(case)/No(control). $\chi_s^2 = 6.72$. H_0 is rejected. There is sufficient evidence ($.001 < P < .01$) to conclude that stroke victims are more likely to be oral contraceptive users. ($p > .5$).

10.66 (a) (i)   1.339

(ii)   1.356

(b) (i)   1.314

(ii)   1.355

10.72 (a) 1.241

(b) (1.036, 1.488)

(c) We are 95% confident that taking heparin increases the odds of a negative response by a factor of between 1.036 and 1.488 when compared to taking enoxaparin.

10.76 (a) H_0: Sex ratio is 1:1 in warm environment ($p_1 = .5$); H_A: Sex ratio is not 1:1 in warm environment ($p_1 \neq .5$), where p_1 denotes the probability of a female in the warm environment. $\chi_s^2 = .18$. H_0 is not rejected. There is insufficient evidence ($P > .20$) to conclude that the sex ratio is not 1:1 in warm environment.

(c) H_0: Sex ratio is the same in the two environments ($p_1 = p_2$); H_A: Sex ratio is not the same in the two environments ($p_1 \neq p_2$), where p denotes the probability of a female and 1 and 2 denote the warm and cold environments. $\chi_s^2 = 4.20$. H_0 is rejected. There is sufficient evidence ($.02 < P < .05$) to conclude that the probability of a female is higher in the cold than the warm environment.

10.80 (a) H_0: Directional choice is random (Pr{toward} = .25, Pr{away} = .25, Pr{right} = .25, Pr{left} = .25); H_A: Directional choice is not random. $\chi_s^2 = 4.88$. H_0 is not rejected. There is insufficient evidence ($.10 < P < .20$) to conclude that the directional choice is not random.

10.89 H_0: Site of capture and site of recapture are independent (Pr{RI|CI} = Pr{RI|CII}); H_A: Flies preferentially return to their site of capture (Pr{RI|CI} > Pr{RI|CII}), where C and R denote capture and recapture and I and II denote the sites. H_0 is rejected. There is sufficient evidence ($.0005 < P < .005$) to conclude that flies preferentially return to their site of capture.

10.94 H_0: The probability of an egg being on a particular type of bean is .25 for all four types of beans; H_A: H_0 is false. $\chi_s^2 = 2.23$. H_0 is not rejected. There is insufficient evidence ($P > .20$) to conclude that cowpea weevils prefer one type of bean over the others.

Chapter 11

11.1 (a) SS(between) = 228, SS(within) = 120

(b) SS(total) = 348

(c) MS(between) = 114, MS(within) = 15, s_{pooled} = 3.87

11.4 (a)

Source	df	SS	MS
Between groups	3	135	45
Within groups	12	337	28.08
Total	15	472	

(b) 4

(c) 16

11.9 (a) H_0: The stress conditions all produce the same mean lymphocyte concentration ($\mu_1 = \mu_2 = \mu_3 = \mu_4$); H_A: Some of the stress conditions produce different mean lymphocyte concentrations (the μ's are not all equal). F_s = 3.84. H_0 is rejected. There is sufficient evidence (.01 < P < .02) to conclude that some of the stress conditions produce different mean lymphocyte concentrations.

(b) s_{pooled} = 2.78 cells/mLi $\cdot 10^{-6}$

11.10 (a) H_0: Mean HBE is the same in all three populations ($\mu_1 = \mu_2 = \mu_3$); H_A: Mean HBE is not the same in all three populations (the μ's are not all equal). F_s = .58. H_0 is not rejected. There is insufficient evidence (P > .20) to conclude that mean HBE is not the same in all three populations.

(b) s_{pooled} = 14.4 pg/mLi

11.18 (a)

Source	df	SS	MS	F ratio
Between species	1	2.19781	2.19781	55.60
Between flooding levels	1	2.25751	2.25751	57.11
Interaction	1	0.097656	0.097656	2.47
Within groups	12	0.47438	.03953	
Total	15	.157468		

(b) F_s = 0.097656/.03953 = 2.47. With df = 1 and 12, Table 10 gives $F_{.20}$ = 1.84 and $F_{.10}$ = 3.18. Thus, .10 < P < .20 and we do not reject H_0. There is insufficient evidence (P > .10) to conclude that there is an interaction present.

(c) F_s = 2.19781/.03953 = 55.60. With df = 1 and 12, Table 10 gives $F_{.0001}$ = 32.43. Thus, P < .0001 and H_0 is rejected. There is strong evidence (P < .0001) to conclude that species affects ATP concentration.

(d) $s_{pooled} = \sqrt{.03953}$ = .199

11.24 (a) 123 mm Hg

(b) 123.2 mm Hg

(d) .851 mm Hg

11.29 .67 < $\mu_E - \mu_S$ < 1.48 g, where $\mu_E = \frac{1}{2}(\mu_{E,Low} + \mu_{E,High})$ and $\mu_S = \frac{1}{2}(\mu_{S,Low} + \mu_{S,High})$

11.30 (b) $L = 3.685 \, \text{nmol}/10^8 \, \text{platelets/hour}; \text{SE}_L = 1.048 \, \text{nmol}/10^8 \, \text{platelets/}$
hour

11.33 The following hypotheses are rejected: $H_0: \mu_C = \mu_D$; $H_0: \mu_A = \mu_D$; $H_0: \mu_B = \mu_D$; $H_0: \mu_C = \mu_E$; $H_0: \mu_A = \mu_E$; $H_0: \mu_B = \mu_E$; $H_0: \mu_B = \mu_C$; $H_0: \mu_A = \mu_C$. The following hypotheses are not rejected: $H_0: \mu_A = \mu_B$; $H_0: \mu_D = \mu_E$. Summary:

$$C \quad \underline{A B} \quad \underline{E D}$$

There is sufficient evidence to conclude that treatments D and E give the largest means, treatments A and B the next largest, and treatment C the smallest. There is insufficient evidence to conclude that treatments A and B give different means or that treatments D and E give different means.

11.37 (b) Treatments 4 and 6 would be declared to be significantly different, as would treatments 4 and 9.

11.38 $.346 < \mu_E - \mu_A < 1.674$

11.40 H_0: The three classes produce the same mean change in fat-free mass ($\mu_1 = \mu_2 = \mu_3$); H_A: At least one class produces a different mean (the μ's are not all equal). $F_s = 0.64$. We do not reject H_0. There is insufficient evidence ($P > .20$) to conclude that the population means differ.

11.42 H_0: The mean refractive error is the same in the four populations ($\mu_1 = \mu_2 = \mu_3 = \mu_4$); H_A: Some of the populations have different mean refractive errors (the μ's are not all equal). $F_s = 3.56$. H_0 is rejected. There is sufficient evidence ($.01 < P < .02$) to conclude that some of the populations have different mean refractive errors.

Chapter 12

12.1 (a) $Y = 1 + 4X$. The $\hat{y}$'s are $13, 17, 5, 9, 21$.
   (c) The residuals ($y_i - \hat{y}_i$) are $0, -2, -1, 2, 1$; SS(resid) $= 10$.

12.4 (b) $Y = -.592 + 7.640X$; $s_{Y|X} = .915°C$

12.7 (a) $Y = 607.3 + 25.01X$
   (b)

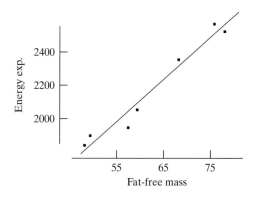

   (c) As fat-free mass goes up by 1 kg, energy expenditure goes up by 25.01 kcal, on average.
   (d) $s_{Y|X} = 64.85$ kcal

12.15 Estimated mean $= 21.1$ mm; estimated SD $= 1.3$ mm

12.19 (a) $.0252 < \beta_1 < .0333$ ng/min

(b) We are 95% confident that the rate at which leucine is incorporated into protein in the population of all *Xenopus* oocytes is between .0252 ng/min and .0333 ng/min.

12.23 (a) $19.4 < \beta_1 < 30.6$ kcal/kg

(b) $20.6 < \beta_1 < 29.4$ kcal/kg

12.24 H_0: There is no linear relationship between respiration rate and altitude of origin $(\beta_1 = 0)$; H_A: Trees from higher altitudes tend to have higher respiration rates $(\beta_1 > 0)$. $t_s = 6.06$. H_0 is rejected. There is sufficient evidence $(P < .0005)$ to conclude that trees from higher altitudes tend to have higher respiration rates.

12.29 (a) $r = .981$

(b) $s_Y = 308.25$, $s_{Y|X} = 64.85$ kcal/kg

12.31 (a) $r^2 = .43^2 = .1849$

(b) We can account for 18.49% of the variation in systolic blood pressure in men by using age in a regression model.

12.32 (a) $.812$

12.34 H_0: There is no correlation between blood urea and uric acid concentration $(\rho = 0)$; H_A: Blood urea and uric acid concentration are positively correlated $(\rho > 0)$. $t_s = 3.953$. H_0 is rejected. There is sufficient evidence $(P < .0005)$ to conclude that blood urea and uric acid concentration are positively correlated.

12.44 .24 g

12.46 (a) Estimated mean $= .85$ kg; estimated SD $= .17$ kg

12.49 (a) $.803$

(b) $s_Y = .210$ cm; $s_{Y|X} = .137$ cm

Chapter 13

13.1 A chi-square test of independence would be appropriate. The null hypothesis of interest is H_0: $p_1 = p_2$, where $p_1 = \Pr\{$clinically important improvement if given clozapine$\}$ and $p_2 = \Pr\{$clinically important improvement if given haloperidol$\}$. A confidence interval for $p_1 - p_2$ would also be relevant.

13.9 A two-sample comparison is called for here, but the data do not support the condition of normality. Thus, the Wilcoxon-Mann-Whitney test is appropriate.

13.11 It would be natural to consider correlation and regression with these data. For example, we could regress $Y =$ forearm length on $X =$ height; we could also find the correlation between forearm length and height and test the null hypothesis that the population correlation is zero.

Index

Index of Examples